ɔr ᵗ ' ᶦᵏ ʳary

Single- and Multi-carrier Quadrature Amplitude Modulation

We dedicate this book to

Rita, Alison and Katie

as well as to the numerous contributors of the field,
many of whom are listed in the acknowledgements and author index sections

Single- and Multi-carrier Quadrature Amplitude Modulation

Principles and Applications for Personal Communications, WLANs and Broadcasting

L. Hanzo
Department of Electronics and Computer Science
University of Southampton, UK

W. Webb
Motorola Arlington Heights, USA
formerly at Multiple Access Communications Ltd, Southampton, UK

T. Keller
Ubinetics, Cambridge Technology Centre, Melbourn, UK
formerly at Department of Electronics and Computer Science, University of Southampton, UK

IEEE PRESS
New York

JOHN WILEY & SONS, LTD
Chichester · New York · Weinheim · Brisbane · Singapore · Toronto

First Editon published in 1994 as Modern Quadrature Amplitude Modulation – Principles and Applications for Fixed and Wireless Channels by Pentech Press.

Baffins Lane, Chichester,
West Sussex, PO19 1UD, England

National 01243 779777
International (+44) 1243 779777

e-mail (for orders and customer service enquiries): cs-books@wiley.co.uk

Visit our Home Page on http://www.wiley.co.uk or http://www.wiley.com

Other Wiley Editorial Offices

New York • Weinheim • Brisbane • Singapore • Toronto

Library of Congress Cataloging-in-Publication Data

Hanzo, Lajos.
Single- and multi-carrier quadrature amplitude modulation : principles and applications for personal communications, WLANs and broadcasting / L. Hanzo, W. Webb, T. Keller.
p.cm.
Includes bibliographical references and index.
ISBN 0-471-49239-6
1. Signal processing—Digital techniques. 2. Amplitude modulation. 3. Curves—Rectification and quadrature. I. Webb, William, 1967- II. Keller, T. (Thomas) III. Title.

TK5102.9 H37 2000
621.382'2—dc21 99-086714

British Library Cataloguing in Publication Data

A catalogue record for this book is available from the British Library

ISBN 0 471 49239 6

Produced from PostScript files supplied by the authors.
Printed and bound in Great Britain by Antony Rowe Ltd, Chippenham, Wiltshire.
This book is printed on acid-free paper responsibly manufactured from sustainable forestry, in which at least two trees are planted for each one used for paper production.

Contents

Preface

Since its discovery in the early 1960s, quadrature amplitude modulation (QAM) has continued to gain interest and practical application. Particularly in recent years many new ideas and techniques have been proposed, allowing its employment over fading mobile channels. This book attempts to provide an overview of most major QAM techniques, commencing with simple QAM schemes for the uninitiated, while endeavouring to pave the way towards complex, rapidly evolving areas, such as trellis-coded pilot-symbol and transparent-tone-in-band assisted schemes, or arrangements for wide-band mobile channels. The second half of the book is targetted at the more advanced reader, providing a research-oriented outlook using a variety of novel QAM-based single- and multi-carrier arrangements.

The book is structured in five parts. Part I - constituted by Chapters 1-4 - is a rudimentary introduction for those requiring a background in the field of modulation and radio wave propagation. Part II is comprised of Chapters 5-9 and concentrates mainly on classic QAM transmission issues relevant to Gaussian channels. Readers familiar with the fundamentals of QAM and the characteristics of propagation channels, as well as with basic pulse shaping techniques may decide to skip Chapters 1-5. Commencing with Chapter 6, each chapter describes individual aspects of QAM. Readers wishing to familiarize themselves with a particular subsystem, including clock and carrier recovery, equalisation, trellis coded modulation, standardised telephone-line modem features, etc. can turn directly to the relevant chapters, whereas those who desire a more complete treatment might like to read all the remaining chapters.

Parts III-V, including Chapters 10-24, are concerned with QAM-based transmissions over mobile radio channels. These chapters provide a research-based perspective and are dedicated to the more advanced reader. Specifically, Chapter 10 concentrates mainly on coherent QAM schemes, including reference-aided transparent-tone-in-band and pilot-symbol assisted modulation arrangements. In contrast, Chapter 11 focuses on low-complexity differentially encoded QAM schemes and on their performance with and without forward error correction coding and trellis coded modulation. Chapter 12 details various timing recovery schemes.

Part IV of the book commences with Chapter 13, which is concerned with variable rate QAM using one- to six-bits per symbol signal constellations. Chapter 14 is dedicated to high-rate wide-band transmissions and proposes a novel equaliser arrangement. Various QAM-related orthogonal signaling techniques are proposed in Chapter 15, while the spectral efficiency of QAM in cellular frequency re-use structures is detailed in Chapter 16. This is followed by Chapter 17, which concentrates on the employment of QAM in a source-matched speech communications system, including various speech codecs, error correction codecs, a voice activity detector and packet reservation multiple access, providing performance figures in contrast to one and two bits per symbol bench-mark schemes.

Part V first appeared in this new edition of the book, concentrating on multi-carrier modulation. Specifically, following a rudimentary introduction to Orthogonal Frequency Division Multiplexing (OFDM) in Chapter 18, Chapters 19-23 detail a range of implementational and performance aspects of OFDM over both Gaussion and wideband fading channels. Lastly, Chapter 24 concentrates on the performance aspects of various standard-compliant and enhanced OFDM-based Digital Video Broadcasting (DVB) systems designed for transmission to mobile receivers.

To the original text of the first edition dealing with many of the fundamentals of single-carrier QAM and QAM-based systems we have added six new chapters dealing with the complexities of the exciting subject of multi-carrier modulation, which has found wide-ranging applications in a past decade, ranging from Wireless Local Area Network (WLAN) to broadcast systems. Whilst the book aims to portray a rapidly evolving area, where research results are promptly translated into products, it is our hope that you will find this second edition comprehensive, technically challenging and above all, enjoyable.

Lajos Hanzo
William Webb
Thomas Keller

Acknowledgements

The authors would like to express their warmest thanks to Prof. Raymond Steele. Without his shrewd long-term vision the research on single-carrier QAM would not have been performed, and without his earnest exhortations a book on the subject would not have been written. Furthermore, Professor Steele has edited some of the chapters and given advice on the contents and style of this book.

Contributions by Dr. P.M. Fortune, Dr. K.H.H. Wong, Dr. R.A. Salami, D. Greenwood, R. Stedman, R Lucas and Dr. J.C.S. Cheung who were formerly with Southampton University are thankfully acknowledged. We thank Multiple Access Communications Ltd. for supporting the work on QAM, particularly in the framework of the DTI LINK Personal Communications Programme, dealing with high data rate QAM transmission over microcellular mobile channels. Special thanks goes to Peter Gould and Philip Evans for the major part they played in the construction of the star QAM test-bed. We are grateful to John Williams of Multiple Access Communications Ltd. for the many simulation results he provided for Chapter 17, the production of many of the figures involving simulated waveforms. Much of the QAM work at Multiple Access Communications Ltd. derives from the support of BT Labs. Martlesham Heath, the DTI and the Radio Communications Agency. Specifically, we thank the latter for the support of the research on spectral efficiency which facilitated Chapter 16.

Much of the results in Chapters 19-22 are based on our work conducted as a sub-contractor of Motorola ECID, Swindon, UK; as part of our involvement in a collaborative Pan-European Wireless Asyncrhonous Transfer Mode (WATM) project known as Median, which was genereously supported by the European Commission (EC), Brussels, Belgium. We would like to acknowledge all our valued friends and colleagues - too numerous to mention individually - who at some stage were associated with the Median consortium and with whom we have enjoyed a stimulating collaboration under the stirling management of IMST, Germany. Our gratitude is due to Andy Wilton and to Paul Crichton of Motorola, who have whole-heartedly sponsored our research. Further thanks are also due to Dr. Joao Da Silva, Bartolome Aroyo, Bernard Barani, Dr. Jorge Pereira, Demosthenes Ikonomou and to the other equally supportive members of the EC's programme management team in Brussels for their enthusiastic support. Furthermore, we enjoyed the valuable support of EPSRC, Swindon UK, and the Mobile VCE, for which we are equally grateful.

Lastly, we express our gratitude for the creative atmosphere to our colleagues Derek Appleby, Steve Braithwaite, Sheng Chen, David Stewart, Jeff Reeve as well as Stephan Weiss at Southampton University, UK and gratefully acknowledge the stimulating embryonic discussions with Prof. G. Gordos (Technical University of Budapest, Hungary), Prof. H.W. Schüssler (University of Erlangen-Nürnberg, Germany) and Dr.Ing. H.J. Kolb as well as the numerous thought-provoking contributions by many established authorities in the field, who appear also in the Author Index Section of the book.

A number of colleagues have influenced our views concerning various aspects of wireless communications and we thank them for the enlightment gained from our collaborations on various projects, papers and books. We are grateful to J. Brecht, Jon Blogh, Marco Breiling, M. del Buono, Clare Brooks, Peter Cherriman, Stanley Chia, Byoung Jo Choi, Joseph Cheung, Peter Fortune, Lim Dongmin, D. Didascalou, S. Ernst, Eddie Green, David Greenwood, Hee Thong How, Thomas Keller, W.H. Lam, C.C. Lee, M.A. Nofal, Xiao Lin, Chee Siong Lee, Tong-Hooi Liew, Matthias Muenster, V. Roger-Marchart, Redwan Salami, David Stewart, Juergen Streit, Jeff Torrance, Spyros Vlahoyiannatos, William Webb, John Williams, Jason Woodard, Choong Hin Wong, Henry Wong, James Wong, Lie-Liang Yang, Bee-Leong Yeap, Mong-Suan Yee, Kai Yen, Andy Yuen and many others with whom we enjoyed an association. Special thanks are due to Dr. Lie-Liang Yang for his insightful contributions on the theory of the coding schemes used in Chapter 23, to Tong-Hooi Liew for his kind assistance in the preparation of a co-authored paper, which was the basis of Chapter 23, which resulted in a joint journal submission. Similarly, the contributions of Matthias Muenster in Sections 22.3 and 22.7 are thankfully acknowledged along with those of Chee-Siong Lee and Spyros Vlahoyiannatos to the papers, which consituted the basis of Chapter 24. Lorenzo Piazzo's permission to expand the material of his Electronics Letter on optimum power- and bit-allocation of OFDM is thankfully acknowledged. We are also grateful to our editors, Mark Hammond and Sarah Hinton and Sarah Lock at Wiley. Finally, the authors warmly thank Rita Hanzo, Denise Harvey and Dr. Peter Cherriman for their dedicated and skilful assistance in typesetting the manuscript in Latex as well as in amalgamating the new material of the second edition with the first edition.

Lajos Hanzo
William Webb
Thomas Keller

Chapter 1

Introduction and Background

This book is concerned with the issues of transmitting digital signals via multilevel modulation. We will be concerned with digital signals originating from a range of sources such as from speech or video encoding, or data from computers. A typical digital information transmission system is shown in Figure 1.1. The source encoder may be used to remove some of the redundancy which occurs in many sources such as speech, typically reducing the transmission rate of the source. The forward error correction (FEC) block then paradoxically inserts redundancy, but in a carefully controlled manner. This redundancy is in the form of parity check bits, and allows the FEC decoder to remove transmission errors caused by channel impairments, but at the cost of an increase in transmission rate. The interleaver systematically rearranges the bits to be transmitted, which has the effect of dispersing a short burst of errors at the receiver, allowing the FEC to work more effectively. Lastly, the modulator generates bandlimited waveforms which can be transmitted over the bandwidth-limited channel to the receiver, where the reverse functions are performed. Whilst we will discuss all aspects of Figure 1.1, it is the generation of waveforms in the modulator in a manner which reduces errors and increases the transmission rate within a given bandwidth, and the subsequent decoding in the demodulator, which will be our main concern in this book.

It is assumed that the majority of readers will be familiar with binary modulation schemes such as binary phase shift keying (BPSK), frequency shift keying (FSK), etc. Those readers who possess this knowledge might like to jump to Section 1.2. For those who are not familiar with modulation schemes we give a short non-mathematical explanation of modulation and constellation diagrams before detailing the history of QAM.

1.1 Modulation Methods

Suppose the data we wish to transmit is digitally encoded speech having a bit rate of 16 kbit/s, and after FEC coding the data rate becomes 32 kb/s. If the radio

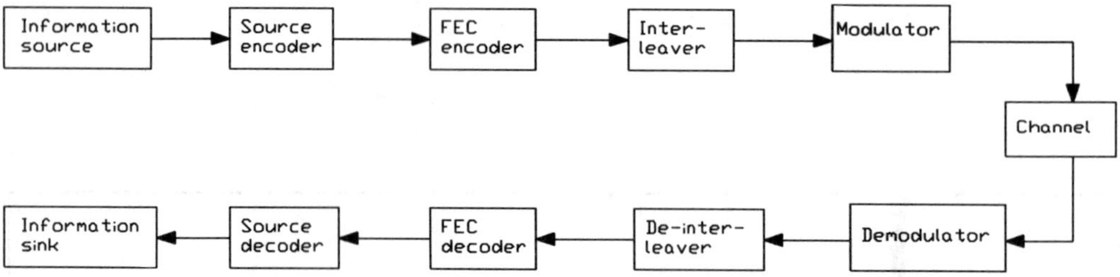

Figure 1.1: Typical digital information transmission system

channel to be used is centred around 1 GHz, then in order to transmit the 32 kb/s we must arrange for some feature of a 1 GHz carrier to change at the data rate of 32 kb/s. If the phase of the carrier is switched at the rate of 32 kb/s, being at 0 deg and 180 deg for bits having logical 0 or logical 1, respectively, then there are $1 \times 10^9/32 \times 10^3 = 31,250$ radio frequency (RF) oscillations per bit transmitted. Figure 1.2(a) and (b) show the waveforms at the output of the modulator when the data is a logical 0 and a logical 1, respectively. On the left is the phasor diagram for a logical 0 and a logical 1 where the logical 0 is represented by a phasor along the positive x-axis, and the logical 1 by a phasor along the negative x-axis. This negative phasor represents a phase shift of the carrier by 180 deg. Figure 1.2(c) shows the modulator output for a sequence of data bits. Note that no filtering is used in this introductory example. Here the waveform can be seen to change abruptly at the boundary between some of the data symbols. We will see later that such abrupt changes can be problematic since they theoretically require an infinite bandwidth, and ways are sought to avoid them. Figure 1.2(d) is called a constellation diagram of phasor points, and as we are transmitting binary data there are only two points. As these two points are at equal distance from the origin we would expect them to represent equal magnitude carriers, and that the magnitude is indeed constant can be seen in Figure 1.2(c). The bandwidth of the modulated signal in this example will be in excess of the signalling rate of 32 kb/s due to the sudden transition between phase states. Later we will consider the bandwidth of modulated signals in-depth. Suffice to say here that if we decreased the signalling rate to 16 kb/s the bandwidth of the modulated signal will decrease. If the data rate is 32 kb/s, and the signalling rate becomes 16 kb/s, then every symbol transmitted must carry two bits of information. This means that we must have four points on the constellation, and clearly this can be done in many ways. Figure 1.3 shows some four-point constellations. The two bits of information associated with every constellation point are marked on the figure. In Figure 1.3(a) and (b) so-called quadrature modulation has been used as the points can only be uniquely described using two orthogonal coordinate axes, each passing through the origin. The orthogonal coordinate axes have a phase rotation of 90 deg with respect to each other, and hence they have a so-called quadrature relationship. The pair of coordinate axes can be associated with a pair of quadrature carriers, normally a sine and a cosine waveform, which can be independently modulated and then transmitted within the same frequency band. Due to their orthogonality they can be separated by the receiver. This implies that whatever symbol is chosen on one

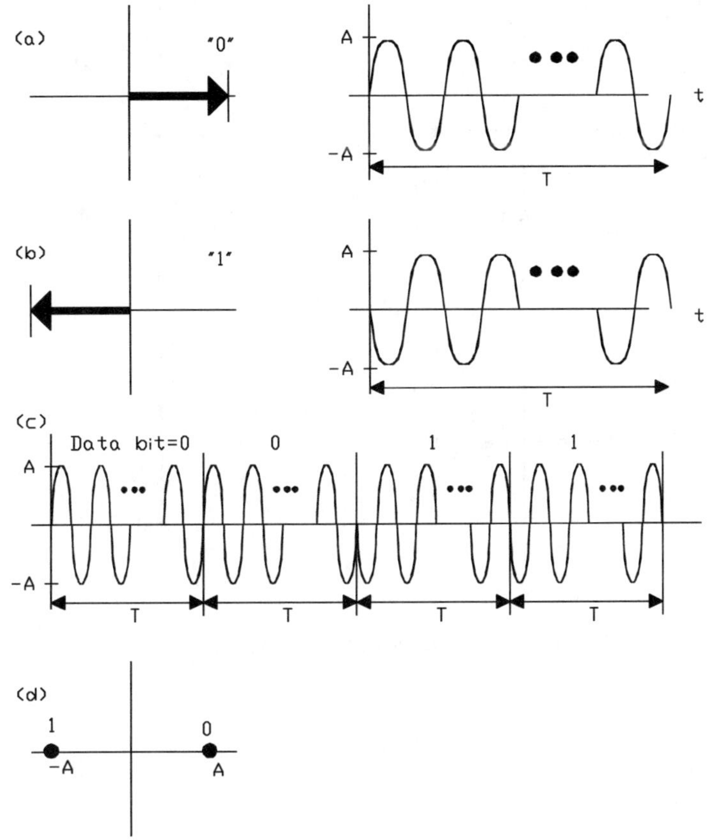

Figure 1.2: Carrier waveform for binary input data

axis (say the x-axis in the case of Figure 1.3(a)) it will have no effect on the data demodulated on the y-axis. Data can therefore be independently transmitted via these two quadrature or orthogonal channels without any increase in error rate, although some increase may result in practice and this is considered in later chapters. We have used I and Q to signify the in-phase and quadrature components, respectively, where in-phase normally represents the x-axis and quadrature the y-axis. Figure 1.3(c) and (d) show constellations where the four points are only on one line. These are not quadrature constellations but actually represent multilevel amplitude and phase modulation where both the carrier amplitude and phase can take two discrete values.

For the constellations in Figure 1.3(a) and (b) we have a constant amplitude signal, but the carrier phase values at the beginning of each symbol period in Figure 1.3(b) would be either $45\,\text{deg}, 135\,\text{deg}, 225\,\text{deg}\,or 315\,\text{deg}$. There are two magnitude values and two phase values for the constellations in Figure 1.3(c) and (d).

In order to reduce the bandwidth of the modulated signal whilst maintaining the same information transmission rate we can further decrease the symbol rate by adding

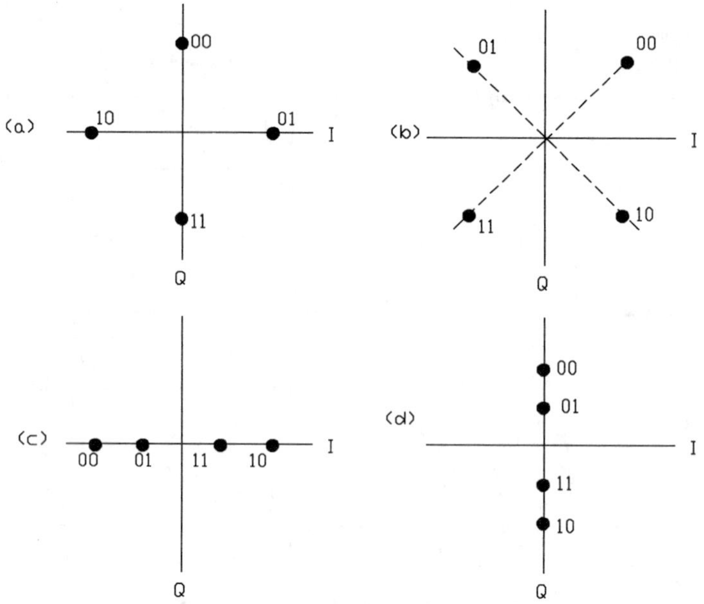

Figure 1.3: Examples of four-level constellations

more points in the constellation. Such a reduction in the bandwidth requirement will allow us to transmit more information in the spectrum we have been allocated. Such capability is normally considered advantageous. If we combine the constellations of Figure 1.3(c) and (d) we obtain the square QAM constellation having four bits per constellation point as displayed in Figure 1.4. We will spend much time in dealing with this constellation in this book. In general, grouping n bits into one signalling symbol yields 2^n constellation points, which are often referred to as phasors, or complex vectors. The phasors associated with these points may have different amplitude and/or phase values, and this type of modulation is therefore referred to as multilevel modulation, where the number of levels is equal to the number of constellation points. After transmission through the channel the receiver must identify the phasor transmitted, and in doing so can determine the constellation point and hence the bits associated with this point. In this way the data is recovered. There are many problems with attempting to recover data transmitted over both fixed channels such as telephone lines and radio channels and many of these problems are given a whole chapter in this book. These problems are generally exacerbated by changing from binary to multilevel modulation, and this is why binary modulation is often preferred, despite its lower capacity. In order to introduce these problems, and to provide a historical perspective to quadrature amplitude modulation (QAM), a brief history of the development of QAM is presented.

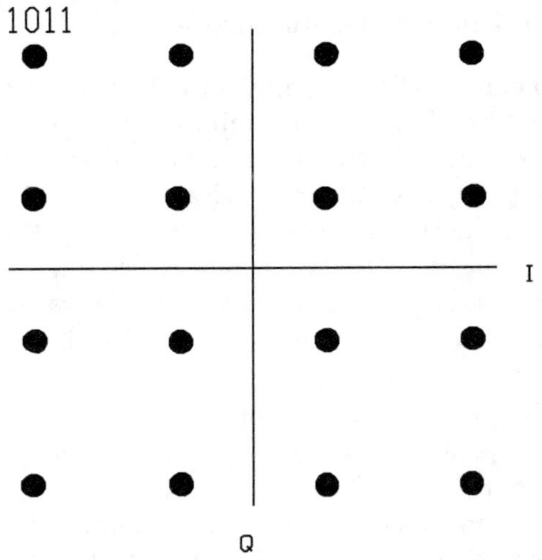

Figure 1.4: Square QAM constellation

1.2 History of Quadrature Amplitude Modulation

1.2.1 Determining the optimum constellation

Towards the end of the 1950s there was a considerable amount of interest in digital phase modulation transmission schemes [13] as an alternative to digital amplitude modulation. Digital phase modulation schemes are those whereby the amplitude of the transmitted carrier is held constant but the phase changed in response to the modulating signal. Such schemes have constellation diagrams of the form shown in Figure 1.3(a). It was a natural extension of this trend to consider the simultaneous use of both amplitude and phase modulation. The first paper to suggest this idea was by C.R. Cahn in 1960, who described a combined phase and amplitude modulation system [14]. He simply extended phase modulation to the multilevel case by allowing there to be more than one transmitted amplitude at any allowed phase. This had the effect of duplicating the original phase modulation or phase shift keying (PSK) constellation which essentially formed a circle. Such duplication led to a number of concentric circles depending on the number of amplitude levels selected. Each circle had the same number of phase points on each of its rings. Only Gaussian channels characteristic of telephone lines impaired by thermal noise were considered. Using a series of approximations and a wholly theoretical approach, he came to the conclusion that these amplitude and phase modulation (AM-PM) systems allowed an increased throughput compared to phase modulation systems when 16 or more states were used and suggested that such a system was practical to construct.

1.2.1.1 Coherent and non-coherent reception

The fundamental problem with PSK is that of determining the phase of the transmitted signal and hence decoding the transmitted information. This problem is also known as carrier recovery as an attempt is made to recover the phase of the carrier. When a phase point at, say, 90 deg is selected to reflect the information being transmitted, the phase of the transmitted carrier is set to 90 deg. However, the phase of the carrier is often changed by the transmission channel with the result that the receiver measures a different phase. This means that unless the receiver knew what the phase change imposed by the channel was it would be unable to determine the encoded information.

This problem can be overcome in one of two ways. The first is to measure the phase change imposed by the channel by a variety of means. The receiver can then determine the transmitted phase. This is known as coherent detection. The second is to transmit differences in phase, rather than absolute phase. The receiver then merely compares the previous phase with the current phase and the phase change of the channel is removed. This assumes that any phase change within the channel is relatively slow. This differential system is known as non-coherent transmission. In his paper, Cahn considered both coherent and non-coherent transmission, although for coherent transmission he assumed a hypothetical and unrealisable perfect carrier recovery device. The process of carrier recovery is considered in Chapter 6, and the details of differential transmission are explained in Chapter 5.

1.2.1.2 Clock recovery

Alongside carrier recovery runs the problem of clock recovery. The recovered clock signal is used to ensure appropriate sampling of the received signal. In Figure 1.2(c) a carrier signal was shown which had vertical lines indicating each bit or symbol period. It was the phase at the start of this period which was indicative of the encoded information. Unfortunately, the receiver has no knowledge of when these periods occur although it might know their approximate duration. It is determining these *symbol periods* which is the task of clock recovery. So carrier recovery estimates the phase of the transmitted carrier and clock recovery estimates the instances at which the data changes from one symbol to another. Whilst the need for carrier recovery can be removed through differential or non-coherent detection, there is no way to remove the requirement for clock recovery.

Clock recovery schemes tend to seek certain periodicities in the received signal and use these to estimate the start of a symbol (actually they often attempt to select the centre of a symbol for reasons which will be explained in later chapters). Clock recovery is often a complex procedure, and poor clock recovery can substantially increase the bit error rate (BER). The issue of clock recovery is considered in Chapter 6. In his work, Cahn overcame the problem of clock recovery by assuming that he had some device capable of perfect clock recovery. Such devices do not exist, so Cahn acknowledged that the error rate experienced in practice would be worse than the value he had calculated, but as he was unable to compute the errors introduced by a practical clock recovery system, this was the only course open to him.

1.2.1.3 The Type I, II and III constellations

A few months later a paper was published by Hancock and Lucky [15] in which they expanded upon the work of Cahn. In this paper they realised that the performance of the circular type constellation could be improved by having more points on the outer ring than on the inner ring. The rationale for this was that errors were caused when noise introduced into the signal moved the received phasor from the transmitted constellation point to a different one. The further apart constellation points could be placed, the less likely this was to happen. In Cahn's constellation, points on the inner ring were closest together in distance terms and so most vulnerable to errors. They conceded that a system with unequal numbers of points on each amplitude ring would be more complicated to implement, particularly in the case of non-coherent detection. They called the constellation proposed by Cahn a Type I system, and theirs a Type II system. Again using a mathematical approach they derived results similar to Cahn's for Type I systems and a 3 dB improvement for the Type II over the Type I system.

The next major publication was some 18 months later, in 1962, by Campopiano and Glazer [16]. They developed on the work of the previous papers but also introduced a new constellation - the square QAM system, which they termed a Type III system. They described this system as "essentially the amplitude modulation and demodulation of two carriers that have the same frequency but are in quadrature with each other" - the first time that combined amplitude and phase modulation had been thought of as amplitude modulation on quadrature carriers, although the acronym QAM was not suggested. They realised that the problem with their Type III system was that it had to be used in a phase coherent mode, that is non-coherent detection was not possible and so carrier recovery was necessary. Again, a theoretical analysis was performed for Gaussian noise channels and the authors came to the conclusion that the Type III system offered a very small improvement in performance over the Type II system, but thought that the implementation of the Type III system would be considerably simpler than that of Types I and II. Examples of the different types of constellation are shown in Figure 1.5.

Three months later another paper was published by Hancock and Lucky [17] in which they were probably unaware of the work done by Campopiano and Glazer. They attempted to improve on their previous work on the Type II system by carrying out a theoretical analysis, supposedly leading to the optimal constellation for Gaussian channels. In this paper they decided that the optimum 16-level constellation had two amplitude rings with eight equispaced points on each ring but with the rings shifted by 22.5 deg from each other. This constellation is shown in Figure 1.6.

Again, they concluded that 16 was the minimum number of levels for AM-PM modulation and that a SNR of at least 11 dB was required for efficient operation with a low probability of bit error.

After this paper there was a gap of nine years before any further significant advances were published. This was probably due to the difficulties in implementing QAM systems with the technology available and also because the need for increased data throughput was not yet pressing. During this period the work discussed in the above papers was consolidated into a number of books, particularly that by Lucky, Salz and Weldon [18]. Here they clearly distinguished between quadrature amplitude

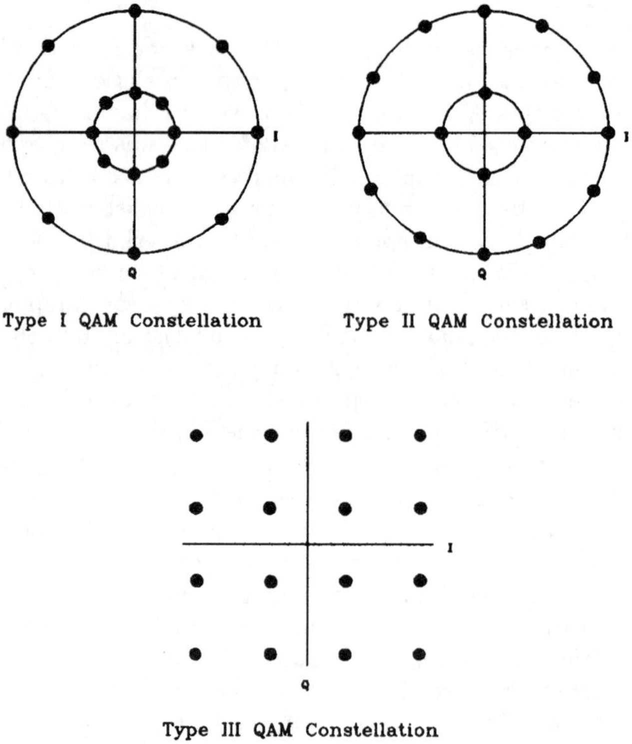

Figure 1.5: Examples of types I, II and III QAM constellations

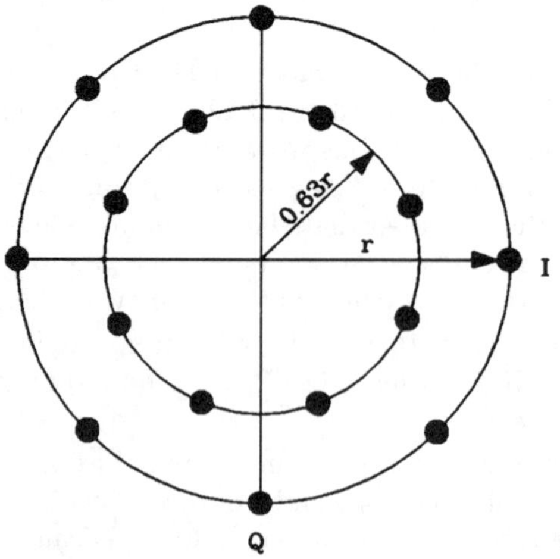

Figure 1.6: "Optimum" 16-level constellation according to Lucky [17] ©IRE, 1962

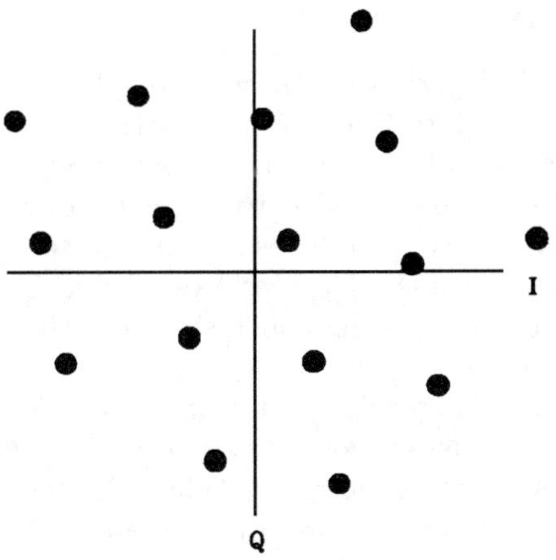

Figure 1.7: "Optimum" 16-level constellation according to Foschini [21] ©IEEE, 1974

modulation (QAM) schemes using square constellations and combined amplitude and phase modulation schemes using circular constellations. It was around this period that the acronym QAM started appearing in common usage along with AM-PM to describe the different constellations.

One of the earliest reports of the actual construction of a QAM system came from Salz, Sheenhan and Paris [19] of Bell Labs in 1971. They implemented circular constellations with 4 and 8 phase positions and 2 and 4 amplitude levels using coherent and non-coherent demodulation. Neither carrier nor clock recovery was attempted. Their results showed reasonable agreement with the theoretical results derived up to that time. This work was accompanied by that of Ho and Yeh [20] who improved the theory of circular AM-PM systems with algorithms that could be solved on digital computers which were by that time becoming increasingly available.

Interest in QAM remained relatively low, however, until 1974. In that year there was a number of significant papers published, considerably extending knowledge about quadrature amplitude modulation schemes. At this time, interest into optimum constellations was revived with two papers, one from Foschini, Gitlin and Weinstein [21] and the other from Thomas, Weidner and Durrani [22]. Foschini *et al.* attempted a theoretical derivation of the ideal constellation using a gradient calculation approach. They came to the conclusion that the ideal constellation was based around an equilateral triangle construction leading to the unusual 16-level constellation shown in Figure 1.7. This constellation has not found favour in practical applications, since the complexities involved in its employment outweigh the associated gains that were claimed for it.

Their conclusions were that this constellation, when limited in terms of power and operated over Gaussian channels, offered a performance improvement of 0.5 dB over square QAM constellations. Meanwhile Thomas *et al.*, working at COMSAT, empirically generated 29 constellations and compared their error probabilities.

1.2.2 Satellite links

In the paper by Thomas *et al.* [22] they also mentioned the first application of QAM, for use in satellite links. Satellite links have a particular problem in that satellites only have a limited power available to them. Efficient amplifiers are necessary to use this power carefully, and these had tended to employ a device known as a travelling wave tube (TWT). Such a device introduces significant distortion in the transmitted signal and Thomas *et al.* considered the effects of this distortion on the received waveform. They came to the interesting conclusion that the Type II constellation (in this case 3 points on the inner ring and 5 on the outer) was inferior to the Type I (4 points on inner and outer rings) due to the increased demand in peak-to-average power ratio of the Type II constellation. Their overall conclusion was that circular Type I constellations are superior in all cases. When they considered TWT distortion they discovered that AM-PM schemes were inferior to PSK schemes because of the need to significantly back off the amplifier to avoid severe amplitude distortion, and concluded that better linear amplifiers would be required before AM-PM techniques could be successfully used for satellite communications. They also considered the difficulties of implementing various carrier recovery techniques, advising that decision directed carrier recovery would be most appropriate, although few details were given as to how this was to be implemented. Decision directed carrier recovery is a process whereby the decoded signal is compared with the closest constellation point and the phase difference between them is used to estimate the error in the recovered carrier. This is discussed in more detail in Chapter 6.

Commensurate with the increasing interest as to possible applications for QAM were the two papers published in 1974 by Simon and Smith which concentrated on carrier recovery and detection techniques. In the first of these [23] they noted the interest in QAM that was then appearing for bandlimited systems, and addressed the problems of carrier recovery. They considered only the 16-level square constellation and noted that the generation of a highly accurate reconstructed carrier was essential for adequate performance. Their solution was to demodulate the signal, quantise it, and then establish the polarity of error from the nearest constellation point, and use it to update the voltage controlled oscillator (VCO) used in the carrier generation section. They provided a theoretical analysis and concluded that their carrier recovery technique worked well in the case of high signal-to-noise (SNR) ratio Gaussian noise, although they noted that gain control was required and would considerably complicate the implementation. They extended their work in Reference [24] where they considered offset QAM or O-QAM. In this modulation scheme the signal to one of the quadrature arms was delayed by half a symbol period in an attempt to prevent dramatic fluctuations of the signal envelope, which was particularly useful in satellite communications. They noted similar results for their decision directed carrier recovery scheme as when non-offset modulation was used.

1.2.2.1 Odd-bit constellations

Despite all the work on optimum constellations, by 1975 interest had centred on the square QAM constellation. The shape of this was evident for even numbers of bits per symbol, but if there was a requirement for an odd number of bits per symbol to

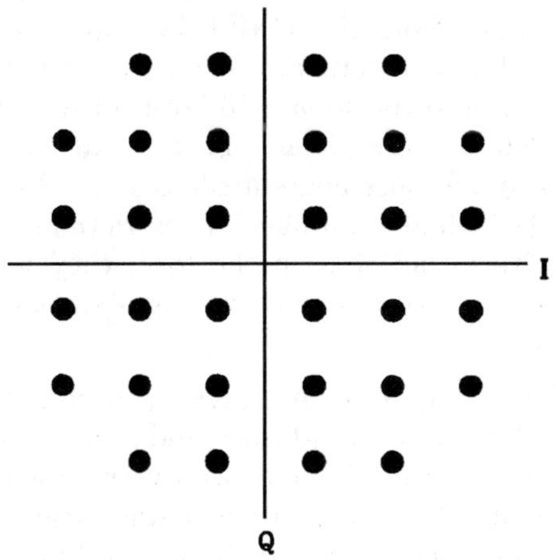

Figure 1.8: "Optimum" 32-level constellation according to Smith [25] ©IEEE, 1975

be transmitted, the ideal shape of the constellation was not obvious, with rectangular constellations having been tentatively suggested. Early in 1975 J.G. Smith, also working on satellite applications, published a paper addressing this problem [25]. He noted that for even numbers of bits per symbol "the square constellation was the only viable choice." In this paper he showed that what he termed "symmetric" constellations offered about a 1 dB improvement over rectangular constellations and he considered both constellations to be of the same implementational complexity. Figure 1.8 shows an example of his symmetric constellation when there are 5 bits per symbol,

1.2.3 QAM modem implementation

About this time, the Japanese started to show interest in QAM schemes as they considered they might have application in both satellite and microwave radio links. In 1976 Miyauchi, Seki and Ishio published a paper devoted to implementation techniques [26]. They considered implementation by superimposing two 4-level PSK modulation techniques at different amplitudes to achieve a square QAM constellation and using a similar process in reverse at the demodulator, giving them the advantage of being able to use existing PSK modulator and demodulator circuits. This method of implementing QAM is discussed in Chapter 5. They implemented a prototype system without clock or carrier recovery and concluded that its performance was sufficiently good to merit further investigation. Further groundwork was covered in 1978 by W. Weber, again working on satellite applications, who considered differential (i.e. non-coherent) encoding systems for the various constellations still in favour [27]. His paper essentially added a theoretical basis to the differential techniques that had been in use at that point, although suggesting that non-coherent demodulation techniques deserved more attention.

In late 1979 evidence of the construction of QAM prototype systems worldwide started to emerge. A paper from the CNET laboratories in France by Dupuis *et al.* [28] considered a 140 Mbit/s modem for use in point-to-point links in the 10-11 GHz band. This prototype employed the square 16-level constellation and included carrier recovery, although no details were given. Theoretical calculations of impairments were presented followed by measurements made over a 58 km hop near Paris. Their conclusions were that QAM had a number of restrictions in its use, relating to its sensitivity to non-linearities, and that in the form they had implemented it, PSK offered improved performance. However, they suggested that with further work these problems might be overcome.

The Japanese simultaneously announced results from a prototype 200 Mbit/s 16-QAM system in a paper by Horikawa, Murase and Saito [29]. They used differential coding coupled with a new form of carrier recovery based on a decision feedback method (detailed in Chapter 6). Their modem was primarily designed for satellite applications and their experiment included the use of TWT amplifiers, but was only carried out back-to-back in the laboratory. Their conclusions were that their prototype had satisfactory performance and was an efficient way to increase bandwidth efficiency.

One of the last of the purely theoretical, as opposed to practical papers on QAM appeared in April 1980, marking the progression of QAM from a technical curiosity into a practical system, some twenty years after its introduction. This came from V. Prabhu of Bell Labs [30], further developing the theory to allow calculation of error probabilities in the presence of co-channel interference. Prabhu concluded that 16-QAM had a co-channel interference immunity superior to 16-PSK but inferior to 8-PSK.

1.2.3.1 Non-linear amplification

In 1982 there came a turning point in the use of QAM for satellite applications when Feher turned his attention to this problem. His first major publication in this field [31] introduced a new method of generation of QAM signals using highly non-linear amplifiers which he termed non-linear amplified QAM (NLA-QAM). Two separate amplifiers for the 16-QAM case were used, one operating with half the output power of the other. The higher power amplifier coded the two most significant bits of the 4-bit symbol only, and the lower power amplifier coded only the two least significant bits. The amplified coded signals were then summed at full output power to produce the QAM signal. Because both amplifiers were therefore able to use constant envelope modulation they could be run at full power with resulting high efficiency, although with increased complexity due to the need for two amplifiers and a hybrid combiner, compared to previous systems which used only a single amplifier. However, in satellite applications, complexity was relatively unimportant compared to power efficiency, and this NLA technique offered a very substantial 5 dB power gain, considerably increasing the potential of QAM in severely power limited applications.

1.2.3.2 Frequency selective fading and channel equalisers

This work was soon followed by a performance study of a NLA 64-state system [32] which extended the NLA scheme to 64 levels by using three amplifiers all operating at different power levels. Performance estimates were achieved using computer simulation techniques which included the effects of frequency selective fading.

Frequency selective fading is essentially caused when there are a number of propagation paths between the transmitter and receiver, e.g. by reflection off nearby mountains. When the time delay of the longest path compared to that of the shortest path becomes comparable to a symbol period, intersymbol interference (ISI) is said to result. As every time domain effect has an equivalent frequency domain effect, this can also be termed frequency selective fading. The two are related through the Fourier transform. This fading caused severe problems, leading Feher to conclude that adaptive equaliser techniques would be required for adequate wideband performance. Equalisers are devices which attempt to remove ISI. They do so by calculating the ISI introduced and then subtracting it from the received signals. They are often extremely complex devices and are considered in detail in Chapter 7.

1.2.3.3 Filtering

In a book published around this time [33], Feher also suggested the use of non-linear filtering (NLF) for QAM satellite communications. Since the I and Q components of the time domain QAM signal change their amplitude abruptly at the signalling intervals their transmission would require an infinite bandwidth. These abrupt changes are typically smoothed by a bandlimiting filter. The design an implementation of such filters is critical, particularly when the NLA-QAM signal is filtered at high power level. In order to alleviate these problems Feher developed the NLF technique along with Huang in 1979 which simplified filter design by simply fitting a quarter raised cosine segment between two initially abruptly changing symbols for both of the quadrature carriers. We have already hinted that filtering is required to prevent abrupt changes in the transmitted signal. This issue is considered in more rigour in Chapter 5. This allowed the generation of jitter-free bandlimited signals, which had previously been a problem, improving clock recovery techniques. Feher's work continued to increase the number of levels used, with a paper on 256-QAM in May 1985 [34] noting the problems that linear group delay distortion caused, and a paper in April 1986 on 512-QAM [35] which came to similar conclusions.

1.2.4 Advanced prototypes

Work was still continuing in France, Japan and also in New Zealand. CNET were continuing their attempt to overcome the problems they had found in their initial trials reported in 1979. A paper published in 1985 by M. Borgne [36] compared the performance of 16, 32, 64 and 128 level QAM schemes using computer simulation with particular interest in the impairments likely to occur over point-to-point radio links. Borgne concluded that non-linearity cancellers and adaptive equalisers would be necessary for this application. Soon after this, the first major paper on adaptive equalisers for QAM was published by Shafi and Moore [37]. Much of this paper was concerned with clock and carrier recovery without going into detail as to how these operations were performed. Details were provided of a fractional decision feedback equaliser (DFE) which they considered suitable for point-to-point radio links. They concluded that carrier recovery and clock timing was critical and likely to cause major problems, which were somewhat ameliorated by their fractionally spaced system.

Although lagging somewhat behind Feher, the Japanese made up for this delay by publishing a very detailed paper describing the development of a 256-level QAM modem in August 1986. In this paper from Saito and Nakamura [38], the authors developed on the work announced in 1979 by Saito *et al.* [26] which was discussed earlier. In this new paper they detailed automatic gain control (which he termed automatic threshold control) and carrier recovery methods. The carrier recovery was a slight enhancement to the system announced in 1979 and the AGC system was based on decision directed methods. Details were given as to how false lock problems were avoided (see Chapter 6) and the back-to-back prototype experiments gave results which the authors considered showed the feasibility of the 256-QAM modem. Evidence of the ever increasing interest in QAM was that in the IEEE special issue on advances in digital communications by radio there was a substantial section devoted to high-level modulation techniques. In a paper by Rustako *et al.* [39] which considered point-to-point applications, the standard times-two carrier recovery method for binary modulation was expanded for QAM. The authors claimed the advantage of not requiring accurate data decisions or interacting with any equaliser. They acknowledged that their system had the disadvantage of slow reacquisition after fades and suggested that it would only be superior in certain situations. This form of carrier recovery is considered in some detail in Chapter 6.

Clearly, until the late 1980s developments were mainly targeted at telephone line and point-to-point radio applications, which led to the definition of the CCITT telephone circuit modem standards V.29 to V.33 based on various QAM constellations ranging from uncoded 16-QAM to trellis coded (TC) 128-QAM. Coded modulation is analysed in Chapter 8, while the CCITT standard modem schemes V.29 to V.33 for telephone lines will be considered in Chapter 9.

1.2.5 QAM for mobile radio

Another major development occurred in 1987 when Sundberg, Wong and Steele published a pair of papers [40,41] considering QAM for voice transmission over Rayleigh fading channels, the first major paper considering QAM for mobile radio applications. In these papers, it was recognized that when a Gray code mapping scheme was used, some of the bits constituting a symbol had different error rates from other bits. Gray coding is a method of assigning bits to be transmitted to constellation points in an optimum manner and is discussed in Chapter 5. For the 16-level constellation two classes of bits occurred, for the 64-level three classes and so on. Efficient mapping schemes for pulse code modulated (PCM) speech coding were discussed where the most significant bits (MSBs) were mapped onto the class with the highest integrity. A number of other schemes including variable threshold systems and weighted systems were also discussed. Simulation and theoretical results were compared and found to be in reasonable agreement. They used no carrier recovery, clock recovery or AGC, assuming these to be ideal, and came to the conclusion that channel coding and post-enhancement techniques would be required to achieve acceptable performance.

This work was continued, resulting in a publication in 1990 by Hanzo, Steele and Fortune [42], again considering QAM for mobile radio transmission, where again a theoretical argument was used to show that with a Gray encoded square constellation, the bits encoded onto a single symbol could be split into a number of subclasses, each subclass having a different average BER. The authors then showed that the difference in BER of these different subclasses could be reduced by constellation distortion at the cost of slightly increased total BER, but was best dealt with by using different error correction powers on the different 16-QAM subclasses. A 16 kbit/s sub-band speech coder was subjected to bit sensitivity analysis and the most sensitive bits identified were mapped onto the higher integrity 16-QAM subclasses, relegating the less sensitive speech bits to the lower integrity classes. Furthermore, different error correction coding powers were considered for each class of bits to optimise performance. Again ideal clock and carrier recovery were used, although this time the problem of automatic gain control (AGC) was addressed. It was suggested that as bandwidth became increasingly congested in mobile radio, microcells would be introduced supplying the required high SNRs with the lack of bandwidth being an incentive to use QAM.

In the meantime, CNET were still continuing their study of QAM for point-to-point applications, and Sari and Moridi published a paper [43] detailing an improved carrier recovery system using a novel combination of phase and frequency detectors which seemed promising. However, interest was now increasing in QAM for mobile radio usage and a paper was published in 1989 by J. Chuang of Bell Labs [44] considering NLF-QAM for mobile radio and concluding that NLF offered slight improvements over raised cosine filtering when there was mild intersymbol interference (ISI).

A technique, known as the transparent tone in band method (TTIB) was proposed by McGeehan and Bateman [9] from Bristol University, UK, which facilitated coherent detection of the square QAM scheme over fading channels and was shown to give good performance but at the cost of an increase in spectral occupancy. This important technique is discussed in depth in Chapter 11. At an IEE colloquium on multilevel modulation techniques in March 1990 a number of papers were presented considering QAM for mobile radio and point-to-point applications. Matthews [45] proposed the use of a pilot tone located in the centre of the frequency band for QAM transmissions over mobile channels.

Huish discussed the use of QAM over fixed links, which was becoming increasingly widespread [46]. Webb *et al.* presented two papers describing the problems of square QAM constellations when used for mobile radio transmissions and introduced the star QAM constellation with its inherent robustness in fading channels [47, 48].

There followed a number of publications by Webb *et al.* describing a number of techniques for enhancing QAM transmissions over mobile radio channels. All of these techniques are described in detail in Chapters 10 to 15 of this book. In December 1991 a paper appeared in the *IEE Proceedings* [49] which considered the effects of channel coding, trellis coding and block coding when applied to the star QAM constellation. This was followed by another paper in the *IEE Proceedings* [50] considering equaliser techniques for QAM transmissions over dispersive mobile radio channels. A review paper appearing in July 1992 [51] considered areas where QAM could be put to most beneficial use within the mobile radio environment, and concluded that its advantages would be greatest in microcells. Further work on spectral efficiency, particularly of multilevel modulation schemes [52] concluded that variable level QAM modulation was substantially more efficient than all the other modulation schemes simulated. Variable level QAM was first discussed in a paper by Steele and Webb in 1991 [53].

Further QAM schemes for hostile fading channels characteristic of mobile telephony can be found in the following recent references [1, 1, 12, 54–86]. If Feher's previously mentioned NLA concept cannot be applied, then power-inefficient class A or AB linear amplification has to be used, which might become an impediment in lightweight, low-consumption handsets. However, the power consumption of the low-efficiency class A amplifier [63, 64] is less critical than that of the digital speech and channel codecs. In many applications 16-QAM, transmitting 4 bits per symbol reduces the signalling rate by a factor of 4 and hence mitigates channel dispersion, thereby removing the need for an equaliser, while the higher SNR demand can be compensated by diversity reception.

Significant contributions were made by Cavers, Stapleton *et al.* at Simon Fraser University, Burnaby, Canada in the field of pre- and post-distorter design. Out-of-band emissions due to class AB amplifier non-linearities and hence adjacent channel interferences can be reduced by some 15-20 dB using Stapleton's adaptive predistorter [1, 1, 65] and a class AB amplifier with 6 dB back-off, by adjusting the predistorter's adaptive coefficients using the complex convolution of the predistorter's input signal and the amplifier's output signal. Further aspects of linearised power amplifier design are considered in references [66] and [67].

A further important research trend is hallmarked by Cavers' work targeted at pilot symbol assisted modulation (PSAM) [12], where known pilot symbols are inserted in the information stream in order to allow the derivation of channel measurement information. The recovered received symbols are then used to linearly predict the channel's attenuation and phase. This arrangement will be considered in Chapter 11. A range of advanced QAM modems have also been proposed by Japanese researchers doing cutting-edge research in the field, including Sampei, Sunaga [68,69], Adachi [70] and Sasaoka [62].

Since QAM research has reached a mature stage, a number of mobile speech, audio and video transmission schemes have been proposed [85–95]. These system design examples demonstrated that substantial system performance benefits accrue, when the entire system is jointly optimised, rather than just a conglomerate of independent system components. Some of these system aspects will be considered in Chapter 17 in the context of voice communications. A range of digital video broadcasting (DVB) schemes will be the topic of Chapter 24.

1.3 Orthogonal Frequency Division Multiplexing Based QAM

1.3.1 History of orthogonal frequency division multiplexing

The first QAM-related so-called orthogonal frequency division multiplexing (OFDM) scheme was proposed by Chang in 1966 [71] for dispersive fading channels, which has also undergone a dramatic evolution due to the efforts of Weinstein, Peled, Ruiz, Hirosaki, Kolb, Cimini, Schüssler, Preuss, Rückriem, Kalet *et al.* [71–84]. OFDM was standardised as the European digital audio broadcast (DAB) as well as digital video broadcast (DVB) scheme. It constituted also a credible proposal for the recent third-generation mobile radio standard competition in Europe. It was recently selected as the high performance local area network (HIPERLAN) transmission technique.

The system's operational principle is that the original bandwidth is divided in a high number of narrow sub-bands, in which the mobile channel can be considered non-dispersive. Hence no channel equaliser is required and instead of implementing a bank of sub-channel modems they can be conveniently implemented by the help of a single fast fourier Transformer (FFT). This scheme will be the topic of Chapters 18-23.

These OFDM systems - often also termed as frequency division multiplexing (FDM) or multi-tone systems - have been employed in military applications since the 1960s, for example by Bello [96], Zimmerman [72], Powers and Zimmerman [97], Chang and Gibby [98] and others. Saltzberg [99] studied a multi-carrier system employing orthogonal time–staggered quadrature amplitude modulation (O-QAM) of the carriers.

The employment of the discrete Fourier transform (DFT) to replace the banks of sinusoidal generators and the demodulators was suggested by Weinstein and Ebert [73] in 1971, which significantly reduces the implementation complexity of OFDM modems. In 1980, Hirosaki [84] suggested an equalisation algorithm in order to suppress both intersymbol and intersubcarrier interference caused by the channel impulse response or timing and frequency errors. Simplified OFDM modem implementations were studied by Peled [77] in 1980, while Hirosaki [78] introduced the DFT based implementation of Saltzberg's O-QAM OFDM system. From Erlangen University, Kolb [79], Schüßler [80], Preuss [81] and Rückriem [82] conducted further research into the application of OFDM. Cimini [74] and Kalet [83] published analytical and early seminal experimental results on the performance of OFDM modems in mobile communications channels.

More recent advances in OFDM transmission were presented in the impressive state-of-the-art collection of works edited by Fazel and Fettweis [100], including the research by Fettweis *et al.* at Dresden University, Rohling *et al.* at Braunschweig University, Vandendorp at Loeven University, Huber *et al.* at Erlangen University, Lindner *et al.* at Ulm University, Kammeyer *et al.* at Brehmen University and Meyr *et al.* [101, 102] at Aachen University, but the individual contributions are too numerous to mention.

While OFDM transmission over mobile communications channels can alleviate the problem of multipath propagation, recent research efforts have focused on solving a set of inherent difficulties regarding OFDM, namely the peak–to–mean power ratio, time and frequency synchronisation, and on mitigating the effects of the frequency selective fading channel. These issues are addressed below in slightly more depth, while a treatment is given in Chapters 18-23.

1.3.2 Peak-to-mean power ratio

It is plausibe that the OFDM signal - which is the superposition of a high number of modulated sub-channel signals - may exhibit a high instantaneous signal peak with respect to the average signal level. Furthermore, large signal amplitude swings are encountered, when the time domain signal traverses from a low instantaneous power waveform to a high power waveform, which may results in a high out-of-band (OOB) harmonic distortion power, unless the transmitter's power amplifier exhibits an extremely high linearity across the entire signal level range (Section 4.5.1). This then potentially contaminates the adjacent channels with adjacent channel interference. Practical amplifiers exhibit a finite amplitude range, in which they can be considered almost linear. In order to prevent severe clipping of the high OFDM signal peaks - which is the main source of OOB emissions - the power amplifier must not be driven into saturation and hence they are typically operated with a certain so-called back-off, creating a certain "head room" for the signal peaks, which reduces the risk of amplifier saturation and OOB emmission. Two different families of solutions have been suggested in the literature, in order to mitigate these problems, either reducing the peak-to-mean power ratio, or improving the amplification stage of the transmitter.

More explicitly, Shepherd [103], Jones [104], and Wulich [105] suggested different coding techniques which aim to minimise the peak power of the OFDM signal by employing different data encoding schemes before modulation, with the philosophy of choosing block codes whose legitimate code words exhibit low so-called Crest factors or peak-to-mean power envelope fluctuation. Müller [106], Pauli [107], May [108] and Wulich [109] suggested different algorithms for post-processing the time domain OFDM signal prior to amplification, while Schmidt and Kammeyer [110] employed adaptive subcarrier allocation in order to reduce the crest factor. Dinis and Gusmão [111–113] researched the use of two-branch amplifiers, while the clustered OFDM technique introduced by Daneshrad, Cimini and Carloni [114] operates with a set of parallel partial FFT processors with associated transmitting chains. OFDM systems with increased robustness to non-linear distortion have been proposed by Okada, Nishijima and Komaki [115] as well as by Dinis and Gusmão [116].

1.3.3 Synchronisation

Time and frequency synchronisation between the transmitter and receiver are of crucial importance as regards to the performance of an OFDM link [117, 118]. A wide variety of techniques have been proposed for estimating and correcting both timing and carrier frequency offsets at the OFDM receiver. Rough timing and frequency acquisition algorithms relying on known pilot symbols or pilot tones embedded into the OFDM symbols have been suggested by Claßen [101], Warner [119], Sari [120], Moose [121], as well as Brüninghaus and Rohling [122]. Fine frequency and timing tracking algorithms exploiting the OFDM signal's cyclic extension were published by

Moose [121], Daffara [123] and Sandell [124].

1.3.4 OFDM/CDMA

Combining OFDM transmissions with code division multiple access (CDMA) allows us to exploit the wideband channel's inherent frequency diversity by spreading each symbol across multiple subcarriers. This technique has been pioneered by Yee, Linnartz and Fettweis [125], by Chouly, Brajal and Jourdan [126], as well as by Fettweis, Bahai and Anvari [127]. Fazel and Papke [128] investigated convolutional coding in conjunction with OFDM/CDMA. Prasad and Hara [129] compared various methods of combining the two techniques, identifying three different structures, namely multi-carrier CDMA (MC-CDMA), multi-carrier direct sequence CDMA (MC-DS-CDMA) and multi-tone CDMA (MT-CDMA). Like non-spread OFDM transmission, OFDM/CDMA methods suffer from high peak-to-mean power ratios, which are dependent on the frequency domain spreading scheme, as investigated by Choi, Kuan and Hanzo [130].

1.3.5 Adaptive antennas

Combining adaptive antenna techniques with OFDM transmissions was shown to be advantageous in suppressing co-channel interference in cellular communications systems. Li, Cimini and Sollenberger [131–133], Kim, Choi and Cho [134], Lin, Cimini and Chuang [135] as well as Münster *et al.* [136] have investigated algorithms for multi-user channel estimation and interference suppression.

1.3.6 OFDM applications

Due to their implementational complexity, OFDM applications have been scarce until quite recently. Recently, however, OFDM has been adopted as the new European digital audio broadcasting (DAB) standard [75, 76, 137–139] as well as for the terrestrial digital video broadcasting (DVB) system [120, 140].

For fixed-wire applications, OFDM is employed in the asynchronous digital subscriber line (ADSL) and high-bit-rate digital subscriber line (HDSL) systems [141–144] and it has also been suggested for power line communications systems [145, 146] due to its resilience to time dispersive channels and narrow band interferers.

More recently, OFDM applications were studied within the European 4th Framework Advanced Communications Technologies and Services (ACTS) programme [147]. The MEDIAN project investigated a 155 Mbps wireless asynchronous transfer mode (WATM) network [148–151], while the Magic WAND group [152, 153] developed a wireless local area network (LAN). Hallmann and Rohling [154] presented a range of different OFDM systems that were applicable to the European Telecommunications Standardisation Institute's (ETSI) recent personal communications oriented air interface concept [155].

1.4 Summary

This concludes our introduction to QAM and the review of publications concerning QAM, spanning a period of over thirty years between the first theoretical study for Gaussian channels through to implementation within microcellular mobile radio environments. We now embark on a detailed investigation of the topics introduced in this chapter.

1.5 Outline of Topics

In Chapter 2 we consider the communications channels over which we wish to send our data. These channels are divided into Gaussian and mobile radio channels, and the characteristics of each are explained.

Chapter 3 forms an introduction to modems, considering the manner in which speech or other source waveforms are converted into a form suitable for transmission over a channel, and introducing some of the fundamentals of modems.

Chapter 4 provides a more detailed description of modems, specifically that of the modulator, considering QAM constellations, pulse shaping techniques, methods of generating and detecting QAM, as well as amplifier techniques to reduce the problems associated with non-linearities.

Chapter 5 provides details of decision theory and highlights the theoretical aspects of QAM transmission, showing how the BER can be mathematically computed for transmission over Gaussian channels.

Chapter 6 considers clock and carrier recovery schemes whilst Chapter 7 looks at equalisers. Chapter 8 considers trellis coded modulation and provides an introduction to Chapter 9 which deals with modems for Gaussian channels such as telephone lines.

Chapter 10 starts the section on mobile radio by providing a theoretical analysis of QAM transmission over Rayleigh fading mobile radio channels. Chapter 11 then considers some of the practicalities of QAM transmission over these links, including constellation distortions and hardware imperfections.

Chapter 12 details clock and carrier recovery schemes for mobile radio, Chapter 13 considers an advanced variable rate QAM scheme, while Chapter 14 proposes various equalisers for wideband transmissions. In Chapter 15 we consider a range of orthogonal transmission and pulse shaping techniques in the context of quadrature-quadrature amplitude modulation.

Chapter 16 considers the spectral efficiency gains that can be achieved when using QAM instead of conventional binary modulation techniques, whilst Chapter 17 provides details of a plethora of practical QAM speech communications systems and compares these with their equivalent binary systems.

In Chapter 18 we focus our attention on Fourier transforms in order to implement orthogonal frequency division multiplexing schemes. In Chapter 19 the BER performance of OFDM modems in AWGN channels is studied for a set of different modulation schemes in the subcarriers. The effects of amplitude limiting of the transmitter's output signal, caused by a simple clipping amplifier model, and of finite resolution D/A and A/D conversion on the system performance are investigated.

Oscillator phase noise is considered as a source of intersubcarrier interference and its effects on the system performance are demonstrated.

In Chapter 20 the effects of time-dispersive frequency-selective Rayleigh fading channels on OFDM transmissions are demonstrated. Channel estimation techniques are presented which support the employment of coherent detection in frequency selective channels. Additionally, differential detection is investigated, and the resulting system performance over the different channels is compared.

Chapter 21 focuses our attention on the time and frequency synchronisation requirements of OFDM transmissions and the effects of synchronisation errors are demonstrated. Two novel synchronisation algorithms for frame and OFDM symbol synchronisation are suggested and compared. The resulting system performance over fading wideband channels is examined.

In Chapter 22, based on the results of Chapter 20, the employment of adaptive modulation schemes is suggested for duplex point-to-point links over frequency-selective time-varying channels. Different bit allocation schemes are investigated and a simplified sub-band adaptivity OFDM scheme is suggested for alleviating the associated signalling constraints. A range of blind modulation scheme detection algorithms are also investigated and compared. The employment of long-block-length convolutional turbo codes is suggested for improving the system's throughput and the turbo coded adaptive OFDM modem's performance is compared using different sets of parameters. Then the effects of using pre-equalisation at the transmitter are examined, and a set of different pre-equalisation algorithms is introduced. A joint pre-equalisation and adaptive modulation algorithm is proposed and its BER and throughput performance is studied.

In Chapter 23 the adaptive OFDM transmission ideas of Chapter 22 are extended further, in order to include adaptive error correction coding, based on redundant residual number system (RRNS) and turbo BCH codes. A joint modulation and code rate adaptation scheme is presented.

Lastly, in Chapter 24 a variety of advanced QAM and OFDM assisted turbo-coded DVB schemes are proposed and analysed. Specifically, it compares the performance of schemes that use blind-equalised QAM modems. It is demonstrated that in comparison to the standard-based DVB systems the employment of turbo coding can provide an extra 5-6 dB channel SNR gain and this can be exploited for example to double the number of transmitted bits in a given bandwidth. This then ultimately allows us to improve for example the associated video quality that can be guaranteed in a given bandwidth at the cost of the associated additional implementational complexity.

Chapter 2

Communications Channels

In this chapter we consider the communication channel which exists between the transmitter and the receiver. Accurate characterisation of this channel is essential if we are to remove the impairments imposed by the channel using signal processing at the receiver. Initially we consider fixed links whereby both terminals are stationary, and then mobile radio communications channels which change significantly with time are investigated.

2.1 Fixed Communication Channels

2.1.1 Introduction

We define fixed communications channels to be those between a fixed transmitter and a fixed receiver. These channels are exemplified by twisted pairs, cables, waveguides, optical fibre and point-to-point microwave radio channels. Whatever the nature of the channel, its output signal differs from the input signal. The difference might be deterministic or random, but it is typically unknown to the receiver. Examples of channel impairments are dispersion, non-linear distortions, delay, and random noise.

Fixed communications channels can often be modelled by a linear transfer function, which describes the channel dispersion. The ubiquitous additive Gaussian noise (AWGN) is a fundamental limiting factor in communications via linear time invariant (LTI) channels. Although the channel characteristics might change due to ageing, temperature changes, channel switching, etc., these variations will not be apparent over the course of a typical communication. It is this inherent time invariance which characterises fixed channels.

An ideal, distortion-free communications channel would have a flat frequency response and linear phase response over the bandwidth (B) of the signals to be transmitted, as seen in Figure 2.1. In this figure $A(\omega)$ represents the magnitude of the channel response at frequency ω, and ωT represents the phase shift at frequency ω. Practical channels always have some linear distortions, i.e. a non-flat frequency re-

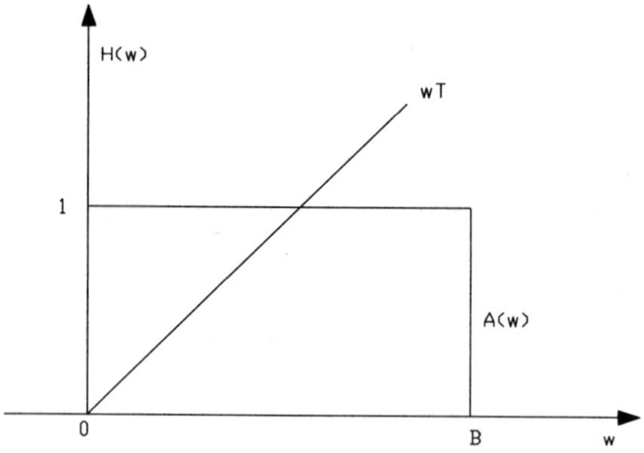

Figure 2.1: Ideal, distortion-free channel model

sponse and a non-linear phase response. In addition, the group delay response of the channel, which is the derivative of the phase response, is often given.

2.1.2 Fixed channel types

Conventional telephony uses twisted pairs to connect subscribers to the local exchange. The bandwidth is approximately 3.4 kHz, and the waveform distortions are relatively benign.

For applications requiring a higher bandwidth coaxial cables can be used. Their attenuation increases approximately with the square root of the frequency. Hence for wide band, long-distance operation they require channel equalisation. Typically, coaxial cables can provide a bandwidth of about 50 MHz, and the transmission rate they can support is limited by the so-called skin effect.

Recent advances in optical communications, such as the invention of the optical amplifiers, render optical fibres more attractive for higher bit rates. Guided optical wave transmission fibres use fine strands of silicon-based, high purity glass. For short-haul systems the so-called step index fibres are used where the higher refractive index internal core is surrounded by a lower refractive index external cladding. Their effective bandwidth is limited by their length and the difference between the refractive index of the core and that of the cladding. For long-distance, high rate applications the so-called graded index fibres are more suitable, where the refractive index gradually reduces from the central core towards the cladding. Their relative loss expressed in dB/km versus wavelength characteristic shows a minimum around 1550 nm. This wavelength also allows the design of low dispersion fibres.

Point-to-point microwave radio channels typically utilise high-gain directional transmit and receive antennae in a line-of-sight scenario, where free space propagation conditions may be applicable.

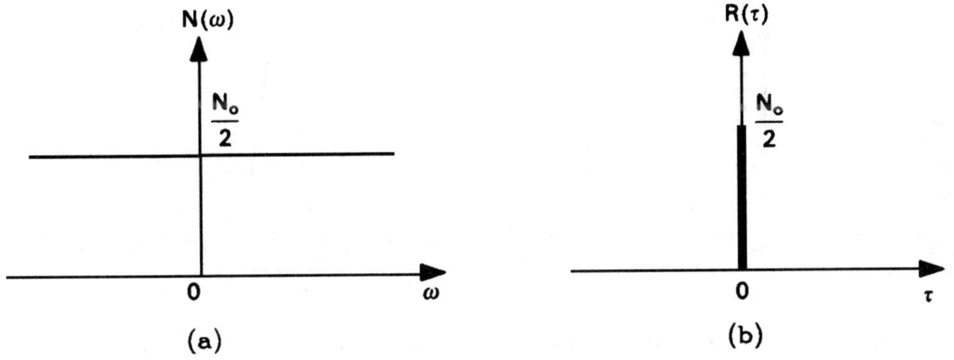

Figure 2.2: Power spectral density and autocorrelation of WN

2.1.3 Characterisation of noise

Irrespective of the communications channel used, random noise is always present. Noise can be broadly classified as natural or man-made. Examples of man-made noise are those due to electrical appliances, lighting, etc., and the effects of those sources can usually be mitigated at the source. Natural noise sources affecting radio transmissions include galactic star radiations and atmospheric noise. There exists a low noise window in the range of 1–10 GHz, where the effects of these sources are minimised.

Natural thermal noise is ubiquitous. This is due to the random motion of electrons and can be reduced by reducing the temperature. Since thermal noise contains practically all frequency components up to some 10^{13} Hz with equal power, it is often referred to as white noise (WN) in an analogy to white light containing all colours with equal intensity. This WN process can be characterised by its uniform power spectral density (PSD) $N(\omega) = N_0/2$ shown together with its autocorrelation function (ACF) in Figure 2.2. The ACF $R(\tau)$ can be computed using the Wiener-Khintchine theorem, which states that $N(\omega)$ and $R(\tau)$ are Fourier transform pairs such that:

$$\begin{aligned} R(\tau) &= \frac{1}{2\pi}\int_{-\infty}^{\infty} N(\omega)e^{j\omega\tau}d\omega = \frac{1}{2\pi}\int_{-\infty}^{\infty}\frac{N_0 e^{j\omega\tau}}{2}d\omega \\ &= \frac{N_0}{2}\frac{1}{2\pi}\int_{-\infty}^{\infty} e^{j\omega\tau}d\omega = \frac{N_0}{2}\delta(\tau), \end{aligned} \tag{2.1}$$

where $\delta(\tau)$ is the Dirac delta function. Clearly, for any time domain shift $\tau > 0$ the noise is uncorrelated.

Bandlimited communications systems bandlimit not only the signal, but the noise as well, and this filtering limits the rate of change of the time domain noise signal, introducing some correlation over the interval of $\pm 1/2B$. The stylised PSD and ACF of bandlimited white noise are displayed in Figure 2.3. With bandlimiting the

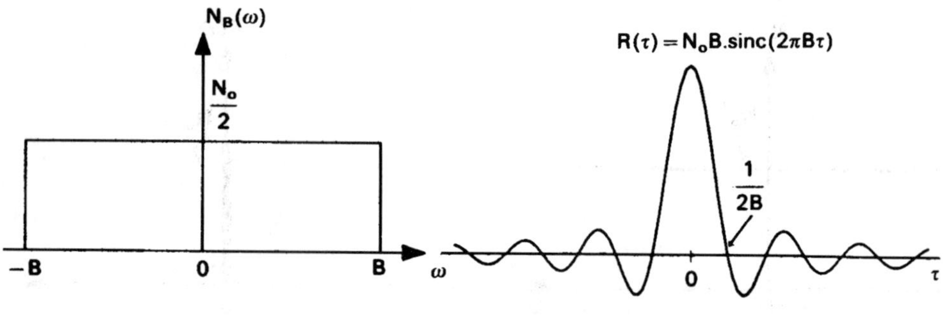

Figure 2.3: Power spectral density and autocorrelation of bandlimited WN

autocorrelation function becomes

$$\begin{aligned} R(\tau) &= \frac{1}{2\pi}\int_{-B}^{B}\frac{N_0}{2}e^{j\omega\tau}d\omega = \frac{N_0}{2}\int_{-B}^{B}e^{j2\pi f\tau}df \\ &= \frac{N_0}{2}\left[\frac{e^{j2\pi f\tau}}{j2\pi\tau}\right]_{-B}^{B} = N_0 B\frac{\sin(2\pi B\tau)}{2\pi B\tau}. \end{aligned} \tag{2.2}$$

In the time domain the amplitude distribution of the white thermal noise has a so-called normal or Gaussian distribution and since it is inevitably added to the received signal, it is usually referred to as additive white Gaussian noise (AWGN). Note that AWGN is therefore the noise generated in the receiver. The probability density function (PDF) is the well-known bell-shaped curve of the Gaussian distribution, given by:

$$p(x) = \frac{1}{\sigma\sqrt{2\pi}}e^{-(x-m)^2/2\sigma^2}, \tag{2.3}$$

where m is the mean and σ^2 is the variance. The effects of AWGN can be mitigated by increasing the transmitted signal power and thereby reducing the relative effects of noise. The signal-to-noise ratio (SNR) at the receiver's input provides a good measure of the received signal quality. This SNR is often referred to as the channel SNR.

If an analogue channel's bandwidth is B and the prevailing SNR is known, then according to the Shannon-Hartley theorem the maximum transmission rate at which information can be sent without errors via this channel is given by:

$$C = B\log_2(1+SNR)\ [\text{bit/sec}], \tag{2.4}$$

where C is referred to as the channel capacity. The Shannon-Hartley law has a number of fundamental consequences, of which we emphasise here only one. Namely, expressing the ratio C/B in terms of SNR we have:

$$\frac{C}{B} = \log_2(1+SNR)\left[\frac{\text{bit/sec}}{\text{Hz}}\right], \tag{2.5}$$

SNR		C/B
Ratio	dB	bit/sec/Hz
1	0	1
3	4.8	2
7	8.5	3
15	11.8	4
31	14.9	5
63	18.0	6
127	21.0	7

Table 2.1: Relative channel capacity versus SNR

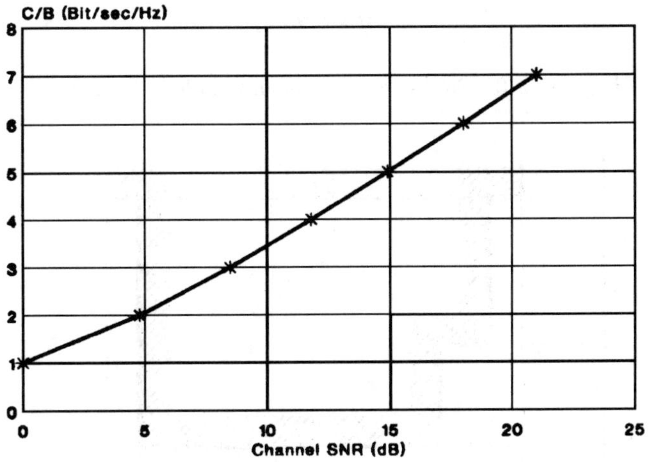

Figure 2.4: Relative channel capacity versus SNR

and selecting the SNR values seen in Table 2.1 we arrive at the near-linear curve seen in Figure 2.4.

It can be seen that the channel capacity for a given bandwidth B is nearly linearly proportional to the SNR. Consequently, using a binary signalling scheme with low C/B ratio when high SNR values are available results in an inefficient use of the channel. This fact provides the main motivation for using the multilevel modulation schemes discussed in this book.

2.2 Telephone Channels

The ubiquitous telephone lines play an important role as data transmission circuits. Although their attenuation and group delay characteristics vary widely, the international leased telephone lines commonly used for multilevel data transmission must comply with the CCITT recommendations, particularly M1020 and M1025. In gen-

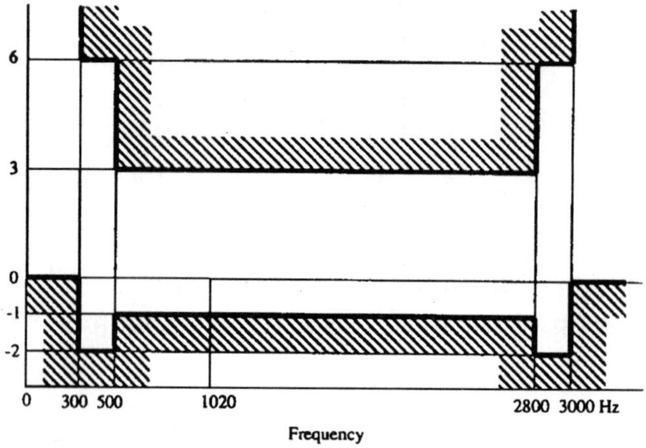

Figure 2.5: Circuit loss as recommended by M1020

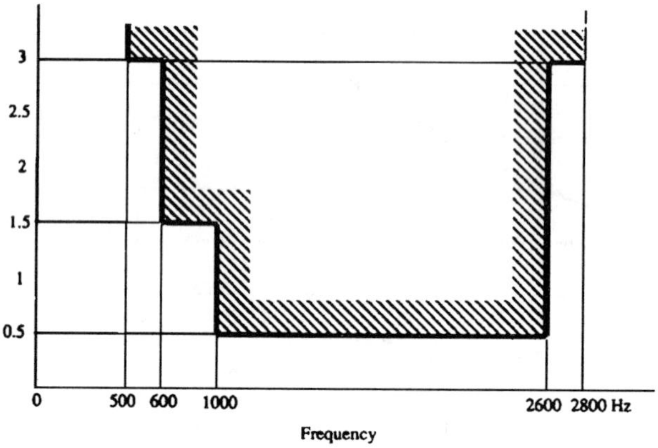

Figure 2.6: Group delay as recommended by M1020

eral, the overall loss is unknown, but should be less than -13 dB relative (dBr) for four-wire circuits, with a variation of less than ±4 dB. The attenuation and group delay characteristics as specified by the M1020 and M1025 recommendations are given in Figures 2.5, 2.6, 2.7 and 2.8.

When multilevel modems are used for data transmission over these lines as described in Chapter 9, the number of sudden amplitude hits greater than ±2 dB should not exceed 10 in any 15 minutes. The psophometrically weighted noise power for lines longer than 10,000 km must be below -38 dBm, where 1 dBm is related to 1 mW measured across a load resistance of 600 ohms. However, for shorter lines the psophometrically weighted noise power must be proportionally lower, e.g. about -50 dBm at 1,000 km. The number of impulse noise peaks exceeding -21 dBm should not be more than 18 in 15 minutes. The maximum allowed phase jitter is limited to 15 deg peak-

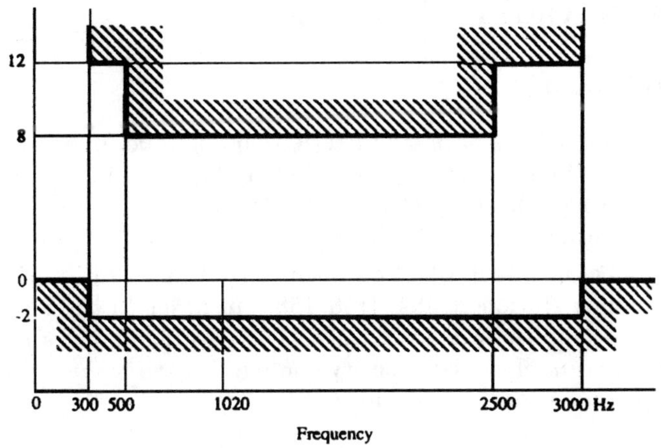

Figure 2.7: Circuit loss as recommended by M1025

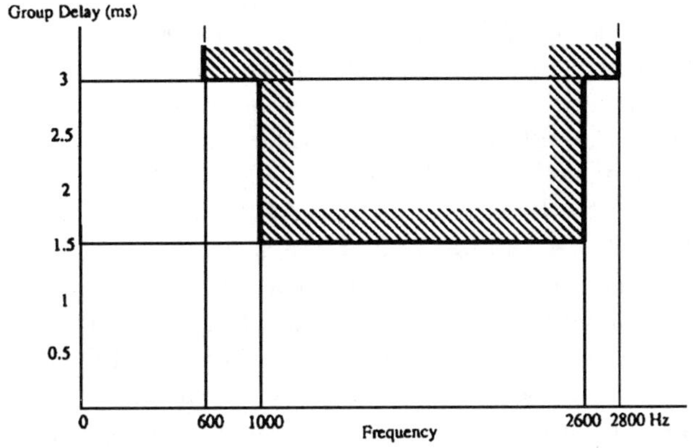

Figure 2.8: Group delay as recommended by M1025

to-peak. The signal-to-total-distortion power ratio should be better than 28 dB when measured with a 1020 Hz -10 dBm sine wave. The level of single tone interference must be at least 3 dB below the random noise level while the maximum frequency error is specified as ± 5 Hz.

Further standards, such as M1030, impose less stringent requirements and are applicable to lower transmission rates or to private switched networks.

Having briefly considered fixed communications channels, we now consider the more complex mobile radio channels.

2.3 Mobile Radio Channels

2.3.1 Introduction

The mobile radio channel has been characterised in a number of excellent treatises by Jakes [156], Lee [157], Steele [158], Parsons [159] and others. In this section we follow a similar approach to Steele [158], Greenwood, Hanzo [160] and Cheung [161], while maintaining the treatment of mobile channels as simple and practical as possible.

Mobile radio links are established between a fixed base station (BS) and a number of roaming mobile stations (MSs) [156–158]. In order to cover a large area it is necessary to have a number of transmitters. The coverage area of each transmitter is defined as the area in which satisfactory communications between the mobile and the transmitter can be achieved, and this is known as a cell. First and second generation mobile systems are using cells with radii up to some 35 km, which implies that the existence of a line-of-sight (LOS) path between transmitter and receiver will be rare. Nearby cells are assigned different frequencies, but as the distance between cells increases, the interference between them reduces, and it becomes possible to reuse frequencies, thus increasing system capacity.

In areas where the generated traffic density is high or when only low power handhelds are available, such as the British Cordless Telecommunications system CT-2 or the Digital European Cordless Telecommunications (DECT) system, the cell size is necessarily reduced. If the BS antenna is mounted below rooftop level of surrounding buildings, i.e. fixed to the side of a building, the resulting radio cell is termed a microcell, and typically has a radius of some 200-400 m [162, 163]. In such cells, due to the proximity to the BS, there is an increased likelihood of a LOS path with the resulting effect that multipath fading is often less severe.

Propagation mechanisms can be divided into three types, distance effects, slow and fast fading. In this section we will neglect the slow fading caused by large obstructions. The distance effect is simply that as the separation between a BS and a MS increases, the received mean signal level tends to decrease. Over distances of a few metres the mean signal level is essentially constant, but the instantaneous signal level can vary rapidly by amounts typically up to 40 dB. These rapid variations are known as fast fading. This is because the MS receives a number of versions of the signal transmitted by the BS, which have been reflected and diffracted by buildings, mountains and other obstructions and have a range of delays, attenuations and frequency shifts. These various signals are all summed at the MS antenna. When this summation is constructive, because all paths are received with the same phase change, the received signal level is enhanced. However, when the multipath signals sum destructively, because of phase changes which cause cancellation, the received signal level is much reduced and the receiver is said to be in a fade. Because differences of 180 deg in the carrier phase could change the interference from constructive to destructive, and at a carrier frequency of 2 GHz this corresponds to some 7 cm, the change in instantaneous signal level can be rapid and so is termed fast fading. As previously suggested, these fades typically occur every half-wavelength.

In mobile communications, digital information is transmitted at a certain bit rate. The propagation phenomena are highly dependent on the ratio of the symbol duration to the delay spread of the time variant radio channel. The delay spread may be defined as the length of the channel's impulse response. We can see that if we transmit data at a slow rate the individual symbols can be resolved at the receiver. This is because the numerous reflected paths all arrive at the receiver before the next symbol is transmitted. However, if we increase the transmitted data rate, a point will be reached where this no longer occurs and each data symbol significantly spreads into adjacent symbols, a phenomenon known as intersymbol interference (ISI). Without the use of channel equalisers to remove the ISI, the bit error rate (BER) may become unacceptably high.

Measurements show that cellular radio networks using large cells, where the excess delay spread may exceed 10 μ s, need equalisers when the bit rate is relatively low, say 64 kb/s, while cordless communications in buildings where the excess delay spread is often significantly below a microsecond may exhibit flat fading when the bit rate exceeds a 1 Mb/s. Very small cells, sometimes referred to as picocells may support many megabits per second without equalisation because the delay spread is only tens of nanoseconds.

2.3.2 Equivalent baseband and passband systems

When modelling mobile radio systems it is advantageous to model them at baseband rather than radio frequencies as this dramatically simplifies the analysis. In this section we show how baseband equivalent models can be developed, following the approach of Proakis [164].

In general the modulated signal $s(t)$ occupies a relatively narrow bandwidth B, when compared to the carrier frequency f_c, where $f_c \gg B$, and can be represented by:

$$s(t) = a(t)\cos[2\pi f_c t + \Theta(t)] = \Re\{a(t)e^{j[\omega_c t + \Theta(t)]}\}, \tag{2.6}$$

where $\Re$ represents the real part of a complex expression. The modulated signal's amplitude is represented by $a(t)$ and its phase by $\Theta(t)$. The carrier f_c is amplitude modulated by the signal $a(t)$ and phase modulated by the signal $\Theta(t)$. It is common practice to refer to the modulated radio signal $s(t)$ as a narrowband bandpass signal. In computer simulations, however, it is impractical to directly model the above equation. This can be appreciated by the following example.

If we have a signal bandwidth of 16 kHz transmitted at 2 GHz, then for every cycle of our baseband signal, the RF signal goes through $(2 \times 10^9)/(16 \times 10^3) = 1.25 \times 10^5$ cycles. To simulate each cycle of the RF carrier, we must oversample by a minimum of a factor of 2. This means that an order of a quarter of a million operations are required for every baseband symbol transmitted, in addition to the simulation of RF mixers, filters, etc.

Fortunately, we can translate the bandpass signal to an equivalent baseband signal with no loss of accuracy. Expanding Equation 2.6 leads to

$$\begin{aligned} s(t) &= a(t)\cos\Theta(t)\cos(2\pi f_c t) - a(t)\sin\Theta(t)\sin(2\pi f_c t) \\ &= u_I\cos(2\pi f_c t) - u_Q\sin(2\pi f_c t) \end{aligned} \tag{2.7}$$

where u_I and u_Q are termed the quadrature components of $s(t)$ and modulate quadra-

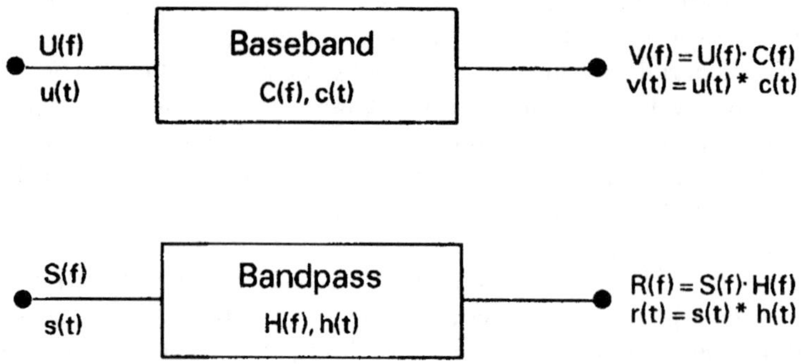

Figure 2.9: Equivalent baseband and passband channels

ture carriers, and therefore are defined as

$$\begin{aligned} u_I &= a(t)\cos\Theta(t) \\ u_Q &= a(t)\sin\Theta(t). \end{aligned} \tag{2.8}$$

This is a Cartesian or rectangular representation of the baseband signal $u(t)$ with u_I and u_Q corresponding to in-phase (I) and quadrature (Q) signals. There is an equivalent polar representation which is often convenient, given by

$$u(t) = a(t)e^{j\Theta(t)}\,, \tag{2.9}$$

where $u(t) = u_I(t) + ju_Q(t)$ is the complex baseband envelope or complex lowpass equivalent of $s(t)$. When combining Equations 2.6 and 2.9, this leads to

$$s(t) = \Re\left[u(t)e^{j2\pi f_c t}\right] \tag{2.10}$$

where $\Re[\;]$ represents the real part of $[\;]$. This equation implies that the knowledge of $u(t)$ and f_c uniquely describes the modulated signal $s(t)$, where all the useful information is conveyed by $u(t)$.

In order to establish the validity of the complex baseband model for studying bandpass systems, we will show that the response of a bandpass system to a bandpass signal is identical to the response of the equivalent baseband system to the equivalent baseband signal. This confirms that any operations such as filtering, which are carried out at RF in hardware, can be represented as baseband operations in software, and thus simulations can be carried out at baseband. In order to show this, we take a bandpass modulated signal $s(t)$ with its equivalent lowpass signal $u(t)$ as defined in Equation 2.9. We then allow $s(t)$ to excite a narrowband bandpass system having impulse response $h(t)$ and equivalent baseband impulse response $c(t)$. The baseband equivalent response $c(t)$ is derived from $h(t)$ in the same way as $u(t)$ was derived from $s(t)$. The notation is summarised in Figure 2.9.

The time domain response $r(t)$ of the bandpass system is given by:

$$r(t) = \int_{-\infty}^{\infty} s(\tau)h(t-\tau)\, d\tau. \tag{2.11}$$

This is the well-known convolution integral and is often simply written as

$$r(t) = s(t) * h(t), \tag{2.12}$$

where * indicates convolution. When translating Equation 2.12 to the frequency domain we have:

$$R(f) = S(f)H(f), \tag{2.13}$$

where $S(f)$ is the Fourier transform of $s(t)$, given by:

$$S(f) = \int_{-\infty}^{\infty} s(t)e^{-j2\pi ft}dt. \tag{2.14}$$

Upon substituting $s(t)$ from Equation 2.10 into Equation 2.14 we can write:

$$S(f) = \int_{-\infty}^{\infty} \left\{\Re\left[u(t)e^{j2\pi f_c t}\right]\right\} e^{-j2\pi ft}dt. \tag{2.15}$$

Noting the identity

$$\Re(z) = \frac{1}{2}(z + z^*), \tag{2.16}$$

where the asterisk represents the complex conjugate and taking into account that $\mathcal{F}\{u^*(t)\} = U^*(-f)$, where $\mathcal{F}$ represents the Fourier transform, we can rewrite Equation 2.15 as:

$$\begin{aligned} S(f) &= \frac{1}{2}\int_{-\infty}^{\infty} \left[u(t)e^{j2\pi f_c t} + u^*(t)e^{-j2\pi f_c t}\right] e^{-j2\pi ft}dt \\ &= \frac{1}{2}\left[U(f-f_c) + U^*(-f-f_c)\right], \end{aligned} \tag{2.17}$$

where $U(f)$ is the Fourier transform of $u(t)$. The spectra $U(f)$ of $u(t)$ and $S(f)$ of $s(t)$ are shown in Figure 2.10. Observe that the multiplication of $u(t)$ by $e^{j2\pi f_c t}$ in Equation 2.15 shifts the baseband spectrum $U(f)$ to $\pm f_c$ and its conjugate complex symmetry implies that the modulated signal $s(t)$ is real. It is worth noting that since $f_c \gg B$, the energy is concentrated in a tight band around the carrier frequency.

In a similar fashion, the conjugate complex symmetric frequency response $H(f)$ of the bandpass system can be described by the help of shifted versions of the equivalent baseband system's frequency response. Since the bandpass system's impulse response $h(t)$ is real, $H(f)$ is conjugate complex symmetric, i.e. $H(f) = H^*(-f)$. Defining

$$C(f-f_c) = \begin{cases} H(f) & \text{if } f > 0 \\ 0 & \text{if } f \leq 0 \end{cases} \tag{2.18}$$

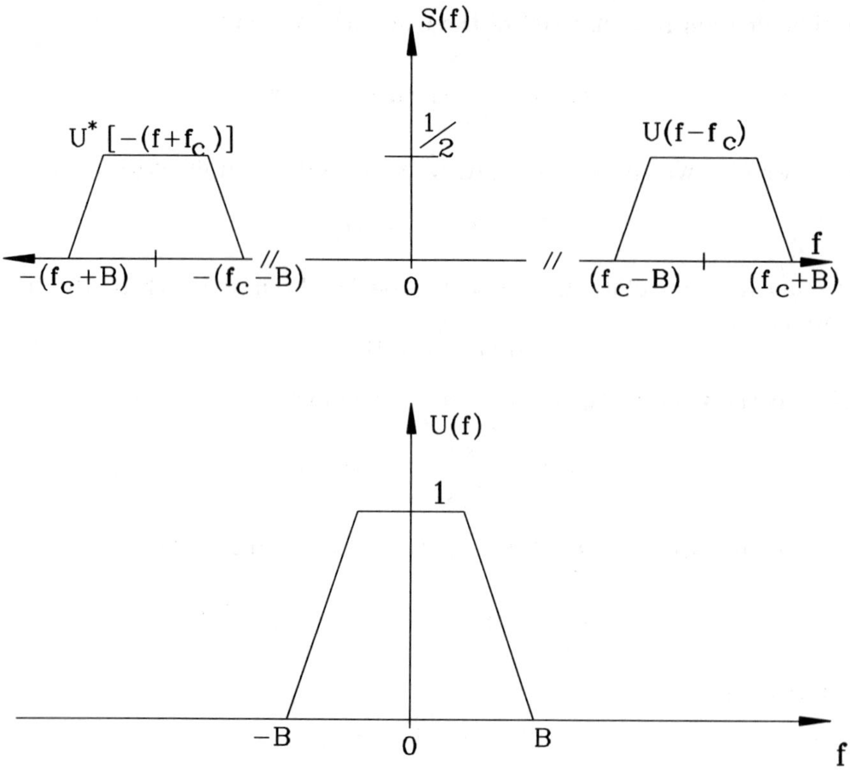

Figure 2.10: Relationship between the baseband spectrum $U(f)$ and the bandpass spectrum $S(f)$

and

$$C^*[-(f+f_c)] = \begin{cases} 0 & \text{if } f \geq 0 \\ H^*(-f) & \text{if } f < 0 \end{cases}, \tag{2.19}$$

as seen in Figure 2.11 allows us to express $H(f)$ as

$$H(f) = C(f - f_c) + C^*(-f - f_c). \tag{2.20}$$

When returning to the time domain the relationship of the baseband and passband impulse responses is given by:

$$\begin{aligned} h(t) &= c(t)e^{j\omega_c t} + c^*(t)e^{-j\omega_c t} \\ &= 2\Re\left\{c(t)e^{j\omega_c t}\right\}, \end{aligned} \tag{2.21}$$

indicating that the complex baseband equivalent impulse response $c(t)$ forms the time domain envelope of $h(t)$, where $c(t)$ can be described by its in-phase and quadrature

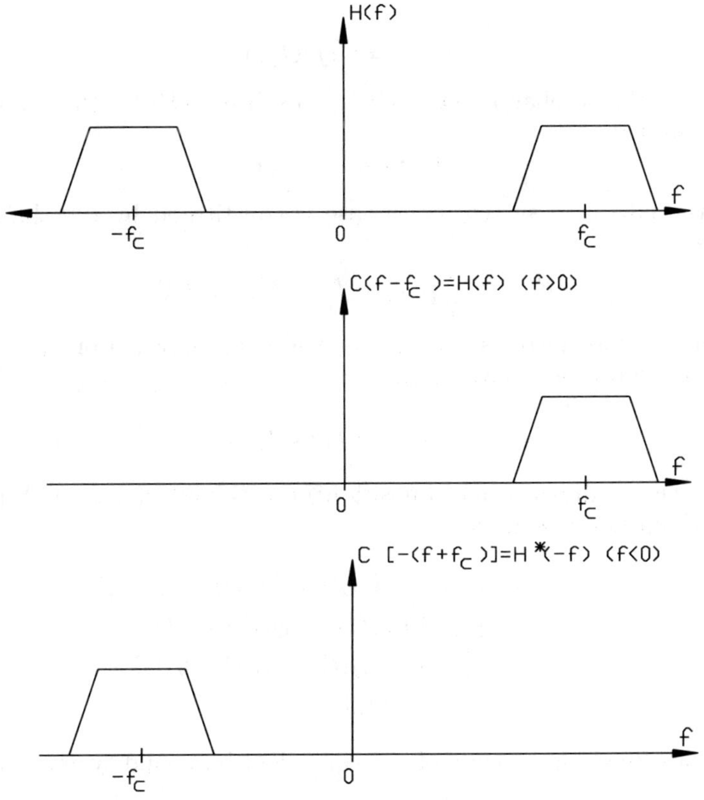

Figure 2.11: Spectral relationship of bandpass and lowpass equivalent systems

phase components as follows:

$$c(t) = c_I(t) + jc_Q(t). \tag{2.22}$$

The received passband signal $r(t)$ or $R(f)$ can now be expressed also by the help of the baseband equivalents $U(f)$ and $C(f)$ or $u(t)$ and $c(t)$. Substituting Equations 2.17 and 2.20 into Equation 2.13 yields:

$$R(f) = \frac{1}{2}\left[U(f-f_c) + U^*(-f-f_c)\right]\left[C(f-f_c) + C^*(-f-f_c)\right]. \tag{2.23}$$

Figures 2.10 and 2.11 reveal that $U(f-f_c)\cdot C^*(f-f_c) = 0$ if $f \leq 0$ and $U^*(-f-f_c)\cdot C(-f-f_c) = 0$ if $f \geq 0$, hence Equation 2.23 simplifies to:

$$\begin{aligned} R(f) &= \frac{1}{2}\left[U(f-f_c)C(f-f_c) + U^*(-f-f_c)C^*(-f-f_c)\right] \\ &= \frac{1}{2}\left[V(f-f_c) + V^*(-f-f_c)\right], \end{aligned} \tag{2.24}$$

where

$$V(f) = U(f)C(f) \tag{2.25}$$

is the response of the baseband system $C(f)$ to excitation $U(f)$. But for a real impulse response $v(t)$ we have

$$V(f) = V^*(-f) \tag{2.26}$$

giving that the passband system's response is constituted by the shifted baseband responses, i.e.:

$$R(f) = \frac{1}{2}\left[V(f - f_c) + V(f + f_c)\right]. \tag{2.27}$$

The baseband systems' response $v(t)$ to the baseband information signal $u(t)$ can be expressed by the following convolution:

$$v(t) = u(t) * c(t). \tag{2.28}$$

Substituting the quadrature decomposition of $u(t)$ and $c(t)$ from Equations 2.22 and 2.8 into Equation 2.28 gives

$$\begin{aligned} v(t) &= [u_I(t) + ju_Q(t)] * [c_I(t) + jc_Q(t)] \qquad (2.29) \\ &= [u_I(t) * c_I(t) - u_Q(t) * c_Q(t)] + \\ &\quad j[u_I(t) * c_Q(t) + u_Q(t) * c_I(t)] \\ &= v_I(t) + jv_Q(t), \end{aligned}$$

where the quadrature components of the equivalent baseband system's response are:

$$\begin{aligned} v_I(t) &= u_I(t) * c_I(t) - u_Q(t) * c_Q(t) \\ v_Q(t) &= u_I(t) * c_Q(t) + u_Q(t) * c_I(t). \end{aligned} \tag{2.30}$$

These operations are portrayed in Figure 2.12.

In summary, in this section we revealed the equivalence of bandpass and lowpass systems. The relationships portrayed allow us to study bandpass systems in the baseband both by computer simulation and analytical methods. The relevant equivalences are summarised in Table 2.2.

2.3.3 Gaussian mobile radio channel

The simplest type of channel is the Gaussian channel as described in Section 2.1.3. Basically it is a linear time invariant transmission system impaired by the thermal noise generated in the receiver. The noise is assumed to have a constant power spectral density (PSD) over the channel bandwidth, and a Gaussian amplitude probability density function (PDF). This type of channel is occasionally realised in digital mobile radio communications mainly in microcells where it is possible to have a line of sight (LOS) with essentially no multipath propagation. Even when there is multipath propagation, but the mobile is stationary and there are no other moving objects such in its vicinity, as vehicles, the mobile channel may be thought of as Gaussian. This is because, due to the zero mobile speed, the fading is represented by a local path loss.

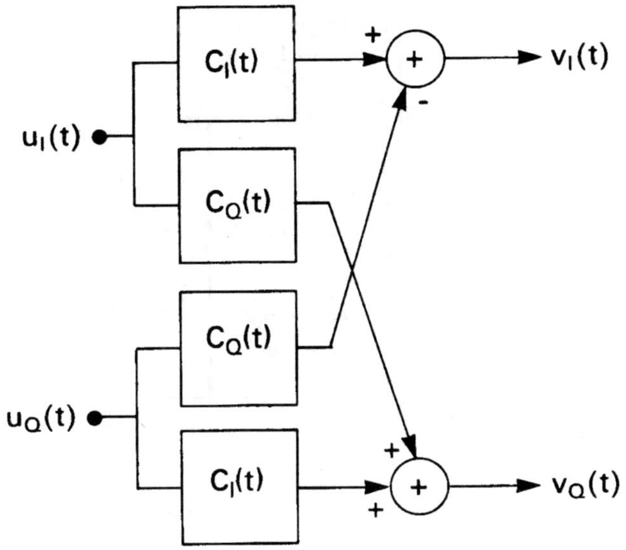

Figure 2.12: Complex convolution in the baseband

Type	Equations
Inputs	$s(t) = \Re\left\{u(t)e^{j2\pi f_c t}\right\}$ $S(f) = \frac{1}{2}[U(f - f_c) + U^*(-f - f_c)]$
System Description	$r(t) = s(t) * h(t)$ $R(f) = S(f)H(f)$ $v(t) = u(t) * c(t)$ $V(f) = U(f)C(f)$ $h(t) = 2\Re\{c(t)e^{j2\pi f_c t}\}$ $H(f) = C(f - f_c) + C^*(-f - f_c)$ $c(t) = c_I(t) + jc_Q(t)$
Outputs	$r(t) = \Re\left\{v(t)e^{j2\pi f_c t}\right\}$ $R(f) = \frac{1}{2}[V(f - f_c) + V^*(-f - f_c)]$ $u(t) = u_I(t) + ju_Q(t)$ $v(t) = v_I(t) + jv_Q(t)$

Table 2.2: Summary of baseband and bandpass system equations

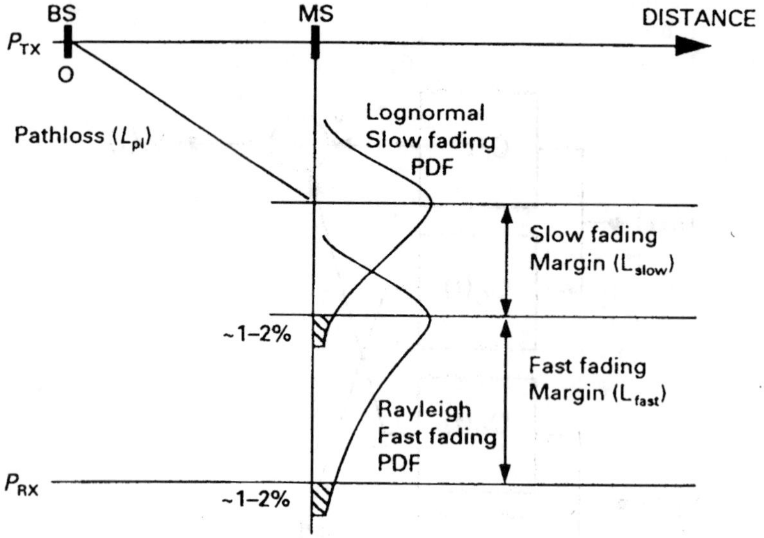

Figure 2.13: Pathloss, slow-fading and fast-fading [158]©1992, Steele

The Gaussian channel is also important for providing an upper bound on system performance. For a given modulation scheme we may calculate or measure the BER performance in the presence of a Gaussian channel. When multipath fading occurs the BER will increase for a given channel SNR. By using techniques to combat multipath fading, such as diversity, equalisation, channel coding, data interleaving, and so forth, that are to be described throughout the book, we can observe how close the BER approaches that for the Gaussian channel.

2.3.4 Narrow band fading channels

Propagation in narrowband mobile radio channels can be divided into three phenomena. These are:

(1) Propagation path loss law

(2) Slow fading statistics

(3) Fast fading statistics

These all vary with the propagation frequency, surrounding natural and man-made objects, vehicular speed, etc. Consequently a deterministic treatment is not possible and so statistical methods are used instead. The calculation of a power budget for a mobile radio link considering the propagation path loss law, the slow fading and fast fading margins is shown in Figure 2.13.

Since most existing public mobile radio systems use the 0.45-1.8 GHz band, we concentrate on channels in this range. These frequencies fall in the so-called ultra high frequency (UHF) band. At these frequencies the signal level drops quickly once over the radio horizon, limiting co-channel interference when the frequencies are reused in neighbouring cell clusters. At these frequencies, even if there is no line-of-sight path between transmitter and receiver, sufficient signal power may be received by means of wave scattering, reflection and diffraction to allow communications.

The prediction of the expected mean or median received signal power plays a crucial role in planning the coverage area of a specific base station and in determining the closest acceptable reuse of the propagation frequency deployed. For high antenna elevations and large rural cells a more slowly decaying power exponent is expected than for low elevations and densely built-up urban areas. As suggested by Figure 2.13, the received signal is also subjected to slow or shadow fading which is mainly governed by terrain and topographical features in the vicinity of the mobile receiver such as small hills and tall buildings. When designing the system's power budget and coverage area pattern, the slow fading phenomenon is taken into account by including a shadow fading margin as demonstrated by Figure 2.13.

Slow fading is compensated by increasing the transmitted power P_{tx} by the slow fading margin L_{slow}, which is usually chosen to be the $1 - 2\%$ quantile of the slow fading probability density function (PDF) to minimise the probability of unsatisfactorily low received signal power P_{rx}. Additionally, the short-term fast signal fading due to multipath propagation is taken into account by deploying the so-called fast fading margin L_{fast}, which is typically chosen to be also a few percent quantile of the fast fading distribution. There remains a certain probability that both of these fading margins are simultaneously exceeded by the superimposed slow and fast envelope fading. This situation is often referred to as "fading margin overload", and results in a very low received signal level which may cause call dropping in mobile telephony. The worst-case probability of these events can be taken to be the sum of the individual fading margin overload probabilities.

Mathematical models can be derived, or measurements made, to determine appropriate margins, which can then be used to analyse likely system performance. For practicality, any models used must consider signals to have been transmitted at baseband through a channel which is equivalent to the mobile radio channel.

2.3.4.1 Propagation path loss law

When considering the propagation path loss, the parameter of prime concern is normally the distance from the BS. Path loss increases with distance due to the increasing area of the circular wavefront expanding from the BS. More details about path loss calculations can be found in [156, 157, 159]. Path loss calculations become significantly more complex when there is any form of obstruction between the transmitter and receiver. In large cells these obstructions will often be hills or undulating terrain. In modelling tools these obstacles are often considered to be knife-edges at the point of the summit of the obstruction, and knife-edge diffraction techniques can then be used [159, 165] to predict the path loss. The analysis becomes increasingly complex if there are a number of hills between the transmitter and receiver. An alternative approach is to determine which Fresnel zone the hill obstructs and perform the prediction accordingly [166].

In urban areas modelling becomes more complex, and in the case of large cells it is rarely possible to model each individual building, although this can be achieved in microcells [167]. Typically this situation is resolved by adding a clutter loss dependent on the density of the buildings in the urban area, and also allowing for a shadowing loss behind buildings when computing the link budget. This approximation leads to sub-optimal network design, with overlapping cells and inefficient frequency reuse. Nevertheless, it seems unlikely that more accurate tools will become available in the near future.

In our probabilistic approach it is difficult to give a worst-case path loss exponent for any mobile channel. However, it is possible to specify the most optimistic scenario. This is given by propagation in free space. The free space path loss, L_{pl} is given by [159]:

$$L_{pl} = -10\log_{10} G_T - 10\log_{10} G_R + 20\log_{10} f^{Hz} + 20\log_{10} d^m - 147.6 \text{ dB}, \quad (2.31)$$

where G_T and G_R are the transmitter and receiver antenna gains, f^{Hz} is the propagation frequency in hertz and d^m is the distance from the BS antenna in meters. Observe that the free space path loss is increased by 6 dB every time, the propagation frequency is doubled or the distance from the mobile is doubled. This corresponds to a 20 dB/decade decay and at $d = 1$ km, $f = 1$ GHz and $G_T = G_R = 1$ a path loss of $L_F = 92.4$ dB is encountered. Clearly, not only technological difficulties, but also propagation losses discourage the deployment of higher frequencies. Nevertheless, spectrum is usually only available in these higher frequency bands.

In practice, for UHF mobile radio propagation channels of interest, the free space conditions do not apply. There are, however, a number of useful path loss prediction models that can be adopted to derive other prediction bounds. One such case is the "plane earth" model. This is a two-path model constituted by a direct line-of-sight path and a ground-reflected path which ignores the curvature of the earth's surface. Assuming transmitter base station (BS) and receiver mobile station (MS) antenna heights of $h_{BS}, h_{MS} \ll d$, the plane earth path loss formula [159] can be written as follows:

$$\begin{aligned} L_{pl} = -10\log_{10} G_T - 10\log_{10} G_R - 20\log_{10} h^m_{BS} - \\ 20\log_{10} h^m_{MS} - 40\log_{10} d^m, \end{aligned} \quad (2.32)$$

where the dependence on propagation frequency is removed. Observe that a 6 dB path loss reduction is the result, when doubling the transmitter or receiver antenna elevations, and there is an inverse fourth power law decay when increasing the BS-MS distance d. In the close vicinity of the transmitter antenna, where the condition h_{BS} or $h_{MS} \ll d$ does not hold, Equation 2.33 is no longer valid.

Hata [168] developed three path loss models for large cells which are widely used, forming the basis for many modelling tools. These were developed from an extensive database derived by Okumura [169] from measurements in and around Tokyo. The typical *urban* Hata model is defined as:

$$L_{Hu} = 69.55 + 26.16\log_{10} f - 13.82\log_{10} h_{BS} - a(h_{MS}) +$$

$$(44.9 - 6.55 \log_{10} h_{BS}) \log_{10} d \, [dB], \tag{2.33}$$

where f is the propagation frequency in MHz, h_{BS} and h_{MS} are the BS and MS antenna elevations (m), $a(h_{MS})$ is a terrain dependent correction factor, while d is the BS-MS distance in (km). The correction factor $a(h_{MS})$ for small and medium-sized cities was found to be

$$a(h_{MS}) = (1.1 \log_{10} f - 0.7) h_{MS} - (1.56 \log_{10} f - 0.8), \tag{2.34}$$

while for large cities it is frequency-parametrised:

$$a(h_{MS}) = \begin{cases} 8.29[\log_{10}(1.54 h_{MS})]^2 - 1.1 & \text{if } f \leq 200MHz \\ 3.2[\log_{10}(11.75 h_{MS})]^2 - 4.97 & \text{if } f \geq 400MHz \end{cases}. \tag{2.35}$$

The typical *suburban* Hata model applies a correction factor to the urban model, and it yields:

$$L_{Hsuburban} = L_{Hu} - 2[\log_{10}(f/28)]^2 - 5.4 \, [dB]. \tag{2.36}$$

The *rural* Hata model modifies the urban formula like this

$$L_{Hrural} = L_{Hu} - 4.78 (\log_{10} f)^2 + 18.33 \log_{10} f - 40.94 \, [dB]. \tag{2.37}$$

The limitations of its parameters as listed by Hata are as follows:

$$\begin{array}{ll} f: & 150 - 1500MHz \\ h_{BS}: & 30 - 200m \\ h_{MS}: & 1 - 10m \\ d: & 1 - 20km. \end{array}$$

For a 900 MHz public land mobile radio (PLMR) system these conditions can usually be satisfied, but for a typical 1.8 GHz personal communications network (PCN) urban microcell all these limits have to be slightly stretched.

Microcellular channels can be described by a four-path model including two more reflected waves from building walls along the streets [170, 171]. In this scenario it is assumed that the transmitter antenna is below the characteristic urban skyline. The four-path model used by Green [170] assumed smooth reflecting surfaces yielding specular reflections with no scattering, finite permittivity and conductivity, vertically polarised waves and half-wave dipole antennas. The resultant path loss profile vs. distance becomes rather erratic with received signal level variations in excess of 20 dB, which renders path loss modelling by a simple power exponent rather inaccurate, however attractive it would appear due to its simplicity.

Figure 2.14 also allows a comparison of the two-path model, the inverse second law and the inverse fourth law characteristics.

2.3.4.2 Slow fading statistics

Having studied a few propagation path loss prediction models we briefly focus our attention on the characterisation of the slow fading phenomena, which constitutes

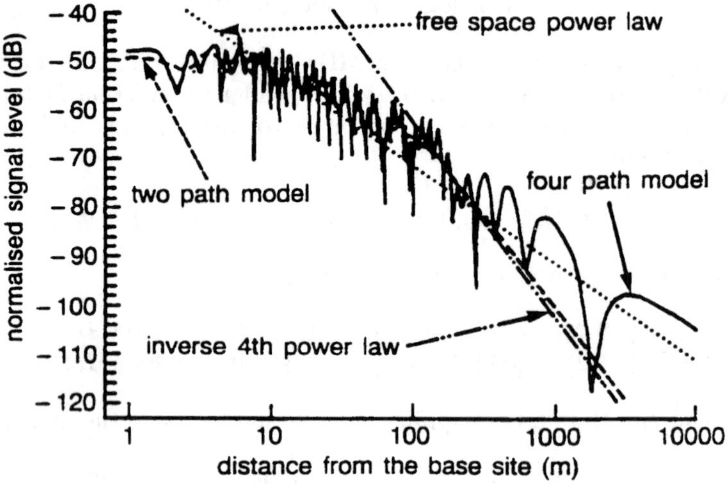

Figure 2.14: Signal level profiles for the two- and four-path models. Also shown are the free space and inverse fourth power laws [170] ©1990, Green

the second component of the overall power budget design of mobile radio links, as portrayed in Figure 2.13. In slow fading analysis the effects of fast fading and path loss have to be removed. The fast fading fluctuations are removed by averaging the signal level over a distance of typically some 20 wavelengths. The slow fading fluctuations are separated by subtracting the best-fit path loss regression estimate from each individual 20 wavelength spaced averaged received signal value. A slow fading histogram derived this way is depicted in Figure 2.15.

The figure suggests a lognormal distribution in terms of dBs caused by normally distributed random shadowing effects. Indeed, when subjected to rigorous distribution fitting using the lognormal hypothesis, the hypothesis is confirmed at a high confidence level. The associated standard deviation in this particular case is 6.5 dB.

2.3.4.3 Fast fading statistics

Irrespective of the distribution of the numerous individual constituent propagation paths of both quadrature components (a_i, a_q) of the received signal, their distribution will be normal due to the central limit theorem. Then the complex baseband equivalent signal's amplitude and phase characteristics are given by:

$$a(k) = \sqrt{a_i^2(k) + a_q^2(k)} \quad (2.38)$$

$$\phi(k) = \arctan[a_q(k)/a_i(k)]. \quad (2.39)$$

Our aim is now to determine the distribution of the amplitude $a(k)$, if $a_i(k)$ and $a_q(k)$ are known to have a normal distribution.

In general, for n normally distributed random constituent processes with means

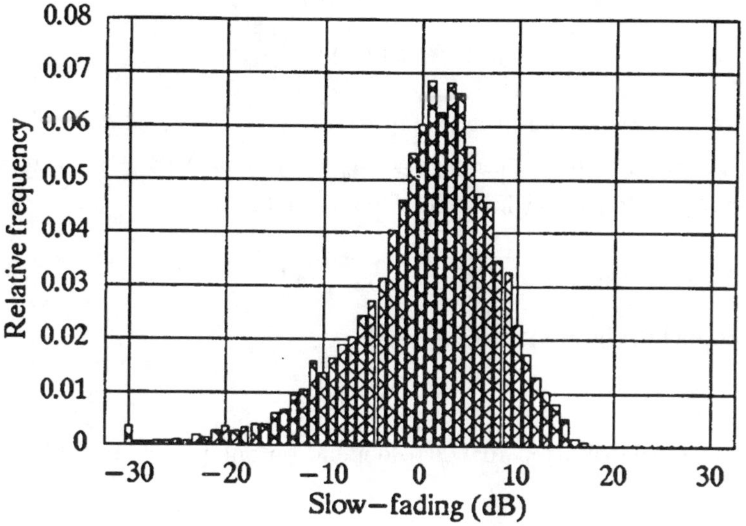

Figure 2.15: Typical microcellular slow fading histogram [158] ©1992, Steele

$\overline{a_i}$ and identical variances σ^2, the resultant process $y = \sum_{i=1}^{n} a_i^2$ has a so-called χ^2 distribution with the PDF given below [164]:

$$p(y) = \frac{1}{2\sigma^2}\left(\frac{y}{s^2}\right)^{(n-2)/4} \cdot e^{-(s^2+y)/2\sigma^2} \cdot I_{(n/2)-1}\left(\sqrt{y}\frac{s}{\sigma^2}\right) \tag{2.40}$$

where

$$y \geq 0 \tag{2.41}$$

and

$$s^2 = \sum_{i=1}^{n}(\overline{a_i})^2 \tag{2.42}$$

is the so-called non-centrality parameter computed from the first moments of the $a_1, a_2, \cdots, a_n$component processes . If the constituent processes have zero means, the χ^2 distribution is central, otherwise non-central. Each process has a variance of σ^2 and $I_k(x)$ is the modified k th order Bessel function of the first kind, given by

$$I_k(x) = \sum_{j=0}^{\infty}\frac{(x/2)^{k+2j}}{j!\Gamma(k+j+1)}, \quad x \geq 0. \tag{2.43}$$

The Γ function is defined as

$$\Gamma(p) \quad = \quad \int_0^{\infty} t^{p-1}e^{-t}dt \quad \text{if } p > 0$$

$$\Gamma(p) = (p-1)! \quad \text{if } p > 0 \text{ integer} \tag{2.44}$$
$$\Gamma(\frac{1}{2}) = \sqrt{\pi} \ , \ \Gamma(\frac{3}{2}) = \frac{\sqrt{\pi}}{2}.$$

In our case we have two quadrature components, i.e. $n = 2$, $s^2 = (\overline{a_i})^2 + (\overline{a_q})^2$, the envelope is computed as $a = \sqrt{y} = \sqrt{a_i^2 + a_q^2}, a^2 = y, p(a)\ da = p(y)\ dy$, and hence $p(a) = p(y)\ dy/da = 2ap(y)$ yielding the Rician PDF.

$$p_{\text{Rice}}(a) = \frac{a}{\sigma^2} e^{-(a^2+s^2)/2\sigma^2} I_o\left(\frac{as}{\sigma^2}\right) \quad a \geq 0. \tag{2.45}$$

Formally introducing the Rician K-factor as

$$K = s^2/2\sigma^2 \tag{2.46}$$

renders the Rician distribution's PDF to depend on one parameter only:

$$p_{\text{Rice}}(a) = \frac{a}{\sigma^2} \cdot e^{-\frac{a^2}{2\sigma^2}} \cdot e^{-K} \cdot I_o\left(\frac{a}{\sigma} \cdot \sqrt{2K}\right), \tag{2.47}$$

where K physically represents the ratio of the power received in the direct line-of-sight path, to the total power received via indirect scattered paths. Therefore, if there is no dominant propagation path, $K = 0$, $e^{-K} = 1$ and $I_0(0) = 1$ yielding the worst-case Rayleigh PDF:

$$p_{\text{Rayleigh}}(a) = \frac{a}{\sigma^2} e^{-\frac{a^2}{2\sigma^2}}. \tag{2.48}$$

Conversely, in the clear direct line-of-sight situation with no scattered power, $K = \infty$, yielding a "Dirac delta shaped" PDF, representing a step-function-like cumulative distribution function (CDF). The signal at the receiver then has a constant amplitude with a probability of one. Such a channel is referred to as a Gaussian channel. This is because although there is no fading present, the receiver will still see the additive white Gaussian noise (AWGN) referenced to its input. If the K-factor is known, the fast fading envelope's distribution is completely described. The set of Rician PDFs for $K = 0$, 1, 2, 4, 10 and 15 is plotted in Figure 2.16.

Note that the relationship between the variances and means of the Rayleigh and the component Gaussian distributions are given as follows [164]:

$$\sigma_R^2 = (2 - \pi/2) \cdot \sigma^2$$

and

$$m_R = \sqrt{2} \cdot \sigma \cdot \sqrt{\pi}/2 = \sigma\sqrt{\pi/2},$$

Clearly, the Rayleigh distribution's variance is not twice that of the composite Gaussian processes, since we are adding the squares of the Gaussian processes, not the processes themselves.

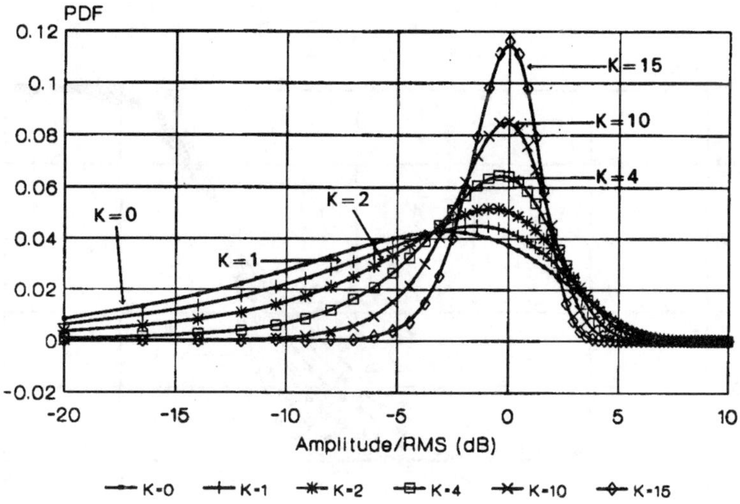

Figure 2.16: Rician PDFs

The Rician CDF takes the shape of [164]:

$$\begin{aligned} C_{\text{Rice}}(a) &= 1 - e^{-\left(K+\frac{a^2}{2\sigma^2}\right)} \sum_{m=0}^{\infty} \left(\frac{s}{a}\right)^m \cdot I_m\left(\frac{a\,s}{\sigma^2}\right) \\ &= 1 - e^{-\left(K+\frac{a^2}{2\sigma^2}\right)} \sum_{m=0}^{\infty} \left(\frac{\sigma\sqrt{2K}}{a}\right)^m \cdot I_m\left(\frac{a\sqrt{2K}}{\sigma}\right). \end{aligned} \tag{2.49}$$

Clearly, this formula is more difficult to evaluate than the PDF of Equation 2.47 due to the summation of an infinite number of terms, requiring double or quadruple precision and it is avoided in numerical evaluations, if possible. However, in practical terms it is sufficient to increase m to a value where the last term's contribution becomes less than 0.1%.

A range of Rician CDFs evaluated from Equation 2.49 are plotted on a linear probability scale in Figure 2.17 for $K = 0, 1, 2, 4, 10$ and 15. Figure 2.18 shows the same Rician CDFs plotted on a more convenient logarithmic probability scale, which reveals the enormous difference in terms of deep fades for the K values considered. When choosing the fading margin overload probability, Figure 2.18 is more useful as it expands the tails of the CDFs, where for a Rician CDF with $K = 1$ the 15 dB fading margin overload probability is approximately 10^{-2}.

2.3.4.4 Doppler spectrum

Having described the fading statistics let us now concentrate on the effects of the Doppler shift and consider again the transmission of an unmodulated carrier frequency

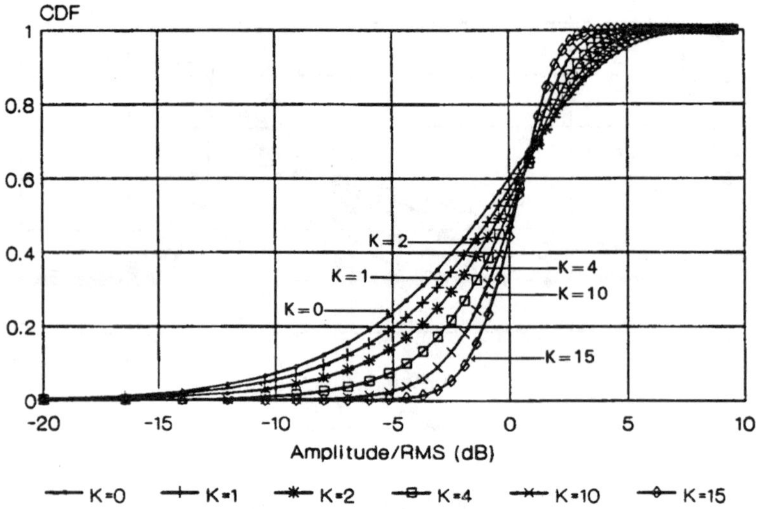

Figure 2.17: Rician CDFs

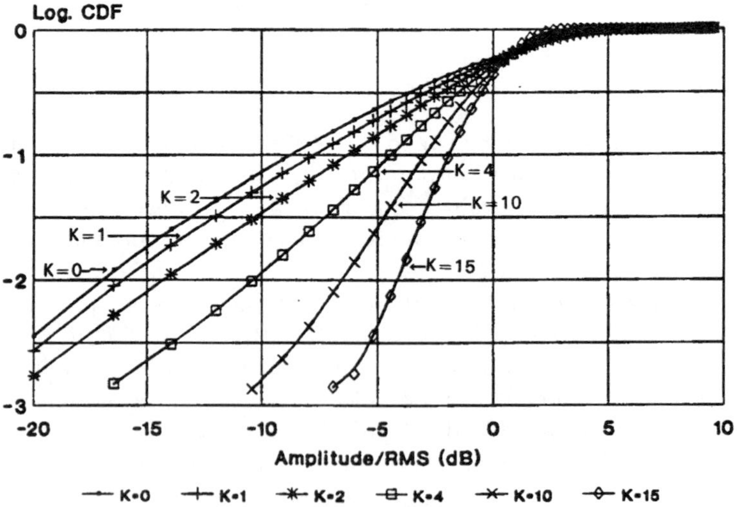

Figure 2.18: Rician logarithmic CDFs

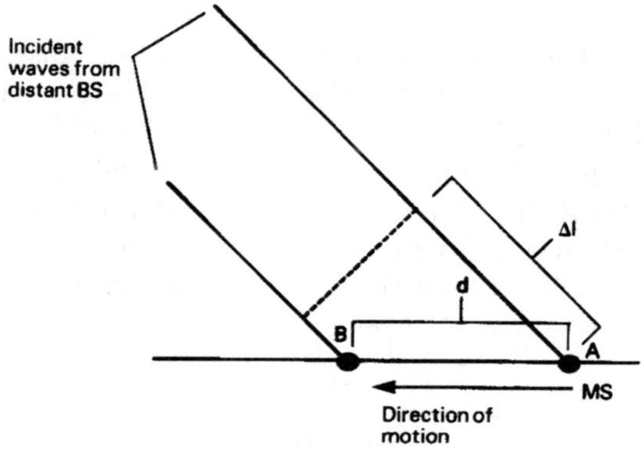

Figure 2.19: Relative path-length change due to the MS's movement

f_c from a BS. A MS travelling in a direction making an angle α_i with respect to the signal received on the ith path as seen in Figure 2.19 advances a distance of $d = v \cdot \Delta t$ during Δt, when travelling at a velocity of v. This introduces a relative carrier phase change of $\Delta\Phi = 2\pi$, if the flight of the wave is shortened by an amount given by $\Delta l = \lambda$, where the wavelength can be computed from $\lambda = c/f_c$ with c being the velocity of light.

For an arbitrary Δl value we have $\Delta\Phi = -2\pi \cdot \Delta l/\lambda$, where the negative sign implies that the carrier wave's phase delay is reduced if the MS is travelling towards the BS. From the simple geometry of Figure 2.19 we have $\Delta l = d \cdot \cos \alpha_i$ and therefore the phase change becomes:

$$\Delta\Phi = -\frac{2\pi v \Delta t \cos \alpha_i}{\lambda}. \tag{2.50}$$

The Doppler frequency can be defined as the phase change due to the movement of the MS during the infinitesimal interval Δt:

$$f_D = -\frac{1}{2\pi}\frac{\Delta\Phi}{\Delta t}. \tag{2.51}$$

When substituting Equation 2.50 into 2.51 we get:

$$f_D = \frac{v}{\lambda}\cos \alpha_i = f_m \cos \alpha_i, \tag{2.52}$$

where $f_m = v/\lambda = v f_c/c$ is the maximum Doppler frequency deviation from the transmitted carrier frequency due to the MS's movement. Notice that a Doppler frequency can be positive or negative depending on α_i, and that the maximum and minimum Doppler frequencies are $\pm f_m$. These extreme frequencies correspond to the $\alpha_i = 0°$ and $180°$, when the ray is aligned with the street that the MS is travelling along, and corresponds to the ray coming towards or from behind the MS, respectively. It is analogous to the change in the frequency of a whistle from a train perceived by a person standing on a railway line when the train is bearing down or receding from the person, respectively.

According to Equation 2.52 and assuming that α_i is uniformly distributed, the Doppler frequency has a so-called random cosine distribution. The received power in an angle of $d\alpha$ around α_i is given by $p(\alpha_i)\ d\alpha$, where $p(\alpha_i)$ is the probability density function (PDF) of the received power, which is assumed to be uniformly distributed over the range of $0 \leq \alpha_i \leq 2\pi$, giving $p(\alpha_i) = 1/2\pi$. The Doppler power spectral density $S(f_D)$ can be computed using Parseval's theorem by equating the incident received power $p(\alpha_i)\ d\alpha$ in an angle $d\alpha$ with the Doppler power $S(f_D)\ df_D$, yielding $S(f_D) = d\alpha_i/(2\pi \cdot df_D)$. Upon expressing α_i from $f_D = f_m \cos\alpha_i$ and exploiting that

$$\frac{d\cos^{-1} x}{dx} = -\frac{1}{\sqrt{1-x^2}}$$

we then have:

$$S(f_D) = \frac{d\alpha_i}{2\pi\ df_D} = \frac{d(\cos^{-1} f_D/f_m)}{2\pi\ df_D} = -\frac{1}{2\pi f_m\sqrt{1-(f_D/f_m)^2}}. \tag{2.53}$$

The incident received power at the MS depends on the power gain of the antenna and the polarisation used. Thus the transmission of an unmodulated carrier is received as a "smeared" signal whose spectrum is not a single carrier frequency f_c, but contains frequencies up to $f_c \pm f_m$. In general we can express the received RF spectrum $S(f_D)$ for a particular MS speed, propagation frequency, antenna design and polarisation as,

$$S(f_D) = \frac{C}{\sqrt{1-(f_D/f_m)^2}} \tag{2.54}$$

where C is a constant that absorbs the $1/2\pi f_m$ multiplier in Equation 2.53. Notice that the Doppler spectrum of Equation 2.54 becomes $S(f_D = 0) = C$ at $f_D = 0$, while $S(f_D = f_m) = \infty$, when $f_D = f_m$. Between these extreme values $S(f)$ has a U-shaped characteristic, as it is portrayed in a stylised form in the simulation model of Figure 2.20, which will be developed in the next section.

2.3.4.5 Simulation of narrowband channels

The Rayleigh fading channel model used in the previous subsection can be represented by the quadrature arrangement shown in Figure 2.20, where the distribution of both received quadrature components (a_i, a_q) is normal due to the central limit theorem. These can be modelled as uncorrelated normally distributed AWGN sources. The outputs from the AWGN sources are applied to lowpass filters having U-shaped frequency domain transfer functions that represent the effects of Doppler frequency shifts, as will be demonstrated.

Observe that the maximum Doppler frequency $f_m = v/\lambda = v \cdot f_c/c$ depends on the product of the speed v of the MS and the propagation frequency f_c. Therefore, the higher the speed or the propagation frequency, the wider the frequency band over which the received carrier is smeared. The effect of deploying the Doppler filter in Figure 2.20 is that the originally uncorrelated quadrature components (a_i, a_q) are

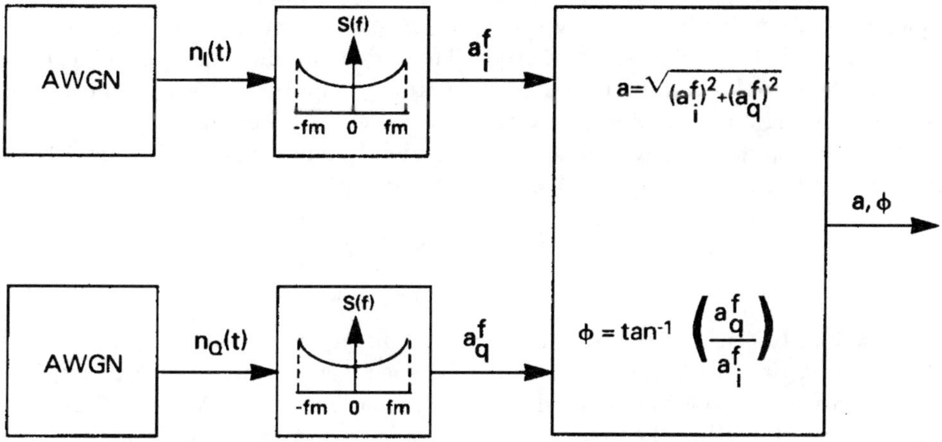

Figure 2.20: Baseband Rayleigh fading simulation model

now effectively lowpass filtered, which introduces some intersample correlation by restricting the maximum rate of change of (a_i, a_q), without altering the Rayleigh distributed envelope fading. This is true, because passing an AWGN process through a linear system, such as a lowpass filter, will reduce the variance of the signal but retain the Gaussian nature of the distribution for both quadrature components. Having no lowpass Doppler filter at all would be equivalent to $f_m = \infty$, implying an infinitely high vehicular speed hence allowing the reception of arbitrary uncorrelated frequencies when transmitting the unmodulated carrier frequency of f_c.

The simulation of the Rayleigh distributed fast fading envelope is based on the concept portrayed in Figure 2.20, but it can be implemented in both the time and the frequency domain.

2.3.4.5.1 Frequency domain fading simulation In the frequeny domain approach one could exploit the fact that the Fourier transform of an AWGN process is another AWGN process. Hence the frequency domain AWGN sequence generated by the Box-Müller algorithm to be described below can be multiplied by the frequency domain Doppler transfer function of Equation 2.54 and then transformed back by IFFT to the time domain in order to generate the I and Q components (a_i^f, a_q^f) of the filtered complex Gaussian process. Then the Rayleigh distributed magnitude a is given by

$$a = \sqrt{(a_i^f)^2 + (a_q^f)^2},$$

while the uniformly distributed phase ϕ is computed as

$$\phi = \arctan\left[\frac{a_q^f}{a_i^f}\right].$$

The main difficulty associated with this frequency domain approach is that the Doppler filter's bandwidth f_m is typically much lower than the sampling frequency at which the faded samples are produced. The faded samples have to be generated exactly at the signalling rate at which modulation symbols are transmitted in order to supply a fading envelope and phase rotation value for their corruption.

A typical Doppler bandwidth f_m at a carrier frequency of $f_c = 1.8$ GHz and vehicular speed of $v = 30$ mph $= 13.3$ m/s is given by

$$f_m = \frac{v \cdot f_c}{c} = \frac{13.3\frac{m}{s} \cdot 1.8 \cdot 10^9 \frac{1}{s}}{3 \cdot 10^8 \frac{m}{s}} \approx 80Hz,$$

where $c = 3 \times 10^8$ m/s is the speed of light. This f_m value becomes about 40 Hz at $f_c = 900$ MHz and 8 Hz at a pedestrian speed of 3 mph. Typical multi-user signalling rates in state-of-art time division multiple access (TDMA) systems are in excess of $f_s = 40$ KHz. Therefore the pedestrian relative Doppler bandwidth D_r becomes $D_r = 8$ Hz/40 KHz $= 2 \times 10^{-4}$ suggesting that only a negligible proportion of the input samples will fall in the Doppler filter's passband. This fact may severely affect the statistical soundness of this approach, unless a very long IFFT is invoked, which will retain a sufficiently high number of frequency domain AWGN samples after removing those outside the Doppler filter's bandwidth.

2.3.4.5.2 Time domain fading simulation When using the time domain approach the AWGN samples generated by the Box-Müller algorithm must be convolved with the impulse response $h_D(t)$ of the Doppler filter $S(f)$ of Equation 2.54 in both quadrature arms of Figure 2.20. Similarly to the frequency domain method, we are faced with a difficult notch filtering problem. Since the filter bandwidth is typically very narrow in comparision to the sampling frequency, the impulse response $h_D(t)$ will be very slowly decaying, requiring tens of thousands of tap values to be taken into account in the convolution in case of a low Doppler bandwidth. This problem is aggravated by the non-decaying U-shaped nature of the Doppler's transfer function $S(f)$. The consequence of these difficulties is that the confidence measures delivered by rigorous goodness-of-fit techniques such as the χ^2 test or the Kolmogorov-Smirnov test described for example in Chapter 2 of Reference [158] might become low, although the resulting fading envelope and phase trajectories are appropriate for simulation studies of mobile radio systems.

2.3.4.5.3 Box-Müller algorithm of AWGN generation The previously mentioned Box-Müller algorithm is formulated as follows:

(1) Generate two random variables u_1, u_2 and let:

$$s = u_1^2 + u_2^2.$$

(While $s \geq 1$, discard s and re-compute u_1, u_2 and s.

(2) If $s < 1$ is satisfied, compute the I and Q components of the noise as follows:

$$\begin{aligned} u_I &= u_1\sqrt{-\frac{2\sigma^2}{s}\cdot\log_e s} \\ u_Q &= u_2\sqrt{-\frac{2\sigma^2}{s}\cdot\log_e s}, \end{aligned}$$

where σ is the standard deviation of the AWGN. This algorithm can be used both for the generation of the Rayleigh-faded signal envelope as well as for generating the thermal AWGN.

2.3.5 Wideband channels

2.3.5.1 Modelling of wideband channels

The impulse response of the flat Rayleigh fading mobile radio channel consists of a single delta function whose weight has a Rayleigh PDF. This occurs because all the multipath components arrive almost simultaneously and combine to have a Rayleigh PDF. If the signal's transmission bandwidth is narrower than the channel's so-called coherence bandwidth $B_c = 1/(2 \cdot \pi \cdot \Delta)$, where Δ represents the time dispersion interval over which significant multipath components are received, then all transmitted frequency components encounter nearly identical propagation delays. Therefore the so-called narrowband condition is met and the signal is subjected to non-selective or flat envelope fading. The channel's coherence bandwidth (B_c) is defined as the frequency separation, where the correlation of two received signal components' attenuation becomes less than 0.5.

The effect of multipath propagation is to spread the received symbols. If the path delay differences are longer than the symbol duration, several echoes of the same transmitted modulated symbol are received over a number of symbol periods. This is equivalent to saying that in wideband channels the symbol rate is sufficiently high that each symbol is spread into adjacent symbols, causing intersymbol interference (ISI). In order for the receiver to remove the ISI and regenerate the symbols correctly it must determine the impulse response of the mobile radio channel by channel sounding. This response must be frequently measured, since the mobile channel may change rapidly both in time and space.

The magnitude of a typical impulse response is a continuous waveform when plotting received amplitude against time delay. If we partition the time delay axis into equal delay segments, usually called delay bins, then there will generally be, a number of received signals in each bin corresponding to the different paths whose times of arrival are within the bin duration. These signals when vectorially combined can be represented by a delta function occurring in the centre of the bin having a magnitude that is Rayleigh distributed. Impulses which are sufficiently small that they are of no significance to the receiver can then be discarded.

As an example here we describe a set of frequently used typical wideband channel impulses specified by the Group Speciale Mobile committee during the definition of the pan-European mobile radio system known as GSM. These impulse responses describe typical urban, rural and hilly terrain environments, as well as an artificially contrived equaliser test response.

The wideband propagation channel is the superposition of a number of dispersive fading paths, suffering from various attenuations and delays, aggravated by the phenomenon of Doppler shift caused by the MS's movement. The maximum Doppler shift (f_m) is given by $f_m = v/\lambda = v \cdot f_c/c$, where v is the vehicular speed, λ is the wavelength of the carrier frequency f_c and c is the velocity of light. The momentary Doppler shift f_D depends on the angle of incidence α_i, which is uniformly distributed, i.e. $f_D = f_m \cdot \cos \alpha_i$, hence it has a so-called random cosine distribution with a Doppler spectrum limited to $-f_m < f_D < f_m$. Due to time-frequency duality, this "frequency dispersive" phenomenon results in "time selective" behaviour and the wider the Doppler spread, i.e. the higher the vehicular speed, the faster the time domain impulse response fluctuation.

The set of 6-tap GSM impulse responses is depicted in Figure 2.21. In simple terms the wideband channel's impulse response is measured by transmitting an impulse and detecting the received echoes at the channel's output in every D-spaced so-called delay bin. In some bins no delayed and attenuated multipath component is received, while in others significant energy is detected, depending on the typical reflecting objects and their distance from the receiver. The path delay can be easily related to the distance of the reflecting objects, since radio waves are travelling at the speed of light. For example, at a speed of 300 000 km/s a reflecting object at a distance of 0.15 km yields a multipath component at a round-trip delay of 1 μs.

The typical urban (TU) impulse response spreads over a delay interval of 5 μs, therefore it definitely spills energy into adjacent signalling intervals for signalling rates in excess of 200 ksymbols/s, resulting in serious ISI. In practical terms the transmissions can be considered non-dispersive, if the so-called excess path delay does not exceed 10% of the signalling interval, which in our example would correspond to 20 ksymbols/s or 20 kBaud. The hilly terrain (HT) model has a sharply decaying short-delay section due to local reflections and a long-delay part around 15 μs due to distant reflections. Therefore in practical terms it can be considered a two- or three-path model having reflections from a distance of $3 \cdot 10^8$ m/s $\cdot$ 15 μs $=$ 2.25 km. The rural area (RA) response seems the least hostile amongst all standardised responses, decaying fast within 1 μs and hence up to signalling rates of 100 kBaud can be treated as a flat-fading narrowband channel.

The last standardised GSM impulse response is artificially contrived to test the channel equaliser's performance and is constituted by six equidistant unit-amplitude impulses representing six equal-powered independent Rayleigh fading paths with a delay spread of over 16 μs. With these impulse responses in mind the required channel is simulated by summing the appropriately delayed and weighted received signal components. In all but one case the individual components are assumed to have Rayleigh amplitude distribution. In the RA model the main tap at zero delay is supposed to have Rician distribution with the presence of a dominant path.

We can model a wideband impulse response using the algorithm of Figure 2.22.

The in-phase modulated signal $s_I(t)$ is applied to a series of delays equal to the width of a delay bin. At each delay it is multiplied by the magnitude of the wideband channel at that delay. However, the wideband channel will be changing as the mobile moves, so it is necessary to superimpose fading on each of the wideband responses. This is done using a noise generator and Doppler filter as described earlier. An

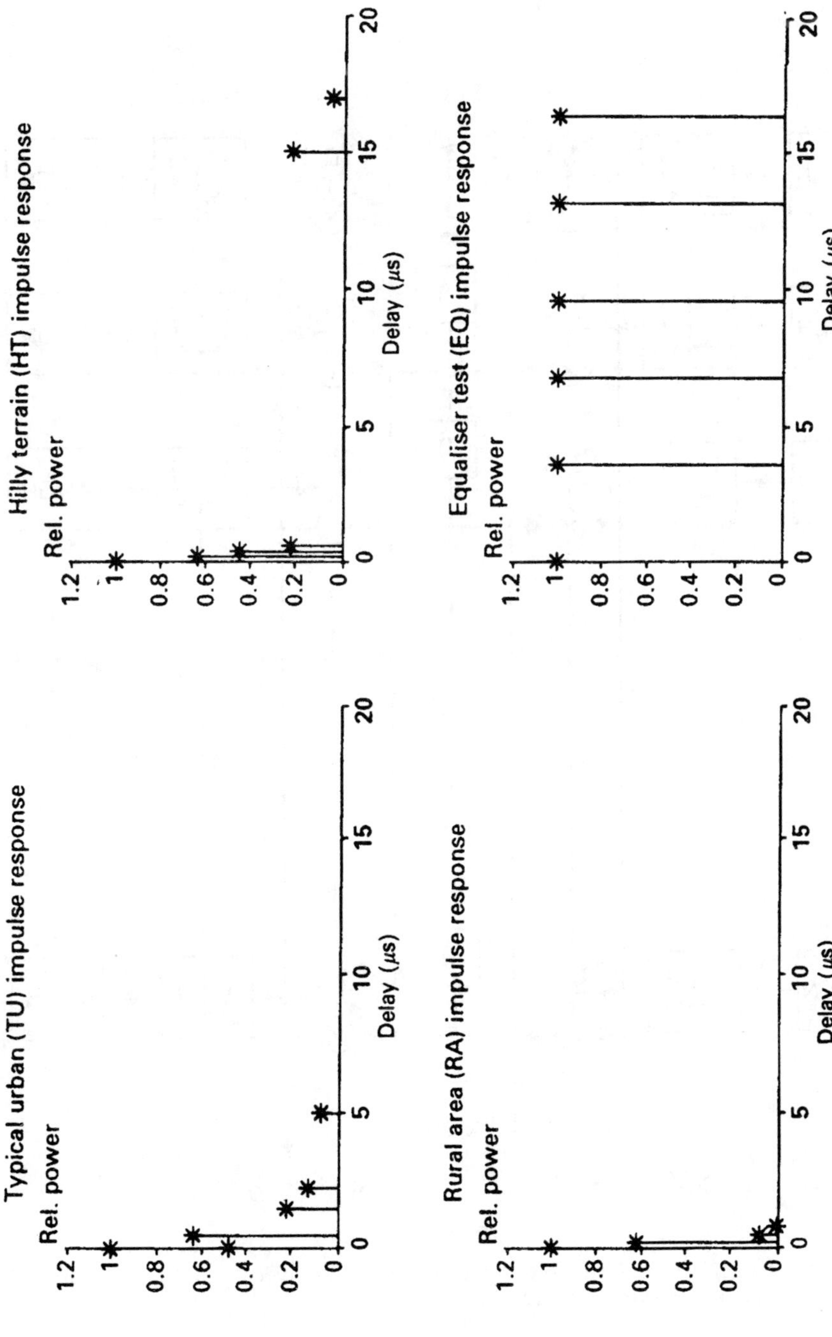

Figure 2.21: GSM specified impulse responses

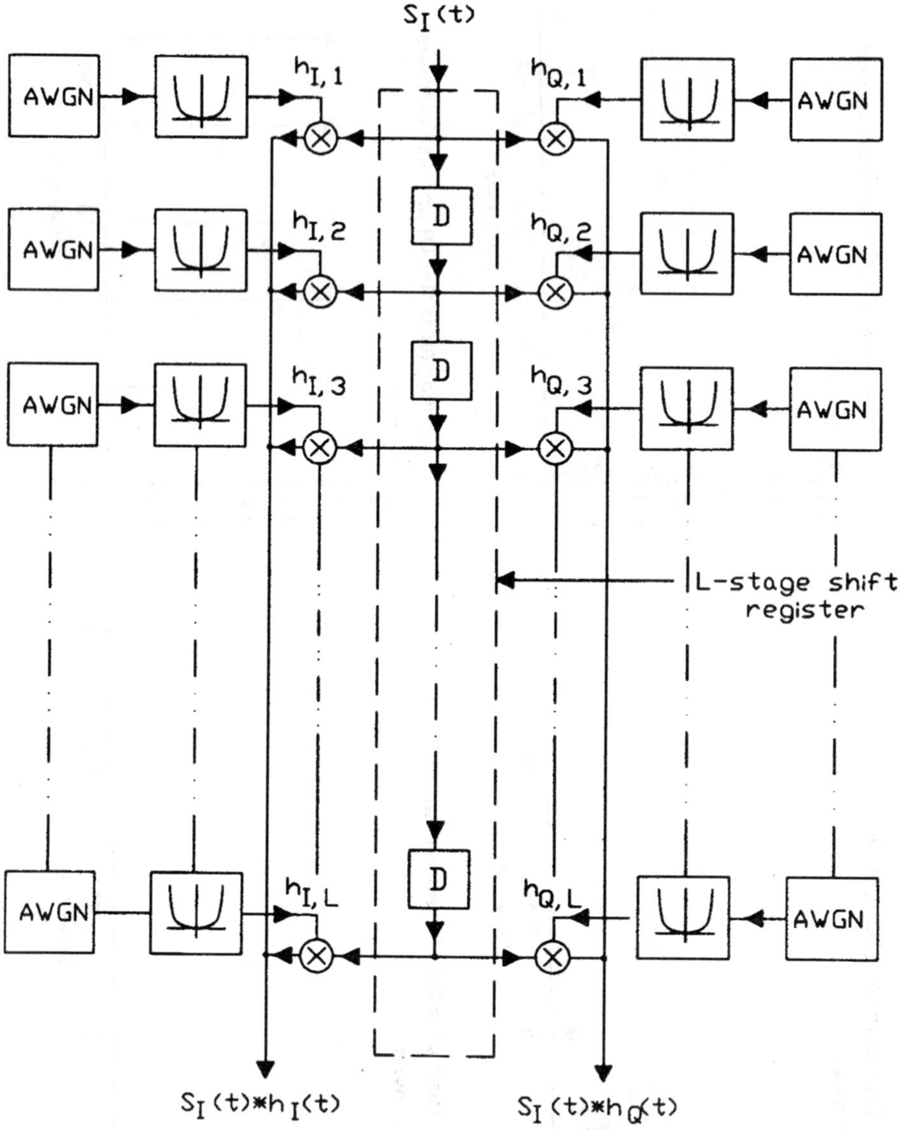

Figure 2.22: Generation of the baseband in-phase received signal via wideband channels [158] ©1992, Steele

identical arrangement to that in Figure 2.22 is used for the quadrature component $s_Q(t)$, and the appropriate convolutional terms are combined to obtain the received signal's quadrature components $r_I(t)$ and $r_Q(t)$.

2.4 Mobile Satellite Propagation

2.4.1 Fixed-link satellite channels

These links are normally between geostationary satellites and fixed, dish-type receivers on the ground. With the spread of satellite broadcasting systems to home users, such links have become widespread. Because of the stationary nature of both ends of the link, the channel itself is normally modelled by a stationary process. Furthermore, because of the high directivity of the antennas used, there is little likelihood of reflected rays being received. To first order, they can be modelled as Gaussian channels.

More precise channel models take into account rain fading. This is a reduction in signal level caused by the signal being absorbed by moisture present in the atmosphere. It is dependent on the meterological conditions at the time of the transmission. Rain fading is also dependent on the transmission frequency. A general guideline is that rain fading starts to become significant at frequencies above 10 GHz, with a typical fade margin at 20 GHz being 10 dB [172].

Satellite channels are often power limited as the satellite only has a limited amount of power available to it. This has often discouraged the use of power inefficient modulation schemes such as QAM. However, techniques are continually emerging to increase the power efficiency of QAM, making it increasingly attractive for satellite applications.

2.4.2 Satellite-to-mobile channels

Satellite-to-mobile communications is still relatively little used due to the difficulties of this particular environment [173–176]. Mobiles are generally unable to deploy highly directional dish antennas, and satellites are normally unable to transmit at high power levels. Some satellite communication systems are restricted to LOS transmission. Because of the elevation of the satellite, LOS transmission will be more frequent than in a conventional cellular system, especially in suburban and rural areas. However, LOS is often masked by trees in rural areas, and by tall buildings in urban areas. Furthermore, the user may be inside a building, further increasing the path loss of the communications link. Once the LOS path is obscured, the signal may arrive at the receiver via a number of reflected paths. The differing phase of these paths will ensure Rician fading.

Few models have been suggested for this channel. Loo [177] suggests that the channel should be modelled as a LOS component with lognormally distributed shadow fading and a multipath component with Rayleigh distribution. This leads to the channel model that

$$p(r) \quad = \quad r/(b_0\sqrt{2\pi d_0}) \int_o^{\infty} 1/z I_0(rz/b_0) \qquad (2.55)$$

$$\cdot \quad \exp[-(ln\ z - mu)^2/2d_0 - (r^2 + z^2)/2b_0)]dz$$

where b_0 is the average scattered power due to multipath, d_0 and μ are the variance and mean due to shadowing, respectively, I_0 is the modified Bessel function, and r is the received signal envelope. The shadowing parameters depend not only on the local environment, but also the satellite elevation, with higher elevations tending to produce lower shadowing losses due to the reduced likelihood of having an obscured LOS path between mobile and satellite.

Because of the hostility of these channels, it is unlikely that the use of multilevel modulation will be widespread.

2.5 Summary

In this chapter we have briefly reviewed the properties of various fixed and mobile communications channels and have shown the equivalence of baseband and passband channels. This allows us to carry out simulations and theoretical analysis in the baseband. In the case of fixed channels the dominant impairment is typically AWGN and linear distortions, as we have seen for example in case of the CCITT standard channels M1020 and M1025 in Figures 2.5 to 2.8. Narrowband mobile radio channels were characterised in terms of path loss, lognormal slow fading and Rician fast fading, as seen in Figure 2.13. The best- and worst-case Rician channels, respectively associated with $K = \infty$ and $K = 0$, are the Gaussian and Rayleigh fading channels, characteristic of the strong LOS and no LOS scenarios. The phenomenon of Doppler shift was introduced, which yielded a smeared spectrum when a single tonal frequency corresponding to an unmodulated carrier was transmitted. A simple simulation model was derived for narrowband channels, which was later extended to dispersive channels. The concept of dispersive wideband channels was portrayed in the context of the GSM wideband channels.

Having considered the non-ideal transmission medium we now embark upon a rudimentary introduction to modems, and review how modems can counteract the channel impairments and facilitate reliable information transmission in a finite bandwidth.

Chapter 3

Introduction to Modems

In this chapter we expand upon our introduction in Chapter 1 by looking in more detail at the basic constituent parts of a modem and considering filtering in more detail.

A typical communications system will input data, perform some form of processing and frequency translation, transmit the data, and perform the converse operations at the receiver. The basic block diagram of such a system employing QAM is shown in Figure 3.1. In a digital communications system the modem's input signal will be a digital signal stream from a digital source or channel encoder. If, however, the modem's input signal is generated by an analogue information source, the signal must be first bandlimited to a bandwidth of B before sampling takes place. According to Nyquist's fundamental theory [178] the sampling frequency must be equal to or higher than twice the bandwidth B, i.e. $f_s \geq 2B$. If this condition is met, the original bandlimited signal can be recovered from its $1/(2B)$ spaced sampled representation with the aid of a lowpass filter having a cut-off frequency of B.

For example, most of the energy of a voice signal is concentrated at frequencies below 4 kHz, hence speech signals are typically lowpass filtered to 4 kHz. As a result, a sampling rate of 8 kHz or higher would be required in order to accurately reconstruct such a signal. In practice, most voice communication systems use a sampling rate of 8 kHz.

We now describe the operation of each of the blocks in the block diagram of Figure 3.1.

3.1 Analogue-to-Digital Conversion

The analogue-to-digital convertor (ADC) takes the bandlimited signal to be transmitted and digitises it by converting the analogue level of each sample to a discrete level. For example, in an 8-bit ADC each discrete level is represented by eight binary output bits. This provides a resolution of 256 distinct digital levels. The action of the ADC is shown in Figure 3.2, where a typical analogue signal is depicted along with

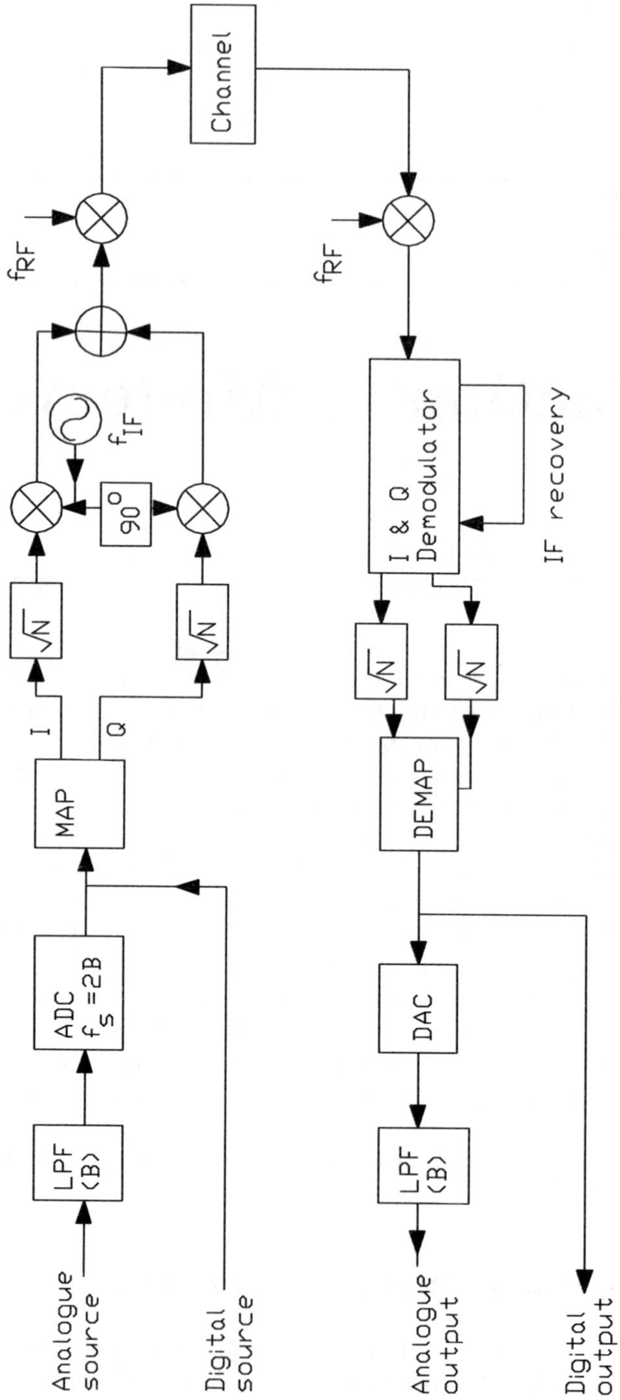

Figure 3.1: Basic QAM modem schematic

its digitised counterpart using only a small number of quantised bits.

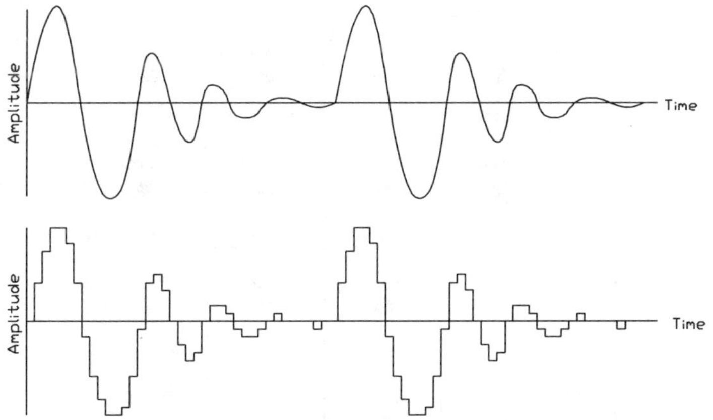

Figure 3.2: Typical ADC input and output waves

The key difference between digital communication systems and the more traditional analogue systems is the signal transmission techniques employed. In an analogue radio, the signal to be transmitted is modulated directly, often by simple multiplication, onto the carrier. On the other hand, digital systems must incorporate modulation schemes which can map the input data onto the carrier in an efficient and readily decoded manner. The benefit gained from using the more sophisticated digital techniques is that the digital link can be rendered error-free by means of signal processing, while in analogue systems the inherent thermal device noise always causes some impairment and hence loss of information.

3.2 Mapping

The process of mapping the information bits onto the streams modulating the I and Q carriers plays a fundamental role in determining the properties of the modem and this is discussed in Chapter 5. The mapping can be represented by the so-called constellation diagram. A constellation is the resulting two-dimensional plot when the amplitudes of the I and Q levels of each of the points which could be transmitted (the constellation points) are drawn in a rectangular coordinate system. For a simple binary amplitude modulation scheme, the constellation diagram would be two points both on the positive x-axis. For a binary PSK (BPSK) scheme the constellation diagram would consist again of two points on the x-axis, but both equidistant from the origin, one to the left and one to the right, as we have seen in Figure 1.2. The "negative amplitude" of the point to the left of the origin represents a phase shift in the transmitted signal of 180 deg. If we allow phase shifts of angles other than 0 and 180 deg, then the constellation points move off the x-axis. They can be considered to possess an amplitude and phase, the amplitude representing the magnitude of the transmitted carrier, and the phase representing the phase shift of the carrier relative to

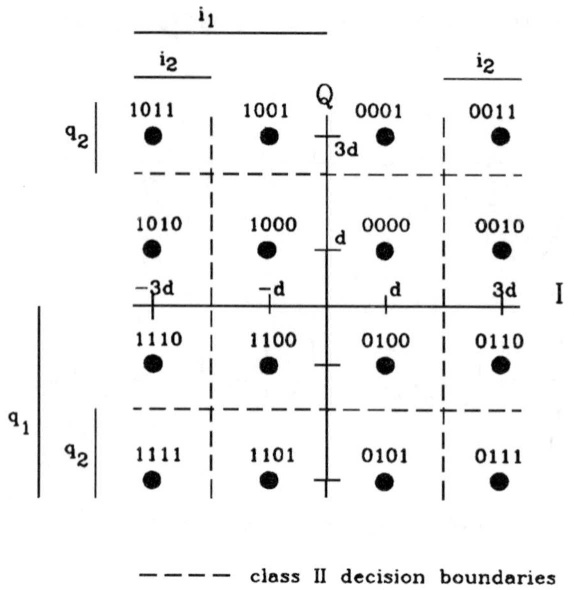

Figure 3.3: 16-QAM square constellation

the local oscillator in the transmitter. The constellation points may also be considered to have Cartesian, or complex coordinates, which are normally referred to as in-phase (I) and quadrature (Q) components corresponding to the x and y axes, respectively.

In the 16-QAM square constellation of Figure 3.3 each phasor is represented by a 4-bit symbol, constituted by the in-phase bits i_1, i_2 and quadrature bits q_1, q_2, which are interleaved to yield the sequence i_1, q_1, i_2, q_2. The quaternary quadrature components I and Q are Gray encoded by assigning the bits 01, 00, 10 and 11 to the levels $3d, d, -d$ and -$3d$, respectively. This constellation is widely used because it has equidistant constellation points arranged in a way that the average energy of the phasors is maximised. Using the geometry of Figure 3.3 the average energy is computed as

$$E_0 = (2d^2 + 2 \times 10d^2 + 18d^2)/4 = 10 \times d^2. \tag{3.1}$$

For any other phasor arrangement the average energy will be less and therefore, assuming a constant noise energy, the signal-to-noise ratio required to achieve the same bit error rate (BER) will be higher.

Notice from the mapping in Figure 3.3 that the Hamming distance amongst the constellation points, which are "closest neighbours" with a Euclidean distance of $2d$ is always 1. The Hamming distance between any two points is the difference in the mapping bits for those points, so points labelled 0101 and 0111 would have a Hamming distance of 1, and points labelled 0101 and 0011 would have a Hamming distance of 2. This is a fundamental feature of the Gray coding process and ensures that whenever a transmitted phasor is corrupted by noise sufficiently that it is incorrectly identified as a neighbouring constellation point, the demodulator will choose a phasor with a

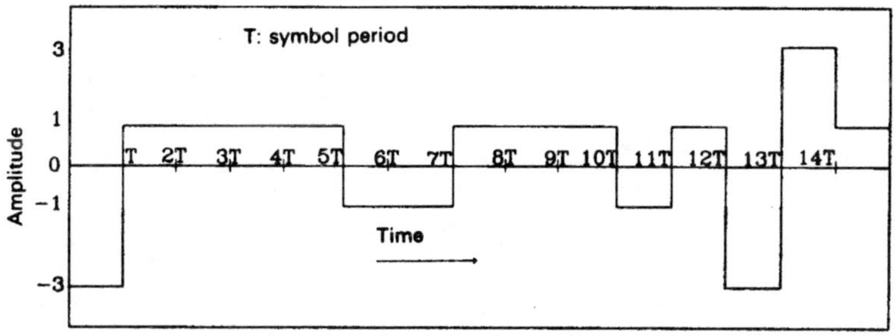

Figure 3.4: Typical I or Q component time-domain representation

single bit error. This minimises the error probability.

In Figure 3.4 we plot a typical quaternary I component sequence generated by the mapper. Because of the instantaneous transitions in the time domain this I sequence has an infinite bandwidth and hence would require an infinite channel bandwidth for transmission. The Q component has similar time and frequency domain representations. These signals must be bandlimited before transmission in order to contain the spectrum within a limited band and so minimise interference with other users or systems sharing the spectrum.

3.3 Filtering

An ideal linear phase lowpass filter (LPF) with a cut-off frequency of $f_N = f_s/2$, where $f_s = 1/T$ is the signalling frequency, T is the signalling interval duration and f_N is the so-called Nyquist frequency, would retain all the information conveyed by the quadrature components I and Q within a compact frequency band. Due to the linear phase response of the filter all frequency components would exhibit the same group delay. Because such a filter has a sinc function shaped impulse response with equidistant zero-crossings at the sampling instants, it does not result in intersymbol interference (ISI). This ideal transfer function is referred to as a Nyquist characteristic and is shown in Figure 3.5 along with its impulse response. However, such a filter is unrealisable, as all practical lowpass (LP) filters exhibit amplitude and phase distortions, particularly towards the transition between passband and stopband. Conventional Butterworth, Chebyshev or inverse Chebyshev LP filters have impulse responses with non-zero values at the equispaced sampling instants and hence introduce ISI. They therefore degrade the bit error rate (BER) performance.

Nyquist's fundamental theoretical work [178] suggested that special pulse shaping filters must be deployed, ensuring that the total transmission path, including the channel, has an impulse response with a unity value at the correct signalling instant and zero-crossings at all other sampling instants. He showed that any odd-symmetric frequency domain extension characteristic fitted to the ideal LPF amplitude spectrum yields such an impulse response, and is therefore free from ISI.

A practical odd-symmetric extension characteristic is the so-called raised cosine

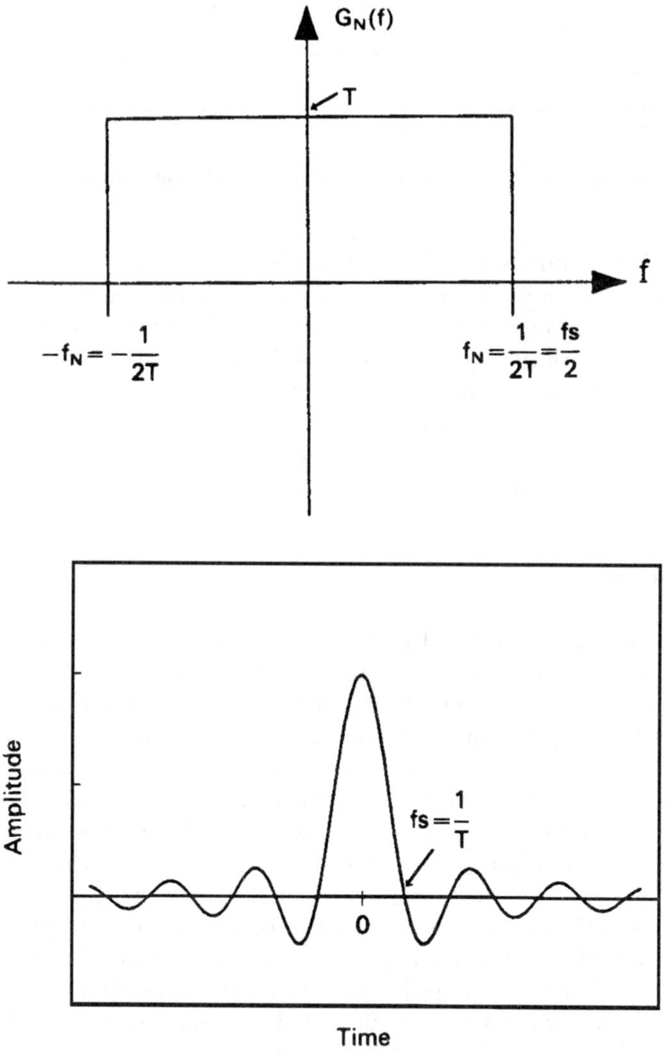

Figure 3.5: Ideal Nyquist filter characteristic and its ISI-free impulse response

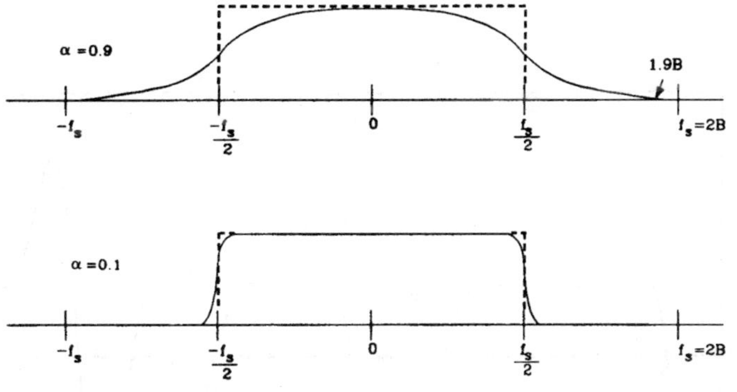

Figure 3.6: Stylised frequency response of two filters with $\alpha = 0.9$ and 0.1

(RC) characteristic fitted to the ideal LPF of Figure 3.5. The parameter controlling the bandwidth of the Nyquist filter is the so-called roll-off factor α; which is one if the ideal LPF bandwidth is doubled by the extension characteristic. If $\alpha = 0.5$ a total bandwidth of $1.5 \times B$ results, and so on. The lower the value of the roll-off factor, the more compact the spectrum becomes but the higher the complexity of the required filter. The frequency response of these filters is shown in Figure 3.6 for $\alpha = 0.9$ and $\alpha = 0.1$.

Observe in Figures 3.7 and 3.8 how the roll-off factor's value affects the time domain signal in case of 16-QAM for $\alpha = 0.9$ and $\alpha = 0.1$, respectively. With the signal is less sharply filtered and in the time domain it strongly resembles the unfiltered I component of Figure 3.4. The high frequency signal components are removed and hence the sharp time domain transitions are smoothed. The spectrum extends to $1.9 \times B$, i.e. it is virtually doubled when compared to the minimum requirement of B. Waveforms for $\alpha = 0.1$ are shown in Figure 3.8 where it can be seen that the spectrum is confined to $1.1 \times B$, and due to the sharper filtering the time domain waveform is highly smoothed, exhibiting a lesser resemblance to the unfiltered I component of Figure 3.4.

In case of additive white Gaussian noise (AWGN) with a uniform power spectral density (PSD) the noise power admitted to the receiver is proportional to its bandwidth. Therefore it is also necessary to limit the received signal bandwidth at the receiver to a value close to the transmitter's bandwidth. Optimum detection theory [179] shows that the SNR is maximised, if so-called matched filtering is used, where the Nyquist characteristic is divided between two identical filters, each characterised by the square root of the Nyquist shape, as suggested by the filters $\sqrt{N}$ in Figure 3.1.

3.4 Modulation and Demodulation

Once the analogue I and Q signals have been generated and filtered, they are modulated by an I-Q modulator as shown in Figure 3.1. This modulator essentially consists

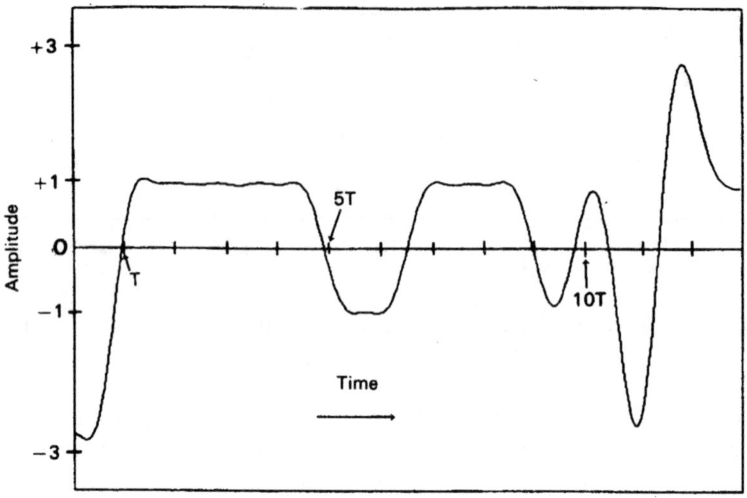

Figure 3.7: Typical filtered I component for $\alpha = 0.9$

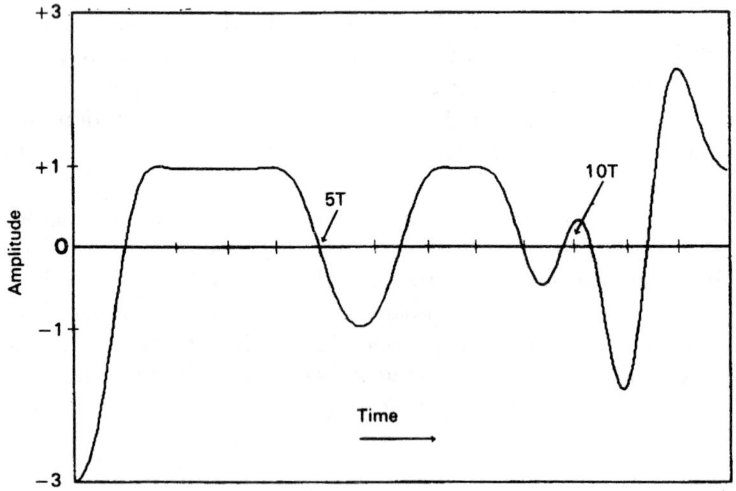

Figure 3.8: Typical filtered I component for $\alpha = 0.1$

of two mixers, one for the I channel and another for the Q channel. The I channel is mixed with an intermediate frequency (IF) signal that is in phase with respect to the carrier, and the Q channel is mixed with an IF that is 90 deg out of phase. This process allows both signals to be transmitted over a single channel within the same bandwidth using quadrature carriers. In a similar fashion, the signal is demodulated at the receiver. Provided signal degradation is kept to a minimum, the orthogonality of the I and the Q channels will be retained and their information sequences can be demodulated.

Following I-Q modulation, the signal is modulated by a radio frequency (RF) mixer, increasing its frequency to that used for transmission. Since the IF signal occurred at both positive and negative frequencies, it will occur at both the sum and difference frequencies when mixed up to the RF. Since there is no reason to transmit two identical sidebands, one is usually filtered out, as seen plotted in dashed lines in Figure 3.9.

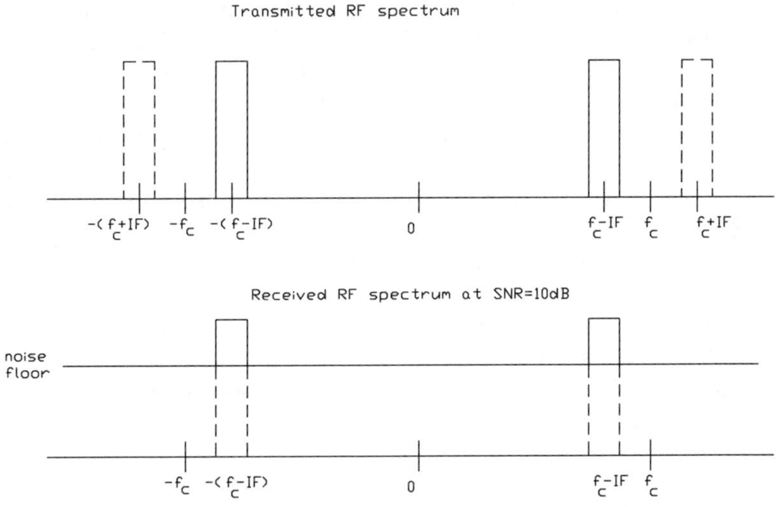

Figure 3.9: Typical transmitted and received RF spectra

The transmission channel is often the most critical factor influencing the performance of any communications system. Here we consider only the addition of noise based on the signal to noise ratio (SNR). The noise is often the major contributing factor to signal degradation and its effect exhibits itself as a noise floor, as portrayed in the received RF spectrum of Figure 3.9.

The RF demodulator mixes the received signal down to the IF for the I-Q demodulator. In order to accurately mix the signal back to the appropriate intermediate frequency, the RF mixer operates at the difference between the IF and RF frequencies. Since the I-Q demodulator includes IF recovery circuits, the accuracy of the RF oscillator frequency is not critical. However, it should be stable, exhibiting a low phase noise, since any noise present in the downconversion process will be passed on to the detected I and Q baseband signals, thereby adding to the possibility of bit errors. The recovered IF spectrum is similar to the transmitted one but with the additive noise floor seen in the RF spectrum of Figure 3.9.

Returning to Figure 3.1, I-Q demodulation takes place in the reverse order to the modulation process. The signal is split into two paths, with each path being mixed down with IFs that are 90 deg apart. Since the exact frequency of the original reference must be known to determine absolute phase, IF carrier recovery circuits are used to reconstruct the precise reference frequency at the receiver. The recovered I component should be near identical to that transmitted, with the only differences being caused by noise.

3.5 Data Recovery

Once the analogue I and Q components have been recovered, they must be digitised. This is done by the bit detector. The bit detector determines the most likely bit transmitted by sampling the I and Q signals at the correct sampling instants and comparing them to the legitimate I and Q values of $-3d$, $-d$, d, $3d$ in the case of a square QAM constellation. From each I and Q decision two bits are derived, leading to a 4-bit 16-QAM symbol. The four recovered bits are then passed on to the DAC. Although the process might sound simple, it is complicated by the fact that the "right time" to sample is a function of the clock frequency at the transmitter. The data clock must be regenerated upon recovery of the carrier. Any error in clock recovery will increase the BER.

If there is no channel noise or the SNR is high, the reconstructed digital signal is identical to the original input signal. Provided the DAC operates at the same frequency and with the same number of bits as the input ADC, then the analogue output signal after lowpass filtering with a cut-off frequency of B, is also identical to the output signal of the LPF at the input to the transmitter. Hence it is a close replica of the input signal.

3.6 Summary

In this short chapter we have described the fundamental structure of modems using the basic schematic shown in Figure 3.1. A rudimentary introduction to Gray coding and Nyquist filtering was offered. Nyquist filtering is an essential prerequisite for intersymbol interference (ISI) free communications over bandlimited channels, as we will show in the next chapter.

Accordingly, Chapter 4 will focus on a range of further QAM techniques, including the design of various phasor constellations for channels having a range of impairments. Then an in-depth treatment of ISI-free signalling using Nyquist filtering will be given, which is followed by a detailed discussion on various QAM detection methods, such as threshold detection, matched filtering and correlation receivers. Finally, Chapter 4 will incorporate a section on the linearisation problems of power amplifiers and it will be concluded by a discussion on non-differential versus differential coding of QAM signals.

Chapter 4

Basic QAM Techniques

In this chapter we consider some of the details of QAM modems in more depth than Chapter 3. We are specifically concerned with constellation types for various channels having different dominant impairments, encoding techniques and suitable forms of pulse shaping for QAM transmission. We will also present a short theoretical discourse on the necessary conditions for interference-free communications over bandlimited channels and derive the optimum transmitter and receiver filters. Methods of QAM detection, such as threshold detection, matched filtering and correlation receivers are also the subject of this chapter, which is concluded with a brief discussion on differential versus non-differential coding of QAM signals.

4.1 Constellations for Gaussian Channels

As discussed in Chapter 1, a large number of different constellations have been proposed for QAM transmissions over Gaussian channels. However, the three constellations shown in Figure 4.1 are often preferred. The essential problem is to maintain a high minimum distance, d_{min}, between points whilst keeping the average power required for the constellation to a minimum. Calculation of d_{min} and the average power is a straightforward geometric procedure, and has been performed for a range of constellations by Proakis [164]. The results show that the square constellation is optimal for Gaussian channels. The Type I and Type II constellations require a higher energy to achieve the same d_{min} as the square constellation and so are generally not preferred for Gaussian channels. However, there may be implementational reasons for favouring circular constellations over the square ones.

When designing a constellation, consideration must be given to:

(1) The minimum Euclidean distance amongst phasors, which is characteristic of the noise immunity of the scheme.

(2) The minimum phase rotation amongst constellation points, determining the phase jitter immunity and hence the scheme's resilience against clock recovery

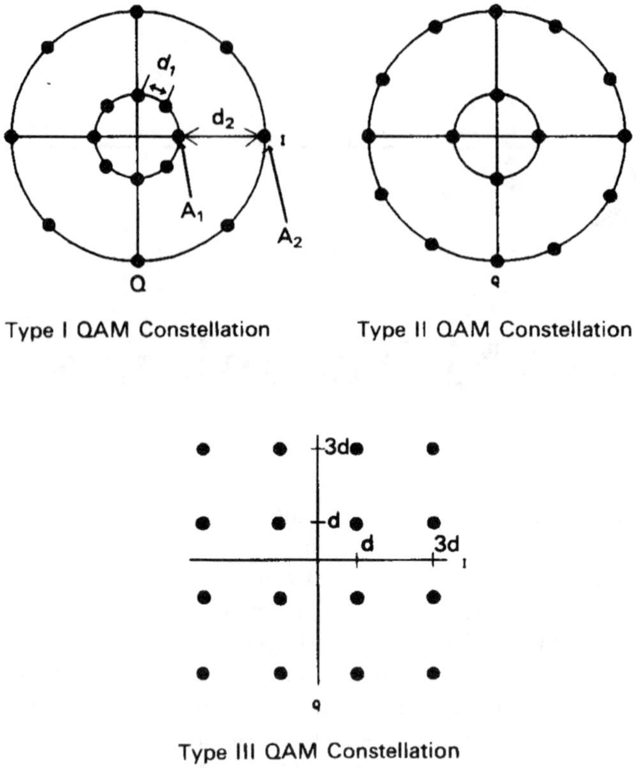

Figure 4.1: Variety of QAM constellations

imperfections and channel phase rotations.

(3) The ratio of the peak-to-average phasor power, which is a measure of robustness against non-linear distortions introduced by the power amplifier.

It is quite instructive to estimate the optimum ring ratio for the Type I constellation of Figure 4.1 in AWGN under the constraint of a constant average phasor energy E_0. Accordingly, a high ring ratio value implies that the Euclidean distance amongst phasors on the inner ring is reduced, while the distance amongst phasors on different rings is increased. In contrast, upon reducing the ring ratio the cross-ring distance is reduced and the distances on the inner ring become larger.

Intuitively, one expects that there will be an optimum ring ratio, where the overall bit error rate (BER) constituted by detection errors on the same ring plus errors between rings is minimised. This problem will be revisited in Figure 11.9 for transmissions over Rayleigh fading channels using simulation experiments. Suffice to say here that the minimum Euclidean distance amongst phasors is maximised if $d_1 = d_2 = A_2 - A_1$ in the Type I constellation of Figure 4.1. Using the geometry of

Figure 4.1 we can write that:

$$\begin{aligned}\cos 67.5^o &= \frac{d_1}{2} \cdot \frac{1}{A_1} \\ d_1 &= 2 \cdot A_1 \cdot \cos 67.5^o\end{aligned}$$

and hence

$$A_2 - A_1 = d_1 = d_2 = 2 \cdot A_1 \cdot \cos 67.5^o.$$

Upon dividing both sides by A_1 and introducing the ring ratio RR we arrive at:

$$\begin{aligned}RR - 1 &= 2 \cdot \cos 67.5^o \\ RR &\approx 1.77.\end{aligned}$$

Simulation results using 1.5 $<$ RR $<$ 3.5 both over Rayleigh and AWGN channels showed [180] that the BER depends on the channel SNR and has a very flat minimum in the above range, hence from Chapter 11 onwards over Rayleigh fading channels we often opted for RR = 3.

Under the constraint of having identical distances amongst constellation points, when $d_1 = d_2 = d$, the average energy E_0 of the Type I constellation can be computed as follows:

$$E_0 = \frac{8 \cdot A_1^2 + 8 \cdot A_2^2}{16} = \frac{1}{2}(A_1^2 + A_2^2)$$

where

$$A_1 = \frac{d}{2 \cdot \cos 67.5^o} \approx \frac{d}{0.765} \approx 1.31d$$

and

$$A_2 \approx 1.77 \cdot A_1 \approx 2.3d$$

yielding

$$E_0 \approx 0.5 \cdot (5.3 + 1.72)d^2 \approx 3.5d^2.$$

The minimum distance of the constellation for an average energy of E_0 becomes:

$$d_{min} \approx \sqrt{E_0/3.5} \approx 0.53 \cdot \sqrt{E_0},$$

while the peak-to-average phasor energy ratio is:

$$r \approx \frac{(2.3d)^2}{3.5d^2} \approx 1.5.$$

The minimum phase rotation θ_{min}, the minimum Euclidean distance d_{min} and the peak-to-average energy ratio r are summarised in Table 4.1 for both the Type I and Type III constellations.

Let us now derive the above characteristic parameters for the Type III constellation. Observe from Figure 4.1 that $\theta_{min} < 45$ deg, while the distance between phasors

Type	θ_{min}	d_{min}	r
I	45 deg	$0.53\sqrt{E_0}$	1.5
III	$<$ 45 deg	$0.63 \cdot \sqrt{E_0}$	1.8

Table 4.1: Comparison of Type I and Type III constellations

is $2 \cdot d$. Hence the average phasor energy becomes:

$$\begin{aligned} E_0 &= \frac{1}{16}\left[4 \cdot (d^2 + d^2) + 8(9d^2 + d^2) + 4 \cdot (9d^2 + 9d^2)\right] \\ &= \frac{1}{16}(8d^2 + 80 \cdot d^2 + 72d^2) \\ &= 10d^2. \end{aligned}$$

Hence, assuming the same average phasor energy E_0 as for the Type I constellation we now have a minimum distance of

$$d_{min} = 2d = 2 \cdot \sqrt{E_0/10} = \sqrt{E_0/2.5} \approx 0.63 \cdot \sqrt{E_0}.$$

Lastly, the peak-to-average energy ratio r is given by:

$$r = \frac{18d^2}{10d^2} = 1.8.$$

The Type III constellation's characteristics are also summarised in Table 4.1. Observe that the Type I constellation has a higher jitter immunity and a slightly lower peak-to-average energy ratio than the Type III scheme. However, the Type III phasor constellation has an almost 50% higher minimum distance at the same average phasor energy and hence it is very attractive for AWGN channels, where noise is the dominant channel impairment.

The Type III square constellation can only be implemented, when there are N bits per symbol, where $N = 2M$ and M is an integer. If an odd number of bits is to be encoded then the optimum constellation shape is less obvious. This issue was investigated by Smith [25] who determined that the best constellations were those which have come to be known as cross-QAM. An example of such a constellation is shown in Figure 4.2. It is not possible to Gray encode such a cross constellation so that all nearest neighbour points separated by d_{min} have a Hamming distance of one, and so a small Gray coding penalty must be accepted along with an increased encoding and decoding complexity. For these reasons, even-bit square constellations are significantly more common than odd-bit constellations.

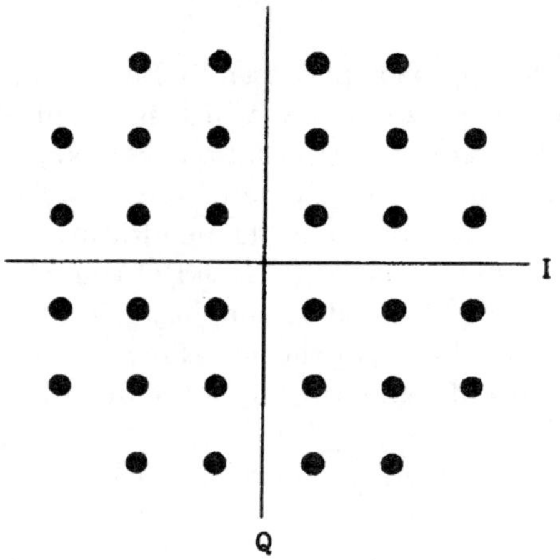

Figure 4.2: 5-bit QAM cross constellation

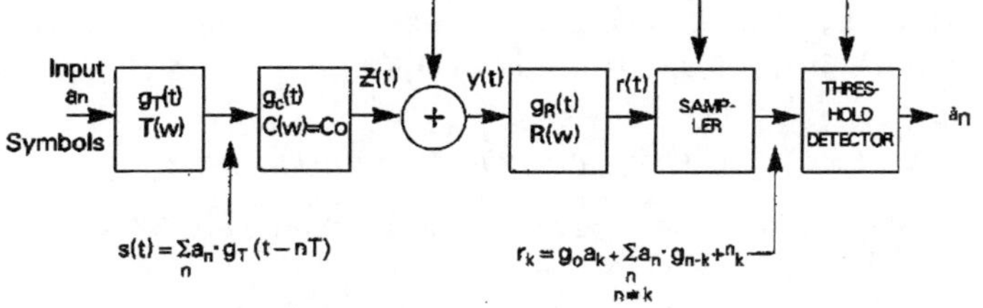

Figure 4.3: Baseband equivalent system model for the I or Q component of square 16-QAM

4.2 General Pulse Shaping Techniques

4.2.1 Baseband equivalent system

In this section we follow the approach proposed by Lucky, Salz and Weldon [18] using the simplified system model of Figure 4.3, which is essentially a simplified version of Figure 3.1, ignoring the frequency translation or mixing sections of the modem schematic, which do not influence the shape of the signal's spectrum or the signal detection process. In the figure only one of the quadrature arms is portrayed, but we consider a 2^m-ary QAM system, where the I and Q components are $2^{m/2}$-ary. The information to be conveyed is encoded as amplitude values of the signalling pulse shape $g_T(t)$, and these values change at the signalling intervals nT, $n = 1, 2, \ldots, \infty$. Thus the baseband transmitted signal on one of the quadrature channels is

$$s(t) = \sum_n a_n g_T(t - nT) \tag{4.1}$$

where $g_T(t)$ is the transmitter filter's impulse response, implying that the information symbols a_n are idealised zero-width pulses. The demodulator has to infer the transmitted information a_n from the received signal.

We continue to use the baseband equivalent system model introduced in Chapter 2, and try to find that transmitter and receiver filter pair $T(\omega)$ and $R(\omega)$, respectively, which minimises the intersymbol interference (ISI). The legitimate QAM levels are $\pm d$, $\pm 3d$, $\ldots$, $\pm(2^{m/2} - 1)$, as seen, for example, in the Type III phasor constellation of Figure 4.1. The transmitted signal is contaminated by AWGN $n(t)$ and may suffer multipath dispersion (see Chapter 2) via the channel whose transfer function is $C(\omega)$. The receiver filters the signal and after sampling at instants $kT + \tau$, where τ is the channel delay, infers the transmitted information by threshold detection. The signal $r(t)$ at the output of the receiver filter for one quadrature arm is given by:

$$r(t) = \sum_n a_n g(t - nT) + n(t), \tag{4.2}$$

where $g(t)$ represents the impulse response of the total transmission path constituted by the transmitter, receiver and channel transfer functions $T(\omega)$, $R(\omega)$ and $C(\omega)$, respectively. Using their impulse responses of $g_T(t)$, $g_R(t)$ and $g_C(t)$, respectively, we have

$$g(t) = g_T(t) * g_C(t) * g_R(t), \tag{4.3}$$

where * means convolution. The received signal at the receiver's sampling instants is written as:

$$r(kT + \tau) = \sum_n a_n g(kT - nT + \tau) + n(kT + \tau), \tag{4.4}$$

or using a convenient shorthand we have:

$$r_k = \sum_n a_n g_{k-n} + n_k, \tag{4.5}$$

where the system delay τ has been absorbed in the signalling interval index. Equation 4.5 can be rewritten to emphasise the effect of the current symbol of index $k = n$ as:

$$r_k = g_0 a_k + \sum_{\substack{n \\ n \neq k}} a_n g_{n-k} + n_k, \tag{4.6}$$

where g_0 is the main tap of the system's impulse response, while the second and third terms represent the ISI and AWGN, respectively. The receiver compares the signal r_k to the appropriately scaled decision levels of $0, \pm 2g_0 d, \ldots, \pm(2^{m/2} - 2)g_0 d$, and decides which of the legitimate a_k values is most likely to have been transmitted. An erroneous decision occurs when the (ISI + AWGN) term becomes larger than $g_0 d$, i.e.:

$$| \sum_{\substack{n \\ n \neq k}} a_n g_{n-k} + n_k| > g_0 d\,. \tag{4.7}$$

Minimising the left-hand side (LHS) term of Equation 4.7 will minimise the BER. Since the channel's transfer function may be varying with time, we assume a linear phase, distortion-free, equalised channel with $C(\omega) = C_0$ and $\phi_C = \omega T$ within the transmission band. Then the problem is that of finding the filter characteristics $T(\omega)$ and $R(\omega)$ minimising the LHS of Equation 4.7.

4.2.2 Nyquist filtering

From Equation 4.7 the BER performance of one channel of the QAM system depends on the samples g_{n-k}, $n \neq k$ of the system's impulse response and the noise n_k. It is Nyquist's sampling theorem [178] that enables us to relate these time domain samples to the spectral domain filter characteristic $G(\omega) = T(\omega)R(\omega)$, where $G(\omega)$ is the Fourier transform pair of g_{n-k}. If the system is bandlimited to $[-B, B]$, its behaviour is determined by its samples spaced apart by $T = (1/2B)$. Hence

$$g(nT) = g\left(\frac{n}{2B}\right). \tag{4.8}$$

The system's transfer function $G(\omega)$ is given in terms of these samples by the Fourier series expansion of

$$G(\omega) = \begin{cases} \frac{1}{2B}\sum_n g\left(\frac{n}{2B}\right) e^{-j\omega n/2B} & \text{if } |\omega| \leq 2\pi B \\ 0 & \text{otherwise.} \end{cases} \tag{4.9}$$

Nyquist's sampling theorem also states that the original signal $g(t)$ can be perfectly recovered from its samples by lowpass filtering (LPF) the signal with a filter having a cut-off frequency of B. In time domain this filtering corresponds to convolution with the impulse response

$$h_{LPF}(t) = 2B\frac{\sin(2\pi Bt)}{2\pi Bt} = 2Bsinc(2\pi Bt) \tag{4.10}$$

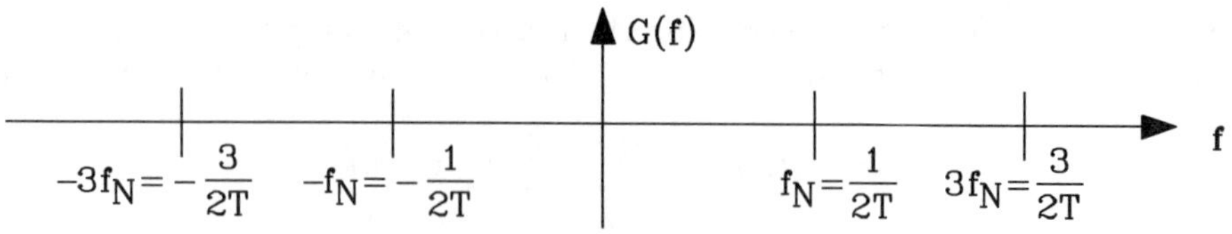

Figure 4.4: Nyquist frequency segments

of the LPF. The sampling interval given by $T = 1/2B$ is often referred to as Nyquist interval and the associated frequency of $f_N = 1/2T$ is known as the Nyquist frequency.

The ISI term of Equation 4.7 can only be eliminated if $g_{n-k} = 0$ is met for $n \neq k$, which requires g_n to have zero-crossings at $nT \neq 0$. Since we are only using the T-spaced samples of $g(n)$, this requirement can be formulated by the help of the Fourier transform as:

$$g_n = \int_{-\infty}^{\infty} G(f) e^{j2\pi f nT} df = \delta_n. \tag{4.11}$$

where δ_n is the Kronecker delta. The above integration can be carried out in segments of Nyquist bandwidths given by $f_N = 1/2T$, as seen in Figure 4.4. Then we have:

$$g_n = \sum_k \int_{(2k-1)/2T}^{(2k+1)/2T} G(f) e^{j2\pi f nT} df = \delta_n\,, \tag{4.12}$$

Instead of integrating $G(f)$ piecewise over the Nyquist frequency segments, as k is increased we can integrate over $[-1/2T, 1/2T]$ and shift $G(f)$ along the frequency axis by $2f_N = 1/T$, as k is increased, giving:

$$g_n = \sum_k \int_{-1/2T}^{1/2T} G\left(f + \frac{k}{T}\right) e^{j2\pi f nT} df = \delta_n\,. \tag{4.13}$$

Upon interchanging the summation with the integration, g_n can be expressed by the equivalent Nyquist characteristic $G_N(f)$ as follows:

$$g_n = \int_{-1/2T}^{1/2T} G_N(f) e^{j\omega nT} df = \delta_n\,, \tag{4.14}$$

where

$$G_N(f) = \begin{cases} \sum_k G\left(f + \frac{k}{T}\right) & \text{if } |f| \leq \frac{1}{2T} \\ 0 & \text{otherwise.} \end{cases} \tag{4.15}$$

Equation 4.15 reveals that the equivalent Nyquist characteristic $G_N(f)$ is the superposition of all the $1/T$ wide frequency domain slices of $G(f)$ within the band $[-1/2T, 1/2T]$, while it is zero outside this band. Since Equation 4.14 fully specifies $g_n = \delta_n$ in terms of its $T = 1/2B$ spaced samples represented by the Kronecker delta, it also describes $G_N(f)$. The *sinc* function in Equation 4.10 fulfils the criterion of ISI-free signalling and constrains $G_N(f)$ to be an ideal LPF characteristic with zero phase, as seen in Figure 4.5 and given by

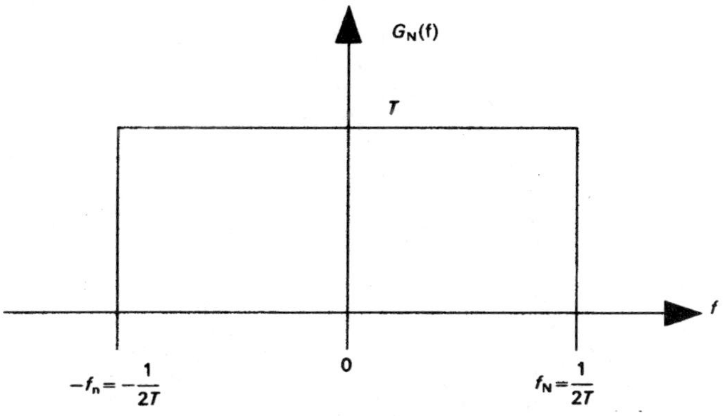

Figure 4.5: Baseband equivalent Nyquist characteristic for ISI-free transmission

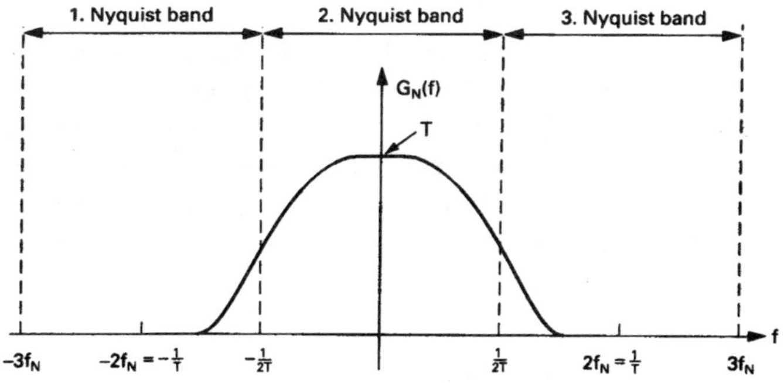

Figure 4.6: Sketch of a practical Nyquist filter characteristic

$$G_N(f) = \begin{cases} T & \text{if } |f| \le \frac{1}{2T} \\ 0 & \text{if } |f| > \frac{1}{2T}. \end{cases} \tag{4.16}$$

The simplest implementation of the ISI-free Nyquist equivalent characteristic would then be the ideal LPF of Equation 4.16 plotted in Figure 4.5. Unfortunately, this filter characteristic is unrealisable and any approximation to it results in serious ISI penalty. Any practical Nyquist filter must have a wider bandwidth than $f_N = 1/2T$, but a bandwidth in excess of $2f_N$ is usually unacceptable in terms of bandwidth efficiency. This situation is depicted in Figure 4.6, where $G(f)$ is limited to $[-2f_N, 2f_N]$ and hence has only three Nyquist segments. Then Equation 4.15 has only three terms as seen below:

$$G_N(f) = G\left(f - \frac{1}{T}\right) + G(f) + G\left(f + \frac{1}{T}\right) \tag{4.17}$$

Observe from Figure 4.6 that the Nyquist segment overlaying specified by Equation 4.17 is equivalent to folding 1 and 3 Nyquist segments back in segment 2 around

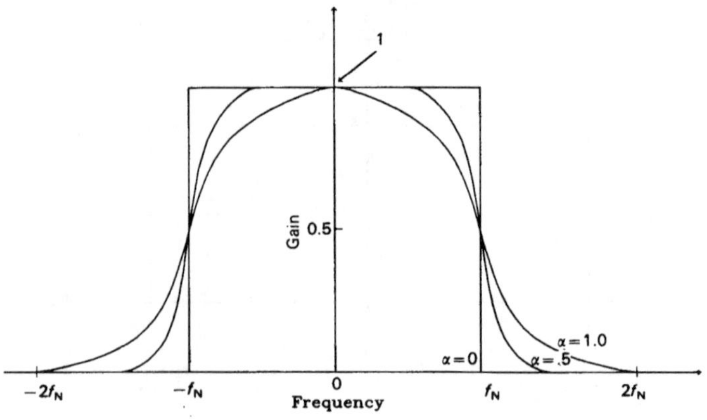

Figure 4.7: Sketch of raised-cosine Nyquist filter characteristics for various roll-off factors

the frequencies $-f_N$ and f_N, respectively. This is due to the symmetry of the positive and negative wings of the spectrum. However, if the phase characteristic is non-zero, the spectrum's conjugate complex symmetry must be taken into account.

Equivalently, the folding process can be stated as follows. The Nyquist equivalent characteristic is constituted by an ideal LPF with a cut-off frequency of $f_N = 1/2T$ plus an overlaid characteristic having odd symmetry around f_N, but not extending beyond $2f_N$.

4.2.3 Raised cosine Nyquist filtering

From the above discussion we know that there is a plethora of filters fulfilling the Nyquist criterion, having bandwidths between $f_N = 1/2T$ and $2f_N = 1/T$. In general, the higher the bandwidth efficiency the better. A practical set of convenient Nyquist filters are the so-called raised cosine (RC) filters, where the characteristic fits a quarter period of a frequency domain cosine-shaped curve to the ideal LPF transfer characteristic, as seen in Figure 4.7.

The bandwidth is controlled by the so-called roll-off factor α, defined as the ratio of excess bandwidth above f_N to f_N. The RC characteristic is then defined [18]as

$$G(f) = \begin{cases} T & \text{for } 0 \leq |f| \leq \frac{1-\alpha}{2T} \\ \frac{T}{2}\left(1 - \sin\left[\frac{\pi T}{\alpha}\left(f - \frac{1}{2T}\right)\right]\right) & \text{for } \frac{1-\alpha}{2T} < |f| \leq \frac{1+\alpha}{2T}. \end{cases} \quad (4.18)$$

4.2.4 The choice of roll-off factor

The impulse response of the RC characteristic plays an important role in deciding upon the choice of α in a particular system. Namely, when the channel becomes non-ideal and hence the ISI is non-zero, then the decay of the impulse response must be as rapid as possible to minimise the duration over which a previous transmitted symbol can influence the received signal.

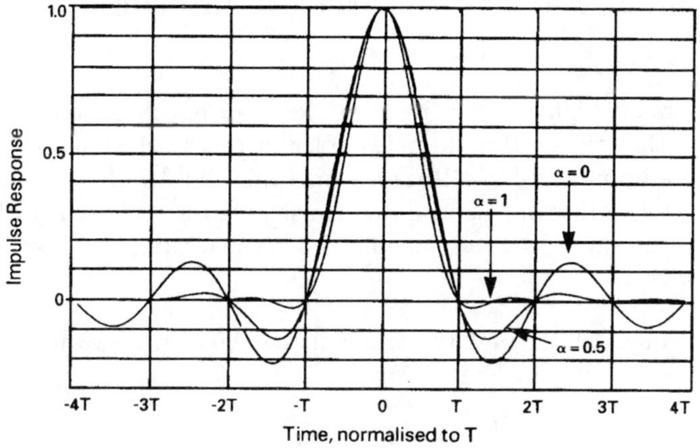

Figure 4.8: Raised-cosine Nyquist filter impulse responses for various roll-off factors

The set of impulse responses corresponding to the RC curves in Equation 4.18 is given [18] by:

$$g(t) = \frac{\sin(\pi t/T)}{\pi t/T} \frac{\cos(\alpha \pi t/T)}{1 - 4\alpha^2 t^2/T^2}. \tag{4.19}$$

These impulse responses are plotted in Figure 4.8 for roll-off factors of $\alpha = 0, 0.5$ and 1. Observe in Equation 4.19 that for $\alpha = 0$ we have:

$$g(t) = \frac{\sin(\pi t/T)}{\pi t/T}, \tag{4.20}$$

associated with the lowest bandwidth occupancy but also with the most slowly decaying impulse response. The fastest decay is ensured, when $\alpha = 1$, at the price of doubled bandwidth occupancy. In this case the impulse response exhibits an additional set of zero-crossings between those stipulated by the Nyquist condition. A good compromise is represented by $\alpha = 0.5$ in terms of both bandwidth occupancy and impulse response decay. Notice that the impulse response decay plays a crucial role in determining the system's robustness against ISI, because for slowly decaying responses the channel's ISI has a prolonged influence on future and past sampling instants. Therefore higher roll-off factors must be used in high-ISI environments , despite the lower bandwidth efficiency.

In summary, we determined the required transfer characteristic of the total transmission system, which must have an odd-symmetric roll-off around $f_N = 1/2T$. Lower roll-off factors yield spectral compactness, but they are also more vulnerable to ISI via non-ideally equalised channels. This makes clock recovery much more difficult to achieve.

4.2.5 Optimum transmit and receive filtering

In this subsection we highlight briefly how the transfer characteristic $G_N(f)$ must be split between the transmitter filter $G_T(f)$ and the receiver filter $G_R(f)$ in order to maximise noise immunity [18]. Recall from Equation 4.7 that we have minimised the effects of ISI by selecting an overall transfer characteristic $G_N(f)$. The BER due to AWGN is minimised by maximising the SNR at sampling instants for a given transmitted power.

Recall from Equation 4.1 that the transmitted signal was given by

$$s(t) = \sum_{n=-\infty}^{\infty} a_n g_T(t - nT),$$

where a_n represents the Dirac-shaped information symbols, while $g_T(t)$ is the transmitter filter's impulse response. Then the average transmitted power over an interval of $2 \cdot N \cdot T$ is given by the expectation value:

$$P_T = \langle \lim_{N\to\infty} \frac{1}{2NT} \int_{-NT}^{NT} \left[\sum_{n=-N}^{N} a_n g_T(t - nT) \right]^2 dt \rangle. \tag{4.21}$$

Upon interchanging the expectation and limit computation we have:

$$P_T = \lim_{N\to\infty} \frac{1}{2NT} \sum_{n=-N}^{N} \sum_{m=-N}^{N} \langle a_n a_m \rangle \int_{-NT}^{NT} g_T(t - nT) g_T(t - mT) dt. \tag{4.22}$$

For uncorrelated input symbols we get:

$$\langle a_n a_m \rangle = \begin{cases} 0 & n \neq m \\ \bar{a^2} & n = m \end{cases} \tag{4.23}$$

and hence

$$\begin{aligned} P_T &= \lim_{N\to\infty} \frac{\bar{a^2}}{2NT} \sum_{n=-N}^{N} \int_{-NT}^{NT} g_T^2(t - nT) dt \\ &= \frac{\bar{a^2}}{T} \int_{-\infty}^{\infty} g_T^2(t) dt. \end{aligned} \tag{4.24}$$

By exploiting Parseval's theorem:

$$\int_{-\infty}^{\infty} g_T^2(t) dt = \int_{-\infty}^{\infty} |G_T(f)|^2 df \tag{4.25}$$

the channel's input power can also be expressed in terms of the spectral domain transmitter transfer function $G_T(f)$:

$$P_T = \frac{\bar{a^2}}{T}\int_{-\infty}^{\infty}|G_T(f)|^2 df = \frac{\bar{a^2}}{T}\int_{-\infty}^{\infty}\left|\frac{G_N(f)}{G_R(f)}\right|^2 df, \tag{4.26}$$

where we have used the fact that $G_N(f) = G_T(f)\cdot G_R(f)$. The received noise has a spectral density of $N(f)$ and after passing through the receiver filter it has a power of

$$P_N = \int_{-\infty}^{\infty} N(f)|G_R(f)|^2 df. \tag{4.27}$$

The noise power P_N has to be minimised under the constraint of constant channel input power P_T so as to maximise the SNR. Using the technique of variational calculus [18] the functional F defined below must be minimised:

$$F = \lambda P_T + P_N\,, \tag{4.28}$$

where λ is the so-called Lagrange multiplier.

The result of this minimisation problem is that for a flat noise spectral density N_0 and for a real Nyquist equivalent characteristic, the transmitter and receiver filters must be identical in order to maximise the SNR, i.e.:

$$G_T(f) = G_R(f) = \sqrt{G_N(f)}. \tag{4.29}$$

This important result was already implicitly exploited in Figure 3.1, where we represented the square root of the Nyquist filter by $\sqrt{N}$.

4.2.6 Characterisation of ISI by eye diagrams

When a dispersive channel introduces ISI, previously transmitted symbols will cause non-zero contributions at subsequent sampling instants. The so-called eye diagram is a simple and convenient tool to study the effects of ISI and other channel impairments both by simulation and in hardware experiments.

Let us imagine an ideal noiseless and ISI-free, infinite bandwidth communications system transmitting a binary data sequence having values of ± 1, at a bit rate of $R = 1/T$, as shown in Figure 4.9.

If this system's output signal is displayed on an analogue oscilloscope screen with persistence, when using external synchronisation having a frequency of $R = 1/T$, the beam starts its travel across the screen every $1/T$ second and draws the adjacent signal sections on top of each other, as also shown in Figure 4.9.

If this PRBS is now transmitted via a Nyquist-filtered, bandlimited, but ISI-free channel, the corresponding signals are portrayed in Figure 4.10.

Observe that since no ISI and no noise are present, the eye is "clean" or "fully open" at the sampling instants and the zero-crossings occur always at multiples of T, appearing at the same instant in the eye diagram.

Let us now introduce ISI. The signalling waveforms will have typically non-zero values at all sampling instants and the received PRBS and its eye diagram will typically resemble those in Figure 4.11.

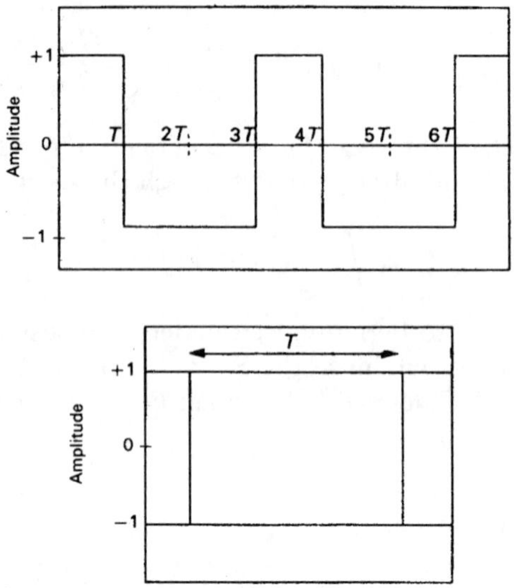

Figure 4.9: A typical binary data sequence and its eye diagram

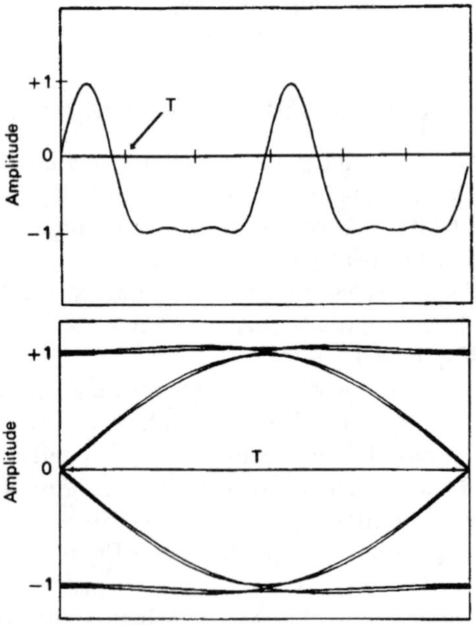

Figure 4.10: Received signal and its eye diagram via bandlimited channel with no noise and ISI

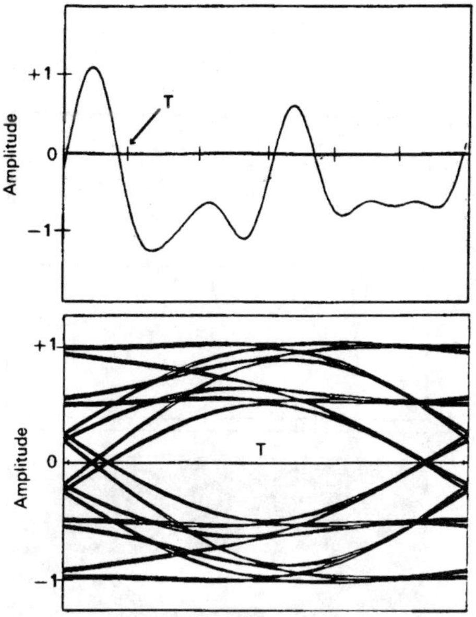

Figure 4.11: Received signal and eye diagram impaired by ISI

When AWGN is also superimposed, we expect waveforms similar to those in Figure 4.12.

As demonstrated by Figure 4.13, nearly all typical channel impairments become recognisable in the eye diagrams, and hence eye diagrams are very useful in quick overall system evaluation. Observe that imperfectly timed sampling has a similar effect to increasing the AWGN. The zero-crossing jitter becomes very crucial in case of clock recovery circuits that operate by detecting zero-crossings (see Chapter 6) and using the time at which these occur to derive the optimal sampling point. When multilevel transmissions are used, the eye diagram becomes much more complicated due to the increased number of possible transitions. The eye opening is narrower, as demonstrated by Figure 4.14, where the transmitter's eye pattern is seen for a 16-QAM Nyquist-filtered modem having a roll-off factor of $\alpha = 1$.

This Nyquist filter with $\alpha = 1$ has the sharpest possible impulse response decay and hence minimises the effects of ISI. Lower α values result in smoother time domain transitions, yielding a more closed eye pattern. This increased sensitivity is the price paid for the higher spectral compactness of narrower Nyquist filters.

4.2.7 Non-linear filtering

In 1982 Feher [181] suggested the use of non-linear filtering (NLF) for QAM satellite communications. This was a technique he had developed along with Huang in 1979 which simplified filter design by simply fitting a time domain quarter period of a sine wave between two symbols for both of the quadrature carriers. This allowed the

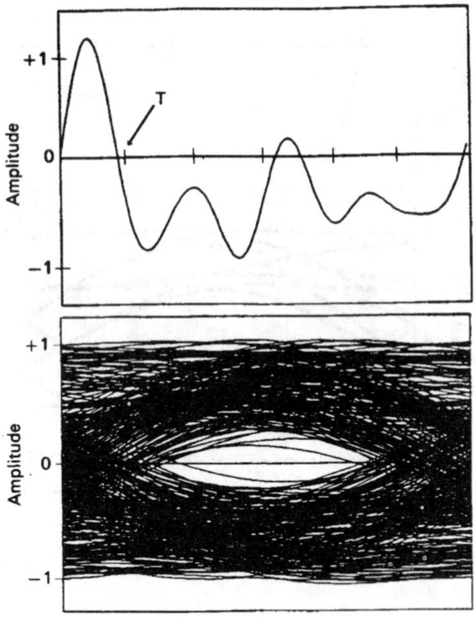

Figure 4.12: Received signal and eye diagram with ISI and AWGN

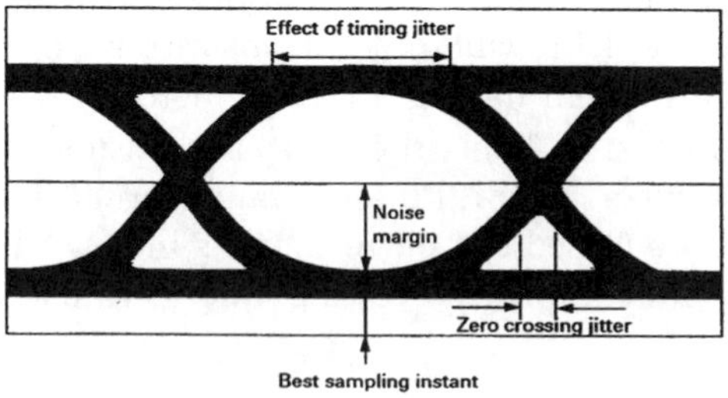

Figure 4.13: Effect of channel impairments on the eye diagram

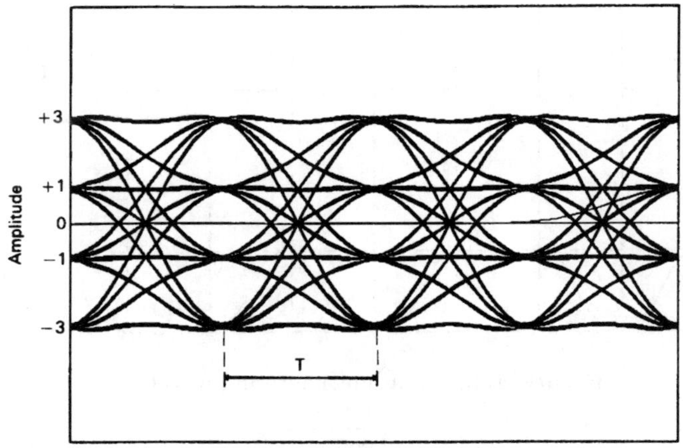

Figure 4.14: 16-QAM transmitter eye diagram using $\alpha = 1$

generation of jitter-free bandlimited signals, which had previously been a problem, improving clock recovery techniques. Furthermore, unlike partial response filtering techniques which intentionally introduce ISI, NLF produces an ISI-free waveform since there is no contribution from previous symbols at any sample point, which is advantageous when complex high-level QAM constellations are transmitted. The disadvantage of this form of filtering is that it is less spectrally efficient than optimal partial response filtering schemes. We performed some simulation work to compare their spectral efficiency, and our results suggested that whilst optimal partial response schemes [182] necessitated an excess bandwidth of about 20% above the Nyquist frequency $f_N = 1/(2 \cdot T)$, the NLF scheme required an excess bandwidth of about 40%. Nevertheless, its implementational advantages often render this loss of efficiency acceptable. The power spectrum of a NLF signal is given [181] by

$$S(f) = T\left(\frac{\sin 2\pi fT}{2\pi fT}\frac{1}{1-4(fT)^2}\right)^2 \tag{4.30}$$

and some sample waveforms are given in Figure 4.15 for one quadrature arm. The other quadrature arm is identical.

4.3 Methods of Generating QAM

4.3.1 Generating conventional QAM

As discussed in Chapter 3 and as shown in Figure 4.16,

in a typical QAM modem the incoming binary data stream is passed through a serial-to-parallel convertor of an appropriate width. The data is then passed through a logic block which performs the translation between the binary input data and the constellation points, to allow Gray coding. Typically such a block would be imple-

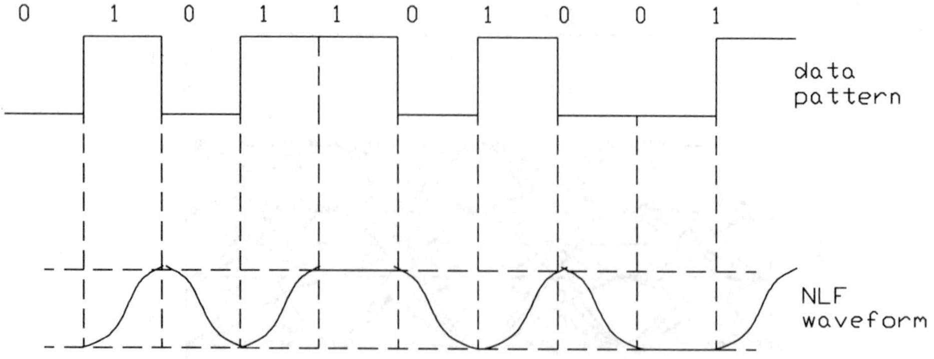

Figure 4.15: Non-linear filtering waveforms

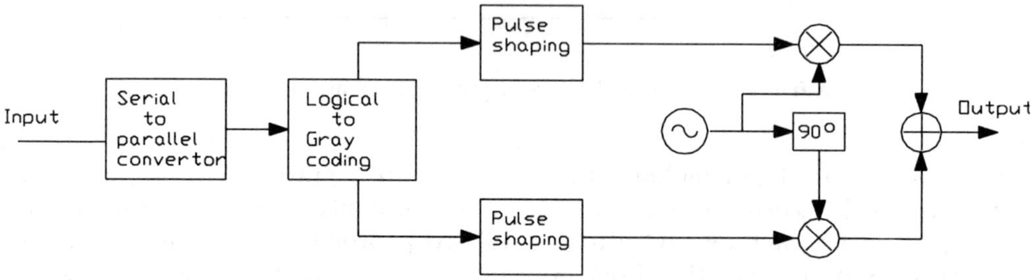

Figure 4.16: Standard QAM modulator

mented as a look-up table using a programmable memory device. In the case of square QAM the output from the serial-to-parallel convertor can be partitioned into two subsets, and each encoded onto a separate axis using identical mapping blocks. Digital-to-analogue (D/A) conversion is then required, followed by pulse shaping, before upconverting the signal to the carrier frequency. It is often advantageous to perform the pulse shaping prior to the D/A conversion by oversampling the transmitted data and using further look-up tables containing the pulse shaping law in order to produce a digital representation of the smoothed signal. This will still need to be filtered after D/A conversion in order to remove aliasing errors caused by the oversampling ratio, but this filtering is substantially simpler than would otherwise be required.

In order to demodulate the signal, the reverse process is followed. The incoming waveform is sampled at the correct instants provided by the clock recovery circuit, A/D converted and a decision is made as to the most likely constellation point to have been transmitted. The position of this constellation point is then passed through a logic block which performs the reverse operation to the one in the transmitter, leading to the encoded information. This is passed through a parallel-to-serial convertor leading to the output data stream.

Whilst this is by far the most common method for generating QAM signals, an alternative suggestion referred to as superposed QAM was made by Miyauchi *et al.* [26],

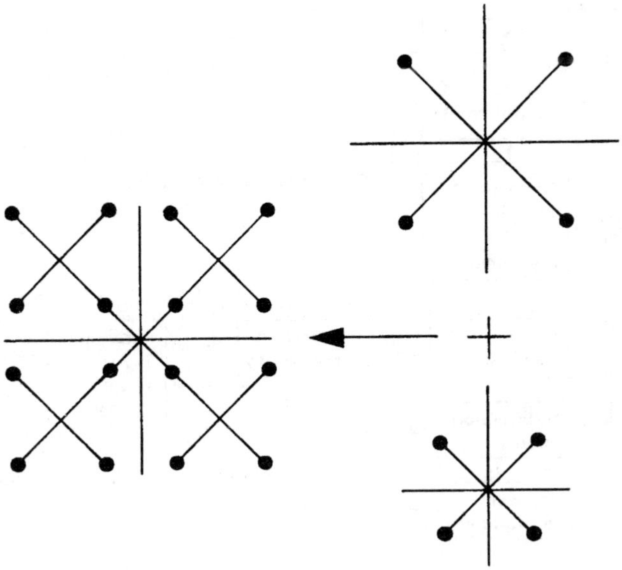

Figure 4.17: Constructing QAM with two PSK modulators

which will be considered in the next section.

4.3.2 Superposed QAM

In superposed QAM modems the approach is to use a number of PSK modems in tandem as explained below. This had the advantage of being able to use circuitry already developed for PSK modems, allowing ease of implementation. This technique has not gained much popularity, since the conventional QAM generation method described in the previous section has generally proved more efficient. Nevertheless, a brief description is included for completeness.

A diagram showing how a 16-level square QAM constellation could be constructed by this method is given in Figure 4.17. Essentially, the first two bits of the 4-bit symbol are passed to a QPSK modulator operating at a certain power and the second two bits are passed to a second QPSK modulator operating at one-quarter of the power of the first. The two QPSK modulated signals are then superimposed before transmission. A similar philosophy can be employed at the receiver.

Block diagrams of transmitter and receiver structures are given in Figure 4.18. At the receiver the signal is passed to a QPSK decoder. This essentially determines the quadrant of transmission, or the signal level produced by the first QPSK modulator within the transmitter. The decision is remodulated and passed to a subtraction circuit. This forms the difference between the incoming signal and the remodulated signal, giving the signal produced by the second QPSK modulator. This can also be demodulated to produce the remaining output bits.

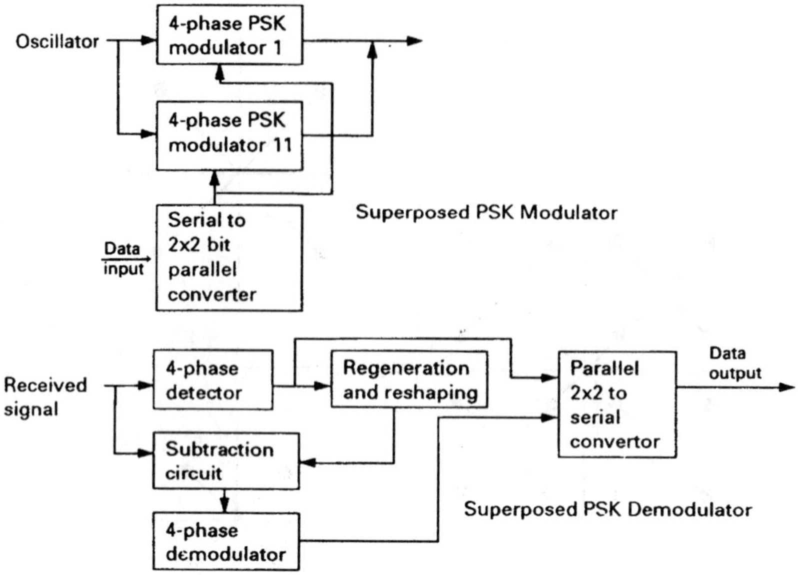

Figure 4.18: Superposed PSK modulator and demodulator

4.3.3 Offset QAM

Offset, or staggered modulation, is a technique which evolved with PSK transmissions in an effort to reduce the envelope variations of the transmitted signal, and in particular to prevent the envelope of the transmitted signal passing through zero during transitions between symbols as this required the power amplifier to maintain linearity across a wide amplitude range. This was problematic as non-linearities in amplifiers, particularly those used for satellite applications, could cause considerable distortion of the transmitted waveform. Furthermore, large variations in the envelope could result in a requirement for tight filtering if out-of-band spillage was to be avoided. Offset modulation involves delaying one of the quadrature arms by half a symbol period with respect to the other arm. Therefore, when the maximum rate of change of level is taking place on one arm (i.e. in the centre of a transition between levels) then the other arm is near stationary. Offset QAM (O-QAM) can be expressed as:

$$\begin{aligned} s(t) &= \left[\sum_k a_{2k} h(t-2kT)\right] \cos(2\pi f_c t) \\ &- \left[\sum_k a_{2k-1} h(t-(2k-1)T)\right] \sin(2\pi f_c t) \end{aligned} \tag{4.31}$$

where f_c is the carrier frequency, $1/T$ is the symbol rate, a_k is the kth data symbol, for 16-level square QAM taking on values ± 1 and ± 3, and $h(t)$ is the signalling waveform corresponding to the impulse response of the transmitter's pulse shaping filter. A

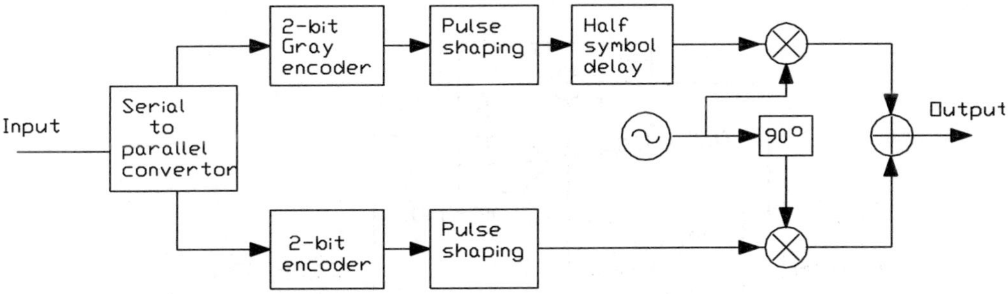

Figure 4.19: O-QAM modulator structure

C/I_c dB	S_r ($\alpha = 2$)	S_r ($\alpha = 4$)	S_r ($\alpha = 6$)
-9	1.07	1.07	1.06
-15	1.05	1.04	1.03
-20	1.04	1.03	1.03

Table 4.2: Relative efficiency of O-QAM compared to QAM

diagram showing the structure of an O-QAM modulator is shown in Figure 4.19.

Offsetting the modulation can make clock recovery more problematic, and decoding more complex. Further uncertainty can exist within the receiver concerning which axis is delayed with respect to the other, and we consider differential coding to overcome this ambiguity in Section 4.7. Claims have been made for less out-of-band spillage for O-QAM compared to QAM [60], with a resulting increase in spectral efficiency due to tighter packing of adjacent channels. Malkamaki [60] performed a number of simulations where he compared the spectral efficiency of QAM with O-QAM. Spectral efficiency is defined in many ways, and we consider some of these definitions in Chapter 17 of this book. In the meantime, we use Malkamaki's definition that the spectral efficiency is given by

$$S_r = \frac{R_m}{B_m}\left(\frac{C}{I_c}\right)^{-2/\alpha} \tag{4.32}$$

where R_m is the modulation rate, B_m is the channel spacing, C is the received carrier power, I_c is the interference power, and α is the assumed propagation exponent. Malkamaki performed a number of simulations and reported the relative efficiency results displayed in Table 4.2.

The results suggest that in terms of spectral efficiency O-QAM is only marginally better than QAM, and implementation difficulties may reduce this slight advantage. Furthermore, O-QAM requires a less efficient differential coding system than QAM, as discussed in Section 4.7, and it is not clear whether Malkamaki has taken this into account. It would appear that O-QAM is only advantageous where large envelope variations must be avoided due to amplifier design considerations.

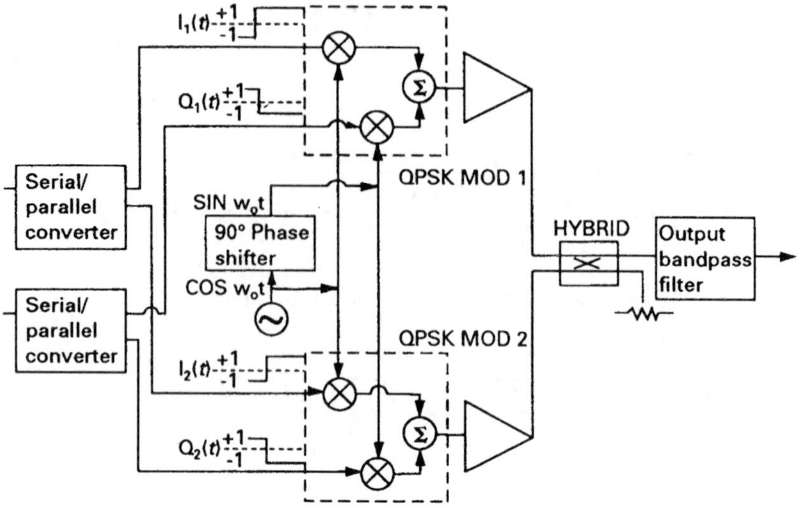

Figure 4.20: NLA structure

4.3.4 Non-linear amplification

QAM had been proposed for satellite applications in the late 1960s but a major implementational stumbling block was that of amplification. Because of limited power within satellites, power efficient non-linear class C amplifiers were preferred. QAM, being a non-constant envelope modulation, requires linear amplifiers. The increase in power required when using low-efficiency linear class A amplifiers precluded QAM from most satellite applications. Feher [31] suggested a new method for generating QAM signals using highly non-linear amplifiers, which he termed non-linearly amplified QAM (NLA-QAM), in order to overcome this problem. Two separate amplifiers for the 16-QAM case were used, one operating with a quarter the output power of the other. This is shown in Figure 4.20.

The higher power amplifier processed the two most significant bits (MSBs) of the 4-bit symbol only, and the lower power amplifier processed only the two least significant bits (LSBs). The two MSBs and LSBs represented the high power and low power phasors, respectively, which always maintained a constant magnitude and hence were class C amplified. The amplified processed signals were then summed at full output power to produce the QAM signal.

Again, in this arrangement each amplifier is producing a QPSK signal, therefore the scheme has certain similarities with that proposed by Miyauchi, except the latter performs summation of QPSK signals prior to amplification, whereas Feher sums the signals after amplification. Because both amplifiers were able to use constant envelope modulation they could be operated at a constant level with resulting high power efficiency. However, Feher's NLA scheme resulted in an increased complexity due to the need for two amplifiers and a hybrid combiner, compared to previous

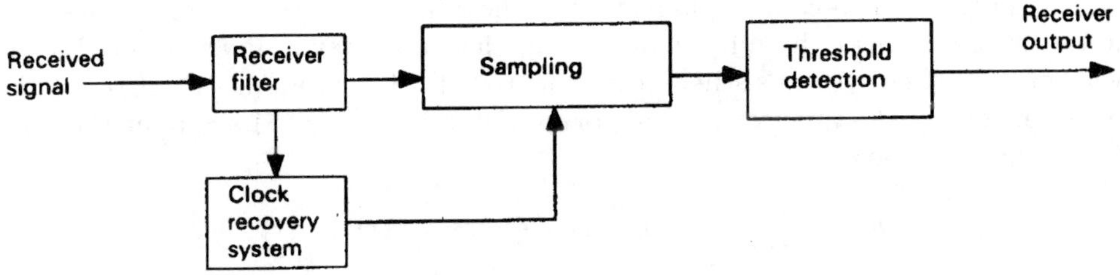

Figure 4.21: Receiver structure using threshold detection

systems which used only a single amplifier. Nevertheless, in satellite applications, complexity was relatively unimportant compared to power efficiency, and this NLA technique offered a substantial 5 dB power gain, considerably increasing the potential of QAM in severely power-limited applications. This work was soon followed by a performance study of a NLA 64-state system [32] which extended the NLA scheme to 64 levels by using three amplifiers all operating at different power levels.

Non-linear amplification is rarely used in other applications due to the increase in hardware complexity.

4.4 Methods of Detecting QAM Signals

4.4.1 Threshold detection of QAM

In most of this book we will be assuming that the filtered received waveform is sampled at the correct point by use of a clock recovery circuit (see Chapter 6) and subjected to threshold detection in order to derive the received information symbol for that symbol instant. A generic receiver designed around these principles is shown in Figure 4.21. The optimum receiver filter maximising the signal-to-noise ratio (SNR) and hence minimising the bit error rate (BER) was derived in Section 4.2, while the BER using a perfectly coherent threshold detector will be analytically derived for 16-QAM and 64-QAM over AWGN channels in Chapter 5.

Unless stated to the contrary, throughout this book we will be considering the receiver of Figure 4.21, which is based on threshold detection.

4.4.2 Matched filtered detection

In the previous section we have briefly summarised the salient features of the low complexity threshold detector receiver, which is constituted by a receiver filter followed by a sampler and threshold detector. In Section 4.2 we optimised the receiver filter under the constraint of constant channel input power or transmitted power P_T and minimised the amount of noise power P_N admitted to the threshold detector by appropriately designing the receiver filter. The result was that the Nyquist characteristic had to be split symmetrically between the transmitter and receiver filter, while the channel was assumed to be equalised.

When designing a so-called matched filter, the approach is to maximise the signal-to-noise ratio (SNR) at the output of the matched filter for a given received signal waveform [183]. The system's schematic is identical to that in Figure 4.3. Considering for example one of the quadrature components of a 2^m-ary QAM signal at the input of the receiver we have:

$$y(t) = \sum_n a_n \cdot h(t - nT) + n(t) = z(t) + n(t)$$

where a_n represents the $M = 2^{m/2}$-ary I or Q components, $n(t)$ is the AWGN and $h(t)$ is given by the convolution:

$$h(t) = g_T(t) * g_c(t),$$

with $g_T(t)$ and $g_c(t)$ being the transmitter filter's and the channel's impulse response, respectively. Clearly, signalling is carried out using the signalling waveforms $h(t)$ weighted by the $M = 2^{m/2}$-ary I and Q components. Assuming that there is no intersymbol interference (ISI) the sampling circuit's output signal at time $t = T$ is the sum of a signal contribution r_k and a noise contribution n_k.

Our goal is then to maximise the following SNR term [164, 183, 184]:

$$SNR_T = \frac{r_k^2}{\sigma_0^2},$$

at $t = T$, where σ_0^2 is the noise variance, as a function of the filter transfer function $R(f)$. The information signal $r(t)$ at the filter's output can be formulated by the help of the Fourier transform as:

$$r(t) = \int_{-\infty}^{\infty} R(f)Z(f) \cdot e^{j2\pi ft} df,$$

where $Z(f)$ is the frequency domain representative of the received signal $z(t)$ and $R(f)$ is the transfer function of the filter $g_R(t)$ at the receiver's input.

Following Sklar's approach [183] and assuming a double-sided noise spectral density of $N_0/2$, the filter's output noise power is given by:

$$\sigma_0^2 = \frac{N_0}{2} \int_{-\infty}^{\infty} |R(f)|^2 df.$$

Then the SNR is expressed as:

$$SNR_T = \frac{|\int_{-\infty}^{\infty} R(f) \cdot Z(f) e^{j2\pi fT} df|^2}{N_o/2 \int_{-\infty}^{\infty} |R(f)|^2 df}. \quad (4.33)$$

In order to find the optimum filter $R(f) = R_{opt}(f)$, Schwartz's inequality can be invoked, which is formulated in general as:

$$|\int_{-\infty}^{\infty} g_1(x) \cdot g_2(x) dx|^2 \leq \int_{-\infty}^{\infty} |g_1(x)|^2 dx \cdot \int_{-\infty}^{\infty} |g_2(x)|^2 dx.$$

The equality is satisfied, if $g_1(x) = c \cdot g_2^*(x)$, where c is a constant and $*$ denotes the complex conjugate. If we let $g_1(x) = R(f)$ and $g_2(x) = Z(f) \cdot e^{j2\pi fT}$, we arrive at:

$$|\int_{-\infty}^{\infty} R(f) \cdot Z(f) e^{j2\pi fT} df|^2 \leq \int_{-\infty}^{\infty} |R(f)|^2 df \cdot \int_{-\infty}^{\infty} |Z(f)|^2 df, \tag{4.34}$$

where equality is satisfied if

$$R(f) = c \cdot Z^*(f) e^{-j2\pi fT}. \tag{4.35}$$

Upon substituting Equation 4.34 into Equation 4.33 we have

$$SNR_T \leq \frac{\int_{-\infty}^{\infty} |R(f)|^2 df \cdot \int_{-\infty}^{\infty} |Z(f)|^2 df}{N_0/2 \int_{-\infty}^{\infty} |R(f)|^2 df}$$

giving

$$SNR_T \leq \frac{2}{N_0} \int_{-\infty}^{\infty} |Z(f)|^2 df.$$

If we denote the received signal's energy by E_z and exploit the Parseval theorem, we arrive at:

$$E_z = \int_{-\infty}^{\infty} |z(t)|^2 dt = \int_{-\infty}^{\infty} |Z(f)|^2 df$$

and hence the maximum SNR is given by:

$$SNR_T^{max} = \frac{2E_z}{N_0}. \tag{4.36}$$

Observe that the received signal's waveform shape does not appear in the above equation, indicating that the SNR depends on the energy of $z(t)$, rather than its shape. Recall from Equations 4.35 and 4.36 that the equality condition ensured the maximum SNR and from Equation 4.35 this requires in the time domain that

$$\begin{aligned} g_R(t) &= \mathcal{F}^{-1}\{R(f)\} \\ &= \mathcal{F}^{-1}\{c \cdot Z^*(f) e^{-j2\pi fT}\}, \end{aligned}$$

where $\mathcal{F}$ represents the Fourier transform. When taking into account that:

$$\mathcal{F}^{-1}\{Z^*(f)\} = z(-t)$$

and that the factor $e^{-j2\pi fT}$ corresponds to a time domain delay of T, we arrive at

$$g_R(t) = \begin{cases} c \cdot z(T-t) & 0 \leq t \leq T \\ 0 & \text{otherwise.} \end{cases} \tag{4.37}$$

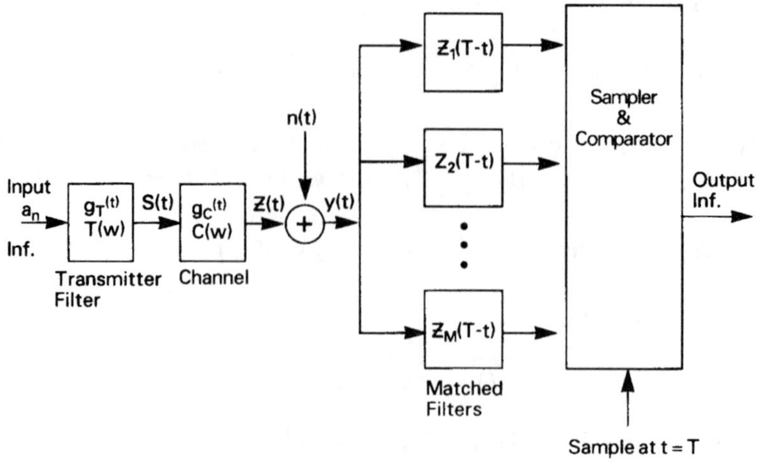

Figure 4.22: Matched filtered receiver

Explicitly, the impulse response of the filter maximising the SNR at $t = T$ must be the reflected and shifted version of the received signal $z(t)$. Observe from Equation 4.37 that the delay of T renders the impulse response causal or realisable, since $g_R(t) = z(-t)$ would not be realisable.

The matched filtered receiver will then have a bank of filters, each of which is matched to one of the possible $M = 2^{m/2}$ received I or Q waveforms $z_1(t), z_2(t), \ldots, z_M(t)$, as portrayed in Figure 4.22. The received signal is filtered through all matched filters and the output signal of that specific filter will be the largest, which is adequately matched to the received signal. The output signal of the remaining filters will be significantly lower, and hence the comparator finally infers, which a_n $n = 1, \ldots, M$ has been transmitted.

Having described the matched filtered receiver we now briefly focus our attention on an alternative implementation of it, the so-called *correlation receiver.*

4.4.3 Correlation receiver

Observe from Figure 4.22 that the causal filter's output signal can also be described by the help of the convolution:

$$r(t) = y(t) * g_R(t) = \int_0^t y(\tau) \cdot g_R(t-\tau)d\tau \tag{4.38}$$

and setting arbitrarily $c = 1$ in Equation 4.37 as well as exploiting that:

$$g_R(t-\tau) = c \cdot z[T-(t-\tau)]$$

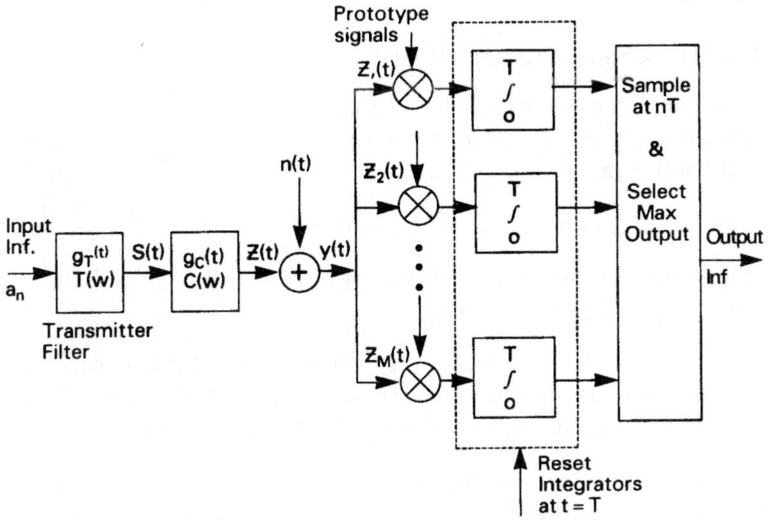

Figure 4.23: Correlation receiver

yields:

$$r(t) = \int_0^t y(\tau) \cdot z(T - t + \tau) d\tau.$$

For $t = T$ this can be written as:

$$r(t = T) = \int_0^{t=T} y(\tau) \cdot z(\tau) d\tau,$$

which allows us to interpret the matched filtered receiver as a *correlation receiver.*

Explicitly, in the correlation receiver of Figure 4.23 the received signal $y(t)$ is correlated with all legitimate prototype signals $z_n(t), n = 1, \ldots, M$ using a bank of M correlators and one of them is assumed to have been transmitted which exhibits the highest correlation at $t = T$. It is important to emphasise, however, that the output of the matched filter is only equal to that of the correlator at $t = T$, which can be easily demonstrated using simple specific received signal waveforms.

Lastly, it is important to note that the channel's impulse response $g_c(t)$ can be determined by using channel sounding methods, briefly explained later in Section 7.6. If $g_c(t)$ is known, then the prototype signals $z_n(t), n = 1, \ldots, M$ can be determined. Alternatively, the channel can be equalised initially in order to render $C(f) = C_0$, where C_0 is a constant. This constant transfer function is associated with an ideal Diracdelta impulse response $g_c(t)$, in which case we have for the prototype signals $z_n(t) = s_n(t), n = 1, \ldots, M = 2^{m/2}$ given by $z_n(t) = s_n(t) = a_n g_T(t - nT)$.

4.5 Linearisation of Power Amplifiers

4.5.1 The linearisation problem

A major problem with bandwidth efficient QAM schemes is that their performance is strongly dependent on the linearity of the transmission system. The system designer is faced with the following choices:

(1) Using a low-efficiency linear class A amplifier.

(2) Opting for Feher's non-linear amplification principle [185] relying on separate amplifiers for all amplitude levels.

(3) Deploying a power efficient, weakly non-linear (NL) amplifier and using linearisation techniques to minimise the power amplifier's out-of-band emission and produce a waveform which can be accurately demodulated.

Techniques for achieving (3) include active biasing [186], the feedforward method [187], the LINC technique [188], negative feedback [189, 190], predistortion [1, 65, 191] or postdistortion [192].

4.5.2 Linearisation by predistortion [1]

4.5.2.1 The predistortion concept

This subsection concentrates on linearisation by predistortion following the approach proposed by Stapleton *et al.* [1, 65, 191, 193–197] whereby a slowly adapting predistorter is used [1] to minimise the out-of-band spectral spillage due to power amplifier non-linear (NL) distortions. The simplest method would be to use a non-linear device in front of the power amplifier to produce intermodulation products in antiphase to those of the amplifier in an open-loop configuration. However, this will not adapt to varying transmitter characteristics and needs to be set individually for each amplifier. Instead we use a slowly adapting feedback system, which is shown in Figure 4.24. This method monitors the out-of-band power produced by the non-linear amplifier (NLA) and adjusts the predistorter's parameters to minimise it.

As seen in this figure, the predistorter creates the distorted signal $v_d(t)$ from the undistorted modulated signal $v_m(t)$. Upon inputting this signal to the amplifier A, its output signal is given by $v_a(t)$.The adaptive feedback path downconverts the bandpass amplifier's output signal, which is then bandpass filtered (BPF) to separate the out-of-band (OOB) signal power from the wanted signal. The OOB signal power is then averaged by the power detector and used by the controller to adjust the predistorter's complex transfer characteristic to minimise the OOB power.

4.5.2.2 Predistorter description

Based on the above concept Stapleton *et al.* have shown [1] that using the in-phase and quadrature-phase second-order non-linear functions, F_1 and F_2, which are derived from the envelope of the predistorter's input signal $v_m(t)$, the inverse functions of the power amplifier's amplitude and phase responses can be interpolated. The

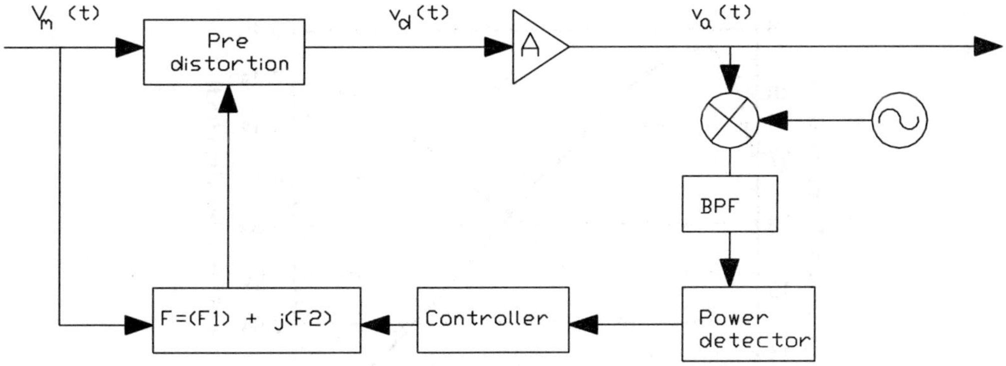

Figure 4.24: Basic predistorter block diagram [1] ©IEEE, 1992, Stapleton *et al.*

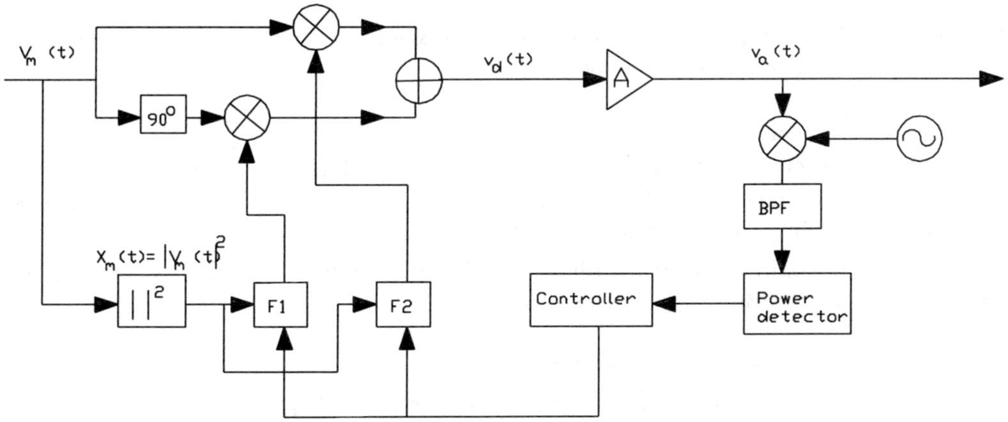

Figure 4.25: Detailed predistorter block diagram [1] ©IEEE, 1992, Stapleton *et al.*

predistorted signal $v_d(t)$ is derived by multiplying the I and Q components of the modulated signal $v_m(t)$ with the second-order non-linear functions F_1 and F_2, as portrayed in Figure 4.25. The magnitude and phase characteristics of a typical class AB amplifier are depicted in Figure 4.26 as a function of the input power. Observe the diagrammatic gain and phase variations due to the input power fluctuations. These amplitude characteristics, which depend exclusively on the input power level, must be linearised. It follows that the second-order functions F_1 and F_2 will also only have to depend on the magnitude of the modulated signal, i.e. on $x_m(t) = |v_m(t)|^2$, as shown below:

$$\begin{aligned} F_1\{x_m(t)\} &= a_{11} + a_{13}x_m(t) + a_{15}{x_m}^2(t) \\ F_2\{x_m(t)\} &= a_{21} + a_{23}x_m(t) + a_{25}{x_m}^2(t). \end{aligned} \quad (4.39)$$

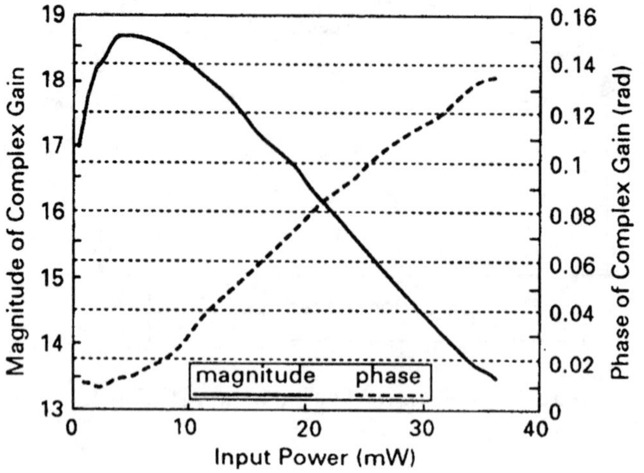

Figure 4.26: Magnitude and phase versus input power for a typical class AB amplifier [1] ©IEEE, 1992, Stapleton *et al.*

Note that the coefficient indices have been chosen for later notational convenience. Observe also that the scalar gains a_{11} and a_{21} linearly weight the I and Q components, without introducing any dependency on $v_m(t)$.

The in-phase and quadrature phase weighting functions F_1 and F_2 allow us to express the predistorter's complex gain as:

$$\begin{aligned} F\{x_m(t)\} &= F_1\{x_m(t)\} + jF_2\{x_m(t)\} \\ &= a_{11} + ja_{21} + (a_{13} + ja_{23})x_m(t) + (a_{15} + ja_{25}){x_m}^2(t) \\ &= a_1 + a_3x_m(t) + a_5{x_m}^2(t), \end{aligned} \tag{4.40}$$

where the complex coefficients are given below:

$$\begin{aligned} a_1 &= a_{11} + ja_{21} \\ a_3 &= a_{13} + ja_{23} \\ a_5 &= a_{15} + ja_{25}. \end{aligned} \tag{4.41}$$

Using the predistorter's complex transfer function F in Equation 4.40, the predistorted signal $v_d(t)$ is given by:

$$\begin{aligned} v_d(t) &= v_m(t)F\{x_m(t)\} \qquad (4.42) \\ &= a_1v_m(t) + a_3v_m(t)x_m(t) + a_5v_m(t){x_m}^2(t) \\ &= a_1v_m(t) + a_3v_m(t)|v_m(t)|^2 + a_5v_m(t)|v_m(t)|^4. \end{aligned}$$

Observe in the above Equation that $v_d(t)$ has a first-, a third- and a fifth-order $v_m(t)$-dependent term, which explains the choice of coefficient indices. Therefore, by

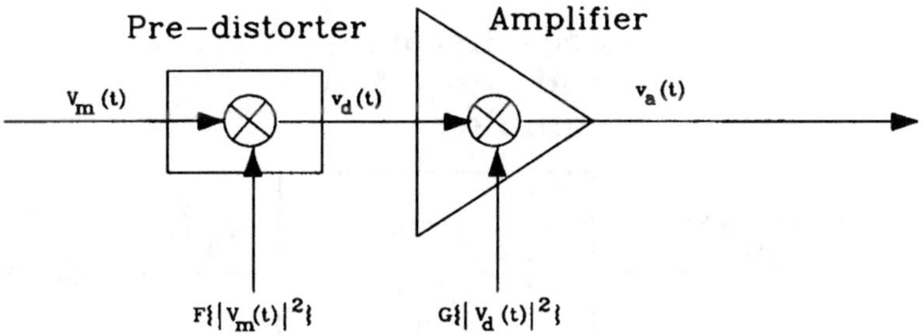

Figure 4.27: Simplified predistorter schematic

appropriately choosing the coefficients a_3 and a_5 the amplifier's third- and fifth-order intermodulation distortion can be reduced.

In a similar approach to our previous deliberations we introduce $x_d(t) = |v_d(t)|^2$ for the envelope of the predistorted signal and model the power amplifier also by a simple complex gain expression of the form:

$$G\{x_d(t)\} = G_1\{x_d(t)\} + jG_2\{x_d(t)\} = G\{|v_d(t)|^2\}, \tag{4.43}$$

giving the amplifier's output signal $v_a(t)$ in the following form:

$$v_a(t) = v_d(t)G\{|v_d(t)|^2\}. \tag{4.44}$$

The complex amplifier's gain can be modelled by the following complex power series:

$$G\{x_d(t)\} = g_1 + g_3 x_d(t) + g_5 {x_d}^2(t). \tag{4.45}$$

This simple model allows us to portray the whole system as shown in Figure 4.27. Observe that both $F\{|v_m(t)|^2\}$ and $G\{|v_d(t)|^2\}$ depend only on the power of their input signals, but not on their phases. The complex coefficients of G in Equation 4.45 describe both the amplitude modulation to amplitude modulation (AM-AM) and the amplitude modulation to phase modulation (AM-PM) conversion characteristics of the power amplifier.

The system model displayed in Figure 4.27 can be further simplified, as depicted in Figure 4.28, where the overall complex system gain $K\{|v_m(t)|^2\}$ can be derived as follows. First the amplifier's output signal is expressed as:

$$\begin{aligned} v_a(t) &= v_d(t).G\{x_d(t)\} \\ &= v_m(t).F\{x_m(t)\}.G\{x_m(t).|F[x_m(t)]|^2\}, \end{aligned} \tag{4.46}$$

where we exploited the relationships that $x_d(t) = x_m(t).|F\{x_m(t)\}|^2$ and $v_d = v_m.F\{x_m(t)\}$. From Equation 4.46 the complex system gain is given by:

$$K\{|v_m(t)|^2\} = \frac{v_a(t)}{v_m(t)} = F\{x_m(t)\}.G\{x_m(t).|F[x_m(t)]|^2\}. \tag{4.47}$$

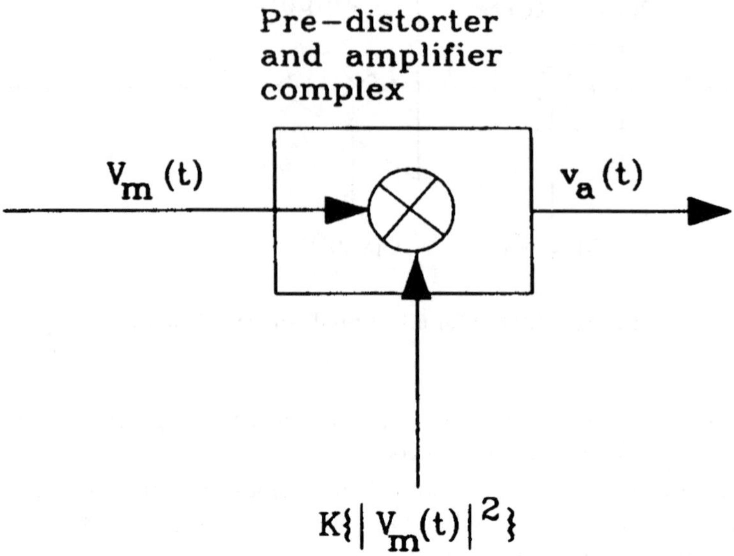

Figure 4.28: Block diagram of predistorter and amplifier complex

Similarly to the amplifier, the whole system can be modelled by a complex power series, truncated after the fifth-order term, given by:

$$\begin{aligned} K\{x_m(t)\} &= K_1\{x_m(t)\} + jK_2\{x_m(t)\} \\ &= k_1 + k_3 x_m(t) + k_5 {x_m}^2(t). \end{aligned} \quad (4.48)$$

Then, in accordance with Equation 4.42 we can write:

$$\begin{aligned} v_a(t) &= v_m(t) K\{x_m(t)\} \\ &= k_1 + k_3 v_m(t) x_m(t) + k_5 v_m(t) {x_m}^2(t) \\ &= k_1 + k_3 v_m(t)|v_m(t)|^2 + k_5 v_m(t)|v_m(t)|^4 , \end{aligned} \quad (4.49)$$

which shows the third- and fifth-order dependence of $v_a(t)$ on $v_m(t)$.

The computation of the coefficients k_1, k_3 and k_5 proceeds by equating the right hand sides of Equations 4.47 and 4.48 and substituting the appropriate expressions for $F\{\ \}$ and $G\{\ \}$ to yield:

$$\begin{aligned} & k_1 + k_3 x_m(t) + k_5 {x_m}^2(t) \\ = & [a_1 + a_3 x_m(t) + a_5 {x_m}^2(t)][g_1 + g_3 x_m(t)|F\{x_m(t)\}|^2 \\ & + g_5 {x_m}^2(t)|F\{x_m(t)\}|^4]. \end{aligned} \quad (4.50)$$

Performing the required multiplications gives:

$$\begin{aligned} &k_1 + k_3 x_m(t) + k_5 {x_m}^2(t) \\ &= [a_1 + a_3 x_m(t) + a_5 {x_m}^2(t)][g_1 + g_3 x_m(t)|a_1 + a_3 x_m(t) + \\ &\quad a_5 {x_m}^2(t)|^2 + g_5 {x_m}^2(t)|a_1 + a_3 x_m(t) + a_5 {x_m}^2(t)|^4]. \end{aligned} \tag{4.51}$$

After collecting the zeroth-, first- and second-order terms of $x_m(t)$ on the right hand side of Equation 4.51 we get the following relationships:

$$\begin{aligned} k_1 &= a_1 g_1 \\ k_3 &= a_3 g_1 + a_1 g_3 |a_1|^2 \\ k_5 &= a_5 g_1 + a_3 g_3 |a_1|^2 + a_1 g_5 |a_1|^4 + 2 a_1 g_3 Re\{a_1 {a_3}^*\}. \end{aligned} \tag{4.52}$$

Observe that the third-order distortion depends on the coefficients a_1 and a_3 of the predistorter (PD), while the fifth-order distortion can be controlled by a_1, a_3 and a_5. This suggests that both the third- and fifth-order intermodulation products can be efficiently reduced by the appropriate choice of the coefficients a_1, a_3 and a_5.

4.5.2.3 Predistorter coefficient adjustment

The multiplicative second-order non-linear functions F_1 and F_2 to be used by the predistorter are derived from the amplifier's AM-AM and AM-PM characteristics. A specific 5 W class AB amplifier's characteristics operating in the 800 MHz frequency band have been measured by Stapleton *et al.* [1] and are shown in Figure 4.29. The inverses of these characteristics are the required F_1 and F_2 non-linear functions, which can be inspected in Figure 4.30 along with the minimum mean squared error (MMSE) fitted second-order curves. The exact coefficients for the initial F_1 and F_2 functions for this particular amplifier were provided by Stapleton *et al.* [1]:

$$\begin{aligned} F_1\{x_m(t)\} &= 0.58315 - 0.08498 x_m(t) + 0.05169 {x_m}^2(t) \\ F_2\{x_m(t)\} &= -0.00395 + 0.0214 x_m(t) - 0.01887 {x_m}^2(t). \end{aligned} \tag{4.53}$$

These coefficients represent the initial values for the adaptive system. The functions F_1 and F_2 can be implemented for example by the help of operational amplifiers and linear four-quadrant multipliers, using the block diagram of Figure 4.31, where $F\{{x_m}^2(t)\} = a_1 + {x_m}^2(t)[a_3 + a_5 {x_m}^2(t)]$. The function $x_m(t) = |v_m(t)|^2$ can be realised by means of an envelope detector, which must exhibit good linearity across the range of input signal magnitudes.

The adaptive PD coefficient adjustment is based on measuring the OOB signal power at the frequency of the third and fifth upper harmonics. To render this power measurement practical, the output signal of the power amplifier (PA) must be down-converted to a frequency where the third and fifth harmonics can be filtered out. The

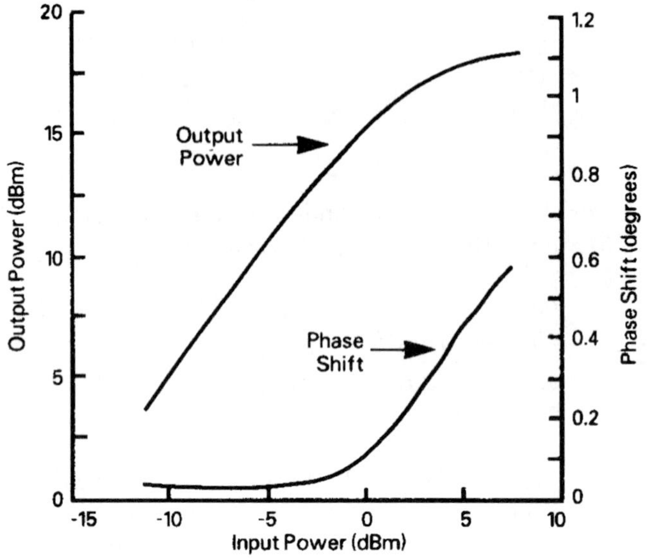

Figure 4.29: AM-AM and AM-PM characteristics of class AB power amplifier [1] ©IEEE 1992, Stapleton *et al.*

power of these components must be measured by a peak power detector and then averaged. As seen in Figure 4.24, the power detector's output drives the predistorter's controller, which adjusts the coefficients a_1, a_3 and a_5 adaptively to minimise the OOB signal power and hence minimise the detrimental effects of the non-linearity of the power amplifier (PA).

4.5.2.4 Predistorter performance

In general the choice of a specific minimisation algorithm will depend on the initial convergence, tracking accuracy and computational complexity requirements. In our application the initial convergence requirement is not particularly crucial, but low complexity and good tracking accuracy are desirable. Direct search algorithms, such as the Hooke-Jeeve method [198] are attractive in terms of both low complexity and high noise immunity, but converge generally more slowly than gradient techniques.

Stapleton *et al.* [199] studied the performances of three different coefficient PD adaptation methods, namely, the Hook-Jeeve direct search technique, the steepest descent (SD) gradient algorithm [198] and the Davidson–Fletcher–Powell (DFP) surface fit method [198]. Their convergence results are displayed in Figure 4.32.

As demonstrated by the figure, the DFP method has the fastest convergence at somewhat higher computational complexity. The initial convergence of the SD algorithm appears to be identical to that of the DFP method. However, after the rapid initial intermodulation decay its performance gradually tails off, finally exhibiting similar tracking performance to its other two counterparts.

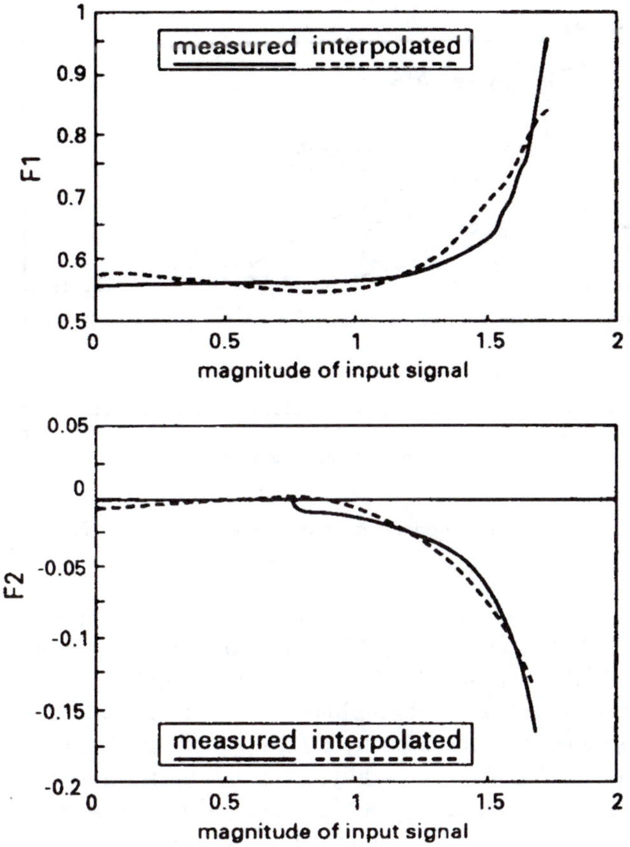

Figure 4.30: Interpolated F_1 and F_2 functions [1] ©IEEE 1992, Stapleton *et al.*

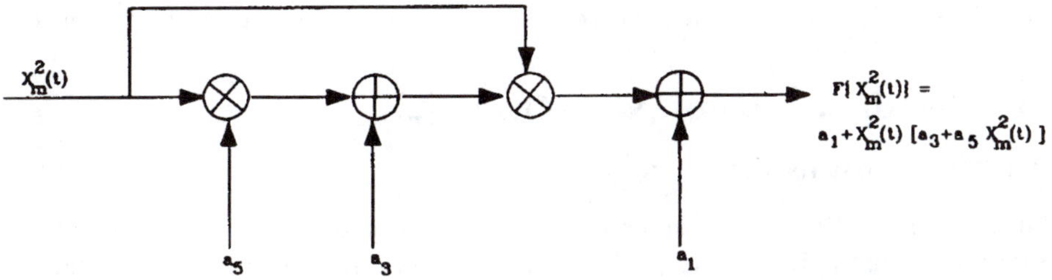

Figure 4.31: *F*-function implementation [1] ©IEEE, 1992, Stapleton *et al.*

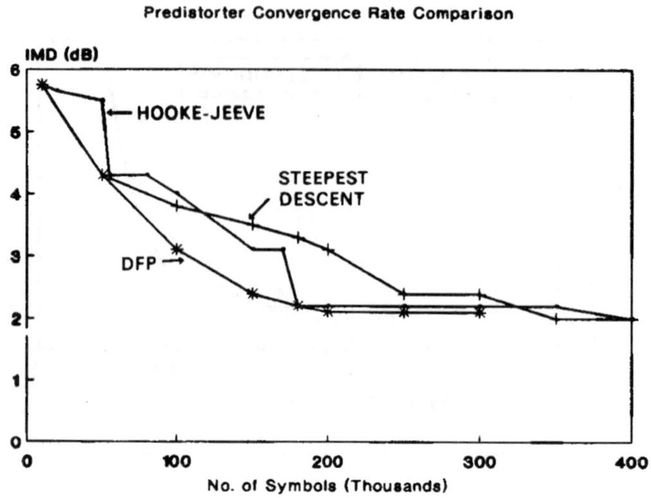

Figure 4.32: Predistorter convergence rate comparison [199] ©IEEE 1992, Stapleton *et al.*

The overall performance of the predistorter can be characterised by the spectral plot of Figure 4.33 [199], where the PA's input spectrum was plotted along with its output spectrum both with and without the PD. In the underlying experiment the previously portrayed 5 W class AB amplifier was used with a back-off power of 6 dB. The modulated signal was generated by a 64 kbps 16–QAM modem having a tight roll-off factor of $\alpha = 0.33$. The spectrally compact sharp roll-off yields slow impulse response decay and increased ISI sensitivity. The PD coefficients were adjusted using the DFP method and the spectrum depicted was generated after a convergence time of 9.6 sec.

In conclusion, predistortion is an efficient technique to reduce the non-linear power amplifier's intermodulation distortion, and hence out-of-band spectral spillage, which appears in adjacent channels in frequency division multiplex systems. Stapleton *et al.* demonstrated that the out-of-band emissions of a specific 5 W class AB amplifier deployed in a 16–QAM system can be reduced by some 10 dB, while maintaining high power efficiency in the amplifier. The complexity of the predistorter is moderate.

4.5.3 Postdistortion of NLA-QAM [2]

4.5.3.1 The postdistortion concept

In the previous section we considered cancelling the third- and fifth-order OOB intermodulation distortions (IMD) of non-linearly amplified (NLA) QAM signals by predistortion. The OOB spectral spillage was reduced by about 15 dB, when using a power efficient class AB power amplifier. However, the predistorter's complexity and power consumption are not negligible.

It may prove more attractive to relax the OOB emission requirement, for instance,

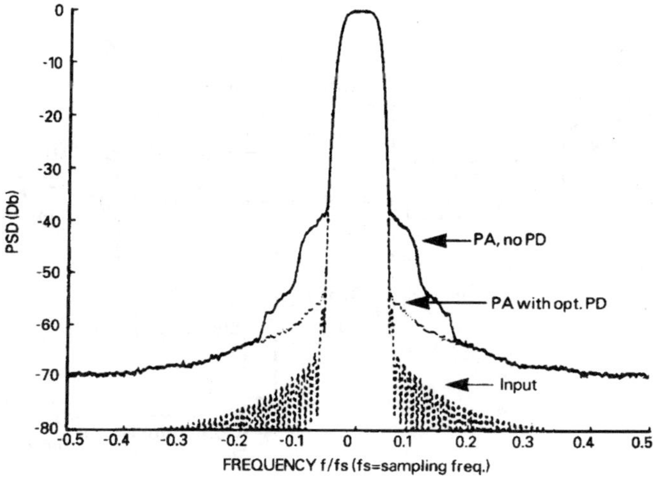

Figure 4.33: Simulated power spectrum after DFP optimisation of PD coefficients [199] ©IEEE 1992

from -60 dBc to -45 dBc and remove the resulting unwanted adjacent channel interference at the base station (BS). This eliminates the predistorter in front of the QAM modulator and hence reduces the complexity of the hand-held mobile station (MS). The undesirable IMD is reduced at the BS by a postdistorter. The BS transmits the signal to the receiving MS using power inefficient class A linear amplification, since the BS's complexity and power consumption constitute a less severe limitation than that of the MS, and so no postdistortion is required in the mobile.

Best IMD cancellation is achieved using an adaptive postdistorter, which accounts for time variant IMD caused by amplifier ageing, supply voltage variations, temperature fluctuations or other slowly drifting time variant IMD sources. The basic principle of the postdistorter is fairly similar to that of the predistorter, in that the AM-AM and AM-PM non-linearities of the power amplifier are adaptively measured and compensated. The postdistorter measures the envelope of the received signal and generates a non-linear signal, which models the MS power amplifier's third- and fifth-order IMD. Then the postdistorter's output signal is subtracted from the distorted QAM signal in order to cancel the IMD and hence the adjacent channel interference is significantly reduced.

The block diagram of a three-channel, QAM-based, frequency division multiple access (FDMA) scheme is depicted in Figure 4.34. The non-linearly amplified (NLA) QAM signals of three adjacent channels having carrier frequencies of f_1, f_2 and f_3 are received by the BS. Due to the power efficient class AB or class B amplification at the MSs, channels 1 and 3 spill their third- and fifth-order intermodulation distortions into the adjacent channel 2. The postdistorter's coefficients are initially adjusted to cancel the typical average third- and fifth-order IMD. Then their spectral spillage into channel 2 is measured by the IMD detector during channel 2's idle periods and the

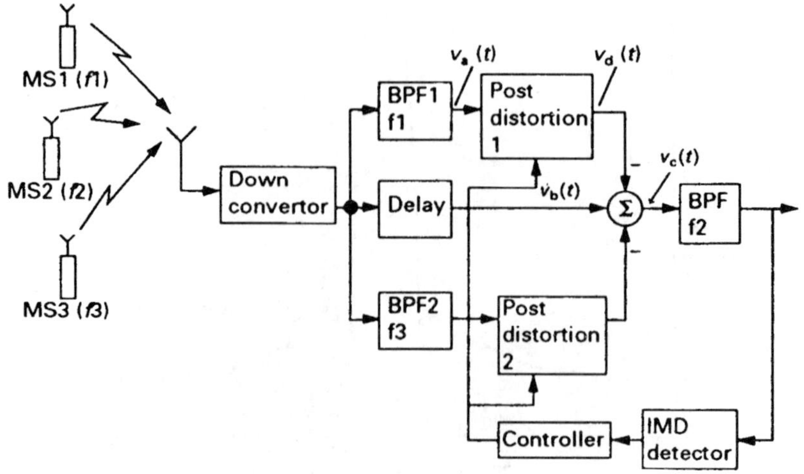

Figure 4.34: QAM-based three-channel FDM mobile radio system using postdistortion [2] ©IEEE, 1993, Quach *et al.*

postdistorter coefficients are adaptively adjusted to minimise the IMD.

In this simple three-channel scheme only two adjacent channels can spill IMD into channel 2's band. If, however, there are more channels and their third- or fifth-order IMD falls into channel 2, the schematic of Figure 4.34 should be extended to reduce their spillage as well. The downconverted received signal is filtered by the band-pass filters BPF1 and BPF2 to pass only channel 1's signal to postdistorter 1 and that of channel 3 to postdistorter 2. The delay block is provided to account for the processing delays of the BPFs and ostdistorters and to pass the received composite signal to the "distortion canceller" block, which is denoted by Σ. The postdistorted channel 1 and channel 3 signals are subtracted from the received composite signal and the "clean" channel 2 signal is bandpass filtered to remove any vestige of its OOB components.

Adaptive coefficient update can be introduced to counteract the slow IMD variations due to ageing, temperature fluctuations, etc. The IMD detector senses the OOB spillage of channel 1 and channel 3 into channel 2 during idle intervals of channel 2 and the controller adaptively adjusts the postdistorter's coefficients to minimise the IMD.

4.5.3.2 Postdistorter description

Similarly to the previously described predistorter, the postdistorter uses a memoryless, non-linear function to model the power amplifier's non-linearity. Using the postdistorter's input and output signal, $v_a(t)$ and $v_d(t)$, respectively, we have:

$$v_d(t) = v_a(t)F(|v_a(t)|^2), \tag{4.54}$$

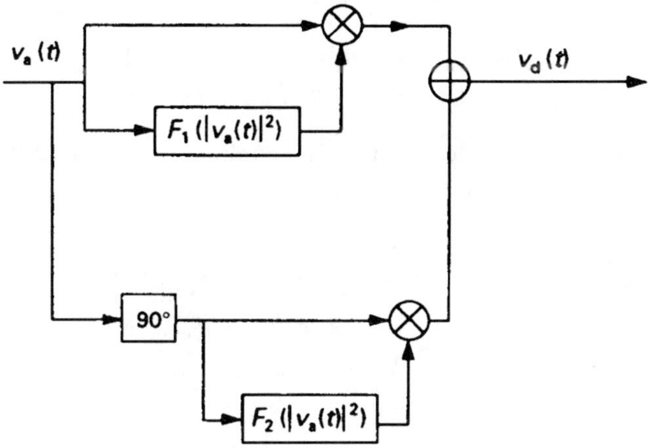

Figure 4.35: Postdistorter block diagram [2] ©IEEE, 1993, Quach *et al.*

where $F(|v_a(t)|^2)$ is the postdistorter's transfer function. Since we want to cancel the third- and fifth-order IMDs, we assume that $v_d(t)$ is a fifth-order function of the input signal $v_a(t)$. The postdistorter's transfer function $F(\ )$ is a second-order function of the envelope of $v_a(t)$, i.e. that of $x_a(t) = |v_a(t)|^2$:

$$\begin{aligned} F(|v_a(t)|^2) &= a_1 + a_3|v_a(t)|^2 + a_5(|v_a(t)|^2)^2 \\ &= a_1 + a_3x_a(t) + a_5{x_a}^2(t)\,, \end{aligned} \tag{4.55}$$

where the coefficients a_1, a_3 and a_5 are complex. Then the postdistorter's output signal $v_d(t)$ will be a fifth-order function of $v_a(t)$:

$$\begin{aligned} v_d(t) &= v_a(t)F(|v_a(t)|^2) \\ &= a_1v_a(t) + a_3v_a(t)|v_a(t)|^2 + a_5v_a(t)|v_a(t)|^4\,. \end{aligned} \tag{4.56}$$

Again, similarly to the predistorter, the postdistorter's complex gain $F(\ )$ can be expressed as a function of the in-phase and quadraturephase distortion functions as follows:

$$F[x_a(t)] = F_1[x_a(t)] + jF_2[x_a(t)]. \tag{4.57}$$

The in-phase and quadrature phase components of the amplified signal $v_a(t)$ are multiplied by the corresponding $v_a(t)$-dependent non-linear functions $F_1(|v_a(t)|^2)$ and $F_2(|v_a(t)|^2)$, respectively, in order to form the complex signal $v_d(t)$, which is also shown in Figure 4.35. The non-linearly class AB or class B amplified signal's expression is given by

$$v_a(t) = v_m(t)G[|v_m(t)|^2], \tag{4.58}$$

where $v_m(t)$ is the modulated signal and $G[|v_m(t)|^2]$ is the non-linear class AB power amplifier's complex gain, which is assumed to be a quadratic non-linear function of the envelope $x_m(t) = |v_m(t)|^2$. Then we have:

$$\begin{aligned} G[x_m(t)] &= g_1 + g_3 x_m(t) + g_5 {x_m}^2(t) \\ &= g_1 + g_3|v_m(t)|^2 + g_5|v_m(t)|^4 \,. \end{aligned} \tag{4.59}$$

Similarly to the predistorter we can define the complex system transfer function

$$\begin{aligned} K[|v_m(t)|^2] &= \frac{v_d(t)}{v_m(t)} \\ &= K_1[|v_m(t)|^2] + jK_2[|v_m(t)|^2] \\ &= k_1 + k_3 x_m(t) + k_5 {x_m}^2(t) \\ &= k_1 + k_3|v_m(t)|^2 + k_5|v_m(t)|^4. \end{aligned} \tag{4.60}$$

Using Equations 4.56 and 4.58 we then get:

$$\begin{aligned} K[|v_m(t)|^2] &= \frac{v_a(t)F[|v_a(t)|^2]}{v_m(t)} \\ &= G[|v_m(t)|^2].F\left\{|v_m(t)G[|v_m(t)|^2]|^2\right\} \\ &= G[x_m(t)]F\{x_m(t).|G[x_m(t)]|^2\}, \end{aligned} \tag{4.61}$$

which has the reciprocal structure of the predistorter's complex transfer function given in Equation 4.47. Therefore the coefficients k_1, k_3 and k_5 of $K[|v_m(t)|^2]$ are obtained by interchanging a_i with g_i, $i = 1, 3, 5$ in Equation 4.52 yielding:

$$\begin{aligned} k_1 &= g_1 a_1 \\ k_3 &= g_3 a_1 + g_1 a_3 |g_1|^2 \\ k_5 &= g_5 a_1 + g_3 a_3 |g_1|^2 + g_1 a_5 |g_1|^4 + 2 g_1 a_3 Re\{g_1 {g_3}^*\}, \end{aligned} \tag{4.62}$$

where $*$ means the complex conjugate.

From Figure 4.34 we see that the IMD cancellation is carried out by subtracting the OOB IMD components of channel 1 and channel 3 from the delayed composite signal $v_b(t)$. Considering channel 1's IMD only, we obtain the "cleaned" signal $v_c(t)$ as follows:

$$v_c(t) = v_b(t) - v_d(t) = v_a(t) - v_d(t), \tag{4.63}$$

where we have exploited that, apart from a delay component, $v_b(t) = v_a(t)$. Cancelling the IMD produced by channel 3 can be carried out in the same way, as specified by Equation 4.63. When substituting $v_a(t)$ from Equations 4.58 and $v_d(t) = K[|v_m(t)|^2].v_m(t)$ from Equation 4.60 into Equation 4.63 we have:

$$\begin{aligned} v_c(t) &= G[|v_m(t)|^2]v_m(t) - K[|v_m(t)|^2]v_m(t) \\ &= \left\{G[|v_m(t)|^2] - K[|v_m(t)|^2]\right\} v_m(t) \\ &= [(g_1 - k_1) + (g_3 - k_3)|v_m(t)|^2 + (g_5 - k_5)|v_m(t)|^4]v_m(t). \end{aligned}$$

$$(4.64)$$

Substituting the coefficients k_1, k_3 and k_5 from Equation 4.62 into Equation 4.64, the OOB spectral spillage into channel 2's band due to channel 1's IMD can be reduced by minimising the third- and fifth-order coefficients of Equation 4.64, which are given below:

$$\begin{aligned}(g_3 - k_3) &= g_3 - (g_3 a_1 + g_1 a_3 |g_1|^2) \\ (g_5 - k_5) &= g_5 - (g_5 a_1 + g_3 a_3 |g_1|^2 + g_1 a_5 |g_1|^4 + \\ &\quad 2 g_1 a_3 Re\{g_1 g_3{}^*\})\end{aligned} \tag{4.65}$$

Since the linear term $(g_1 - k_1)v_m(t)$ is due to channel 1's desired signal, it is outside of the bandwidth of BPF2, and therefore $(g_1 - k_1)$ can be set to any arbitrary value; it will be removed by BPF2. The remaining two terms, $(g_3 - k_3)$ and $(g_5 - k_5)$, must be minimised by appropriately choosing the postdistorter coefficients a_1, a_3 and a_5.

4.5.3.3 Postdistorter coefficient adaptation

Stapleton and Le Quach have shown [192] that the average OOB IMD power is a parabolic function of the postdistorter coefficients a_3 and a_5 and hence it has a global minimum, which can be located by optimisation algorithms. The initial setting of the postdistorter coefficients can be obtained by least squares fitting (LSF) of the second-order non-linear functions F_1 and F_2 to model the power amplifier's AM-AM and AM-PM characteristics, as described earlier in Section 4.5.2.3, in the context of the predistorter. This initial set-up can be significantly improved by optimisation methods, such as the Hook-Jeeve algorithm [198], as also proposed for the predistorter's optimisation in Section 4.5.2.3.

4.5.3.4 Postdistorter performance

Stapleton and Le Quach [192] evaluated the adjacent channel's IMD rejection of a three-channel 16-QAM system with and without postdistortion by means of simulations. Their results are presented in Figure 4.36 for the following scenarios:

(1) Without postdistortion.

(2) With postdistortion using LSF coefficients.

(3) With Hook-Jeeve optimisation

Observe the significant OOB IMD reduction in the idle channel 2 band from an average of about -40 dB to about -55 dB.

In summary, similarly to the predistorter, the postdistorter also offers significant out-of-band intermodulation distortion reduction, when using non-linear power amplifiers for QAM transmissions. Its performance and complexity are comparable to that of the predistorter, but it is incorporated into the BS, where its additional cost, complexity and power consumption are more readily justified than in the portable station.

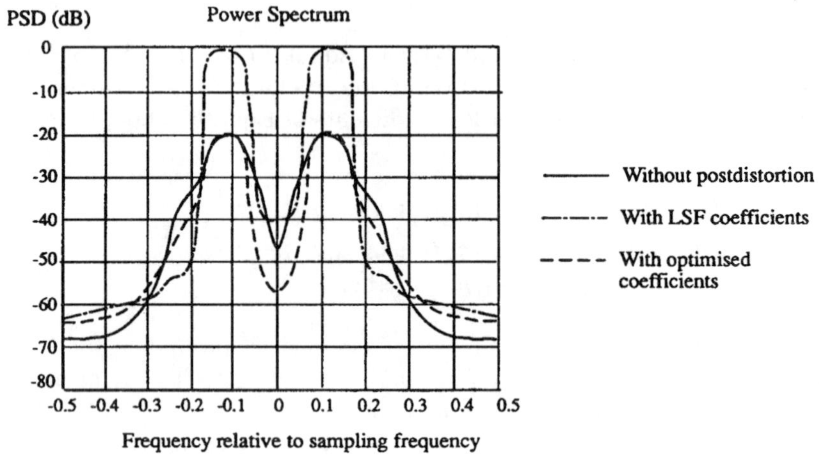

Figure 4.36: Power spectra of a postdistortion system with optimal coefficients, channel 2 is Idle [192]©IEEE 1992, Stapleton *et al.*

Having considered a variety of QAM constellations, along with some pulse shaping and amplification techniques, we now focus our attention on differential versus non-differential QAM transmissions.

4.6 Non-differential Coding for Square QAM

When using a multilevel modulation scheme, it is necessary to map the binary bit stream to be transmitted onto the constellation in an optimal manner. Typically, each of the M constellation points is assigned an N-bit code, where $M = 2^N$. The binary input stream is passed through an N-bit serial-to-parallel convertor and the appropriate constellation point transmitted. The mapping of the data onto the constellation is generally not made in a haphazard fashion, but so as to minimise the number of bit errors given a symbol error. That is, if an incorrect symbol is received, composed of N bits, it is required that the number of bits in error be as small as possible. Since errors are most likely to be made to neighbouring symbol points, this suggests that Gray coding should be employed, whereby the code for each constellation point only differs by one bit from any of its closest neighbours.

An example of an optimally coded 16-level QAM constellation is given in Figure 4.37, where two bits are Gray encoded onto the I-axis and two onto the Q-axis. Assuming that only one bit error occurs per symbol, which is the most likely form of error, we have that

$$P_{E^B} = \frac{1}{N} P_{E^S} \tag{4.66}$$

where P_{E^B} and P_{E^S} are the probability of bit error and of symbol error, respectively. With some constellations, it is not possible to have perfect Gray coding, (e.g. circular

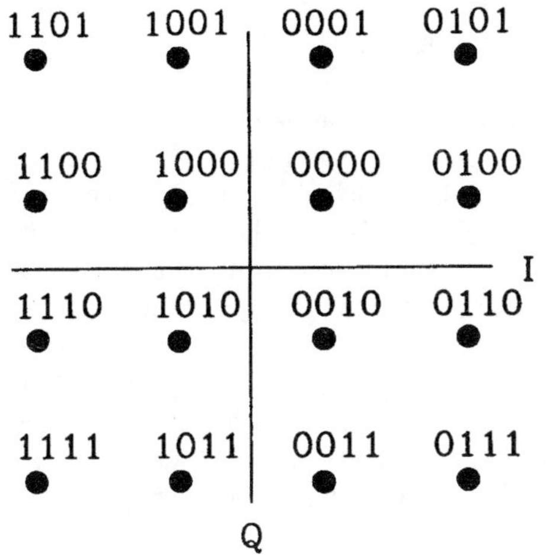

Figure 4.37: Perfect Gray coded 16-level QAM constellation

QAM with four points on the inner ring and eight on the outer [27]), and in this case the right-hand side of the above equation must be multiplied by the Gray coding penalty, $g > 1$, to reflect this coding problem.

4.7 Differential Coding for Square QAM

A problem which arises in conjunction with most practical QAM constellations is that the receiver is unable to resolve a number of possible carrier lock positions without additional information from the transmitter. This problem was considered by Weber [27] and we follow his approach here. We define an L-fold rotationally symmetric constellation as a signal set for which the phasor constellation maps onto itself for a rotation of $\pm K \cdot (2\pi/L)$ radians where K and L are integers. Then, given $M = 2^N$ constellation points, the receiver cannot uniquely assign the N bits to each constellation point without resolving the L-fold ambiguity.

Weber suggested that a general algorithm for differentially encoding an arbitrary constellation, where $L = 2^I$, I integer, was to:

(1) Divide the signal space into L equal pie-shaped sectors, and differentially Gray encode the identifiers of these sectors in a rotational manner. For example, in case of square 16-QAM the $L = 2^2 = 4$ pie-shaped sectors have a 90 deg angle, and are encoded by the identifiers 00, 01, 11, 10 sequentially when rotating around the constellation.

(2) Gray encode the remaining $(N - I)$ bits within each sector. For our 16-QAM example $(N - I) = 2$ bits belonging to the same quarter of the constellation will be Gray coded by assigning codes of 00 and 11 as well as 01 and 10 to opposite corners of a square. This will allow closest neighbours to differ only in one bit position.

This process has the advantage of differentially encoding as few of the bits as possible, which results in the best performance, by keeping error propagation to a minimum.

If it is assumed that the SNR is sufficiently high that errors are only made to neighbouring symbols then the penalty through using Gray coding can be readily computed. There are now two forms of errors, those within sectors and those between symbols in different sectors. For the former there will be no differential coding penalty, as differential coding has not been used within a sector. There may be a Gray coding penalty, depending on whether perfect Gray coding was possible. When a sector boundary is crossed, a minimum of two bit errors, one in each of two consecutive symbols due to the comparisons used in the differential decoding process, will result. Further errors may result if the sector identifier's Gray coding was such that it was not perfect across boundaries. This can be represented by a Gray coding penalty for boundary symbols. In order to derive a general formula representing this we form the following equation:

$$P_{E^B}(diff\ encoded) = F.P_{E^B}(non\ diff) \tag{4.67}$$

where F is the relative performance of the differentially coded system to the non-differentially coded system. For the non-differentially coded case the general relationship between bit and symbol errors is given by

$$P_{E^B}(non\ diff) = \frac{g}{N} P_{E^S} \tag{4.68}$$

where g is the Gray coding penalty representing the average number of bits in error per symbol error, where only errors between adjacent symbols are considered. A similar expression for the differentially encoded non-ideal Gray coding case is

$$P_{E^B}(diff\ encoded) = \frac{f}{N} P_{E^S} \tag{4.69}$$

where f is the differentially encoded Gray penalty. The penalty for differentially encoding a particular signal set is then given by

$$F = \frac{f}{g}. \tag{4.70}$$

Weber suggests the following algorithm for finding f, which is demonstrated with reference to the square QAM constellation of Figure 4.38:

(a) Within a single sector draw lines between symbol point pairs separated by δ_{min} and let the total number of these lines be N_1.

(b) Write next to each line the Hamming distance, i.e. the number of bits which differ between the two signal points, and let the sum of all the Hamming distances within one sector be H_1.

(c) Draw lines between adjacent constellation points separated by δ_{min} and lying on opposite sides of the sector border, and let the number of these lines be N_2

and the sum of their Hamming distances be H_2.

(d) The differential coding penalty is then given by

$$f = \frac{H_1 + H_2 + 2N_2}{N_1 + N_2} \tag{4.71}$$

where the weighting factor of 2 in the third term of the numerator is due to error propagation in the differentially encoded bits. Note that $g = H_1/N_1$ is the Gray coding penalty within a sector, whereas H_2/N_2 is the Gray coding penalty across a sector.

In order to augment our exposition, an example showing the application of these steps to rectangular 16-level QAM is given in Figure 4.38. From the figure it can be seen that $N_1 = 4$ and $H_1 = 4$ whilst $N_2 = 2$ and $H_2 = 2$, implying that $g = 1$ and therefore there is no Gray coding penalty with this constellation. Because of this $F = f$ and we have that $F = (4 + 2 + 4)/(4 + 2) = 1.67$, or 2.22 dB. For the general M-level square constellation it can be shown [27] that

$$\begin{aligned} N_1 &= 2^{N-1} - 2^{N/2} \\ H_1 &= 2^{N-1} - 2^{N/2} \\ N_2 &= 2^{N/2-1} \\ H_2 &= \left(\frac{N}{2} - 1\right) 2^{(N/2-1)} \end{aligned} \tag{4.72}$$

and since $g = 1$ we have that

$$F = 1 + \frac{N/2}{2^{n/2} - 1}. \tag{4.73}$$

Because only two bits are ever required for differential coding of the four sectors, the coding penalty drops as the number of bits increases, from 2 (3 dB) for N=2 (QPSK) to nearly 1 (0 dB) for higher-order systems.

A further problem arises when offset QAM (O-QAM) is employed as not only must the L-fold rotational ambiguity be resolved, but it is also necessary to determine which axis is delayed relative to the other. Weber proposed overcoming this ambiguity by encoding the I and Q channel transitions independently using separate differential coding systems so that now all bits are being differentially encoded. This will lead to a pattern of I bits and then Q bits, alternating. It will therefore be of no importance which axis is staggered with relation to the other as they are now treated independently. This will increase the differential coding penalty to the full 3 dB, and is an argument against using O-QAM as compared to QAM.

4.8 Summary

In this chapter we have deepened our knowledge as regards to general QAM signal processing techniques. After a brief discourse on various QAM constellations for different channels we have analytically derived the transfer functions of the optimum

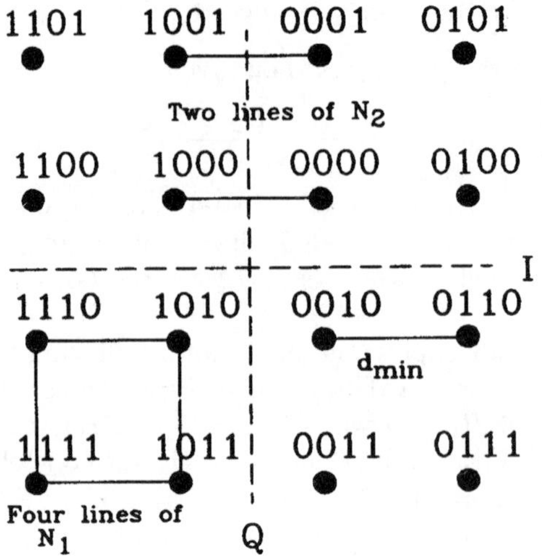

Figure 4.38: Calculating differential coding penalty with a square constellation

transmitter and receiver filters. Both of these filters must retain a square root Nyquist shape in order to minimise the ISI and maximise the noise immunity. Various methods of generating QAM signals have been proposed and techniques of QAM detection, such as threshold detection, matched filtering and the correlation receiver have been reviewed. The chapter was concluded by a short discussion on differential versus non-differential coding.

Equipped with these QAM techniques in the next chapter we now consider the bit error rate versus channel SNR performance of 16-QAM and 64-QAM over AWGN channels.

Chapter 5

Square QAM

In this chapter we consider the transmission and demodulation of square QAM signals over Gaussian channels. The theory of QAM transmissions over Rayleigh fading channels is given in Chapter 10. In order to arrive at explicit BER versus SNR formulae, let us first review a few fundamental results of decision theory.

5.1 Decision Theory

The roots of decision theory stem from Bayes' theorem formulated as follows:

$$P(X/Y) \cdot P(Y) = P(Y/X) \cdot P(X) = P(X, Y), \tag{5.1}$$

where the random variables X and Y have probabilities of $P(X)$ and $P(Y)$, their joint probability is $P(X, Y)$ and their conditional probabilities are given by $P(X/Y)$ and $P(Y/X)$.

In decision theory the above theorem is invoked in order to infer from the noisy analogue received sample y, what the most likely transmitted symbol was, assuming that the so-called *a priori* probability $P(x)$ of the transmitted symbols $x_n, n = 1, \ldots, M$ is known. Given that the received sample y is encountered at the receiver, the conditional probability $P(x_n/y)$ quantifies the chance that x_n has been transmitted:

$$P(x_n/y) = \frac{P(y/x_n) \cdot P(x_n)}{P(y)}, \qquad n = 1, \ldots, N \tag{5.2}$$

where $P(y/x_n)$ is the conditional probability of the continuous noise sample y, given that $x_n, n = 1, \ldots, N$ was transmitted. The probability of encountering a specific y value will be the sum of all possible combinations of receiving y and $x_n, n = 1, \ldots, N$ simultaneously, which is given by:

$$P(y) = \sum_{n=1}^{N} P(y/x_n) \cdot P(x_n) = \sum_{n=1}^{N} P(y, x_n). \tag{5.3}$$

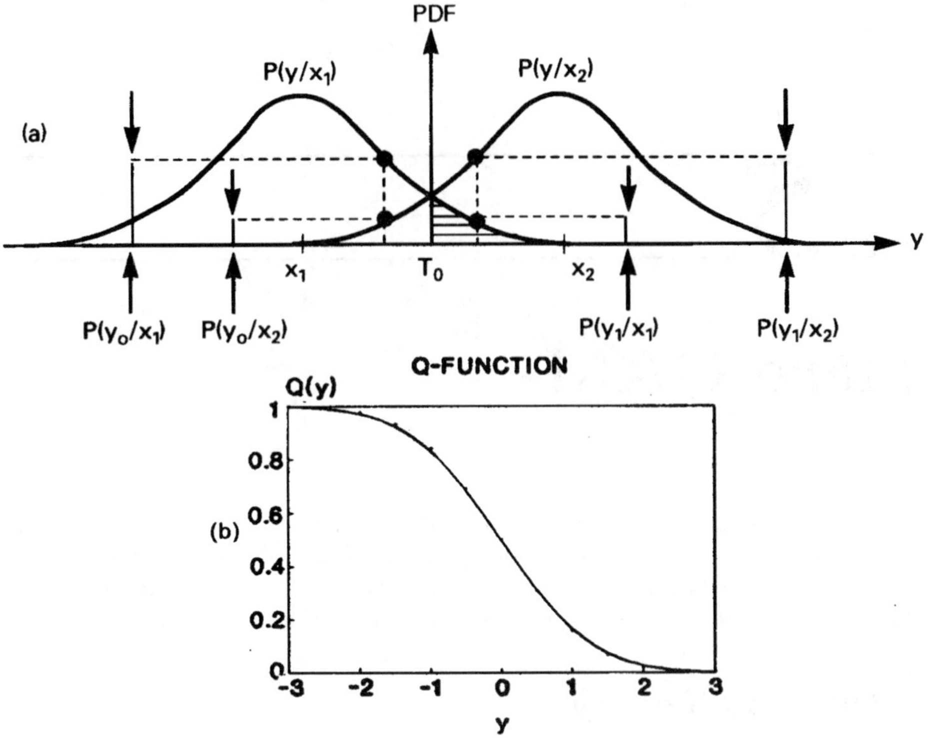

Figure 5.1: (a) Transmitted phasors and noisy received phasors for BPSK, (b) Gaussian Q-function

Let us now consider the case of binary phase shift keying (BPSK), where there are two legitimate transmitted waveforms, x_1 and x_2, which are contaminated by noise, as portrayed in Figure 5.1(a). The conditional probability of receiving any particular analogue sample y given that x_1 or x_2 was transmitted is quantified by the Gaussian probability density functions (PDFs) seen in Figure 5.1(a), which are described by:

$$P(y/x) = \frac{1}{\sigma\sqrt{2\pi}} e^{\frac{-(y-x)^2}{2\sigma^2}}, \tag{5.4}$$

where $x = x_1$ or x_2 is the mean and σ^2 is the variance.

Observe from the figure that the shaded area represents the probability of receiving values larger than the threshold T_0, when x_1 was transmitted and this is equal to the probability of receiving a value below T_0, when x_2 was transmitted. As displayed in the figure, when receiving a specific $y = y_0$ sample, there is an ambiguity, as to which symbol was transmitted. The corresponding conditional probabilities are given by $P(y_0/x_1)$, and $P(y_0/x_2)$ and their values are also marked on Figure 5.1(a). Given the knowledge that x_1 was transmitted, we are more likely to receive y_0 than on condition that x_2 was transmitted. Hence, upon observing $y = y_0$, statistically speaking it is advisable to decide that x_1 was transmitted. Following similar logic, when receiving y_1 as seen in Figure 5.1(a), it is logical to conclude that x_2 was transmitted.

Indeed, according to optimum decision theory [183], the optimum decision threshold above which x_2 is inferred is given by:

$$T_0 = \frac{x_1 + x_2}{2} \tag{5.5}$$

and below this threshold x_1 is assumed to have been transmitted. If $x_1 = -x_2$ then $T_0 = 0$ is the optimum decision threshold minimising the bit error probability.

In order to compute the error probability in case of transmitting x_1, the PFD $P(y/x_1)$ of Equation 5.4 has to be integrated from x_1 to ∞, which gives the shaded area under the curve in Figure 5.1(a). In other words, the probability of a zero-mean noise sample exceeding the magnitude of x_1 is sought, which is often called the *noise protection distance*, given by the so-called Gaussian Q-function:

$$Q(x_1) = \frac{1}{\sigma\sqrt{2\pi}} \int_{x_1}^{\infty} e^{\frac{-y^2}{2\sigma^2}} dy, \tag{5.6}$$

where σ^2 is the noise variance. Notice that since $Q(x_1)$ is the probability of exceeding the value x_1, it is actually the complementary cumulative density function (CDF) of the Gaussian distribution.

Assuming that $x_1 = -x_2$, the probability that the noise can carry x_1 across $T_0 = 0$ is equal to that of x_2 being corrupted in the negative direction. Hence, assuming that $P(x_1) = P(x_2) = 0.5$, the overall error probability is given by:

$$\begin{aligned} P_e &= P(x_1) \cdot Q(x_1) + P(x_2) \cdot Q(x_2) \\ &= \frac{1}{2}Q(x_1) + \frac{1}{2}Q(x_1) = Q(x_1) \end{aligned} \tag{5.7}$$

The values of the Gaussian Q-function plotted in Figure 5.1(b) are tabulated in many textbooks [183], along with values of the Gaussian PDF in case of zero-mean, unit-variance processes. For abscissa values of $y > 4$ the following approximation can be used:

$$Q(y) \approx \frac{1}{y\sqrt{2\pi}} e^{\frac{-y^2}{2}} \quad \text{for} \quad y > 4. \tag{5.8}$$

Having provided a rudimentary introduction to decision theory, let us now focus our attention on the demodulation of 16-QAM signals in AWGN.

5.2 QAM Modulation and Transmission

In general the modulated signal can be represented by

$$s(t) \;=\; a(t)\cos[2\pi f_c t \;+\; \Theta(t)] \;=\; Re(a(t)e^{j[w_c t + \Theta(t)]}) \tag{5.9}$$

where the carrier $\cos(w_c t)$ is said to be amplitude modulated if its amplitude $a(t)$ is adjusted in accordance with the modulating signal, and is said to be phase modulated if $\Theta(t)$ is varied in accordance with the modulating signal. In QAM the amplitude of the baseband modulating signal is determined by $a(t)$ and the phase by $\Theta(t)$. The in-phase component I is then given by

$$I = a(t)\cos\Theta(t) \tag{5.10}$$

and the quadrature component Q by

$$Q = a(t)\sin\Theta(t). \tag{5.11}$$

This signal is then corrupted by the channel. Here we will only consider AWGN. The received signal is then given by

$$r(t) = a(t)\cos[2\pi f_c t + \Theta(t)] + n(t) \tag{5.12}$$

where $n(t)$ represents the AWGN, which has both an in-phase and quadrature component. It is this received signal which we will attempt to demodulate.

5.3 16-QAM Demodulation in AWGN

The demodulation of the received QAM signal is achieved by performing quadrature amplitude demodulations using the decision boundaries seen in Figure 3.3 for the I and Q components, as shown below for the bits i_1 and q_1:

$$\begin{array}{llll} \text{if} & I,Q \geq 0 & \text{then} & i_1,q_1 = 0 \\ \text{if} & I,Q < 0 & \text{then} & i_1,q_1 = 1 \end{array} \tag{5.13}$$

The decision boundaries for the third and fourth bits i_2 and q_2, respectively, are again shown in Figure 3.3, and thus:

$$\begin{array}{llllllll} \text{if} & & & I,Q & \geq & 2d & \text{then} & i_2,q_2 = 1 \\ \text{if} & -2d & \leq & I,Q & < & 2d & \text{then} & i_2,q_2 = 0 \\ \text{if} & -2d & > & I,Q & & & \text{then} & i_2,q_2 = 1. \end{array} \tag{5.14}$$

We will show that in the process of demodulation the positions of the bits in the QAM symbols associated with each point in the QAM constellation have an effect on the probability of them being in error. In the case of the two most significant bits (MSBs) of the 4-bit symbol i_1, q_1, i_2, q_2, i.e. i_1 and q_1, the distance from a demodulation decision boundary of each received phasor in the absence of noise is $3d$ for 50% of the time, and d for 50% of the time; if each phasor occurs with equal probability. The average protection distance for these bits is therefore $2d$ although the bit error probability for a protection distance of $2d$ would be dramatically different from that calculated. Indeed, the average protection distance is never encountered, we only use this term to aid our investigations. The two least significant bits (LSBs), i.e. i_2 and q_2, are always at a distance of d from the decision boundary and consequently the average protection distance is d. We may consider our QAM system as a class 1 ($C1$) and as a class 2 ($C2$) sub-channel, where bits transmitted via the $C1$ sub-channel are received with a lower probability of error than those transmitted via the $C2$ sub-channel.

Observe in the phasor diagram of Figure 3.3 that upon demodulation in the $C2$ sub-channel, a bit error will occur if the noise exceeds d in one direction or $3d$ in the opposite direction, where the latter probability is insignificant. Hence the $C2$ bit error probability becomes

$$P_{2G} = Q\left\{\frac{d}{\sqrt{N_0/2}}\right\} = \frac{1}{\sqrt{2\pi}}\int_{\frac{d}{\sqrt{N_0/2}}}^{\infty} \exp\left(-x^2/2\right)dx \tag{5.15}$$

where $N_0/2$ is the double-sided spectral density of the AWGN, $\sqrt{N_0/2}$ is the corresponding noise voltage, and the $Q\{\}$ function was given in Equation 5.6 and Figure 5.1. As the average symbol energy of the 16-level QAM constellation computed for the phasors in Figure 3.3 is

$$E_0 = 10d^2, \tag{5.16}$$

we have that

$$P_{2G} = Q\left\{\sqrt{\frac{E_0}{5N_0}}\right\}. \tag{5.17}$$

For the $C1$ sub-channel data the bits i_1, q_1 are at a protection distance of d from the decision boundaries for half the time, and their protection distance is $3d$ for the remaining half of the time. Therefore the probability of a bit error is

$$\begin{aligned} P_{1G} &= \frac{1}{2}Q\left\{\frac{d}{\sqrt{N_0/2}}\right\} + \frac{1}{2}Q\left\{\frac{3d}{\sqrt{N_0/2}}\right\} \\ &= \frac{1}{2}\left[Q\left\{\sqrt{\frac{E_0}{5N_0}}\right\} + Q\left\{3\sqrt{\frac{E_0}{5N_0}}\right\}\right]. \end{aligned} \tag{5.18}$$

The probabilities P_{1G} and P_{2G} as a function of E_0/N_0 are given by Equations 5.17 and 5.18 and displayed in Figure 5.2.

Also shown is the average probability P_{AV} of bit error for the 16-level QAM system as

$$P_{AV} = (P_{1G} + P_{2G})/2. \tag{5.19}$$

Our simulation results gave practically identical curves to those in Figure 5.2, where the AWGN performance is seen to have only a small advantage in using the $C1$ sub-channel over using the $C2$ sub-channel.

5.4 64-QAM Demodulation in AWGN

Each constellation point in 64-level QAM (64-QAM) systems is represented by a unique 6-bit symbol, which is Gray coded to minimise the decoded error probability, as seen in Figure 5.3.

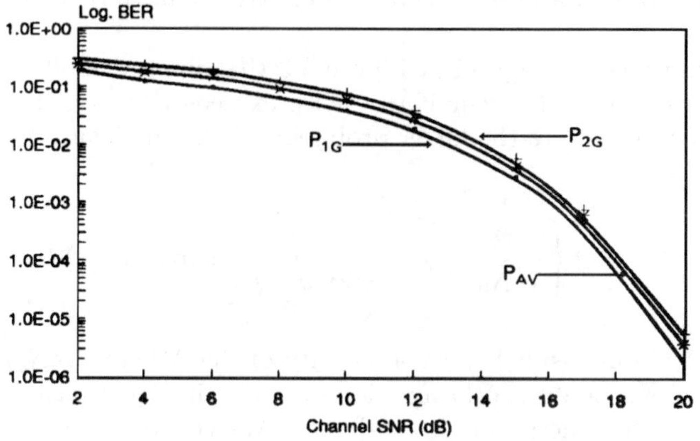

Figure 5.2: 16-QAM BER versus channel SNR curves over AWGN channel

The complex phasors of the constellation are decomposed into the 8-level I and Q components. The amplitudes $7d$, $5d$, $3d$, d, $-d$, $-3d$, $-5d$ and $-7d$ of the I and Q AM signals are assigned the 3-bit Gray codes 011, 010, 000, 001, 101, 100, 110 and 111, respectively. The three I and Q bits are denoted by i_1, i_2, i_3 and q_1, q_2, q_3, respectively. These bits are interleaved to give a 6-bit QAM symbol represented by i_1, q_1, i_2, q_2, i_3, q_3. The 64-level QAM phasors are transmitted over the channel, where they become corrupted. They are demodulated using the decision boundaries shown in Figure 5.3. Observe that the construction of the signal constellation is similar to that of a Karnaugh table, where in the left half-plane of the coordinate axis i_1 is always logical 1 and in the bottom half-plane q_1 is logical 1, etc. The bits i_1, q_1, i_2, q_2, i_3, q_3, i_4, q_4 are recovered as follows:

$$\begin{array}{llllll} \text{if} & I,Q & \geq & 0 & \text{then} & i_1,q_1 = 0 \\ \text{if} & I,Q & < & 0 & \text{then} & i_1,q_1 = 1, \end{array} \tag{5.20}$$

for the most significant bits, and

$$\begin{array}{lllllll} \text{if} & & & I,Q & \geq & 4d & \text{then} \quad i_2,q_2 = 1 \\ \text{if} & -4d & \leq & I,Q & < & 4d & \text{then} \quad i_2,q_2 = 0 \\ \text{if} & -4d & > & I,Q & & & \text{then} \quad i_2,q_2 = 1. \end{array} \tag{5.21}$$

for the next most significant bits, while finally for the least significant bits

$$\begin{array}{lllllll} \text{if} & & & I,Q & \geq & 6d & \text{then} \quad i_3,q_3 = 1 \\ \text{if} & 2d & \leq & I,Q & < & 6d & \text{then} \quad i_3,q_3 = 0 \\ \text{if} & -2d & \leq & I,Q & < & 2d & \text{then} \quad i_3,q_3 = 1 \\ \text{if} & -6d & \leq & I,Q & < & -2d & \text{then} \quad i_3,q_3 = 0 \\ \text{if} & -6d & > & I,Q & & & \text{then} \quad i_3,q_3 = 1. \end{array} \tag{5.22}$$

Similarly to 16-QAM the position of the bits in the 6-bit QAM symbol has an

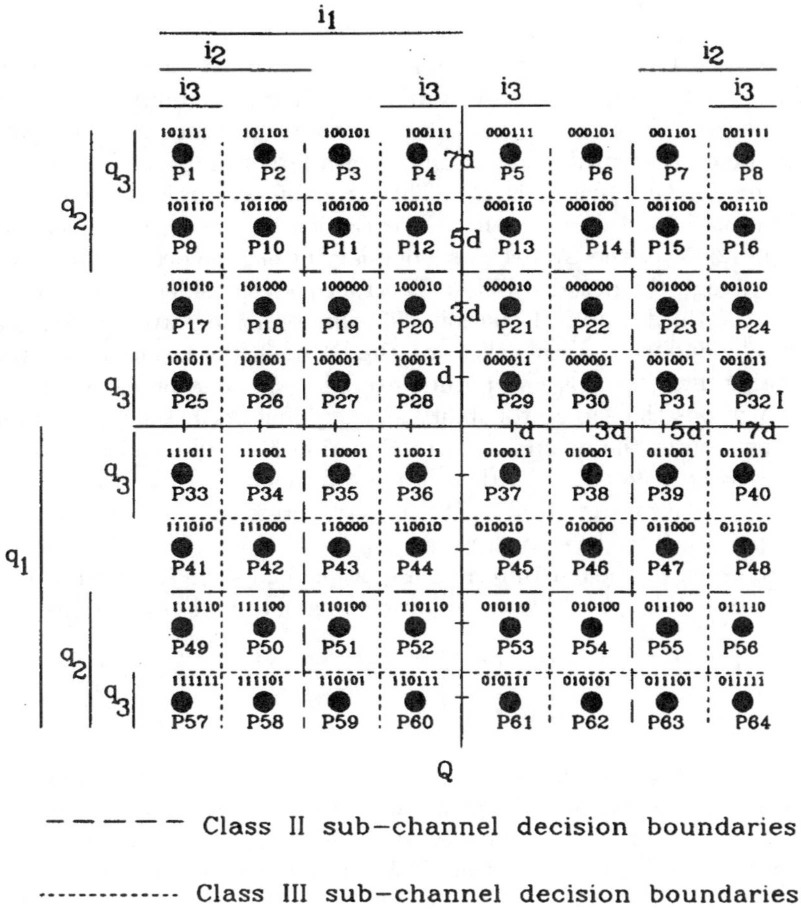

Figure 5.3: Square 64-QAM phasor constellation

effect on their error probabilities. In case of the i_1, q_1 bits, for example, the phasor can be at a distance d, $3d$, $5d$ or $7d$ from the decision boundary represented by the coordinate axis. Therefore their "average protection distance" is $16d/4 = 4d$. Note again that this average protection distance is never encountered and due to the non-linear nature of the $Q\{\}$ function the actual BER will be quite different from that of a $4d$ average distance PSK constellation. Figure 5.3 reveals that this average protection distance for the i_2, q_2 bits is $2d$, while for the i_3, q_3 bits it is d. We may view the bits (i_1, q_1), (i_2, q_2) and (i_3, q_3) as three sub-channels each having different integrities. These channels will be referred to as $C1$, $C2$ and $C3$, respectively.

We assume that an independent random data sequence is conveyed to each of the $C1$, $C2$ and $C3$ sub-channels, and that perfect coherent detection is carried out at the receiver using Equations 5.20 to 5.22. Let us determine the BER performance of the I component, as the Q component being in quadrature will support data that is subjected to the same BER.

Let us consider first the $C3$ sub-channel (i_3, q_3) with its decision boundaries of -$6d$, -$2d$, $2d$ and $6d$. Each phasor has a different error probability, depending on its position in the constellation of Figure 5.3. The bit i_3 of phasor P_5 (a logical 1) will, for example, be corrupted, if a noise sample with larger amplitude than d is added to it when the decoded phasor is P_6. This probability is represented by the first term of Equation 5.23. However, when P_5 is carried by a noise vector having larger than $5d$ amplitude over the $C3$ decision boundary at $6d$, the decision becomes error-free again, although the decoded phasor is erroneously considered to be P_8. This is quantified by the third term in Equation 5.23, where the negative sign takes account of the favourable influence of this event on the BER. In the negative direction there is no corruption until the noise amplitude exceeds $-3d$ (phasor P_3) in the negative direction, which has the same probability as exceeding $3d$ in the positive direction, and hence contributes the second term to Equation 5.23. For noise vectors having amplitudes between $-3d$ and $-7d$ there are erroneous decisions. However, once the $-6d$ decision boundary is exceeded, i.e. the negative noise value is larger than $-7d$, the received sample falls into the error-free domain again, although the decoded phasor is P_1 which is reflected by the fourth term of Equation 5.23. The i_3 bit error probability of the P_5 phasor is then given by the following summation of Q-functions:

$$P_{e5} = Q\left[\frac{d}{\sqrt{N_0/2}}\right] + Q\left[\frac{3d}{\sqrt{N_0/2}}\right] - Q\left[\frac{5d}{\sqrt{N_0/2}}\right] - Q\left[\frac{7d}{\sqrt{N_0/2}}\right], \tag{5.23}$$

where $N_0/2$ denotes the double-sided Gaussian noise spectral density and $\sqrt{N_0/2}$ is the corresponding noise voltage. After averaging the powers of the individual phasors $P_1, \ldots, P_{64}$ the average symbol energy for $64-QAM$ is found to be $E = 42d^2$. Substituting $d = \sqrt{E/42}$ into Equation 5.23, and introducing the average SNR $\gamma = E/N_0$, we get:

$$P_{e5} = Q\left[\sqrt{\frac{\gamma}{21}}\right] + Q\left[3\sqrt{\frac{\gamma}{21}}\right] - Q\left[5\sqrt{\frac{\gamma}{21}}\right] - Q\left[7\sqrt{\frac{\gamma}{21}}\right]. \tag{5.24}$$

For the phasor P_8 the situation is different, as its $C3$ bit i_3 is not corrupted by positive noise samples of arbitrarily large amplitudes. For negative noise levels it cycles in and out of error as the noise increases past -d, -$5d$, -$9d$ and -$13d$, as seen in Figure 5.3. Therefore the associated error probability is given by

$$P_{e8} = Q\left[\sqrt{\frac{\gamma}{21}}\right] + Q\left[9\sqrt{\frac{\gamma}{21}}\right] - Q\left[5\sqrt{\frac{\gamma}{21}}\right] - Q\left[13\sqrt{\frac{\gamma}{21}}\right]. \tag{5.25}$$

Following a similar argument the error probability P_{e6} of the i_3 bit of phasor P_6 is

$$P_{e6} = Q\left[\sqrt{\frac{\gamma}{21}}\right] + Q\left[3\sqrt{\frac{\gamma}{21}}\right] - Q\left[5\sqrt{\frac{\gamma}{21}}\right] + Q\left[9\sqrt{\frac{\gamma}{21}}\right], \tag{5.26}$$

while

$$P_{e7} = Q\left[\sqrt{\frac{\gamma}{21}}\right] + Q\left[3\sqrt{\frac{\gamma}{21}}\right] - Q\left[7\sqrt{\frac{\gamma}{21}}\right] + Q\left[11\sqrt{\frac{\gamma}{21}}\right]. \quad (5.27)$$

The error probabilities P_{e1}, P_{e2}, P_{e3}, P_{e4} are equivalent to those given by P_{e8}, P_{e7}, P_{e6}, P_{e5}, respectively, and the same holds for all corresponding phasors in the columns of the phasor diagram of Figure 5.3. Furthermore, the q_3 bit error probability is identical to that of i_3, if independent random sequences are transmitted. Averaging the $C3$ bit error probabilities yields:

$$\begin{aligned} P_{C3}(\gamma) = {} & Q\left[\sqrt{\frac{\gamma}{21}}\right] + \frac{3}{4}Q\left[3\sqrt{\frac{\gamma}{21}}\right] - \frac{3}{4}Q\left[5\sqrt{\frac{\gamma}{21}}\right] - \frac{1}{2}Q\left[7\sqrt{\frac{\gamma}{21}}\right] \\ & + \frac{1}{2}Q\left[9\sqrt{\frac{\gamma}{21}}\right] + \frac{1}{4}Q\left[11\sqrt{\frac{\gamma}{21}}\right] - \frac{1}{4}Q[13\sqrt{\frac{\gamma}{21}}]. \end{aligned} \quad (5.28)$$

The last three terms in Equation 5.28 represent extremely unlikely events as the Gaussian noise sample must exceed the $9d$ protection distance of the best-protected $C1$ bit. Consequently we neglect these terms.

Applying similar arguments to those used in the formulation of $P_{C3}(\gamma)$, we have the bit error probabilities for $C2$ and $C1$ as

$$P_{C2}(\gamma) = \frac{1}{2}Q\left[\sqrt{\frac{\gamma}{21}}\right] + \frac{1}{2}Q\left[3\sqrt{\frac{\gamma}{21}}\right] + \frac{1}{4}Q\left[5\sqrt{\frac{\gamma}{21}}\right] + \frac{1}{4}Q\left[7\sqrt{\frac{\gamma}{21}}\right], \quad (5.29)$$

and

$$P_{C1}(\gamma) = \frac{1}{4}Q\left[\sqrt{\frac{\gamma}{21}}\right] + \frac{1}{4}Q\left[3\sqrt{\frac{\gamma}{21}}\right] + \frac{1}{4}Q\left[5\sqrt{\frac{\gamma}{21}}\right] + \frac{1}{4}Q\left[7\sqrt{\frac{\gamma}{21}}\right]. \quad (5.30)$$

The P_{C1}, P_{C2} and P_{C3} error probability curves evaluated using Equations 5.28-5.30 are displayed in Figure 5.4.

Our simulation results differed by less than a decibel from these curves. Observe from the figure the consistent BER advantage obtained using the $C1$ sub-channel. All sub-channels have BERs$< 10^{-3}$ for SNR values in excess of 23 dB. They are therefore suitable for digitised speech transmission. Notice that for SNRs above 23 dB there is only a modest BER advantage in using the $C1$ sub-channel.

As we have seen, both 64-QAM and 16-QAM possess sub-channels with different BERs. The sub-channel integrities are consistently different and these differences can be used advantageously when transmitting speech and video signals. The most sensitive speech or image bits can be transmitted via the lower BER sub-channel, while the more robust source coded bits can be sent over the higher BER sub-channel. In certain applications this property is very advantageous, because it might remove the need for complex forward error correction codecs with unequal error protection [42, 85–87].

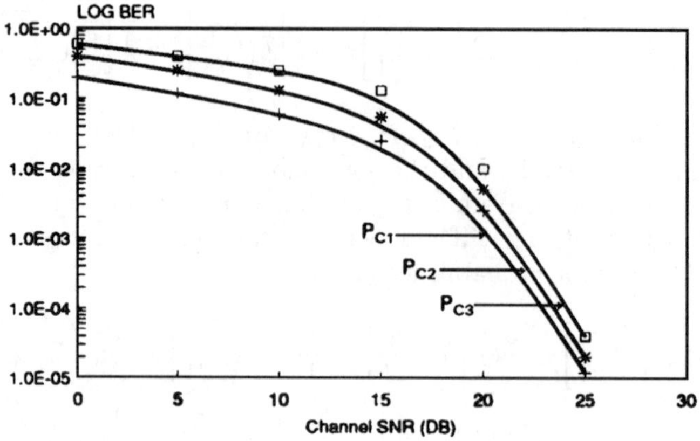

Figure 5.4: Square 64-QAM BER versus channel SNR curves over AWGN channel

5.5 Summary

In this chapter we have examined the theoretical performance of square QAM transmissions over Gaussian channels. Theoretical calculations showed the presence of different classes of bits with different performance within the constellation, and we will see in Chapter 10 how the relative performance of these bits diverges even further when the constellation is transmitted over mobile radio channels.

In the next chapter we consider some of the practical difficulties of receiving modulated signals by examining clock and carrier recovery techniques. These are needed to recover the symbol timing and the carrier phase, respectively.

Chapter 6

Clock and Carrier Recovery Systems for QAM

6.1 Introduction

Both clock and carrier recovery systems attempt to derive information about timing from the received signal, often in a similar manner. Whilst carrier recovery is only necessary in a coherent demodulation system, clock recovery is required in all schemes, and accurate clock recovery is essential for reliable data transmission. Confusion often exists between clock and carrier recovery. Clock recovery attempts to synchronise the receiver clock with the baseband symbol rate transmitter clock, whereas carrier recovery endeavours to align the receiver local oscillator with the transmitted carrier frequency.

It is not our intention here to give a detailed account of all clock and carrier recovery schemes, such background information can be found in the literature [183, 200–202]. Instead we describe the particular clock and carrier recovery schemes which have proved suitable for QAM over Gaussian channels. In many cases these are either the same, or modified versions of timing recovery systems which have been used for binary modulation, and in these instances we describe how these schemes have been adapted for QAM. We consider the issue of timing recovery systems for mobile radio transmissions in Chapter 12.

6.2 Clock Recovery

There exist a plethora of different clock recovery techniques. However, many apparently different clock recovery techniques are equivalent, and this is not surprising as they must all use the properties of the received waveform. Basic clock recovery systems which can be used successfully in conjunction with QAM include times-two, early-late, zero-crossing and synchroniser clock recovery systems [200].

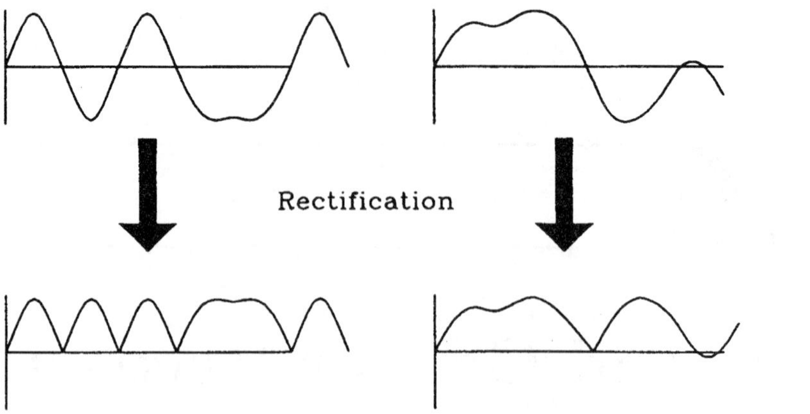

Figure 6.1: Effect of squaring waveforms

We now consider each of these clock recovery systems. We have evaluated the performance of all these systems using computer simulation and found that they all offer similar performance in a Gaussian environment.

6.2.1 Times-two clock recovery

The most fundamental of all clock recovery schemes is the times-two, or squaring system [200]. Typical BPSK data signals often contain sections of repetitive strings of $-1, +1, -1, +1, \ldots$ which after bandlimiting and pulse shaping has a waveform similar to that shown on the left-hand side of Figure 6.1. Observe that this signal has periodic sections having a frequency of half the signalling clock rate. Therefore, if the received demodulated signal is squared, or passed through a non-linear rectifier, then it will possess a periodic frequency domain component at the symbol rate. A bandpass filter tuned close to the symbol rate will extract this periodic signal, allowing derivation of the required timing information. This is shown diagrammatically at baseband for both a binary and a multilevel modulation scheme in Figure 6.1. The times-two clock recovery works best for binary modulation schemes where the $-1, +1, -1, +1, \ldots$ sections are frequent and hence the energy of the clock frequency component is high, but not so well for multilevel sections. The reason for this is that the multilevel scheme has a reduced component at the symbol frequency due to the increased possibility of non-zero-crossing transitions such as $+3, +1, -1, \ldots$ A schematic of a simple baseband times-two clock recovery system is shown in Figure 6.2 where the clock pulse regenerator could be implemented with a saturating amplifier, leading to a rectangular pulse shape.

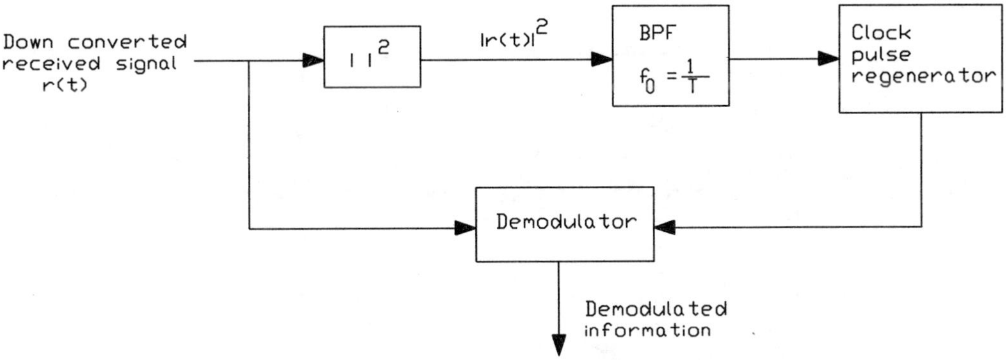

Figure 6.2: Times-two clock recovery schematic

6.2.2 Early-late clock recovery

Another well-known form of clock recovery is early-late clock recovery [203], which is portrayed in Figure 6.3. Whilst times-two clock recovery exploits the whole of the incoming waveform, early-late clock recovery works on the peaks in the received waveform. The basic assumptions made by the early-late method are that the peaks in the incoming waveform are at the correct sampling points and that these peaks are symmetrical. This is often true, but for some modulation schemes, such as partial response raised cosine arrangements [182], neither of these assumptions is valid. It may still be possible to use early-late clock recovery with these modulation schemes but the timing jitter will be increased.

The early-late scheme of Figure 6.3 firstly squares the incoming signal in order to render all peaks positive, and then takes two samples of the received waveform, both equispaced around the predicted sampling instant. If the predicted sampling instant is aligned with the correct sample point, and the assumptions made above are correct, then the sample taken just prior to the sample point - the early sample - will be identical to the sample taken just after the sampling instant - the late sample. If the early sample is larger than the late sample, this indicates that the recovered clock is sampling too late, and if the early sample is smaller than the late sample this indicates that the recovered clock is sampling too early. It is common practice to lowpass filter (LPF) the difference of each pair of samples to reduce the effect of random noise on the system. This filtered difference signal adjusts the frequency of a voltage controlled oscillator (VCO) in order to delay or advance the arrival of the next clock impulse, as required by the early and late samples.

The early-late clock recovery works well in conjunction with binary modulation schemes which have peaks in most of the symbol periods, but as with times-two clock recovery, less satisfactorily with multilevel schemes because there are fewer distinctive peaks.

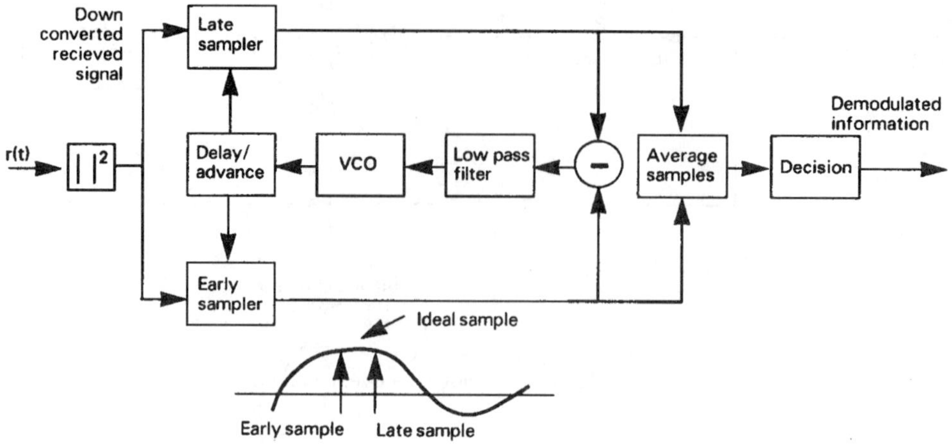

Figure 6.3: Early-late clock recovery schematic

6.2.3 Zero-crossing clock recovery

The zero-crossing clock recovery principle is similar to early-late clock recovery, in that it looks for a specific feature in the received waveform. This works on the premise that with symmetrical signalling pulses, the received waveform will pass through zero exactly midway between the sampling points. The receiver detects a change in the polarity of the received signal, and if this does not occur midway between predicted sampling points, it speeds up or slows down its reconstructed clock appropriately. Again, the zero-crossing clock recovery is a scheme which works well for binary modulation, but the assumption that the waveform will pass through zero midway between sampling instances is not always valid with multilevel schemes. For example, if the received waveform moves between amplitudes of $+1$ and -3 in a smooth manner, it will pass through zero before the middle of the sample period.

There are two solutions to allow the use of zero-crossing with multilevel modulation. The first is to ignore this problem, as on average the crossing will be in the middle of the symbol period, and to lowpass filter or integrate the observed zero-crossing point in order to produce this average. This increases the timing jitter compared to binary modulation schemes. The second is only to observe the zero-crossings when we expect them to fall midway between sampling instants [204]. If we detect a transition between two symbols of equal magnitude but opposite polarity then we can expect the zero-crossing associated with this transition to be in the middle of a symbol period and can use this to update the timing. If only these transitions are used for timing updates then, because of the accuracy of the zero-crossing detection, less averaging needs to be performed to remove any noise and so the filtering requirements are reduced, leading to a lower jitter in the recovered clock.

In the case of binary modulation, a similar schematic to that of Figure 6.3 is applicable to zero-crossing clock recovery. However, in the case of multilevel modulation a control logic block must be included in order to enable or disable sampling adjustments, depending on whether zero-crossings fall midway between sampling instances.

6.2.4 Synchroniser

A completely different clock recovery system is the so-called synchroniser. This is the scheme used by the pan-European mobile radio system known as GSM. In this system the transmitter periodically sends a sounding sequence. The receiver searches for this sequence by performing an autocorrelation at a rate significantly faster than the symbol rate (typically four times). The oversampling point at which the maximum correlation occurs is deemed to be the correct sampling point, and is assumed to remain correct until the next sounding sequence is received. This scheme is simple, but requires a sounding sequence to be periodically inserted into the data stream, increasing the bandwidth requirements. It also requires that the transmitter and receiver clocks do not drift significantly over a synchronisation block, which may place constraints on the circuitry that can be employed. This synchronisation scheme has the advantage that it performs equally well regardless of the modulation scheme. The simplified block diagram of such a scheme is shown in Figure 6.4.

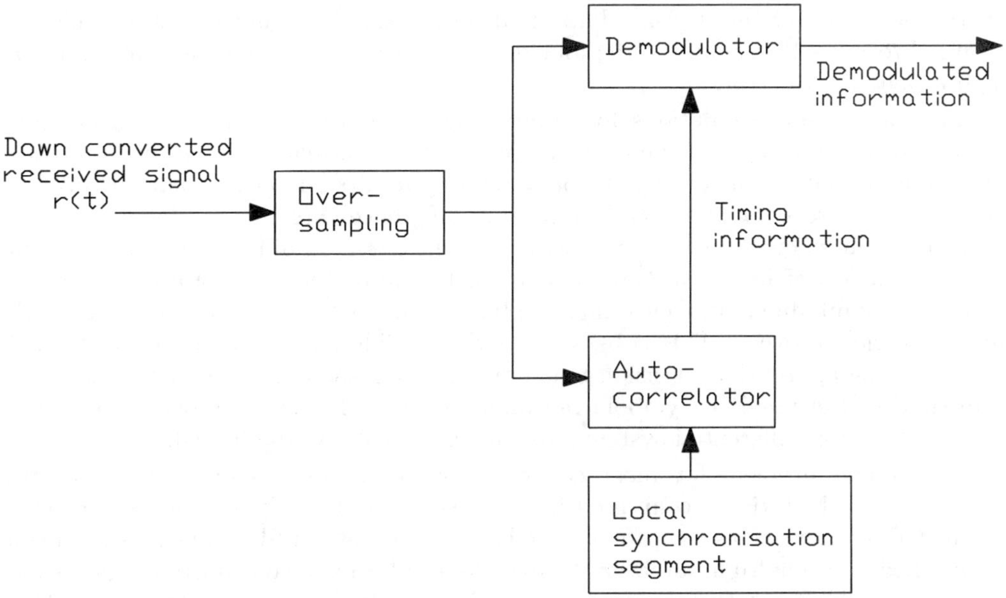

Figure 6.4: Synchroniser clock recovery schematic

6.3 Carrier Recovery

Carrier recovery has received more attention than clock recovery [205,206] but is not always required as non-coherent detection can often be applied, whereas all modulation schemes require accurate clock recovery. Carrier recovery is always required for square QAM constellations, even when differential coding is employed. For an explanation of differential coding see Chapter 4. This is because the square constellation consists of discrete information encoded on the I and Q axes separately. In order to decode this information it is essential to be able to separate the effects of the I and Q modulation and this can only be performed if a coherent source exists within the receiver.

The effect of coherent detection suffering from inaccurate carrier recovery in the case of a non-differential square 16-QAM system is shown in Figure 6.5, and that on a differential system is shown in Figure 6.6. With the non-differential system a constellation point with a magnitude of (3, 1) on the I and Q axes, respectively, is shown to be transformed to another one with magnitude of approximately (2.2, 2.2) due to the channel-induced carrier phase rotation. When demodulated with reference to the original coordinate system drawn in dashed lines this could easily be decoded as an incorrect symbol. For the differential case shown in Figure 6.6 let us suppose that the information to be transmitted requires that the constellation point be moved positively by one quadrant and positively by 90 deg within that quadrant. At the receiver when demodulating using the original unrotated coordinates shown in dashed lines we will perceive the correct quadrant move but the received point is on the boundary of the decision bars, and only a small amount of noise would be required to move it into any of the four decision areas within the quadrant, giving a high probability of error.

Circular QAM constellations do not need carrier recovery if differential encoding is used. This is because, with these constellations, the information is encoded onto the phase and amplitude of the signal, corresponding to a polar rather than rectangular encoding, and so only phase and amplitude differences are required which can be obtained even when the receiver's carrier reference is at a different frequency from the transmitter reference. In Gaussian channels, the performance of a non-coherent system using full differential encoding will be 3 dB worse than a coherent system with *perfect* carrier recovery. This is because, with the differential coding, if one symbol is outside the appropriate decision boundary it will cause two errors, one when it is compared to the previous symbol, and another when the next symbol is compared with it. In a non-differential system only one error will be experienced.

A common problem for most carrier recovery schemes is an inability to resolve phase ambiguities encountered in the case of rotationally symmetric, or near-symmetric constellations. So for constellations such as QPSK, phase lock can be established at all multiples of 90 deg, but without additional transmitted information it is not possible to resolve the angle at which phase lock has been established. This problem can be overcome by sending a sounding sequence known to the receiver, or through the use of differential coding. The decision as to whether to employ coherent non-differential, coherent differential, or non-coherent differential modulation depends on the particular environment and the QAM constellation chosen. If the channel changes only very slowly, which would be the case for a Gaussian channel with a high SNR, then coherent non-differential transmission is to be preferred, with a training sequence in order to resolve the false locking problem. If the square constellation is employed in channels whose phase change can be relatively rapid then coherent differential modulation will prevent error bursts caused by the possible false lock-

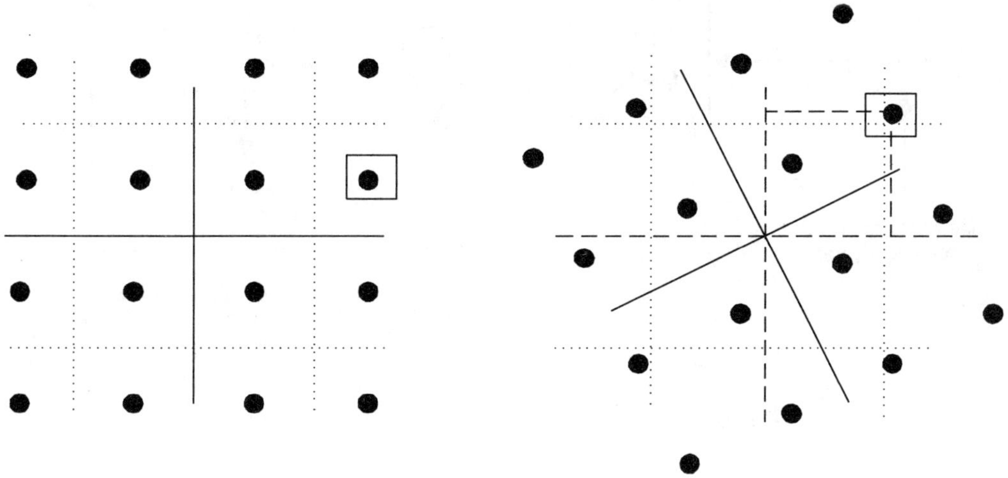

Figure 6.5: Coherent detection suffering from inaccurate carrier recovery in the case of non-differential QAM

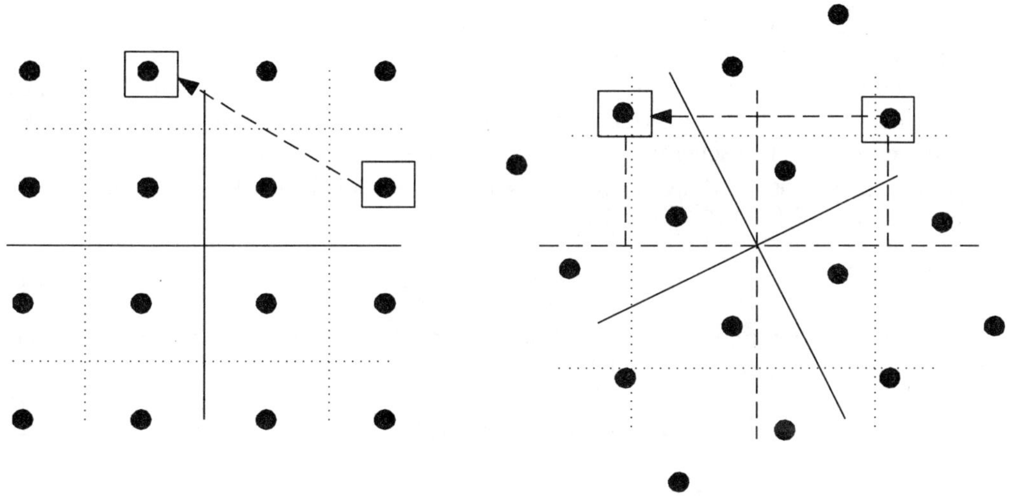

Figure 6.6: Coherent detection suffering from inaccurate carrier recovery in the case of differential QAM

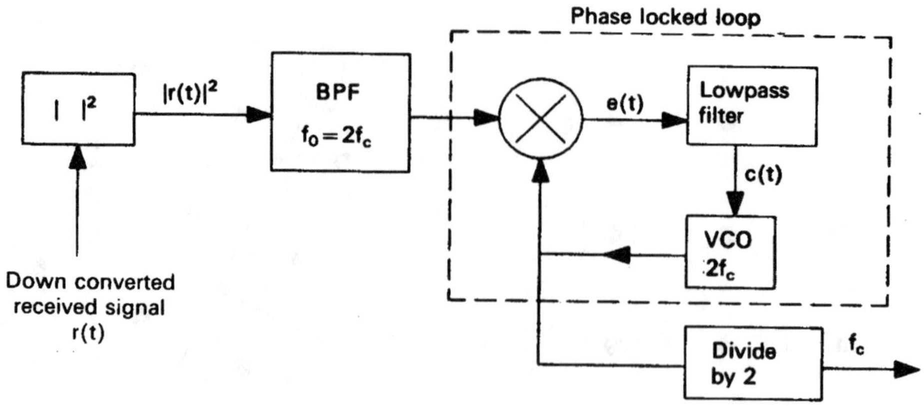

Figure 6.7: Times-two carrier recovery schematic

ing. If acceptable carrier recovery performance cannot be achieved then non-coherent differential modulation coupled with a circular QAM constellation is preferable.

After this introduction we now describe some of the more common carrier recovery systems and examine techniques which have been proposed for QAM systems.

6.3.1 Times-n carrier recovery

As with clock recovery, a common form of carrier recovery for binary schemes is the times-two or the squared carrier recovery. In this scheme the incoming carrier wave is squared, leading to a periodic component at the carrier frequency which can be isolated using a filter and phase locked loop. Times-two carrier recovery can be implemented using the schematic of Figure 6.7 [164].

Assuming BPSK modulation and noiseless communications for the sake of simplicity, the received signal is given by

$$r(t) = a(t).\cos(w_c t + \psi), \tag{6.1}$$

where $a(t) = \pm 1$ represents the modulating signal, and ψ the phase of the carrier. After squaring the received signal we get

$$\begin{aligned} r^2(t) &= a^2(t)\cos^2(w_c t + \psi) \qquad (6.2) \\ &= \frac{1}{2}[1 + \cos[2(w_c t + \psi)]]. \end{aligned}$$

Upon removing the DC component and bandpass filtering, the remaining signal $\cos[2(w_c t + \psi)]$ can be used to drive a phase locked loop (PLL) constituted by a lowpass filter, a multiplier and a voltage controlled oscillator (VCO).

Assuming that the VCO oscillates at $2w_c$ and has an initial phase of $\hat{\psi}$, its output signal of $\sin[2(w_c t+\hat{\psi})]$ is multiplied by $\cos[2(w_c t+\psi)]$ in order to generate the signal

$$\begin{aligned} e(t) &= \cos[2(w_c t+\psi)].\sin[2(w_c t+\hat{\psi})] \\ &= \frac{1}{2}\sin[4w_c t+2(\psi+\hat{\psi})]+\frac{1}{2}\sin[2(\hat{\psi}-\psi)]. \end{aligned} \quad (6.3)$$

The first term of the above equation is removed by the lowpass loop filter, while the second term of

$$c(t) = \frac{1}{2}\sin[2(\hat{\psi}-\psi)] \approx \Delta\psi \; if \; \Delta\psi \ll 1 \quad (6.4)$$

is used to drive the VCO. Finally, the output from the VCO is divided by 2.

Unfortunately, according to Franks [200] the squaring approach does not work with quadrature modulation. This is because when a quadrature signal with equal power on average in each quadrature branch is squared, the signal power in each of the branches tends to cancel, leaving a low signal level on which to perform carrier recovery. However, this can be overcome through the use of a fourth-power device which can generate a significant component at four times the carrier frequency. This can then be divided by 4 to provide carrier recovery. Passing the complex received signal through a fourth-power device has the effect of raising the magnitude of the received phasor to the fourth power and multiplying the phase angle by 4. Times-four carrier recovery was investigated by Rustako *et al.* [39].

Rustako notes that decision directed carrier recovery (see Section 6.3.2) suffers when the occasional receiver outage destroys the accuracy of the data detections. This causes a loss of carrier recovery and considerable time may be required to reacquire the carrier. Furthermore, when adaptive equalisers are used, which update their tap coefficients using a decision directed algorithm, then the equaliser and carrier recovery system can interact in unwanted manner. He notes that times-four carrier recovery does not depend on data decisions or interact with the equaliser. Higher powers than 4 increase carrier jitter due to the increased effects of the data pattern.

A schematic diagram of their proposed carrier recovery circuit is shown in Figure 6.8. The received RF signal is downconverted by mixing with the local oscillator (LO). Selectivity and automatic gain control are achieved at the IF stage. The input is raised to the fourth power and input to the PLL. The function of the PLL is to extract the line component at four times the carrier frequency, and to filter out the random component due to the data. Observe that in Figure 6.8 the PLL has an identical structure to that of Figure 6.7, except that it is tuned to f_c rather than $4f_c$, and hence it delivers the required output frequency. The carrier frequency f_c must then be multiplied by four as shown in the $\times 4$ block. This could be achieved by raising the signal to the fourth power. Rustako noted that the times-four function could be performed by a frequency doubler which by virtue of its design also produces frequencies at a fourth power. With suitable filtering such a device could be readily used. In his paper, Rustako gave a thorough mathematical derivation of the theory of times-four carrier recovery. Rustako then went on to construct a times-four carrier recovery and test it using a fading channel simulator. From his experiment Rustako noted that using this form of carrier recovery synchronisation can be maintained even when severe fading causes high error rates. The drawback is that in order to keep

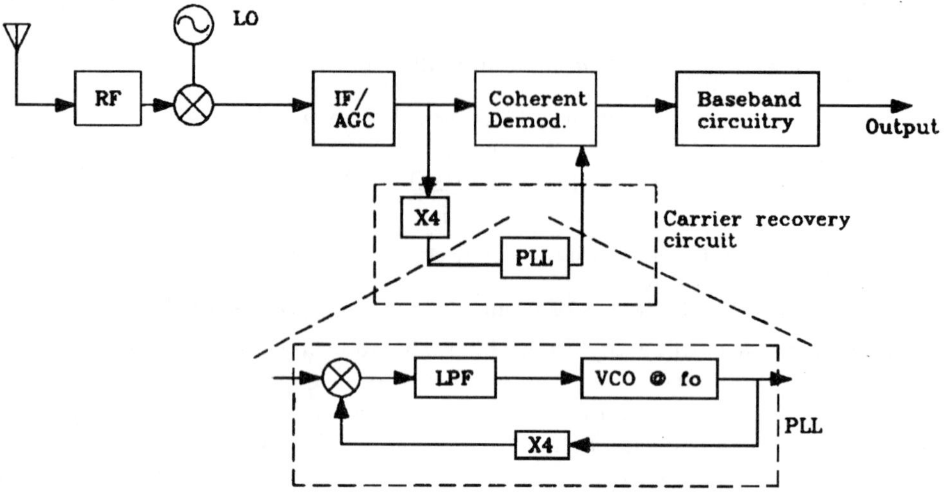

Figure 6.8: Times-four carrier recovery block diagram

pattern-related phase jitter under control, the PLL noise bandwidth must be made small. This will slow down acquisition when loss of carrier does occur. Therefore, times-four carrier recovery will be most suitable in situations where fades are benign and hence loss of lock is not a frequent event.

6.3.2 Decision directed carrier recovery

A carrier recovery scheme which is completely different from all the carrier and clock recovery systems discussed so far is decision directed carrier recovery. In fixed link QAM systems this has proved to be one of the most popular carrier recovery schemes. In this scheme, a decision is made as to which was the most likely constellation point transmitted given the received symbol. This is normally the constellation point closest to the received symbol. It is then assumed that the phase difference between the received symbol and the constellation point is due to carrier recovery drift, and the receiver carrier recovery is updated accordingly. Decision directed carrier recovery has the advantage that it can work with all constellations; however, there is typically a BER threshold. If the prevailing channel BER is less than this threshold then such methods operate extremely well as they effectively remove additive noise from the input phasor. If the BER is higher than this threshold then cumulative catastrophic failure can result whereby the constellation point selected in the receiver is the wrong one. The update signal to the carrier recovery system is therefore erroneous, driving the carrier phase further away from the correct value and further increasing the chance of error, etc. Whether decision directed carrier recovery will be suitable for a specific application will depend on this BER threshold, which itself is dependent on the number of modulation levels used.

Whilst simple in principle, decision directed carrier recovery can often require a

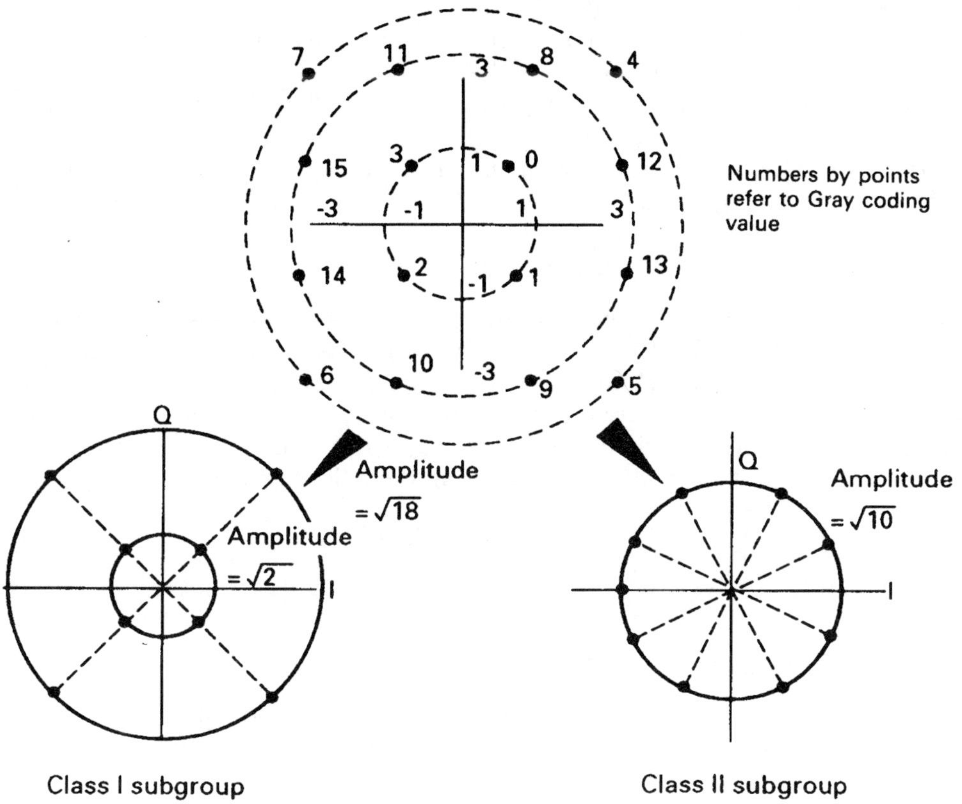

Figure 6.9: 16-level QAM subgroups

moderately complex implementation such as that proposed by Horikawa *et al.* [29] who produced an elegant arrangement for decision directed carrier recovery combined with QAM. The basic premise of the circuit is that the quality of carrier recovery will be acceptable if only half the input signals are used for carrier recovery update.

The 16-QAM phasors can be divided into two subgroups called class I and class II phasors. Note these are not related to the class 1 ($C1$) and class 2 ($C2$) sub-channels derived in Chapter 5. This is shown in Figure 6.9.

Class I consists of the eight phasors lying on the angles:

$$\theta_I = \pi/2 \times (n + 1/2) \quad n = 0, 1, 2, 3 \tag{6.5}$$

and class II consists of all the remaining phasors:

$$\theta_{II} \approx \pi/2 \times (n \pm 1/5) \quad n = 0, 1, 2, 3 \tag{6.6}$$

where the subscripts refer to class I and class II respectively, and $(\pi/2) \cdot (1/5) = (\pi/10) = 18\,\text{deg} \approx \tan^{-1}(1/3)$ from the geometry of the figure. In the square 16-

QAM constellation shown in Figure 6.9 the class I phasors have $|I| = |Q|$, where $|x|$ means magnitude of x, and I and Q refer to the received voltage on the real and imaginary axes, respectively. The class II signals have $|I| = 3|Q|$ or $|Q| = 3|I|$. The normalised amplitudes for class I phasors are $\sqrt{18}$ and $\sqrt{2}$, and for the class II phasors $\sqrt{10}$.

Horikawa's circuit operates only on class I phasors. It does so by comparing the values of the incoming I and Q signals and if they satisfy the relationship

$$|I|/m < |Q| < n \times |I|, \tag{6.7}$$

where m and n are typically 2 then the circuit concludes that a class I phasor has been sent. For a class I signal received with perfect carrier recovery over a perfect channel we would have $|I| = |Q|$. By subtracting $|I|$ from $|Q|$ we have a signal whose polarity indicates the direction of carrier drift. Determining the quadrant of transmission allows deduction of the required signal to be applied to the VCO. This computation can be carried out by a simple algorithm using the following quantities:

$$\begin{aligned} a &= pol(I)\ XOR\ pol(Q) \\ b &= pol(I+Q) \\ c &= pol(I-Q) \\ d &= (a\ XOR\ b)\ XOR\ c \end{aligned} \tag{6.8}$$

where pol (x) is a function which returns 0 if x is negative and 1 if x is positive. The value of d is then used to drive the VCO so that if $d = 1$ the VCO runs slower and if $d = 0$ the VCO runs faster.

For example, say a phasor at 135 deg had been sent but because of an increase in the receiver carrier frequency this had been received with a phase advance at 140 deg. This would give $I = -1.08$ and $Q = 0.91$. So $a =$ (0 XOR 1) $= 1$, $b =$pol$(-0.17) = 0$, $c =$pol$(-1.99) = 0$ and $d =$ [(1 XOR 0) XOR 0] $= 1$. Thus the VCO would be instructed to run slower as required. Examples taken from any quadrant can be seen to function correctly.

The circuit is shown in Figure 6.10. The input phasor is split into real and imaginary components using the receiver's recovered clock. The quadrature components are lowpass filtered (LPF) to remove the unwanted frequency components of the mixing operation. As mentioned, the decision as to whether a class I signal has been received is based on evaluating whether $|I|/2 < |Q| < 2 \cdot |I|$. Both quadrature components are full-wave rectified and input to a comparator which determines whether the signal level is within the limits expected for a class I signal. The XORed outputs from the two comparators are fed into the control input (c) of an enabling gating arrangement which allows the updating of the VCO if the I and Q amplitudes are within a ratio of 2. The central part of the circuit performs the algorithm described above by computing the parameters a, b, c, d in Equation 6.8, with the loop filter and VCO acting as a phase locked loop (PLL).

This circuit suffers from possible false locking as 90 deg rotations of the constellation will not change the parameters $a \ldots d$. This must be resolved through the use of a training sequence or some other ambiguity resolving technique.

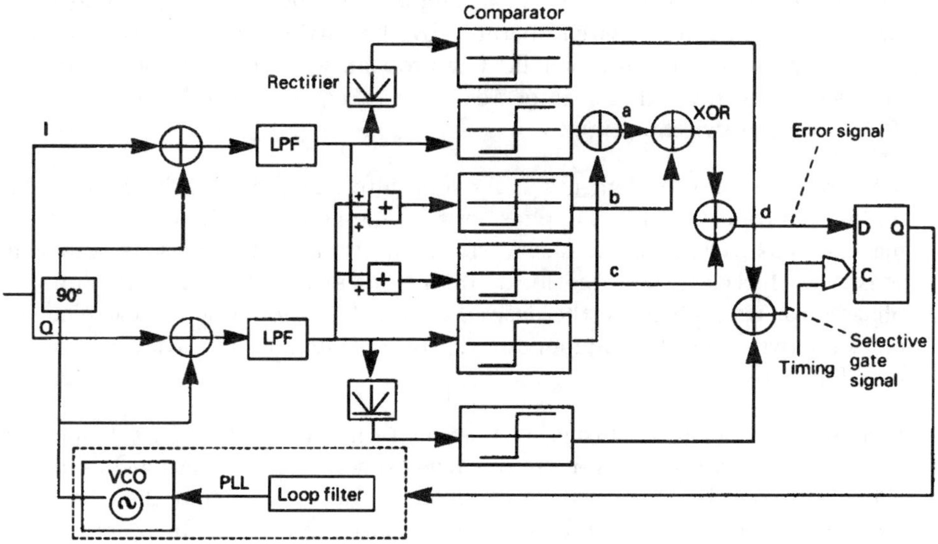

Figure 6.10: Implementation of a decision directed carrier recovery system for 16-level square QAM [43] ©IEEE, 1988, Sari *et al.*

6.3.2.1 Frequency and phase detection systems

Within amplitude modulation schemes where the phase shift imposed by the modulating signal on the carrier remains constant it is possible to use phase locked loop (PLL) systems. In PLLs phase detectors (PDs) are deployed in order to detect the phase difference between the recovered carrier and the internal reference frequency of the receiver which is used to update the internal reference frequency appropriately. Such a system is referred to as a phase locked loop, which is described in detail in Reference [207].

Such simple systems cannot be used in conjunction with phase or quadrature modulation schemes due to the variation in the carrier phase imposed by the encoded information. These modulation-induced carrier phase variations can either be ignored with the assumption that the average of this phase deviation will be zero, or attempts can be made to remove the phase variation using decision directed methods such as those described in the previous subsection.

Further problems associated with PLL methods for point-to-point microwave links are discussed by Sari and Moridi [43]. Sari notes that the high number of modulation levels used in point-to-point links requires a low phase jitter, which demands narrow PLL bandwidths. However, the frequency uncertainty of the RF oscillators used in the up and down conversion process was often greater than the affordable PLL bandwidth. Possible solutions to this problem include:

(1) Switching of the loop filter. Here a large filter bandwidth is used during the initial phase acquisition process and then a narrow bandwidth is invoked once

lock has been achieved. In its most simple form such a system would include two loop filters, and a switch operated by the carrier recovery circuit once the error signal was sufficiently small. The problem with this method is that some means of detecting lock, i.e. a lock detector is required in order to activate and deactivate the switch optimally.

(2) The use of non-linear elements in the loop filter. Due to this the error signal is no longer proportional to the phase error. The non-linear function is chosen to increase sensitivity (i.e. the phase error seen by the PLL) near lock, and reduce sensitivity further away from lock. Taking the square root or logarithm of the phase error would achieve the desired effect. The chosen non-linear element is placed between the PD and loop filter. Appropriate non-linear devices can often be difficult to construct.

(3) Frequency sweeping. This is a common technique within digital microwave systems whereby the receiver's frequency is slowly changed until lock is achieved using a narrow bandwidth detector. Such a detector should then be able to maintain accurate lock if the channel is relatively benign. In practice the IF oscillator frequency is swept over the uncertainty interval by adding a periodic signal to the filtered PD output. Frequency sweeping is halted once phase lock has been achieved . This process has the disadvantage of taking some time due to theoretical constraints as regards to the speed at which the sweep can take place.

(4) Frequency detectors (FDs). These are devices which produce a signal proportional to the frequency difference between the received signal and local reference carriers. Hence they can readily determine the frequency offset between the received signal and the local reference but as the frequency error does not depend on the phase they cannot be used for phase tracking. They are therefore often used in parallel with PDs, the FD being used during initial acquisition. When the FDs give a near-zero output then a switch is operated such that the FDs are deactivated and the PDs are activated. Sari notes [43] that such a system requires a substantial amount of circuitry.

In his paper, Sari [43] describes an architecture for a new implementation of a combined phase and frequency detector which he refers to as a PFD. A diagram showing the architecture of such a system is given in Figure 6.11. The top PD will be converted into an FD by appropriately manipulating its output. The downconverted received signal is input to a pair of sinusoidal phase detectors (PD) operating in quadrature with each other so that their outputs are $sin\ \phi(t)$ and $cos\ \phi(t)$, where $\phi(t)$ is the phase error, as shown in Figure 6.12. Note that the required stable lock points for a rectangular QAM constellation are at $n \cdot (\pi/2), n = 1, 2, 3, 4$, and these points can be identified as those at the zero crossings of $sin\ \phi$ in Figure 6.12, where $cos\ \phi$ is simultaneously positive. During steady state, the phase error $\phi(t)$ fluctuates around a stable lock point where $sin\ \phi \ll 1$, and only the upper PD and the VCO are being used in a normal PLL configuration. However, during the initial acquisition phase the phase error $\phi(t)$ significantly deviates from zero, and changes at a rate proportional

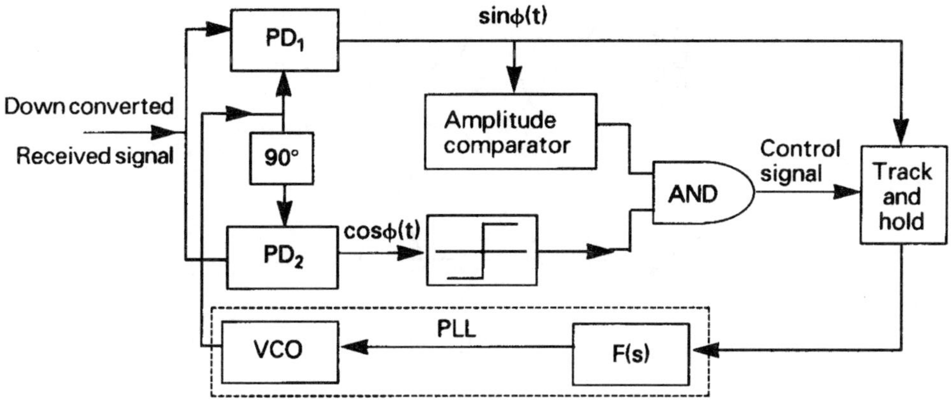

Figure 6.11: PFD block diagram

to the frequency offset between the received signal and VCO reference according to $\phi(t) = \Delta wt + \phi(0)$. Clearly, the PD output is a sinusoid with a frequency equal to the difference Δw between the received signal and VCO reference, and therefore has a zero DC content delivering no information about the polarity of the frequency error. In contrast, an FD would provide a DC signal proportional to the frequency offset Δw.

The circuit of Figure 6.11 overcomes this problem by the use of a track and hold device. The device tracks the upper PD when it is in phase lock and so its output is in the range $k\pi/2 \pm \theta$ where $k = 1, 2, 3, 4$ is an integer representing phase lock in four different quadrants, and θ is the maximum allowed phase deviation. In order to track the phase error, the circuitry also requires that the output from the lower quadrant PD is positive, which pinpoints the stable locking domains. Should, however, the phase error $|\phi(t) - n\pi/2|$ be outside the range $\pm\theta$, the circuitry operates in a hold mode and the track and hold device retains the last valid level. The control signal $sin\ \phi$ originating from the upper PD determines whether the phase error $\phi(t)$ is within a predetermined interval $\pm\theta$, and $cos\ \phi$ originating from the lower PD allows recognition of the stable lock points as seen in Figure 6.12. The operation of the device is now explained with reference to Figure 6.13.

The operation of the PFD of the circuit of Figure 6.11 for positive and negative frequency offsets is shown in Figures 6.13(a) and (b), respectively. In these figures which show the situation during initial phase acquisition (i.e. less lock has not been acquired) the dashed line represents the error signal from the upper PD and the solid line represents the output of the track and hold device (and thus the input to the VCO). With a positive frequency offset as shown in Figure 6.13(a) the upper PD tracks the offset until it falls outside the range θ. At this point the amplitude comparator detects that the PD output has moved outside the allowed range and switches the track and hold device to hold mode. It is maintained in this configuration until the output from the upper PD falls back within the allowed range. Note that the continuous

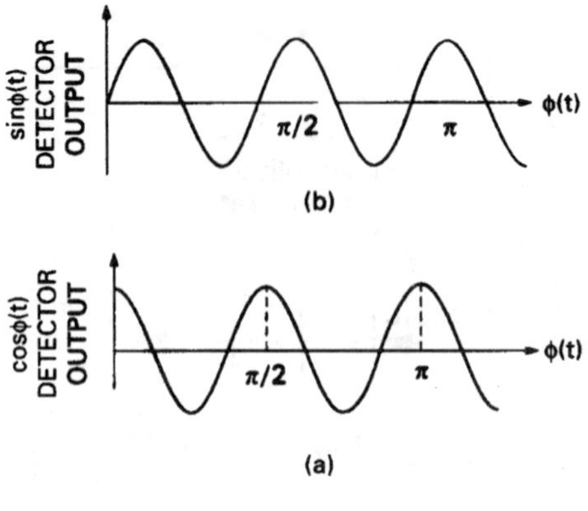

Figure 6.12: In-phase and quadrature phase PD output signal [43] ©IEEE, 1988, Sari *et al.*

line does not fall back to the dashed line as the latter passes through the axis moving downwards. This is because at this point the track and hold device is maintained in the hold position by the output of the lower PD, which precedes that of the upper by 90 deg as demonstrated in Figure 6.12. When $cos\ \phi$ is negative, so the upper branch is moving from +90 deg to −90 deg, the track and hold device is maintained at hold. As can be seen, in the positive case portrayed in Figure 6.13(a), the control signal to the VCO is predominantly positive, ensuring that there is sufficient information to allow acquisition. This condition is detected by the comparator at the output of the lower PD. Once lock is acquired, the system will automatically revert to a tracking role as the upper PD output falls within its allowed range. A similar effect can be seen for a negative frequency offset as shown in Figure 6.13(b). The parameter θ should be chosen to maximise the DC at the detector output. For a sinusoidal PD this means that θ should be approximately 16 deg.

In the case of QAM, according to Sari [43], the phase error associated with each incoming symbol is assessed, and the error angle becomes the PD's output. This is then passed to the track and hold block, which performs its function as before. However, with 16-level square QAM, when the receiver's local oscillator is unsynchronised with that of the transmitter, three circles are apparent, as shown in Figure 6.14. On the inner and outer circles there are four equispaced points forming a QPSK signal set. On the middle ring there are 8 non-regularly spaced points, which can lead to phase ambiguity when used in a circuit of the form proposed by Sari as there will now be 8 possible lock positions, but only in 4 of these can data be correctly demodulated. To overcome this problem, Sari, like Horikawa, proposed to invoke only the class I QAM signals, using a windowing system in order to determine whether a symbol is suitable for carrier recovery. A block diagram of such a system is shown in Figure 6.15.

In this figure the input signal is quadrature downconverted, lowpass filtered (LPF) and analogue-to-digital converted (ADC). It is then input to a practical version of the PFD shown in Figure 6.11. In this scheme the track and hold function is performed by a D-type flip-flop which will only pass the input to the output when clocked. The

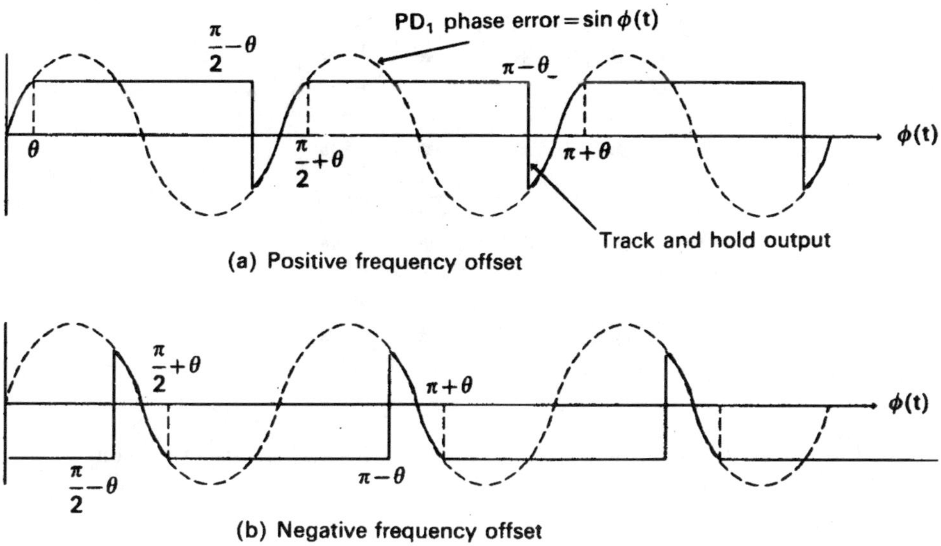

Figure 6.13: PFD waveforms for positive and negative frequency offsets

flip-flop is normally clocked regularly by the clock recovery system (not shown) but can be inhibited by the control logic device. This latter device determines whether the incoming signal is within one of the windows shown as square boxes on the QAM constellation in Figure 6.14. If the signal is within these boxes then the PD signal is used for carrier recovery purposes, otherwise the input to the loop filter is maintained at the previous level.

Sari claimed a substantially reduced acquisition time for his system compared to a discrete FD and PD system, and a reduced complexity architecture.

6.4 Summary

In this chapter we have considered clock and carrier recovery techniques designed for fixed link QAM systems. Clock recovery techniques used in this environment have tended to be similar or identical to those used for other modulation schemes, with zero-crossing techniques often finding favour. That clock recovery has not been a particular problem for QAM is evidenced by a paucity of publications on the topic.

A considerably more problematic issue has been that of carrier recovery schemes for QAM. Many of the existing schemes proved unsuitable due to the phase-related content of the encoded information and also because a more accurate lock was required due to the higher number of levels used compared to binary modulation. The carrier recovery techniques employed essentially were either times-N or decision directed. Decision directed methods generally were more suitable, where the BER was low and where there were no problems inflicted by a decision feedback equaliser's interactions with a decision directed carrier recovery system.

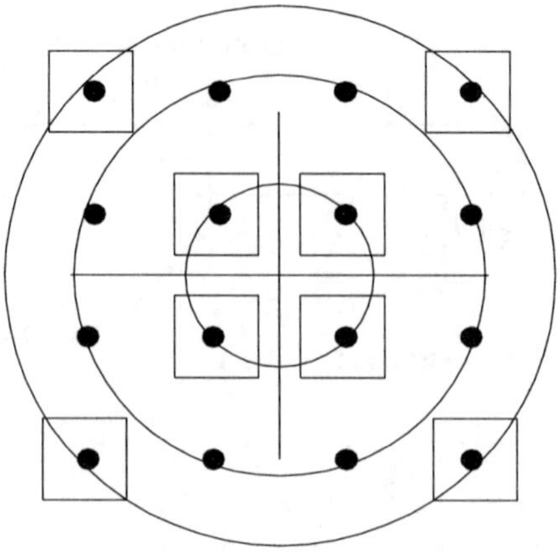

Figure 6.14: Circles and windows within a 16-level square QAM constellation

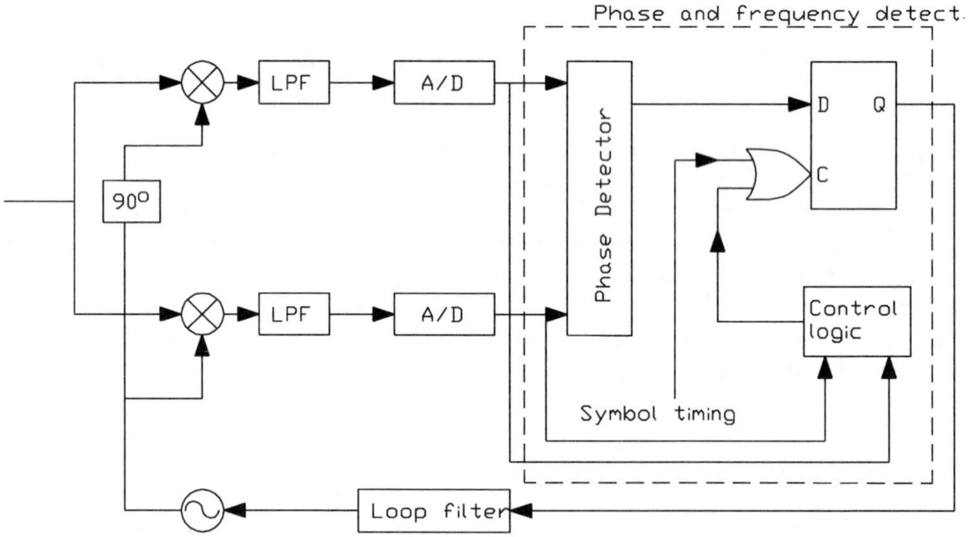

Figure 6.15: General block diagram of a PFD system

Of the decision directed carrier recovery schemes, the one proposed by Sari and Moridi addressed the issue of both frequency and phase acquisition, whereas that of Horikawa looked at practical implementation of the discrete time phase detector required. Both decided to use only the class I QAM symbols for carrier recovery, although for different reasons - Sari to prevent possible false locking errors and Horikawa to simplify the decoding circuitry required. In the next chapter we consider the important topic of channel equalisation.

Chapter 7

Trained and Blind Equaliser Techniques

7.1 Introduction

In Chapter 4 we considered the transmission of information via bandlimited channels. It was shown that all bandlimited channels have impulse responses of infinite duration which may result in intersymbol interference (ISI). However, if the Nyquist criterion is met then the impulse response has zeros at adjacent signalling instants and a unity main tap. Therefore, no ISI is introduced.

In contrast to the approach pursued in Chapter 4, where it was assumed that the channel is distortion-free and hence has a constant frequency domain transfer function and linear phase within the passband, here we consider the more realistic scenario of transmission media having linear distortions. If we characterise the channel by its baseband impulse response $g(t)$, then the baseband received signal $y(t)$ is given by the following convolution:

$$y(t) = x(t) * g(t).$$

A typical continuous channel impulse response is depicted in Figure 7.1. Note that it is no longer symmetric, its zero-crossings are no longer uniformly spaced and this channel introduces linear distortions, which can be seen in the eye pattern.

The aim of channel equalisation is to remove the distortions introduced by the channel. This can be carried out either in the frequency domain by filtering the received signal through the inverse filter of the channel, or in the time domain by convolving it with the appropriate equaliser impulse response.

Linear distortions are introduced, for example, by wideband radio channels described in Chapter 2, when there are a multitude of radio paths between the transmitter and the receiver, and the delay on some of these paths is significantly greater than on others. In extreme cases the longer paths can result from reflections off mountains, more often they occur due to reflections off nearby buildings and some-

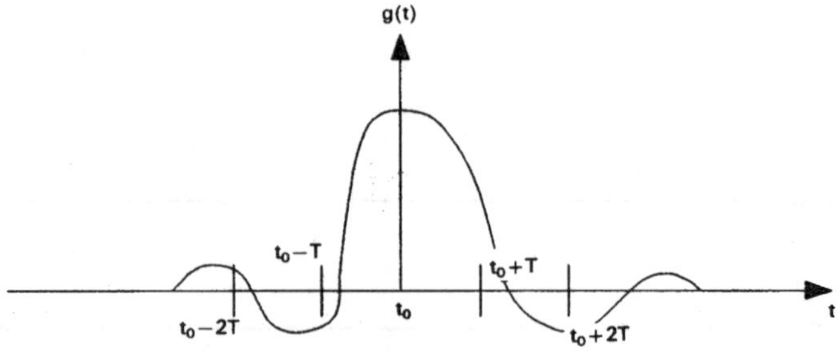

Figure 7.1: Typical channel impulse response

times from vehicles such as buses. Wideband channels can be characterised by their impulse response. This is the signal that would be received were an impulse transmitted (i.e. a pulse of extremely short duration). The impulse response is composed of the superposition of all the paths arriving at the receiver, each path suffering a certain attenuation and delay. Textbooks often show example impulse responses as exponentially decaying curves falling off steeply at first and then becoming more gently undulating. This is because the responses that are significantly delayed are often highly attenuated due to the extra distance which they have travelled.

Historically, wideband channels have been rare in QAM implementations. The only areas where significant multipath fading has been encountered are on high transmission rate (typically $>$ 140 Mbit/s) point-to-point radio links. These are characterised by slowly varying impulse responses extending over a few symbol periods only. Furthermore, the high transmission rate has implied relatively simple equalisers if they are to be implemented with existing circuitry. Two equalisers which are frequently used are the linear and the decision feedback equaliser, while the Viterbi equaliser has rarely been used for QAM mainly due to its high complexity. For that reason we do not describe the Viterbi equaliser here, but refer the reader to Proakis [164]. In the following two sections we consider linear and decision feedback equalisers. In the final section we consider aspects of equalisers specific to QAM, and consider some actual implementations. For more in-depth treatment of equalisation the interested reader is referred to the literature [3, 18, 158, 161, 185, 208].

7.2 Linear Equalisers

7.2.1 Zero-forcing equalisers

The basic structure of a linear equaliser (LE) system is given in Figure 7.2.

It is a linear transversal filter with an impulse response of

$$c(t)_{t=jT} = c(jT) = \sum_{j=-N}^{N} c(j) z^{-jT}, \tag{7.1}$$

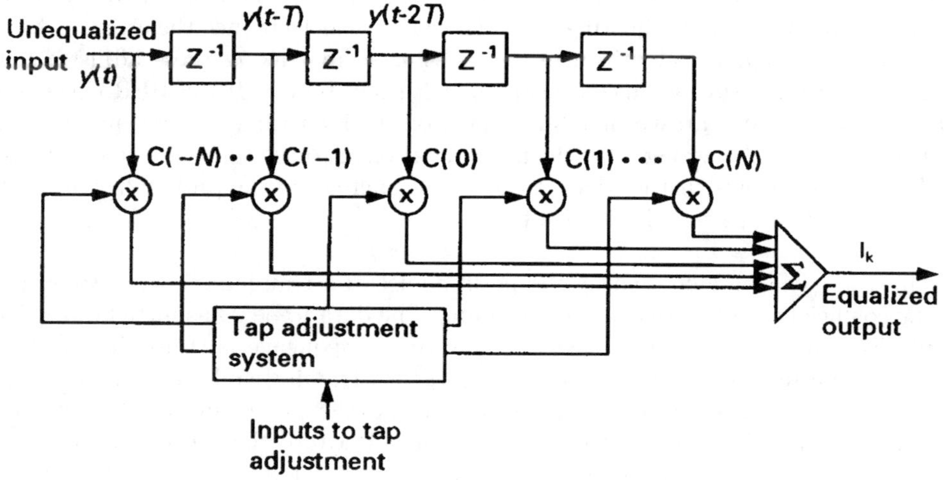

Figure 7.2: General linear equaliser system

where $c(t)$ is represented by its T-spaced samples. Equalisation is performed by convolving the T-spaced samples of the received signal, $y(t) = y(kT) = y_k$, with $c(t)$. In practice it is performed by passing y_k through a delay chain and the equalised output I_k is derived by a multiplication using taps with variable coefficients $c(j)$. Thus

$$I_k = \sum_{j=-N}^{N} c(j)y(k-j), \tag{7.2}$$

where there are $(2N+1)$ taps.

In order to satisfy the Nyquist criterion, the equaliser taps can be chosen to drive the T-spaced samples of the combined impulse response of the concatenated channel and equaliser towards zero, allowing only for the main tap to be non-zero. Due to its operating principle this scheme is referred to as a zero-forcing equaliser. Naturally, zero-forcing is only possible over the length of the transversal filter's memory. Nevertheless, as the filter length tends to infinity, the ISI at the equaliser's output tends to zero for all values of jT, $j = 1, 2, \ldots, \infty$. The fundamental limitation of the zero-forcing equaliser is due to the fact that the channel transfer function's inverse filter $C(f)$ constituted by the equaliser when cascaded with the channel transfer function $G(f)$ enhances the channel noise in those frequency intervals where $G(f)$ has a high attenuation. This is because the flat noise spectrum $N(f)$ is boosted along with the signal, where the channel inflicted a high attenuation, which becomes particularly adverse for transmissions over frequency selective fading radio channels.

There are numerous ways of deriving the coefficients $c(j)$. If the channel impulse response can be considered time invariant as a leased telephone line, then the coefficients are constant. In most cases, however, the channel impulse response is time variant and so the tap coefficients need to be variable. A large class of linear equalisers adjust these coefficients by comparing the equalised output with the nearest constellation point and adjusting the coefficients so as to reduce the error. A form of recursive least squares algorithm is normally used in making these adjustments. This is adequate for slowly varying channels with low BERs. If the BER exceeds a certain threshold then catastrophic failure will occur due to the wrong symbol being decoded and hence the wrong adjustment being applied to the tap coefficients. These maladjusted coefficients further distort the signal, increasing the probability of error, and so on, until complete failure ensues.

Robustness can be increased by the periodic use of a training sequence to ensure that the coefficient values are close to optimum. The data to be transmitted is divided into blocks, and at the start of each block a known sequence is transmitted. This method can be further subdivided into convergence training and channel sounding, depending on the known sequence transmitted and the processing at the receiver. In the former, any known sequence can be sent. Because the receiver knows this sequence there can be no decoding errors during this period and so the coefficients can be driven towards the optimal value with little risk of catastrophic failure. During normal data transmission the coefficients can either be left unchanged, or updated according to a more cautious update rule compared to that used during the training period.

Another way that the transmission of a known sequence can be used is to periodically sound the channel and then set the coefficients accordingly, eliminating the need for complicated convergence algorithms. The channel may be sounded by sending a maximal length pseudorandom binary sequence (PRBS) having a flat, almost "white" spectrum as the known sequence, and correlating the received sequence against a replica of itself held in the receiver. This correlation process is invoked for each of the symbol intervals over which the expected delay spread of the incoming signal occurs. Maximal length PRBSs have the property that the peak amplitude of the autocorrelation function with zero delay is equal to the length of the code, but for any offset delay the autocorrelation function is equal to -1. The autocorrelation is performed with the XOR function giving a result of either 1 or -1 (rather than 1 or 0). This method is used for example in the CCITT V.27, V.29, V.32 and V.33 standard high speed data modems proposed for telephone lines having linear distortions. These schemes will be described in Chapter 9. Once we have sounded the channel and its impulse response is known, we can calculate the tap coefficients for the LE.

The optimal coefficients for an infinite length LE are the T-spaced samples of the inverse filter. Again, if the baseband equivalent transmitted signal is $x(t)$, and the baseband impulse response of the channel is $g(t)$, then the baseband equivalent input to our receiver is

$$y(t) = g(t) * x(t) \tag{7.3}$$

where * symbolises convolution. An equaliser at the receiver having an impulse response $c(t)$ results in an equalised output

$$h(t) = c(t) * y(t) = c(t) * g(t) * x(t) \tag{7.4}$$

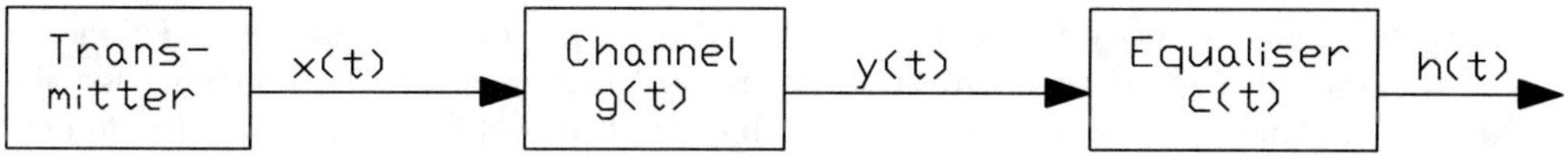

Figure 7.3: System model with equaliser

which, to be equal to the input $x(t)$, requires

$$c(t) * g(t) = \delta(t) \tag{7.5}$$

where $\delta(t)$ is the unit pulse sequence [1, 0, 0, ..]. This system is shown in Figure 7.3. Applying the z-domain transfer function to Equation 7.5 gives the z-transform of the filter required to equalise the channel having transfer function $G(z)$ as:

$$C(z) = \frac{1}{G(z)}. \tag{7.6}$$

In practice the channel response is of finite length and we represent it by

$$G(z) = g(0) + g(1)z^{-1} + \ldots + g(n)z^{-n} \tag{7.7}$$

where $g(0)$, $g(1)$, ..., $g(n)$ are the coefficients of the finite impulse response (FIR) filter modelling $G(z)$. Its inverse filter has a z-transform computed as follows:

$$\begin{aligned} C(z) &= \frac{1}{g(0) + g(1)z^{-1} + \ldots + g(n)z^{-n}} \\ C(z) &= c(0) + c(1)z^{-1} + c(2)z^{-2} + \ldots \end{aligned} \tag{7.8}$$

where $c(0)$, $c(1), \ldots$ form an infinite length sequence of equaliser coefficients. If the ISI is mild then this sequence will decay sharply and the equaliser coefficients, i.e. $c(n)$, are insignificant for large n. Then we can set the coefficients, $c(n)$, of the LE by the long division of $1/G(z)$, which requires little computational effort and can easily be implemented in real time. Thus, if our channel filter has a response of

$$G(z) = g(0) + g(1)z^{-1} + \ldots + g(n)z^{-n} \tag{7.9}$$

where $g(n)$ is a complex number then the complex equaliser coefficients are given by

$$\begin{aligned} c(0) &= \frac{1}{g(0)} \\ c(1) &= \frac{[-c(0) \times g(1)]}{g(0)} \\ c(2) &= \frac{[(-c(0) \times g(2)) + (-c(1) \times g(1))]}{g(0)} \end{aligned} \tag{7.10}$$

and generally

$$c(n) = \frac{\sum_{i=0}^{n-1}(-c(i) \times g(n-i))}{g(0)}. \tag{7.11}$$

If there is no single dominant tap in the impulse response then the coefficients of the LE may fail to decay for a reasonable value of n. In this case the coefficients calculated by this long division method are no longer the optimal coefficients for the finite length equaliser. Although optimal values for the coefficients of the finite length linear equaliser can be found by more complex processes [164], the linear equaliser is no longer suitable.

7.2.2 Least mean squared equalisers

Least mean squared (LMS) equalisers are considered superior to zero-forcing (ZF) equalisers, because they can minimise the joint effects of ISI and noise at the equaliser's output over their memory length. In our analysis, however, we assume a high signal-to-noise ratio and neglect the effect of noise for the sake of simplicity. We also note that LMS equalisers in general can tolerate higher ISI and have faster convergence than ZF equalisers.

Lucky *et al.* [18] define the mean squared distortion (MSD) as

$$MSD = \frac{1}{h^2(0)} \sum_{\substack{n=-\infty \\ n \neq 0}}^{\infty} h^2(n), \tag{7.12}$$

which is simply the normalised sum of the square of symbol period spaced system response samples at the equaliser's output. Recall from Equation 7.4 that $h(n)$ is the convolution of the received signal $y(n)$ and the equaliser's impulse response, giving:

$$h(n) = y(n) * c(n) = \sum_{j=-N}^{N} c(j)y(n-j), \tag{7.13}$$

and hence

$$h(0) = \sum_{j=-N}^{N} c(j)y(-j). \tag{7.14}$$

Minimising the effects of ISI requires minimising the mean squared error (MSE) term

$$\epsilon = \sum_{\substack{n=-\infty \\ n \neq 0}}^{\infty} h^2(n) = \left(\sum_{n=-\infty}^{\infty} h^2(n) \right) - h^2(0), \tag{7.15}$$

which due to Equation 7.13 and Equation 7.14 is a quadratic function of the equaliser coefficients $c(j)$ and hence it has a minimum. Then from Equation 7.15 we get:

$$\frac{\partial \epsilon}{\partial c(j)} = \left(\sum_{k=-\infty}^{\infty} 2h(n) \frac{\partial h(n)}{\partial c(j)} \right) - 2h(0) \frac{\partial h(0)}{\partial c(j)} = 0, \tag{7.16}$$

and exploiting Equations 7.13 and 7.14 yields:

$$\sum_{n=-\infty}^{\infty} h(n)y(n-j) = h(0)y(-j), \; j = -N \dots N. \tag{7.17}$$

The main tap value $h(0)$ does not play any role in minimising the (MSE) of the system response's ISI contributions. Hence for the sake of simplicity we set the main tap to $h(0) = 1$ and fulfil this condition by appropriately scaling all equaliser taps by a constant at a later stage.

Substituting Equation 7.13 into Equation 7.17 gives:

$$\sum_{n=-\infty}^{\infty} \sum_{k=-N}^{N} c(k)y(n-k)y(n-j) = y(-j). \tag{7.18}$$

After exchanging the order of summations we arrive at:

$$\sum_{k=-N}^{N} c(k) \sum_{n=-\infty}^{\infty} y(n-k)y(n-j) = y(-j), \tag{7.19}$$

where

$$b(j,k) = \sum_{n=-\infty}^{\infty} y(n-k)y(n-j) \tag{7.20}$$

are recognised as the correlation coefficients of the received signal $y(n)$. Therefore Equation 7.19 simplifies to:

$$\sum_{k=-N}^{N} c(k)b(j,k) = y(-j) \quad j = -N \dots N. \tag{7.21}$$

This is a set of $(2N+1)$ simultaneous linear equations, from which the $(2N+1)$ equaliser coefficients $c(k)$, $k = -N \; \dots \; N$, minimising the (MSE) term of Equation 7.12 can be determined. Subsequently all coefficients must be scaled by a constant in order to arrive at $h(0) = 1$, which does not affect the shape of the equaliser transfer function. In a more convenient matrix form Equation 7.21 is expressed as:

$$\mathbf{B} \cdot \mathbf{c} = \mathbf{y} \tag{7.22}$$

where $\mathbf{B}$ is a $[2N+1] \times [2N+1]$ dimensional matrix of correlations, while $\mathbf{c}$ and $\mathbf{y}$ are $[2N+1]$ dimensional column vectors. Assuming a stationary received signal $y(n)$, its correlation coefficients will not be dependent on the actual values of the indices

(j, k), only their differences, hence Equation 7.22 can be written as follows:

$$\begin{bmatrix} b(0) & b(1) & \dots & b(2N) \\ b(1) & b(0) & \dots & b(2N-1) \\ b(2) & b(1) & \dots & b(2N-2) \\ \vdots & \vdots & & \vdots \\ b(2N) & b(2N-1) & \dots & b(0) \end{bmatrix} \begin{bmatrix} c(-N) \\ c(-N+1) \\ c(-N+2) \\ \vdots \\ c(N) \end{bmatrix} = \begin{bmatrix} y(N) \\ y(N-1) \\ y(N-2) \\ \vdots \\ y(-N) \end{bmatrix}. \quad (7.23)$$

Observe that the correlation matrix in Equation 7.23 is not only symmetric but all the elements along the main diagonals are identical, hence it has a so-called Töplitz structure. This set of equations are often referred to as the Wiener–Hopf equations in the literature of LMS spectral estimation.

An important consequence of Equation 7.21 can be inferred as follows [18]. The ideal system response has a unity value at $n = 0$, i.e. $h(0) = 1$ and $h(n) = 0$ otherwise. Any deviation from these desired values is actually the error sample $e(n)$ associated with sampling instant n, therefore

$$\begin{aligned} e(0) &= h(0) - 1 \text{ for } n = 0 \\ e(n) &= h(n) \quad \text{for } n \neq 0. \end{aligned} \quad (7.24)$$

From Equation 7.17 we infer that

$$\sum_{\substack{n=-\infty \\ n\neq 0}}^{\infty} h(n)y(n-j) + h(0)y(-j) = h(0)y(-j), \quad (7.25)$$

and hence

$$\sum_{\substack{n=-\infty \\ n\neq 0}}^{\infty} h(n)y(n-j) = \sum_{\substack{n=-\infty \\ n\neq 0}}^{\infty} e(n)y(n-j) = 0 \quad j = -N \dots N. \quad (7.26)$$

Equation 7.26 implies that the optimum equaliser setting will result in an error sequence $e(n)$ which is uncorrelated with the equaliser's input signal $y(n)$ within its range of $[-N, \dots, N]$.

Similar sets of simultaneous equations are frequently encountered in various spectral estimation problems and a plethora of computationally efficient recursive algorithms are available for their solution. For example, when computing the optimum predictor coefficients in linear predictive speech coding, a similar set of equations must be solved, which is often carried out using the Levinson–Durbin algorithm [158]. Another example is that of designing an optimum autoregressive filter in the decoding of Bose–Chaudhuri–Hocquenghem block codes, which generates the required sequence of syndromes and facilitates the computation of the error locations. In this application the Berlekamp–Massey algorithm is normally employed [158]. The specific solution to be used will depend on the final formulation of Equation 7.21 as regards to the computation of $b(j, k)$ in Equation 7.20. Namely, the error minimisation in Equation 7.15 extends over $[-\infty, \dots, \infty]$, but in practical terms it must be set to zero outside a finite interval length. Depending on how this windowing is carried out, different solutions arise, because the matrix $\mathbf{B}$ containing the elements $b(j,k)$ will have different properties. Efficient solutions avoid straight matrix inversion, because their complexity is usually proportional to N^3. Instead, recursive solutions exploiting the symmetry of $\mathbf{B}$ or the Töplitz structure are generally favoured due to their lower complexity, which is typically proportional to N^2.

The solution of Equation 7.22 in terms of matrix inversion is given in the form

$$\mathbf{c}^{\text{opt}} = \mathbf{B}^{-1}\mathbf{y} \tag{7.27}$$

but as mentioned, iterative solutions are preferable.

A frequently used straightforward iterative procedure is the *steepest descent or gradient algorithm*, where an arbitrary initial coefficient vector $\mathbf{c}_0$ is assumed. As seen from Equation 7.13 and Equation 7.15, the (MSE) is a quadratic function of the equaliser coefficients comprised by $\mathbf{c}$, and hence exhibits a minimum. The initial $\mathbf{c}_0$ equaliser setting corresponds to an arbitrary momentary MSE ϵ_0 on a $(2N+1)$ dimensional paraboloid surface. The $(2N+1)$ dimensional gradient vector $\mathbf{g}_0$ of the MSE at this instant is given by:

$$\mathbf{g}_0 = \frac{\partial \epsilon}{\partial \mathbf{c}_0}. \tag{7.28}$$

With the $(2N+1)$ gradient components $g_{-N}, \ldots, g_N$ known, every equaliser coefficient is modified by a step value $d_0(n)$, $n = -N, \ldots, N$, proportional to its magnitude in the direction opposite to its gradient. For example, if the point $\mathbf{c}_0$ was at a location of negative gradient on the $(2N+1)$ dimensional paraboloid in terms of all $(2N+1)$ components, all components must be increased by the common step size d and vice versa, giving the relationship

$$c_{k+1}(n) = c_k(n) - d \cdot g_k(n), \quad n = -N \ldots N. \tag{7.29}$$

Observe that the convergence speed and accuracy of the algorithm depends on the choice of the step size $d_k(n)$ that was fixed to d in our approach. In Section 7.4 we will show that faster convergence can be achieved using a technique referred to as Kalman filtering [164]. In general, the larger the step size, the faster the initial convergence and the larger the residual convergence error. The minimum MSE cannot be attained with this algorithm, but due to its simplicity the method is quite popular.

7.2.3 Decision directed adaptive equalisers

The simplified schematic of a decision directed automatic adaptive equaliser is portrayed in Figure 7.4, where the term "decision" directed implies that after a training period the equaliser is required to update its coefficients on the basis of the signal at the output of the receiver's decision device (D), which is not always an error-free signal. Initially a training sequence known to both the transmitter and the receiver is sent, allowing the equaliser to derive the error signal $e(n) = x(n) - \hat{x}(n)$ from the transmitted signal $x(n)$ and the equalised received signal $\hat{x}(n)$. Observe that in this

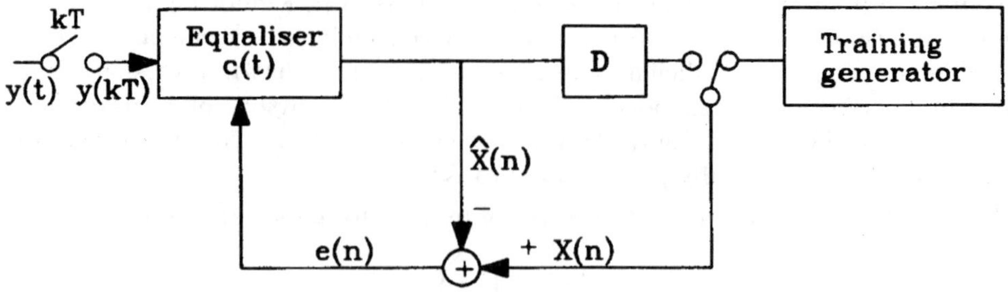

Figure 7.4: Adaptive equaliser schematic

formulation $e(n)$ can now contain both ISI and noise. Then the MSE term of

$$\epsilon = \sum_{n=-\infty}^{\infty} e^2(n) \tag{7.30}$$

can be used to adjust the equaliser's coefficients. After the training segment has elapsed, the transmitted signal $x(n)$ becomes unknown. However, assuming that no transmission errors have occurred, the output signal of the decision device (D) in Figure 7.4 is the exact replica of the transmitted signal $x(n)$. Therefore flipping the switch in Figure 7.4 to the output of the decision device (D) allows decision directed equalisation to continue unhindered. Using this approach the error signal is given by:

$$e(n) = x(n) - \sum_{k=-N}^{N} c(k)y(n-k). \tag{7.31}$$

Then the MSE is yielded as:

$$\epsilon = \sum_{n=-\infty}^{\infty} \left[x(n) - \sum_{k=-N}^{N} c(k)y(n-k) \right]^2 \tag{7.32}$$

which we want to minimise as a function of the equaliser coefficients $c(k)$. Upon setting the partial derivatives to zero we arrive at:

$$\frac{\partial \epsilon}{\partial c(i)} = -2 \sum_{n=-\infty}^{\infty} \left\{ \left[x(n) - \sum_{k=-N}^{N} c(k)y(n-k) \right] y(n-i) \right\} = 0. \tag{7.33}$$

After rearranging Equation 7.33 we get:

$$\sum_{n=-\infty}^{\infty} x(n)y(n-i) = \sum_{n=-\infty}^{\infty} \sum_{k=-N}^{N} c(k)y(n-k)y(n-i) \tag{7.34}$$

which on exchanging the order of summations yields:

$$\sum_{n=-\infty}^{\infty} x(n)y(n-i) = \sum_{k=-N}^{N} c(k) \sum_{n=-\infty}^{\infty} y(n-k)y(n-i). \tag{7.35}$$

In Equation 7.32 it was implicitly assumed that the error is minimised over the interval $-\infty < n < \infty$, which cannot be implemented in practice. Depending on how the input data undergoing spectral estimation is windowed to a finite interval, two frequently used methods arise, the so-called autocorrelation method and the covariance method [158]. Pursuing the former we implicitly suppose that the data outside the equaliser's range $[-N, \ldots, N]$ is set to zero. Then recognising that in Equation 7.35 for a stationary received signal $y(n)$

$$r(i,k) = r(|i-k|) = \sum_{n=-N}^{N} y(n-k)y(n-i) \tag{7.36}$$

is the autocorrelation of the received samples, while

$$\psi(-i) = \sum_{n=-N}^{N} x(n)y(n-i) \tag{7.37}$$

is the cross-correlation of the received samples with the transmitted samples, which in the case of error-free decisions are known to the receiver, allows us to express Equation 7.35 as follows:

$$\sum_{k=-N}^{N} c(k)r(|i-k|) = \psi(-i), \;\; i = -N \ldots N. \tag{7.38}$$

Similarly to Equation 7.21, Equation 7.38 also represents a set of $(2N+1)$ simultaneous equations, from which the equaliser coefficients $c(k)$ can be determined. The sole difference arises on the right-hand side of Equation 7.38, where now we have the cross-correlation between the received samples $y(n)$ and the transmitted samples $x(n)$ that are assumed to be locally known. This implies either that the received signal $y(n)$ is an error-free replica of $x(n)$ or that an equaliser training pattern is transmitted, which is known to both the transmitter and the receiver. The solution of Equation 7.38 ensues similarly to that of Equation 7.21, i.e. either by Gauss-Jordan elimination or by some recursive method.

For the success of the equaliser training process it is vital to establish initial synchronisation before the training segment is received in order to ensure that the error signal $e(n)$ is derived from the appropriately synchronised received and locally stored training patterns. If the bit error rate (BER) is low, the decision directed equaliser is capable of tracking slowly varying channel characteristics, such as slow drift in telephone lines, etc. High BERs, however, result in catastrophic error precipitation. In order to attain fast initial convergence, usually a larger step size is used during the training process, which is reduced to a smaller one during decision directed operation

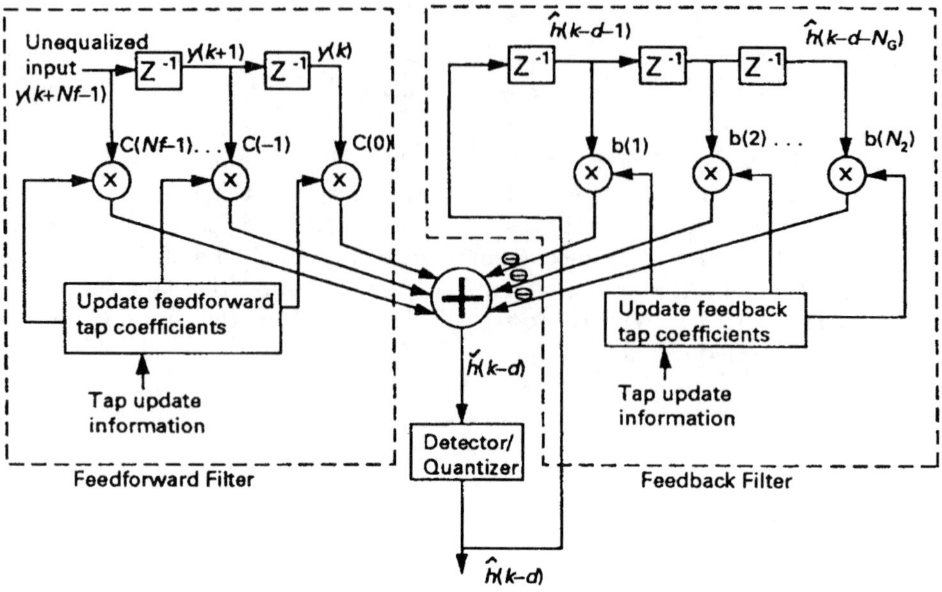

Figure 7.5: General decision feedback equaliser Structure

to maintain a low tracking error.

7.3 Decision Feedback Equalisers

The structure of a general decision feedback equaliser (DFE) is shown in Figure 7.5.

The equaliser is composed of a feedforward section and a feedback section. The feedforward section is identical to the linear equaliser but the feedback section differs in its input. Instead of using the received input signal, a decision is made as to the transmitted symbol, and this value is passed into the feedback section. This leads to a number of advantages. The data passed into the feedback section is the output signal of the detector and hence no longer contains any noise, thus increasing the accuracy of the interference cancellation. The structure of the feedback system implies that the length of the feedback register need only be as long as the delay spread of the received signal. This is in contrast to the linear equaliser where the equaliser length is governed by the effective length of the infinite impulse response inverse filter, and can easily be an order of magnitude greater than the delay spread.

From Figure 7.5, following Cheung's approach [161], we infer the equalised signal at the detector's input as:

$$\check{h}(k-d) = \sum_{m=0}^{N_f-1} c(m)y(k+m) - \sum_{p=1}^{N_b} b(p)\hat{h}(k-d-p), \tag{7.39}$$

which has been delayed by d sampling intervals. The error signal $e(k-d)$ between the transmitted signal $x(k-d)$ and the equalised signal $\check{h}(k-d)$ can be expressed by the help of Equation 7.39 as follows:

$$\begin{aligned} e(k-d) &= x(k-d) - \check{h}(k-d) \\ &= x(k-d) - \sum_{m=0}^{N_f-1} c(m)y(k+m) + \sum_{p=1}^{N_b} b(p).\hat{h}(k-d-p) \end{aligned} \tag{7.40}$$

Using the minimum MSE criterion requires minimisation of the expected value

$$\epsilon = E\{|e(k-d)|^2\}. \tag{7.41}$$

As we have seen in the previous section, this is achieved by that equaliser setting which renders the error signal $e(k-d)$ uncorrelated with the equaliser's input signal. As regards to the forward filter this requires

$$E\{e(k-d) \cdot y^*(k+m)\} = 0 \quad m = 0 \ldots (N_f - 1) \tag{7.42}$$

and for the backward filter it requires

$$E\{e(k-d) \cdot \hat{h}^*(k-d-q)\} = 0 \quad q = 1 \ldots N_b \tag{7.43}$$

where the superscript * represents conjugate complex computation. Supposing that no decision errors have been encountered, we have

$$\hat{h}(k-d) = x(k-d) \tag{7.44}$$

i.e. the detector's output signal is an identical replica of the transmitted signal. Upon substituting Equation 7.40 into Equation 7.43 we arrive at:

$$\begin{aligned} E\{\left[x(k-d) - \sum_{m=0}^{N_f-1} c(m)y(k+m) + \sum_{p=1}^{N_b} b(p)\hat{h}(k-d-p)\right] \\ \cdot\hat{h}^*(k-d-q)\} = 0 \quad q = 1 \ldots N_b \end{aligned} \tag{7.45}$$

and taking into account Equation 7.44 yields:

$$\begin{aligned} E\{\left[x(k-d) - \sum_{m=0}^{N_f-1} c(m)y(k+m) + \sum_{p=1}^{N_b} b(p)x(k-d-p)\right] \\ \cdot x^*(k-d-q)\} = 0 \quad q = 1 \ldots N_b. \end{aligned} \tag{7.46}$$

For the random uncorrelated transmitted signal samples we have:

$$E\{x(k-d-p) \cdot x^*(k-d-q)\} = \begin{cases} 1 & \text{for } p = q \\ 0 & \text{otherwise} \end{cases} . \tag{7.47}$$

The received signal $y(k)$ is the convolution of the L_c sample's long channel impulse response $g(k)$ and the transmitted signal $x(k)$ contaminated by the additive noise $n(k)$, as shown below:

$$y(k) = \sum_{i=0}^{L_c-1} g(i) \cdot x(k-i) + n(k). \tag{7.48}$$

Taking Equations 7.47 and 7.48 into account, Equation 7.46 can be written as:

$$\begin{aligned}\sum_{m=0}^{N_f-1} c(m) \sum_{i=0}^{L_c-1} E\{[g(i)x(k+m-i) + n(k+m)] \cdot x^*(k-d-q)\} \\ = \sum_{p=1}^{N_b} b(p)E\{x(k-d-p) \cdot x^*(k-d-q)\} \quad q = 1 \ldots N_b\end{aligned} \tag{7.49}$$

Exploiting the orthogonality of the additive noise $n(k+m)$ and transmitted signal $x(k-d-q)$ we have:

$$E\{n(k+m) \cdot x^*(k-d-q)\} = E\{n^*(k+m)x(k-d-p)\} = 0 \tag{7.50}$$

and therefore Equation 7.50 simplifies to:

$$\sum_{m=0}^{N_f-1} c(m)g(m+d+q) = b(q) \quad q = 1 \ldots N_b. \tag{7.51}$$

Explicitly, Equation 7.51 is a set of N_b simultaneous equations delivering the feedback filter's coefficients $b(q)$, $q = 1, \ldots, N_b$ as a function of the feedforward filter coefficients $c(m)$, $m = 0, \ldots, (N_f - 1)$, and that of the channel impulse response $g(m+d+q)$. The feedforward coefficients can be determined using Equation 7.21, while the channel impulse response is derived by means of channel sounding methods.

7.4 Fast Converging Equalisers

As we have seen in Section 7.2.2, the stochastic gradient method is a low complexity technique for adjusting the equaliser coefficients and it exhibits slow convergence properties. Considerably faster convergence can be attained using a computationally more demanding recursive coefficient adjustment technique, which is referred to as Kalman filtering [209].

Kalman filtering is a fast-converging spectral estimation algorithm for the solution of Wiener–Hopf type normal equations and it has been well documented in references on adaptive filtering and channel equalization [3, 164, 185] . In our deliberations we follow the approach of Proakis [164] and Haykin [3].

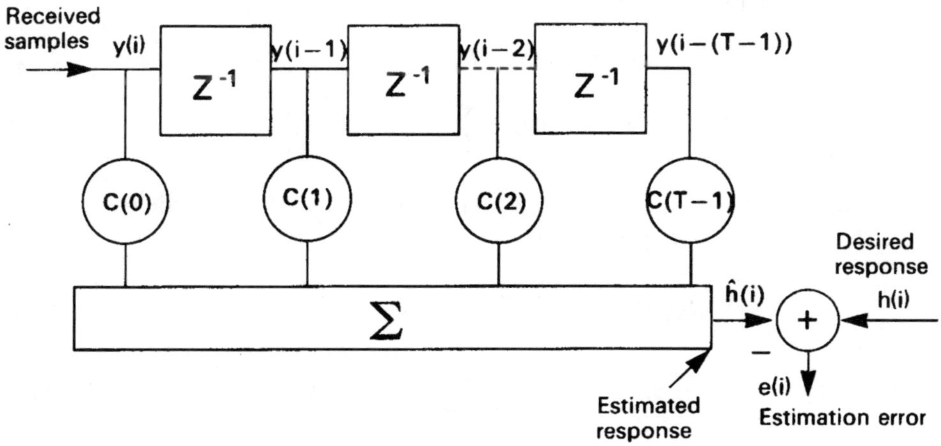

Figure 7.6: LS equaliser structure

7.4.1 Least squares method

The minimum mean squared estimation method used in Sections 7.2.2 and 7.2.3 was based on a statistical ensemble average minimisation of the square of the error signal, which led to the standard Wiener–Hopf equations given in Equation 7.21 or Equation 7.38. The set of simultaneous equations was then solved by matrix inversion, recursive solutions or by the computationally simple steepest descent algorithm, which yielded slow convergence due to applying a common step size d, as seen in Equation 7.29.

In contrast, the least squares (LS) method's optimisation criterion is expressed in terms of a time average, rather than ensemble average. The LS estimation problem is then formulated as estimating the required equaliser output sequence $h(i)$ by $\hat{h}(i)$ from its input sequence $y(i)$ by the help of the equaliser coefficients $c(t)$, using a linear relationship similar to that in Equation 7.31,

$$h(i) = \sum_{t=0}^{T-1} c(t)y(i-t) + e(i) \tag{7.52}$$

where the index t is used to emphasise the notion of time dependence and $e(i)$ represents the random estimation error. This scheme is depicted in Figure 7.6. In other words, Equation 7.52 states that the required equaliser output sample $h(i)$ can be estimated from the previous T received samples by a linear combination, where the random estimation error $e(i)$ has zero mean and a variance of σ^2. This formulation of the estimation problem is identical to the linear regression model often used in source encoding theory.

Our approach is now to find the equaliser coefficients $c(t)$, which minimise the cost function defined as the sum of error squares over a finite interval. The estimation error is expressed from Equation 7.52 as follows:

$$\begin{aligned} e(i) &= h(i) - \hat{h}(i) \\ &= h(i) - \sum_{t=0}^{T-1} c(t)y(i-t). \end{aligned} \tag{7.53}$$

Then the optimisation cost function ϵ depends on the equaliser coefficients $c(t)$, $t = 0, \ldots, T-1$ and can be expressed as:

$$\begin{aligned} \epsilon &= \sum_i |e(i)|^2 \\ &= \sum_i |h(i) - \sum_{t=0}^{T-1} c(t)y(i-t)|^2, \end{aligned} \tag{7.54}$$

where the interval over which minimisation is carried out depends on the data windowing method invoked. Observing that our cost function ϵ to be minimised in Equation 7.54 is similar to that of Equation 7.32 in Section 7.2.3 and setting the range over which optimisation is carried out in accordance with the covariance method to $i = 1, \ldots, I$, the equaliser coefficients can be computed.

Setting the partial derivatives of Equation 7.54 with respect to the equaliser coefficients $c(t)$, $t = 0, \ldots, T-1$ to zero gives a similar set of equations to Equation 7.38, as shown below:

$$\sum_{t=0}^{T-1} c(t)\Phi(t,k) = \psi(-k), \quad k = 0\ldots(T-1) \tag{7.55}$$

Here

$$\psi(-k) = \sum_{i=1}^{I} h(i)y(i-k), \quad k = 0\ldots(T-1) \tag{7.56}$$

is the time-averaged cross-correlation between the equaliser's input signal $y(i)$ and its desired output signal $h(i)$, $c(t)$ represents the equaliser coefficients, while

$$\phi(t,k) = \sum_{i=1}^{I} y(i-k)y(i-t), \quad 0 \leq t, \; k \leq T-1 \tag{7.57}$$

is the time-averaged correlation of $y(i)$. In matrix form we have:

$$\mathbf{\Phi} \cdot \mathbf{c} = \mathbf{\Psi} \tag{7.58}$$

or explicitly:

$$\begin{pmatrix} \phi(0,0) & \phi(1,0) & \phi(2,0) & \dots & \phi(T-1,0) \\ \phi(0,1) & \phi(1,1) & \phi(2,1) & \dots & \phi(T-1,1) \\ \phi(0,2) & \phi(1,2) & \phi(2,2) & \dots & \phi(T-1,2) \\ \vdots & \vdots & \vdots & & \vdots \\ \phi(0,T-1) & \phi(1,T-1) & \phi(2,T-1) & \dots & \phi(T-1,T-1) \end{pmatrix} \begin{bmatrix} c(0) \\ c(1) \\ c(2) \\ \vdots \\ c(T-1) \end{bmatrix} = \begin{bmatrix} \psi(0) \\ \psi(-1) \\ \psi(-2) \\ \vdots \\ \psi(-T+1) \end{bmatrix}. \tag{7.59}$$

The elements of ϕ can be computed from Equation 7.57 as follows:

$$\begin{aligned} \phi(0,0) &= y(1)y(1) + \ldots + y(I)y(I) \\ &= \sum_{i=1}^{I} y(i)y(i) \\ \phi(0,1) &= y(0)y(1) + \ldots + y(I-1)y(I) \\ &= \sum_{i=1}^{I} y(i-1)y(i) \\ &\vdots \\ \phi(0,T-1) &= y(2-T)y(1) + \ldots + y(I-T+1)y(I)y(I) \\ &= \sum y(i-T+1)y(i) \\ \phi(1,0) &= y(1)y(0) + \ldots + y(I)y(I-1) \\ &= \sum_{i=1}^{I} y(i)y(i-1) \\ \phi(1,1) &= y(0)y(0) + \ldots + y(I-1)y(I-1) \\ &= \sum_{i=1}^{I} y(i-1)y(i-1) \\ &\vdots \\ \phi(1,T-1) &= y(2-T)y(0) + \ldots + y(I-T+1)y(I-1) \\ &= \sum_{i=1}^{I} y(i-T+1)y(i-1) \\ &\vdots \\ \phi(T-1,0) &= y(1)y(2-T) + \ldots + y(I)y(I-T+1) \end{aligned} \tag{7.60}$$

$$= \sum_{i=1}^{I} y(i)y(i-T+1)$$

$$\vdots$$

$$\begin{aligned}\phi(T-1,T-1) &= y(2-T)y(2-T)+\ldots+y(I-T+1)y(I-T+1)\\ &= \sum_{i=1}^{I} y(i-T+1)y(i-T+1)\end{aligned}$$

The diagonal elements of the matrix $\mathbf{\Phi}$ are now not identical, so it does not have a Töplitz structure. The equaliser coefficients optimised for the interval $i = 1, \ldots, I$ can be found from Equation 7.58 by simply inverting the matrix $\mathbf{\Phi}$ or using more efficient recursive algorithms. Having presented the fundamental principles of LS estimation, in the next section we present a recursive LS (RLS) algorithm for equaliser coefficient adjustment.

7.4.2 Recursive Least Squares Method [3]

7.4.2.1 Cost function weighting

The principle of the RLS [3] algorithm is that given the current LS equaliser coefficient set, it is possible to update this set, when new equaliser input samples arrive. The RLS algorithm converges quickly, but has a high complexity, as we will see. The cost function ϵ of RLS estimation usually contains an exponentially decaying multiplicative weighting factor β^{n-i} in order to reduce the effect of "old" error terms and emphasise more recent ones:

$$\epsilon = \sum_{i=1}^{n} \beta^{n-i}|e(i)|^2, \tag{7.61}$$

where the error term $e(i)$ was given in Equation 7.53. The factor β is typically close to, but less than unity, and the number of error terms n in the cost function above is variable. This is because in the RLS method the computations commence from a known initial state determined for example by the LS method described in the previous subsection, and new received samples are then used to update previous estimates. For the observation interval n the equaliser coefficients $c(t), t = 0, \ldots, (T-1)$ remain unchanged.

For our further deliberations it is useful to introduce the vector of equaliser input samples, which can be complex valued:

$$\mathbf{y}(i) = [y(i), y(i-1), y(i-2)\ldots y(i-(T-1))]^T, \tag{7.62}$$

where the superscript T means transposition. Then we can express the correlation matrix $\mathbf{\Phi(n)}$ for the observation interval n as follows:

$$\mathbf{\Phi(n)} = E\left\{\mathbf{y}(i)\cdot\mathbf{y}^{*T}(i)\right\} = \sum_{i=1}^{n}\mathbf{y}(i)\cdot\mathbf{y}^{*T}(i) =$$

$$\begin{bmatrix} \sum_{i=1}^{n} y(i)y^*(i) & \dots & \sum_{i=1}^{n} y(i)y^*(i-T+1) \\ \sum_{i=1}^{n} y(i-1)y^*(i) & \dots & \sum_{i=1}^{n} y(i-1)y^*(i-T+1) \\ \vdots & & \vdots \\ \sum_{i=1}^{n} y(i-T+1)y^*(i) & \dots & \sum_{i=1}^{n} y(i-T+1)y^*(i-T+1) \end{bmatrix}. \tag{7.63}$$

Observe that apart from the upper summation limit n this matrix has the same elements as in the previous section.

The key equation in Equation 7.55 or 7.58 is now defined in terms of the modified correlation matrix incorporating the exponential scaling factor β^{n-i}:

$$\mathbf{\Phi}(n) = \sum_{i=1}^{n} \beta^{n-i} \mathbf{y}(i) \cdot \mathbf{y}^{*T}(i), \tag{7.64}$$

while the cross-correlation $\mathbf{\Psi}(n)$ takes the form

$$\mathbf{\Psi}(n) = \sum_{i=1}^{n} \beta^{n-i} \mathbf{y}(i) \cdot h^*(i). \tag{7.65}$$

7.4.2.2 Recursive correlation update

Since we intend to derive a recursive update formula for $\mathbf{\Phi}(n)$ based on its previous value, we separate the last term from the summation in Equation 7.64 yielding:

$$\mathbf{\Phi}(n) = \beta \left[\sum_{i=1}^{n-1} \beta^{n-i-1} \mathbf{y}(i) \cdot \mathbf{y}^{*T}(i) \right] + \mathbf{y}(n)\mathbf{y}^{*T}(n), \tag{7.66}$$

where the square bracketed term can be recognised as $\mathbf{\Phi}(n-1)$. Hence Equation 7.66 can be simplified to the required recursive relationship

$$\mathbf{\Phi}(n) = \beta\mathbf{\Phi}(n-1) + \mathbf{y}(n)\mathbf{y}^{*T}(n), \tag{7.67}$$

where the matrix given by the outer product $\mathbf{y}(n)\mathbf{y}^{*T}(n)$ constitutes the correction term in the update process. Analogously, the update formula for $\mathbf{\Psi}(n)$ accrues as follows:

$$\mathbf{\Psi}(n) = \beta\mathbf{\Psi}(n-1) + \mathbf{y}(n)h^*(n). \tag{7.68}$$

Again, similarly to the LS method, the normal equations given by Equation 7.58 have to be solved either by matrix inversion, or preferably by computationally more efficient recursive methods for $n = 1, 2, \dots, \infty$. According to Haykin [3] this can be achieved using Woodbury's identity [208], which is also known as the matrix inversion lemma, if the matrix $\mathbf{\Phi}(n)$ is non-singular, i.e. it has a non-zero determinant and hence it is invertible.

7.4.2.3 The Ricatti equation of RLS estimation

Upon exploiting Woodbury's identity the following recursive formula accrues for the inverted matrix $\mathbf{\Phi}^{-1}$ [3]:

$$\mathbf{\Phi}^{-1}(n) = \beta^{-1}\mathbf{\Phi}^{-1}(n-1) - \frac{\beta^{-2}\mathbf{\Phi}^{-1}(n-1)\mathbf{y}(n)\mathbf{y}^{*T}(n)\mathbf{\Phi}^{-1}(n-1)}{1+\beta^{-1}\mathbf{y}^{*T}(n)\mathbf{\Phi}^{-1}(n-1)\mathbf{y}(n)} \quad (7.69)$$

This is already an explicit recursive expression for updating $\mathbf{\Phi}^{-1}(n-1)$, when new received signal samples $y(n)$ become available, which can be further streamlined using the following shorthand:

$$\mathbf{k}(n) = \frac{\beta^{-1}\mathbf{\Phi}^{-1}(n-1)\mathbf{y}(n)}{1+\beta^{-1}\mathbf{y}^{*T}(n)\mathbf{\Phi}^{-1}(n-1)\mathbf{y}(n)}, \quad (7.70)$$

where $k(n)$ is a $T \times 1$ vector, since the numerator is a $T \times 1$ vector, while the denominator is a scalar. Then we have:

$$\mathbf{\Phi}^{-1}(n) = \beta^{-1}\mathbf{\Phi}^{-1}(n-1) - \beta^{-1}\mathbf{k}(n)\mathbf{y}^{*T}(n)\mathbf{\Phi}^{-1}(n-1), \quad (7.71)$$

which is the so-called Ricatti equation of RLS estimation [3]. The vector expression $\mathbf{k}(n)$ in Equation 7.70 can further be simplified by multiplying both sides of it with the denominator, giving:

$$\mathbf{k}(n) + \beta^{-1}\mathbf{k}(n)\mathbf{y}^{*T}(n)\mathbf{\Phi}^{-1}(n-1)\mathbf{y}(n) = \beta^{-1}\mathbf{\Phi}^{-1}(n-1)\mathbf{y}(n)$$

which leads to:

$$\begin{aligned}\mathbf{k}(n) &= \beta^{-1}\mathbf{\Phi}^{-1}(n)\mathbf{y}(n) - \beta^{-1}\mathbf{k}(n)\mathbf{y}^{*T}(n)\mathbf{\Phi}^{-1}(n-1)\mathbf{y}(n) \\ &= \left[\beta^{-1}\mathbf{\Phi}^{-1}(n) - \beta^{-1}\mathbf{k}(n)\mathbf{y}^{*T}(n)\mathbf{\Phi}^{-1}(n-1)\right]\mathbf{y}(n).\end{aligned} \quad (7.72)$$

Closer scrutiny of Equation 7.71 reveals that the square bracketed term of Equation 7.72 is equal to $\mathbf{\Phi}^{-1}(n)$, hence Equation 7.72 simplifies to:

$$\mathbf{k}(n) = \mathbf{\Phi}^{-1}(n)\mathbf{y}(n), \quad (7.73)$$

which defines the vector $\mathbf{k}(n)$ as a transformed version of the received signal vector $\mathbf{y}(n)$, where the transformation is carried out by rotating $\mathbf{y}(n)$ with the aid of the inverse of the correlation matrix.

7.4.2.4 Recursive equaliser coefficient update

Based on the normal equations in Equation 7.58 and the recursive update formula in Equation 7.68 the estimated equaliser coefficients can be expressed by means of $\mathbf{\Phi}^{-1}(n)$ as follows:

$$\hat{\mathbf{c}}(n) = \mathbf{\Phi}^{-1}(n)\mathbf{\Psi}(n)$$

$$= \beta\mathbf{\Phi}^{-1}(n)\mathbf{\Psi}(n-1) + \mathbf{\Phi}^{-1}(n)\mathbf{y}(n)h^*(n). \tag{7.74}$$

Substituting $\mathbf{\Phi}^{-1}(n)$ in the first term of Equation 7.74 by Equation 7.71 leads to:

$$\begin{aligned}\hat{\mathbf{c}}(n) &= \mathbf{\Phi}^{-1}(n-1)\mathbf{\Psi}(n-1) - \mathbf{k}(n)\mathbf{y}^{*T}(n)\mathbf{\Phi}^{-1}(n-1)\mathbf{\Psi}(n-1) \\ &\quad + \mathbf{\Phi}^{-1}(n)\mathbf{y}(n)h^*(n).\end{aligned} \tag{7.75}$$

Upon substituting the vector

$$\mathbf{\Phi}^{-1}(n-1)\mathbf{\Psi}(n-1) = \hat{\mathbf{c}}(n-1) \tag{7.76}$$

into Equation 7.75 we arrive at:

$$\hat{\mathbf{c}}(n) = \hat{\mathbf{c}}(n-1) - \mathbf{k}(n)\mathbf{y}^{*T}(n)\hat{\mathbf{c}}(n-1) + \mathbf{\Phi}^{-1}(n)\mathbf{y}(n)h^*(n). \tag{7.77}$$

Exploiting Equation 7.73 simplifies Equation 7.77 even further to:

$$\begin{aligned}\hat{\mathbf{c}}(n) &= \hat{\mathbf{c}}(n-1) - \mathbf{k}(n)\mathbf{y}^{*T}(n)\hat{\mathbf{c}}(n-1) + \mathbf{k}(n)h^*(n) \\ &= \hat{\mathbf{c}}(n-1) + \mathbf{k}(n)\left[h^*(n) - \mathbf{y}^{*T}(n)\hat{\mathbf{c}}(n-1)\right] \\ &= \hat{\mathbf{c}}(n-1) + \mathbf{k}(n)i(n)\end{aligned} \tag{7.78}$$

where

$$i(n) = h^*(n) - \mathbf{y}^{*T}(n)\hat{\mathbf{c}}(n-1) \tag{7.79}$$

represents an estimation error term given by the difference of the desired equaliser output $h^*(n)$ and its estimate $\mathbf{y}^{*T}(n)\hat{\mathbf{c}}(n-1)$. In this light the estimation error term of Equation 7.79 is weighted in Equation 7.78 by the gain vector $\mathbf{k}(n)$ in order to update the equaliser coefficient set $\hat{\mathbf{c}}(n-1)$ to arrive at the RLS estimate $\hat{\mathbf{c}}(n)$. Since the estimate $\mathbf{y}^{*T}(n)\hat{\mathbf{c}}(n-1)$ of $h^*(n)$ is based on the previous RLS estimate $\hat{\mathbf{c}}(n-1)$, the estimation error term $i(n)$ in Equation 7.79 is an a priori estimation error, as opposed to the a posteriori estimation error of

$$e(n) = h(n) - \hat{\mathbf{c}}^{*T}(n)\mathbf{y}(n). \tag{7.80}$$

In other words, $i(n)$ is the estimation error, which would arise without updating the equaliser coefficients from $\hat{\mathbf{c}}(n-1)$ to $\hat{\mathbf{c}}(n)$. Due to updating the error reduces to $e(n)$, which was our cost function given in Equation 7.53 in the RLS estimation procedure.

In summary, the RLS equaliser coefficient estimation algorithm is specified by Equations 7.70, 7.79, 7.78 and 7.71, repeated here for convenience:

$$\mathbf{k}(n) = \frac{\beta^{-1}\mathbf{\Phi}^{-1}(n-1)\mathbf{y}(n)}{1 + \beta^{-1}\mathbf{y}^{*T}(n)\mathbf{\Phi}^{-1}(n-1)\mathbf{y}(n)}$$

$$i(n) = h^*(n) - \mathbf{y}^{*T}(n)\hat{\mathbf{c}}(n-1)$$

$$\hat{\mathbf{c}}(n) = \hat{\mathbf{c}}(n-1) + \mathbf{k}(n)i(n)$$

$$\mathbf{\Phi}^{-1}(n) = \beta^{-1}\mathbf{\Phi}^{-1}(n-1) - \beta^{-1}\mathbf{k}(n)\mathbf{y}^{*T}(n)\mathbf{\Phi}^{-1}(n-1).$$

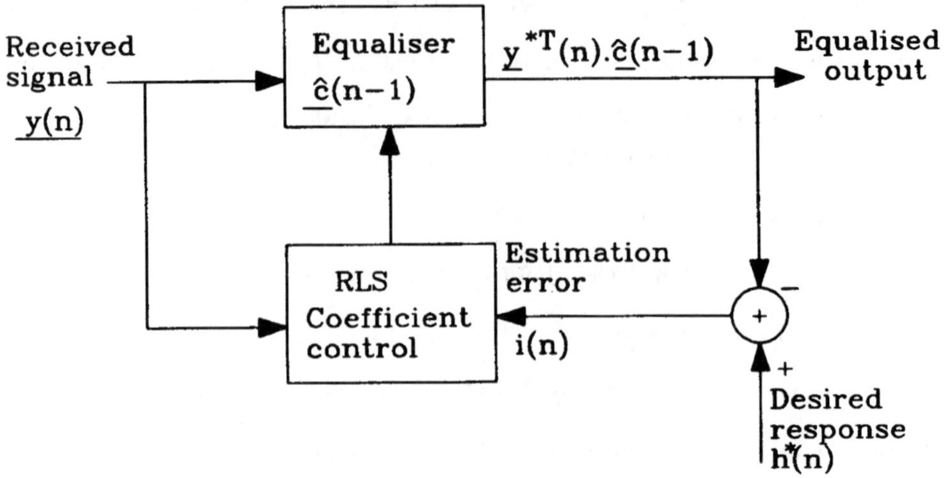

Figure 7.7: Kalman filtering block diagram [3] ©Prentice Hall, 1984, Haykin

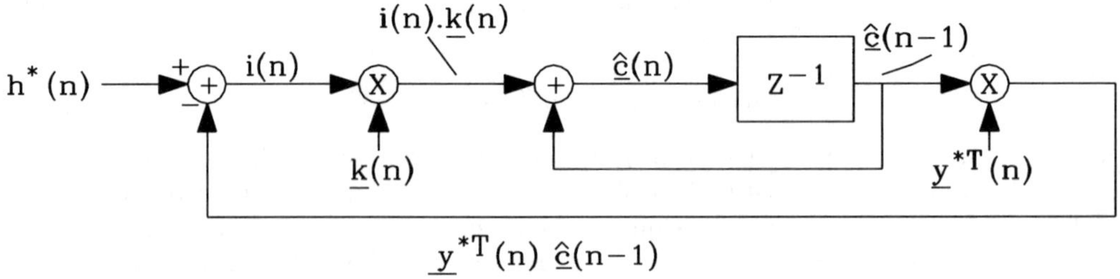

Figure 7.8: Kalman filtering signal flow diagram [3] ©Prentice Hall, 1984, Haykin

Observe that Equation 7.79 describes how the received signal $\mathbf{y}(n)$ is equalised by the coefficient set $\hat{\mathbf{c}}(n-1)$, giving an equalised output of $\hat{\mathbf{c}}(n-1) \cdot \mathbf{y}^{*T}(n)$, as seen in Figure 7.7. The estimation error $i(n)$ is computed according to Equation 7.79 and it is weighted by the so-called Kalman gain factor $\mathbf{k}(n)$ in the computation of the new set of equaliser coefficients, as portrayed by Equation 7.78. Before the recursion enters a new iteration cycle, the inverse of the correlation matrix $\mathbf{\Phi}(n)$ is updated in harmony with Equation 7.71 and used to compute the next value of $\mathbf{k}(n)$ according to Equation 7.70. The sequence of operations is best illustrated by Haykin's signal flow diagram [3] seen in Figure 7.8.

The fast convergence of the RLS Kalman algorithm is particularly important in data transmission modems for telephone links, where the length of the equaliser training pattern is limited. If adequate equalisation is not established within the training window, a new training procedure is initiated, which might be prolonged indefinitely. The Kalman algorithm can also be used in time division multiple access (TDMA) mobile radio systems, where the channel is sounded in every TDMA burst in order

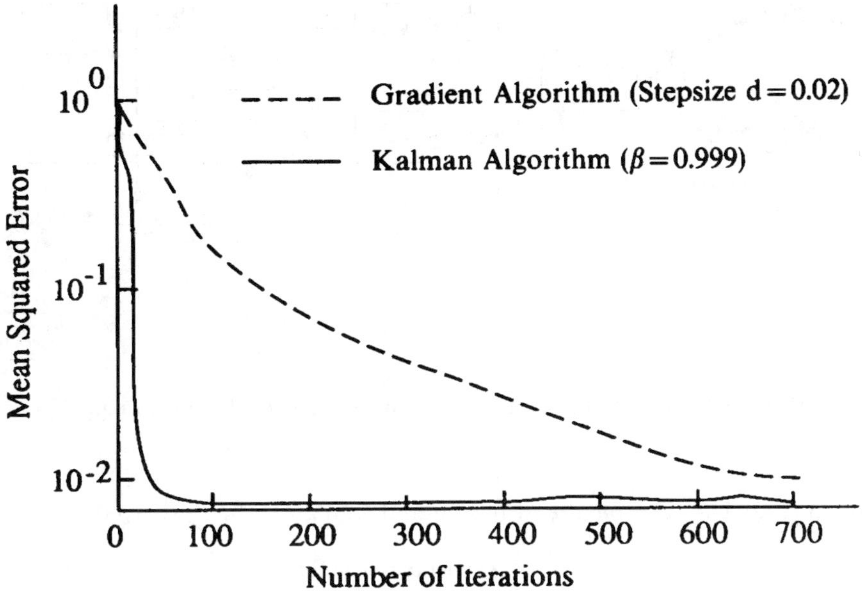

Figure 7.9: Comparison of convergence rate for the Kalman and gradient algorithm [164] ©McGraw-Hill, 1989, Proakis

to acquire an accurate estimate of the non-stationary channel impulse response [161]. In this scenario the Kalman algorithm is fed with the locally stored original and the received sounding sequence in order to derive the estimate of the channel impulse response, which can be used for example in a DFE. Cheung [161] followed this approach in a TDMA mobile scenario, evaluating the channel estimation error for various SNR values and tabulated the computational complexity associated with various steps of the algorithm. The convergence rates of the steepest descent or stochastic gradient and the Kalman algorithm can be characterised by the typical MSE versus iteration index curves [164], as seen in Figure 7.9 where an 11-tap equaliser was used.

7.5 Adaptive Equalisers for QAM

The equalisers discussed so far are suitable in their original form for baseband pulse amplitude modulated (PAM) schemes. If the channel SNR allows the deployment of QAM, higher bit rates can be achieved within the same bandwidth using QAM modems [37,185].

Figure 7.10 shows the schematic diagram of a QAM system with the square root Nyquist filters $\sqrt{N(\omega)}$ and an adaptive, complex baseband equaliser $C(\omega)$. In Chapter 2 on communications channels we have shown the equivalence of passband and

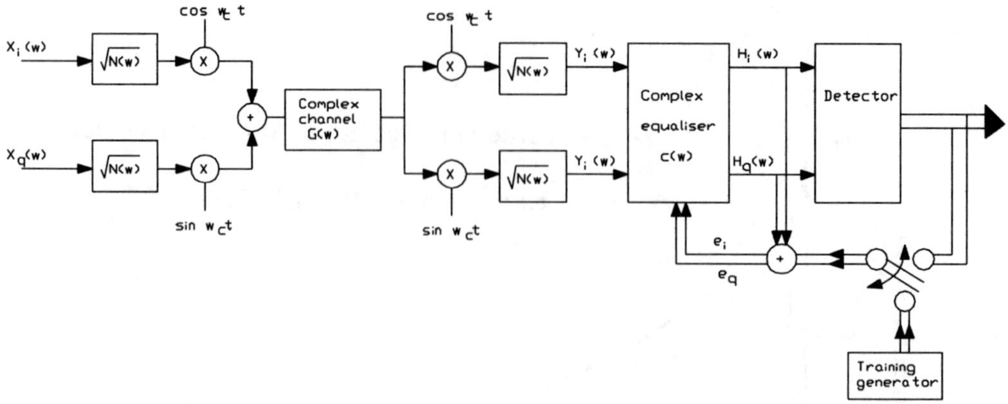

Figure 7.10: QAM system with complex adaptive baseband equaliser

baseband channels. The passband modulated signal can be portrayed as follows:

$$\begin{aligned} x(t) &= A(t)\cos(\omega_c t + \phi(t)) \\ &= x_i(t)\cos(\omega_c t) - x_q(t)\sin(\omega_c t) \end{aligned} \tag{7.81}$$

where the baseband in-phase and quadrature phase components are given by:

$$\begin{aligned} x_i(t) &= A(t)\cos[\phi(t)] \\ x_q(t) &= A(t)\sin[\phi(t)] \end{aligned} \tag{7.82}$$

Exciting the complex lowpass equivalent communication channel represented by its impulse response $g(t)$ with the complex baseband equivalent signal

$$x(t) = x_i(t) + jx_q(t) \tag{7.83}$$

the complex baseband system response becomes:

$$\begin{aligned} y(t) &= x(t) * g(t) \\ &= y_i(t) + jy_q(t) \\ &= [x_i(t) + jx_q(t)] * [g_i(t) + jg_q(t)] \\ &= [x_i(t) * g_i(t) - x_q(t) * g_q(t)] + j[x_i(t) * g_q(t) + g_i(t) * x_q(t)] \end{aligned} \tag{7.84}$$

where the quadrature components of the channel output $y(t)$ are given by:

$$\begin{aligned} y_i(t) &= x_i(t) * g_i(t) - x_q(t) * g_q(t) \\ y_q(t) &= x_i(t) * g_q(t) + g_i(t) * x_q(t). \end{aligned} \tag{7.85}$$

Using the above complex baseband representation the system model seen in Figure 7.3 is redrawn in terms of complex QAM signals in Figure 7.11. As seen in the figure, the real in-phase equaliser paths $c_i(t)$ cancel the ISI in both of the received components

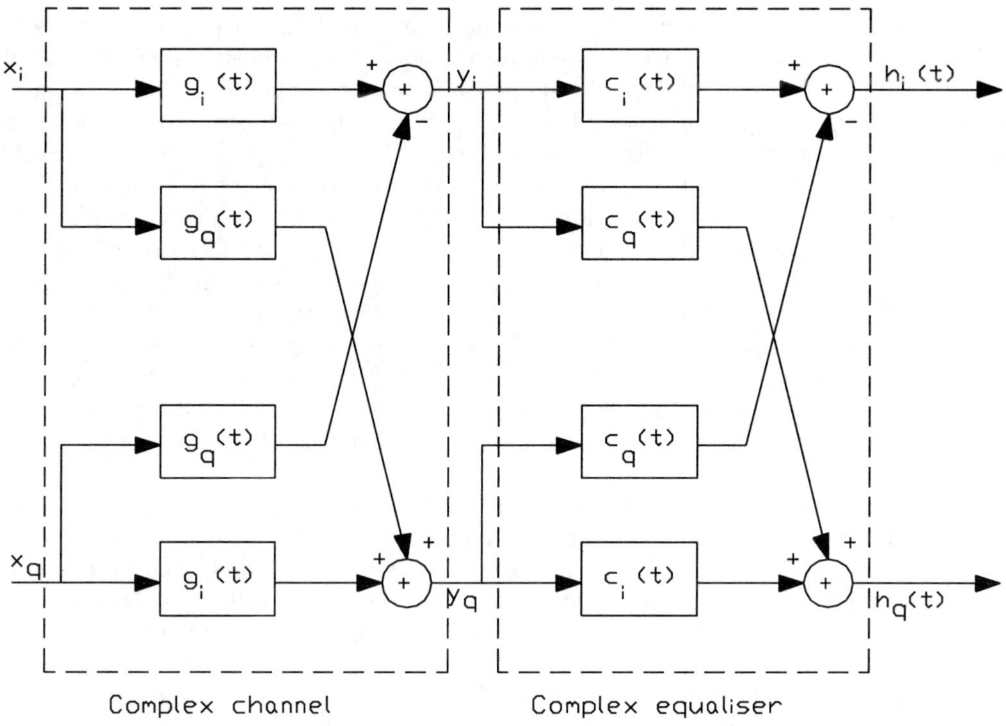

Figure 7.11: Complex baseband channel and equaliser schematic

$y_i(t)$ and $y_q(t)$, while the cross-coupling equaliser paths with impulse responses $c_q(t)$ combat the cross-interference between $y_i(t)$ and $y_q(t)$ caused by channel asymmetries, or imperfections in clock and carrier recovery. The price paid for the higher spectral efficiency of the QAM modem is its quadruple equalisation complexity.

7.6 Viterbi Equalisers for Partial Response Modulation

7.6.1 Partial response modulation

Although the full treatment of partial response (PR) modulation techniques and that of their equalisation is beyond the scope of this book, for the sake of completeness we briefly highlight the principles of maximum likelihood sequence estimation (MLSE) used for the detection and equalisation of PR systems. The most frequently used MLSE scheme is the Viterbi algorithm (VA), which can also be invoked for the maximum likelihood detection of trellis coded modulation (TCM) that will be the subject of the next chapter. For a full treatment of partial response modulation, Viterbi decoding and Viterbi equalisation the reader is referred to Chapters 4 and 6 of Reference [158].

The class of constant envelope, continous phase modulation (CPM) schemes is widely used over fading mobile radio channels due to their robustness against signal fading and interference, while maintaining good spectral efficiency. Many of their attractive properties are attributable to the fact that the transmitted signal has a constant envelope, which facilitates the employment of highly power efficient class C amplifiers. In contrast, QAM systems have to use linearised class AB amplifiers, as we highlighted in Chapter 4. The immunity of CPM to fading is also a consequence of the constant signal envelope, since only the phase changes carry information.

The simplest such constant envelope CPM arrangement is minimum shift keying (MSK), where the modulated signal's phase is constrained to change linearly between adjacent signaling instants over the duration of one signalling interval. Hence MSK is a so-called full-response system. However, the derivative of the phase trajectory has discontinuities at the sampling instants due to its step wise linear nature, indicating that further spectral efficiency gains are possible, when introducing smoother phase changes.

It is plausible that the slower and smoother the phase changes engendered by the modulating signal, the narrower the modulated signal's spectrum and the better the spectral efficiency. Spreading phase changes with maximally flat zero initial and final slopes over several modulation intervals yields a so-called partial response system. Although this PR scheme introduces intersymbol interference (ISI), this can be removed using appropriately designed channel equalisers, which can cope with the additional deliberately introduced controlled ISI (CISI). An attractive representative of this PR CPM family is called Gaussian minimum shift keying (GMSK), where the phase changes are spread typically over two to four signalling intervals using a Gaussian filter [158].

The schematic of the GMSK modulator is portrayed in Figure 7.13. The impulse response $g(t)$ of the Gaussian filter is given [210] by:

$$g(t) = \frac{1}{2T}\left[Q\left(2\pi B_b \frac{t-T/2}{\sqrt{\ell n2}}\right) - Q\left(2\pi B_b \frac{t+T/2}{\sqrt{\ell n2}}\right)\right] \tag{7.86}$$

for

$$0 \leq B_bT \leq \infty$$

where $Q(t)$ is the Q-function

$$Q(t) = \int_t^\infty \frac{1}{\sqrt{2\pi}} \exp(-\tau^2/2)d\tau \tag{7.87}$$

B_b is the 3 dB down bandwidth of the Gaussian phase shaping filter, T is the signalling interval duration and

$$B_N = B_bT \tag{7.88}$$

is the normalised bandwidth.

The Gaussian impulse response $g(t)$ is shown in Figure 7.12 for different normalised bandwidths B_N.

Observe in Figure 7.13 that the Gaussian filter is preceeded by an integrator, which converts the data-induced input frequency changes to phase changes $\phi(t, \alpha_n)$.

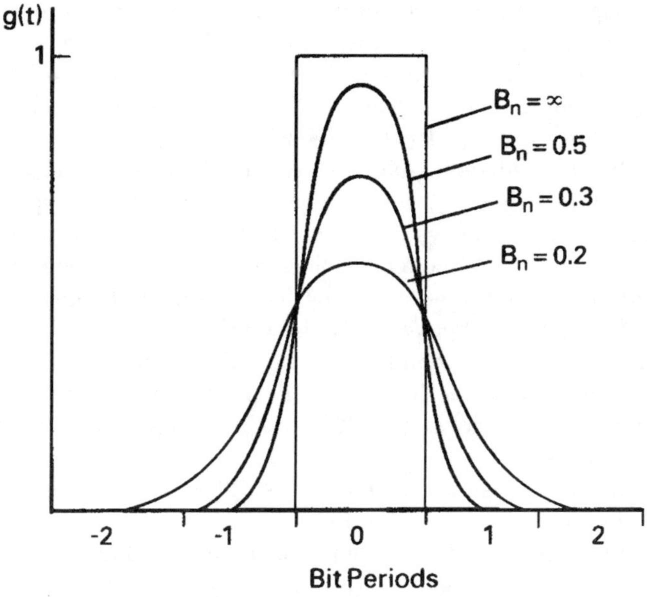

Figure 7.12: Gaussian impulse response $g(t)$ of GMSK for various normalised bandwidths B_N [158] ©Pentech Press, 1992, Steele

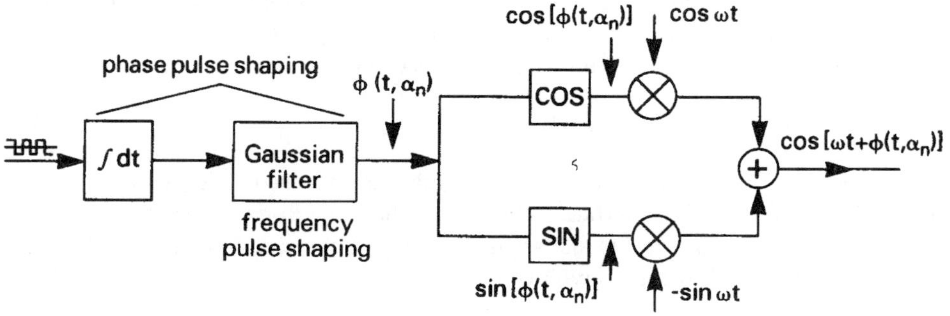

Figure 7.13: GMSK modulator schematic diagram

Taking the sin and cos of this phase yields the base band equavalent in-phase and quadrature phase modulating signals, which modulate two quadrature carriers. Finally, the orthogonal quadrature components are added and transmitted in the same frequency band.

7.6.2 Viterbi equalisation

A set of standardised dispersive wideband mobile radio channels used in the pan-European mobile radio system, known as GSM has been introduced in Figure 2.21.

Since dispersive wideband channels can be described by impulse responses extending over a number of signaling intervals, their uncontrolled ISI must be counteracted by the channel equaliser. Clearly, the equaliser now has to have a longer memory in order to be able to remove the superimposed effects of the channel impulse response plus the CISI introduced by the PR GMSK scheme.

If, however, the transmitted signal's bandwidth is narrow compared to the channel's coherence bandwidth (B_c), all transmitted frequency components encounter nearly identical propagation delays, i.e. the so-called narrowband condition is met and the signal is subjected to non-frequency-selective or flat envelope fading. In this case the channel's impulse response is an ideal Dirac delta, implying no channel-induced ISI and the equaliser has to remove solely the CISI.

Both during and after the definition of the GSM standards a number of Viterbi equaliser (VE) implementations have been proposed in the literature, which have different complexities and performances [211,212]. A simple general VE block diagram is shown in Figure 7.14.

In conventional time division multiple access (TDMA) mobile systems, such as the GSM system, communications is maintained using transmission bursts, incorporating a so-called midamble sequence in the centre of the burst. This periodically transmitted Dirac delta-like sequence has a sharply decaying autocorrelation sequence, which allows the system to measure the channel's impulse response and hence assists in the equalisation process. Furthermore, since the midamble is typically a 16-bit unique word, its detection using autocorrelation facilitates frame synchronisation. This synchroniser concept was briefly introduced already in Section 6.2.4.

Explicitly, the modulated TDMA burst with the channel sounding sequence $s(t)$ in its centre is convolved with the channel's impulse response $h_c(t)$ and corrupted by noise. Neglecting the noise for simplicity, the received sounding sequence becomes:

$$s_r(t) = s(t) * h_c(t). \tag{7.89}$$

Then the received sounding sequence $s_r(t)$ is matched filtered (MF) using the causal impulse response h_{MF}. (For further notes on matched filtering the reader is referred to Section 4.4.2.) In order to derive the estimate $h_e(t)$ of the channel's impulse response we can then use the following convolution:

$$h_e(t) = s_r(t) * h_{MF}(t) = s(t) * h_c(t) * h_{MF}(t) = R_s(t) * h_c(t), \tag{7.90}$$

where $R_s(t)$ is the sounding sequence's autocorrelation function. If $R_s(t)$ is a highly peaked Dirac impulse-like function, then its convolution with $h_c(t)$ becomes $h_e(t) \approx h_c(t)$. With $s(t)$ in the middle of the TDMA burst, the estimated channel impulse response $h_c(t)$ can be considered quasi-stationary for the burst duration, and can be used to equalise the useful information bits at both sides of it. The time variant channel typically precipitates higher bit error rates towards the burst edges, where the impulse response estimate becomes less accurate due to the channel's quasi-stationarity.

Since the complexity of the VE increases exponentially with the number of signalling intervals over which the received signal can be spread due to ISI and CISI, the estimated channel response $h_c(t)$ has to be windowed to a computationally affordable

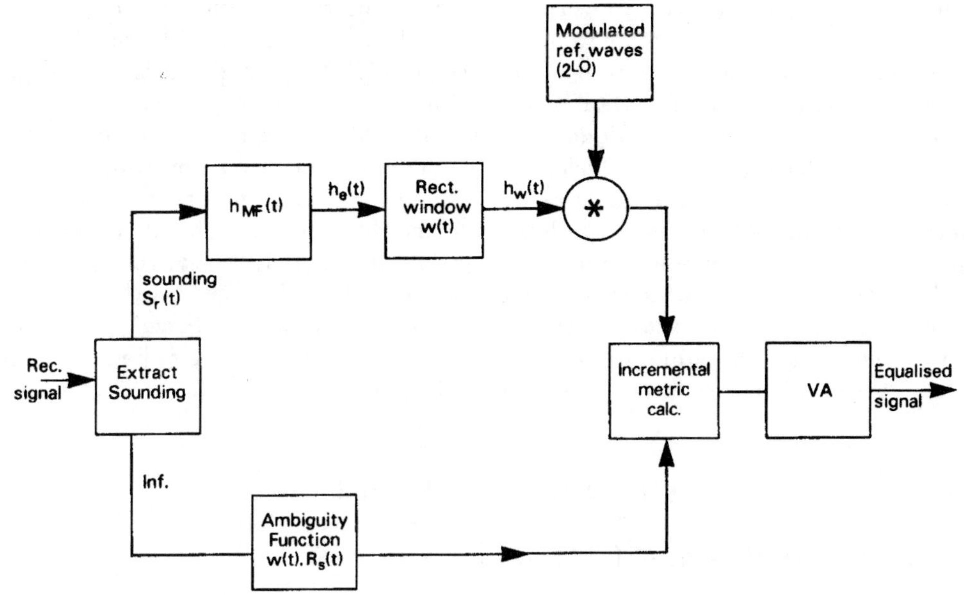

Figure 7.14: MLSE block diagram

length using the rectangular function $w(t)$. However, it has to retain a sufficiently long memory to compensate for the typical channel impulse responses encountered. Specifically, in addition to the duration L_{CISI} of the controlled ISI, the channel's delay spread L_c also has to be considered in calculating the required observation interval $L_o = L_{CISI} + L_c$ of the 2^{L_o-1}-state VE.

In order to augment our exposition, we invoke the GSM system's example. When using a bit interval of 3.69 μs corresponding to the GSM TDMA burst rate of about 270 kbit/s and a maximum channel impulse response duration of around 15–20 μs, as seen in Figure 2.21, a VE with a memory of 4-6 bit intervals is a good compromise. Note that $h_c(t)$ might have an excessively long decay time and hence for practical reasons it is retained over that particular 4-6 bit interval of its total time domain support length, where it is exhibiting the highest energy. In general, L_o consecutive transmitted bits give rise to 2^{L_o} possible transmitted sequences, which are first input to a local modulator to generate the modulated waveforms, and then convolved with the windowed estimated channel response $h_w(t)$ in order to derive the legitimate reference waveforms for metric calculation, as portrayed in Figure 7.14.

Recall that the condition $h_e(t) = h_c(t)$ is met, i.e. the estimated impulse response is identical to the true channel impulse response only, if $R_s(t)$ is the Dirac delta function. This condition cannot be met when finite-length sounding sequences are used. The true channel response $h_c(t)$ could only be computed by deconvolution from Equation 7.90 upon neglecting the rectangular window $w(t)$, which does not give a unique decomposition. Alternatively, the received signal can be convolved for the sake of metric calculation with the known windowed autocorrelation function $w(t) \cdot R_s(t)$ often referred to as ambiguity function, as seen in the lower branch of Figure 7.14 after extracting the sounding sequence from the received TDMA burst. Clearly, this way the received signal is "predistorted" using the ambiguity function, identically to the estimated impulse response in Equation 7.90. This filtered signal is then compared to all possible locally generated reference signals and the incremental metrics m_i, $i = 0, \ldots, (2^{L_o - 1})$ are computed. Lastly, the incremental metrics are utilised by the Viterbi algorithm (VA) in order to determine the maximum likelihood transmitted sequence, as explained in references [158,212]. Suffice to note here that the maximum likelihood transmitted sequence is identified on the basis of the metrics by finding the specific reference sequence that has the lowest deviation from the currently processed received sequence. The same procedure can be used in order to infer the maximum likelihood transmitted trellis coded sequence in the next chapter.

7.7 Overview of Blind Equalizers

S. Vlahoyiannatos and L. Hanzo [1]

7.7.1 Introduction

Blind channel equalization has been attracting scientific interest since 1975, when Sato published his original study of a scheme now known as the "Sato algorithm" [4]. Following Sato's study, the topic of blind equalization attracted more academic than practical attention. In recent years researchers have intensified their efforts in this field due to the definition of a range of high-importance video broadcast standards. The Pan-European Satellite-based Digital Video Broadcast (DVB-S) [213], the terrestrial DVB-T [214] and the cable-based DVB-C [215] systems constitute a family of harmonized systems, where both the DVB-S and the DVB-C schemes recommend blind equalization. In this review, we offer a comprehensive overview of these techniques, focusing on two particular families of methods, namely on the so-called Bussgang algorithms and on the joint channel impulse response (CIR) and data sequence estimation. This section is structured as follows. In Subsection 7.7.2 we provide a brief historical perspective. The basic principles of blind equalization are discussed in Subsection 7.7.3, followed by an overview of the so-called Bussgang equalizers in Subsection 7.7.4. Subsection 7.7.6 is dedicated to convergence results. In Subsection 7.7.7 joint data and channel estimation techniques are discussed. The so-called second–order statistics based algorithms' principles are highlighted in Section 7.7.8. The basic principles of the fourth–order statistics based channel estimation and equalization algorithms using the so-called "polycepstra" are summarized in Subsection 7.7.9. Complexity comparisons are provided in Subsection 7.7.10. Finally, in Subsection 7.7.11 the performance of these equalizers is studied comparatively, while decision–directed improvements are

[1]This section is based on S. Vlahoyiannatos and L. Hanzo; Blind Equalizers: A Comparative Study and Application to Digital Video Broadcasting, submitted to Proceeding of the IEEE

reviewed in Subsection 7.7.12. Let us now commence our discourse with a historical perspective.

7.7.2 Historical background

As mentioned above, blind equalization was originally proposed by Sato [4]. Five years after Sato's publication, the blind equalization problem was further studied for example by Benveniste, Goursat and Ruget in [216], where several blind equalization issues were clarified and a new algorithm was proposed [7]. At the same time Godard [5] introduced a criterion, namely the so-called "constant modulus" (CM) criterion, leading to a new class of blind equalizers. Following Godard's contribution a range of studies were conducted employing the constant modulus criterion. Foschini [217] was the first researcher studying the convergence properties of Godard's equalizer upon assuming an infinite equalizer length. Later, Ding *et al.* continued this study [218] and provided an indepth analysis of the convergence issue in the context of a realistic equalizer. Although numerous researchers studied this issue, nevertheless, a general solution is yet to be found. A plethora of authors have studied Godard's equalizer, rendering it the most widely studied blind equalizer. A well–known algorithm of the so-called Bussgang type [219, 220] was also proposed by Picchi and Prati [8]. Their "Stop-and-Go" algorithm constitutes a combination of the Decision-Directed algorithm [3] with Sato's algorithm [4]. After 1991, a range of different solutions to the blind equalization problem were proposed. Seshadri [221] suggested the employment of the so-called M-algorithm, as a "substitute" for the Viterbi algorithm [222] for the blind scenario, combined with the so-called "least mean squares (LMS)" based CIR estimation. This CIR estimation was replaced by "recursive least squares (RLS)" estimation by Raheli, Polydoros and Tzou [223], combining the associated convolutional decoding with the CIR estimation, leading to what was termed as "*Per-Survivor Processing*". Since then a number of papers have focused on this technique [223–232]. At the same time as Seshadri, Tong *et al.* [233] proposed a different approach to blind equalization, which used oversampling in order to create a so-called "cyclostationary" received signal, and performed CIR estimation by measuring the autocorrelation function of this signal and by exploiting this signal's cyclostationarity. This technique was also applied to the case of 'sampling' the received signals of different antennas (instead of oversampling the signal of a single antenna) and further extended by Moulines *et al.* using a different method of CIR estimation, namely the so–called subspace method in [234]. Furthermore, Tsatsanis and Giannakis suggested that the cyclostationarity can be induced by the transmitter upon transmitting the signal more than once [235]. A number of further contributions have also been published in the context of these techniques [236–251]. Finally, nearly coincidentally with Seshadri [221] and Tong *et al.* [233], Hatzinakos and Nikias [252] proposed a more sophisticated approach to blind equalization by exploiting the so-called "tricepstrum" of the received signal. Until today, the blind equalization problem is an open research topic, attracting significant amount of research. A general answer to the fundamental question "*Under what circumstances is it preferable to use a blind equalizer to a trained–equalizer ?*" is yet to be provided. Despite the scarcity of reviews on the topic, in the context of the Global System of Mobile Communications known as GSM an impressive effort was made by Boss, Kammeyer and Petermann [253], who also proposed two novel blind algorithms. We recommend furthermore the fractionally–spaced equalization review of Endres *et al.* [254] and the Constant Modulus overview of Johnson *et al.* [255] based on a specific type of equalizers, namely on the so-called "fractionally-spaced" equalizers. A review of subspace–ML multichannel blind equalizers was provided by Tong and Perreau [256]. Further important references are the monograph by Haykin [3], the relevant section by Proakis [257] and the blind deconvolution book due to Nandi [258]. Comparative performance studies between various blind equalizers have also been performed. We recommend the second–order statistics-based comparative performance studies of Becchetti *et al.* [259], Kristensson *et al.* [260] and Altuna *et al.* [261], which is based on the mobile environment as well as the second–order statistics and PSP-based comparative study of Skowratanont and Chambers [262]. Furthermore, we recommend the fractionally–spaced Bussgang algorithm based comparative performance study by Shynk *et al.* [263], the CMA comparative performance study of Schirtzinger *et al.* [264] and the comparative convergence study by Endres *et al.* [265]. Let us now review the basic principles of blind equalization in the next section.

7.7.3 Blind equalization principles

In this section we will introduce and discuss a range of blind equalization principles, mainly focusing on the Bussgang techniques [219, 220]. Since Sato's original study in 1975 [4], blind equalization has attracted significant scientific interest due to its potentials in terms of:

- Overhead reduction. Training sequences sacrifice bandwidth in order to assist in determining the CIR, hence assisting equalization. Blind equalizers do not need training sequences and therefore conserve bandwidth.

- Simplification of point to multipoint communications systems or broadcast. When a communications link is reset, equalizer adjustment "from scratch" is necessary. In this case, using a training sequence is inefficient, since the transmitter has to retransmit the training sequence specifically for each receiver, which is reset.

We will now proceed to describe the problem of blind equalization. In our discussion the communication system parameters will be assumed to be time-invariant. This constraint is not necessary in general, although it is beneficial in terms of complexity reduction. Furthermore, the additive noise will be assumed to be Gaussian and white. Let us assume that the input bits, resulting from any encoder prior to modulation, are mapped into complex Quadrature Amplitude Modulation (QAM) [266] input symbols $a(n)$ transmitted at time instant n. These transmitted symbols are filtered by the CIR h_i and then the noise $e(n)$ is added to them, resulting in the received symbols $y(n)$ at time instant n in the form of:

$$y(n) = \sum_{i=-L_1}^{L_2} h_i \cdot a(n-i) + e(n), \tag{7.91}$$

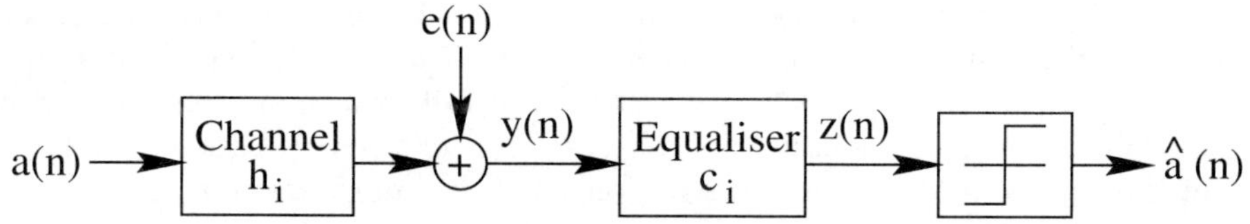

Figure 7.15: Equalized Communications system

where L_1 and L_2 are the length of the CIR's pre- and post-cursor sections surrounding the main tap (measured in terms of the number of transmitted symbols), respectively. An equalizer is typically placed after the channel in order to remove the channel-induced dispersion, as it is shown in Figure 7.15. Blind equalization involves finding the 'best' equalizer filter, which regenerates the input symbols $a(n)$ at the receiver, without any knowledge of the CIR upon exploiting the knowledge of the distribution of the input QAM symbols. If the equalizer has N_1 feedback and N_2 feedforward taps $\{c_i\}$, then the equalized symbols will be of the form:

$$z(n) = \sum_{i=-N_1}^{N_2} c_i \cdot y(n-i) \tag{7.92}$$

and upon using Equation (7.91) we have:

$$z(n) = \sum_{i=-L_1-N_1}^{L_2+N_2} t_i \cdot a(n-i) + \sum_{i=-N_1}^{N_2} c_i \cdot e(n-i), \tag{7.93}$$

where $\{t_i\} = \{h_i\} * \{c_i\}$ is the convolution of the CIR with the equalizer filter, representing the total transfer function of the cascaded system constituted by the channel plus the equalizer. Assuming that the noise power is low compared to the power of the received signal, we can observe that the blind equalization problem corresponds to estimating the suitable equalizer impulse response, which reduces the first term at the right–hand side of Equation (7.93) to only one of the summation terms. In this case, the cascaded system's impulse response tap-vector $\mathbf{t}$ takes the form of:

$$\mathbf{t} = (0, \cdots, 0, A, 0, \cdots, 0), \tag{7.94}$$

where A is a constant. This is the only case that corresponds to zero inter-symbol interference (ISI). Assuming that the strongest signal path is located at time instant 0, Equation (7.93) can also be expressed as:

$$z(n) = t_0 \cdot a(n) + \sum_{i=-L_1-N_1,\ i\neq 0}^{L_2+N_2} t_i \cdot a(n-i) + \sum_{i=-N_1}^{N_2} c_i \cdot e(n-i). \tag{7.95}$$

The first term of Equation (7.95) is the useful one, including the one and only path at time 0. The second term is the ISI term, which is also referred to as *convolutional noise* [267]. This is because this term is a noise term, as far as the receiver is concerned and since it is the result of the convolution of the CIR h_i and the equalizer's impulse response c_i with the input signal $a(n)$. This is usually the main noise contributor at the initialization of the equalization process and it is reduced further during the stages of the equalization process, leaving only the real noise term as the sole signal impairment, when the equalization is perfect. Finally, the third term of Equation (7.95) is the noise term $e(n)$, convolved with the equalizer's impulse response c_i, since the noise has been filtered by the equalizer.

A problem similar to blind equalization is the problem of blind deconvolution [268]. In blind deconvolution the aim is to perform joint channel and data estimation in a non-real time fashion, which is a fundamental difference with respect to the blind equalization problem. The received signal is stored and then deconvolved in order to produce the CIR together with the input signal. This is the case, for example, when seismic signals are considered [269]. We have no knowledge of the signal emerging from the crust of the earth or of the channel that intermediates between the source of this signal and our receiver, hence we attempt to record the signal in order to deconvolve it later. When a channel estimation is available to us, we can estimate the input signal in two ways. A feasible approach is to use a sequence estimation technique, such as the Viterbi algorithm [222]. Another approach is to produce an inverse filter of the channel and to perform filtering with the aid of this inverse filter. We will discuss this technique in more detail in Section 7.7.7.

Let us now point out an important aspect of channel equalization, which was observed by Benveniste *et al.* in [216], noting that if the distribution of the equalizer's input signal is Gaussian, then the equalized signal of Equation (7.93) is also Gaussian distributed. This is true even if there are more than one terms in the sum which are non-zero, i.e. even if ISI exists. Conversely, provided that the distribution of the equalizer's input signal is Gaussian, the equalized signal constituted by a sum of "independent identically distributed (i.i.d.)" Gaussian variables is also a Gaussian variable irrespective of whether there is residual ISI at its output. In this case, equalization is impossible, since the equalizer cannot distinguish the zero–ISI equalized signal sequence from the other possible candidate sequences, which contain ISI. In other words, it is impossible to equalize the received signal, since the zero–ISI candidate received signal sequence is indistinguishable from the other ISI-contaminated candidate sequences. Hence blind equalization cannot be performed, if the distribution of the equalizer's input signal is Gaussian. Fortunately, the uniform distribution of typical QAM data sources is far from the Gaussian, which renders the received signal amenable to blind equalization.

The equalizers alluded to in this section are classified in Figure 7.16. Blind channel identifiability issues have been discussed in [216] and [270]. Following the above general discussion on blind equalizibility, let us now proceed with the overview of the family of Bussgang techniques in the next section.

7.7.4 Bussgang blind equalizers

In this section we set out to characterize a range of basic equalizer schemes involving the class of "*Bussgang*" techniques dating back to 1952 [220]. In our discussion we

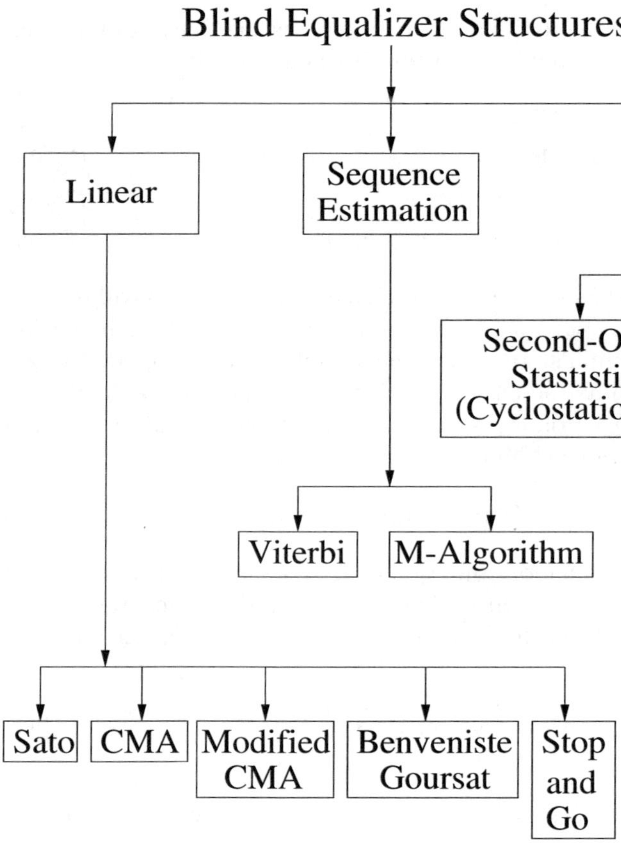

Figure 7.16: Classification of blind equalizers

will consider the following signal vectors, which obey the notations of Figure 7.15:

$$\begin{aligned}
\mathbf{a}(n) &= [a(n+N_1+L_1),\ldots,a(n-N_2-L_2)]^T && (7.96)\\
\mathbf{y}(n) &= [y(n+N_1),\ldots,y(n-N_2)]^T && (7.97)\\
\mathbf{c} &= [c_{-N_1},\ldots,c_{N_2}]^T && (7.98)\\
\mathbf{t} &= [t_{-N_1-L_1},\ldots,t_{N_2+L_2}]^T && (7.99)\\
\mathbf{e}(n) &= [e(n+N_1+L_1),\ldots,e(n-N_2-L_2)]^T && (7.100)
\end{aligned}$$

$$\mathbf{H} = \begin{pmatrix} h_{-L_1} & \ldots & h_{L_2} & 0 & \ldots & 0 & 0 \\ 0 & h_{-L_1} & \ldots & h_{L_2} & 0 & \ldots & 0 \\ \vdots & \ddots & \ddots & \ddots & \ddots & \ddots & \vdots \\ 0 & \ldots & 0 & h_{-L_1} & \ldots & h_{L_2} & 0 \\ 0 & 0 & \ldots & 0 & h_{-L_1} & \ldots & h_{L_2} \end{pmatrix} \qquad (7.101)$$

where ^{T} denotes transpose, * denotes conjugate and ^{H} denotes the Hermitian matrix. From these definitions the following matrix relationships hold:

$$\mathbf{y}(n) = \mathbf{H} \cdot \mathbf{a}(n) \tag{7.102}$$

$$\mathbf{t} = \mathbf{H}^T \cdot \mathbf{c} \tag{7.103}$$

$$z(n) = \mathbf{c}^T \cdot \mathbf{y}(n) \tag{7.104}$$

$$z(n) = \mathbf{t}^T \cdot \mathbf{a}(n) + \mathbf{c}^T \cdot \mathbf{e}(n). \tag{7.105}$$

Explicitly, Equation (7.102) describes the received signal $y(n)$ as the convolution of the transmitted signal $a(n)$ and the CIR h_i. Equation (7.103) reflects the convolution of the CIR with the equalizer's impulse response and Equation (7.104) characterizes the equalizer's output signal $z(n)$ as the convolution of the received signal $y(n)$ with the equalizer's impulse response c_i. Following these definitions, a general form of the Bussgang equalizer update Equations [220] can be expressed as:

$$\mathbf{c}^{(n+1)} = \mathbf{c}^{(n)} - \lambda \cdot \mathbf{y}^*(n) \cdot [z(n) - g\{z(n)\}], \tag{7.106}$$

where $g\{z(n)\}$ is a non–linear zero–memory function of the equalized output $z(n)$ and λ is the so-called "step–size" parameter, controlling the speed and the accuracy of the equalizer's convergence. The condition for attaining convergence in the mean value for these algorithms is [3]:

$$E[z(n) \cdot \mathbf{y}^*(n)] = E[g\{z(n)\} \cdot \mathbf{y}^*(n)] \tag{7.107}$$

or

$$E[z(n) \cdot y^*(n-i)] = E[g\{z(n)\} \cdot y^*(n-i)], \quad i = -N_1, \cdots, N_2. \tag{7.108}$$

Upon multiplying each side of this equation by the relevant equalizer tap coefficient c_i^* and summing the results for $i = -N_1, \cdots, N_2$ as in [3], we obtain:

$$E[z(n) \cdot z^*(n)] = E[g\{z(n)\} \cdot z^*(n)]. \tag{7.109}$$

If instead of multiplying by c_i^* we had multiplied by c_{i-k}^*, assuming that the equalizer has an infinite number of taps, then Equation (7.109) would take the form of:

$$E[z(n) \cdot z^*(n-k)] \simeq E[g\{z(n)\} \cdot z^*(n-k)], \tag{7.110}$$

which reflects the so-called *Bussgang property* that is satisfied by the Bussgang algorithms, when the equalizer length is doubly infinite. When both the feedforward and feedback equalizer lengths are sufficiently high, then the Bussgang property is approximately satisfied. In the strict sense, however, only Equation (7.109) is satisfied. When the equalizer has converged, then the equalized symbols $z(n)$ approximate the transmitted symbols $a(n)$ and Equation (7.109) becomes:

$$E[z(n) \cdot g\{z(n)\}] = E[|a(n)|^2] = 1, \tag{7.111}$$

provided that the input power is normalized.

The problem of finding the optimum Bussgang equalizer corresponds to finding the function $g\{z(n)\}$, which provides the best estimate of the corresponding transmitted symbol $a(n)$ for each equalized symbol $z(n)$, whilst satisfying Equation (7.111). According to the Maximum Likelihood (ML) criterion, this can be achieved by setting $g\{z(n)\} = E[a(n)|z(n)]$, i.e. setting $g\{z(n)\}$ equal to the expected (or most probable) value of the transmitted symbol, given the equalized symbol $z(n)$. In order to find this expected value, we have to estimate the distribution $f_z(z)$ of the equalized symbols $z(n)$. During the equalizer's initialization, in general the equalized signal contains ISI. If during this initialization phase we ignore the channel noise by assuming that the ISI is the main signal impairment at this stage, then this means that the equalized signal consists of a number of transmitted signal replicas, each having a different delay and weight. When the number of these replicas is sufficiently high, according to the central limit theorem we can approximate the distribution of the equalized symbols $z(n)$ with a Gaussian distribution. In our analysis in this section we shall assume that the central limit theorem condition can be invoked since there is a sufficiently high number of ISI terms. Assuming also M-level QAM transmissions, the expected value giving the estimate of $g\{z(n)\}$ obeys the following form, which is similar to the one given in [268] for pulse-amplitude modulation (PAM):

$$g_{ML}\{z(n)\} = E[a(n)|z(n)] = \frac{\sum_{i=1}^{M} A_i e^{-|z(n)-A_i|^2/2\sigma^2}}{\sum_{i=1}^{M} e^{-|z(n)-A_i|^2/2\sigma^2}} \tag{7.112}$$

where the coefficients A_i constitute the signal amplitudes associated with the QAM constellation and σ^2 is the variance of the noise, consisting of two components, namely the convolutional noise and the Additive White Gaussian Noise (AWGN) induced by the channel. An estimate of the noise variance σ^2 must be available for the evaluation of Equation (7.112). However, the function $g\{z(n)\}$ is only the optimum one under the plain assumption of a Gaussian distribution for the above-mentioned composite noise, which is produced by the ISI plus the channel's additive noise. Depending on this distribution, different Bussgang algorithms exist. The well–known Godard (or CMA) [5], Sato [4], Benveniste-Goursat [216] or Stop-and-Go [8] algorithms constitute a few such algorithms. We will describe each of them in the forthcoming paragraphs and discuss their characteristics. As an illustration, in Figure 7.17 we have plotted the real part of the function $g\{(z)\}$, evaluated from Equation (7.112), for the ML algorithm of $g_{ML}\{z(n)\} = E[a(n)|z(n)]$ under the assumption of $-20dB$ additive Gaussian noise power. The real part of a 16-QAM signal can take four discrete values, symmetrically distributed around the origin. The estimated signal, which approximates the most likely transmitted signal, should be close to these legitimate constellation points. We observe that this is the case, when the above ML algorithm is used, under the assumption of low noise ($-20dB$). In this low-noise scenario, the ML algorithm estimates the transmitted signal as the constellation symbol, which is closest to the equalized symbol at time instant n $z(n)$. This is illustrated in Figure 7.17 by the four levels, corresponding to the four discrete values that the real (or the imaginary) part of a 16-QAM signal can assume. If the noise variance is not sufficiently low, however, then the surface of Figure 7.17 loses its resemblance to the 16-QAM constellation. This is shown in Figure 7.18, where the noise variance was assumed to be $-10dB$, i.e. $10dB$ higher than in Figure 7.17. When no CIR

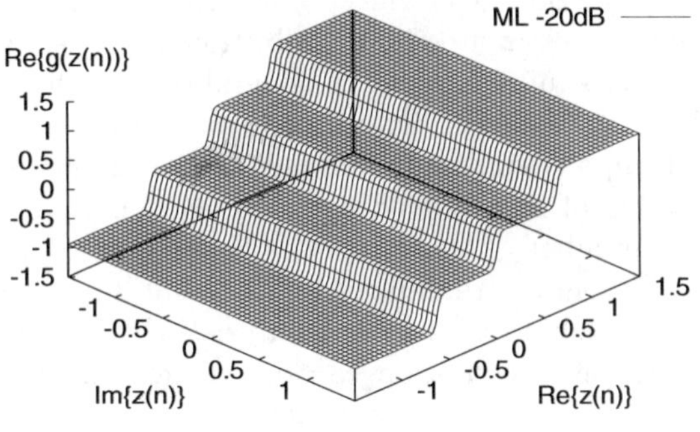

Figure 7.17: The real part of the function $g\{z(n)\}$ plotted against the complex $z(n)-$plane for ML detection assuming Gaussian noise with -20dB variance for 16–QAM.

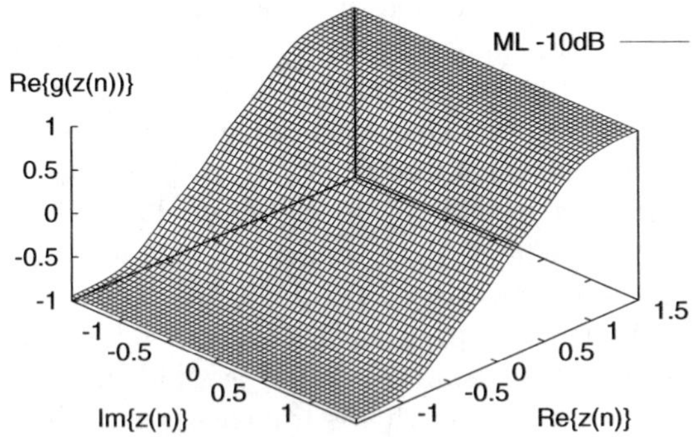

Figure 7.18: The real part of the function $g\{z(n)\}$ plotted against the complex $z(n)-$plane for ML detection assuming Gaussian noise with -10dB variance for 16–QAM.

information is available, it might be extremely optimistic to assume that, even without channel noise, the power of the ISI-induced noise would be as low as $-20dB$. In fact it might be well over $0dB$, thus rendering the 16-QAM pattern unrecognisable in the received signal. The approximation of the estimation function $g\{z(n)\}$ for each Bussgang algorithm is given in a 2D plot in [268] for 8-level PAM, where the signals are real-valued. Similar figures can be generated by extending the approach of [268] to the case of 16-QAM, where the signals assume complex values, as it was seen in the 3D plots of Figures 7.17 and 7.18.
An alternative interpretation of the function $g\{z(n)\}$ can be observed by considering a "cost-function" $J\{z(n)\}$. The minimization of this cost-function leads to the desired equalizer tap values, according to the wide-spread steepest descent algorithm [3]:

$$\mathbf{c}^{(n+1)} = \mathbf{c}^{(n)} - \lambda \cdot \frac{\partial J\{z(n)\}}{\partial \mathbf{c}}, \tag{7.113}$$

which physically implies that the taps $\mathbf{c}^{(n)}$ at instant n are modified by the derivative of the cost-function – after weighting by the step-size λ – in the direction of minimizing the cost-function. According to Equations (7.113) and (7.106) the following relationship holds between $g\{z(n)\}$ and $J\{(z(n)\}$:

$$\frac{\partial J\{z(n)\}}{\partial \mathbf{c}} = \mathbf{y}^*(n) \cdot (z(n) - g\{z(n)\}. \tag{7.114}$$

Again, in simple, but conceptually feasible terms Equations (7.113) and (7.114) can be interpreted as updating each tap of the equalizer on the basis of the gradient of the error term

$$\epsilon(n) = z(n) - g\{z(n)\} \tag{7.115}$$

with respect to (wrt) a specific tap. Depending on the polarity of the cost-function's derivative wrt a specific tap, this tap is updated according to the step-size λ, such that in the next step it reduces $\epsilon(n)$ – hence the negative sign in Equation (7.113). Upon using the differentiation rules with respect to a vector given in Appendix 7.9, we can readily arrive at:

$$g\{z(n)\} = z(n) - J^{'}\{z^*(n)\}, \tag{7.116}$$

where $'$ denotes the derivative and the associated difference quantifies the discrepancy of the equalized output $z(n)$ and the derivative of the cost-function given by Equation (7.114). Equation (7.116) will be useful, when we consider the Bussgang cost-functions individually and derive the corresponding equalizer tap update algorithms.

Again, in all of our discussions, we employ QAM [266], [267] and the general structure of the equalizer is shown in Figure 7.19. As we can see from this figure, the blind equalizer coefficients are updated using the knowledge of the received signal vector $\mathbf{y}(n)$, the equalized signal $z(n)$, the phase–corrected equalized signal $\tilde{z}(n) = z(n)e^{-j\phi}$ and the estimated signal $\hat{a}(n)$. We note, however, that a specific Bussgang equalizer may not make use of all of these signals in order to update its equalizer tap coefficients. It depends on the algorithmic implementation, which of these signals are invoked in the tap update process. What is common, however, to all Bussgang algorithms is

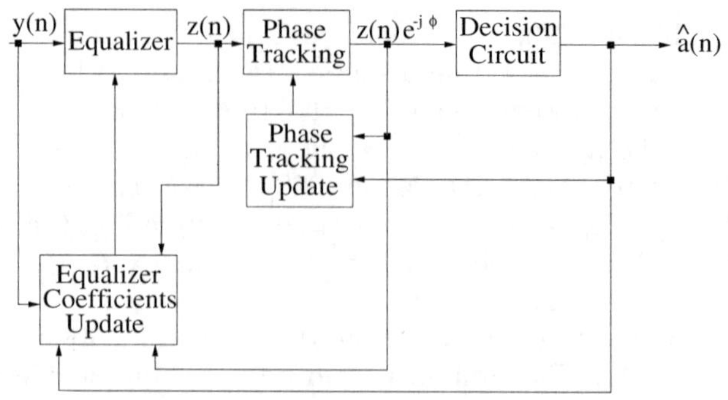

Figure 7.19: Equalizer structure used in Bussgang techniques

that the error estimate in Equation (7.115) will be a function of the equalized symbol $z(n)$ at time n only, i.e. they are based on a zero–memory estimation.

In the equalizers presented in this section the sampling rate is identical to the signalling-rate – or Baud-rate, that is we use only one sample per symbol period. These equalizers are referred to as *symbol-spaced* schemes, as opposed to *fractionally–spaced equalizers*, which use more than one sample per symbol in order to equalize the channel. A typical example of fractionally–spaced equalizers is constituted by the family of second–order cyclostationary statistics based blind channel estimation algorithms [271]. In the context of Bussgang schemes, the extension of these equalizers to fractionally spaced arrangements is relatively straightforward. Such algorithms have been reported in the literature for example by Pei and Shih in [272] or by Dogancay and Kennedy in [273]. They have been further studied for example by Endres, Johnson and Green in [274], by LeBlanc, Fijalkow and Johnson in [275], by Endres, Halford, Johnson and Giannakis in [254], by Magarini *et al.* in [276] and by Papadias and Slock in [277].

Before proceeding to the discussion of Sato's algorithm, we note that the Bussgang zero–memory function $g\{z(n)\}$ of Equation (7.106) can be extended to the non–zero–memory case, if we take into consideration more than one equalized symbols in generating the error function of Equation (7.115). This was proposed by Yang for the CMA in [278]. Let us now consider a range of Bussgang algorithms in a little more depth in the forthcoming subsections.

7.7.4.1 Sato's algorithm [4]

Sato's pioneering contribution in 1975 [4] described the first blind equalization algorithm proposed, which was designed for real–valued signals and PAM. However, its extension to complex–valued signals and QAM is straightforward, especially in the spirit of Godard's publication of the well–known CMA [5], which was derived for complex–valued QAM signals. Sato's algorithm dedicated to real valued signals $z(n)$ uses the following cost-function [4]:

$$J^S(n) = E\left[(|z(n)| - \gamma)^2\right], \tag{7.117}$$

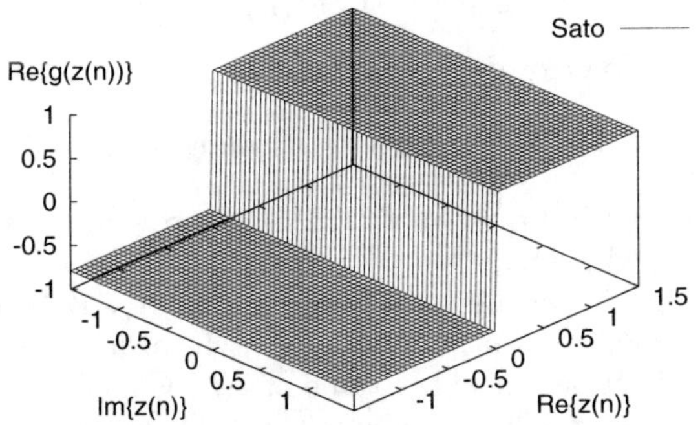

Figure 7.20: The real part of function $g\{z(n)\}$ plotted against the real part of the equalized symbol $z(n)$ for different Sato's algorithm. The modulation used is 16–QAM.

where γ is Sato's scaling coefficient and $E[]$, again, represents the expectation over all possible transmitted data sequences. It is clear that this cost-function is forcing the absolute value of the equalized signal to a fixed value γ. This is a plausible policy to pursue, when Binary Phase Shift Keying (BPSK) is used, but not for any other multilevel PAM scheme. For these multilevel constellations the minimization of the Sato cost-function of Equation (7.117) may not seem to lead to correct update of the equalizer taps at each iteration. Nonetheless, experimental experience shows that the minization of Equation (7.117) may still lead to convergence to the desired zero–ISI equilibrium, although not in all cases, as we shall see in Section 7.7.6. The associated complex Sato cost-function can be defined as in [7]:

$$J^S(n) = E\left[(|Re\{z(n)\}| - \gamma)^2\right] + E\left[(|Im\{z(n)\}| - \gamma)^2\right]. \tag{7.118}$$

The steepest descent algorithm – which results from the cost-function of Equation (7.118) – can be found by determining the gradient of the cost-function with respect to the equalizer tap vector **c**. Alternatively, using Equation (7.116) we obtain:

$$g_{Sato}\{z(n)\} = csgn(z(n)), \tag{7.119}$$

where the complex "signum" function is given by:

$$csgn(z(n)) = sgn(Re\{z(n)\}) + j \cdot sgn(Im\{z(n)\}). \tag{7.120}$$

In Figure 7.20 a plot similar to these in Figures 7.17 and 7.18 is given for Sato's algorithm. Clearly, this surface does not follow the four legitimate values of the 16-QAM constellation pattern of Figure 7.17, it is constituted by two planes. All the other Bussgang algorithms follow a pattern similar to that of Sato's algorithm and none of them follows the specific 16-QAM constellation pattern of the ML algorithm seen in Figure 7.17. Sato's algorithm [4] is thus given, according to Equation (7.106) by:

$$\mathbf{c}^{(n+1)} = \mathbf{c}^{(n)} - \lambda \cdot \mathbf{y}^*(n) \cdot \epsilon^{Sato}(n), \tag{7.121}$$

where $\epsilon^{Sato}(n)$ is the Sato-error defined as:

$$\epsilon^{Sato}(n) = z(n) - \gamma \cdot csgn(z(n)). \tag{7.122}$$

Explicitly, the tap vector **c** in Equation (7.121) is adjusted according to the correction term $\lambda \cdot \mathbf{y}^*(n) \cdot \epsilon^{Sato}(n)$, where Sato's error term $\epsilon^{Sato}(n)$ depends on the cost-function of Equation (7.118). As it can be seen from Equation (7.122), this algorithm uses only the sign of the equalized output values $z(n)$ in order to update the equalizer coefficients. This implies that the exact value of $z(n)$ is ignored. Clearly, an error in the polarity of $z(n)$ is less probable, than an error in its exact value, when compared to the actual transmitted value $a(n)$. Therefore, Sato's algorithm has the advantage that it avoids using the generally error–prone exact value of $z(n)$, in favour of invoking the less spurious polarity of it. In the case of symmetric multilevel PAM transmissions, for which the algorithm was originally proposed, $Re\{z(n)\}$ assumes equi-probable positive and negative values. Using suitable coding and taking into account only the sign of $z(n)$ implies ignoring the fine–resolution channel effects. The same idea can be adopted for QAM transmissions, where $Re\{z(n)\}$ and $Im\{z(n)\}$ can be treated as two independent PAM constellations.

Setting the value of the scaling coefficient γ in Equation (7.121) is very important, since it actually directs the signal $z(n)$ to the point of its convergence, i.e. to the original constellation points. A way of achieving this is by constraining the mean value of the error update term of Equation (7.121) to zero. Therefore, the optimum value is set for γ in the minimum mean squared error sense. This optimum value was set by Sato [4] to:

$$\gamma = \frac{E[a(n)^2]}{E[|a(n)|]} \tag{7.123}$$

for real valued signals. This is the only value of γ, which sets the mean value of the error term in Equation (7.122) to zero. For complex–valued transmitted signals γ is given by a similar relationship:

$$\gamma = \frac{E[Re\{a(n)^2\}]}{E[Re\{|a(n)|\}]} = \frac{E[Im\{a(n)^2\}]}{E[Im\{|a(n)|\}]}. \tag{7.124}$$

Having described Sato's algorithm, let us now consider the so–called "*Constant Modulus Algorithm*", which was proposed by Godard in [5].

7.7.4.2 Constant modulus algorithm [5]

A more general algorithm for blind equalization was proposed by Godard in [5], which was further generalized later by Shalvi and Weinstein in [279]. Its related cost-function is defined as:

$$J^{(p,q)}(n) = \frac{1}{pq} E[||z(n)|^p - R_p|^q], \tag{7.125}$$

which will be elaborated on below. Godard's original algorithm and all other related algorithms use only $q = 2$ and this is the case that we will focus on from now on. One can observe that Sato's algorithm in Equation (7.118) is similar to the CMA of Equation (7.125), when we have $p = 1$ and $q = 2$. The reason for considering the difference betweeen the amplitude of $z(n)$ and a constant R_p instead of the actually transmitted symbol $a(n)$, is that in blind equalization we attempt to match the equalized signal not to the actually transmitted sequence, which is never available, but to its statistics, which of course is known at the receiver. Given this cost-function, the coefficient adjustment algorithm using the steepest descent technique of Equation (7.113) can then be invoked. The value of the parameter R_p has to be matched to the constellation in a way similar to the setting of γ for Sato's algorithm in Equation (7.118). Clearly, this algorithm forces only the amplitude of the received signal to match a desired mean value, but ignores the phase. Consequently, the phase of the received signal might have arbitrary variations. Godard suggested that equalizing only the magnitude of the signal should be adequate. The associated phase ambiguity can be removed by using differential phase encoding, which is congenial to the nature of blind equalization. Indeed, a sign ambiguity is related to any blind equalizer, stemming from the fact that the QAM constellations are symmetric with respect to the x and y-axes. As it is clear from the definition of the CMA's cost-function in Equation (7.125), in the case of pure phase modulation, the equalizer's output $z(n)$ will be constrained to a constant value and the algorithm will readily converge [5]. A variation of this algorithm, which solves the problem of the arbitrary phase rotation is the so–called Modified–CMA, presented in the next section. Using Equations (7.116) and (7.125) we obtain:

$$g_{CMA}(z(n)) = z(n) \cdot \left[1 - |z(n)|^2 + R_2^2\right]. \tag{7.126}$$

Substituting this Equation into Equation (7.106) we readily arrive at the equalizer tap update equation of the CMA [5]:

$$\mathbf{c}^{(n+1)} = \mathbf{c}^{(n)} - \lambda \cdot \mathbf{y}^*(n) \cdot z(n) \cdot \left[|z(n)|^2 - R_2\right]. \tag{7.127}$$

The value of R_2 can be found by constraining the mean value of the update term of Equation (7.127) equal to zero, assuming that the equalized signal $z(n)$ is equal to the transmitted signal $a(n)$ with its phase rotated by a random value, i.e. assuming that the state of perfect equalization has been reached. The procedure of determining R_2 is exactly the same as the one, which was used to compute Sato's scaling coefficient γ, yielding [5] :

$$R_2 = \frac{E[|a(n)|^4]}{E[|a(n)|^2]}. \tag{7.128}$$

A generalization of Godard's algorithm was proposed by Shalvi and Weinstein in [279]. For this algorithm the cost-function to be minimized is the so-called Kurtosis of the equalized signal $z(n)$, defined as [252]:

$$K(z) = E[|z(n)|^4] - 2 \cdot E^2[|z(n)|^2] - |E[z^2(n)]|^2 \tag{7.129}$$

for complex-valued signals $z(n)$. The algorithm resulting from this cost-function is given in [279], where it is shown that this generalized algorithm results in Godard's algorithm as a special case. Following this algorithm, Shalvi and Weinstein proposed their so-called super-exponential algorithm [280], which uses 4-th order cumulants and converges at a nearly super-exponential speed. Cumulant–based blind equalization algorithms have been proposed for example in [281]– [282].

As an important extension of Godard's algorithm, Wesolowsky's modified CMA is presented in the next section.

7.7.5 Modified constant modulus algorithm [6]

Another modified version of Godard's CMA [5] was proposed by Wesolowsky in [6], employing a cost-function, which relies on both the real and imaginary parts of the equalized signal $z(n)$ [5]:

$$\begin{aligned} J(n) &= E[\left(|Re\{z(n)\}|^2 - R_{2,R}\right)^2 + \\ &\quad \left(|Im\{z(n)\}|^2 - R_{2,I}\right)^2]. \end{aligned} \tag{7.130}$$

The idea behind this cost-function, as compared to the CMA cost-function of Equation (7.125) is that both the real and imaginary parts of the signal are forced to a constant value and, therefore, the random phase ambiguity of the CMA now becomes only 90°. This is meaningful in pure phase modulation, in which case the CMA may converge to an arbitrarily phase–shifted solution. For QAM though, the 90° symmetry of the constellation makes it possible for both algorithms to converge to a 90° phase–shifted solution. Following the same procedure as in the context of the other algorithms, based on Equations (7.113) and (7.114) we obtain:

$$\begin{aligned} \mathbf{c}^{(n+1)} &= \mathbf{c}^{(n)} - \\ &\quad -\lambda \cdot \mathbf{y}^*(n) \cdot [Re[z(n)] \cdot \left((Re[z(n)])^2 - R_{2,R}\right) \\ &\quad +j \cdot Im\{z(n)\} \cdot \left((Im\{z(n)\})^2 - R_{2,I}\right)] & (7.131) \\ g_{MCMA}(z) &= z - (Re\{z\} \cdot \left((Re\{z\})^2 - R_{2,R}\right) + \\ &\quad j\left(Im\{z\} \cdot \left((Im\{z\})^2 - R_{2,I}\right)\right)). & (7.132) \end{aligned}$$

The values of $R_{2,R}$ and $R_{2,I}$ can be found using the same method as for the other algorithms, namely by constraining the mean value of the update terms in Equation (7.131), yielding [5]:

$$R_{2,R} = \frac{E\left[(Re\{a(n)\})^4\right]}{E\left[(Re\{a(n)\})^2\right]} \tag{7.133}$$

$$R_{2,I} = \frac{E\left[(Im\{a(n)\})^4\right]}{E\left[(Im\{a(n)\})^2\right]}. \tag{7.134}$$

It was shown by Wesolowsky [6] that the MCMA exhibits slightly faster convergence than the classical CMA, in particular for medium-distortion channels. It also offers slightly better steady–state performance, as it will be shown in Section 7.7.11. Let us now, however, turn our attention to considering the *Benveniste-Goursat* algorithm [7].

7.7.5.1 Benveniste–Goursat algorithm [7]

In Sato's algorithm [4], the error signal was expressed in Equation (7.122), which is repeated here for convenience:

$$\epsilon^{Sato}(n) = z(n) - \gamma \cdot csgn(z(n)).$$

The error term is used to update the equalizer coefficients, according to Equation (7.121). This error signal is however non–zero, when the signal is perfectly equalized, except in the case of QPSK, since γ is a constant value, reflecting the statistics of $a(n)$, while $z(n)$ generally takes its values from a multilevel constellation, when the SNR is sufficiently high and the equalizer has converged. This results in inaccurate steady-state behaviour along with small error fluctuations around the point of equilibrium associated with the minimum error. In other words, even near the optimum equalizer setting, not every tap update drives the equalizer towards the desired equilibrium. In order to remedy these deficiencies, Benveniste and Goursat [7] considered the decision–directed error signal expressed as:

$$\epsilon^{DD}(n) = z(n) - \hat{a}(n), \tag{7.135}$$

which becomes zero, when equalization has been accomplished, giving a good-steady state performance. On the other hand, this error signal cannot be employed during the equalizer's initial convergence phase, since at the beginning of the equalization process the decisions concerning $z(n)$ are often erroneous and this would drive the coefficient update equation to an ill–conditioned state. Combining the error signals in Equation (7.122) and (7.135), each one scaled by a certain weight, Benveniste and Goursat [7] formulated a new error signal as:

$$\epsilon^{BG}(n) = k_1 \cdot \epsilon^{DD}(n) + k_2 \cdot |\epsilon^{DD}(n)| \cdot \epsilon^{Sato}(n), \tag{7.136}$$

where k_1 and k_2 are the corresponding weighting factors. This error signal is zero, when equalization is perfect and, at the same time, it is not as error–prone as a purely blind decision–directed (DD) approach would be at start up, since then the influence of Sato's error term $\epsilon^{Sato}(n)$ in Equation (7.122) offers better error estimation. Using this combined error signal we readily arrive at the Benveniste–Goursat (B-G) algorithm [7], adjusting the equalizer taps according to:

$$\mathbf{c}^{(n+1)} = \mathbf{c}^{(n)} - \lambda \cdot \mathbf{y}^*(n) \cdot \epsilon^{BG}(n). \tag{7.137}$$

A good choice for k_1 and k_2 in Equation (7.136) would be to initialize the algorithm with a large k_2/k_1 ratio and decrease the ratio, when the equalizer is close to convergence in order to render the steady-state equalization more accurate. This philosophy is similar to the idea of switching to decision–directed mode, when the equalizer has converged.

The related $g\{z(n)\}$ function in this case can be found by comparing Equations (7.106) and (7.137), yielding [268]:

$$g(z) = z(n) - k_1 \cdot \epsilon^{DD}(n) + k_2 \cdot |\epsilon^{DD}(n)| \cdot \epsilon^{Sato}(n). \tag{7.138}$$

Having discussed the Benveniste–Goursat algorithm, we will now consider another DD–like algorithm in the next section, namely the *stop-and-go* algorithm by Picchi and Prati [8].

7.7.5.2 Stop-and-Go algorithm [8]

In the previous algorithms the equalizer coefficient update is inevitably occasionally wrong due to the stastistical nature of the algorithms. This leads to a reduced convergence speed and also to a degradation of the steady–state performance of the equalizer. In order to avoid this impediment to some degree, Picchi and Prati [8] suggested an algorithm, which decides whether a specific received symbol should contribute to the update process and updates the equalizer coefficients, only when it has decided that this would bring their values closer to their steady state ones. The algorithm used for updating the coefficients is the classic error feedback algorithm with the decision–directed error expressed as in Equation (7.135), which is repeated here for convenience:

$$\epsilon^{DD}(n) = z(n) - \hat{a}(n).$$

Two variables, namely $f_{n,R}$ and $f_{n,I}$ are introduced in [8], each of which defines a measure of the probability that the update of the real or imaginary part of the equalizer coefficients is correct. Naturally, the actual probability is unknown at the receiver, but it can be estimated using the philosophy of the Sato–type error of Equation (7.122), setting $f_{n,R}$ and $f_{n,I}$ to 1 and 0, depending on our confidence in the success of the update, as [8]:

$$f_{n,R} = \begin{cases} 1 & \text{if} \;\; sgn(Re[\epsilon^{DD}(n)]) = sgn(Re[\epsilon^{Sato}(n)]) \\ 0 & \text{if} \;\; sgn(Re[\epsilon^{DD}(n)] \neq sgn(Re[\epsilon^{Sato}(n)]) \end{cases} \tag{7.139}$$

and

$$f_{n,I} = \begin{cases} 1 & \text{if} \;\; sgn(Im\{\epsilon^{DD}(n)\}) = sgn(Im\{\epsilon^{Sato}(n)\}) \\ 0 & \text{if} \;\; sgn(Im\{\epsilon^{DD}(n)\}) \neq sgn(Im\{\epsilon^{Sato}(n)\}). \end{cases} \tag{7.140}$$

Practically this implies that the update of the real part of the equalizer coefficients only takes place when:

$$\left.\begin{array}{lrcl} & Re[z(n)] > \gamma & \text{and} & Re[\hat{a}(n)] < Re[z(n)] \\ \text{or} & 0 < Re[z(n)] < \gamma & \text{and} & Re[\hat{a}(n)] > Re[z(n)] \\ \text{or} & -\gamma < Re[z(n)] < 0 & \text{and} & Re[\hat{a}(n)] < Re[z(n)] \\ \text{or} & Re[z(n)] < -\gamma & \text{and} & Re[\hat{a}(n)] > Re[z(n)] \end{array}\right\}. \tag{7.141}$$

From this we can suggest that the "correct" values for $z(n)$ are those ones that are expected to bring $\hat{a}(n)$ closer to γ, which was richly illustrated in [8] in geometrical terms. This is a plausible, but certainly imperfect criterion. The probabilities of making a false update decision also are calculated in [8].

With these definitions we can form the algorithm using the classical error feedback algorithm of Equation (7.106), but involving both the real and imaginary parts of the error, enabled or disabled by $f_{n,R}$ and $f_{n,I}$, as follows:

$$\mathbf{c}^{(n+1)} = \mathbf{c}^{(n)} - \lambda \cdot \mathbf{y}^*(n) \cdot \left[f_{n,R} Re\{\epsilon^{DD}(n)\} + j f_{n,I} Im\{\epsilon^{DD}(n)\}\right]. \tag{7.142}$$

This algorithm is expected to have an advantage over the previous ones, since it uses Equations (7.140)-(7.141) to reject unreliable coefficient updates and to render convergence more steady and accurate. Nevertheless, the algorithm's convergence is hampered to a certain degree, since it does not use all the incoming symbols to update the equalizer coefficients.

A modification of this algorithm, suggested by Choi, Hwang and Song in [283], invokes a CMA-type error term, defined as [283]:

$$e^{CMA}(n) = z(n) \cdot (|z(n)|^2 - R_2), \tag{7.143}$$

instead of the Sato error of Equation (7.122), in order to form the decisions concerning the validity of the equalizer update at each symbol, where the symbols $z(n)$ and the constant R_2 were defined in Section 7.7.4.2. This error term corresponds to circular geometric regions in the equalized symbol domain, since it only considers the square of the magnitude of the equalizer's output $z(n)$, ignoring the phase.

Finally, again, we present the symbol estimation function $g\{z(n)\}$ corresponding to this algorithm, which is readily found by observing Equations (7.106) and (7.142):

$$g\{z(n)\} = z(n) - \left[f_{n,R} \cdot Re\{\epsilon^{DD}(n)\} + j f_{n,I} \cdot Im\{\epsilon^{DD}(n)\}\right]. \tag{7.144}$$

Having reviewed a range of Bussgang algorithms, let us now consider some of the associated convergence issues in the next section.

7.7.6 Convergence issues

In this section, the convergence properties of the Bussgang equalizers are discussed and the problems associated with them are explored. These issues attracted the attention of researchers as early as 1980, when Godard studied the convergence of the algorithm he proposed [5]. Later, in 1985, Foschini [217] gave a proof of the convergence of the CMA, when the length of the equalizer is doubly infinite, i.e. when both the number of the feedforward when it has and the feedback taps of the equalizer was infinite. It was not until ten years later that Ding *et al.* [218] proved that when the equalizer is not of infinite length, then there can be undesirable stable local minima, depending on the CIR. Ding *et al.* arrived at this conclusion by considering a special class of channels, namelay the so-called "autoregressive" channels, and by finding the local minima of a CMA equalizer for these channels. Their theory presented in [284] also revealed that the local minima of the CMA based Baud–rate spaced Bussgang equalizer correspond to local minima of all other Baud–rate spaced Bussgang equalizers for the same channel, arising from the fact that the equalizers do not have an infinite length. These minima are thus referred to as length–dependent minima [285]. Algorithm–dependent minima do not exist in the family of CMAs, but do exist in Sato's algorithm [286], in the context of the Benveniste-Goursat algorithm [7] and in conjunction with the Stop-and-Go algorithm [285]. The above two CMAs and also the Shalvi-Weinstein algorithms [280] exhibit only channel–dependent local minima. It has to be mentioned that a general solution for the convergence of the Bussgang equalizers is still an open research issue. The regions around undesirable local minima have also been studied by Ding *et al.* [287, 288], while initialization strategies have been proposed by Li and Ding in [289]. Moreover, it has been indicated that the convergence performances of the CMA and the Shalvi-Weinstein algorithms are similar to each other and also similar to the performance of the LMS (or Wiener) receiver [290–292]. Finally, dynamic convergence issues have been treated, for example, in [293–295]. Similar studies have recently been conducted also for fractionally–spaced equalizers [246, 274, 296, 297]. In this overview we give a basic analysis model for the convergence of Bussgang equalizers and interpret some well–established results.

We commence this analysis with the convergence analysis of the CMA. We consider a noiseless environment, which simplifies our discussion. We recall the error term of the CMA's equalizer tap update formula from Equation (7.127):

$$\epsilon^{\mathbf{CMA}}(n) = \mathbf{y}^*(n) \cdot z(n) \cdot (|z(n)|^2 - R_2). \tag{7.145}$$

This error term is basically the derivative of the CMA's cost-function in Equation (7.125) with respect to the equalizer tap vector. The points at which the mean value of this error term becomes zero define the local minima, maxima and saddle points of this algorithm. Therefore, in order to find the possible local minima, we have to evaluate the local minima of the following equation:

$$\mathbf{y}^*(n) \cdot z(n) \cdot (|z(n)|^2 - R_2) = \mathbf{0}. \tag{7.146}$$

By substituting the vectors from Equation (7.96) to (7.101) into Equation (7.146) and taking the expectation we have:

$$\begin{aligned} & E\left[\mathbf{y}^*(n) \cdot z(n) \cdot (|z(n)|^2 - R_2)\right] \\ &= E\left[\mathbf{H}^* \cdot \mathbf{a}^*(n) \cdot (\mathbf{a}^T(n) \cdot \mathbf{t}) \cdot (z^*(n) z(n) - R_2)\right] \\ &= E\left[\mathbf{H}^* \cdot (\mathbf{a}^*(n) \cdot \mathbf{a}^T(n)) \cdot \mathbf{t} \cdot \left((\mathbf{t}^H \mathbf{a}^*(n)) \cdot (\mathbf{a}^T(n)\mathbf{t}) - R_2\right)\right] \\ &= E\left[\mathbf{H}^* \cdot \left((\mathbf{a}^* \cdot \mathbf{a}^T)(\mathbf{t} \cdot \mathbf{t}^H)(\mathbf{a}^* \cdot \mathbf{a}^T) - R_2 \cdot (\mathbf{a}^* \cdot \mathbf{a}^T)\right) \cdot \mathbf{t}\right] \\ &= \mathbf{H}^* \cdot \mathbf{T} \cdot \mathbf{t} \end{aligned} \tag{7.147}$$

where

$$\begin{aligned} \mathbf{T} \ &= \ \mu_4 diag(|t_i|^2 - 1) + \\ & \quad \ \mu_2^2 \cdot (\mathbf{t}\mathbf{t}^H + \mathbf{t}^T \mathbf{t}^* \mathbf{I} - \\ & \quad -2 \cdot diag(|t_{-K_1}|^2, \cdots, |t_{K_2}|^2)) \end{aligned} \tag{7.148}$$

or

$$[\mathbf{T}]_{i,j} = \begin{cases} \mu_4 \cdot (|t_i|^2 - 1) + \mu_2^2 \sum_{k=-K_1, k \neq i}^{K_2} |t_k|^2 & i = j \\ \mu_2^2 t_i \cdot t_j^* & i \neq j \end{cases} \tag{7.149}$$

and $\mu_j = E[|a(n)|^j]$, while $K_1 = N_1 + L_1$, $K_2 = N_2 + L_2$. The candidate stationary points will satisfy the set of equations:

$$\mathbf{H}^* \cdot \mathbf{T} \cdot \mathbf{t} = \mathbf{0}. \tag{7.150}$$

The resulting equations may have two types of solutions. We can assume that the null–space of matrix $\mathbf{H}^*$ is trivial, i.e. we assume that:

$$\mathbf{H}^* \cdot \mathbf{x} = \mathbf{0} \Rightarrow \mathbf{x} = 0. \tag{7.151}$$

As long as the number of the taps is finite, the channel matrix $\mathbf{H}^*$ is an $(N_1 + N_2 + 1) \times (N_1 + N_2 + L_1 + L_2)$ dimensional matrix, which has less rows than columns. A system characterized by $\mathbf{H}^* \cdot \mathbf{x} = 0$ in conjunction with such a matrix $\mathbf{H}^*$ always has an infinite number of non–trivial solutions. If the equalizer has an infinite number of feedforward and feedback taps though, then the situation changes. We can see that the channel matrix $\mathbf{H}$ of Equation (7.101) now has an infinite number of rows and columns, thus being a square matrix having linearly independent rows, which implies that the system characterized by Equation (7.151) has only the trivial solution. This was shown differently by Ding *et al.* in [284]. For the moment, we will assume that the matrix has a trivial nullspace. In this case we can find the stationary equilibrium points of the algorithm by finding the solution of the following set of equations:

$$\mathbf{T} \cdot \mathbf{t} = \mathbf{0}. \tag{7.152}$$

The solution of these equations can be found to be any vector $\mathbf{t}$, which has some zero entries and some non–zero values, all exhibiting the same magnitude. Since the vector $\mathbf{t}$ is constituted by the convolution of the CIR with the equalizer's impulse response, ideally $\mathbf{t}$ would be a Dirac delta function. The vectors $\mathbf{t}$ having more than one non–zero components represent saddle points, yielding unstable equilibria, which do not affect the equalizer's performance, since they do not constitute convergence points. This can be shown by examining the second derivative of the tap update formulae in Equation (7.127) with respect to the equalizer tap vector $\mathbf{c}$. The second derivatives can be found in the same way as the first derivatives, with the exception that now the derivative is with respect to the conjugate of the equalizer tap vector $\mathbf{c}$, yielding:

$$\frac{\partial^2 J_{CMA}(n)}{\partial \mathbf{c} \partial \mathbf{c}^H} = \mathbf{H}^* \cdot \mathbf{T}' \cdot \mathbf{H}^T, \tag{7.153}$$

where

$$\begin{aligned} \mathbf{T}' &= \mu_4 diag(2|t_i|^2 - 1) + \\ &\quad 2\mu_2^2 \cdot (\mathbf{t}\mathbf{t}^H + \mathbf{t}^T \mathbf{t}^* \mathbf{I} - \\ &\quad -2 \cdot diag(|t_{-K_1}|^2, \cdots, |t_{K_2}|^2)) \end{aligned} \tag{7.154}$$

or

$$[\mathbf{T}^{'}]_{i,j} = \begin{cases} \mu_4 \cdot (2|t_i|^2 - 1) + 2\mu_2^2 \sum_{k=-K_1,k\neq i}^{K_2} |t_k|^2 & i = j \\ 2\mu_2^2 \cdot t_i \cdot t_j^* & i \neq j. \end{cases} \tag{7.155}$$

The positive definiteness of a matrix $\mathbf{A}$ can be verified by considering the term $\mathbf{x}^T \cdot \mathbf{A} \cdot \mathbf{x}^*$. In our case, this term becomes:

$$\begin{aligned} \mathbf{x}^T \cdot \mathbf{H}^* \cdot \mathbf{T}^{'} \cdot \mathbf{H}^T \cdot \mathbf{x}^* &= \\ (\mathbf{H}^H \cdot \mathbf{x})^T \cdot \mathbf{T}^{'} \cdot (\mathbf{H}^H \cdot \mathbf{x})^*. \end{aligned} \tag{7.156}$$

From this relationship we observe that if the matrix $\mathbf{T}^{'}$ is positive (negative) definite, then the matrix $\mathbf{H}^* \cdot \mathbf{T}^{'} \cdot \mathbf{H}^T$ is also positive (negative) definite. Therefore, it is sufficient to estimate the positive definiteness of $\mathbf{T}^{'}$. By examining the sub-determinants of $\mathbf{T}^{'}$, we can easily see that if only one entry of the vector $\mathbf{t}$ is non-zero, then all of the sub-determinants have the form μ_4^i and are positive, which implies that the matrix is positive definite and the associated error surface point is a local minimum. Finally, all the other solutions, for which more than one component of the vector $\mathbf{t}$ is non–zero, have positive and negative subdeterminants for the matrix $\mathbf{T}^{'}$, thus constituting saddle points.

The stationary points of the Modified CMA [6] of Section 7.7.5 have a similar form to the stationary points of the CMA [5] of Section 7.7.4.2, as explored by Wesolowsky in [6], since the cost-functions of these two algorithms are similar.

Above we have found the stationary and saddle points of the CMA under the assumption that the rows of the channel matrix are linearly dependent. In general, this assumtion is not true, unless the number of feedforward and feedback taps of the equalizer is infinite. In practice though, for a high number of taps this is approximately true. Ding *et el.* showed in [218, 284] that if the number of equalizer taps is not infinite, depending on the form of the CIR, undesirable equilibria will be present. The authors considered a specific channel, namely the autoregressive channel and found the undesirable equilibria associated with this channel. These are channel–dependent equilibria, which are likely to exist in a similar form for every other Bussgang equalizer. This can be seen, if we recall Equation (7.147). The multiplicative matrix $\mathbf{H}^*$ in Equation (7.147), which is responsible for the undesirable equilibria explored by Ding *et al.* [218], stems from the $\mathbf{y}^*$ factor multiplying the error term in Equation (7.106). This equation is common to all Bussgang equalizers, which implies that the term $\mathbf{H}^*$ will exist for any other algorithm of this type and therefore all these algorithms will exhibit these unstable equilibria. The stationary points of the error surface in Sato's algorithm were explored in [286], while the stationary points of the modified CMA are similar to those of the CMA, which were explored in [6]. A different type of analysis, suggested by Shynk and Chan in [298], assumed that the convolutional noise is normally distributed and provides similar results. Fractionally–spaced equalizers of this type have also been studied. It has been shown by Ding *et al.* [246, 297] that a sufficient condition for the existence of only one desirable global minimum is that the equalizer's length is at least equal to the channel's delay spread and, at the same time, there are no common zeros between the subchannels created by considering the z-transform of the oversampled channel. In other words this means that the associated subchannels are reasonably diverse. Endres *et al.* [274] have explored the scenario, where the first of these conditions was not met. In closing, we note that the extension of Bussgang equalizers to fast RLS estimation based schemes has been proposed by Douglas *et al.* in [299] as well as by Papadias and Slock in [300]. Having highlighted the most salient Bussgang algorithms, let us now consider joint CIR and data estimation techniques.

7.7.7 Joint channel and data estimation techniques

Joint channel and data detection techniques constitute the blind equivalent of sequence estimation algorithms. These techniques have originally been used for data estimation, when the CIR was known. Here, instead of equalizing the received signal, we estimate the CIR together with the ML data sequence, assuming that the CIR estimate is sufficiently accurate. Originally, Seshadri [221] observed that it may be risky to invoke the Viterbi algorithm [222] for the blind scenario, since by assuming perfect CIR knowledge, we would discard all the surviving paths but one. This would be a bad tactic at the initial stages of the algorithm, when the CIR estimation is poor. In fact, no path should be eliminated in favour of another in the blind detection scenario. This would correspond to a computationally intensive algorithm. Instead, Seshadri suggested that the so-called M–algorithm [301] should be used, retaining M number of surviving paths at each trellis state, rather than just one. He also suggested that the CIR estimation should be initialized and updated at any symbol interval using an LMS estimator. Each survivor path of the trellis should keep its own data plus CIR estimate, as illustrated in Figure 7.21. Here only the stylized trellis states and the associated M number of trellis transitions are indicated. For a deeper exposure to the M-algorithm the interested reader is referred to [301]. The LMS estimation was extended to RLS-based estimation by Xie *et al.* [302] for the case of multiuser communications, where the interference was due to K users, each having p distinct propagation paths of the dispersive channel. Raheli *et al.* [223] introduced the term *Per–Survivor Processing (PSP)* and showed the applicability of this technique to fast fading channels, owing to its fast convergence.

In this section, we give a brief overview of joint CIR and data estimation using the PSP technique. The definitions of the variables are the same as in Figure 7.15 and Section 7.7.4. Additionally, we define $\hat{\mathbf{h}} = [\hat{h}_{-L_1}, \cdots, \hat{h}_{L_2}]^T$ as the estimated CIR. Any transition from state μ_n to state μ_{n+1} in the trellis of Figure 7.21, describing the evolution of the state-machine modelling the channel, is due to the estimated transmitted symbol vector $\hat{\mathbf{a}}(\mu_n \rightarrow \mu_{n+1}) = \hat{\mathbf{a}}(n)$. The estimated received symbol $\hat{y}(n)$ for this transition at time n is then given by the convolution of the tentatively assumed transmitted symbol vector $\hat{\mathbf{a}}(n)$ and the CIR $\hat{\mathbf{h}}^{(n)}$ assumed, which is formulated as $\hat{\mathbf{a}}^T(n) \cdot \hat{\mathbf{h}}^{(n)}$. Each surviving path in the trellis of Figure 7.21 will have its own CIR and data estimate. The LMS CIR estimation algorithm for each surviving path will

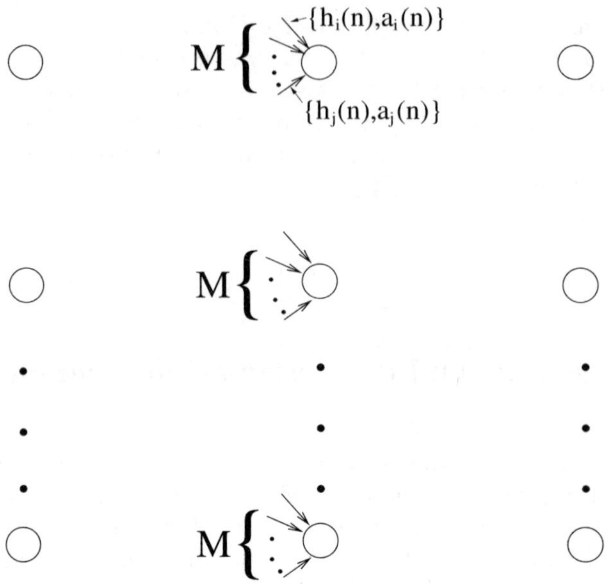

Figure 7.21: Per–survivor processing using the M–algorithm.

then be [221]:

$$\hat{\mathbf{h}}^{(n+1)} = \hat{\mathbf{h}}^{(n)} - \lambda \cdot \hat{\mathbf{a}}^*(n) \cdot \left(y(n) - \underbrace{\sum_{i=0}^{L} \hat{\mathbf{a}}^T(n) \cdot \hat{\mathbf{h}}^{(n)}(n)}_{\hat{y}(n)} \right), \tag{7.157}$$

which is similar in its philosophy to the tap-update formula of Equation (7.106), updating $\mathbf{c}^{(n)}$ using the step-size λ depending on the error term in the round brackets (). The LMS CIR and data estimator can be replaced by an RLS estimator according to the following equations [3, 223]:

$$\begin{aligned}
\mathbf{\Phi}(\mu_n) &= \sum_{i=0}^{n} w^{n-i}\hat{\mathbf{a}}^*(i)\hat{\mathbf{a}}^T(i) = \mathbf{P}^{-1}(\mu_n) \\
\mathbf{k}(\mu_{n+1}) &= \frac{\mathbf{P}(\mu_n)\hat{\mathbf{a}}^*(n)}{w + \hat{\mathbf{a}}^T(n)\mathbf{P}(\mu_n)\hat{\mathbf{a}}^*(n)} \\
\mathbf{P}(\mu_{n+1}) &= \frac{1}{w}\left(\mathbf{P}(\mu_n) - \mathbf{k}(\mu_{n+1})\hat{\mathbf{a}}^T(n)\mathbf{P}(\mu_n)\right) \\
\epsilon(\mu_n \to \mu_{n+1}) &= \sum_{i=0}^{n} w^{n-i}|y(i) - \hat{\mathbf{a}}^T(n)\hat{\mathbf{h}}^{(n)}(i)|^2 \\
\hat{\mathbf{h}}(\mu_{n+1}) &= \hat{\mathbf{h}}(\mu_n) + \mathbf{k}(\mu_{n+1})\epsilon(\mu_n \to \mu_{n+1}).
\end{aligned} \tag{7.158}$$

This CIR and data estimator provides fast adaptation and renders the estimator applicable to fast fading channels.

The blind data and CIR estimator presented in this section has some significant differences wrt the Bussgang blind equalizers of the previous section. Namely:

- In contrast to stand-alone equalization, there is no filter simulating the inverse of the CIR.
- The convergence is significantly faster.
- This technique can be combined with ML sequence estimation techniques and it is also amenable to amalgamation with ML trellis-based channnel decoding.

As for the other blind equalization techniques discussed, numerous studies have provided insight into this method. The associated practical considerations and a range of modifications were studied by Chugg and Polydoros in [224], while in [225] the associated performance was studied both theoretically and by simulations. In [226] and [227], Hidden Markov Model (HMM) theory was applied to the problem of joint data detection as well as CIR estimation and performance results were given for the so–called Gaussian Minimum Shift Keying (GMSK) modulation invoked in a Time Division Multiple Access (TDMA) scenario. In [228] the employment of fuzzy logic was proposed for the calculation of the associated decision metrics. In [229] and [230] the acquisition performance of a PSP detector was evaluated and "smart" initialization strategies were explored. In [231] the algorithm was adapted to make use of soft statistics and to include error prediction. Finally, in [232] a genetic algorithm was applied for estimating the CIR, and the Viterbi algorithm was then invoked.

Let us now consider the family of second–order statistics based techniques.

7.7.8 Blind equalization using second–order cyclostationary statistics

Second–order statistics-based blind channel identifiability

Using second–order statistics [303], in general, is significantly more efficient, than invoking higher–order statistics. This is because less samples are needed, in order to generate the required statistics, implying higher convergence speed and better stability. Nevertheless, second–order statistics are not applicable in all cases of interest. Specifically, in the case of blind equalization, the CIR is not identifiable from second–order statistics, when we sample at the Baud-rate, i.e. at one sample per symbol period. This is because the phase information of the CIR is not contained in the second–order statistics of the received signal sampled at the Baud-rate. This can be readily seen, if we consider the autocorrelation function of the received signal in Figure 7.15, namely:

$$R_{yy}(\tau) = E[y(t+\tau) \cdot y^*(t)] \tag{7.159}$$

and the autocorrelation of the transmitted signal sequence, namely:

$$R_{aa}(\tau) = E[a(t+\tau) \cdot a^*(t)], \tag{7.160}$$

as well as the channel transfer function $H(j\omega)$ in the frequency domain. These considerations assume that the input and the received signals are wide-sense stationary (WSS), implying that the signals' statistical characteristics are insensitive to the instant of capturing them [271]. This is typically true for the input QAM signal. For the received signal though, it is true only when the channel is static; when it is not, then often we can consider it as slowly varying, in which case the above WSS hypothesis can be regarded as met. For example, when TDMA is used, it is adequate to assume that the CIR remains unchanged within one TDMA frame period. It was verified in [304] or [303] that when both the sequences $\{a(n)\}$ and $\{y(n)\}$ are wide–sense stationary, then the Fourier transforms of these autocorrelation functions, namely $\Phi_{yy}(\omega)$ and $\Phi_{aa}(\omega)$, respectively, are related to each other with the aid of:

$$\Phi_{yy}(\omega) = |H(j\omega)|^2 \cdot \Phi_{aa}(\omega), \tag{7.161}$$

indicating that only the CIR's magnitude information can be extracted, when Baud–rate based sampling is employed. In 1991 Tong *et al.* [233] exploited the fact that by employing oversampling, the received signal becomes so-called *cyclostationary*, implying that the associated statistical properties become periodic in time [303]. Exploiting this cyclostationarity, Tong *et al.* proposed a method of CIR estimation, using second–order statistics. More recently, in 1995 Moulines *et al.* [234] observed that the same happens, when we are taking the input of multiple sensors – such as multiple antennas for example – at Baud–rate, i.e. without resorting to oversampling. Finally, in 1997, Tsatsanis and Giannakis [241], [235] proposed another substitute for oversampling at the receiver, namely inducing cyclostationarity at the transmitter, instead of oversampling at the receiver. All these ideas lead to a range of similar methods applicable to CIR estimation, which are based on second–order statistics.

Second–order statistics–based blind CIR estimation algorithms

Let us invoke oversampling and denote the symbol interval by T, while the sampling interval by T_S, so that $T = T_S \cdot \Delta$, where Δ is the oversampling factor. We also assume in the rest of this chapter that the channel's memory extends from $-L_1T$ to L_2T. The channel memory limits do not have to be multiples of the symbol interval T, it is simply notationally convenient. We observe the system over the time interval $[nT, (n+1) \cdot T]$, using the linear model of:

$$\mathbf{y}(n) = \mathbf{H}(n) \cdot \mathbf{a}(n) + \mathbf{e}(n), \tag{7.162}$$

where

$$\mathbf{y}(n) = [y(nT), y(nT+T_S), \ldots, y(nT+(\Delta-1)T_S)]^T \tag{7.163}$$

represents the received signal vector at time instant n,

$$\mathbf{H}(n) = \begin{bmatrix} h_{-L_1T}(n) & \cdots & h_{L_2T}(n) \\ h_{-L_1T+T_S}(n) & \cdots & h_{L_2T+T_S}(n) \\ \vdots & \vdots & \vdots \\ h_{-L_1T+(\Delta-1)T_S}(n) & \cdots & h_{L_2T+(\Delta-1)T_S}(n) \end{bmatrix} \tag{7.164}$$

is constituted by the channel's impulse response taps $\{h_i(n)\}$ for $i = -L_1T, \cdots, L_2T + (\Delta - 1)T_S$ at time nT,

$$\mathbf{a}(n) = [a((n + L_1)T), \ldots, a((n - L_2)T)]^T \tag{7.165}$$

is the transmitted data vector and

$$\mathbf{e}(n) = [e(nT), e(nT + T_S), \ldots, e(nT + (\Delta - 1)T_S)]^T \tag{7.166}$$

is the channel noise vector.

We now point out the equivalence of oversampling the received signal with rendering the transmitted signal cyclostationary and also its equivalence to involving the outputs of Δ number of 'sensors' or antennae. The former [235] involves transmitting the same output Δ times sequentially, instead of just once. This way the transmitted signal becomes cyclostationary and the same happens to the corresponding received signal sampled at the Baud-rate, when the channel is stationary for the duration of processing. If we consider the transmitter's Baud–rate to be Δ-times lower than the receiver's Baud-rate, then the received signal would be the same as in the previous scenario.

Upon using the outputs of multiple sensors or antennae [234], we can observe that the output of each sensor – say for example that of the i–th sensor – is equivalent to the output of oversampling at time instants of $iT_S, i = 0, \cdots, \Delta - 1$ in the $[0, (\Delta - 1)T_S]$ space.

Having shown the above equivalences, we will now proceed to discussing the formation of second–order statistics based algorithms. Let us consider the autocorrelation matrix of the received signal $y(nT + iT_S)$ over the time interval iT at time instant nT:

$$\begin{aligned}
\mathbf{R_{yy}}(i;n) &= E[\mathbf{y}(n) \cdot \mathbf{y}^H(n-i)] \\
&\overset{(7.162)}{=} E[(\mathbf{H}(n) \cdot \mathbf{a}(n) + \mathbf{e}(n)) \cdot \\
&\quad (\mathbf{a}^H(n-i) \cdot \mathbf{H}^H(n-i) + \mathbf{e}^H(n-i))] \\
&= E[\mathbf{H}(n) \left(\mathbf{a}(n)\mathbf{a}^H(n-i)\right) \mathbf{H}^H(n-i) + \\
&\quad + \mathbf{H}(n)\mathbf{a}(n)\mathbf{e}^H(n-i) + \\
&\quad + \mathbf{e}(n)\mathbf{a}^H(n-i)\mathbf{H}^H(n-i) + \\
&\quad + \mathbf{e}(n)\mathbf{n}^H(n-i)],
\end{aligned} \tag{7.167}$$

where $\mathbf{y}(n)$ was defined in Equation (7.163). The two terms in the middle of this expression have a zero mean value, since they are the product of independent variables. Thus, Equation (7.167) becomes:

$$\mathbf{R_{yy}}(i;n) = \mathbf{H}(n) \cdot \mathbf{R_{aa}}(i;n) \cdot \mathbf{H}^H(n-i) + \mathbf{R_{ee}}(i;n). \tag{7.168}$$

The CIR estimation process consists of estimating the CIR matrix $\mathbf{H}$ using measurements of the autocorrelation matrix $\mathbf{R_{yy}}(n)$ according to Equations (7.167) and (7.168), assuming that the autocorrelation of the transmitted signal $\mathbf{R_{aa}}(i;k)$ is known. Tong *et al.* [271] invoked the assumption that the input data source produces completely uncorrelated symbols $\mathbf{a}(n)$, while Hua *et al.* [245] adapted the method to work with correlated input signals. Furthermore, while in Tong's algorithm [271] the noise was assumed to have some known properties, in [236] a variation was proposed, which assumed no knowledge concerning the channel noise. In order to estimate the CIR matrix $\mathbf{H}$ in Equation (7.164), we must also have knowledge of the noise's correlation. The noise samples cannot be assumed to be uncorrelated (white), since they are generated at a higher rate, than the Baud-rate and since the receiver's lowpass filtering imposed autocorrelation upon it. This autocorrelation function has the same shape as the impulse response of the receiver filter, since the noise was white before this filtering and hence its autocorrelation function was a $\delta()$ function.

There are various ways of performing CIR estimation at this stage. Some algorithms, such as that proposed by Tong *et al.*, [233,271], estimate the CIR by assuming that the noise is white – not a valid assumption in general – and the transmitted sequence is also white. In our forthcoming deliberations the channel is assumed to be static. The CIR matrix $\mathbf{H}$ of Equation (7.164) is then estimated by performing the so-called "singular value decomposition (SVD)" [305] of $\mathbf{R_{yy}}(0)$ and $\mathbf{R_{yy}}(1)$ and then by performing a range of further algebraic calculations. In other algorithms [235] iterative procedures have been proposed, exhibiting lower complexity, but having an asymptotically similar performance. After estimating the CIR matrix $\mathbf{H}$ of Equation (7.164) and ignoring the effect of noise in Equation (7.162), we can invert $\mathbf{H}$ in order to extract the original information symbols $\mathbf{a}(n)$ from the received sequence $\mathbf{y}(n)$, using the following LMS estimator:

$$\mathbf{c}_{LMS} = \mathbf{H}^t \cdot \mathbf{y}(n), \tag{7.169}$$

where $\mathbf{H}^t$ satisfies:

$$\mathbf{H}^t \cdot \mathbf{H} = \mathbf{I} \tag{7.170}$$

and it is the "pseudo–inverse" [290] of the CIR matrix. The input signals have been assumed to be i.i.d. variables with a normalized power of unity, while the noise has been neglected.

An important advantage of the second–order statistics based algorithms – compared to the family of the linear equalizers – is that they are asympotically accurate. Simulations presented in [271] demonstrate this accuracy in performance terms and also show that the second–order statistics based algorithms exhibit fast convergence, typically within 100 symbols. This concludes our discussions on the second–order statistics based algorithms. For more detailed discussions on this issue the reader is referred to [233, 271] and also to [236–238]. In [239] a different method of estimating the CIR matrix $\mathbf{H}$ of Equation (7.164) is proposed for the "multiple channel" scenario, using so-called "outer–product matrix decomposition". In [240, 241], the transmitter–induced cyclostationarity algorithms are explored, while in [242] the cyclostationarity–based method is applied to an Orthogonal Frequency Modulation (OFDM) receiver. In [243], a modification of Tong's original method [271] is proposed, while in [244] a general study of the cyclostationary method is given. The effects of non-i.i.d. signal distributions are studied in [245] and in [246]. In [247] the second–order statistics based methods are adapted to the so-called source separation problem, where for example the wanted and interfering signals are separated. In [248] the subspace method is applied to the suppression of both inter-symbol interference and multiple-access interference. In [249] the case of unknown noise distributions is considered. Finally, in [250], the so–called "*Column–Anchored Zeroforcing Equalization (CAZE)*" is proposed and studied. Having reviewed the family of second–order statistics based techniques, let us now consider a range of so-called "polycepstra-oriented" algorithms in the next section.

7.7.9 Blind channel estimation and equalization using polycepstra

The last class of blind equalization algorithms discussed here involves fourth–order – rather than second–order – statistics of the received signal, in order to estimate the inverse of the CIR and to equalize its effects. The algorithms belonging to this class were initially proposed by Hatzinakos and Nikias in [252]. This algorithm, referred to as the *Tricepstrum Equalization Algorithm (TEA)*, employs the complex cepstrum of the so–called fourth–order cumulants (the tricepstrum is defined in Appendix 7.10) of the received signal sequence sampled at the Baud-rate. These algorithms are capable of identifying both so-called "minimum and maximum–phase" channels, which we will characterize more explicitly during our further discourse. Below we briefly introduce the TEA using the fourth–order cumulants and tricepstra as defined in Appendix 7.10.

In this context we aim to equalize a channel having a z–domain channel transfer function of $H(z)$. We consider $H(z)$ as the product of a minimum–phase [3] section – where the z–domain transfer function of the channel has all its zeros inside the unit circle – and a maximum–phase section – where the z–domain transfer function of the channel has all its zeros outside the unit circle. This is formulated as [252]:

$$H(z) = A \cdot I(z^{-1}) \cdot O(z), \tag{7.171}$$

where A is a constant and $I(z^{-1})$ and $O(z)$ are polynomials of the form $\Pi_{i=1}^{M}(1-a_i z^{-1})$ and $\Pi_{j=1}^{K}(1-a_j z)$ respectively, with $|a_i| < 1$, M is the number of channel zeros inside the unit circle and K is the number of channel zeros outside the unit circle. A simple manifestation of a minimum-phase channel exhibits a CIR, where the main tap associated with the time instant 0 is the largest one. The polynomial $I(z^{-1})$ is a minimum-phase polynomial, while $O(z)$ is a maximum-phase polynomial. In order to equalize this channel under the so-called Zero-Forcing (ZF) constraint – implying that the combined CIR and equalizer impulse response is forced to zero at sampling instants $n \neq 0$ – we use an equalizer, having a transfer function of $C(z)$, which is the inverse of that of the channel [252]:

$$C(z) = \frac{1}{H(z)} = \frac{1}{A \cdot I(z^{-1}) \cdot O(z)}. \tag{7.172}$$

Then the cascaded channel and equalizer transfer function constitutes an ideal channel. If instead of the ZF equalizer we use a decision-feedback equalizer (DFE) having a feedforward transfer function of $V(z)$ and a feedback transfer function of $G(z)$, then

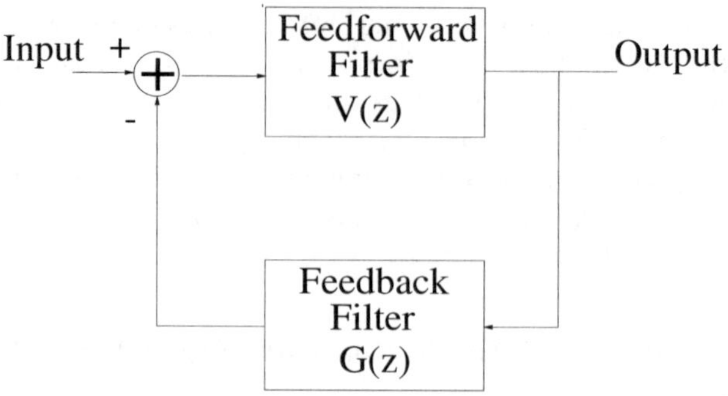

Figure 7.22: The DFE filter structure.

the equalizer's transfer function will have the form [252]:

$$C(z) = \frac{V(z)}{1 + G(z)}, \tag{7.173}$$

where $V(z)$ and $G(z)$ are the feedforward and feedback section's transfer function, respectively, as shown in Figure 7.22. The feedback filter must be realizable. A possible choice for $V(z)$ and $G(z)$ is [252]:

$$V(z) = \frac{O^*(z^{-1})}{AO(z)} \tag{7.174}$$

$$1 + G(z) = I(z^{-1}) \cdot O^*(z^{-1}). \tag{7.175}$$

Then the system $C(z) \cdot H(z)$ is perfectly equalized and the feedforward section $V(z)$ of the DFE is an all–pass filter having zeros inside and poles outside the unit circle, while the feedback filter $G(z)$ is a minimum–phase filter. In order to construct the DFE we have to find an estimate of the coefficients of the feedforward and feedback filters. Equivalently, we have to find an estimate of the poles and zeros of the two equalizer filters. This can be achieved using the fourth–order cumulants of the received signal.

We form the received signal $y(n)$ as the convolution of the transmitted signal $a(n)$ with the CIR $\{h_n\}$, plus a zero–mean additive Gaussian stochastic process $e(n)$, as follows:

$$y(n) = a(n) * h_n + e(n). \tag{7.176}$$

We recall from Appendix 7.10 that the tricepstrum $c_y(m, n, l)$ of the received signal $y(n)$ is related to the tricepstrum $c_h(m, n, l)$ of the CIR by [252]:

$$c_y(m, n, l) = c_h(m, n, l), \qquad (m, n, l) \neq (0, 0, 0). \tag{7.177}$$

This implies that in the tricepstrum domain the received signal is equal to the CIR. Therefore, estimating the received signal's tricepstrum directly gives us the CIR's tricepstrum. The problem then becomes that of estimating the CIR in terms of its tricepstrum representation. It was shown in [252] that it is sufficient to consider tricepstra in the form of $c_y(K,0,0)$ (K integer). We recall from Appendix 7.10 the following relationship:

$$\sum_{I=1}^{p} A^{(I)}[L_y(m-I,n,l) - L_y(m+I,n+I,l+I)] +$$
$$+\sum_{J=1}^{q} B^{(J)}[L_y(m-J,n-J,l-J) - L_y(m+J,n,l)] =$$
$$= -mL_y(m,n,l)$$

where

$$K \cdot c_y(K,0,0) = \begin{cases} -A^{(K)}, & K = 1,\ldots,p \\ B^{(-K)}, & K = -1,\ldots,-q. \end{cases}$$

This relationship is in fact a system of linear equations, which can be solved by iterative methods, as shown in Appendix 7.10. However, the estimation of the CIR and the equalizer filters is quite an elaborate task and hence the associated derivation was relegated to Appendix 7.10, following the approach of [252].

An estimate of these algorithms' complexity compared to the Bussgang algorithms' complexity was also given by Hatzinakos and Nikias in [252], where the complexity of the former appears to be significantly higher, a fact which was mentioned before. As a trade–off, the simulations presented in [252] indicate the superiority of the TEAs in terms of convergence speed, as well as in terms of their ability to equalize non–minimum phase channels, which are often encountered in fading mobile channels. For more detailed investigations of these techniques the reader is referred to [252, 281, 306–309].

Having presented an overview of a range of basic blind equalizer structures, we will now provide a summary of their complexity.

7.7.10 Complexity evaluation

In this section the complexity of the blind equalizers presented is evaluated. We commence our discussions by considering the complexity of the Bussgang algorithms. The complexity of the Bussgang techniques is similar for all of them and it is relatively low. For simplicity, in this section we will assume that the number of equalizer taps is $2N+1$, i.e. $N_1 = N_2 = N$, according to the notation we have used so far. The complexity depends only on the number of equalizer taps, $2N+1$, and it is on the order of N, which is indicated as $O(N)$. An estimate of the number of real additions and multiplications required for each algorithm per equalized symbol interval is presented in Table 7.1. Here, we have assumed that the complex variables are represented in the memory of the associated arithmetic unit in terms of their real and imaginary parts. We have also assumed that the number of equalizer taps $2N+1$ is approximately equal to the channel memory $2L+1$, i.e. $2N+1 \approx 2L+1$ and that the square root evaluation required for the computation of $|\epsilon^{DD}(n)|$ in the Benveniste–Goursat algorithm of Section 7.7.5.1 is performed with the aid of 4 real additions and 2 real multiplications. Finally, in Table 7.1 we have also included an estimate of the memory requirements of each algorithm. In addition to the Bussgang algorithms the table

Algorithm	Additions	Multiplications	Memory
CMA [5]	$16L + 10$	$16L + 13$	$4L + 4$
Sato [4]	$16L + 8$	$16L + 10$	$4L + 4$
B–G [7]	$16L + 17$	$16L + 19$	$4L + 4$
Modified–CMA [6]	$16L + 8$	$16L + 14$	$4L + 4$
Stop-and-Go [8]	$16L + 8$	$20L + 10$	$4L + 4$
Super–Exponential ($p = 2, q = 1$) [280]	$O(L^3)$	$O(L^3)$	$O(L)$

Table 7.1: Complexity estimate of the Bussgang techniques of Section 7.7.4, assuming $2N + 1 \approx 2L + 1$, where L is the channel's memory

	Additions	Mult/ions	Memory
$M \neq 1$	$MQ(8L^2 + 24L + 6Q) + 2Q$	$MQ(16L^2 + 30L + 18) + 4Q$	$4MQ(L + 1)$
$M = 1$	$Q(8L^2 + 24L + 6Q) + 2Q$	$Q(16L^2 + 30L + 18) + 4Q$	$4Q(L + 1)$

Table 7.2: Complexity estimate of the Sequence Estimation algorithms of Section 7.7.7

incorporated the super-exponential algorithm, which does not fall into the Bussgang category, but exhibits similarities with the CMA. A complexity estimate of the PSP-based algorithms of Section 7.7.7 is obtained similarly to the previously introduced Bussgang algorithms of Section 7.7.4 by calculating the total number of additions and multiplications as well as the associated memory requirements. Considering that the convolution of the CIR with the estimated sequence in Equation (7.157) is only calculated once for each survivor transition and stored in memory (thus saving unnecessary calculations), the associated complexity results are summarized in Table 7.2. These complexity figures refer to one symbol interval. The complexity arising from convolutional decoding, if channel coding is used, is ignored in these calculations. In Table 7.2 M is the number of survivors in Figure 7.21 that we retain at each step and Q is the number of possible signal constellation points, i.e. $Q = 2^K$ for K-bit per symbol modulation. By comparing Tables 7.1 and 7.2 we observe that the complexity of the sequence estimation techniques (aided by LMS-based CIR estimation) is an order of magnitude higher, than that of the the Bussgang techniques with respect to the equalizer length, which is $O(L)$. However, the associated complexity also depends on the number of phasors in the modulation constellation as well as on the number

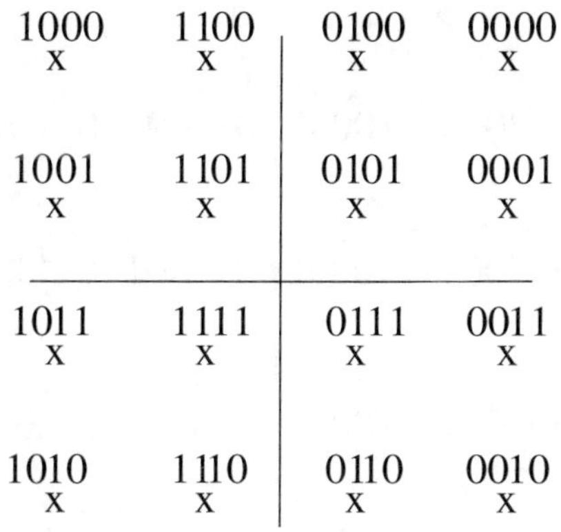

Figure 7.23: 16-QAM constellation

M of survivors per state.

From these tables we can also observe that the Bussgang algorithms are attractive for long CIRs, since in this case the sequence estimation techniques exhibit excessively high computational complexity. By contrast, for short CIRs the sequence estimation techniques offer significant advantages at an affordable complexity. However, as it will be seen in Section 7.7.11, channels exhibiting long CIRs cannot be equalized using Bussgang equalizers due to the enhancement of convolutional noise associated with the large number of equalizer taps required for equalization.

Having presented a rudimentary overview of various blind equalization methods, we note the emergence of a recent approach, based on Neural Networks (NN). For more details concerning NN-based blind methods the reader is referred to [310–315].

7.7.11 Performance results

In this section we present the associated comparative performance results for the algorithms described in Sections 7.7.4 and 7.7.7. Two different types of results are presented, commencing with Bit Error Rate (BER) learning curves, which offer a measure of the algorithms' convergence speed. The second set of results is concerned with the average BER curves of the algorithms over a dispersive Gaussian channel. The modulation schemes involved are 16-QAM and 64-QAM [266,267]. The associated signal constellations are shown in Figures 7.23 and 7.24.

7.7.11.1 Channel models

Three different channels were used in our comparative study. The first one is a typical worst–case Wireless Asynchronous Transfer Mode (WATM) channel, while the second one is a Shortened WATM (SWATM) CIR, both of which are presented in Figure 7.25. These indoor CIRs were generated with the aid of finding the line-of-sight path and the four longest-delay paths in a 100x100x3m^3 hall at a WATM transmission rate of 155 Mbit/s. The third channel used in our comparative study is a simple one–

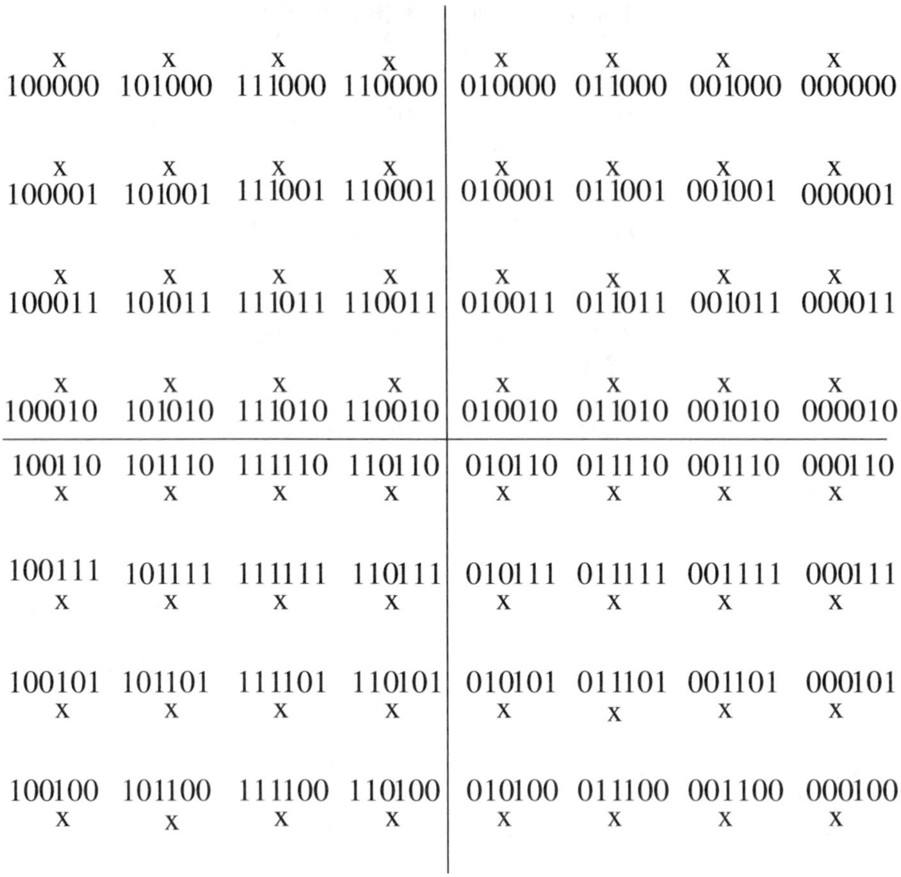

Figure 7.24: 64-QAM constellation

symbol–delay channel, which is shown in Figure 7.26. This channel may characterize a satellite link, where due to the directional parabolic antenna used, only one or two multipath components may arrive at the receiver. The difference between the WATM, SWATM and the one–symbol–delay channel is that while the former two channels carry multipath components at several symbols' delay, the latter exhibites only one additional multipath component at a delay of one symbol. These different CIRs will be used to demonstrate the fact that different equalizers may be appropriate for different channels.

7.7.11.2 Learning Curves

In this section, the associated BER learning curves of the blind equalizers of Sections 7.7.4 and 7.7.7 are presented for the CIRs of Figures 7.25 and 7.26. The step–size parameter λ is common for all the Bussgang algorithms and equal to $5 \cdot 10^{-4}$, while

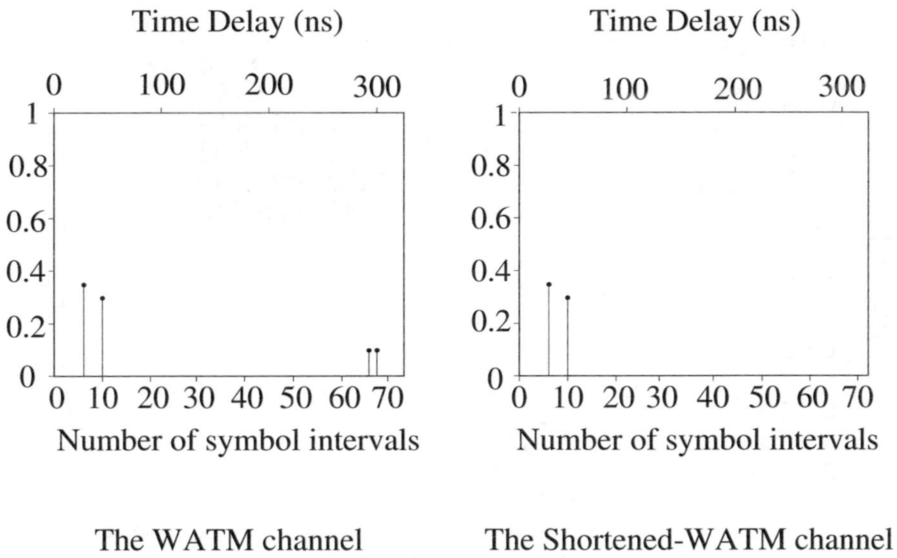

Figure 7.25: (a) The WATM and (b) Shortened WATM channels

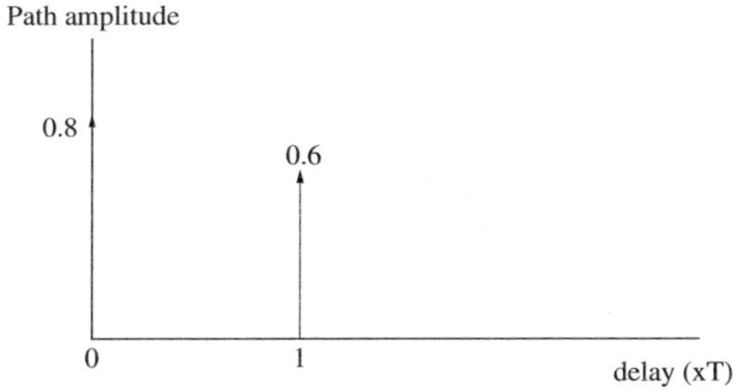

Figure 7.26: The one–symbol–delay channel used in the simulations

the Benveniste–Goursat parameters of Equation (7.137) are $k_1 = 1$, $k_2 = 5$. For the PSP-based sequence estimation algorithm, the step–size in Equation (7.157) is 10^{-2} and only $M = 1$ survivor was retained, owing to the fact that the CIR was time-invariant. The WATM channel simulations associated with the CIR of Figure 7.25(a) are presented in Figure 7.27 for 16-QAM and in Figure 7.30 for 64-QAM. An equalizer length of 68 taps was employed and the Signal–to–Noise ratio (SNR) was $30dB$. The corresponding SWATM and one-symbol-delay CIR based results are presented in Figures 7.28 and 7.29 as well as 7.31 and 7.32 for 16-QAM as well as 64-QAM, respectively. As we see, the CIR spread was gradually shortened from 68 to 10 and then to 2. Observe that for the M-algorithm of Figure 7.21 we only presented BER results in Figures 7.29 and 7.32 over the shortest one-symbol-delay CIR. In order to quantify the BER associated with the learning curves, we have used the values of

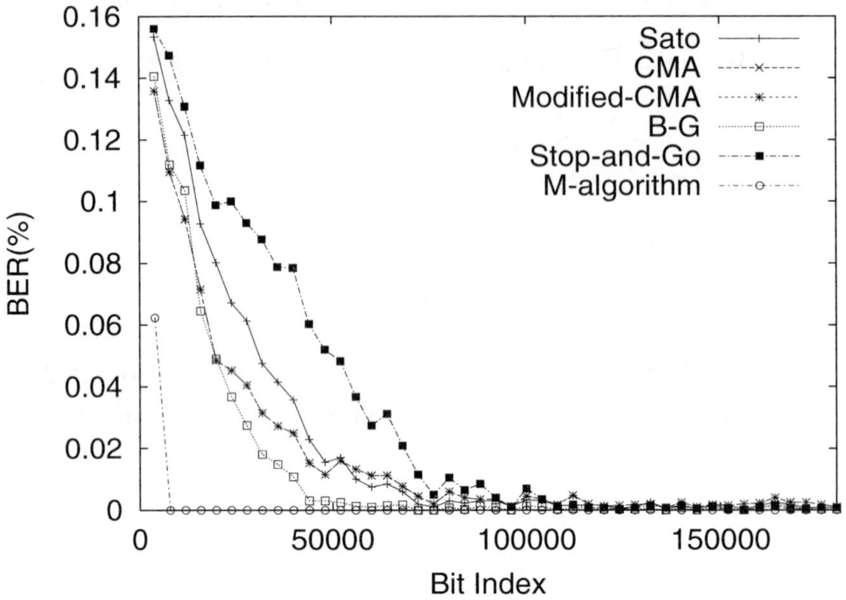

Figure 7.27: BER learning curves over the WATM channel of Figure 7.25(a) for 16-QAM at SNR=30dB using a 68–tap equalizer obeying the schematic of Figure 7.19

BER in variable–length intervals and we have performed smoothing of the curves, which was necessary due to the statistical nature of the blind equalizers of this type; not all of the incoming data symbols drive the equalizer to the point of convergence.

From these curves we can infer some observations concerning the convergence speed of each of the tested algorithms.

- The M–Algorithm converges at a higher speed, than any of the Bussgang algorithms considered. A reason for this is the employment of a larger step–size value, which could not have been used for the Bussgang algorithms, since this would result in poor tracking performance.
- Sato's algorithm converges at a medium to slow speed.
- Godard's algorithm converges with about the same speed as Sato's and its convergence is faster for higher-order QAM.
- The MCMA algorithm converges at a medium speed, but its convergence is faster for higher-order QAM, as we can see from Figures 7.30 and 7.31 for 64-QAM.
- The Benveniste–Goursat algorithm converges rapidly only for low-order QAM, and we can see from Figures 7.27 and 7.28 that for 16-QAM it converges faster, than any of the other Bussgang algorithms. For higher-order constellations, such as 64-QAM, it converges significantly slower.

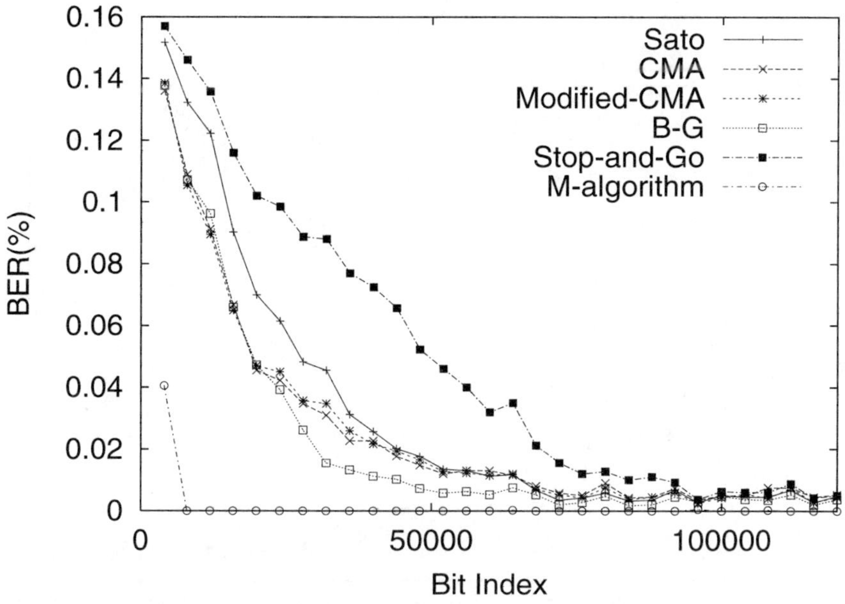

Figure 7.28: BER learning curves over the Shortened WATM channel of Figure 7.25(b) for 16-QAM at SNR=20dB using a 30–tap equalizer obeying the schematic of Figure 7.19

	WATM	S-WATM	TC
M–Algorithm	550	550	3700
Sato	30000	25000	80000
CMA	35000	40000	55000
MCMA	35000	20000	60000
Benveniste–Goursat	20000	15000	-
Stop–and–Go	75000	25000	-

Table 7.3: The number of symbols required for each algorithm to converge for 16-QAM over the WATM and Shortened WATM channels of Figures 7.25(a) and (b) as well as over the simple one–symbol–delay channel of Figure 7.26, which is labelled as TC (Test Channel). The SNR is $30dB$.

- The Stop–and–Go algorithm is definitely the slowest algorithm in all cases, since it is not always enabled to iterate.

In Tables 7.3 and 7.4 an estimate of the number of symbols needed for the convergence of each algorithm is given. The converged state was defined as the state, where the BER has 'just' reached its steady state value and does not change significantly thereafter. The channel SNR is kept at $30dB$. Having presented a comparative

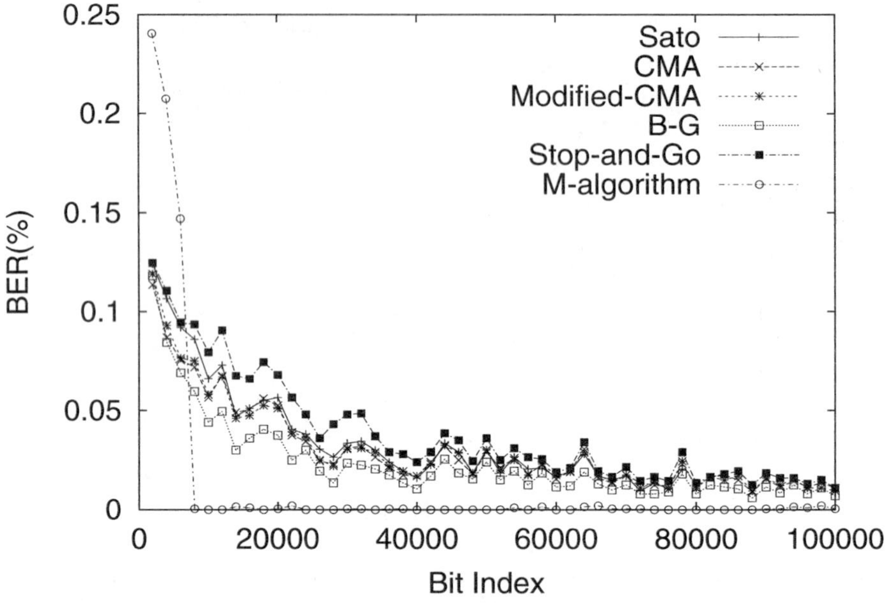

Figure 7.29: BER learning curves over the one–symbol–delay channel of Figure 7.26 for 16-QAM at SNR=20dB using a 10–tap equalizer obeying the schematic of Figure 7.19

	WATM	S-WATM	TC
M–Algorithm	4150	3000	6000
Sato	50000	50000	120000
CMA	55000	35000	110000
MCMA	45000	45000	115000
Benveniste–Goursat	75000	55000	150000
Stop–and–Go	320000	200000	600000

Table 7.4: The number of symbols required for each algorithm to converge for 64-QAM over the WATM and Shortened WATM channels of Figures 7.25(a) and (b) as well as over the simple one–symbol–delay channel of Figure 7.26, which is labelled as TC (Test Channel). The SNR is $30dB$.

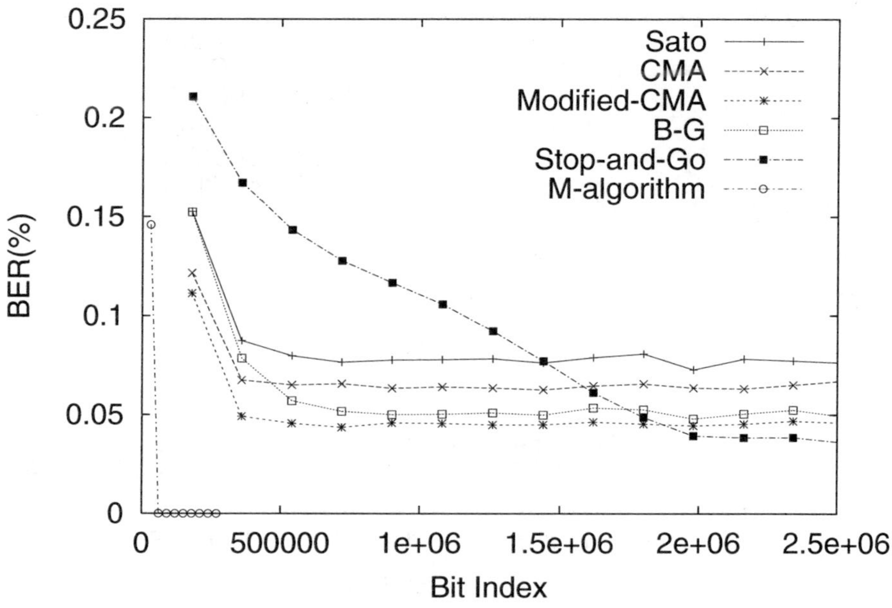

Figure 7.30: BER learning curves over the WATM channel of Figure 7.25(a) for 64-QAM at SNR=30dB using a 68–tap equalizer obeying the schematic of Figure 7.19

simulation study of the convergence speed of the blind equalizers of Sections 7.7.4 and 7.7.7, we will now give an illustration of the blind equalizers' convergence by means of the associated phasor diagrams [266, 267].

7.7.11.3 Phasor diagrams

In this section some phasor diagrams are presented for the various algorithms considered. We are observing the phasor diagram at the equalizer's output at different instants, each giving an idea of how the equalizer is converging. In Figures 7.33 and 7.34, the phasor constellation is plotted in the complex plane for a CMA-based equalizer having 10 taps, equalizing the one–symbol–delay channel of Figure 7.26, when 16-QAM is used. Two snapshots are shown. In the first the equalizer is reaching convergence, but still exhibits residual ISI. In the second snapshot, the equalizer has almost converged. These snapshots demonstrate how the equalizer is adapting and slowly approaching the state of convergence. When it has converged, the residual impairments are the convolutional noise, which is rather small, and the additive channel noise, which has also been chosen to be low, so that the convergence can be better observed. At this state, the equalized signal is confined to small regions around the legitimate QAM constellation points and the diameter of these regions depends basically on the SNR and on the residual ISI, which produces convolutional noise. Finally, in Figures 7.35 and 7.36 the phasor diagram is shown for the case of 64-QAM. Having presented our comparative results for the convergence speed of the

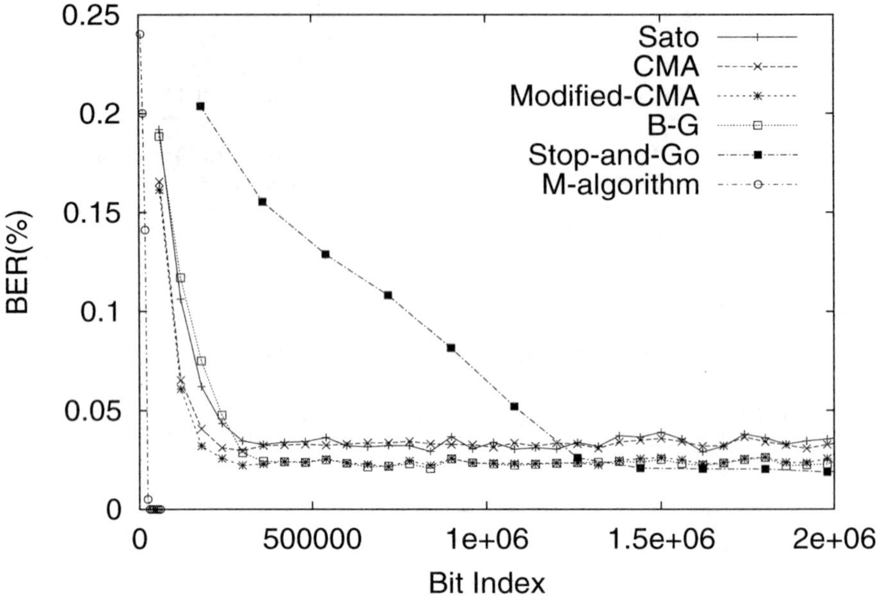

Figure 7.31: BER learning curves over the Shortened WATM channel of Figure 7.25(b) for 64-QAM at SNR=20dB using a 30–tap equalizer obeying the schematic of Figure 7.19

blind equalizers, we will now investigate the accuracy of convergence.

7.7.11.4 Gaussian channel

In this section the steady–state average BER curves are presented as a function of the bit-SNR over the WATM and Shortened WATM channels of Figure 7.25, as well as over the one-symbol delay channel of Figure 7.26 using 16-QAM and 64-QAM. The bit-SNR is defined as the SNR per bit, i.e. the signal power per bit over the noise power per bit. In mathematical terms we have:

$$\text{Bit-SNR} = \frac{SNR}{\text{Number of bits per symbol}}. \tag{7.178}$$

The equalizer characteristics are the same as in Section 7.7.11.2. For the M–algorithm, the channel estimator takes into consideration only the CIR taps that indeed exist, thus reducing the number of calculations. The associated curves are presented in Figures 7.37 to 7.42.

Note here that following the classical initialization strategy – that is initializing the equalizer tap vector to $\mathbf{c}^o = (1.2, 0, \cdots, 0)^T$ as suggested by Godard [5] – is inadequate for the Stop-and-Go equalizer to converge to the desired equilibrium. Instead, the equalizer converges to an undesirable equilibrium associated with the combined CIR

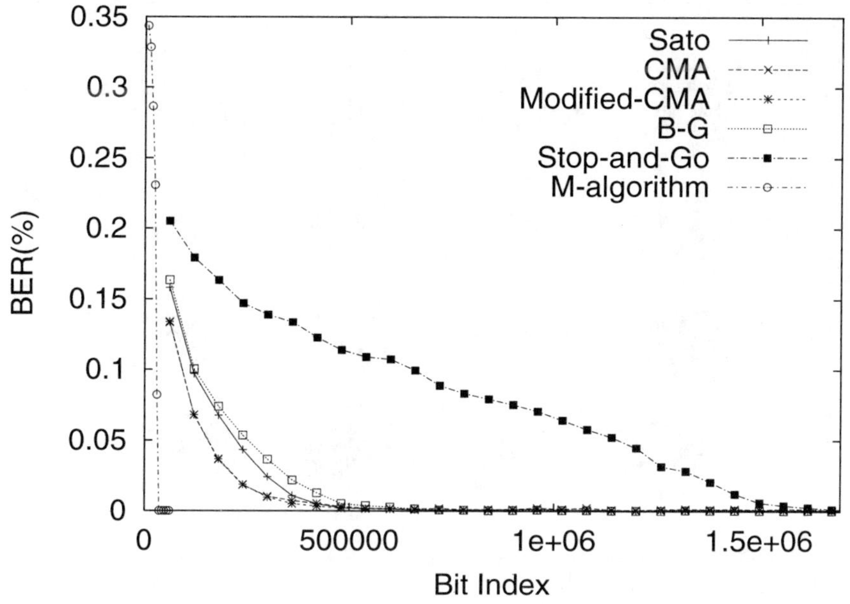

Figure 7.32: BER learning curves over the one–symbol–delay channel of Figure 7.26 for 64-QAM at SNR=30dB using a 10–tap equalizer obeying the schematic of Figure 7.19

plus equalizer tap vector of $\mathbf{t}^1 = (0.74, 0.48, 0, \cdots, 0)^T$. For the equalizer to converge to the correct equilibrium, the initialization has to be closer to the desired equilibrium. For example, initializing the equalizer with the tap vector $(1.2, -0.9, 0.6, 0, \cdots, 0)^T$ is adequate. This phenomenon is directly related to the nature of this algorithm and the undesirable equilibrium is clearly an algorithm-dependent equilibrium. This is why this phenomenon is not observed in the other equalizers. From these curves we can infer some observations concerning the accuracy of convergence for each of the tested algorithms, which exhibits itself in terms of the residual BER at a given SNR after reaching the steady-state. We note furthermore that at the SNR concerned, namely at $30dB$, the converged–state–accuracy of the various techniques becomes explicit also from Figures 7.27 to 7.31.

- The M–algorithm provides the best performance.
- Sato's algorithm converges with a medium accuracy.
- Godard's algorithm converges with a medium convergence accuracy, which is improved for low-order QAM.
- The Modified–CMA algorithm offers a good accuracy, which is even better for higher-order QAM.

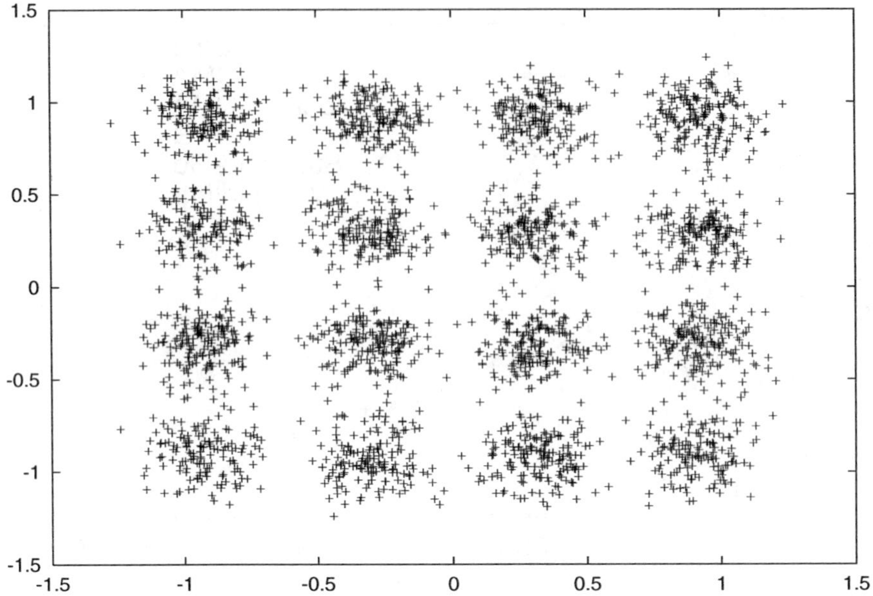

Figure 7.33: The 16-QAM phasor constellation over the one–symbol–delay channel of Figure 7.26 after 40000 symbols at an SNR=30dB using a CMA 10–tap equalizer obeying the schematic of Figure 7.19

- The Benveniste–Goursat algorithm exhibits excellent accuracy, especially in conjunction with higher-order QAM.
- The Stop–and–Go algorithm has a good convergence accuracy, especially in the context of BPSK.
- While for 16-QAM the BER curves do not tend to exhibit residual errors, tending toward BER= 0 for high SNRs, the same does not hold for 64-QAM over the WATM channel. This is, because these channels contain multipath components spread to several symbols' delays. This, in turn, means that the equalizer should also have a high number of taps. However, a blind equalizer having a high number of taps is usually not feasible due to their limited accuracy. We can extend the order of these equalizers and improve the associated BER, but only up to the point, where the equalizer starts to enhance the ISI and the associated convolutional noise.
- The WATM channel is a channel containing multipath components scattered over a delay of 67 symbols. The equalization of this channel would involve a very long equalizer of about 80 taps. In the blind scenario, this is impractical, since the equalizer's resolution is not sufficiently high. Hence beyond a certain length, the taps of these equalizers which are located far from the center tap do not effectively contribute to the equalization process. Furthermore, they

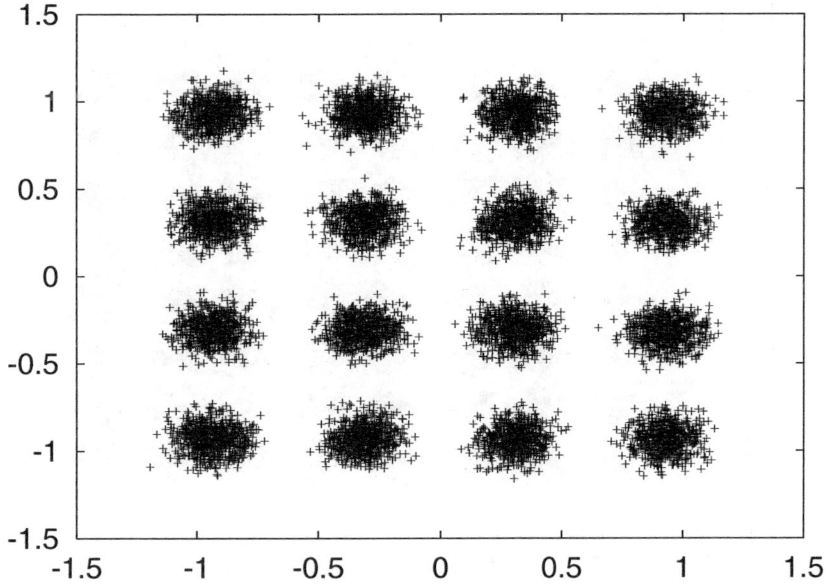

Figure 7.34: The 16-QAM phasor constellation over the one–symbol–delay channel of Figure 7.26 after 60000 symbols at an SNR=30dB using a CMA 10–tap equalizer obeying the schematic of Figure 7.19

enhance the convolutional noise by increasing the ISI. Besides that, the speed of convergence is linearly dependent on the equalizer order and it becomes low for such high-order equalizers. This can be viewed in the Figure 7.43, where the MSE is plotted against the equalizer order for a specific example. In this case, the algorithms used were the CMA and the LMS or Wiener filter, which a trained benchmarker used for comparison. The modulation was 16-QAM and the environment was noiseless.

7.7.12 Simulations with decision–directed switching

In this section, we briefly explore the possibility of switching to decision–directed equalization after the convergence of the blind equalizer. It is expected that for low SNR values this would drive the equalizer away from convergence – since the blind decision-directed equalizer is generally unstable – and therefore, at low SNRs switching to DD mode would be disastrous. Nevertheless, when the SNR is sufficiently high, switching to DD mode is expected to assist the equalizer in converging with a better accuracy. This is indeed what we observed in our investigations. Explicitly, this technique improved the performance of blind equalization and the improvement was higher, when the order of the QAM constellation was higher. This can be explained by the fact that when the SNR is high, the DD technique is powerful. This property

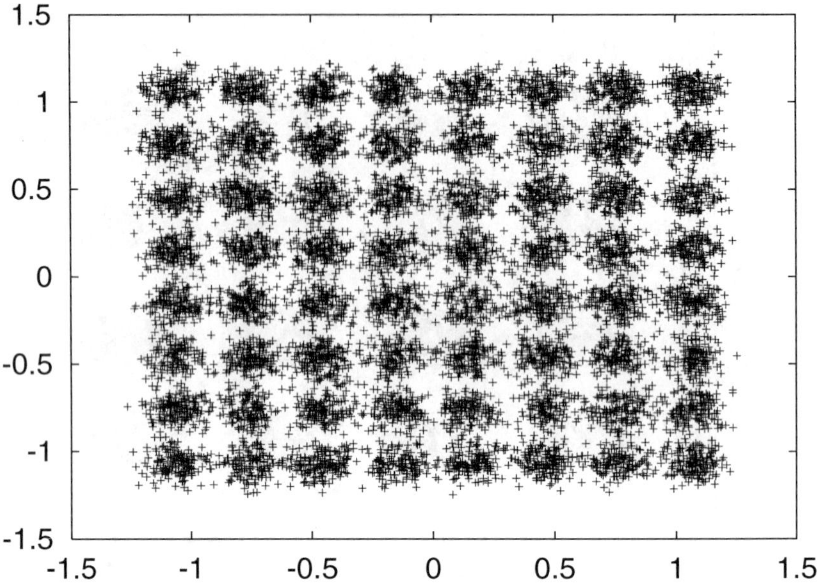

Figure 7.35: The 64-QAM phasor constellation over the one–symbol–delay channel of Figure 7.26 after 70000 symbols at SNR=35dB using a CMA 10–tap equalizer obeying the schematic of Figure 7.19

can be exploited more readily in the context of higher–order QAM constellations, than in lower–order schemes, since the SNR is typically higher for the higher–order QAM schemes, where the distances between constellation points are smaller. In Figures 7.44 and 7.45 the associated improvement is shown for the case of the CMA and for 16-QAM as well as for 64-QAM, respectively. It can be observed that the improvement is modest for 16-QAM, but it is around $1dB$ for 64-QAM at a BER of 10^{-3}.

Having studied the range of blind equalisation solutions in the previous subsections, a prominant application of the algorithms in the context of the Pan-European Satellite-based Digital Video Broadcast (DVB-S) [213] system [316] will be presented in Chapter 24.

7.8 Summary

This chapter has dealt with the issues of channel equalisation for QAM transmissions, commencing with a summary of conventional equalisation techniques. The principles of Kalman filtering were introduced, followed by an overview of the family of blind channel equalisers, which have found wide-ranging applications in data transmission modems and broadcast systems.

The blind equalizers investigated were classified in Figure 7.16. The Busssgang equalizers exhibit a low complexity, but require a considerable number of symbols

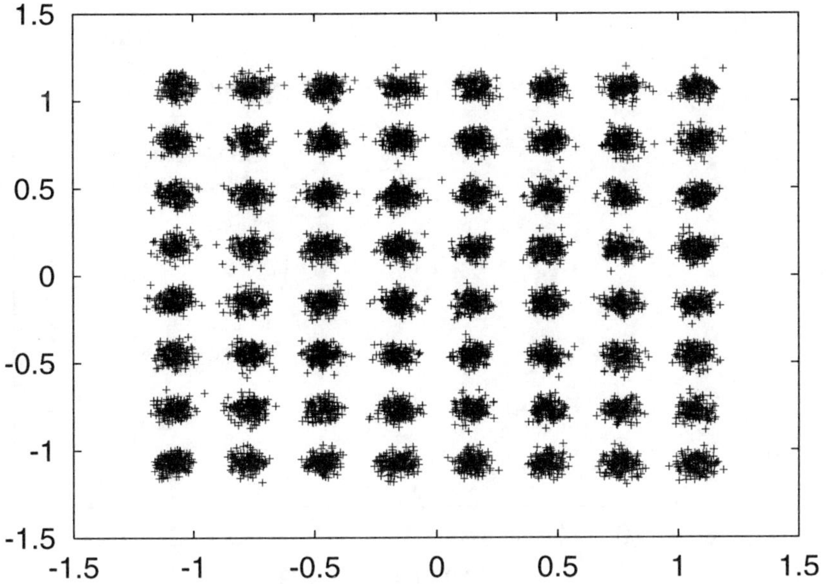

Figure 7.36: The 64-QAM phasor constellation over the one–symbol–delay channel of Figure 7.26 after 180000 symbols at an SNR=35dB using a CMA 10–tap equalizer obeying the schematic of Figure 7.19

to converge. Moreover, their convergence accuracy is mediocre. By contrast, the M–Algorithm appears to be the best choice, giving the best convergence speed and tracking performance at the cost of a higher complexity, depending on the channel's delay spread and the number of survivors M.

For advanced QAM equalisation strategies the reader is referred to Chapter 14 where equalisers for QAM transmissions over dispersive mobile radio channels are considered. Furthermore, a range of blind equalised video broadcast systems will be the topic of Chapter 24.

In the next chapter we consider a transmission technique whereby the design of the constellation used in the modulation scheme and the error correction coding used to improve the channel's bit error rate are combined into one system. Such a system can be shown to have distinct advantages and is known as trellis coded modulation or also often referred to as 'codulation'.

7.9 Appendix: Differentiation with Respect to a Vector

In this appendix we give the definitions of differentiation with respect to complex vectors along with their associated properties. Based on these, we then highlight how

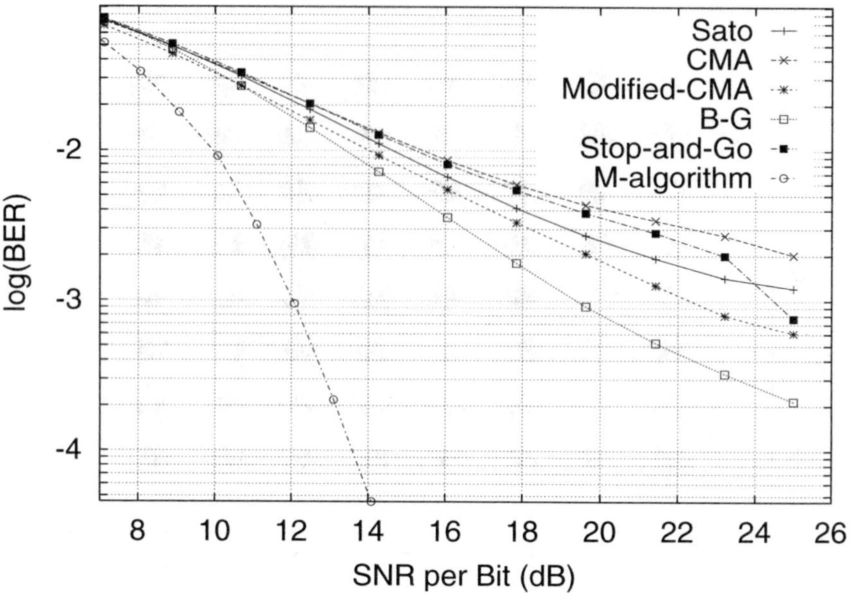

Figure 7.37: Gaussian BER versus SNR curves for the WATM channel of Figure 7.25(a) for 16-QAM using a 68-tap equalizer obeying the schematic of Figure 7.19 and, for the M–Algorithm, a 5–path channel estimation obeying the schematic of Figure 7.21.

we can minimize the cost-functions encountered in our theoretical considerations. First of all we define the gradient vector [3]:

$$\nabla_{\mathbf{z}} = \begin{bmatrix} \frac{\partial}{\partial x_1} + j\frac{\partial}{\partial y_1} \\ \frac{\partial}{\partial x_2} + j\frac{\partial}{\partial y_2} \\ \vdots \\ \frac{\partial}{\partial x_N} + j\frac{\partial}{\partial y_N} \end{bmatrix} \tag{7.179}$$

and

$$\nabla_{\mathbf{z}^*} = \begin{bmatrix} \frac{\partial}{\partial x_1} - j\frac{\partial}{\partial y_1} \\ \frac{\partial}{\partial x_2} - j\frac{\partial}{\partial y_2} \\ \vdots \\ \frac{\partial}{\partial x_N} - j\frac{\partial}{\partial y_N} \end{bmatrix}, \tag{7.180}$$

where $\mathbf{z} = [x_1+jy_1 \cdots, x_N+jy_N]^T$ is a complex vector. The properties of the gradient vectors, which we will exploit are :

$$\nabla_{\mathbf{z}}\mathbf{z} = \mathbf{O} \tag{7.181}$$

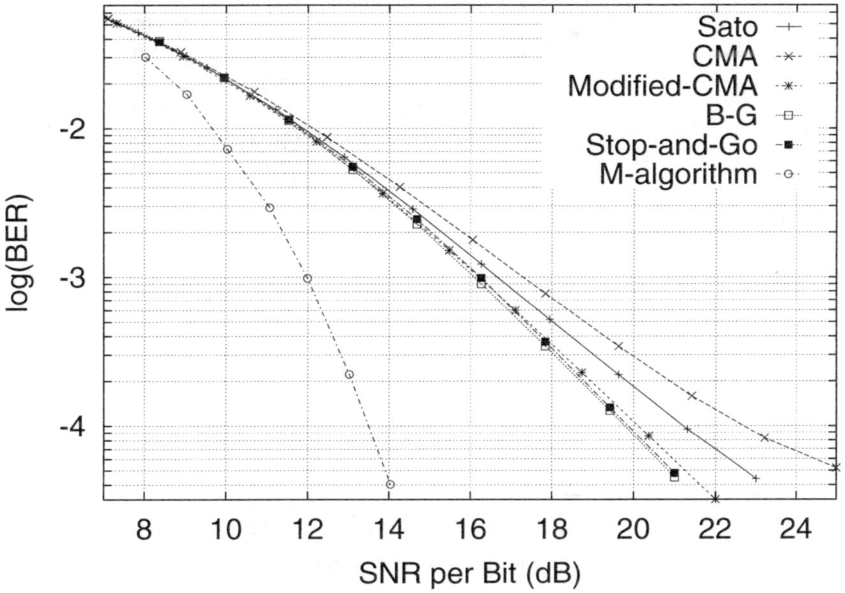

Figure 7.38: Gaussian BER versus SNR curves for the Shortened WATM channel of Figure 7.25(b) for 16-QAM using a 30–tap equalizer obeying the schematic of Figure 7.19 and, for the M–Algorithm, a 3–path channel estimation obeying the schematic of Figure 7.21.

$$\nabla_{\mathbf{z}}\mathbf{z}^* = \mathbf{I} \tag{7.182}$$

$$\nabla_{\mathbf{z}^*}\mathbf{z} = \mathbf{I} \tag{7.183}$$

$$\nabla_{\mathbf{z}^*}\mathbf{z}^* = \mathbf{O} \tag{7.184}$$

$$\nabla_{\mathbf{z}}(\mathbf{a}^T \cdot \mathbf{b}) = \nabla_{\mathbf{z}}\mathbf{b} \cdot \mathbf{a} + \nabla_{\mathbf{z}}\mathbf{a} \cdot \mathbf{b}, \tag{7.185}$$

where $\mathbf{O}$ is the zero matrix and $\mathbf{I}$ is the unit matrix. These properties can be readily verified by applying the definition of the gradient vector of Equations (7.179) and (7.180) and the properties of differentiation of real functions. We will also highlight a few more properties, which we will invoke in order to derive the analytical expressions of the cost-function derivatives with respect to the equalizer tap vector of $\mathbf{c}$:

$$\nabla_{\mathbf{z}}(f(\mathbf{a}))^k = \begin{bmatrix} \frac{\partial}{\partial x_1} + j\frac{\partial}{\partial y_1} \\ \frac{\partial}{\partial x_2} + j\frac{\partial}{\partial y_2} \\ \vdots \\ \frac{\partial}{\partial x_N} + j\frac{\partial}{\partial y_N} \end{bmatrix} (f(\mathbf{a}))^k$$

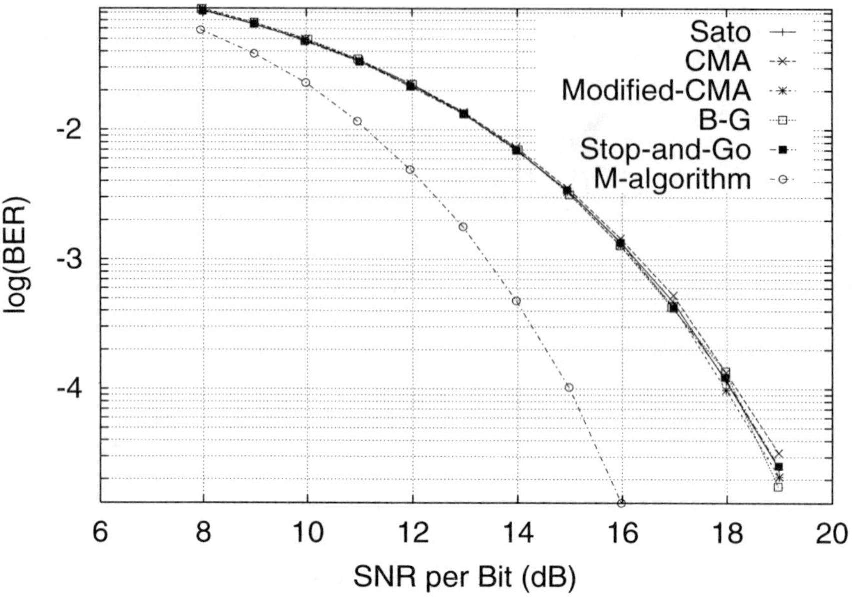

Figure 7.39: Gaussian BER versus SNR curves for the one–symbol–delay channel of Figure 7.26 for 16-QAM using a 10–tap equalizer, obeying the schematic of Figure 7.19 and, for the M–Algorithm, a 2–path channel estimation obeying the schematic of Figure 7.21.

$$
\begin{aligned}
&= \begin{bmatrix} \frac{\partial (f(\mathbf{a}))^k}{\partial x_1} + j\frac{\partial (f(\mathbf{a}))^k}{\partial y_1} \\ \frac{\partial (f(\mathbf{a}))^k}{\partial x_2} + j\frac{\partial (f(\mathbf{a}))^k}{\partial y_2} \\ \vdots \\ \frac{\partial (f(\mathbf{a}))^k}{\partial x_N} + j\frac{\partial}{\partial y_N} \end{bmatrix} \\
&= k \cdot (f(\mathbf{a}))^{k-1} \begin{bmatrix} \frac{\partial f(\mathbf{a})}{\partial x_1} + j\frac{\partial f(\mathbf{a})}{\partial y_1} \\ \frac{\partial f(\mathbf{a})}{\partial x_2} + j\frac{\partial f(\mathbf{a})}{\partial y_2} \\ \vdots \\ \frac{\partial f(\mathbf{a})}{\partial x_N} + j\frac{\partial f(\mathbf{a})}{\partial y_N} \end{bmatrix} \\
&= k \cdot (f(\mathbf{a}))^{k-1} \nabla_{\mathbf{z}} f(\mathbf{a}). \qquad (7.186)
\end{aligned}
$$

These properties are useful for the minimization of the blind equalizer cost-functions given in Equations (7.125), (7.118), (7.130) for the CMA, for Sato's algorithm and for the Modified CMA, respectively. In the next section we will deal with the derivation of the tap-update equations of a specific blind equalization algorithm, namely the CMA and we will give the minimization procedure for this algorithm. Three very useful relationships which we will invoke in the next section come from the direct

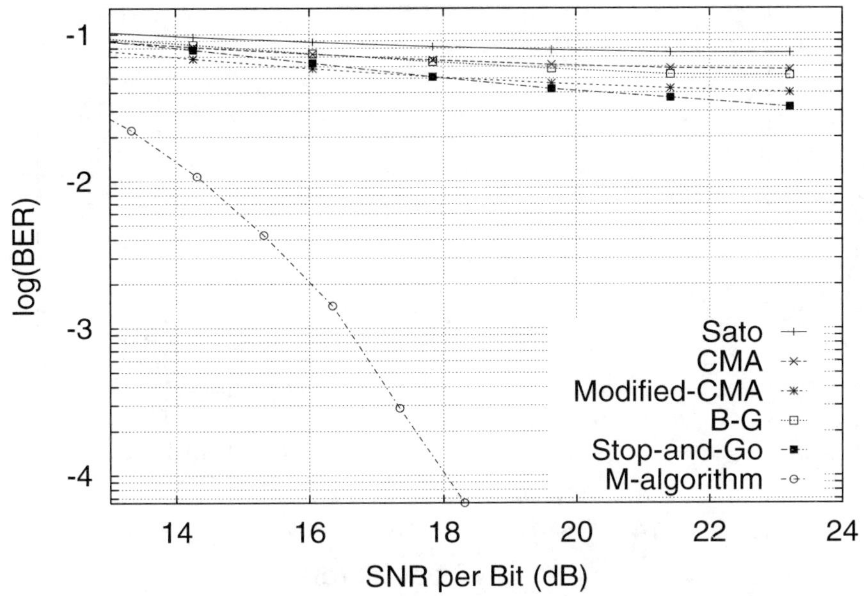

Figure 7.40: Gaussian BER versus SNR curves for the WATM channel of Figure 7.25(a) for 64-QAM using a 68–tap equalizer obeying the schematic of Figure 7.19 and, for the M–Algorithm, a 5–path channel estimation obeying the schematic of Figure 7.21.

application of Equation (7.186), yielding:

$$
\begin{aligned}
\nabla_{\mathbf{c}}|z(n)| &= \nabla_{\mathbf{c}}\sqrt{|z(n)|^2} \\
&= \frac{1}{2\sqrt{|z(n)|^2}}\nabla_{\mathbf{c}}|z(n)|^2 \\
&\overset{(7.186)}{=} \frac{1}{2|z(n)|}\nabla_{\mathbf{c}}(\mathbf{c}^T\mathbf{y}(n))(\mathbf{c}^H\mathbf{y}^*(n)) \\
&= \frac{1}{2|z(n)|}[(\mathbf{c}^T\mathbf{y}(n))\nabla_{\mathbf{c}}(\mathbf{c}^H\mathbf{y}^*(n)) + \\
&\quad +(\mathbf{c}^H\mathbf{y}^*(n))\nabla_{\mathbf{c}}(\mathbf{c}^T\mathbf{y}(n))] \\
&\overset{(7.185)}{=} \frac{1}{2|z(n)|}[(\mathbf{c}^T\mathbf{y}(n))\{(\nabla_{\mathbf{c}}\mathbf{y}^*(n))\cdot\mathbf{c}^* + \\
&\quad +(\nabla_{\mathbf{c}}\mathbf{c}^*)\cdot\mathbf{y}^*(n)\} \\
&\quad +(\mathbf{c}^H\mathbf{y}^*(n))\{(\nabla_{\mathbf{c}}\mathbf{c})\cdot\mathbf{y}(n) + \\
&\quad +(\nabla_{\mathbf{c}}\mathbf{y}^T(n))\cdot\mathbf{c}\}] \\
&\overset{(7.181)-(7.184)}{=} \frac{(\mathbf{c}^T\cdot\mathbf{y}(n))\mathbf{y}^*(n)}{2|z(n)|}
\end{aligned}
$$

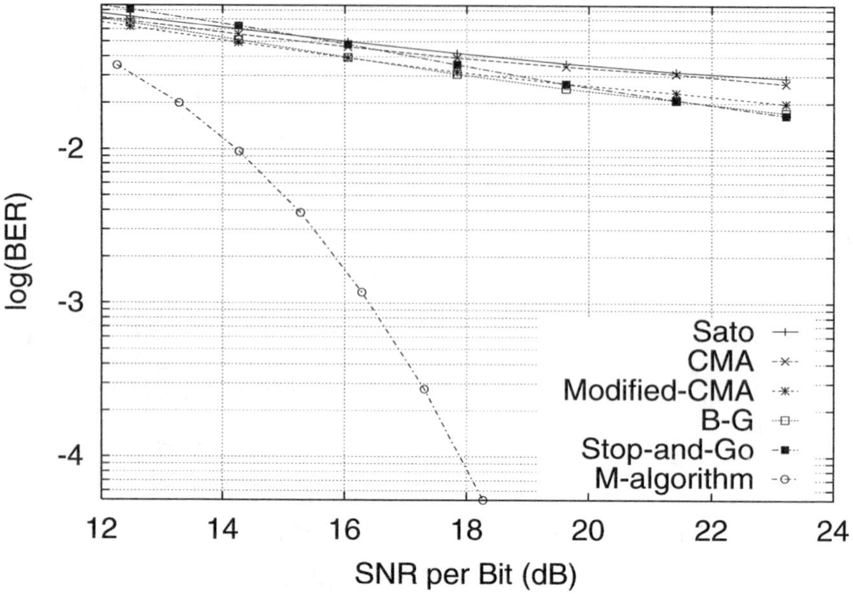

Figure 7.41: Gaussian BER versus SNR curves for the Shortened WATM channel of Figure 7.25(b) for 64-QAM using a 30–tap equalizer obeying the schematic of Figure 7.19.

$$= \frac{z(n) \cdot \mathbf{y}^*(n)}{2|z(n)|} \tag{7.187}$$

$$\begin{aligned}
\nabla_{\mathbf{c}} Re\{z(n)\} &= \nabla_{\mathbf{c}} \frac{z(n) + z^*(n)}{2} \\
&= \nabla_{\mathbf{c}} \frac{\mathbf{c}^T \mathbf{y}(n) + \mathbf{c}^H \mathbf{y}^*(n)}{2} \\
&\overset{(7.185)}{=} \frac{(\nabla_{\mathbf{c}} \mathbf{c}) \cdot \mathbf{y}(n) + (\nabla_{\mathbf{c}} \mathbf{c}^*) \cdot \mathbf{y}^*(n)}{2} \\
&\overset{(7.181)--(7.185)}{=} \frac{1}{2} \cdot \mathbf{y}^*(n)
\end{aligned} \tag{7.188}$$

$$\begin{aligned}
\nabla_{\mathbf{c}} Im\{z(n)\} &= \nabla_{\mathbf{c}} \frac{z(n) - z^*(n)}{2j} \\
&= \nabla_{\mathbf{c}} \frac{\mathbf{c}^T \mathbf{y}(n) - \mathbf{c}^H \mathbf{y}^*(n)}{2j} \\
&\overset{(7.185)}{=} \frac{(\nabla_{\mathbf{c}} \mathbf{c}) \cdot \mathbf{y}(n) - (\nabla_{\mathbf{c}} \mathbf{c}^*) \cdot \mathbf{y}^*(n)}{2j}
\end{aligned}$$

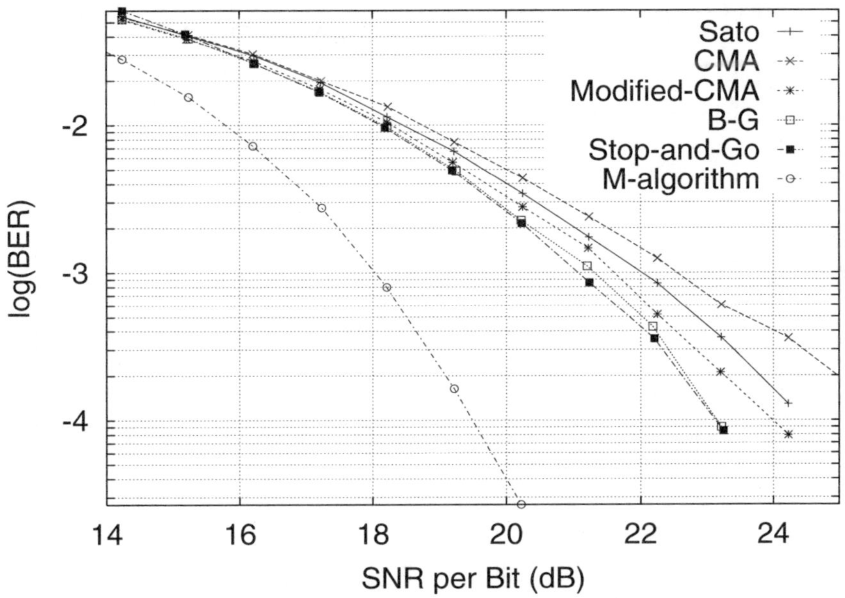

Figure 7.42: Gaussian BER versus SNR curves for the one–symbol–delay channel of Figure 7.26 for 64-QAM using a 10–tap equalizer, obeying the schematic of Figure 7.19 and, for the M–Algorithm, a 2–path channel estimation obeying the schematic of Figure 7.21.

$$\overset{(7.181)--(7.185)}{=} \quad \frac{j}{2} \cdot \mathbf{y}^*(n), \tag{7.189}$$

where $z(n) = \mathbf{c}^T \mathbf{y}(n)$ is the equalized signal in Figure 7.15, expressed as the product of the equalizer tap vector $\mathbf{c}$ and the received signal vector $\mathbf{y}(n)$, which forms the convolution of the received signal with the equalizer's impulse response.

7.9.1 An illustrative example: CMA cost-function minimization

The CMA's cost-function was defined in Equation (7.125), which is repeated here for convenience:

$$J(n) = E[(|z(n)|^2 - R_2)^2]$$

If we omit the expectation, then its gradient vector, with respect to $\mathbf{c}$, is :

$$\begin{aligned}
\nabla_{\mathbf{c}} J(n) &= \nabla_{\mathbf{c}} \left(|z(n)|^2 - R_2\right)^2 \\
&\overset{(7.186)}{=} 2 \cdot \left(|z(n)|^2 - R_2\right) \cdot \nabla_{\mathbf{c}} \left(|z(n)|^2 - R_2\right) \\
&= \left(|z(n)|^2 - R_2\right) \cdot \nabla_{\mathbf{c}} z(n) \cdot z^*(n)
\end{aligned}$$

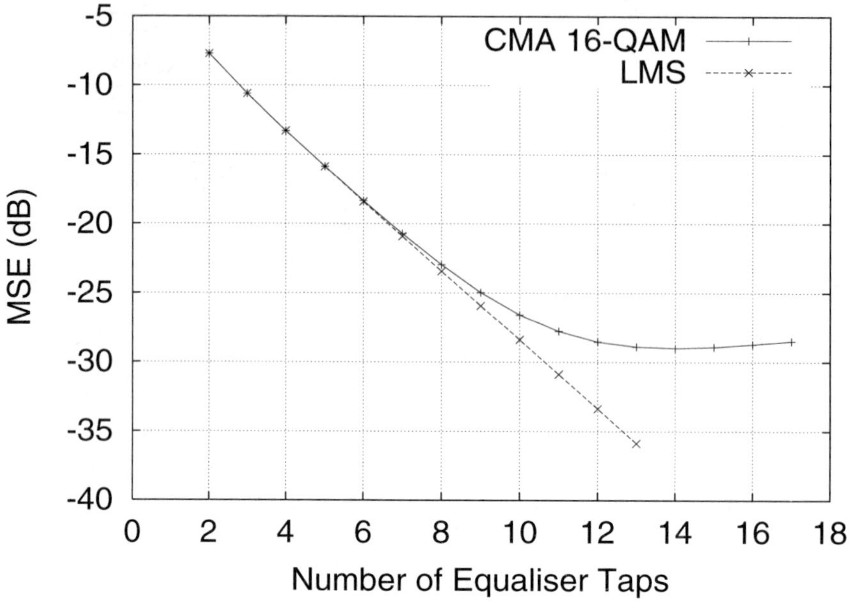

Figure 7.43: MSE as a function of the equalizer order for 16–QAM using the CMA-based equalizer and a noiseless scenario. The channel model used is the one–symbol delay channel as given in Figure 7.26. The LMS-based MSE is given as a lower limit, as this is a non–blind equalizer which gives a benchmark performance. Clearly, for small number of taps the blind equalizer curve coincides with the LMS benchmark curve, while for larger number of taps the CMA produces large MSE, attaining a local minimum at about 14 taps, which is the ideal setting for this scenario.

$$
\begin{aligned}
&= 2 \cdot \left(|z(n)|^2 - R_2\right) \cdot z(n) \cdot \nabla_{\mathbf{c}} z^*(n) \\
&= 2 \cdot \left(|z(n)|^2 - R_2\right) \cdot z(n) \cdot \nabla_{\mathbf{c}} \mathbf{c}^H \mathbf{y}^*(n) \\
&\overset{(7.185)}{=} 2 \cdot \left(|z(n)|^2 - R_2\right) \cdot z(n) \cdot \mathbf{y}^*(n). \qquad (7.190)
\end{aligned}
$$

This is exactly the equalizer coefficients' update term of Equation (7.127).

7.10 Appendix: Polycepstra definitions

Based on [3,252,268,280], in this appendix we give some fourth–order statistics definitions and properties, which are useful for the blind equalization algorithms of Section 7.7.9. The definitions are based on [252], in which the polycepstra–based algorihms first appeared and on [280], where the super–exponential algorithm was proposed, using fourth–order cumulants. In general, the joint cumulant of the random variables

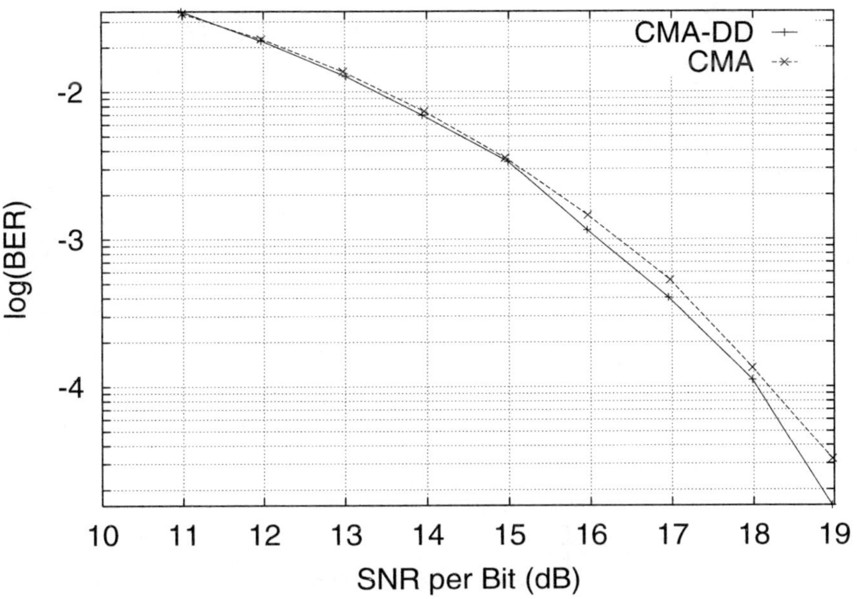

Figure 7.44: Gaussian BER versus SNR curves for the one–symbol–delay channel of Figure 7.26 for 16-QAM using a 10–tap equalizer obeying the schematic of Figure 7.19 for the CMA using switching to decision–directed equalization after convergence

$x_{n_1}, x_{n_2}, \ldots x_{n_m}$ is given by [280]:

$$L(x_{n_1}; x_{n_2}; \cdots; x_{n_m}) = (-j)^m \frac{\partial^m ln\phi(\omega)}{\partial\omega_{n_1} \ldots \partial\omega_{n_m}|_{\omega=0}}, \tag{7.191}$$

where $\phi(\omega)$ is the so-called joint characteristic function of the random variables x_i, defined as [303]:

$$\phi(\omega) = E\left[e^{j\omega\mathbf{x}}\right] = E\left[e^{j\sum_{i=1}^{m}\omega_i x_i}\right]. \tag{7.192}$$

For zero–mean random variables and cumulants up to fourth–order, the definition is equivalent to [252]:

$$\begin{aligned}
L(x_1) &= 0 && (7.193)\\
L(x_1; x_2) &= E[x_1 \cdot x_2] && (7.194)\\
L(x_1; x_2; x_3) &= E[x_1 \cdot x_2 \cdot x_3] && (7.195)\\
L(x_1; x_2; x_3; x_4) &= E[x_1 \cdot x_2 \cdot x_3 \cdot x_4] - \\
&\quad E[x_1 \cdot x_2] \cdot E[x_3 \cdot x_4] - \\
&\quad E[x_1 \cdot x_3] \cdot E[x_2 \cdot x_4] -
\end{aligned}$$

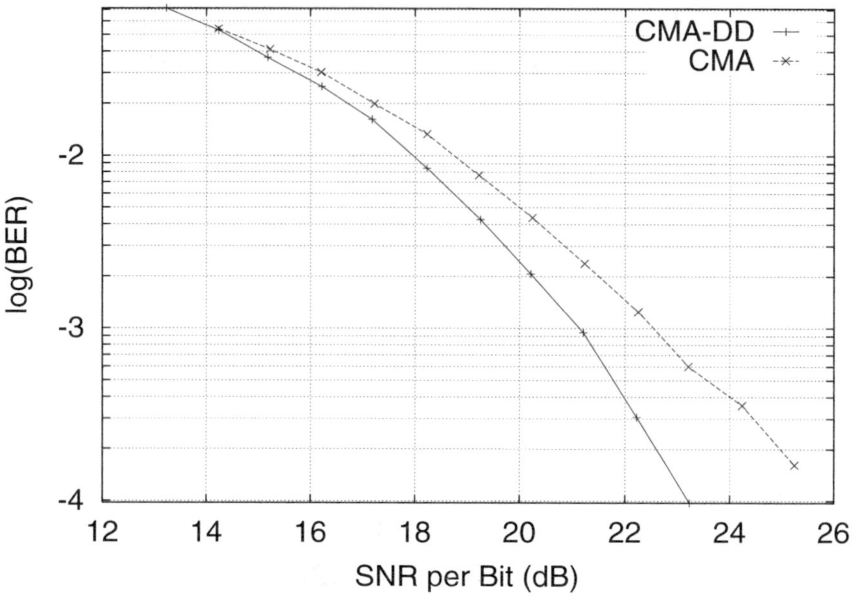

Figure 7.45: Gaussian BER versus SNR curves for the one–symbol–delay channel of Figure 7.26 for 64-QAM using a 10–tap equalizer obeying the schematic of Figure 7.19 for the CMA using switching to decision–directed equalization after convergence

$$E[x_1 \cdot x_4] \cdot E[x_2 \cdot x_3] \tag{7.196}$$

When the random variables are actually samples from a zero–mean stochastic process, such as the received signal $y(k)$, then the fourth–order cumulant is given by:

$$\begin{aligned} L_y(m,n,l) &= E[y(k)y(k+m)y(k+n)y(k+l)] - \\ &\quad E[y(k)y(k+m)] \cdot E[y(k)y(k+l-n)] - \\ &\quad E[y(k)y(k+n)] \cdot E[y(k)y(k+l-m)] - \\ &\quad E[y(k)y(k+l)] \cdot E[y(k)y(k+m-n)]. \end{aligned} \tag{7.197}$$

Following the above definition, we will now summarize some of the associated cumulant properties. First of all, the cumulant of the sum of two independent and zero–mean stationary signals $x_1(k)$ and $x_2(k)$ is equal to the sum of the cumulants of $x_1(k)$ and $x_2(k)$, i.e.:

$$L_{x_1+x_2}(m,n,l) = L_{x_1}(m,n,l) + L_{x_2}(m,n,l). \tag{7.198}$$

Furthermore, the cumulant of the convolution of two zero–mean stationary signals corresponds to the convolution of their cumulants in the cumulant domain:

$$L_{x_1 * x_2}(m,n,l) = L_{x_1}(m,n,l) * L_{x_2}(m,n,l), \tag{7.199}$$

where $*$ stands for convolution. Two further properties related to the signal distributions:

- If $x(k)$ is Gaussian, then the cumulant of $x(k)$ is 0
- If $x(k)$ is i.i.d., then the cumulant of $x(k)$ is given by:

  $$L_x(m,n,l) = K(x(k)) \cdot \delta(m,n,l), \tag{7.200}$$

  where $K(x(k))$ is the Kurtosis of $x(k)$ defined as:

  $$K(x(k)) = E[|x(k)|^4] - 3\left(E[x^2(k)]\right)^2 \tag{7.201}$$

  for real-valued signals Equation (7.129) defined the Kurtosis for complex-valued signals, which is repeated here:

  $$K(z) = E[|z(n)|^4] - 2 \cdot E^2[|z(n)|^2] - |E[z^2(n)]|^2.$$

Finally, below we will now give a cumulant property of stochastic signals passing through linear systems. We assume that the i.i.d. signal $x(k)$ is input to a system having an impulse response h_k. The output of the system is given by:

$$y(k) = \sum_i h_i \cdot a(k-i). \tag{7.202}$$

The cumulant of the output is then given by:

$$L_y(n,k,l) = K(a(k)) \cdot L_h(m,n,l), \tag{7.203}$$

where $L_h(m,n,l)$ is the fourth–order moment of the CIR. Here, the moment has replaced the cumulant because $h(k)$ is a deterministic signal. We can now form the cumulant of the received signal $y(k)$, which is given by the convolution of the i.i.d. input signal $a(k)$ with the CIR h_k plus the additive Gaussian noise:

$$\begin{aligned} y(k) &= h_k * x(k) + e(k) \Rightarrow & (7.204)\\ L_y(m,n,l) &= K(x(k)) \cdot L_h(m,n,l), & (7.205) \end{aligned}$$

where the Gaussian noise term vanishes, since its fourth–order cumulant is zero. We will now define the tricepstrum of a signal $x(k)$ as:

$$c_x(m,n,l) = Z_{(3)}^{-1}\{ln(Z_{(3)}\{L_x(m,n,l)\})\}, \tag{7.206}$$

where $Z_{(3)}$ stands for the 3D z–transform. Complicated as this definition may seem, it simplifies the cumulant–based relationship of Equation (7.205) to the tricepstrum domain as:

$$c_y(m,n,l) = c_h(m,n,l) \qquad (m,n,l) \neq (0,0,0), \tag{7.207}$$

which means that the tricepstrum of the received signal is the same as the tricepstrum of the CIR, provided that the input is i.i.d. and the noise is Gaussian distributed. By measuring the tricepstrum of the received signal, we can now have an estimate of the tricepstrum of the CIR. It can be shown that the following identity holds [252]:

$$\begin{aligned} &\sum_{I=1}^{p} A^{(I)}[L_y(m-I,n,l) - L_y(m+I,n+I,l+I)] + \\ &+ \sum_{J=1}^{q} B^{(J)}[L_y(m-J,n-J,l-J) - L_y(m+J,n,l)] = \\ &\qquad\qquad = -mL_y(m,n,l), \end{aligned} \tag{7.208}$$

where

$$K \cdot c_y(K,0,0) = \begin{cases} -A^{(K)}, & K = 1,\ldots,p \\ B^{(-K)}, & K = -1,\ldots,-q \,. \end{cases} \tag{7.209}$$

This relationship holds for any (m,n,l). Nonetheless, a rule of thumb, suggested in [252] is to define $w = max(p,q)$, $z \leq w/2$, $s \leq z$ and then select $m = -w,\ldots,-1,1,\ldots,w$, $n = -z,\ldots,0,\ldots,z$ and $l = -s,\ldots,0,\ldots,s$. It is shown in [252] that $A^{(I)}, B^{(J)}$ are the minimum– and maximum–phase differential cepstrum coefficients related to the zeros of the CIR. Thus, estimating $A^{(I)}, B^{(J)}$ from the system of Equations (7.208) means that we can invert the CIR apart from an uncertainty of the magnitude A. Theoretically, it is required that $p, q \to \infty$, but in practice we can use a sufficiently large discrete value for each of them. Equations (7.208) can then be rewritten in matrix form as:

$$\mathbf{P} \cdot \mathbf{a} = \mathbf{p}, \tag{7.210}$$

where $\mathbf{P}$ is the matrix containing the L_y–elements of the left–hand side of Equation (7.208), $\mathbf{a}$ is the vector containing $A^{(I)}$ and $B^{(J)}$ and $\mathbf{p}$ is the vector containing the L_y– elements of the right–hand side of Equation (7.208). We have to estimate $\mathbf{a}$, in order to invert the CIR. This can be carried out upon invoking the LMS solution of Equation (7.210), as in [252], yielding:

$$\hat{\mathbf{a}}^{(i+1)} = \hat{\mathbf{a}}^{(i)} + \mu^{(i)} \cdot (\hat{\mathbf{P}}^{(i)})^H \cdot (\hat{\mathbf{p}}^{(i)} - \hat{\mathbf{P}}^{(i)} \cdot \hat{\mathbf{a}}^{(i)}) \tag{7.211}$$

with

$$0 < \mu^{(i)} < 2/tr\{(\hat{\mathbf{P}}^{(i)})^H \cdot \hat{\mathbf{P}}^{(i)}\},$$

where $tr()$ stands for the trace of a matrix. When $\mu^{(i)}$ is in this region, stability is ensured, as shown in [3].

The exact relationship of the coefficients $A^{(I)}$ and $B^{(I)}$ with the poles and zeros of the equalizer filters' polynomials is as follows. If the z–domain channel transfer function polynomial $H(z)$ is given by $H(z) = I(z^{-1}) \cdot O(z)$, as in Equation (7.171), where

$$I(z^{-1}) = \frac{\Pi_{i=1}^{L_1}(1 - a_i z^{-1})}{\Pi_{i=1}^{L_3}(1 - c_i z^{-1})} \tag{7.212}$$

with $|a_i| < 1$ and $|c_i| < 1$ and

$$O(z) = \Pi_{i=1}^{L_2}(1 - b_i z) \tag{7.213}$$

with $b_i < 1$, then the coefficients $A^{(I)}$ and $B^{(I)}$ are given by:

$$A^{(I)} = \sum_{i=1}^{L_1} a_i^I - \sum_{i=1}^{L_3} c_i^I \tag{7.214}$$

$$B^{(I)} = \sum_{i=1}^{L_2} b_i^I \tag{7.215}$$

Having estimated $A^{(I)}$ and $B^{(I)}$ we can now use Equation (7.215) to estimate the time–domain coefficients $\hat{i}_{inv}^{(i)}(k)$ and $\hat{o}_{inv}^{(i)}(k)$, which are the coefficients of $1/I(z^{-1})$ and $1/O(z)$, respectively, as in [252] :

$$\hat{i}_{inv}^{(i)}(k) = -\frac{1}{k}\sum_{n=2}^{k+1}[-A^{(n-1),(i)}] \cdot \hat{i}_{inv}^{(i)}(k-n+1), \quad k = 1, \ldots, N_1 \tag{7.216}$$

$$\hat{o}_{inv}^{(i)}(k) = \frac{1}{k}\sum_{n=k+1}^{0}[-B^{(1-n),(i)}] \cdot \hat{o}_{inv}^{(i)}(k-n+1), \quad k = -1, \ldots, -N_2, \tag{7.217}$$

where the superscript i denotes the i-th iteration of the algorithm. The equalizer's impulse response is finally given by the convolution $\hat{i}_{inv}^{(i)}(k) * \hat{o}_{inv}^{(i)}(k)$. In the case of a DFE we also have to estimate the inverse z–transforms of $I(z^{-1})$ and $O^*(z^{-1})$, which is given by [252]:

$$\begin{aligned} \hat{i}_o^{(i)}(k) &= \frac{1}{k}\sum_{n=2}^{k+1}[A^{(n-1),(i)} + (B^{(n-1),(i)})^*]\hat{i}_o^{(i)}(k-n+1) \\ & \qquad k = 1, \ldots, N_3 \end{aligned} \tag{7.218}$$

$$\hat{o}_o^{(i)}(k) = -\frac{1}{k}\sum_{n=2}^{k+1}[B^{(n-1),(i)}]^* \cdot \hat{o}_o^{(i)}(k-n+1), \quad k = 1, \ldots, N_4. \tag{7.219}$$

Finally, the feedforward and feedback equalizer impulse responses are now extracted as $\hat{o}_o^{(i)}(k) * \hat{o}_{inv}^{(i)}(k)$ and $\hat{i}_{inv}^{(i)}(k) * \hat{o}_o^{(i)}(k) - \delta(k)$, respectively. The channel attenutation parameter A can be estimated using Automatic Gain Control (AGC). A synopsis of the equations involved in the TEA is given in Table 7.10.

$\sum_{I=1}^{p} A^{(I)}[L_y(m-I,n,l) - L_y(m+I,n+I,l+I)]+$ $+\sum_{J=1}^{q} B^{(J)}[L_y(m-J,n-J,l-J) - L_y(m+J,n,l)] =$ $= -mL_y(m,n,l)$
$\hat{\mathbf{a}}^{(i)} + \mu^{(i)} \cdot (\hat{\mathbf{P}}^{(i)})^H \cdot (\hat{\mathbf{p}}^{(i)} - \hat{\mathbf{P}}^{(i)} \cdot \hat{\mathbf{a}}^{(i)}))$ $\mathbf{a} = (A^{(1)}, \ldots, A^{(p)}, B^{(1)}, \ldots, B^{(q)})^T$ $[\mathbf{p}]_i = -m \cdot L_y(m,n,l)$ $[\mathbf{P}]_{ij} = L_y(m-I,n,l) - L_y(m+I,n+I,l+I)$ $i = 1, \ldots, p$ $[\mathbf{P}]_{ij} = L_y(m-J,n-J,l-J) - L_y(m+J,n,l)$ $i = p+1, \ldots, p+q$ $w = max(p,q),\ z \leq w/2,\ s \leq z$ $m = -w, \ldots, -1, 1, \ldots, w$ $n = -z, \ldots, 0, \ldots, z$ $l = -s, \ldots, 0, \ldots, s$
$\hat{i}_{inv}^{(i)}(k) = -\frac{1}{k}\sum_{n=2}^{k+1}[-A^{(n-1),(i)}] \cdot \hat{i}_{inv}^{(i)}(k-n+1)$ $k = 1, \ldots, N_1$
$\hat{o}_{inv}^{(i)}(k) = \frac{1}{k}\sum_{n=k+1}^{0}[-B^{(1-n),(i)}] \cdot \hat{o}_{inv}^{(i)}(k-n+1)$ $k = -1, \ldots, -N_2$
$\hat{i}_{o}^{(i)}(k) = \frac{1}{k}\sum_{n=2}^{k+1}[A^{(n-1),(i)} + (B^{(n-1),(i)})^*]\hat{i}_{o}^{(i)}(k-n+1)$ $k = 1, \ldots, N_3$
$\hat{o}_{o}^{(i)}(k) = -\frac{1}{k}\sum_{n=2}^{k+1}[B^{(n-1),(i)}]^* \cdot \hat{o}_{o}^{(i)}(k-n+1)$ $k = 1, \ldots, N_4$

Table 7.5: A synopsis of the equations related to the TEA

Chapter 8

Trellis Coded Modulation

8.1 Introduction

Historically, modulation and forward error correction (FEC) have been treated as distinct subjects, where FEC is typically invoked in order to lower the BER of the modem. However, in his visionary paper of 1974 Massey [317] surmised that treating modulation and FEC as an integrated entity would allow significant coding gains to be achieved over power and bandlimited transmission media. The first practical coded modulation scheme was proposed by Imai and Hirakawa [318], shortly followed by Ungerboeck's more frequently referenced papers [319, 320], where trellis coded modulation (TCM) is introduced.

During the 1980s TCM has reached a state of maturity and has been incorporated into the CCITT V. series multilevel modems designed for Gaussian telephone channels (see Chapter 9). With the emergence of widespread digital communications via hostile fading mobile radio channels, researchers turned their attention towards adapting TCM principles for these applications, which led to the invention of block coded modulation (BCM) [321]. The importance of coded modulation is evidenced in the large number of conferences [321], special issues [322], and books [323] devoted to this subject.

The motivation for TCM is based on the disadvantages of the conventional approach of using separate modulation and coding blocks. In power-limited applications, such as satellite channels, FEC codes can be invoked in order to improve power efficiency at the cost of expanding the bandwidth required. In a bandwidth-limited scenario, more bits/symbol can be transmitted but at the cost of increased power requirements. In trellis coded modulation the advantages of higher-level modulation are combined with those of FEC, where modulation/encoding and demodulation/FEC decoding form an integral operation.

TCM and BCM were initially deployed over telephone channels. The telephone channel has historically been the scene of the earliest applications of many efficient modulation systems. This is because this ubiquitous medium constitutes a benign channel, being relatively time invariant with a high SNR, normally modelled as 28 dB. However, the channel is severely bandlimited over the range 300-3000 Hz. Calculations of channel capacity [324] showed that uncoded modulation techniques were not realising the full potential of the channel, as also suggested by Figure 2.4. For example, if one-dimensional pulse amplitude modulation (PAM) is transmitted with m bits per dimension using one of the 2^m equally likely levels, normally ± 1, ± 3, ± 5, ..., $\pm(2^m - 1)$, then the average energy of each coordinate is given by

$$S_m = (4^m - 1)/3 \tag{8.1}$$

and it then follows that

$$S_{m+1} = 4S_m + 1. \tag{8.2}$$

Therefore, it takes approximately four times as much energy (6 dB) to send a further 1 bit/dimension, or 2 bits/symbol. The maximum possible information transmission rate over a noisy channel was determined by Shannon [325] by defining the so-called channel capacity C (in bits per dimension) as

$$C = \frac{1}{2}\log_2(1 + S/N) \ \ bits/dimension, \tag{8.3}$$

where S is the signal energy, N is the noise energy, and so S/N is the SNR. When $S/N \gg 1$ it takes approximately four times as much energy to increase the capacity by 1 bit/dimension. Therefore, the ratio of bits/dimension to channel capacity, which is a good measure of how efficiently the channel is being exploited, approaches 1 as S becomes large. How this high efficiency is achieved is not stated by Shannon, but adding more modulation levels until the maximum channel capacity is reached would initially seem the best approach. Unfortunately, high-level modulation schemes are prone to error and hence joint modulation and coding might contribute better towards approaching the predicted maximum channel efficiency.

However, many authors [317, 326] regard a parameter R_0, defined below, as a better estimate of the maximum rate that is practically achievable [327] where

$$R_0 = \frac{1}{2}\log_2(1 + S/2N) \ \ bits/dimension. \tag{8.4}$$

According to Equation 8.4 it therefore takes 3 dB more power to signal at a certain number of bits/dimension. The practical potential saving using coding is therefore probably about 3 dB/dimension, or 6 dB for two-dimensional QAM.

This gain was not achievable using standard coding techniques. It was this which led workers to investigate alternative coding techniques that would actually realise this coding gain.

8.2 TCM Fundamentals

Whilst considerable work has been performed on channel coding [328–332], here we consider a different approach where coding implies the introduction of interdependancies between sequences of signal points such that not all sequences are possible. The

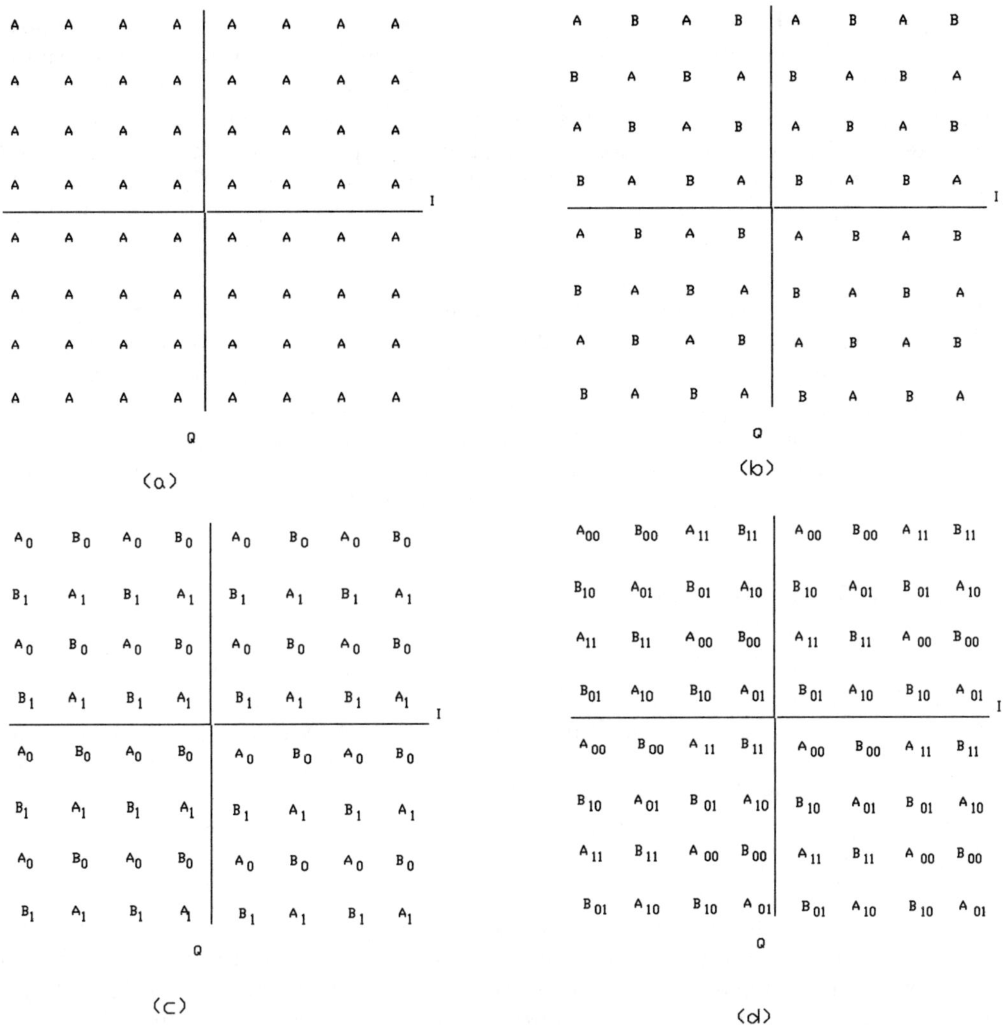

Figure 8.1: Mapping by set partition

surprising consequence of this is that the minimum distance d_0 between two uncoded sequences in two-dimensional space is increased to d_{min} in an expanded N-dimensional space when coding is used. Use of maximum likelihood sequence detection at the receiver yields a coding gain factor of d_{min}^2/d_0^2 in energy efficiency, less the amount of extra energy needed for the transmission of the additional redundant coding bits. Conventional channel coding cannot realise this gain. Instead it is necessary to use a principle described by Ungerboeck [320] as "mapping by set partition".

The basis of set partitioning is shown in Figure 8.1 and described below. The original constellation shown in Figure 8.1(a), which is 64-level square QAM in this example, can be initially divided into two subsets (Figure 8.1(b)) by assigning alternate points A and B to each subset. This has the result that the minimum squared distance between points within a subset is twice that between points in the original constellation given by d_0^2. Observe that the points in each subset now lie on a rectangular grid rotated by 45 deg with respect to the original grid. This procedure can be repeated to yield 4, 8, 16, ... subsets with similar properties and intrasubset squared distances of 4, 8, 16, ... times d_0^2. This is illustrated in Figure 8.1 for 2 subsets (A, B) of 32 points, 4 subsets (A_0, A_1, B_0, B_1) of 16 points and 8 subsets (A_{00}, ..., A_{11}, B_{00}, ..., B_{11}) of 8 points. A relatively simple coding technique can then be employed. Some of the incoming six data bits of a modulation symbol are encoded in a channel coder resulting in a larger number of bits due to the coding process. The additional protection bits are used to select which subset the phasor representing the original input bits will be assigned to. The remaining uncoded bits select the specific constellation point to be transmitted within the subset selected. The coding can be either block or convolutional, although convolutional coding has consistently proved both more effective and simpler to implement for Gaussian channels, and so we concentrate our efforts on convolutionally coded TCM.

8.3 8-PSK TCM

It is not our intention here to describe TCM in detail as Ungerboeck has written an excellent tutorial paper [320] which fully describes TCM, and which this section is based upon. Ungerboeck suggested that, for an AWGN channel, significant gains could be achieved by expanding the symbol set size and using the extra bits accommodated by a higher-level modulation scheme for channel coding, as explained above. In its most common form this is known as trellis coded modulation (TCM).

TCM schemes employ redundant non-binary modulation in combination with a finite state FEC encoder which governs the selection of the modulation signals to generate coded signal sequences. Essentially the expansion of the original symbol set gives more bits per symbol than required by the data rate, and these extra bits are used by a convolutional encoder which restricts the possible state transitions. In the receiver, the noisy signals are decoded by a soft-decision maximum-likelihood sequence decoder. This takes the incoming data stream and attempts to fit it onto each of the legitimate phasor sequences allowed by the constraints imposed by the encoder. The best-fit sequence having the minimum Euclidean distance from the received sequence is used as the most likely estimate of the transmitted sequence.

Simple four-state TCM schemes, where four-state refers to the number of possible states that the encoder can be in, can improve the robustness of digital transmission against additive noise by 3 dB compared to conventional uncoded modulation. With more complex TCM systems the coding gain can reach 6 dB. As opposed to traditional error correction schemes, these gains are obtained without bandwidth expansion, or reduction of the effective information rate. This is because the FEC encoder's parity bits are absorbed by expanding the signal constellation in order to transmit a higher number of bits per symbol. The term "trellis" is used because these schemes can be described by a state transition diagram similar to the trellis diagrams of binary convolutional codes [332]. The difference is that in the TCM scheme the trellis branches

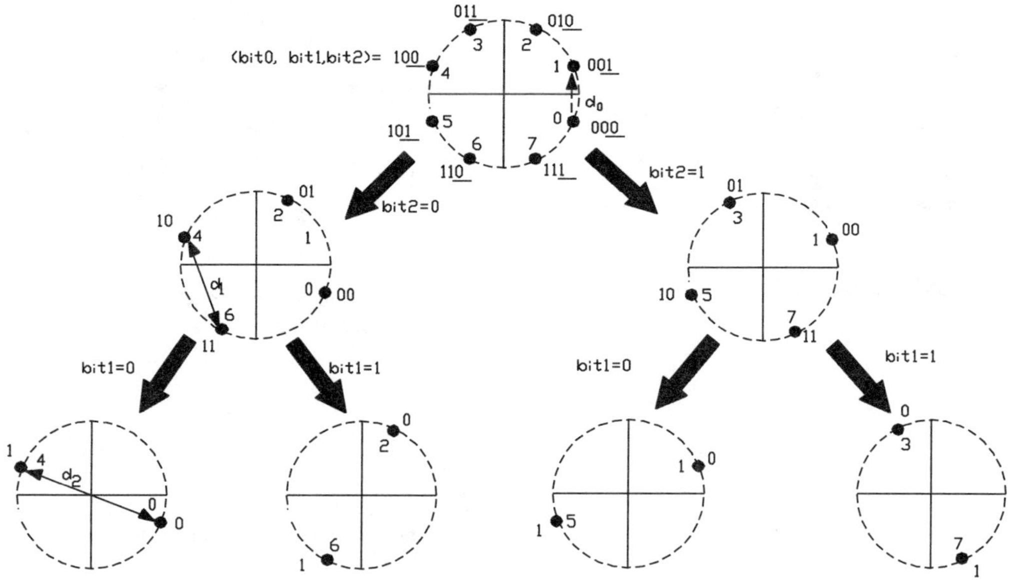

Figure 8.2: 8-PSK set partitioning

are labelled with redundant non-binary modulation phasors rather than with binary code symbols.

We now illustrate the principle of TCM using the example of a four-state trellis code for 8-PSK modulation, as this relatively simple case helps understanding of the principles involved. We will then go on to consider coding schemes suitable for QAM.

The partitioned signal set proposed by Ungerboeck [320] is shown in Figure 8.2 where the binary phasor identifiers are now not Gray encoded. Observe in the figure that the Euclidean distance amongst constellation points is increased at every partitioning step. The underlined last two bits, bit 1 and bit 2, are used to identify one of the four partitioned sets, while bit 0 finally pinpoints a specific phasor in each partitioned set.

Signal sets and state transition diagrams for (a) uncoded 4-PSK modulation and (b) coded 8-PSK modulation with four trellis states are given in Figure 8.3 while the corresponding four-state encoder modulator structure is shown in Figure 8.4. Observe that after differential encoding bit 0 is fed direct to the 8-PSK signal mapper, whilst bit 1 is half-rate convolutionally encoded by a two-stage four-state linear circuit. The convolutional encoder adds the parity bit, bit 2, to the sequence, and these two protected bits are used to identify which constellation subset the bits will be assigned to, whilst the more widely spaced constellation points will be selected according to the unprotected bit 0.

The trellis diagram for 4-PSK is a trivial one-state trellis which serves to indicate uncoded 4-PSK from the viewpoint of TCM. Every connected path through the trellis represents an allowed signal sequence. In both systems, starting from any state, four transitions can occur, as required to encode 2 bits/symbol. The four parallel

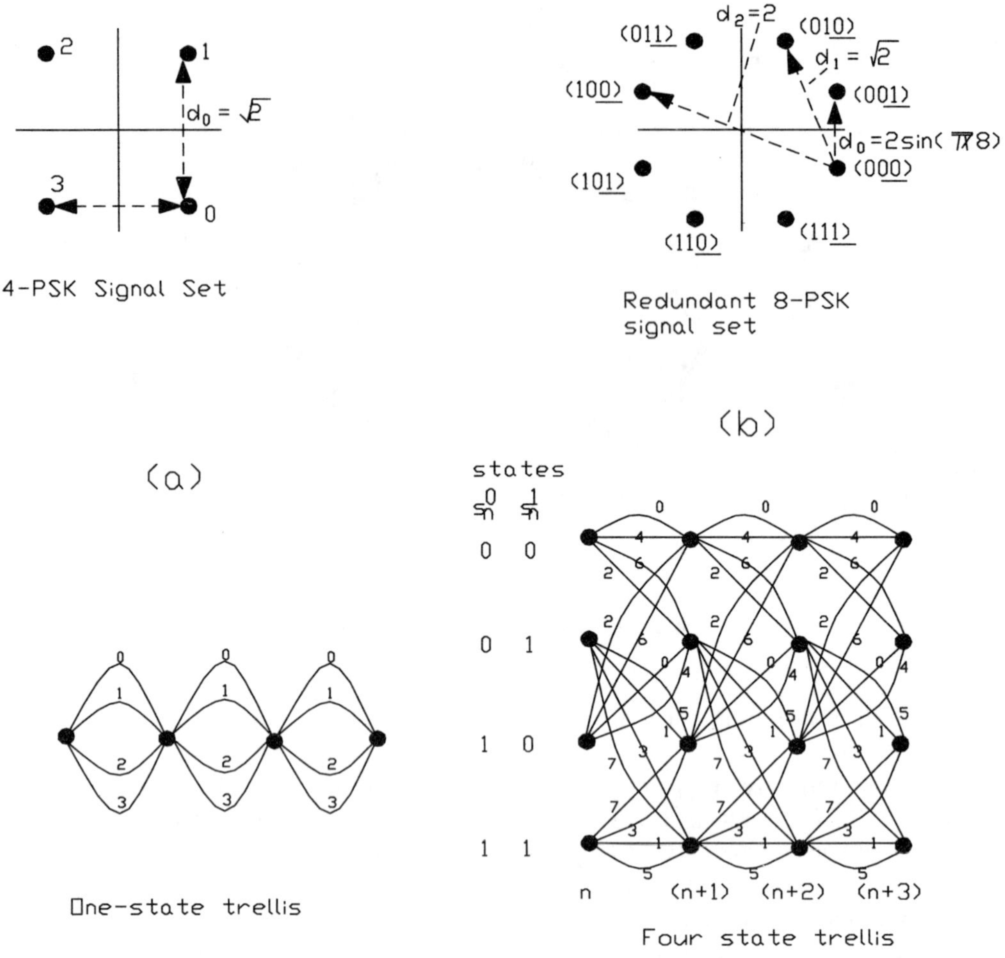

Figure 8.3: Constellation and trellis for 4-PSK and 8-PSK [320] ©IEEE, 1982, Ungerboeck

transitions in the one-state trellis diagram of Figure 8.3(a) do not restrict the sequence of 4-PSK symbols that can be transmitted, i.e. there is no symbol coding and therefore all trellis paths are legitimate. Hence the optimum detector can only make nearest signal decisions for each individual symbol received. The smallest distance between the 4-PSK signals is $\sqrt{2}$, denoted as d_0, and this is termed the free distance of uncoded 4-PSK. Each 4-PSK symbol has two nearest neighbours at this distance. Each phasor is represented by a 2-bit symbol and transitions from any state to any other are legitimate.

The situation for 8-PSK TCM is a little less simplistic. The trellis diagram of Figure 8.3(b) is constituted by four states according to the four possible states of the shift register of Figure 8.4 which we represent by the four vertically stacked bold nodes. Following the elapse of a signal period a new 2-bit input symbol arrives and

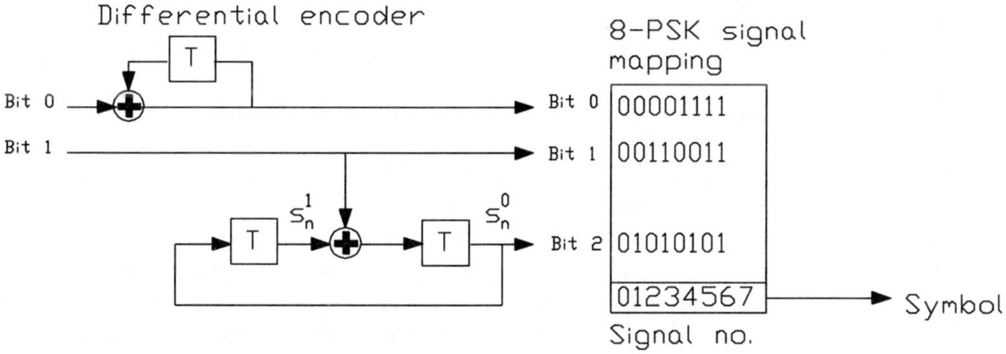

Figure 8.4: Encoder for the four-state 8-PSK trellis [320] ©IEEE, 1982, Ungerboeck

the convolutional encoder's shift register is clocked. This event is characterised by a transition in the trellis from state S_n to state S_{n+1}, tracking one of the four possible paths corresponding to the four possible input symbols.

In the four-state trellis of Figure 8.3(b) for the coded 8-PSK scheme, the transitions occur in pairs and the states corresponding to the bold nodes are represented by the shift register states S_n^0 and S_n^1 in Figure 8.4. Due to the limitations imposed by the convolutional encoder of Figure 8.4 only a limited set of state transitions associated with certain phasor sequences are possible. These limitations allow us to detect and reject illegitimate decoded sequences. For example, when the shift register of Figure 8.4 is in state (0, 0), only transitions to phasor points (0, 2, 4, 6) are legitimate whilst those to phasor points (1, 3, 5, 7) are illegitimate. Observe in the 8-PSK constellation of Figure 8.3 that the underlined bit 1 and bit 2 identify four twin-phasor subsets whose phasors are opposite to each other and hence have a high intra subset separation. The unprotected bit 0 is then invoked to select the required phasor point within the subset. Since the redundant bit 2 is also one of the shift register state bits, namely S_n^0, assuming initial states of $(S_n^0, S_n^1) = (0, 0)$ or $(0, 1)$ only phasors (0, 2, 4, 6) can emerge as also seen in Figure 8.3(b). Similarly, if we have $(S_n^0, S_n^1) = (1, 0)$ or $(1, 1)$ then the branches emerging from these lower two states of the trellis in Figure 8.3(b) can only be associated with phasors (1, 3, 5, 7).

There are other possible codes, for example four distinct transitions from each state to all possible successor states, but the one selected here proved to be the most effective [320]. Within the 8-PSK constellation we have the following distances: $d_0 = 2\sin(\pi/8)$, $d_1 = \sqrt{2}$ and $d_2 = 2$. The 8-PSK signals are assigned to the transitions in the four-state trellis in accordance with the following rules.

(1) Parallel transitions are associated with phasors having a maximum distance of d_2 between them, which is characteristic of phasor points in the subsets (0, 4), (1, 5), (2, 6) and (3, 7). Since these parallel transitions belong to the same subset and are controlled by the unprotected bit 0, symbols associated with them should be as far apart as possible.

(2) All four-state transitions originating from, or merging into any one state are

labelled with phasors having a distance of at least $d_1 = \sqrt{2}$ between them. These are the phasors belonging to subsets (0, 2, 4, 6) or (1, 3, 5, 7).

(3) All 8-PSK signals are used in the trellis diagram with equal frequency.

Observe that the assignment of bits to the 8-PSK constellation of Figure 8.3 does not obey Gray coding; adjacent phasors can have arbitrary Hamming distances between them. However, the bit mapping and encoding process employed is designed to exploit the high Euclidean distances between sets of points in the constellation. The underlined bit 1 and bit 2 representing the convolutional codec's output are identical for all parallel branches of the trellis. For example, the branches labelled with phasors 0 and 4 between states (0, 0) and (0, 0) are associated with bit 1 = 0 and bit 2 = 0, while the uncoded bit 0 can be either 0 or 1. However, due to appropriate code design this unprotected bit has a maximum protection distance of $d_2 = 2$, requiring the corruption of phasor 0 into phasor 4 in order to inflict a single bit 0 error.

The effect of channel errors exhibits itself at the decoder by diverging from the trellis path in the encoder. Let us consider the example of Figure 8.5, where the encoder traversed through the phasors 0-0-0 from state (0, 0), but due to channel errors the decoder's path was different from this, with the phasor sequence 2-1-2 being encountered. The so-called free distance of a TCM scheme can be computed as the lower one of two distances. Namely, the Euclidean distances between parallel branches in the trellis associated with the uncoded bits, which is $d_2 = 2$ in our example, as well as the distances between diverging and remerging trellis paths. The lower of these two distances characterises the error behaviour of the underlying TCM scheme, since the error event associated with it will be the one most frequently encountered. If the received phasors are at a Euclidean distance larger than half the code's free distance from the transmitted phasor, an erroneous decision will be made. It is essential to ensure that by appropriate code design the number of decoded bit errors is minimised in the most likely error events, and this is akin to Gray coding a non-trellis-coded constellation.

The Euclidean distance between parallel branches is $d_2 = 2$ in our example.

The distance between the diverging phasor sequences of 0-0-0 and 2-1-2 portrayed in Figure 8.5 is inferred from Figure 8.3 as d_1-d_0-d_1. By inspecting the trellis in Figure 8.3 we infer that this diverging path is the shortest one that can be found, all other diverging paths have higher free distances from the error-free 0-0-0 path. Furthermore, this is the only path having this minimum free distance. The free distance of this TCM sequence is given by

$$\begin{aligned} d_{free} &= min\{d_2; \sqrt{d_1^2 + d_0^2 + d_1^2}\} \\ &= min\{2; \sqrt{2 + (2.\sin\frac{\pi}{8})^2 + 2}\}. \end{aligned} \tag{8.5}$$

Clearly, the free distance of this TCM scheme is given by the Euclidean distance between parallel trellis branches associated with the uncoded bit 0, i.e.

$$d_{free} = 2. \tag{8.6}$$

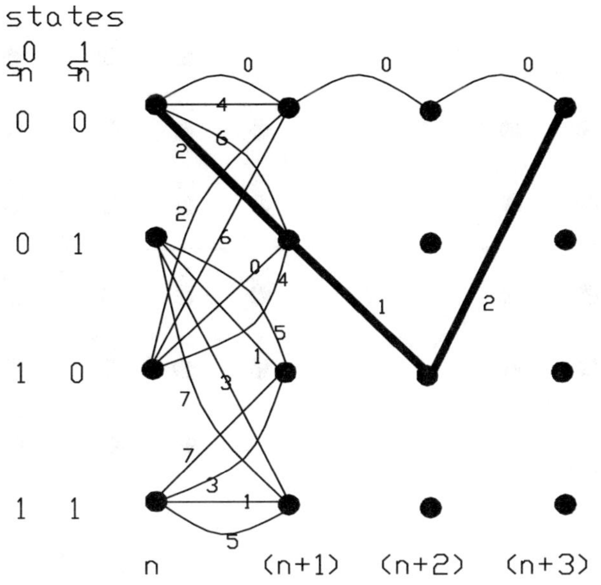

Figure 8.5: Diverging trellis paths for the computation of d_{free}

The free distance of the uncoded 4-PSK constellation of Figure 8.3 was $\sqrt{2}$ and hence the use of TCM has given a coding gain of $g = 2^2/(\sqrt{2})^2 = 2$, which corresponds to 3 dB. There is only one nearest neighbour phasor at $d_{free} = 2$, corresponding to the π-rotated phasor in Figure 8.3. Consequently the phasor arrangement can be rotated by π whilst retaining all of its properties, but other rotations are not admissible.

The number of erroneous decoded bits induced by the diverging path 2-1-2 is seen from the phasor constellation of Figure 8.3 to be 1-1-1, yielding a total of three bit errors. The more likely event of a bit 0 error which is associated with a Euclidean distance $d_2 = 2$ yields only a single bit error.

The soft decision decoding is accomplished in two steps. The first step is known as subset decoding, where within each phasor subset assigned to parallel transitions, i.e. the uncoded bits, the phasor closest to the received channel output in terms of Euclidean distance is determined. In the second step, the Viterbi algorithm is used to find a signal path through the trellis with the minimum sum of squared Euclidean distances from the sequence of noisy channel outputs received. Only the signals already selected by the subset decoding are considered. For a description of the Viterbi algorithm the reader is referred to the literature [328–332].

8.4 16-QAM TCM

For the case of, say, 16-level QAM, TCM causes an expansion of the constellation to a minimum of 32 levels. We then have 5 bits per symbol while previously we had 4, and the extra bit is used for FEC encoding. We partition the constellation

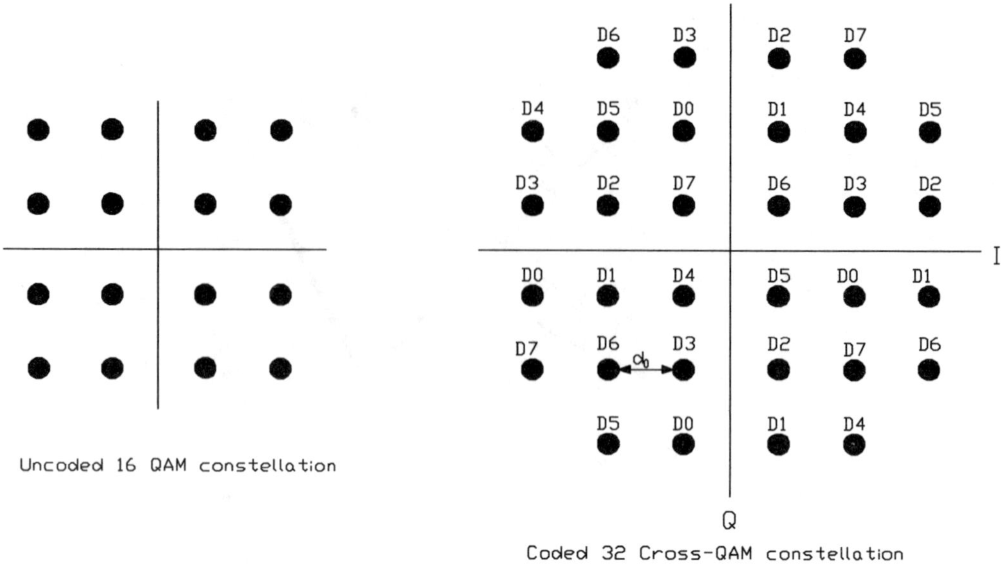

Figure 8.6: Set partitioning of a 32-level QAM constellation

into eight groups, known as D-groups, numbered from 0 to 7, each consisting of four points as portrayed in Figure 8.6. The partitioning is performed in such a way that the four uncoded constellation points representing two uncoded bits of any of the eight coded groups are widely spaced across the constellation, thus decreasing the probability that one of these uncoded four points will be mistakenly decoded for any of the other three points within that subset. Such an error would not be corrected as these bits are unprotected. This procedure is adopted for each of the eight sets of four bits. This has the effect that originally neighbouring constellation points reside within different D-groups. An error to a neighbouring point will therefore result in an incorrect D-group being detected. However, due to the FEC coding employed with these groups, this error may be corrected.

Essentially, of the five bits which now constitute a symbol, two bits have a low error probability, and three bits a higher one. The three bits, though, are protected by a 2/3 rate FEC code which has the effect of decreasing the error probability of the two protected bits to near that of the uncoded two bits. It should be noted that no two points of the same D-group are in close proximity. Of the incoming data, two bits are directly mapped onto one of the four states in each D-group, and the other two bits are passed to the convolutional encoder which generates a third parity bit. This is shown in Figure 8.7. These three bits are mapped onto one of the eight D-groups.

For this constellation the minimum distance between any two points is d_0. Between phasor points of the subsets forming the first division, i.e. between any point in the set (D0, D2, D4, D6) and any point in the set (D1, D3, D5, D7) the minimum distance will be $\sqrt{2}d_0$ as can be inferred from the geometry of Figure 8.6. Between the points of the subsets forming the next division, i.e. between any point in the set (D0, D4) and any point in the sets (D1, D5), (D2, D6) or (D3, D7), the minimum distance

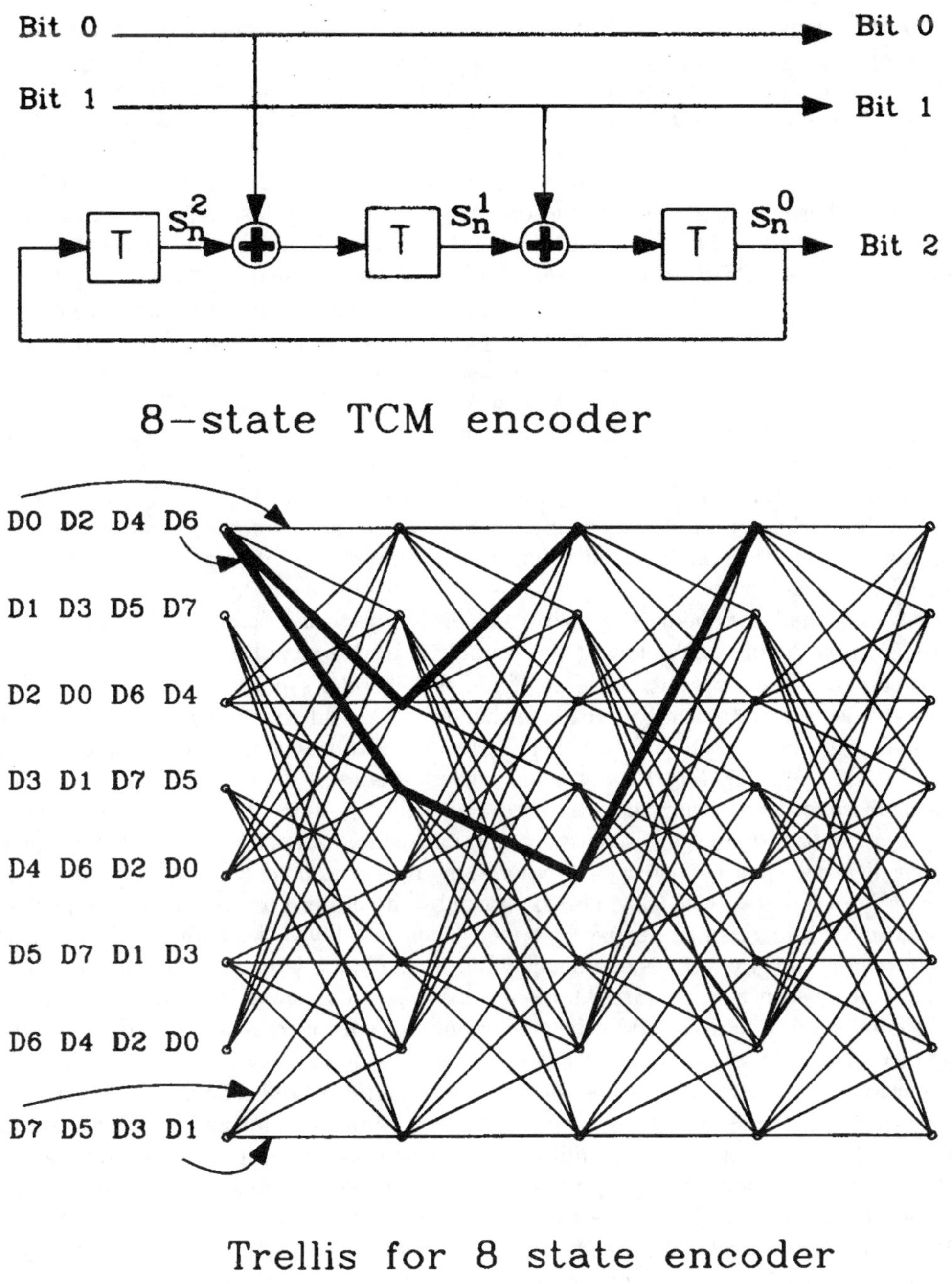

Figure 8.7: 16-QAM TCM encoder and trellis [320] ©IEEE, 1982, Ungerboeck

will be $\sqrt{4}d_0$. Finally, between any two points of any subset in the final partition the minimum distance will be $\sqrt{8}d_0$.

Since the convolutional encoder of Figure 8.7 now has three state bits (S_n^0, S_n^1, S_n^2), the trellis representing this TCM scheme has eight states. The redundant bit, bit 2, is also called S_n^0. In the eight-state trellis shown, four transitions diverge from and merge into each state. The overall signal set of Figure 8.6 is denoted by A0, which contains 2^{m+1} levels, where $m = 4$ is the original number of bits/symbol required to be sent per modulation interval. Each of the subsets, or D-groups, then contains $2^{m-2} = 4$ phasor points. The transitions in the trellis therefore represent $2^{m-2} = 4$ parallel transitions. The assignment of signal subsets to transitions satisfies the same rules as the coded PSK, appropriately adapted to the cross constellation.

The four transitions to or from the same state are always assigned either the set (D0, D2, D4, D6) or (D1, D3, D5, D7) according to the value of the redundant bit 2. Since each D-group hosts four phasors, this allows 16 legitimate phasors to be transmitted from each node. By consulting Figure 8.6 it can be seen that the intrasubset distances are always larger than the minimum distances between points of different D-groups. The order of assignment of the even or odd transition sets (D0, D2, D4, D6) or (D1, D3, D5, D7) against each of the states in Figure 8.7 can be found by examining the action of the encoder in Figure 8.7. For example, in state (0, 0, 1), the next to top state on the trellis diagram, an input pair of (bit 1, bit 0) = (0, 0) changes the encoder to (1, 0, 0) and leads to an output of (0, 0, 0) thus moving from state 1 to 4, the latter being the fifth from top on the trellis diagram. The input pair of (0, 0) corresponds to the lowest path number, in this case D1.

When the decoder's path diverges from the encoder's path at any state in Figure 8.7 this can happen only within either even or odd indexed groups, i.e. from any path associated with (D0, D2, D4, D6) to any other within this set, or from any path in (D1, D3, D5, D7) to any other within this set. This guarantees a squared phasor distance of at least $2d_0^2$ when sequences diverge plus the same amount when they immediately remerge. If paths remerge after two transitions then the squared phasor distance must be at least $4d_0^2$ when they diverge and $2d_0^2$ when they reconverge, making a total distance of $6d_0^2$. This is shown by the uppermost of the bold sequences in Figure 8.7 which is erroneously decoded when 0-0-0 was transmitted. In the first symbol period a D4 phasor is erroneously decoded when a D0 was transmitted. These points can be seen from Figure 8.6 to be separated by a squared distance of $4d_0^2$. In the second symbol period a D0 subset is erroneously decoded as a D2, with a squared distance between them of $2d_o^2$.

For longer divergent sequences, such as the lowermost bold sequence of Figure 8.7, at least one intermediate transition must contribute an additional squared signal distance d_0^2. This means that the minimum differing path length involves a divergence within a four-member subgroup giving $2d_0^2$, a transition giving d_0^2 and a reconvergence giving $2d_0^2$, totalling $5d_0^2$. This is smaller than the $8d_0^2$ squared intrasubset distance. The minimum distance is therefore $\sqrt{5}d_0$. Observe that although this lower bold divergent path has remerged one signalling interval later than the upper one, it had a lower minimum distance from the transmitted path.

Two further differing code paths with this distance are shown in Figure 8.8, where the upper bold path indicates the tranmitted sequence, while the lower one indicates

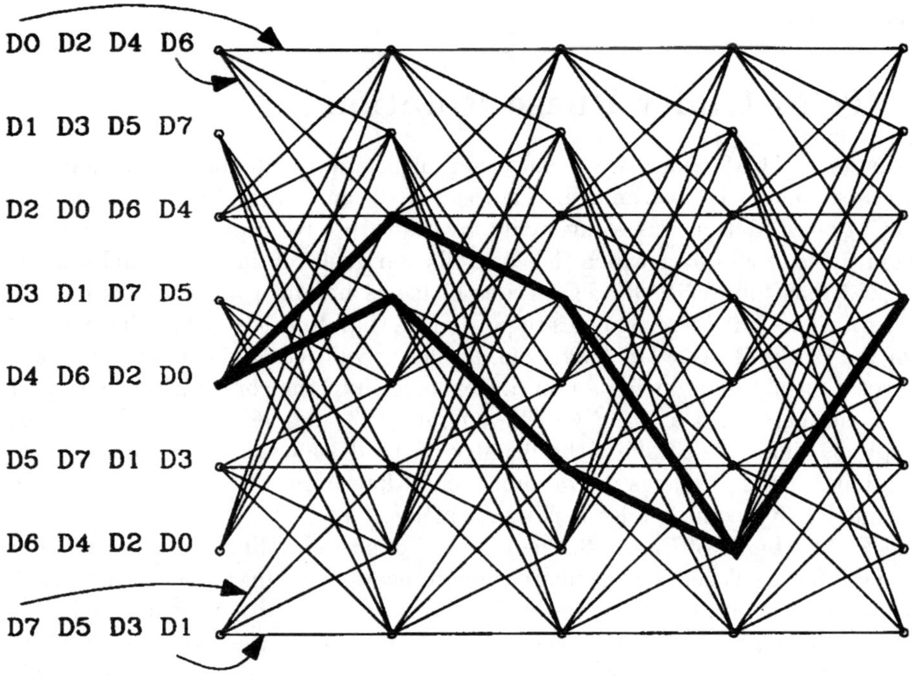

Figure 8.8: 16-QAM TCM path differing by the minimum distance

the erroneous received one. The divergence at the first node is caused by a D2 transition being erroneously decoded as a D0 transition with a squared Euclidean distance of $2d_0^2$ as can be seen from Figure 8.6. At the next node D4 is erroneously decoded as D1 with a squared Euclidean distance of d_0^2, and at the final node before remerging D7 is erroneously decoded as D1 encountering a squared Euclidean distance of $2d_0^2$. Therefore, the total squared Euclidean distance by which these two sequences differ is $5d_0^2$.

The coding gain for this scheme is given by

$$G_c = 10\log_{10}\left[\frac{d^2_{free,coded}}{d^2_{free,uncoded}}\frac{E_{uncoded}}{E_{coded}}\right] \tag{8.7}$$

where $E_{uncoded}$ and E_{coded} are the average signal energies for the uncoded 16-level constellation and the coded 32-level constellation, respectively. The increase in terms of squared distance is a factor of 5, and upon adapting Equation 8.1 for the 16-QAM/32-TCM case the increase in signal energy will always be approximately 2. Hence the coding gain $G_c = 10\log_{10}(5/2) \approx 4$ dB.

Ungerboeck surmises [320] that in general it is possible to gain 3 dB when using a two-stage shift register associated with 4 trellis states, 4 dB with 8 states, 5 dB with 16 states and up to 6 dB with 128 or more states. In his paper he publishes code tables giving optimal codes for several different modulation schemes and numbers of encoder states.

8.5 TCM Under Phase Rotation

Unfortunately, TCM codes are usually not invariant to all phase rotations under which the signal set is invariant. In Figures 8.3(b) and 8.2 we have seen that the 8-PSK TCM constellation was rotationally invariant to a rotation of π, but not for the other multiples of $\pi/4$ over which the constellation itself retains rotational symmetry. Oeder [333] has found 8-PSK TCM constellations having rotational symmetry for $(n.\pi)/4$ rotations but these schemes require more than four states to achieve a coding gain of 3 dB. Similarly, Wei [334] found that a code could be made invariant to 90 deg rotations by introducing a non-linear device into the convolutional coder. This system is actually used in the CCITT standard V.32 modem and will be discussed in Chapter 9. This code suffers the penalty of 0.8 dB lower coding gain than linear convolutional codes. The system designer must determine whether such a penalty is worthwhile for overcoming false locking problems.

The TCM scheme of Figure 8.3 requires coherent demodulation if low error rates are to be achieved. Where loss of phase synchronisation is unlikely this is not problematic, as by inspection of the metrics in the Viterbi decoder it is possible to determine that an incorrect phase lock has been achieved, and rotate the received constellation appropriately. Alternatively, differential encoding can be used, as discussed in Chapter 11, to overcome this problem, albeit at the cost of a differential encoding SNR or BER penalty.

8.6 Summary

TCM has proved to be a useful coding technique over high SNR bandlimited channels, in particular telephone lines. Ungerboeck's original TCM schemes led to a power efficiency improvement of some 4 dB over uncoded schemes at a reasonable decoder complexity. Modern advances have improved this gain slightly at the cost of a substantial increase in decoder complexity. TCM is a rapidly evolving area, in particular for fading mobile channels, which was detailed for example in [321,323]. In this chapter our intention has been to introduce the reader to its principles. The decoding and demodulation of TCM schemes is typically performed using maximum likelihood sequence estimation (MLSE) based on the Viterbi algorithm. A few associated issues of MLSE for equalisers have been touched upon in Section 7.6, but for a full treatment the reader is referred to the literature [327–332].

In the next chapter we detail the first and most widespread use of TCM, the high bit rate transmission of information over telephone lines, and we show that the combination of quadrature modulation and advanced TCM schemes with channel equalisers has dramatically increased the amount of data which can be transmitted via such a channel.

Chapter 9

Classic QAM Modems for Bandlimited AWGN Channels

9.1 Introduction

The most widely used AWGN communication channel is the conventional telephone line. Since its amplitude and group delay characteristics vary slowly, mainly due to ageing, humidity, etc., telephone lines can be considered as linear and time invariant. They are typically bandlimited to 300-3000 Hz and have SNR values around 30 dB. The international lines must comply with the CCITT M1020 Recommendation, the amplitude and group delay characteristics of which were shown in Figure 2.5.

The early history of modems for AWGN channels is comprehensively documented in Reference [324]. Early data transmission modems designed in the 1950s used frequency shift keying (FSK) at 300 bps for switched telephone lines and at 1200 bps for private or leased lines. From the 1960s onward, in conjunction with the upgrading of the telephone network, the equalised bandwidth was gradually extended to 1200 Hz, 1600 Hz and 2400 Hz. This allowed signalling rates to increase from 1200 Hz to 2400 Hz, and the higher SNR values facilitated the transmission of several bits per sample.

An important milestone was the launch of the first four-phase 2400 bps modem in 1962, which was later standardised in the V.26 CCITT Recommendation. This was followed in 1967 by the introduction of the eight-phase differential phase shift keying (8-DPSK), 4800 bps modem using a signalling rate of 1600 Bd and 3 bits/sample. To achieve this signalling rate over leased lines a manual channel equaliser (EQ) was needed, as specified by the V.27 CCITT Recommendation. However, with rapid advances in digital equalisation, the V.27bis and V.27ter Recommendations quickly followed, proposing adaptive channel equalisation, which allowed operation via switched telephone lines.

In 1971 the first 4 bits/symbol 16-QAM modem was developed, which deployed an advanced channel equaliser to achieve a signalling rate of 2400 Bd. This modem trans-

Rate (bps)	Baud rate (Bd)	Constellation	Standard	Features
2400	1200	4-DPSK	V.26, 1968	fixed EQ
4800	1600	8-DPSK	V.27, 1972	manual EQ
9600	2400	16-QAM	V.29, 1976	adapt. EQ
9600	2400	32-QAM, TC	V.32, 1984	adapt. EQ
14 400	2400	128-QAM, TC	V.33, 1988	adapt. EQ

Table 9.1: Summary of modem features

mitted at 9600 bps over point-to-point four-wire leased lines and it was standardised in 1976 in the V.29 CCITT Recommendation. This development was further improved to facilitate 9600 bps transmissions via two-wire switched telephone lines and it was standardised in the V.32 Recommendation in 1984. Although the basic configuration uses a square-shaped 16-QAM constellation, it also has a 32-point trellis coded option to improve its robustness against channel errors. The most advanced QAM scheme at the time of writing, standardised in 1988 in the V.33 Recommendation, is the 14 400 bps point-to-point four-wire modem, which is recommended for leased lines. It uses trellis coded (TC) 128-QAM for 14 400 bps transmissions, but it also allows trellis coded 64-QAM transmissions at a fallback rate of 12 000 bps over lower quality channels. The history of point-to-point multilevel modems is summarised in Table 9.1.

9.2 Transmission Bit Rate Limits

As we have seen before, in case of square QAM constellations both the I and Q components assume the equiprobable, equispaced transmission levels of $\pm d$, $\pm 3d$, $\ldots$, $\pm(2^m - 1)d$, where m is the number of bits per quadrature component. The average energy per quadrature component [324] is then

$$S_m = (4^m - 1)/3, \tag{9.1}$$

which gives for 16-QAM $S_2 = 5d^2$ and for 64-QAM $S_3 = 21d^2$. Then the total constellation energy is $E_{16\text{-QAM}} = 10d^2$ and $E_{64\text{-QAM}} = 42d^2$. From Equation 9.1 it follows that

$$S_{m+1} = 4S_m + 1, \tag{9.2}$$

and hence when increasing the number of bits per quadrature component by one, the average energy is increased by about a factor of 4, or 6 dB.

The channel capacity C of the AWGN channel bandlimited to B Hz was determined by Shannon [325] as follows:

$$C = B\log_2(1 + S/N) \text{ bit/sec.} \tag{9.3}$$

Observe that for high SNR $= S/N$ values and constant AWGN power N, the channel capacity doubles when the signal power S is quadrupled by transmitting an additional bit per quadrature component, for example when transmitting 64-QAM in

Absolute phase	Q_1	Relative signal element amplitude
0 deg, 90 deg, 180 deg, 270 deg	0	3
	1	5
45 deg, 135 deg, 225 deg, 315 deg	0	$\sqrt{2}$
	1	$3\sqrt{2}$

Table 9.2: Phasor amplitude mapping for the V.29 modem

place of 16-QAM. In other words, an additional 6 dB channel SNR is required, if the channel capacity is to be doubled by transmitting two extra bits per complex QAM phasor. This fact manifests itself, for example, in an approximately 6 dB shift of the 4 bits/symbol 16-QAM BER curve to higher SNRs, when using 6 bits/symbol 64-QAM transmissions. Assuming that sampling takes place at the Nyquist rate of $f_s = 2B$, the maximum information conveyed by a QAM channel symbol per quadrature component is

$$\frac{C}{2B} = \frac{1}{2}\log_2(1 + \frac{S}{N}) \text{ bit/quad.comp..} \tag{9.4}$$

Using a lower number of bits/quadrature component clearly does not fully exploit the maximum capacity of the channel. Accordingly, in the bandlimited telephone channel the achievable transmission rate depends on the number of bits per symbol, which is governed by the channel SNR available.

In the following sections we consider how modems have evolved towards high capacity systems by considering recent CCITT recommendations for modems [335], commencing with the V.29 modem.

9.3 V.29 Modem

The 9600 bps V.29 modem was recommended for use over point-to-point four-wire leased telephone lines, but its users are not precluded from using it over lower quality links. There are provisions for optional fallback rates of 7200 and 4800 bps when the channel conditions are hostile. A generic block diagram for all the modems we consider in this chapter is shown in Figure 9.1.

9.3.1 Signal constellation

The 9600 bps input data stream is "scrambled" using a linear feedback shift register in order to assist the carrier and clock recovery operation by preventing a long sequence of logical 1's or 0's. The scrambled data sequence is then converted into 4 bits/symbol called "quadbits", which are Gray coded onto the signal constellation depicted in Figure 9.2. The first bit (Q_1) of every symbol determines the amplitude of the signal element to be transmitted, but the amplitude is also dependent on the instantaneous absolute phase, and the two different sets of amplitudes used are given in Table 9.2. The remaining three bits Q_2, Q_3 and Q_4 are differentially Gray coded onto the phase positions, as summarised in Table 9.3. After differential mapping of the bits, 16-QAM

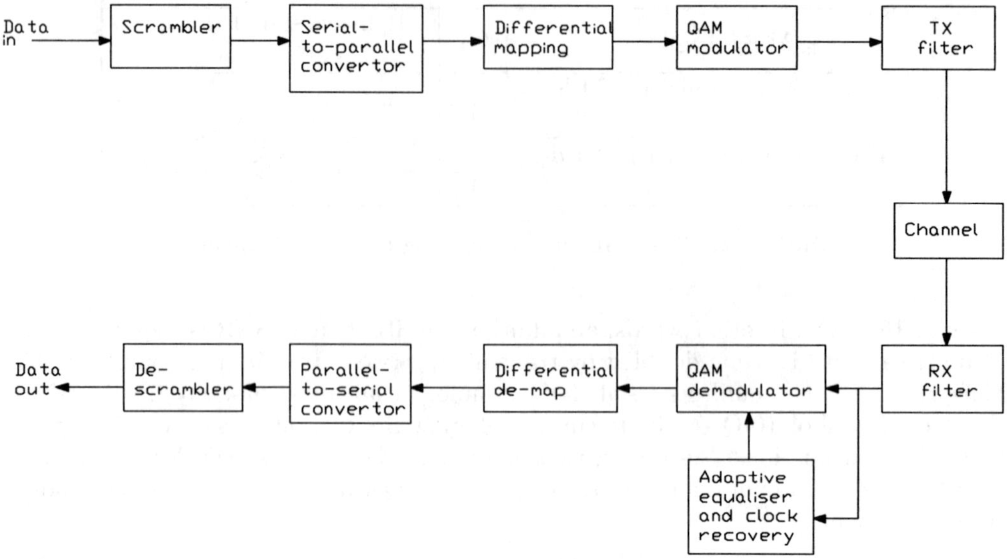

Figure 9.1: General modem schematic for bandlimited AWGN channels

Q2	Q3	Q4	Phase change
0	0	1	0 deg
0	0	0	45 deg
0	1	0	90 deg
0	1	1	135 deg
1	1	1	180 deg
1	1	0	225 deg
1	0	0	270 deg
1	0	1	315 deg

Table 9.3: Phasor phase mapping for the V.29 modem

modulation is carried out using the signal constellation of Figure 9.2 at a signalling rate of 2400 Bd.

At the fallback rate of 7200 bps the bit Q_1 is permanently set to zero, limiting the amplitude to either $\sqrt{2}$ or 3, depending on the phase information, as seen in Figure 9.3. Otherwise the mapping is identical to that of the 9600 bps constellation. The bit rate is 2400 $Bd \times 3$ bits/symbol=7200 bps. In case of 4800 bps transmissions $Q_1 = 0$ and $Q_4 = \overline{(Q_2 \oplus Q_3)}$ are set, where $\oplus$ means modulo 2 addition and $\overline{(\ )}$ is the inverse of $(\ )$, yielding the mapping summarised in Table 9.4. The resulting constellation is shown in Figure 9.4 and the 2 bits/symbol transmissions at 2400 Bd yield a bit rate of 4800 bps.

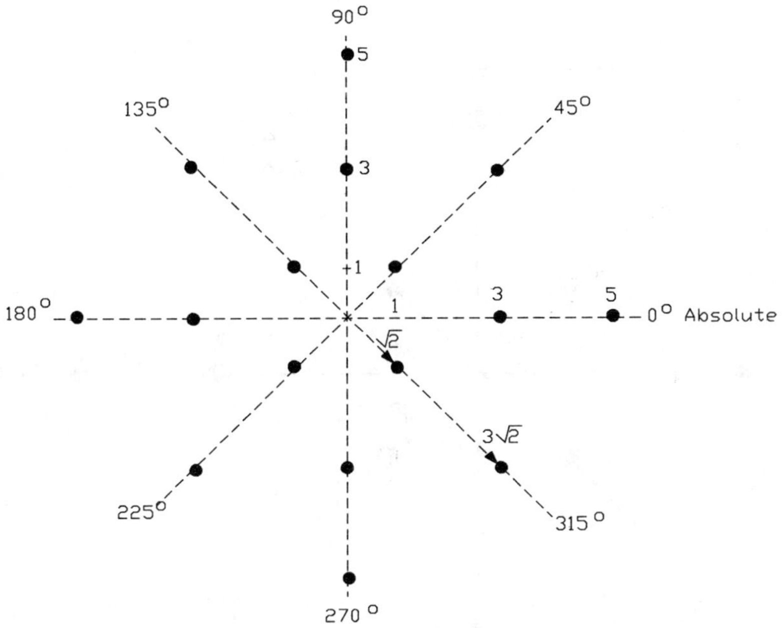

Figure 9.2: Phasor constellation for 9600 bps operation

Data bits		Quadbits				Phase change
		Q1	Q2	Q3	Q4	
0	0	0	0	0	1	0 deg
0	1	0	0	1	0	90 deg
1	1	0	1	1	1	180 deg
1	0	0	1	0	0	270 deg

Table 9.4: Mapping table for 4800 bps operation

9.3.2 Training signals

The operation of the adaptive channel equaliser, carrier recovery and timing recovery circuits is aided by the transmission of specific training signals which allow characterisation of the telephone line in terms of its impulse response, and the establishment of timing references. These are summarised in Table 9.5 for the V.29 modem. The training sequence is split into four segments, each containing different signals. Segment 1 contains no energy and forms a so-called precursor. Segment 2 consists of the alternate sequence of phasors A and B for the duration of 128 symbol intervals, where phasors A and B are depicted in Figure 9.5 for the transmission rates of 9600, 7200 and 4800 bps, respectively. Observe in the figure that symbol A determines the absolute phase reference of 180 deg. The segment 2 sequence toggles between two distinct phasor states at a signalling rate of 2400 Bd, hence it is periodic at 1200 Hz, therefore having a strong 1200 Hz frequency domain component which assists timing

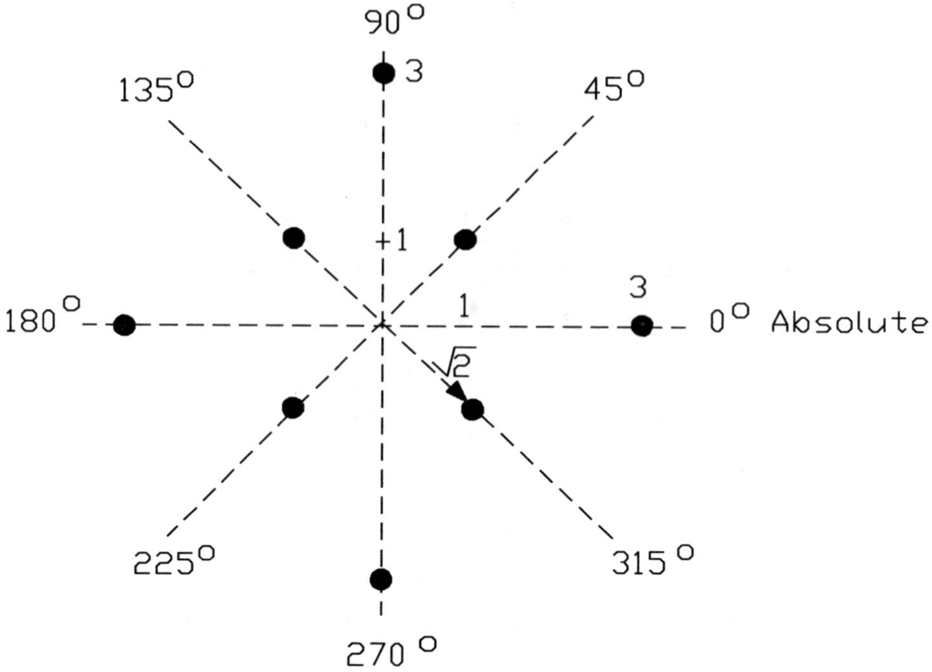

Figure 9.3: Signal space diagram at 7200 bps

Training segments	Type of line signal	No. sym. intervals	Approx. time (ms)
Segment 1	No transmitted energy	48	20
Segment 2	Alternations	128	53
Segment 3	Equaliser conditioning pattern	384	160
Segment 4	Scrambled all binary 1's	48	20
Total signal	Total synchronising signal	608	253

Table 9.5: Training signals for the V.29 modem

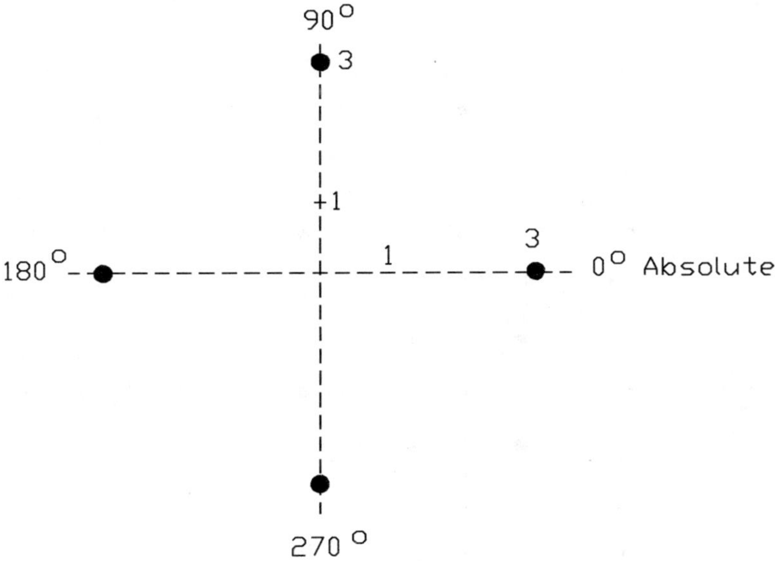

Figure 9.4: Signal space diagram at 4800 bps

recovery, as explained in Chapter 6. By the time that segment 2 is finished the timing recovery must be established.

Segment 3 of the training pattern is designed to initialise the adaptive channel equaliser by transmitting a pseudorandom (PR) sequence of phasors C and D. This pseudorandom sequence can be used to sound the channel. Observe in Figure 9.5 that the phasor D depends on the transmission rate. The PR sequence generator uses the polynomial $p(x) = 1 + x^{-6} + x^{-7}$, and is shown in Figure 9.6, where $\oplus$ represents modulo 2 addition. Every time the PR sequence contains a logical 0, phasor C is transmitted; for a logical 1, phasor D is sent. The seven-stage PR sequence generator is initialised to 0101010, yielding an initial sequence of CDCDCDC and continues for 384 symbol durations. Since these phasors are widely spaced in the signal constellation, the transmitted signal becomes similar to that of the 2400 Bd BPSK transmissions. These phasors have large Euclidean distances between them and hence are unlikely to be corrupted by AWGN. The same phasors can be used to assist initial equaliser convergence as explained in Chapter 7.

In segment 4 of the training sequence, 48 scrambled binary 1's are transmitted for further updating and fine tuning of the clock and carrier recovery circuits. This is because the initial clock and carrier recovery during segment 2 was performed without equalisation. Now that equalisation is present it is possible to fine-tune these timing recovery circuits. The total training lasts 608 symbol durations, amounting to about 253 ms.

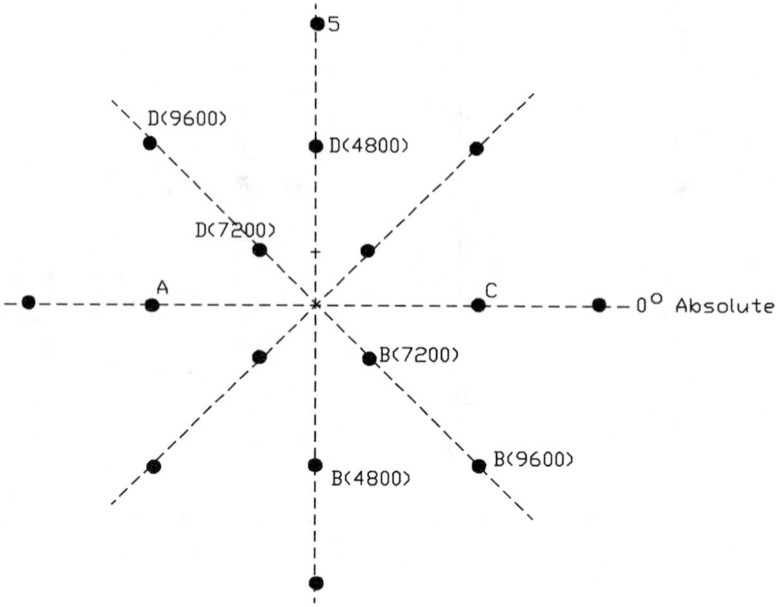

Figure 9.5: Signal space diagram showing synchronising signal points

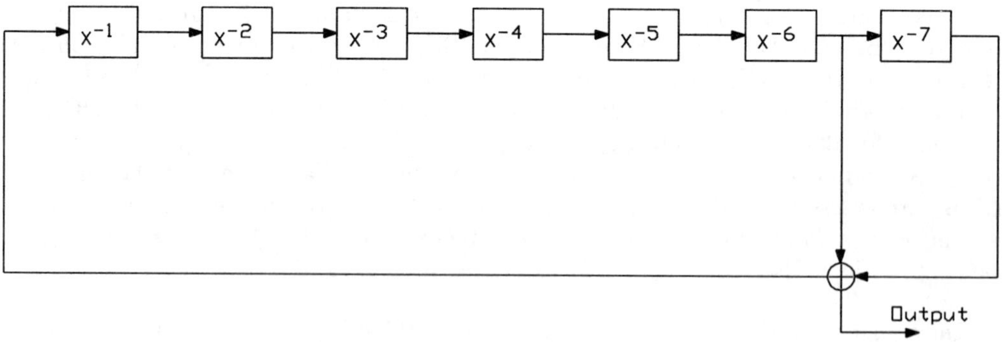

Figure 9.6: Pseudorandom sequence generator

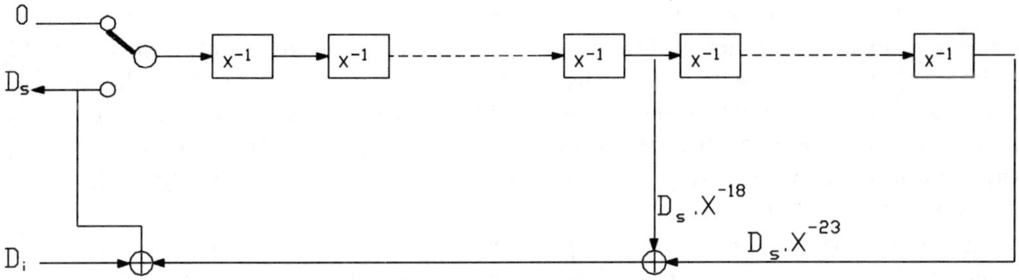

Figure 9.7: V.29 scrambler

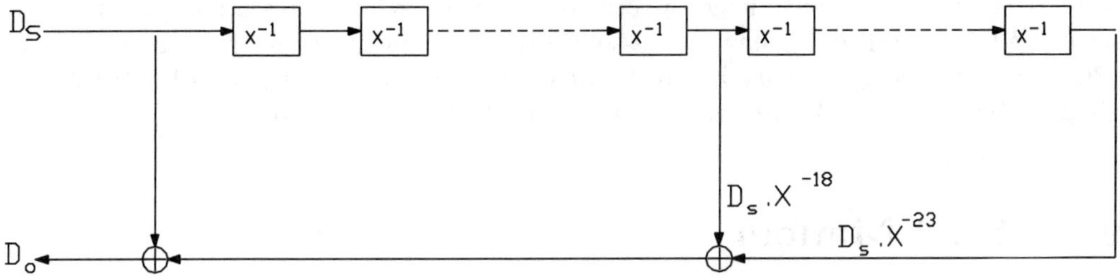

Figure 9.8: V.29 descrambler

9.3.3 Scrambling and descrambling

Data scrambling and descrambling is invoked in order to assist the clock and carrier recovery circuits by preventing the transmission of long sequences of all 1's or all 0's. Its operation is based on the generator polynomial

$$g(x) = 1 + x^{-18} + x^{-23}. \tag{9.5}$$

The scrambler performs a logical operation on the data bits to be transmitted using the generator polynomial $g(x)$. Figure 9.7 shows a suitable circuit structure where D_i are the input data bits and D_s the output data stream. At the beginning of transmissions the switch on the scrambler is set to 0, causing the scrambler to be filled with logical 0's. During this period the input data bits D_i are part of the training sequence and can be sent unscrambled since the input 0's do not affect the sequence D_s. During normal data transmission the scrambled data is fed back to the shift register by setting the switch to D_s. Descrambling is carried out by the circuit of Figure 9.8 by performing the inverse operation using the generator polynomial of Equation 9.5.

9.3.4 Channel equalisation and synchronisation

The use of channel equalisation is compulsory in the modem and the receiver has to be provided with appropriate means of detecting whether the channel has changed significantly since the equaliser training sequence was sent. In this instance the equalisation will no longer be correct, and the receiver sends a training sequence to the original transmitter which indicates to this transmitter that in response it should send another training sequence to the receiver. Upon resynchronisation the modem also has to be able to detect the four-segment initial synchronisation signal sequence described earlier in Table 9.5. When detecting the reception of the training sequence, say from modem A to modem B, then modem B has to initiate the transmission of its own training pattern.

If modem A does not receive a synchronising signal from modem B within the interval of the maximum expected turnaround delay, which is typically 1.2 sec, it sends another synchronising signal. Also, if modem A fails to synchronise on its received sequence, it transmits another four-segment sequence. The carrier frequency is 1700 Hz and the transmitted spectrum measured when transmitting continuous scrambled logical 1 must have an essentially linear phase over the range of 200–2700 Hz. The energy density has to be attenuated at the frequencies of 500 Hz and 2900 Hz by 4.5 ± 2.5 dB with respect to its level at the carrier frequency.

9.4 V.32 Modem

9.4.1 General features

The V.32 Recommendation [335] proposes a family of two-wire duplex modems for transmission rates of up to 9600 bps for the general switched telephone network as well as for point-to-point leased telephone circuits. The signalling rate is 2400 Bd and transmission rates of 9600 bps as well as 4800 bps are supported using a variety of QAM constellations. At 9600 bps either square QAM or trellis coded 32-QAM are used. There is provision for manual as well as for automatic transmission rate selection. The carrier frequency is 1800 Hz and the spectrum, when transmitting logical 1's, has to be attenuated by 4.5 ± 2.5 dB at frequencies of 600 and 3000 Hz, when compared to the carrier frequency.

9.4.2 Signal constellation and bitmapping

9.4.2.1 Non-redundant 16-QAM

The 9600 bps non-redundant square 16-QAM uses a partially differential bit mapping onto the phasor constellation. The scrambled data stream is divided into groups of four bits and the first two bits $Q1_n$ and $Q2_n$ of the nth symbol are differentially encoded to generate the bits $Y1_n$ and $Y2_n$ according to Table 9.6 where the signal state applies to 4800 bps. Following this the bits $Y1_n$, $Y2_n$, $Q3_n$ and $Q4_n$ are then mapped onto the square 16-QAM constellation, as shown in Figure 9.9. In case of 4800 bps transmissions the 2-bit symbols are generated by differentially encoding $Q1_n$ and $Q2_n$ into $Y1_n$ and $Y2_n$, as specified in Table 9.6. The corresponding phasor subset designated by states A, B, C and D coincides with four of the 16-QAM phasors and is also shown in Figure 9.9.

Inputs		Previous outputs		16-QAM outputs		32-QAM outputs		Signal state
$Q1_n$	$Q2_n$	$Y1_{n-1}$	$Y2_{n-1}$	$Y1_n$	$Y2_n$	$Y1_n$	$Y2_n$	
0	0	0	0	0	1	0	0	B
0	0	0	1	1	1	0	1	C
0	0	1	0	0	0	1	0	A
0	0	1	1	1	0	1	1	D
0	1	0	0	0	0	0	1	A
0	1	0	1	0	1	0	0	B
0	1	1	0	1	0	1	1	D
0	1	1	1	1	1	1	0	C
1	0	0	0	1	1	1	0	C
1	0	0	1	1	0	1	1	D
1	0	1	0	0	1	0	1	B
1	0	1	1	0	0	0	0	A
1	1	0	0	1	0	1	1	D
1	1	0	1	0	0	1	0	A
1	1	1	0	1	1	0	0	C
1	1	1	1	0	1	0	1	B

Table 9.6: Differential coding for the V.32 modem

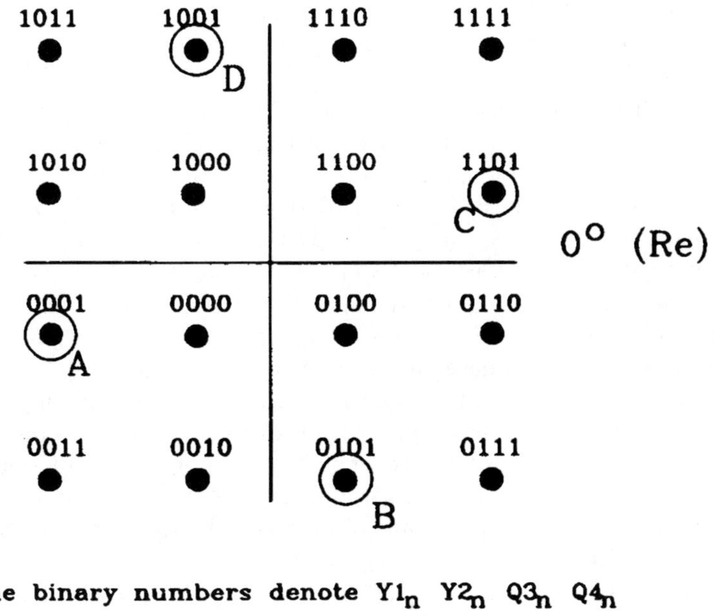

Figure 9.9: V.32 16-point phasor constellation with non-redundant coding for 9600 bps and subset A B C D of states used at 4800 bps and for training

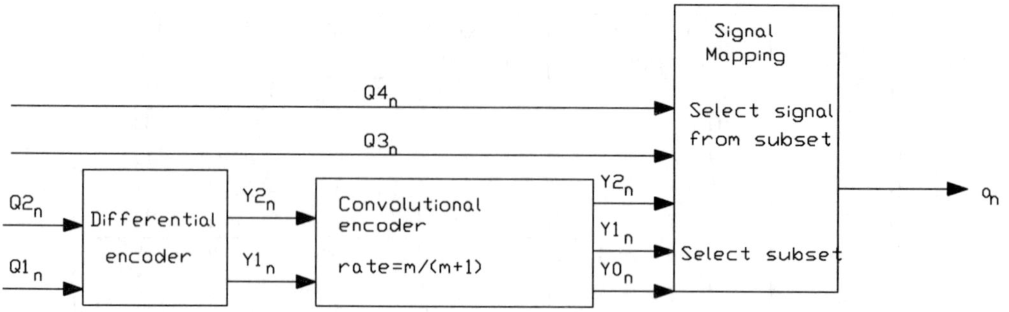

Figure 9.10: General structure of encoder modulator for trellis coded 32-QAM modulation [320] ©IEEE, 1987, Ungerboeck

9.4.2.2 Trellis coded 32-QAM

The fundamentals of trellis coded modulation (TCM) were discussed in Chapter 8; here we focus our attention on its employment in the V.32 modem. The general structure of a trellis coded 32-QAM modulator is shown in Figure 9.10 [320]. When using $(m + 1) = 5$ bits per QAM symbol, the signal constellation is constituted by $2^{m+1} = 32$ phasors as displayed in Figure 9.11.

In our specific case the 32 phasors are subdivided into eight subsets $D0, \ldots, D7$, each constituted by four phasors, which are as far apart from each other, as possible. This is shown in Figure 9.12, where the three most significant bits $Y0_n$, $Y1_n$ and $Y2_n$ are the subset identifiers. Phasors of a subset are encoded by the two least significant bits $Q3_n$ and $Q4_n$.

Phasors of different subsets might be close to each other, hence they are easily corrupted and therefore they need error correction coding. The two most significant bits $Q1_n$ and $Q2_n$ of the 4-bit input symbol are therefore $R = 2/3$ rate, constraint length $\nu = 3$ convolutionally coded to yield three bits $Y0_n$, $Y1_n$, and $Y2_n$, which select a specific subset, i.e. D-group, of the 32-QAM constellation. Accordingly, the constraint length $\nu = 3$ code has $\nu = 3$ storage elements and $2^\nu = 8$ trellis states.

TCM schemes have been shown to have significant coding gains over uncoded constellations [320], but Ungerboeck also demonstrated (see Section 8.5) that they are more sensitive to carrier phase offsets than the less dense uncoded constellations. Therefore, in general it is advantageous to design TCM codes having as many rotationally invariant, phase symmetric phasor positions as possible. This would ensure shorter bursts of errors and faster recovery after temporary loss of synchronisation. However, the overall TCM scheme must be rendered transparent to the data, i.e. insensitive to phase rotations, by means of differential encoding and decoding which typically incurs a BER penalty.

A TCM scheme was found by Wei [336] for the 32-QAM constellation of Figure 9.11 that could achieve a coding gain of 4 dB using an eight-state, constraint length $\nu = 3$ code, while retaining 90 deg phase symmetry. This convolutional encoder contained non-linear elements, as seen in Figure 9.13.

The 32-QAM trellis coded scheme generates the bits $Y1_n$ and $Y2_n$ from the first

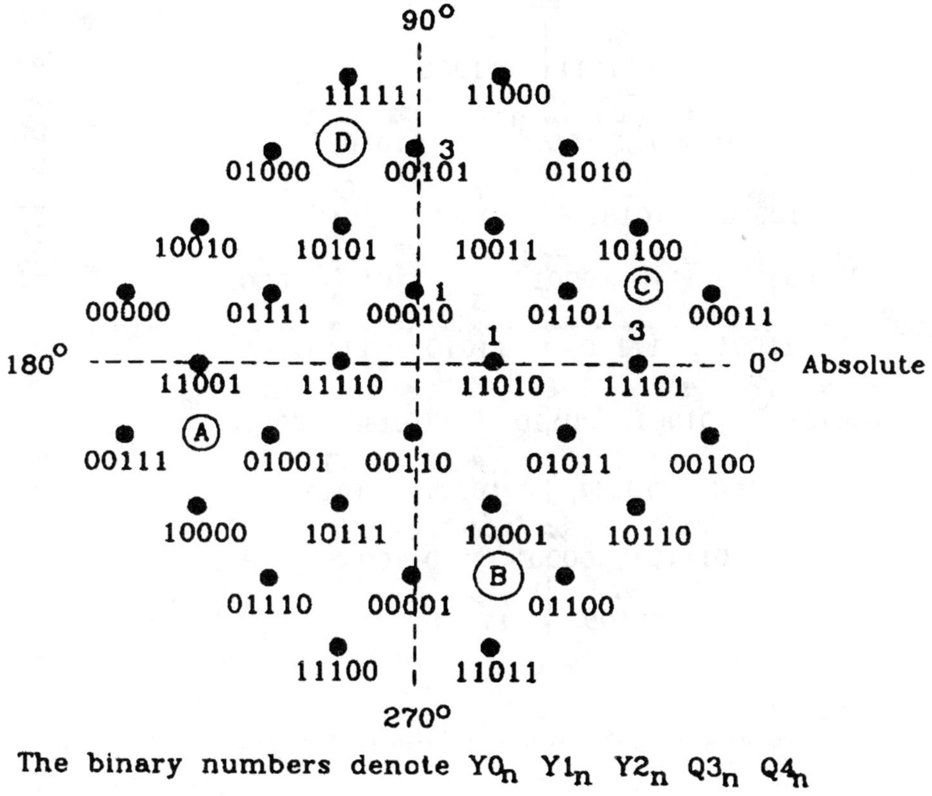

Figure 9.11: V.32 32-point signal structure with trellis coding for 9600 bps and states A, B, C, D Used at 4800 bps and for training

two input bits $Q1_n$ and $Q2_n$ by differential coding as seen in Table 9.6. The differentially encoded bits $Y1_n$ and $Y2_n$ are then input to the $R = 2/3$ rate systematic convolutional encoder depicted in Figure 9.13. The figure also shows the differential encoding and the mapping onto the 32-QAM signal constellation.

Since the convolutional encoder is systematic, its input bits $Y1_n$ and $Y2_n$ are copied to its output and the simple Boolean logic seen in Figure 9.13 specifies the necessary operations to generate the parity bit $Y0_n$. Finally, the bits $Y0_n$, $Y1_n$, $Y2_n$, $Q3_n$ and $Q4_n$ are mapped onto the trellis coded 32-QAM constellation displayed in Figure 9.11. Observe that at the fallback rate of 4800 bps the same states A, B, C and D are used, as in the non-redundant constellation of Figure 9.9 but these now do not coincide with any of the legitimate 32-QAM phasors. The I and Q quadrature component coordinates of these states are identical to those of the corresponding states in Figure 9.9 and hence they are distinctly different from the 32-QAM constellation points.

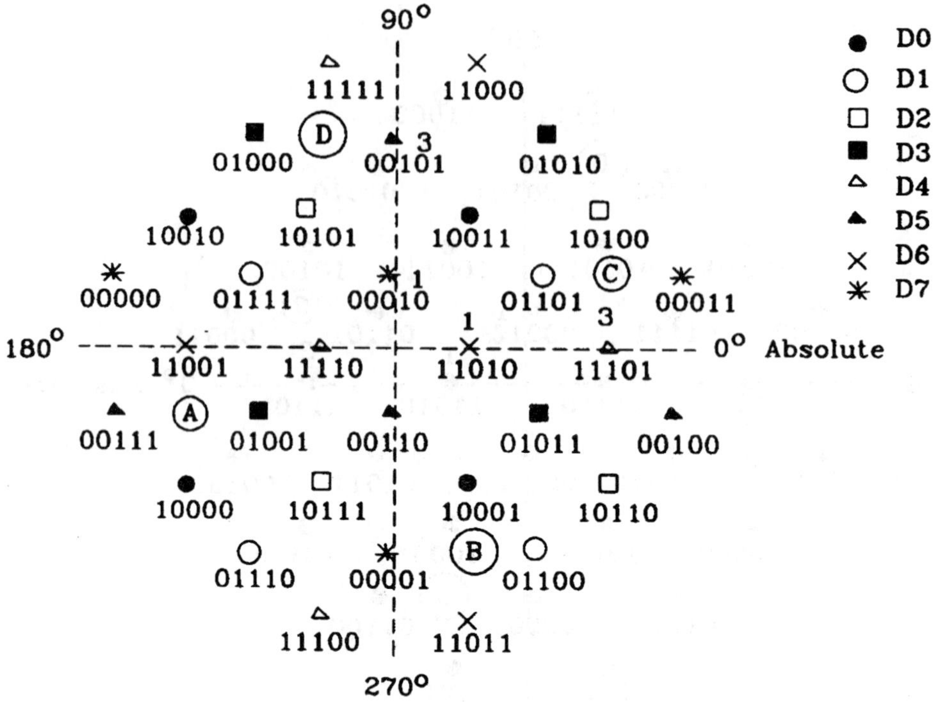

Figure 9.12: V.32 Trellis coded signal subsets $D0, \ldots, D7$

9.4.3 Scrambler and descrambler

In the V.32 modem each transmission direction uses a different scrambler. The calling modem uses the same scrambler as the V.29 Recommendation shown in Figure 9.7, specified by the generator polynomial of Equation 9.5, namely $g_1(x) = 1 + x^{-18} + x^{-23}$. At the calling modem, scrambling is performed by dividing the message data polynomial by the generator polynomial, and the coefficients of the quotient constitute the scrambled data. The answering modem computes the descrambled data signal by multiplying the received data with the generator polynomial $g_1(x)$, while scrambling is carried out using:

$$g_2(x) = 1 + x^{-5} + x^{-23} . \tag{9.6}$$

This long division can be carried out by means of a similar circuit to that of Figure 9.7; however, the feedback taps of the linear shift register are at the positions given by the generator polynomial $g_2(x)$.

The initial set-up and training procedures of the V.32 modem are much more complicated than for the V.29 modem, which were briefly described in the previous section. These training procedures tend to use the signalling states A, B, C and D or some subset of them to assist the initial set-up of the carrier and clock recovery circuitries, as well as that of the adaptive channel equaliser.

Since this book is concerned with QAM modulation, we will not consider further

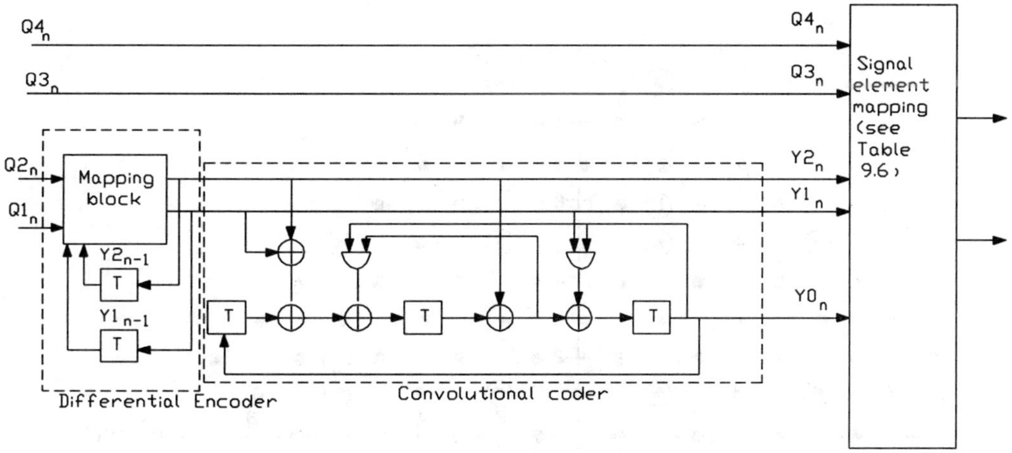

Figure 9.13: Trellis coding at 9600 bps [336] ©IEEE, 1984, Wei

standardised features of the V.32 modem. The interested reader is referred to the newest edition of the CCITT V-Series Recommendations [335].

9.5 V.33 Modem

9.5.1 General features

The 14 400 bps V.33 modem is designed primarily for special quality four-wire leased telephone lines, which comply with the CCITT M.1020 Recommendation shown in Chapter 2. It has a 12 000 bps fallback rate and includes an eight-state trellis -coded modulator and an optional multiplexer to combine data rates of 2400 bps, 4800 bps, 7200 bps, 9600 bps and 12 000 bps. The carrier frequency is 1800 Hz and the signalling rate is 2400 Baud. The generator polynomial for the scrambler and descrambler is identical to that of the V.29 modem, namely $g(x) = 1 + x^{-18} + x^{-23}$.

9.5.2 Signal constellations and Bitmapping

The 14 400 bps scrambled data stream is divided into groups of six consecutive bits and the first two bits $Q1_n$ and $Q2_n$ are differentially encoded into $Y1_n$ and $Y2_n$, as we have seen in Table 9.6 for the V.32 modem. The differentially encoded bits $Y1_n$ and $Y2_n$ are convolutionally encoded using the same systematic, non-linear convolutional encoder as in case of the V.32 modem, which was portrayed in Figure 9.13. The $R = 2/3$ rate systematic convolutional encoder generates the parity bit $Y0_n$, which is concatenated to the differential encoder's output bits $Y1_n$ and $Y2_n$ and to the uncoded input bits $Q3_n$, $Q4_n$, $Q5_n$ and $Q6_n$. This 7-bit sequence $Q6_n$, $Q5_n$, $Q4_n$, $Q3_n$, $Y2_n$, $Y1_n$, $Y0_n$ is then mapped onto the 128-QAM signal constellation seen in Figure 9.14.

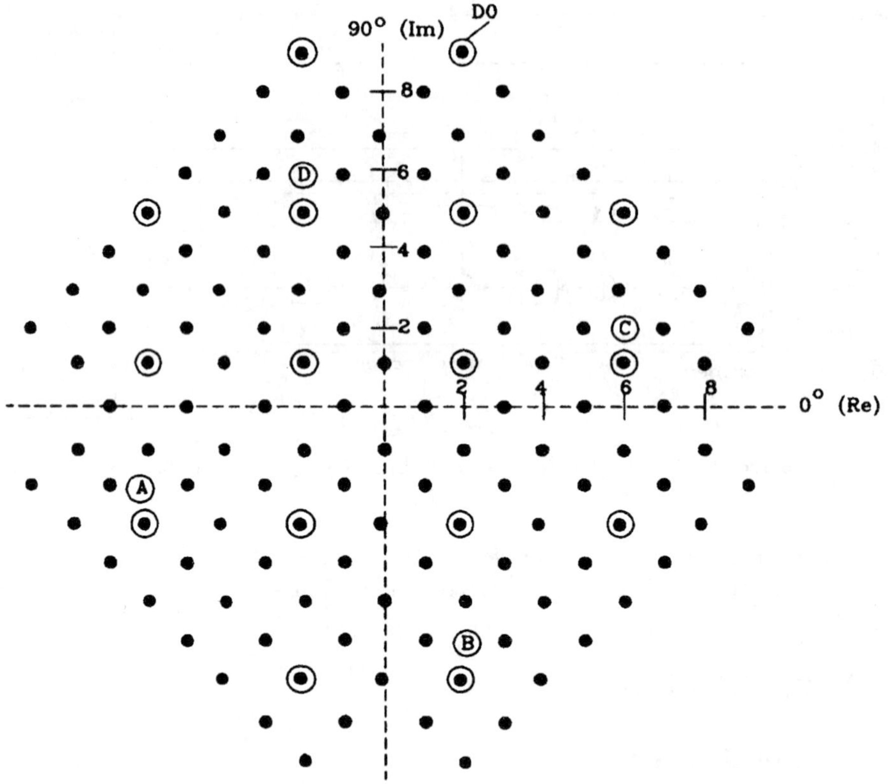

Figure 9.14: Signal space diagram and mapping for trellis coded modulation at 14 400 bps

Similarly to the V.32 trellis coded constellation, the three least significant protected bits $Y2_n$, $Y1_n$ and $Y0_n$ are used as subset identifiers to select one of the eight trellis subsets. For example, in Figure 9.14 the circled phasors belong to the subset identified by $[Y2_n, Y1_n, Y0_n] = [110]$. The remaining four unprotected bits are used to encode the 16 phasor positions within the trellis coded subset. Since these phasors are sufficiently far apart in the signal constellation, they are less prone to noise corruption and hence they can be transmitted unprotected.

At the 12 000 bps fallback rate the modem operates using 6 bits/symbol and it has the phasor constellation depicted in Figure 9.15. The input data rate of 12 000 bps would only require 5 bits/symbol at 2400 Bd, but the previously described trellis coding scheme generates and attaches the parity bit $Y0_n$ to protect the differentially coded subset partitioning bits $Y2_n$ and $Y1_n$. Apart from removing the bit $Q6_n$ and using the constellation of Figure 9.15, the 12 000 bps scheme is identical to the 14 400 bps scheme, but with its increased noise protection distances amongst constellation points allows the modem to operate at lower signal-to-noise ratios. The 12 000 bps set partitioning scheme is based again on the three convolutionally coded

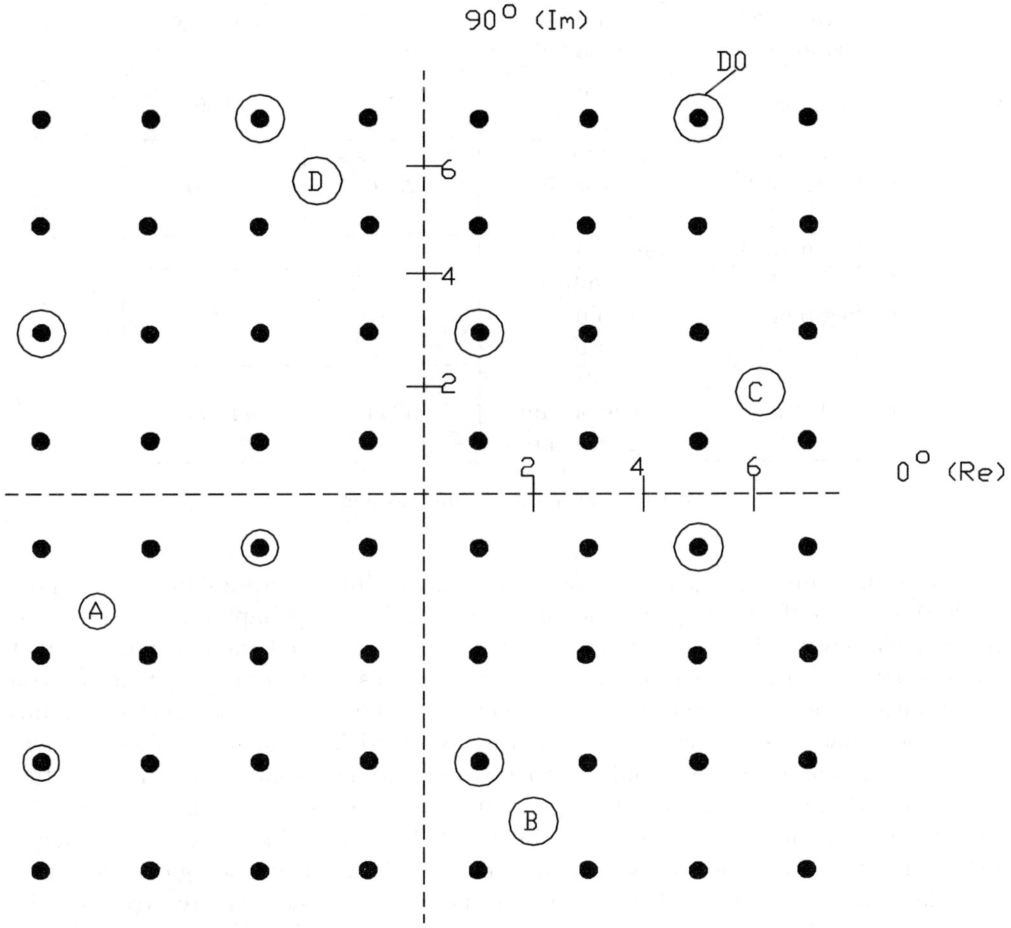

Figure 9.15: Signal space diagram and mapping for trellis coded modulation at 12 000 bps

least significant bits $Y2_n$, $Y1_n$ and $Y0_n$ of the circled phasors in Figure 9.15.

9.5.3 Synchronising signals

Before the transmission of useful data, the modem requires a variety of training signals in order to condition its carrier and clock recovery circuits, as well as the adaptive channel equaliser. The standardised initial synchronising signals are summarised in Table 9.7. Segment 1 of the training pattern is constituted by 256 alternations between the signal constellation states A and B, depicted in Figures 9.14 and 9.15, which occupy the same positions as in the V.32 Recommendations and hence coincide with four of the legitimate non-redundant 16-QAM phasors. This segment assists the initial carrier and clock recovery, as detailed in Chapter 6. This is followed by segment 2, consisting of a scrambled pseudorandom sequence of the states A, B, C and D, shown

Training segments	Type of line signal	No. sym. intervals	Approx. time (ms)
Segment 1	Alternations ABAB	256	106
Segment 2	Equaliser conditioning A,B,C,D	2976	1240
Segment 3	Rate sequence	64	27
Segment 4	Scrambled all binary 1's	48	20
Total	Total synchronising signal time	3344	1393

Table 9.7: V.33 training sequence

in Figures 9.14 and 9.15, which is transmitted at 4800 bps and designed to support the initial set-up of the adaptive equaliser as highlighted in Chapter 7. Segment 3 is designed to convey a 16-bit binary sequence, which is scrambled and transmitted eight times at 4800 bps using differentially encoded bits. This sequence carries transmission rate, multiplex set-up or other configuration information. Finally, segment 4 contains scrambled binary 1's transmitted at the rate specified in segment 3. This segment is used to fine-tune the clock and carrier recovery circuits because the linear distortions of the channel should now be equalised and so the clock and carrier references established over the unequalised channel can now be more reliably set. Scrambling is invoked in order to introduce signal transitions into long strings of logical 1's or 0's.

V.33 modems are provided with adaptive channel equalisers. The receiver must be able to detect loss of equalisation and initiate a resynchronisation procedure. Upon initiating a resynchronisation sequence the initiating modem expects to receive a synchronising signal from the remote modem within the maximum turnaround delay.

9.6 Summary

In this chapter we have examined a range of classic QAM schemes designed for AWGN telephone lines, which were standardised by the CCITT - now known as the ITU. The more recent modems use high-order constellations, complex trellis coded systems and long training sequences in order to maximise the transmission rate facilitated by a bandlimited transmission medium.

This consideration of one of the most widespread uses of QAM modems concludes our section on QAM for Gaussian channels. In the remaining chapters we concentrate on transmissions over hostile fading mobile channels. We commence the next section by considering the theoretical performance of certain QAM constellations over mobile radio channels and then continue to investigate means means of mitigating the impairments imposed by the channel.

Chapter 10

Square QAM for Rayleigh Fading Channels

10.1 Square 16-QAM Performance over Rayleigh Fading Channels

Having considered the BER versus channel SNR performance of QAM transmissions via AWGN channels in Chapter 5, we now turn to the analysis of the square 16-QAM constellation of Figure 3.3 over Rayleigh fading channels [42,85,86]. Our discussion in this chapter assumes that the reader has already consulted Chapter 5. Assuming a so-called non-frequency-selective, narrowband fading channel, where the signal bandwidth is much lower than the coherence bandwidth of the channel, all frequency components of the transmitted signal undergo the same attenuation and phase shift. The multipath components in the received signal are not resolvable and the channel transfer function is given by

$$c(t) = \alpha(t) \cdot e^{j\Phi(t)}, \tag{10.1}$$

where $\alpha(t)$ represents the Rayleigh fading envelope and $\Phi(t)$ the phase of the channel, which is uniformly distributed over $[-\pi, \pi]$. If additionally the fading is slow, so that $\alpha(t) = \alpha$ and $\Phi(t) = \Phi$ for the duration of one signalling interval, the received signal $r(t)$ at the channel's output is:

$$r(t) = \alpha \cdot e^{j \cdot \Phi} \cdot m(t) + n(t), \tag{10.2}$$

where $m(t)$ is the transmitted modulated signal and $n(t)$ is an AWGN process.

In Chapter 5 we derived the following $C1$ and $C2$ BER results for the AWGN channel using the Gaussian Q-function [337]:

$$P_{1G} = \frac{1}{2} Q \left\{ \frac{d}{\sqrt{N_0/2}} \right\} + \frac{1}{2} Q \left\{ \frac{3d}{\sqrt{N_0/2}} \right\}$$

$$= \frac{1}{2}\left[Q\left\{\sqrt{\frac{E_0}{5N_0}}\right\} + Q\left\{3\sqrt{\frac{E_0}{5N_0}}\right\}\right]. \tag{10.3}$$

and

$$P_{2G} = Q\left\{\sqrt{\frac{E_0}{5N_0}}\right\}. \tag{10.4}$$

where $E_0 = 10d^2$ is the average constellation energy.

Assuming perfectly phase coherent reception, no fade compensation, and rewriting Equations 10.3 and 10.4 in terms of the instantaneous SNR γ via Rayleigh fading channels, we get:

$$P_{1R}(\gamma) = \frac{1}{2}\left[Q\left\{\sqrt{\gamma/5}\right\} + Q\left\{3\sqrt{\gamma/5}\right\}\right], \tag{10.5}$$

and

$$P_{2R}(\gamma) = Q\left\{\sqrt{\gamma/5}\right\}. \tag{10.6}$$

As the fading envelope fluctuates, the instantaneous SNR γ is also changing. The average BER is found by calculating the bit error probability at a given instantaneous SNR γ and averaging it over all possible SNRs. This is carried out by multiplying the bit error probability $P(\gamma)$ at the instantaneous SNR γ with the occurrence probability of the specific SNR γ expressed in terms of its probability density function (PDF) $C(\gamma)$ and then averaging, i.e. integrating, this product over all possible values of γ:

$$P_{be} = \int_0^{\infty} P(\gamma) \cdot C(\gamma) d\gamma. \tag{10.7}$$

The random variable $\gamma = \alpha^2 \cdot E_0/N_0$ is a transformed form of the fading envelope α and the PDF of α, $C(\alpha)$ is known to be Rayleigh [158, 338]:

$$C(\alpha) = \left(\frac{\alpha}{\alpha_0{}^2}\right) e^{-\alpha^2/2\alpha_0{}^2}, \tag{10.8}$$

where α_0^2 is the identical variance of the independent quadrature components constituting the fading envelope α, while the second moment of α is given by $E(\alpha^2) = \overline{\alpha^2} = 2\alpha_0{}^2$ giving the variance of the envelope itself.

Let us now compute the PDF of γ, $C(\gamma)$, given that $C(\alpha)$ is Rayleigh distributed, and $\gamma = \alpha^2.E_o/N_o$ is a second-order function of α. The following theorem can be invoked [337].

<u>Theorem</u>

Let α be a continuous random variable with the PDF $C(\alpha)$, and let $\gamma = \alpha^2.E_o/N_o$ be a transformation of α. In order to determine the PDF of γ, $C(\gamma)$, we solve the equation $\gamma = \alpha^2.E_o/N_o$ for α in terms of γ. If α_1 and α_2 are the only real solutions then

$$C(\gamma) = C(\alpha_1)\left|\frac{d\alpha_1}{d\gamma}\right| + C(\alpha_2)\left|\frac{d\alpha_2}{d\gamma}\right|. \tag{10.9}$$

As $\gamma = \alpha^2 \cdot E_0/N_0$ has two real roots, α_1 and α_2, but the Rayleigh distribution does not exist in the negative amplitude domain, α_2 can be eliminated and hence $C(\gamma)$ is given by

$$C(\gamma) = C(\alpha_1) \cdot |\frac{d\alpha_1}{d\gamma}|. \tag{10.10}$$

When taking into account that

$$\alpha = \sqrt{\frac{N_0}{E_0}} \cdot \gamma^{\frac{1}{2}} \tag{10.11}$$

and that

$$\frac{d\alpha_1}{d\gamma} = \frac{1}{2}\sqrt{\frac{N_0}{E_0\gamma}} \tag{10.12}$$

we arrive at

$$C(\gamma) = \frac{1}{2\alpha_0{}^2}\frac{N_0}{E_0\gamma} \cdot e^{\frac{-N_0\gamma}{2E_0\alpha_0^2}}. \tag{10.13}$$

If we define the average SNR Γ as:

$$\Gamma = \overline{\gamma} = E\{\alpha^2\frac{E_0}{N_0}\} = \overline{\alpha^2}\frac{E_0}{N_0} = 2\alpha_0^2\frac{E_0}{N_0}, \tag{10.14}$$

where we exploited the fact that $E(\alpha^2) = \overline{\alpha^2} = 2\alpha_0{}^2$ for a Rayleigh PDF, then from Equations 10.13 and 10.14 the transformed PDF $C(\gamma)$ is given by

$$C(\gamma) = \frac{1}{\Gamma}e^{-\gamma/\Gamma}, \tag{10.15}$$

which is known as the chi-square distribution [158], with the instantaneous SNR $\gamma \geq 0$.

The $C1$ BER over Rayleigh fading channels is found by substituting into Equation 10.7 the PDF $C(\gamma)$ from Equation 10.15 and the instantaneous bit error probability $P_{1R}(\gamma)$ over Rayleigh fading channels from Equation 10.5, yielding:

$$P_{1R}(\Gamma) = \frac{1}{2\Gamma}\int_0^\infty \left[Q\left\{\sqrt{\gamma/5}\right\} + Q\left\{3\sqrt{\gamma/5}\right\}\right] \cdot e^{-\gamma/\Gamma}d\gamma. \tag{10.16}$$

The $C2$ BER cannot be formulated without considering the effect of a particular level of channel attenuation on the demodulation process. The demodulation is now carried out on an attenuated signal constellation. Therefore we multiply the decision boundaries specified in Equation 4.6 for the $C2$ bits in the I and Q components by the average attenuation $\overline{\alpha}$:

$$\begin{array}{lcccccl} \text{if} & & I,Q & \geq & 2\overline{\alpha}d & \text{then} & i_2, q_2 = 1 \\ \text{if} & -2\overline{\alpha}d & \leq & I,Q & < 2\overline{\alpha}d & \text{then} & i_2, q_2 = 0 \\ \text{if} & -2\overline{\alpha}d & > & I,Q & & \text{then} & i_2, q_2 = 1. \end{array} \tag{10.17}$$

Considering the case when the transmitted bit i_2 or q_2 is a logical 0 and assuming that $\alpha < 2\overline{\alpha}$, the protection distance between the received phasor αd and the decision boundary $2\overline{\alpha}d$ is $d_1 = (2\overline{\alpha}d - \alpha d)$. Upon substituting d_1 into Equation 4.7 instead of

d, we get the $C2$ BER for the transmission of a logical 0 as

$$\begin{aligned} P_{2,0,<2} &= Q\left\{\frac{d_1}{\sqrt{N_0/2}}\right\} \\ &= Q\left\{\frac{d(2\overline{\alpha}-\alpha)}{\sqrt{N_0/2}}\right\}. \end{aligned} \tag{10.18}$$

From $E_0 = 10d^2$ we have $d = \sqrt{E_0/10}$, and whence:

$$P_{2,0,<2} = Q\left\{(2\overline{\alpha}-\alpha)\sqrt{\frac{E_0}{5N_0}}\right\}. \tag{10.19}$$

From $\gamma = \alpha^2 E_0/N_0$ we get $E_0/N_0 = \gamma/\alpha^2$, and therefore:

$$\begin{aligned} P_{2,0,<2} &= Q\left\{\left(\frac{2\overline{\alpha}-\alpha}{\alpha}\right)\sqrt{\frac{\gamma}{5}}\right\} \\ &= Q\left\{\left(\frac{2\overline{\alpha}}{\alpha}-1\right)\sqrt{\frac{\gamma}{5}}\right\}. \end{aligned} \tag{10.20}$$

Upon substituting Equations 10.20 and 10.15 into Equation 10.7 we get one component of the total $C2$ error probability. First the upper integration boundary of Equation 10.7 is derived for $\alpha \leq 2\overline{\alpha}$ as

$$\gamma = \alpha^2 E_0/N_0 \leq 4(\overline{\alpha})^2 E_0/N_0\,. \tag{10.21}$$

For a Rayleigh distribution we have [338] that $\overline{\alpha} = \alpha_0\sqrt{\pi/2}$, and this means we arrive at:

$$\gamma = 4\left(\alpha_0\sqrt{\frac{\pi}{2}}\right)^2\frac{E_0}{N_0} = 2{\alpha_0}^2\frac{E_0}{N_0}\pi = \Gamma\pi, \tag{10.22}$$

and

$$P_{2,0,<2}(\Gamma) = \int_0^\infty P(\gamma)C(\gamma)d\gamma = \int_0^{\Gamma\pi} Q\left\{(\frac{2\overline{\alpha}}{\alpha}-1)\sqrt{\frac{\gamma}{5}}\right\}\cdot\frac{1}{\Gamma}e^{-\gamma/\Gamma}d\gamma. \tag{10.23}$$

Similarly, for $d_1 = (2\overline{\alpha}d - \alpha d) < 0$, i.e. for $\alpha > 2\overline{\alpha}$, we have

$$P_{2,0,>2} = 1 - \left[1 - Q\left\{-\frac{2\overline{\alpha}d-\alpha d}{\sqrt{N_0/2}}\right\}\right] = 1 - Q\left\{\frac{d(2\overline{\alpha}-\alpha)}{\sqrt{N_0/2}}\right\}. \tag{10.24}$$

Exploiting $d = \sqrt{E_0/10}$ we have

$$P_{2,0,>2} = 1 - Q\left\{\sqrt{\frac{E_0}{5N_0}}(2\overline{\alpha}-\alpha)\right\}, \tag{10.25}$$

and by taking into account that $E_0/N_0 = \gamma/\alpha^2$ we arrive at:

$$P_{2,0,>2} = 1 - Q\left\{(\frac{2\overline{\alpha}}{\alpha} - 1)\sqrt{\frac{\gamma}{5}}\right\}. \tag{10.26}$$

The lower integration limit is now identical to the previously computed upper limit, i.e. $\gamma = \pi\Gamma$, and the total probability of receiving a logical 1 in the $C2$ subchannel instead of the transmitted logical 0 is given by:

$$P_{2,0}(\Gamma) = P_{2,0,<2}(\Gamma) + P_{2,0,>2}(\Gamma). \tag{10.27}$$

Upon substituting Equations 10.23 and 10.26 into Equation 10.27 we get:

$$\begin{aligned} P_{2,0}(\Gamma) &= \frac{1}{\Gamma}\int_0^{\pi\Gamma} Q\left\{(\frac{2\overline{\alpha}}{\alpha} - 1)\sqrt{\frac{\gamma}{5}}\right\} e^{-\gamma/\Gamma} d\gamma \\ &+ \frac{1}{\Gamma}\int_{\pi\Gamma}^{\infty}\left[1 - Q\left\{(\frac{2\overline{\alpha}}{\alpha} - 1)\sqrt{\frac{\gamma}{5}}\right\}\right] e^{-\gamma/\Gamma} d\gamma. \end{aligned} \tag{10.28}$$

In the reverse situation, when a transmitted $C2$ logical 1 is corrupted into a logical 0, the protection distance is $d_1 = 3\alpha d - 2\overline{\alpha}d$, which is positive if $3\alpha d > 2\overline{\alpha}d$, i.e. $\alpha/\overline{\alpha} > 2/3$. The AWGN has to overcome this protection distance to cause an error, and hence the error probability for a positive protection distance d_1 is

$$P_{2,1,>2/3} = Q\left\{\left(\frac{3\overline{\alpha}}{\alpha} - 2\right)\sqrt{\frac{\gamma}{5}}\right\}. \tag{10.29}$$

If the protection distance d_1 is negative, then $3\alpha d < 2\overline{\alpha}d$, i.e. $\alpha/\overline{\alpha} < 2/3$. In this situation the decisions are always erroneous when the noise resides above the level $-(3\alpha d - 2\overline{\alpha}d)$, which is expressed with the aid of the Gaussian Q-function as:

$$P_{2,1,<2/3} = Q\left\{\frac{-(3\alpha d - 2\overline{\alpha}d)}{\sqrt{N_0/2}}\right\} = Q\left\{\frac{-d(3\alpha - 2\overline{\alpha})}{\sqrt{N_0/2}}\right\}$$

or

$$P_{2,1,<2/3} = 1 - Q\left\{(\frac{3\overline{\alpha}}{\alpha} - 2)\sqrt{\frac{\gamma}{5}}\right\}. \tag{10.30}$$

Before computing the total $C2$ bit error probability for the transmission of a logical 1, we determine the integration limit for Equation 10.7. Since $\alpha > 2\overline{\alpha}/3$,

$$\gamma = \alpha^2\frac{E_0}{N_0} = \left(\frac{2}{3}\overline{\alpha}\right)^2\frac{E_0}{N_0} = \frac{4}{9}(\overline{\alpha})^2\frac{E_0}{N_0}. \tag{10.31}$$

For a Rayleigh distribution we have [338] that $\overline{\alpha} = \alpha_0\sqrt{\pi/\gamma}$, so

$$\gamma = \frac{4}{9}{\alpha_0}^2\frac{\pi}{\gamma}\frac{E_0}{N_0} = \Gamma\frac{\pi}{9}. \tag{10.32}$$

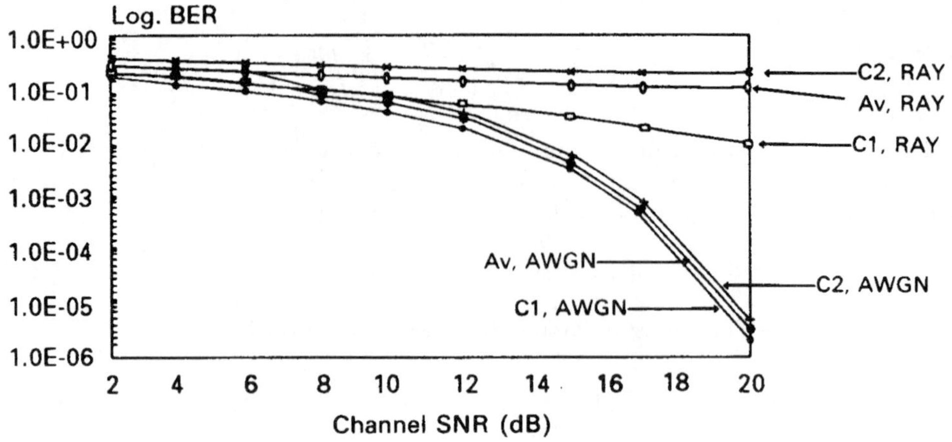

Figure 10.1: Square 16-QAM BER versus channel SNR curves for AWGN and Rayleigh channels

Therefore:

$$P_{2,1}(\Gamma) = P_{2,1,<2/3}(\Gamma) + P_{2,1,>2/3}(\Gamma), \tag{10.33}$$

or

$$\begin{aligned} P_{2,1}(\Gamma) &= \frac{1}{\Gamma}\int_0^{\pi\Gamma/9} Q\left\{(\frac{3\overline{\alpha}}{\alpha} - 2)\sqrt{\frac{\gamma}{5}}\right\} e^{-\gamma/\Gamma} d\gamma \\ &+ \frac{1}{\Gamma}\int_{\pi\Gamma/9}^{\infty}\left[1 - Q\left\{(\frac{3\overline{\alpha}}{\alpha} - 2)\sqrt{\frac{\gamma}{5}}\right\}\right] e^{-\gamma/\Gamma} d\gamma. \end{aligned} \tag{10.34}$$

For random transmitted data the overall $C2$ sub-channel error probability is given by:

$$P_{2R}(\Gamma) = \frac{1}{2}[P_{2,1}(\Gamma) + P_{2,0}(\Gamma)]. \tag{10.35}$$

When we evaluated the $C1$ and $C2$ bit error probabilities from Equations 10.16 and 10.35, respectively, we received nearly identical curves to those in Figure 10.1 denoted by C1,Ray and C2,Ray, for our simulations. The average $C1$ and $C2$ BER curve is represented by Av,Ray in the figure.

For comparison we also included the corresponding AWGN curves denoted by C1,AWGN and C2,AWGN as well as the average BER Av,AWGN, of the two sub-channels. We noted in Chapter 5 that the $C1$ and $C2$ BER differences are consistent but not drastic when transmitting square 16-QAM signals via the AWGN channel. The discrepancy becomes more dramatic over the Rayleigh fading channel. While the $C1$ BER is monotonically decreasing, the $C2$ curve shows a high residual BER. This is because due to the received signal envelope fluctuations the multilevel QAM constellation collapses towards the origin of the coordinate system, causing excessive BERs for the low protection $C2$ sub-channel. The high protection $C1$ sub-channel withstands the channel fading better and hence its behaviour is mostly governed by the AWGN. Automatic gain control (AGC) techniques with and without channel sounding side information for fading compensation are reported in references [42, 85, 87] allowing the reduced residual BERs to be efficiently combated by appropriate FEC methods.

In the final part of this section, we briefly note that there exists a closed-form solution for the integral of the product of the Gaussian Q-function and the Rayleigh PDF. The closed-form $C2$ bit error probability can be derived as follows:

$$\begin{aligned} P_{2R}(\alpha_0) &= \int_0^\infty Q\left\{\frac{\alpha d}{\sqrt{N_o/2}}\right\} C(\alpha)d\alpha \\ &= \int_0^\infty Q\left\{\frac{\alpha d}{\sqrt{N_o/2}}\right\} \frac{\alpha}{\alpha_0^2} \cdot e^{\frac{-\alpha^2}{2\alpha_0^2}} d\alpha. \end{aligned} \tag{10.36}$$

We exploit the following fact [339]:

$$\int_0^\infty [1 - erf(\beta x)]e^{-\mu x^2} \cdot x dx = \frac{1}{2\mu}\left(1 - \frac{\beta}{\sqrt{\mu + \beta^2}}\right) \tag{10.37}$$

where the relationship between the Gaussian error function $erf(x)$ and the Q-function is given as follows [337]:

$$1 - erf(x) = 2 \cdot Q\left\{\sqrt{2}x\right\}. \tag{10.38}$$

Upon rewriting Equation 10.36 in a more convenient form we have

$$P_{2R}(\alpha_0) = \frac{1}{2\alpha_0^2}\int_0^\infty 2 \cdot Q\left\{\sqrt{2}\frac{d}{\sqrt{N_o}}\alpha\right\} \cdot \alpha \cdot e^{\frac{-\alpha^2}{2\alpha_0^2}} d\alpha \tag{10.39}$$

and exploiting Equation 10.38 gives

$$P_{2R}(\alpha_0) = \frac{1}{2\alpha_0^2}\int_0^\infty 2 \cdot \left(1 - erf\left\{\frac{d}{\sqrt{N_o}}\alpha\right\}\right) \cdot \alpha \cdot e^{\frac{-\alpha^2}{2\alpha_0^2}} d\alpha \tag{10.40}$$

Assigning $\beta = d/\sqrt{N_o}$ and $\mu = 1/2\alpha_0^2$ in order to enable us to use the closed-form formula of Equation 10.37 we get

$$P_{2R}(\alpha_0) = \frac{1}{2\alpha_0^2}\frac{1}{2\mu}\left(1 - \frac{\beta}{\sqrt{\mu + \beta^2}}\right) \tag{10.41}$$

and after substituting back $\beta = d/\sqrt{N_0}$ and $\mu = 1/2\alpha_0^2$ we arrive at

$$P_{2R}(\alpha_0) = \frac{\alpha_0^2}{2\alpha_0^2}\left(1 - \frac{\frac{d}{\sqrt{N_o}}}{\sqrt{\frac{1}{2\alpha_0^2} + \frac{d^2}{N_o}}}\right)$$

$$= \frac{1}{2}\left[1 - \sqrt{\frac{2\alpha_0^2 \frac{d^2}{N_o}}{1 + 2\alpha^2 \frac{d^2}{N_o}}}\right]. \tag{10.42}$$

The derivation of the $C1$ bit error probability follows similar lines but is beyond the scope of this book.

10.2 Square 64-QAM Performance over Rayleigh Fading Channels

Adopting an approach similar to that used for the 16-QAM in Section 10.1, the formula for the $C1$ BER can be obtained by substituting the Gaussian channel result of Equation 4.22, repeated here for convenience:

$$P_{C1}(\gamma) = \frac{1}{4}Q\left[\sqrt{\frac{\gamma}{21}}\right] + \frac{1}{4}Q\left[\sqrt{\frac{3\gamma}{21}}\right] + \frac{1}{4}Q\left[\sqrt{\frac{5\gamma}{21}}\right] + \frac{1}{4}Q\left[\sqrt{\frac{7\gamma}{21}}\right] \tag{10.43}$$

into Equation 10.7 along with the transformed PDF $C(\gamma)$ in Equation 10.15, in order to arrive at:

$$P_{C1}(\Gamma) = \frac{1}{4\Gamma}\int_0^{\infty}\left\{Q\left[\sqrt{\frac{\gamma}{21}}\right] + Q\left[3\sqrt{\frac{\gamma}{21}}\right] + Q\left[5\sqrt{\frac{\gamma}{21}}\right] + Q\left[7\sqrt{\frac{\gamma}{21}}\right]\right\} e^{-\gamma/\Gamma} d\gamma. \tag{10.44}$$

This simple derivation is achieved because the $C1$ decision boundaries constituted by the coordinate axis and shown in Figure 4.2 are unaltered by the fading. However, the probability of bit error for the $C2$ and $C3$ sub-channels cannot be derived without considering the effects of fading on each phasor and decision boundary individually.

When calculating the probability of bit error for $C2$, we acknowledge that although the received phasors are attenuated by the instantaneous α value, the receiver is arranged to demodulate with reference to the expected value $2\alpha_0$. The effect of fading must be computed individually for each phasor. For P_6 the modified protection distance d_1 measured from the decision boundary $4\overline{\alpha}d$ is seen from the phasor constellation depicted in Figure 4.2 to be $d_1 = 4\overline{\alpha}d - 3\alpha d$. If the modified protection distance is overcome by noise, erroneous decisions occur. If the modified protection distance $d_1 < 0$ then in the absence of noise the decisions are consistently erroneous. In the presence of noise the phasors can be carried back into the error-free decision domain by sufficiently large negative noise samples. In both cases the error probability is equivalent to the noise samples exceeding the protection distance, irrespective of its actual value. In terms of the $Q[\,]$ function we have

$$P_{e6}(\gamma) = Q\left[\left(\frac{4\overline{\alpha}}{\alpha} - 3\right)\sqrt{\frac{\gamma}{21}}\right]. \tag{10.45}$$

For the phasor P_5 the modified protection distance from the $C2$ decision boundary $4\overline{\alpha}d$ is $d_1 = 4\overline{\alpha}d - \alpha d$. Observe that the effect of the instantaneous attenuation α is now less profound than for P_6, where it is multiplied by a factor of 3. Using this d_1

value in the argument of the $Q[\,]$ function gives

$$P_{e5}(\gamma) = Q\left[\left(\frac{4\overline{\alpha}}{\alpha} - 1\right)\sqrt{\frac{\gamma}{21}}\right]. \tag{10.46}$$

The modified protection distance from the $C2$ decision boundary at $4\overline{\alpha}d$ for P_7 is seen from the phasor constellation of Figure 4.2 to be $d_1 = 4\overline{\alpha}d - 5\alpha d$, which yields an error probability of

$$P_{e7}(\gamma) = Q\left[(\frac{4\overline{\alpha}}{\alpha} - 5)\sqrt{\frac{\gamma}{21}}\right]. \tag{10.47}$$

The phasor P_8 has the highest distance from the coordinate axis, therefore it is profoundly affected by the instantaneous fading. It has an associated protection distance of $d_1 = 4\overline{\alpha}d - 7\alpha d$ which results in an error probability of

$$P_{e8}(\gamma) = Q\left[(\frac{4\overline{\alpha}}{\alpha} - 7)\sqrt{\frac{\gamma}{21}}\right]. \tag{10.48}$$

The partial $C2$ error probabilities from Equations 10.45 to 10.48 have to be substituted in Equation 10.7 along with Equation 10.15 in order to deliver the average square 64-QAM $C2$ BER for Rayleigh fading channels:

$$\begin{aligned} P_{C2}(\Gamma) &= \frac{1}{4\Gamma}\left\{\int_0^\infty Q\left[(\frac{4\overline{\alpha}}{\alpha} - 1)\sqrt{\frac{\gamma}{21}}\right] e^{-\gamma/\Gamma} d\gamma \right. \\ &+ \int_0^\infty Q\left[(\frac{4\overline{\alpha}}{\alpha} - 3)\sqrt{\frac{\gamma}{21}}\right] e^{-\gamma/\Gamma} d\gamma \\ &+ \int_0^\infty Q\left[(\frac{4\overline{\alpha}}{\alpha} - 5)\sqrt{\frac{\gamma}{21}}\right] e^{-\gamma/\Gamma} d\gamma \\ &+ \left. \int_0^\infty Q\left[(\frac{4\overline{\alpha}}{\alpha} - 7)\sqrt{\frac{\gamma}{21}}\right] e^{-\gamma/\Gamma} d\gamma \right\}. \end{aligned} \tag{10.49}$$

The $C3$ sub-channel performance is computed similarly, but for each phasor there are two modified protection distances, d_1 and d_2. For the phasor P_6 in Figure 4.2 we find that $d_1 = 6\overline{\alpha}d - 3\alpha d$ and $d_2 = 3\alpha d - 2\overline{\alpha}d$, representing the distances from the $C3$ decision boundaries at $6\overline{\alpha}d$ and $2\overline{\alpha}d$, respectively. When exposed to AWGN, the decisions are erroneous if the modified protection distance d_1 is exceeded by the noise samples. This holds for negative d_1 values as well. Erroneous decisions are also encountered if the phasor P_6 is carried across the $C3$ decision boundary at $2\overline{\alpha}d$ by sufficiently large negative noise samples. This happens whenever the noise value resides below the level $-d_2$. Accordingly, the $C3$ bit error probability for P_6 is

$$P_{e6}(\gamma) = Q\left[(\frac{6\overline{\alpha}}{\alpha} - 3)\sqrt{\frac{\gamma}{21}}\right] + Q\left[(3 - \frac{2\overline{\alpha}}{\alpha})\sqrt{\frac{\gamma}{21}}\right]. \tag{10.50}$$

Considering the case of the phasor P_5 the modified protection distances are $d_1 = 2\overline{\alpha}d - \alpha d$ and $d_2 = 6\overline{\alpha}d - \alpha d$, respectively. Erroneous decisions are engendered if the protection distance d_1 is exceeded by noise, but noise samples larger than d_2 carry the phasors into another error-free decision domain. Whence the error probability is given by

$$P_{e5}(\gamma) = Q\left[(\frac{2\overline{\alpha}}{\alpha} - 1)\sqrt{\frac{\gamma}{21}}\right] - Q\left[(\frac{6\overline{\alpha}}{\alpha} - 1)\sqrt{\frac{\gamma}{21}}\right] \quad (10.51)$$

When the phasor P_7 is transmitted the situation is characterised by the distances $d_1 = 6\overline{\alpha}d - 5\alpha d$ and $d_2 = 5\alpha d - 2\overline{\alpha}d$, and hence

$$P_{e7}(\gamma) = Q\left[(\frac{6\overline{\alpha}}{\alpha} - 5)\sqrt{\frac{\gamma}{21}}\right] + Q\left[(5 - \frac{2\overline{\alpha}}{\alpha} - 1)\sqrt{\frac{\gamma}{21}}\right]. \quad (10.52)$$

Finally, for the phasor P_8 we find $d_1 = 7\alpha d - 6\overline{\alpha}d$ and $d_2 = 7\alpha d - 2\overline{\alpha}d$, where errors are caused by noise samples below the level $-d_1$. Should, however, the noise vector fall below $-d_2$, the phasors are switched to another error-free decision domain across the decision boundary at $2\overline{\alpha}d$, therefore

$$P_{e8}(\gamma) = Q\left[(7 - \frac{6\overline{\alpha}}{\alpha})\sqrt{\frac{\gamma}{21}}\right] - Q\left[(7 - \frac{2\overline{\alpha}}{\alpha})\sqrt{\frac{\gamma}{21}}\right]. \quad (10.53)$$

Let us assume again that random sequences are sent, i.e. the occurrence probability of each phasor $P_1, \ldots, P_{64}$ in Figure 4.2 is identical, and further, let us exploit the symmetry of the signal constellation. On substituting Equations 10.50 to 10.53, along with Equation 10.15, into Equation 10.7 we obtain the average $C3$ bit error probability as a function of the average SNR Γ:

$$\begin{aligned} P_{C3}(\Gamma) &= \frac{1}{4\Gamma}\{\int_0^\infty [Q[(\frac{6\overline{\alpha}}{\alpha} - 3)\sqrt{\frac{\gamma}{21}}] + Q[(3 - \frac{2\overline{\alpha}}{\alpha})\sqrt{\frac{\gamma}{21}}]]e^{-\gamma/\Gamma}d\gamma \\ &+ \int_0^\infty [Q[(\frac{2\overline{\alpha}}{\alpha} - 1)\sqrt{\frac{\gamma}{21}}] - Q[(\frac{6\overline{\alpha}}{\alpha} - 1)\sqrt{\frac{\gamma}{21}}]]e^{-\gamma/\Gamma}d\gamma \\ &+ \int_0^\infty [Q[(\frac{6\overline{\alpha}}{\alpha} - 5)\sqrt{\frac{\gamma}{21}}] + Q[(5 - \frac{2\overline{\alpha}}{\alpha} - 1)\sqrt{\frac{\gamma}{21}}]]e^{-\gamma/\Gamma}d\gamma \\ &+ \int_0^\infty [Q[(\frac{7 - 6\overline{\alpha}}{\alpha})\sqrt{\frac{\gamma}{21}}] - Q[(7 - \frac{2\overline{\alpha}}{\alpha})\sqrt{\frac{\gamma}{21}}]]e^{-\gamma/\Gamma}d\gamma\}. \end{aligned} \quad (10.54)$$

In summary, the square 64-QAM $C1$, $C2$ and $C3$ sub-channel performances over Rayleigh fading channels are given by Equations 10.44, 10.49 and 10.54, respectively. The equivalent BER vs. channel SNR curves are plotted in Figure 10.2, which again coincide within a fraction of a dB with our simulation results. The BER differences between the three classes become more profound in Rayleigh channels than in the AWGN channel. This is in harmony with our experience for the $16-QAM$ BER curves displayed in Figure 10.1. The BER of the $C1$ sub-channel is well suited for FEC-coded speech or image transmission, while the $C2$ and $C3$ performances have to be improved by fading-compensating automatic gain control and optional diversity reception for FEC techniques to become efficient [42, 85, 87].

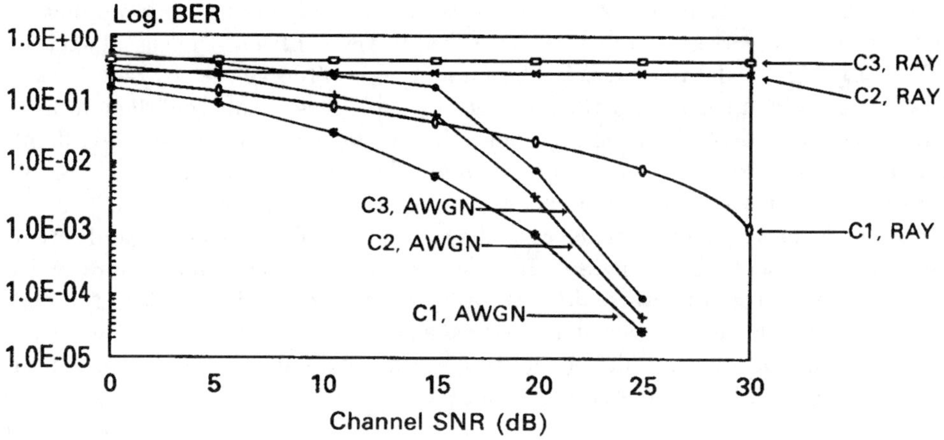

Figure 10.2: Square 64-QAM BER versus channel SNR curves for AWGN and Rayleigh channels

10.3 Reference Assisted Coherent QAM

Two powerful methods have been proposed in order to ensure coherent QAM operation in fading environments. Both these techniques deliver channel measurement information in terms of attenuation and phase shift due to fading. The first is transparent tone in band (TTIB) assisted modulation, where a pilot carrier is inserted typically in the centre of the modulated spectrum. At the receiver the signal is extracted and used to estimate the channel-induced attenuation and phase rotation.

The other technique is called pilot symbol assisted modulation (PSAM) where known channel sounding phasors are periodically inserted into the transmitted time domain signal sequence. Similarly to the frequency domain pilot tone, these known symbols deliver channel measurement information.

10.3.1 Transparent tone in band modulation [9]

10.3.1.1 Introduction

Transparent tone in band (TTIB) assisted modulation has been proposed for various mobile radio applications by McGeehan, Bateman *et al.* [9, 340–346]. TTIB schemes place a pilot tone within the spectrum of the transmitted signal which can be used for a range of applications from in-band signalling in telephone systems, through fading compensation in mobile radio systems [346], to coherent data transmissions [344, 347] to aid the operation of direct conversion receivers. Further related contributions can be found in the literature [10, 11, 348–352].

In the following discussion of TTIB schemes we follow McGeehan and Bateman's approach [9] and concentrate on the use of TTIB in a mobile radio environment. TTIB schemes constitute a frequency domain alternative to pilot symbol assisted modulation (PSAM) [10] (see Section 10.3.2.1). In TTIB systems for mobile radio applications a spectral gap of a certain bandwidth is created in the centre of the signal spectrum to allow insertion of the pilot tone. The pilot tone is often chosen to be the carrier frequency as this can then assist the receiver in the process of coherent demodulation. The use of coherent demodulation overcomes the carrier recovery problems highlighted in Chapter 6 and allows the use of the maximum least square distance constellations discussed in the previous two sections over mobile radio channels.

Further advantages can be achieved if the separation of the two sub-bands created by dividing the spectrum is made to be a subharmonic of the symbol timing. This allows relatively straightforward clock recovery procedures to be employed. Finally, since the amplitude and phase of the transmitted pilot symbol are known to the receiver, extrication of this pilot symbol allows an estimate of the complex fading channel envelope, which removes the need for AGC. However, this will only be possible if the channel is exhibiting flat fading, i.e. all the other frequency components of the signal suffer the same amplitude and phase shift.

The disadvantages of this system are that the bandwidth requirements are slightly increased, and higher transmitter and receiver complexity are required to perform the necessary spectral operations. Furthermore, increased transmitted power is required for the transmission of the pilot tone, and although this is typically at a lower level than the power level of the transmitted signal, at very low levels the pilot signal can become corrupted by noise, reducing the reliability of the fading estimates. Furthermore, the inclusion of the pilot tone increases the signal envelope fluctuations, which results in increased out-of-band emissions due to the amplifier non-linearities, as discussed in Chapter 5.

10.3.1.2 Principles of TTIB

The basic principles of TTIB transmission are shown in Figure 10.3, while the corresponding frequency domain manipulations are portrayed in Figure 10.4, with the 10 stages in processing shown in each of the figures.

The input waveform is firstly filtered by the LPF in order to remove high frequencies which might lead to aliasing. In Figure 10.4 both the lower sub-band (L) and upper sub-band (U) occupy half the total bandwidth B, but there is no reason why the signal could not be split into a number of unequal sidebands apart from the implementational practicalities. The ideal LPF, LPF2 removes U and passes L to the transmitter. Then U is frequency translated using a carrier frequency f_T leading to the spectrum shown at stage 3. This is then bandpass filtered by BPF1 before being recombined with L for transmission. At stage 5, the frequency gap between L and U is f_G, extending the total transmission bandwidth to $B + f_G$. Next the pilot carrier frequency f_P is added to the composite signal. If the total bandwidth $(B + f_G)$ is less than the coherence bandwidth of the channel then both the received spectrum and the received pilot tone suffer the same attenuation and phase shift, so the same correction factor can be used to remove the fading from both the pilot and the signal spectrum.

At the receiver, the pilot signal is extricated using BPF3, while L is retained by

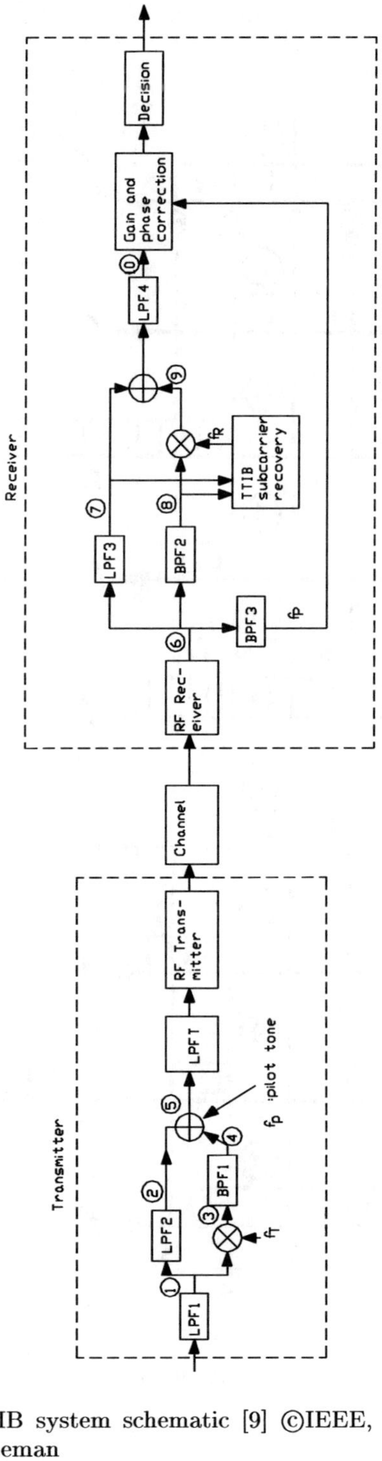

Figure 10.3: TTIB system schematic [9] ©IEEE, 1984, McGeehan, Bateman

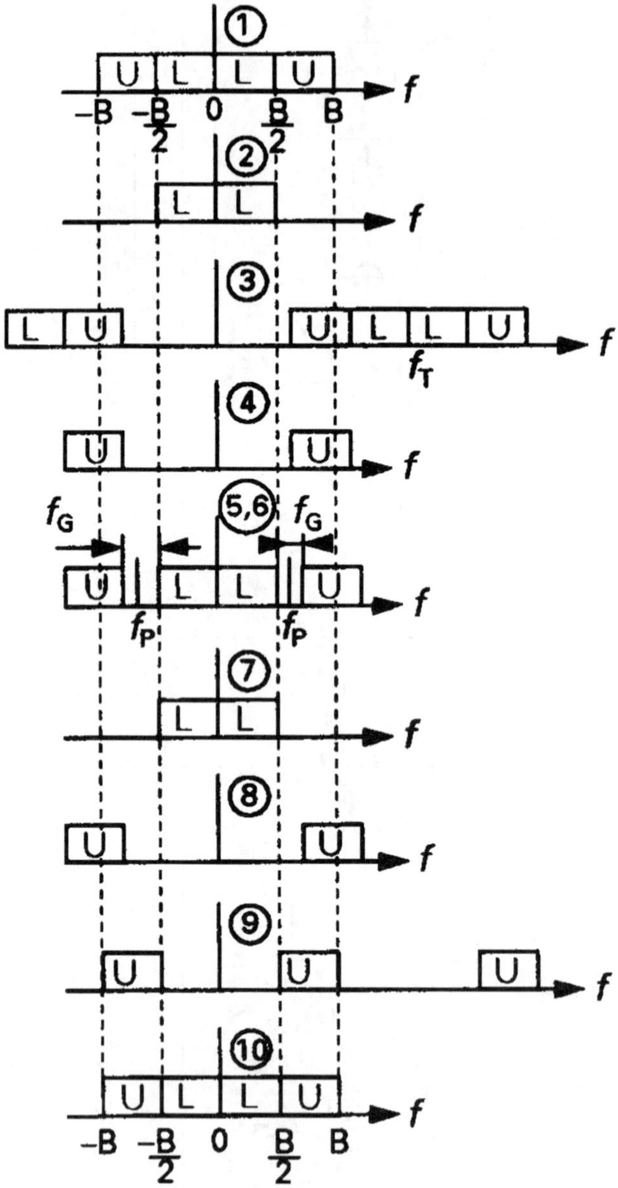

Figure 10.4: TTIB processed spectra [9] ©IEEE, 1984, McGeehan, Bateman

LPF3 and U by BPF2. If the frequency f_T, is known to the receiver, and there is perfect alignment between nominal and actual frequencies giving $f_R = f_T$ then the original signal spectrum can be recovered at stage 10. Before a decision is made as to the transmitted signal, the extricated pilot tone is used to remove the estimated effect of the complex channel fading by deriving an amplitude and phase correction factor. The bandwidth of BPF3 is an important system parameter; if it is made too small then Doppler shift and oscillator drift may prevent recovery of the pilot tone; if it is too large then it will process an excessive amount of channel noise, degrading the reliability of the channel estimate and hence the system performance.

The TTIB receiver typically relies on signals 7 and 8 in order to recover the TTIB subcarrier f_T used by the transmitter in the frequency translation process. The manner in which this is performed is discussed in the next section.

In practice there is usually a frequency offset Δw between the TTIB subcarrier f_T at the transmitter, and that at the receiver, denoted f_R. The effect of this frequency offset of $\Delta w = 2\pi\Delta f = 2\pi(f_T - f_R)$ will also be addressed in the next section, while a practical sub-band splitting technique known as quadrature mirror filtering (QMF) which does not require infinitely steep cut-off filters will be discussed in the next but one section.

10.3.1.3 TTIB subcarrier recovery

In order to eliminate interferences between the L and U sub-bands and achieve perfect signal reconstruction, it is crucial to combine their frequency and phase coherently. Because the received pilot tone has been subject to up and down conversions in the transmitter and receiver, respectively, and the multipath channel might have introduced some random frequency offset, the pilot will be prone to some frequency error and so cannot be relied upon in producing the receiver's TTIB subcarrier f_R. The channel effects and RF up and down conversions have an identical influence over L and U, whereas the TTIB subcarrier offset affects only U, as shown in Figure 10.4.

To overcome this problem the TTIB subcarrier recovery method shown in Figure 10.5 is used [9].

Its operation relies on the characteristics of the signals labelled 7 and 8, which are identical to those shown in Figures 10.3 and 10.4. Observe in Figure 10.5 that U is phase shifted by 90 deg and then both signals 7 and 8 are input to a limiter or null comparator in order to remove any modulation or channel imposed phase variations. The two signals are then multiplied together to create the difference and sum of their frequency components. The loop filter removes the sum of the frequencies but it retains the difference, which is then input to a voltage controlled oscillator (VCO) that generates the frequency and phase coherent local TTIB carrier $f_R = f_T$.

The operation of the TTIB subcarrier recovery system is best understood if a single tone of

$$i(t) = a_i \cos(w_i t + \alpha_i) \tag{10.55}$$

is used to excite the transmitter, where a_i, w_i and α_i are the amplitude, frequency and initial phase of the tone, respectively. We also assume that overlapping regions of the L and U sub-bands do contain some signal energy. We choose w_i specifically to satisfy this condition and assume that the L and U sub-band filters have attenuations

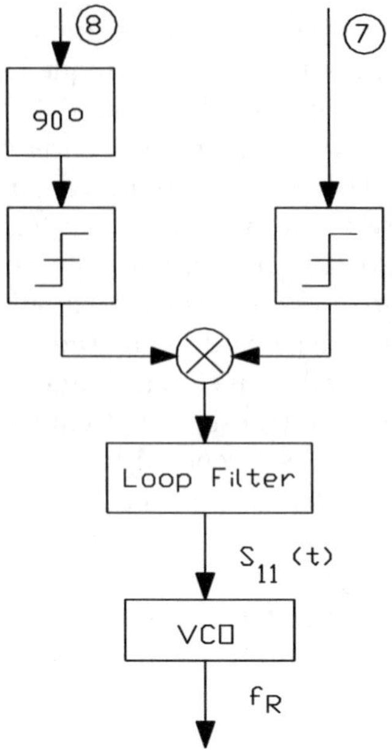

Figure 10.5: TTIB subcarrier recovery [9] ©IEEE, 1984, McGeehan, Bateman

of A_1 and A_2, respectively, at this transition band frequency.

Initially, the U sub-band tone is frequency translated using the TTIB subcarrier of

$$C_T(t) = \cos(w_T t + \alpha_T) \tag{10.56}$$

having a frequency of w_T and initial phase of α_T. It is then added to the L tone giving a signal at point 5 of Figure 10.3 as follows:

$$S_5(t) \;=\; A_1 a_i \cos(w_i t + \alpha_i) \;+\; A_2 a_i \cos[(w_i + w_T)t + \alpha_i + \alpha_T] \tag{10.57}$$

where due to the filter delay τ_1 introduced by both LPF2 and BPF1, and the delay τ_2 introduced by the transmitter filter LPFT in Figure 10.3, the time axis must be translated such that t becomes $t - \tau_1 - \tau_2$. Whilst it might seem unlikely that LPF2 and BPF1 would have identical delays, we will show how this might be arranged using QMF in the next section.

The received signal is corrupted by the complex fading introducing an envelope factor of $\alpha(t)$ and a phase factor of $\psi(t)$. It also has a residual oscillator error of Δw due to up and down conversion. The signal received at stage 6 of Figure 10.3 is therefore given by

$$\begin{aligned} S_6(t) \;=\; & A_1 a_i \alpha(t) \cos[w_i t + \alpha_i + \psi(t) + \Delta w t] \\ & + A_2 a_i \alpha(t) \cos[(w_i + w_T)t + \alpha_i + \alpha_T + \psi(t) + \Delta w t]. \end{aligned} \tag{10.58}$$

The TTIB subcarrier at the receiver is given by

$$C_R(t) = \cos(w_R t + \alpha_R) \tag{10.59}$$

where α_R is its initial phase and w_R is its angular frequency. Then the signal at stage 7 in Figure 10.3, the output of LPF3, becomes

$$S_7(t) = A_3 A_1 a_i \alpha(t) \cos[w_i t + \alpha_i + \psi(t) + \Delta w t] \tag{10.60}$$

where A_3 is the attenuation of LPF3 at $(w_i + \Delta w)$ and the time t is now redefined as $t - \tau_1 - \tau_2 - \tau_3$, with τ_3 being the identical delay of LPF3 and BPF2. After passing $S_6(t)$ through BPF2 we get

$$S_8(t) = A_4 A_2 a_i \alpha(t) \cos[(w_i + w_T)t + \alpha_i + \alpha_T + \psi(t) + \Delta w t] \tag{10.61}$$

where A_4 is the attenuation of BPF2 at $(w_i + w_T + \Delta w)$. By mixing $S_8(t)$ with the output signal from the VCO having a free running frequency of w_R and an initial phase of α_R, components are generated at the difference frequency of $(w_i + w_T - w_R)$ and the sum frequency $(w_i + w_T + w_R)$, with the latter term being removed by LPF4. Ignoring, the latter term, therefore, the difference frequency term at position 9 is then given by

$$\begin{aligned} S_9^{diff}(t) = {} & A_4 A_2 a_i \alpha(t) \cos[(w_i + w_T - w_R)t + \alpha_i + \alpha_T - \alpha_R \\ & + \psi(t) + \Delta w t - w_R(\tau_1 + \tau_2 + \tau_3)]. \end{aligned} \tag{10.62}$$

The last term of $w_R(\tau_1 + \tau_2 + \tau_3)$ in Equation 10.62 is introduced now because the time variable t is defined for the received components as $[t - (\tau_1 + \tau_2 + \tau_3)]$ due to the filtering delays incurred, but the time associated with the receiver's frequency is not delayed. To obtain coherent phase and frequency combination for L and U given by $S_7(t)$ and $S_9^{diff}(t)$ then $w_T = w_R$ and $\alpha_T = \alpha_R - w_R(\tau_1 + \tau_2 + \tau_3) = 0$ must be satisfied. As portrayed in Figure 10.5, initially $S_8(t)$ is phase shifted by 90 deg in order to rotate the cos() function and generate a sin() function. Then both $S_7(t)$ and the phase-shifted $S_8(t)$ are input to a limiter which removes any envelope fluctuations, and the amplitude-limited signals are multiplied together. Both the sum and differences of their arguments are generated, but the loop filter of Figure 10.5 admits only their difference $S_11(t)$ to the VCO, hence we arrive at:

$$S_{11}(t) = \sin[(w_T - w_R)t + (\alpha_T - \alpha_R) - w_R(\tau_1 + \tau_2 + \tau_3)] \tag{10.63}$$

Observe that $S_{11}(t)$ is independent of any channel fading since $\psi(t)$ was cancelled, but retains the frequency and phase error terms between the transmitter and receiver.

After the addition of $S_7(t)$ and $S_9(t)$ and the removal of the unwanted mixer output component by LPF4, the recovered signal can be derived from Equations 10.60 and 10.62 as

$$S_{10}(t) = (A_1 A_3 + A_2 A_4)\alpha(t)\ a_i \cos(w_i t + \alpha_i + \Delta w t). \tag{10.64}$$

If appropriate filter design is able to maintain $A_1 A_3 + A_2 A_4 = 1$, the recovered signal is corrupted by the channel amplitude fading $\alpha(t)$ and the frequency offset Δw only. However, the TTIB processing itself appears perfectly transparent. Since the pilot signal and the information signal suffer the same attenuation in a flat fading narrowband channel, the channel sounding information derived by the pilot can be used to remove the effects of the fading from the information sequence.

10.3.1.4 TTIB schemes using quadrature mirror filters

Quadrature mirror filters (QMFs) [353] allow relatively straightforward TTIB implementations. A variety of suitable QMF designs have been proposed by Johnston [354]. The fundamental property of QMFs is that the lower (L) and upper (U) sidebands are symmetrical around the joint cut-off frequency of the LPF and HPF stages. Hence, despite their overlapping transition region, they do not introduce aliasing distortions and automatically ensure identical signal delays for both sub-band signals. Furthermore, since both stages retain signal energy from the transition band, their output signals in this frequency band will be highly correlated with each other. Through the use of correlation techniques we can combine these two correlated sub-bands perfectly after the removal of the pilot tone at the receiver.

The number of sub-bands is not limited to 2, m cascaded stages of QMFs can be used to provide 2^m sub-bands. With an increased number of sub-bands the system becomes more complex, but as each sub-band becomes narrower in comparison to the coherence bandwidth of the mobile radio channel, and the number of pilots which can be inserted increases, the channel is effectively sounded across the signal's bandwidth, making it possible to dispense with a channel equaliser. However, the increased number of pilots reduces the spectral efficiency of the transmission. In Chapter 15 we will introduce a technique based on an orthogonal transformation, and this band splitting might employ 128 sub-channels and a large number of pilot tones.

In this section we consider the twin-band, QMF-based TTIB processing shown in Figure 10.6. Here a complex baseband spectrum is split by the QMF HP and LP stages. In part (c) the subbands are split apart to make room for the pilot tone.

In order for distortion-free signal reconstruction to occur using QMFs, the QMF structure must have an all-pass transfer function given by

$$|H_{LP}(f)|^2 + |H_{HP}(f)|^2 = 1. \tag{10.65}$$

In order to maintain linear phase both the HP and LP impulse responses must be symmetric. Finally, the alias, free condition will be met if the impulse responses satisfy the condition

$$h_{LP}(n) = (-1)^n h_{HP}(n). \tag{10.66}$$

Assuming a complex baseband input spectrum of $I(f)$ the QMF outputs can be written as

$$\begin{aligned} I_{LP}(f) &= I(f) \cdot H_{LP}(f) \\ I_{HP}(f) &= I(f) \cdot H_{HP}(f). \end{aligned} \tag{10.67}$$

In order to shift $I_{LP}(f)$ and $I_{HP}(f)$ sufficiently far apart to create a spectral gap for the pilot symbol, we multiply both transfer functions by $e^{\pm jw_T t}$, where $2w_T$ is the

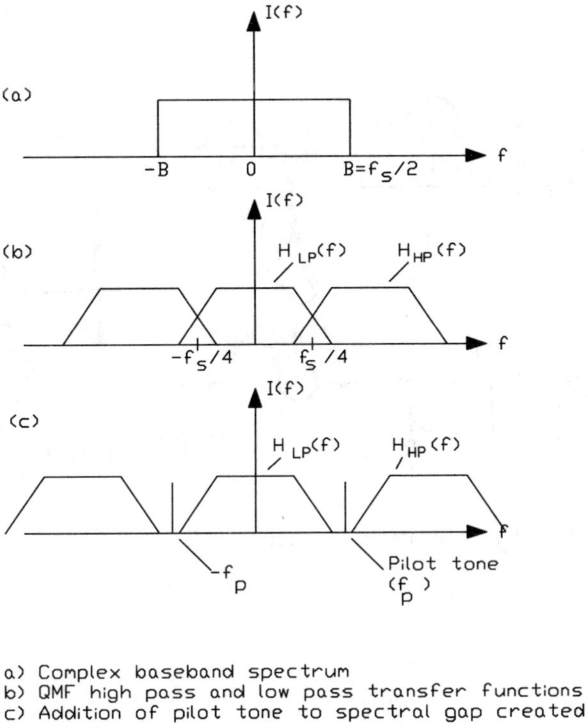

Figure 10.6: QMF splitting of a TTIB signal

required width of the spectral notch. This leads to a transmitted signal of

$$T(f) \;=\; I(f)[H_{LP}(f)e^{-jw_T t} + H_{HP}(f)e^{jw_T t}]. \tag{10.68}$$

These operations are summarised in Figure 10.7 where the complex baseband spectra are also shown at the various stages of the processing.

Note that, due to the half-band splitting, the sub-band signals must be decimated by 2 and in general the connections between blocks represent complex signals. Bateman [347] suggested a QMF-based TTIB receiver structure, but Cavers [10] proposed an improved alternative. Here we explain Caver's structure, but use Bateman's frequency domain approach in our descriptive analysis.

As previously detailed, the transmitted signal $T(f)$ is subjected to amplitude fading and phase rotation by the mobile radio channel. However, in this analysis we neglect these effects since they are considered in the previous section, and concentrate purely on the signal reconstruction. For perfect, alias-free signal regeneration, the sidebands must be shifted back to their original position, and this operation is typically performed in two steps. This is shown in Figure 10.8 where the QMF-based TTIB receiver's schematic is shown along with the typical baseband spectral representation of the various stages of processing.

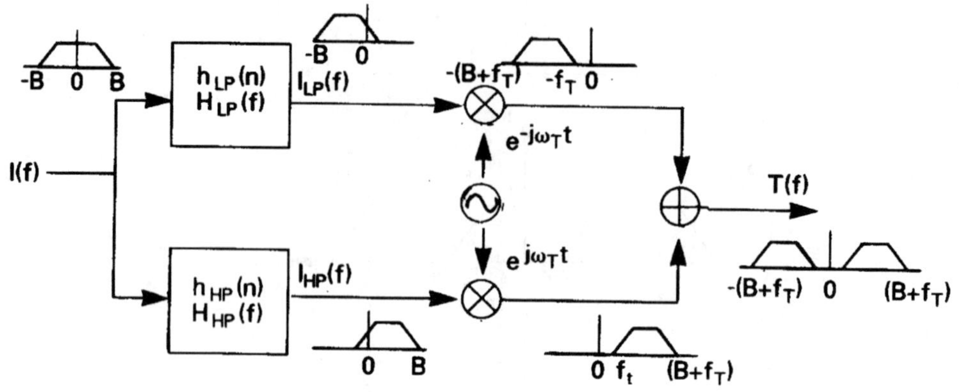

Figure 10.7: TTIB transmitter scheme [347] ©IEEE, 1990, Bateman

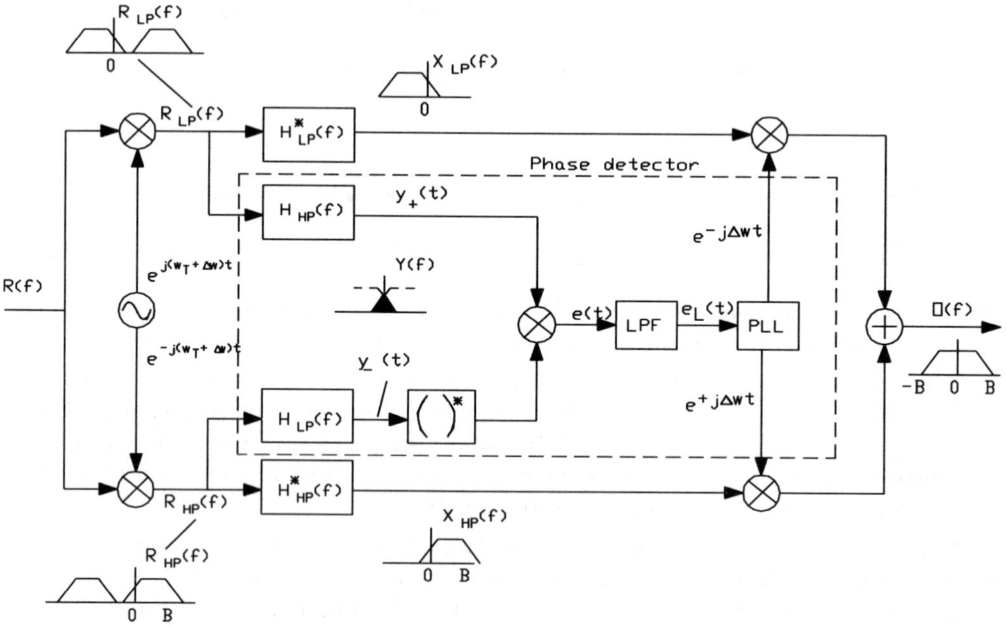

Figure 10.8: QMF-based TTIB receiver structure [347] ©IEEE, 1990, Bateman

Initially, after removing the pilot tone, the sub-band components of the received signal $R(f)$ are translated back close to their original positions by being shifted using a frequency of $\pm w_R = \pm(w_T + \Delta w)$, where Δw represents the frequency difference $(w_R - w_T)$ due to misalignment between the transmitter and receiver oscillators. These translated signals can be expressed in terms of the transmitted signal $T(f)$ and noise $N(f)$ using

$$R(f) = T(f) + N(f) \tag{10.69}$$

so that

$$\begin{aligned} R_{LP}(f) &= R(f).e^{j(w_T+\Delta w)t} &= [T(f) + N(f)]e^{j(w_T+\Delta w)t} \\ R_{HP}(f) &= R(f).e^{-j(w_T+\Delta w)t} &= [T(f) + N(f)]e^{-j(w_T+\Delta w)t}. \end{aligned} \tag{10.70}$$

Upon substituting $T(f)$ from Equation 10.68 into Equation 10.70 we obtain

$$\begin{aligned} R_{LP}(f) &= I(f).H_{LP}(f)e^{j\Delta wt} + I(f).H_{HP}(f).e^{j(2w_T+\Delta w)t} + \\ &\quad N(f)e^{j(w_T+\Delta w)t} \\ R_{HP}(f) &= I(f).H_{LP}(f).e^{-j(2w_T+\Delta w)t} + I(f).H_{HP}(f)e^{-j\Delta wt} + \\ &\quad N(f)e^{-j(w_T+\Delta w)t}. \end{aligned} \tag{10.71}$$

It can be seen in Figure 10.8 that one of the side lobes of each of the spectra $R_{LP}(f)$ and $R_{HP}(f)$ is now shifted back to their original positions, apart from a small frequency error Δw. Filtering $R_{LP}(f)$ and $R_{HP}(f)$ with the filter characteristics $H_{LP}(f)$ and $H_{HP}(f)$, respectively, removes the incorrectly positioned side lobes and retains the required ones, yielding

$$\begin{aligned} X_{LP} &= R_{LP}(f).H_{LP}^*(f) \\ &= I(f).|H_{LP}(f)|^2 e^{j\Delta wt} + N(f).H_{LP}^*(f)e^{j(w_T+\Delta w)t} \\ X_{HP} &= R_{HP}(f).H_{HP}^*(f) \\ &= I(f).|H_{HP}(f)|^2 e^{-j\Delta wt} + N(f).H_{HP}^*(f)e^{j(w_T+\Delta w)t}. \end{aligned} \tag{10.72}$$

The remaining operations in Figure 10.8 are concerned with the removal of the residual frequency error Δw and the reconstruction of the original full-band signal. Assuming that the frequency error can be extricated from $R_{LP}(f)$ and $R_{HP}(f)$ by exploiting the correlations inherent in their overlap transition bands (which we will address later) then the output signal $O(f)$ can be written as

$$O(f) = X_{LP}(f).e^{-j\Delta wt} + X_{HP}(f).e^{j\Delta wt}. \tag{10.73}$$

Substituting Equation 10.72 into Equation 10.73 yields

$$\begin{aligned} O(f) &= I(f)\left[|H_{LP}(f)|^2 + |H_{HP}(f)|^2\right] + \\ &\quad N(f).H_{LP}^*(f).e^{jw_Tt} + N(f).H_{HP}^*(f).e^{-jw_Tt} \end{aligned} \tag{10.74}$$

where the filtered and phase-shifted noise term of

$$N_F(f) = N(f).H_{LP}^*(f).e^{jw_Tt} + N(f).H_{HP}^*(f).e^{-jw_Tt} \tag{10.75}$$

is identical to $N(f)$ since it has the same total bandwidth, the filters $H_{LP}(f)$ and $H_{HP}(f)$ have complementary transfer functions and the noise in the two sidebands is uncorrelated due to their narrow overlapping regions. Taking the all-pass nature of the QMFs given in Equation 10.65 into account, Equation 10.74 can be written as

$$O(f) = I(f) + N(f) \tag{10.76}$$

which is the original input signal plus the additive noise. Therefore, the QMF-based TTIB transceiver ensures perfectly transparent baseband processing.

10.3.1.5 Residual frequency error compensation [10]

As mentioned earlier, the compensation of the residual frequency error Δw is based on the correlation of the spectral components of the HP and LP transition regions [10]. If this region does not contain any significant energy, some deterministic signal can be added, although this reduces the overall power efficiency [341].

The filters $H_{HP}(f)$ and $H_{LP}(f)$ in the phase detector section of Figure 10.8 extract the transition band components of the L and U band signals $R_{LP}(f)$ and $R_{HP}(f)$, respectively, which were given in Equation 10.71. Due to the antisymmetric filter arrangement in the phase detector compared to the filters in the corresponding branches of the transmitter, the combined frequency response $H(f)$ of the TTIB transmitter and the receiver's concatenated phase detector becomes identical in both branches, and is described by

$$H(f) = H_{LP}(f) \cdot H_{HP}(f). \tag{10.77}$$

This antisymmetric filter arrangement blocks all frequency components outside the transition band, and the input signal with spectrum $I(f)$ generates an output having a spectrum of

$$\begin{aligned} Y(f) &= I(f) \cdot H(f). \\ &= I(f).H_{LP}(f) \cdot H_{HP}(f). \end{aligned} \tag{10.78}$$

In the time domain this corresponds to

$$y(t) = i(t) * h(t) = \int_{-\infty}^{\infty} i(t-\tau)h(\tau)d\tau \tag{10.79}$$

where $y(t)$, $i(t)$ and $h(t)$ are the time domain representations of $Y(f)$, $I(f)$ and $H(f)$, respectively.

If the transmission channel is non-ideal, the received signal $r(t)$ differs from the transmitted signal $t(t)$ which is described by the following relationship

$$r(t) = \alpha(t) \cdot e^{j\psi(t)} \cdot t(t) + n(t) \tag{10.80}$$

where $\alpha(t)$, $\psi(t)$ and $n(t)$ represent the fading envelope, the fading phase, and the additive noise, respectively. These quantities are considered to be slowly varying with time, which allows us to drop their dependency on t within a symbol period. The signal $y(t)$ in Equation 10.79 is subjected to the channel effects portrayed in Equation 10.80, but due to the oscillator frequency error $\Delta w = (w_T - w_R)$ between the transmitter and receiver, a further phase component Δwt must be taken into account, yielding the phase-shifted signals $y_+(t)$ and $y_-(t)$ as follows:

$$\begin{aligned} y_+(t) &= \alpha \cdot e^{j(\psi+\Delta wt)} \cdot y(t) + n_+(t) \\ &= \alpha \cdot e^{j\psi} \cdot e^{j\Delta wt} \cdot y(t) + n_+(t) \\ y_-(t) &= \alpha \cdot e^{j(\psi-\Delta wt)} \cdot y(t) + n_-(t) \\ &= \alpha \cdot e^{j\psi} \cdot e^{-j\Delta wt} \cdot y(t) + n_-(t) \end{aligned} \tag{10.81}$$

where $n_\pm(t)$ now represents the phase shifted additive noise components. After taking the conjugate of $y_-(t)$ in Equation 10.81 and multiplying $y_+(t)$ and $y_-^*(t)$ as shown in Figure 10.8 we derive the error signal

$$\begin{aligned} e(t) &= y_+(t) \cdot y_-^*(t) \\ &= \alpha^2 \cdot e^{j2\Delta wt} \cdot |y(t)|^2 + \alpha \cdot e^{j\psi} \cdot e^{j\Delta wt} \cdot y(t) \cdot n_-^*(t) + \\ & \alpha \cdot e^{-j\psi} \cdot e^{j\Delta wt} \cdot y^*(t) \cdot n_+(t) + n_+(t) \cdot n_-^*(t). \end{aligned} \tag{10.82}$$

In the case of low channel noise we can approximate this equation by

$$e(t) \approx \alpha^2 \cdot e^{j2\Delta wt} \cdot |y(t)|^2 \tag{10.83}$$

which upon evaluating the phase $\Phi = 2\Delta wt$ and halving the frequency allows the receiver to coherently recombine the L and U sub-bands by shifting the signals $X_{LP}(f)$ and $X_{HP}(f)$ in Figure 10.8 back to their original positions. Note, however, that there is a phase ambiguity of π in the phase detection process since shifts of both Φ and $\pi+\Phi$ produce the same signal $e(t)$. In this situation the TTIB receiver could erroneously invert the polarity of the output signal $O(t)$. This problem can be overcome through the use of differential coding, although this does incur a performance penalty as discussed in Chapter 4.

10.3.1.6 TTIB system parameters [11]

The performance of a TTIB-based 16-level QAM scheme has been reported in Reference [11] where initially an AWGN channel was assumed and the hardware was constructed using a TMS320C25 digital signal processing device. With a perfectly coherent, synchronous implementation the SNR degradation due to device imperfections was about 0.5 dB when compared to the theoretical performance. When TTIB notch-width-based timing recovery was used along with perfect carrier recovery, a further 0.5 dB degradation was incurred.

When evaluating the TTIB performance using pilot-based carrier recovery, further degradations were introduced which were dependent on the level of the pilot signal transmitted. At higher pilot levels the system is less prone to channel noise, but also less power efficient. If a constant total power level is assumed, there is an optimum trade-off between pilot and information signal power. Furthermore, by narrowing the pilot filter bandwidth, less noise is introduced and the pilot signal's SNR is improved, but the maximum fading rate which can be followed is reduced. In their implementation, Martin and Bateman [11] found a bandwidth of 300 Hz to be most appropriate.

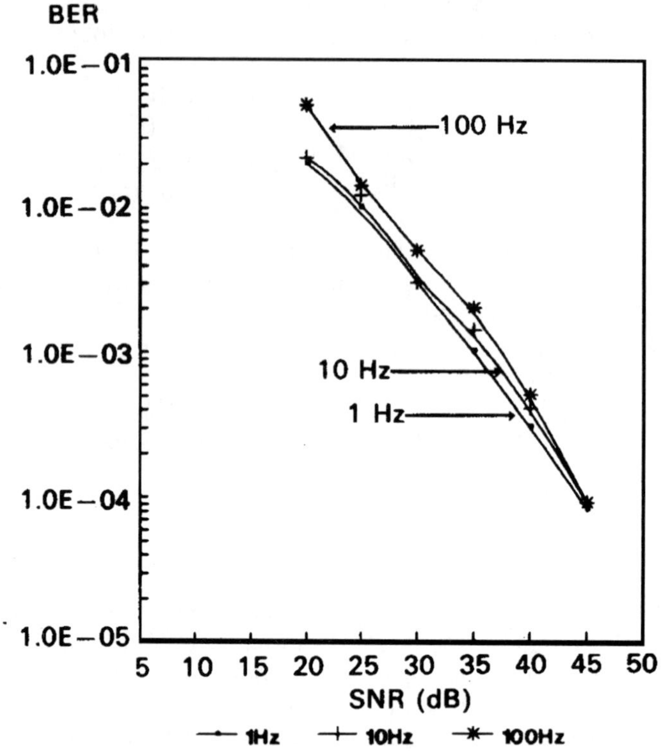

Figure 10.9: BER versus channel SNR performance of TTIB-assisted 16-QAM with f_P=70 MHz, $f_{Doppler}$=1, 10 and 100 Hz. [11] ©IEEE, 1991, Martin, Bateman

They also found that at a pilot power level of -3 dB relative to the signal, the 16-QAM BER against SNR performance was very similar to the perfectly coherent scheme, if the pilot power was not included when calculating the E_b/N_0 ratio. When the pilot power was included, a performance penalty of about 2.5 dB was incurred. At a pilot level of -6 dB relative to the mean information power a further 1 dB loss in SNR was added, which was deemed acceptable by the authors.

When the performance of the TTIB modem was evaluated over a fading channel at a frequency of 70 MHz, the pilot level of -3 dB was still adequate and yielded a 2.5 dB E_b/N_0 penalty for fading rates between 1 and 100 Hz. However, for higher fading rates the pilot filter bandwidth limited the possible range of fading fluctuations and impaired the performance unless a higher notch width was introduced. The BER versus channel SNR performance of the TTIB assisted 16-QAM modem proposed by Martin and Bateman [11] is shown in Figure 10.9 for Rayleigh fading channels exhibiting Doppler frequencies of 1, 10 and 100 Hz.

Clearly, TTIB assisted coherent QAM ensures high-integrity information transmission via fading channels by providing both amplitude and phase references for the receiver. These references assist the receiver in attaining coherent demodulation against a pilot scaled set of demodulation thresholds at a moderate complexity. Having discussed the insertion of channel sounding information in the spectral domain we now focus our attention on a time domain technique called pilot symbol assisted modulation.

10.3.2 Pilot symbol assisted modulation [12]

10.3.2.1 Introduction

As we have seen in Chapter 5, the BER of the square constellation 16-QAM and 64-QAM is sufficiently low for speech transmission over AWGN channels, but as shown in Chapter 10 it becomes prohibitively high for transmission over Rayleigh fading channels. This is particularly so for the lower integrity $C2$ sub-channel in case of 16-QAM and for the $C2$ and $C3$ sub-channels of 64-QAM, resulting in the average BER becoming unacceptably high.

As shown in the previous section, one way to improve the BER of the square constellation over fading channels is to use transparent tone in band (TTIB) transmission. This results in higher out-of-band emissions and hence higher adjacent channel interferences, whilst also increasing complexity. An alternative technique which is analogous to TTIB but, instead of transmitting constantly in part of the available spectrum, transmits periodically in all the available spectrum is refered to as pilot symbol assisted modulation (PSAM) [85,355–357]. This method relies upon the insertion of known phasors into the stream of useful information symbols for the purpose of channel sounding. These pilot symbols allow the receiver to extract channel attenuation and phase rotation estimates for each received symbol, facilitating the compensation of the fading envelope and phase. Closed-form formulae for the BER of PSAM were provided by Cavers [12] for binary phase shift keying (BPSK), and quadrature phase shift keying (QPSK), while for 16-QAM he derived a tight upper bound of the symbol error rate (SER).

10.3.2.2 PSAM system description

Following Caver's approach [12], the block diagram of a general PSAM scheme is depicted in Figure 10.10, where the pilot symbols p are cyclically inserted into the data sequence prior to pulse shaping, as demonstrated by Figure 10.11. A frame of data is constituted by M symbols, and the first one in every frame is assumed to be the pilot symbol $b(0)$, followed by $(M-1)$ useful data symbols $b(1), b(2), \ldots, b(M-1)$.

Detection can be carried out by matched filtering, and the output of the matched filter is split into data and pilot paths, as seen in Figure 10.10. The set of pilot symbols can be extracted by decimating the matched filter's sampled output sequence using a decimation factor of M. The extracted sequence of pilot symbols must then be interpolated in order to derive a channel estimate $v(k)$ for every useful received information symbol $r(k)$. Decision is carried out against a decision level reference grid, scaled and rotated according to the instantaneous channel estimate $v(k)$.

Observe in Figure 10.10 that the received data symbols must be delayed according to the interpolation and prediction delay incurred. This delay becomes longer

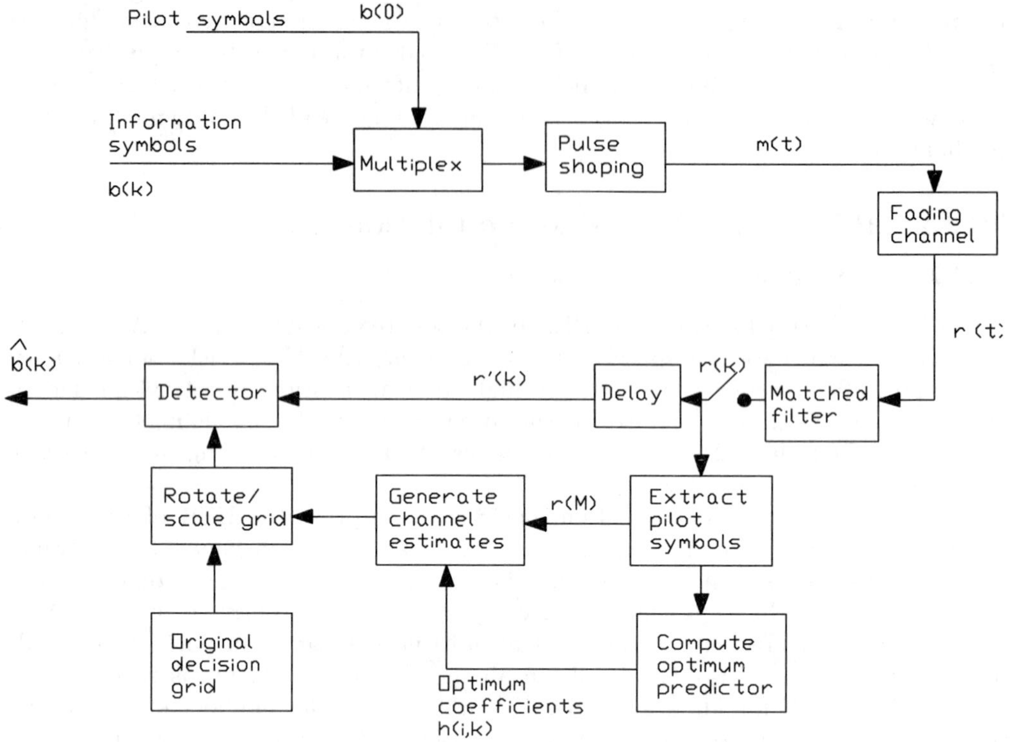

Figure 10.10: PSAM schematic diagram [12] ©IEEE, 1991, Cavers

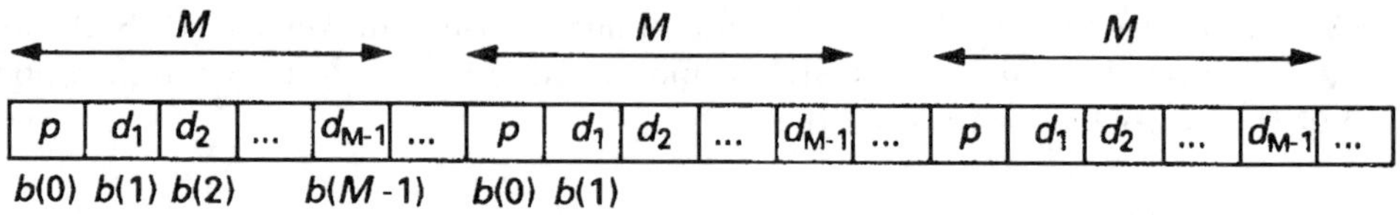

Figure 10.11: Insertion of pilot symbols

if interpolation is carried out using a longer history of the received signal to yield better channel estimates. Consequently, there is a trade-off between processing delay and accuracy. The interpolation coefficients can be kept constant over a whole pilot period of length M, but better channel estimates can be obtained if the interpolator's coefficients are optimally updated for every received symbol.

The complex envelope of the modulated signal can be formulated as:

$$m(t) = \sum_{k=-\infty}^{\infty} b(k)p(t - kT), \tag{10.84}$$

where $b(k) = -3, -1, 1$ or 3 represents the quaternary I or Q components of the 16-

QAM symbols to be transmitted, T is the symbol duration and $p(t)$ is a bandlimited unit-energy signalling pulse, for which we have:

$$\int_{-\infty}^{\infty} |p(t)|^2 dt = 1. \tag{10.85}$$

The value of the pilot symbols $b(kM)$ can be arbitrary, although sending a sequence of known pseudorandom symbols avoids the transmission of a periodic tone, which would increase the detrimental adjacent channel interference [356].

The narrowband Rayleigh channel is assumed to be "flat" fading, which implies that all frequency components of the transmitted signal suffer the same attenuation and phase shift. This condition is met if the transmitted signal's bandwidth is much lower than the channel's coherence bandwidth. The received signal is then given by:

$$r(t) = c(t) \cdot m(t) + n(t), \tag{10.86}$$

where $n(t)$ is the AWGN and $c(t)$ is the channel's complex gain. Assuming a Rayleigh fading envelope $\alpha(t)$, uniformly distributed phase $\phi(t)$ and a residual frequency offset of f_0, we have:

$$c(t) = \alpha(t)e^{j\phi(t)} \cdot e^{j\omega_0 t} \tag{10.87}$$

The matched filter's output symbols at the sampling instant kT are then given as:

$$r(k) = b(k) \cdot c(k) + n(k). \tag{10.88}$$

Without imposing limitations on the analysis, Cavers [12] assumed that in every channel sounding block $b(0)$ was the pilot symbol and considered the detection of the useful information symbols in the range $\lfloor -M/2 \rfloor \leq k \leq \lfloor (M-1)/2 \rfloor$, where $\lfloor \bullet \rfloor$ is the integer of $\bullet$. Optimum detection is achieved if the corresponding channel gain $c(k)$ is estimated for every received symbol $r(k)$ in the above range. The channel gain estimate $v(k)$ can be derived as a weighted sum of the surrounding K received pilot symbols $r(iM)$, $\lfloor -K/2 \rfloor \leq i \leq \lfloor K/2 \rfloor$, as shown below:

$$v(k) = \sum_{i=\lfloor -K/2 \rfloor}^{\lfloor K/2 \rfloor} h(i,k) \cdot r(iM), \tag{10.89}$$

and the weighting coefficients $h(i,k)$ explicitly depend on the symbol position k within the frame of M symbols.

The estimation error $e(k)$ associated with the gain estimate $v(k)$ is computed as:

$$e(k) = c(k) - v(k). \tag{10.90}$$

10.3.2.3 Channel gain estimation

While previously proposed PSAM schemes used either a lowpass interpolation filter [356] or an approximately Gaussian filter [357], Cavers deployed an optimum Wiener filter to minimise the channel estimation error variance ${\sigma^2}_e(k) = E\{e^2(k)\}$, where $E\{\ \}$ represents the expectation. This well-known estimation error variance minimisation problem can be formulated as follows:

$$\begin{aligned}\sigma_e^2(k) &= E\{e^2(k)\} = E\{[c(k) - v(k)]^2\} \\ &= E\left\{\left[c(k) - \sum_{i=\lfloor -K/2 \rfloor}^{\lfloor K/2 \rfloor} h(i,k) \cdot r(iM)\right]^2\right\}.\end{aligned} \tag{10.91}$$

In order to find the optimum interpolator coefficients $h(i,k)$, minimising the estimation error variance $\sigma^2{}_e(k)$ we consider estimating the kth sample and set:

$$\frac{\partial \sigma_e^2(k)}{\partial h(i,k)} = 0 \quad \text{for} \quad \lfloor -K/2 \rfloor \leq i \leq \lfloor K/2 \rfloor. \tag{10.92}$$

Then using Equation 10.91 we have:

$$\frac{\partial \sigma^2{}_e(k)}{\partial h(i,k)} = E\left\{2\left[c(k) - \sum_{i=\lfloor -K/2 \rfloor}^{\lfloor K/2 \rfloor} h(i,k) \cdot r(iM)\right] \cdot r(jM)\right\} = 0. \tag{10.93}$$

After multiplying both square bracketed terms with $r(jM)$, and computing the expected value of both terms separately, we arrive at

$$E\{c(k) \cdot r(jM)\} = E\{\sum_{i=\lfloor -K/2 \rfloor}^{\lfloor K/2 \rfloor} h(i,k) \cdot r(iM) \cdot r(jM)\}. \tag{10.94}$$

Observe that

$$\Phi(j) = E\{c(k) \cdot r(jM)\} \tag{10.95}$$

is the cross-correlation of the received pilot symbols and complex channel gain values, while

$$R(i,j) = E\{r(iM) \cdot r(jM)\} \tag{10.96}$$

represents the pilot symbol autocorrelations, hence Equation 10.94 yields:

$$\sum_{i=\lfloor -K/2 \rfloor}^{\lfloor K/2 \rfloor} h(i,k) \cdot R(i,j) = \Phi(j), \quad j = \lfloor -\frac{k}{2} \rfloor \ldots \lfloor \frac{k}{2} \rfloor. \tag{10.97}$$

If the fading statistics can be considered stationary, the autocorrelations $R(i,j)$ will only depend on the difference $|i-j|$, giving $R(i,j) = R(|i-j|)$. Therefore Equation 10.97 can be written as:

$$\sum_{i=\lfloor -K/2 \rfloor}^{\lfloor K/2 \rfloor} h(i,k) \cdot R(|i-j|) = \Phi(j), \quad j = \lfloor -K/2 \rfloor \ldots \lfloor K/2 \rfloor, \tag{10.98}$$

which is a form of the well-known Wiener-Hopf equations [3], often used in estimation and prediction theory [158].

This set of K equations contains K unknown prediction coefficients $h(i,k)$, $i =$

$\lfloor -K/2 \rfloor, \ldots, \lfloor K/2 \rfloor$, which must be determined in order to arrive at a minimum error variance estimate of $c(k)$ by $v(k)$. First the correlation terms $\Phi(j)$ and $R(|i-j|)$ must be computed and to do this the expectation value computations in Equations 10.95 and 10.96 need to be restricted to a finite duration window. This approach is referred to as the *autocorrelation method* [158].

The set of Equations 10.98 can also be expressed in a convenient matrix form as:

$$\begin{bmatrix} R(0) & R(1) & R(2) & \ldots & R(K) \\ R(1) & R(0) & R(1) & \ldots & R(K-1) \\ R(2) & R(1) & R(0) & \ldots & R(K-2) \\ \vdots & \vdots & \vdots & \ldots & \vdots \\ R(K) & R(K-1) & R(K-2) & \ldots & R(0) \end{bmatrix} \cdot \begin{bmatrix} h\left(\lfloor -\frac{K}{2} \rfloor, k\right) \\ h\left(\lfloor -\frac{K}{2}+1 \rfloor, k\right) \\ h\left(\lfloor -\frac{K}{2}+2 \rfloor, k\right) \\ \vdots \\ h\left(\lfloor \frac{K}{2} \rfloor, k\right) \end{bmatrix} = \begin{bmatrix} \Phi\left(\lfloor -\frac{K}{2} \rfloor\right) \\ \Phi\left(\lfloor -\frac{K}{2}+1 \rfloor\right) \\ \Phi\left(\lfloor -\frac{K}{2}+2 \rfloor\right) \\ \vdots \\ \Phi\left(\lfloor \frac{K}{2} \rfloor\right) \end{bmatrix}, \tag{10.99}$$

which can be solved for the optimum predictor coefficients $h(i,k)$ by matrix inversion using Gauss-Jordan elimination or any appropriate recursive algorithm.

Once the optimum predictor coefficients $h(i,k)$ are known, the minimum error variance channel estimate $v(k)$ can be derived from the received pilot symbols using Equation 10.89, as also demonstrated by Figure 10.10.

10.3.2.4 PSAM parameters

Cavers [12] provided a range of results for a variety of PSAM parameters. As regards to the pilot spacing M, the optimum value is determined by the Nyquist rate of the fading envelope. Inserting pilot symbols less frequently results in a significant increase in BER due to the inability to track the fading, while an unnecessarily frequent insertion wastes energy without significantly improving the BER performance.

The typical energy loss and Doppler frequency interdependence can be highlighted by the following simple example. Let us assume a fast-moving vehicle with a Doppler frequency of $f_D = 100$ Hz, and a TDMA multi-user signalling rate of 100 ksymbols/s, yielding a normalised Doppler shift of $f_d \cdot T = 10^{-3}$. This would require a Nyquist frequency of $f_N = 2 \cdot 10^{-3}$, implying a pilot spacing of $M = 500$, which results in an energy loss of only 0.2%. If the signalling rate reduces to 10 ksymbols/s, the relative frequency of the pilot symbol insertion must be increased tenfold in order for the pilot symbols to appear in the same positions in terms of absolute time, facilitating equally frequent channel sounding, as in the case of 100 ksymbols/s signalling. The corresponding pilot insertion frequency of $M = 50$ implies an energy loss of about 2%. However, when using the PSAM scheme in a single-user scenario with a low signalling rate of, say, 2 ksymbols/s, the energy loss due to five times more frequent channel sounding associated with $M = 10$ becomes 10%.

As seen in Figure 10.10, the operation of the PSAM scheme is prone to a delay of

$KM/2$ symbols, which might become an impediment in case of speech transmissions. Due to this start-up delay the first $KM/2$ modulation symbols are corrupted by fading and hence must be discarded. This might affect, for example, some $7 \cdot 10/2 = 35$ start-up symbols for the specific case of $K = 7$ and $M = 10$.

Clearly, the design of the predictor coefficients $h(i, k)$ depends on a variety of PSAM system parameters, such as K and M, as well as on channel conditions, such as the prevailing SNR and normalised Doppler shift. The symbol error rate performance upper bounds derived by Cavers [12] for 16-QAM cannot be approached if the above conditions are drastically different from those for which the predictor coefficients were optimised. Best results can be attained using adaptive predictors updating their parameters online, as signalling and channel conditions vary.

A further advantage of PSAM is that it generates an absolute phase reference and hence differential encoding of the data is not required. In contrast, TTIB transmissions require a perfect phase lock to be maintained [10], which assumes differential coding, associated with some SNR penalty. In a more recent paper [358] Cavers investigated the performance of PSAM schemes under dispersive channel conditions introduced by frequency selective fading channels in case of BPSK and QPSK. He found that the performance gains are sensitive to delay spread, but PSAM always ensured a better performance than differential coding. Recently Seymour and Fitz [359] recognised that ignoring the transmitted information symbols in the predictor is wasteful. This is because the modulated received information symbols also carry channel-specific information. Exploiting this property allows for more sparse pilot insertion, thereby saving modulation energy, or achieving slightly better BER performance.

10.3.2.5 PSAM performance

In this section we embarked on the performance evaluation of two low-complexity square-constellation pilot-assisted modems, namely a 16-QAM/PSAM scheme and a 64-QAM/PSAM scheme. The pilot separation of the 16-QAM scheme was $M = 10$, while that of the 64-QAM modem was $M = 5$ at a pedestrian speed of 4 mph, propagation frequency of 1.9 GHz and signalling rate of 100 kBd. The channel magnitude and phase fluctuation was linearly interpolated between two adjacent pilot symbols. The performance of these scheme is displayed in Figures 10.12. and 10.13. This low complexity approach typically failed to track the amplitude and phase trajectory upon emerging from a deep fade, which manifests itself in a residual bit error rate, but overall provided a significant performance improvement when compared to the scenarios portrayed earlier in Figures 10.1 and 5.3 without PSAM.

The performance of a combined 16-QAM/PSAM/TCM scheme over Rayleigh fading channels is reported in Figure 11.8, where three PSAM modulation schemes having a net throughput of 2 bits/symbol are studied comparatively. Specifically, uncoded QPSK, 2/3 rate coded 8-PSK/TCM and 1/2 rate square 16-QAM/TCM are compared and the 16-QAM/TCM scheme has the best performance, since it has no unprotected bits.

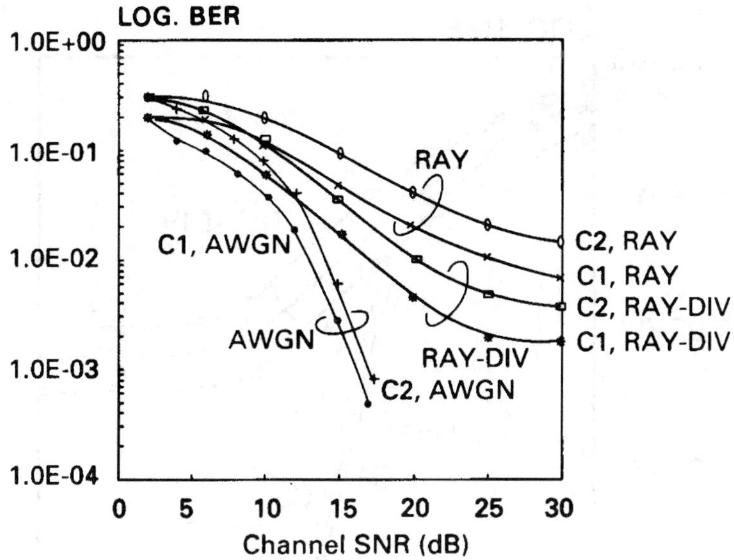

Figure 10.12: Square 16-QAM/PSAM BER versus channel SNR curves for AWGN and Rayleigh channels with and without diversity at 4 mph, 1.9 GHz, and 100 kBd

10.4 Summary

In this chapter we started by considering the theoretical performance of 16-level and 64-level square QAM over mobile radio channels. The results showed a significant drop in performance compared to transmission over the Gaussian channels considered earleir in this book.

We then continued to consider two particular schemes of QAM transmission over mobile radio channels whereby an attempt was made to transmit some information to the receiver concerning the instantaneous state of the mobile radio channel. Transparent tone in band attempted to split the transmitted spectrum in two, to increase the spacing between these two parts, and to insert a continuous tone which the receiver could use to estimate the channel response. Such a scheme was shown to be complex and a number of problems with it were detailed. Pilot symbol assisted modulation inserted a known symbol periodically into the data stream. Since the receiver had been pre-programmed as to what this symbol was and when it would appear, it could use this information to estimate the channel response at that particular instant. Advanced extrapolation techniques could then be used to predict the channel response between any two of these periodic symbols. Both schemes were shown to have some success in overcoming the worst problems of the mobile radio channel although both had the disadvantage of increasing the required bandwidth due to the transmission of redundant information.

In the next chapter we consider a completely different means of overcoming the

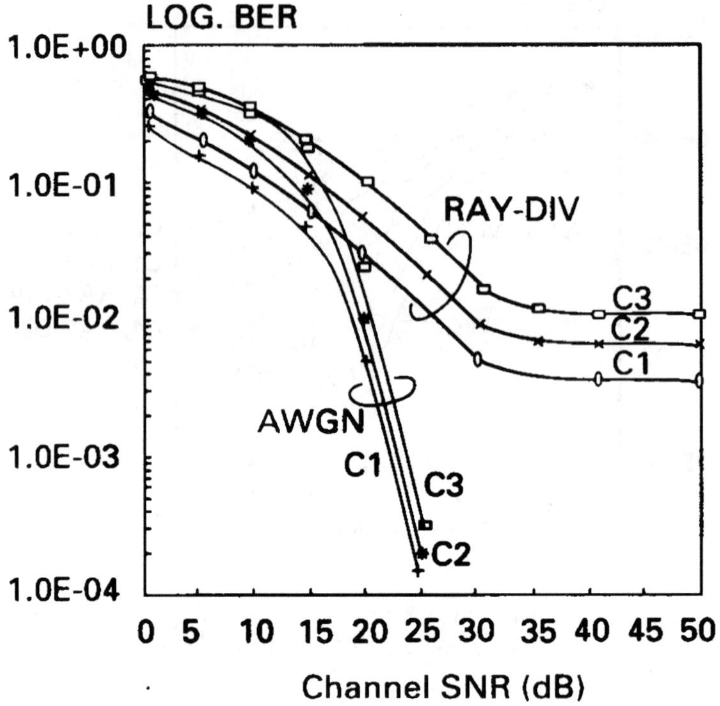

Figure 10.13: Square 64-QAM/PSAM BER versus channel SNR curves for AWGN and Rayleigh channels with diversity at 4 mph, 1.9 GHz and 100 kBd

problem of a rapidly changing channel. This is through the use of a different QAM constellation known as star QAM which, through its symmetry and use of differential coding, removes the need for the receiver to estimate the channel response at all.

Chapter 11

Star QAM Schemes for Rayleigh Fading Channels

11.1 Introduction

In Chapter 10 we considered the transmission of coherently detected maximum least distance square QAM signals over Rayleigh fading channels. We have shown that adequate BER performance can only be achieved through deploying either frequency domain pilots (TTIB) or using time domain pilot signals (PSAM). The channel attenuation and phase rotation information derived using these techniques facilitates fading compensation and coherent detection at moderate implementational complexity. In many applications, such as light weight hand-held portable telephones, power consumption, weight and low cost construction are crucial issues, and hence lower complexity modem schemes are desirable. In this case it may be preferable to reduce the system performance slightly in order to be able to employ a low-complexity non-coherent differentially encoded QAM constellation, such as the circular or star QAM scheme proposed in this chapter. We can then dispense with TTIB, PSAM, or AGC reference schemes, and no longer require carrier recovery. Consequently, the result is an attractive low complexity system. In this chapter we investigate the properties and performance of a variety of star QAM systems and then consider constellation distortion and practical issues.

11.2 Star 16-QAM Transmissions over Mobile Channels

In Section 4.1 we presented a brief discussion on the most appropriate choice of constellation for a range of channel conditions, and showed the superiority of the Type III rectangular constellation of Figure 4.1 over the Type I circular or star constellation for AWGN channels. The QAM constellation which we advocate for Rayleigh fading

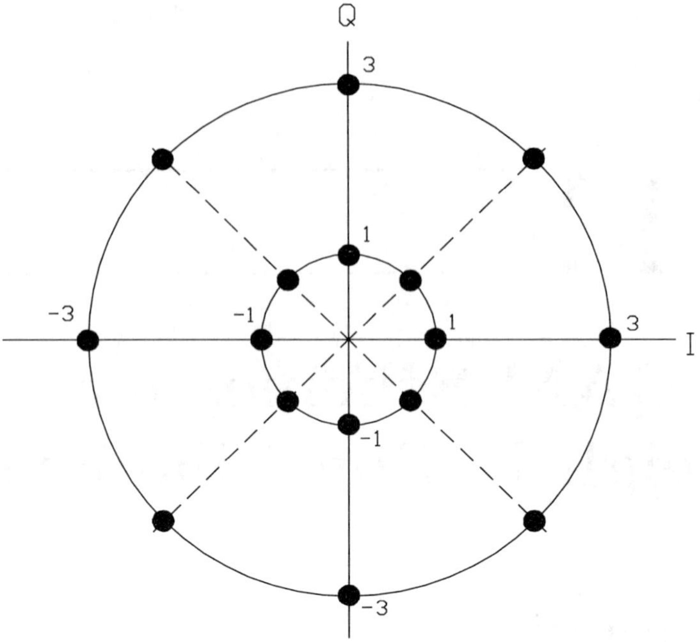

Figure 11.1: Transmitted differential star QAM constellation

channels is shown in Figure 11.1. This constellation does not have a minimum least free distance between points in the manner of the square constellation of Chapter 10, but does allow efficient differential encoding and decoding methods to be used which remove the need for AGC and carrier recovery and thus go some way towards mitigating the effects of Rayleigh fading. The structure of this constellation will be further investigated throughout this section.

11.2.1 Differential coding

In the case of PLL type carrier tracking, some form of differential encoding is essential as the Rayleigh fading channel can introduce phase shifts in excess of 50 deg between consecutive symbols in the case of slow signalling rates, making it extremely difficult to establish an absolute phase reference.The encoding method used in the star QAM system is as follows. Of the four bits in each symbol, b_1, b_2, b_3 and b_4, the first is differentially encoded onto the QAM phasor amplitude so that a 1 causes a change to the amplitude ring which was not used in the previous symbol, and a 0 causes the current symbol to be transmitted at the same amplitude as the previous symbol. The remaining three bits are differentially Gray encoded onto the phase so that, for example,000 would cause the current symbol to be transmitted with the same phase as the previous one, 001 would cause a 45 deg phase shift relative to the previous symbol, "011" a 90 deg shift relative to the previous symbol, and so on.

11.2.2 Differential decoding

Using the above differential encoding technique, decoding data is now reduced to a comparison test between the previous and current received symbols. Suppose we fix the transmitted rings at amplitude levels A_1 and A_2. Let the received phasor amplitudes be Z_t and Z_{t+1} at times t and $t+1$, respectively. The demodulator must identify whether there has been a significant change in amplitude in order to regenerate a logical 1. This is equivalent to asking whether the amplitude of the phasor at the transmitter changed rings at time $t+1$ compared to time t. The algorithm employed at the demodulator uses two adaptive thresholds to carry out its decisions. If

$$Z_{t+1} \geq \left(\frac{A_1 + A_2}{2}\right) Z_t \tag{11.1}$$

or if

$$Z_{t+1} < \left(\frac{2}{A_1 + A_2}\right) Z_t \tag{11.2}$$

then a significant change in amplitude is deemed to have occurred and bit b_1 is set to logical 1 at time $t+1$. Notice that the thresholds are dependent on Z_t, and as the amplitudes of the phasors change in fading conditions so do the thresholds. Should both of Inequalities 11.1 and 11.2 fail to be satisfied, b_1 is assigned logical 0.

If the received symbol phases are θ_t and θ_{t+1} at time t and t+1, respectively, the demodulated angle is

$$\theta_{dem} = (\theta_{t+1} - \theta_t) \; mod \; 2\pi. \tag{11.3}$$

This angle is then quantised to the nearest multiple of 45 deg and a lookup table is consulted in order to derive the remaining three output bits, b_2, b_3 and b_4. This differential system can considerably improve the BERs compared to the square constellation because it eliminates the long error bursts that occur when a false lock was inflicted. It is considerably important that with differential amplitude encoding there is no longer any need for AGC. This not only simplifies the system, but also removes errors caused by an inability of the AGC to follow the fading envelope.

11.2.3 Effect of oversampling

In mobile radio channels the fading continues to cause problems because there are changes in channel amplitude and phase during symbol periods, which in general move the differentially decoded phasors nearer to the decision boundaries. The most likely cause of error in the star QAM system is when both the noise, and the change in the phasor's amplitude or phase due to fading, combine to drive the incoming signal level over a decision boundary. As the Rayleigh fading envelope is bandlimited by the normalised Doppler frequency, it is highly correlated and predictable, particularly at low vehicular speeds. Hence a correction factor may be applied to the incoming signal in order to compensate for the changes in the fading envelope during the last symbol period. Such a system must be fast acting so that the sudden change from an amplitude decrease to an amplitude increase experienced at the bottom of a fade can immediately be detected and compensated. Methods such as using the differential of the PLL error signal in order to change the step size of the phase correction signal tend

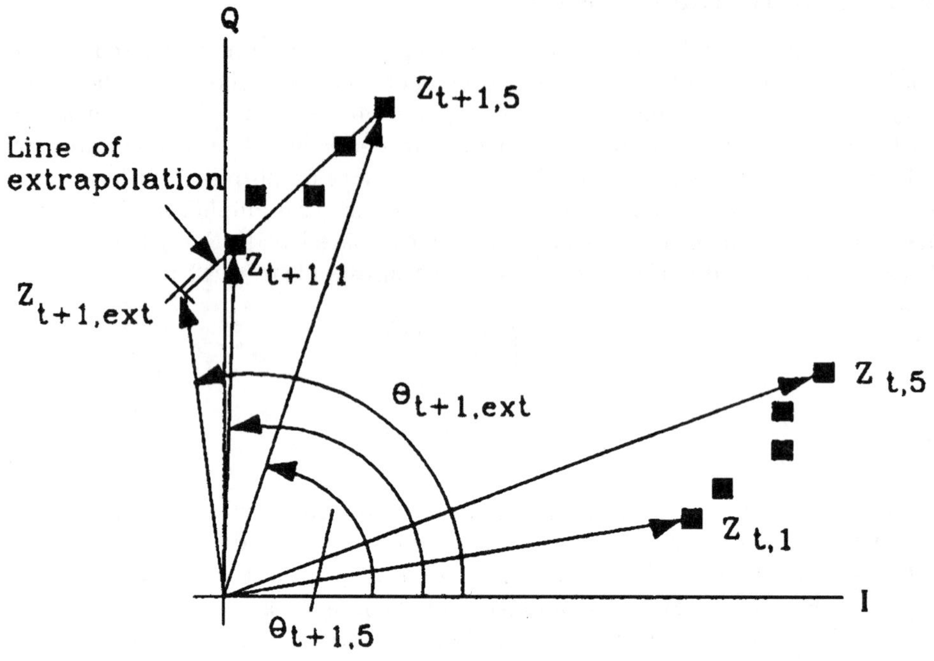

Figure 11.2: Correction for changes in fading envelope

to overshoot at sudden changes and exhibit damped second-order system behaviour.

In order to overcome this problem a simple oversampling receiver can be used. In this system n observations equally spaced in time are made per symbol period. At SNRs above 30 dB the current phasor can be modified to compensate for the fading. This is done by finding the change in the incoming symbol phase and amplitude over the current symbol period. We do this by subtracting the phase at the end of the symbol from the phase at the beginning of the symbol, and subtracting the amplitude at the end of the symbol from the amplitude at the beginning of the symbol.

The situation is shown in Figure 11.2.

We use the same notation as before, but with the addition of subscript n to the symbols, to signify the nth observation of the symbol at time t. Thus

$$Z_{t+1,diff} = Z_{t+1,n} - Z_{t+1,1} \tag{11.4}$$

where *diff* means differential, and

$$\theta_{t+1,diff} = \theta_{t+1,n} - \theta_{t+1,1}. \tag{11.5}$$

These changes can be used to extrapolate from the first observation in the current symbol back to the time when the last observation from the previous symbol was made.

With the aim of improving the accuracy of Inequalities 11.1 and 11.2 we replace Z_t by its last oversampled value, namely $Z_{t,n}$ and we use an estimate of Z_{t+1} which is formed by extrapolation as

$$Z_{t+1,ext} = Z_{t+1,1} - \frac{Z_{t+1,diff}}{n} \tag{11.6}$$

rather than Z_t. Notice that $Z_{t+1,1}$ is the first sample of Z_{t+1} and is closest in time to $Z_{t,n}$. By subtracting the average change in Z_{t+1} over a symbol period, i.e. $\frac{Z_{t+1,diff}}{n}$ from $Z_{t+1,1}$, we obtain an estimate of Z_{t+1} had it been transmitted at time t. This is beneficial as there may have been significant amplitude and phase changes between times t and $t+1$ as can be seen in Figure 11.2.

After determining bit b_1 with the aid of Inequalities 11.1 and 11.2 using the modified Z_t and Z_{t+1} values, we determine bits b_2, b_3 and b_4 by formulating

$$\theta_{dem} = (\theta_{t+1,ext} - \theta_{t,n}) \; mod \; 2\pi. \tag{11.7}$$

where $\theta_{t+1,ext}$ and $\theta_{t,n}$ are the phase angles associated with $Z_{t+1,ext}$ and $Z_{t,n}$. Again θ_{dem} is quantised and used to address a look-up table which provides values of b_2, b_3 and b_4.

In the above description it has been assumed that no pulse shaping or filtering was employed so that square pulse shapes result. This is impractical because of the spectral spillage caused by such modulation. The oversampling system can work with practical pulse shaping systems. With pulse shaping the oversampling system needs to perform an inverse operation before predicting the change undergone in the channel. At the receiver an attempt is made to reconstruct the transmitted waveform. This is done by passing the estimated received and regenerated data in square waveform through an identical filter to the transmitter filter. The difference between this reconstructed waveform and the received waveform can then be used to estimate the average change that has taken place over the symbol period, and thus form some estimate of the Rayleigh fading. This estimate can then be used in the decoding of the next symbol, with extrapolation techniques instigated as required. Simulations showed the degradation in BER over unfiltered transmissions to be negligible.

11.2.4 Star 16-QAM performance

Previous QAM systems tended to exhibit a residual BER at high SNRs due to the rapidly changing Rayleigh channel, rather than the additive noise. With this system the residual BER is reduced by approximately two orders of magnitude. The simulation results of this system when the 16-QAM carrier was 1.9 GHz, the symbol rate was 16kSym/s, the mobile's speed was 30 mph and the channel exhibited Rayleigh fading with additive white Gaussian noise (AWGN) are given in Figure 11.3. The receiver operated as described above, with the oversampling ratio set at $n = 8$. Non-linear filtering as introduced in Chapter 5 was employed at the transmitter. Both transmitter and receiver used a fourth-order Butterworth lowpass filter with a 3dB point at 1.5 times the Baud rate, i.e. 24 kHz.

The variation of BER versus channel SNR is shown in curve (c) in Figure 11.3.

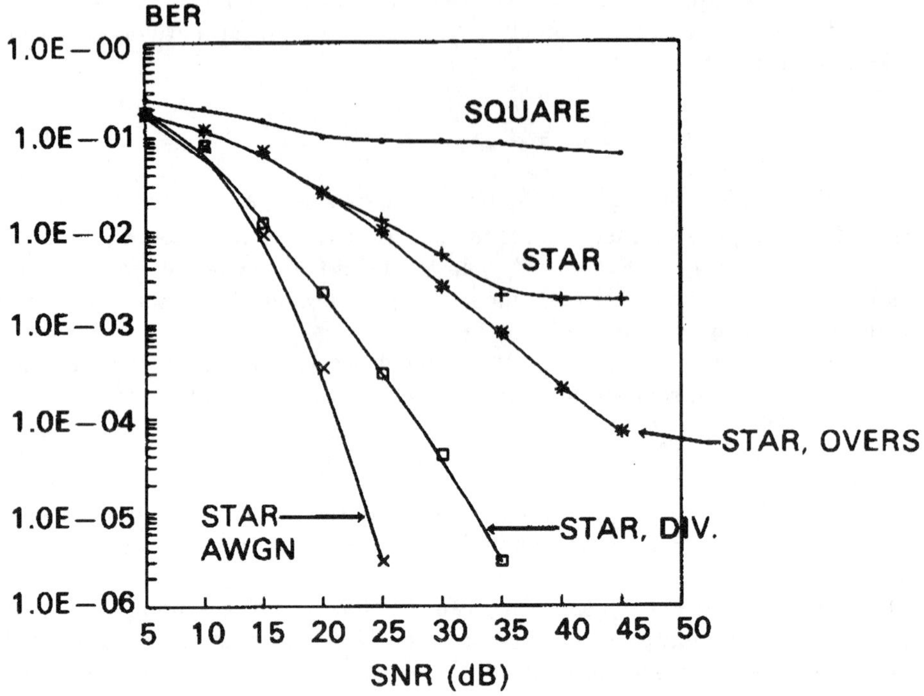

Figure 11.3: BER versus channel SNR results for various 16-QAM systems

Also shown as benchmarks are the results for the similar complexity coherent non-differential square QAM constellation using no fading compensation AGC, TTIB or PSAM (curve (a)) and the star 16-QAM with differential encoding (curve (b)) both over a Rayleigh fading channel, and the star 16-QAM with differential encoding over an AWGN channel (curve (e)). The performance of the system having differential encoding and oversampling (curve(c)) can be considerably enhanced by the use of spatial diversity, where two antennas and receiver circuits are used. For these simulations, switched diversity was used, whereby for each phasor received, the receiver with the incoming phasor of largest magnitude was selected. Both receivers must have their own differential decoders. Curve (d) shows the performance of this second, order diversity assisted system.

As can be seen, a very substantial improvement in BER has been obtained over the coherent non-compensated square 16-QAM scheme having similar complexity by using star QAM, and at high SNRs oversampling gave a further significant gain. By introducing second-order switched diversity the system operated with a channel SNR of only 5 dB above that for an AWGN channel for a BER of 10^{-3}.

11.3 Trellis Coded Modulation for QAM

In an attempt to further reduce the BER achieved by the star 16-QAM schemes proposed in the previous section, we turned our attention towards various FEC coding schemes. Papers published by Ungerboeck and others [320] suggest that, for an AWGN channel, significant coding gains can be achieved by expanding the symbol set size and using the extra bit(s) gained for channel coding. In its most common form this is known as trellis coded modulation (TCM). TCM schemes employ redundant non-binary modulation in combination with a finite state encoder which governs the selection of the modulation signals to generate coded signal sequences. This process is described in Chapter 8.

Historically TCM was conceived to operate with modems over AWGN channels. Here it could be assumed that an accurate phase reference was established and maintained and so there was no problem with false locking. One of the properties of TCM systems is that they exhibit invariance to 180 deg rotations, i.e. if lock is established 180 deg from the correct lock position their operation will be unaffected. Systems have been suggested [360] having non-linear convolutional encoders which exhibit invariance to all 90 deg rotations, but no systems exhibiting invariance to 45 deg rotations have been published. This means that TCM cannot be used as it stands with the star 16-QAM system, as half of the possible lock points will lead to incorrect system operation.

In order to maintain phase invariance to 45 deg rotations it is necessary to introduce differential coding whereby we associate each TCM word with a phase and amplitude change, rather than to an absolute phase and amplitude as is normally done. The constellation we opt for is a four-ring star scheme with 8 points on each level, i.e. the 32-level star QAM shown in Figure 11.4.

This arrangement does not have a constant free distance amongst constellation points, but it does enable the TCM codewords to be optimally and differentially mapped onto the constellation. Based on our experiments we discovered that suboptimal mappings involving constellations, for example, that have 4, 6, 10, and 12 points on the first, second, third and fourth rings, respectively, where the rings are numbered starting with the inner ring, have poorer results than a constellation having 8 points on each of its four rings. Diagonal interleaving over approximately three fades was used in order to randomise errors. Simulations of TCM revealed marginal improvements in BER at very high SNRs compared to 16-level star QAM, although significant degradation in BER was experienced at the lower SNRs encountered in practice. This degradation at lower SNRs was because the BERs on which the convolutional decoder was operating were so high that the decoder often chose the incorrect path through the trellis and thus increased the number of errors. It was not until the BER fell below a certain level that the convolutional coding system was able to reduce the number of errors. TCM systems only proved beneficial when switched diversity was used, achieving a residual BER of 5×10^{-5} at SNRs above 35 dB. This was because current TCM systems have been optimised for AWGN.

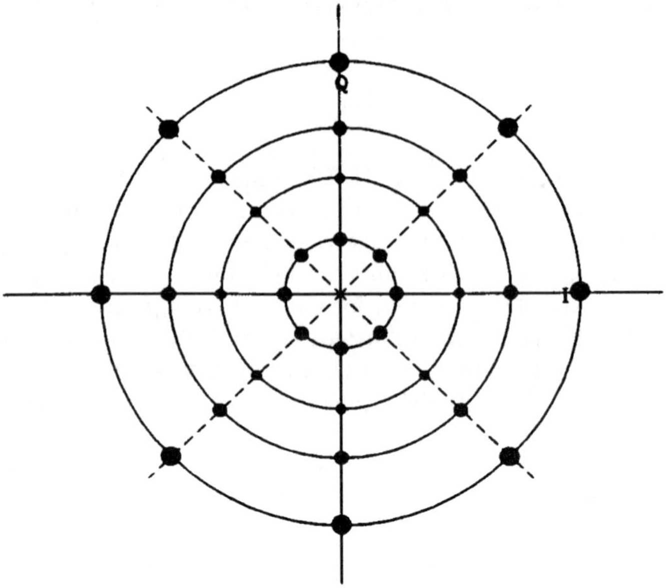

Figure 11.4: Star 32-TCM constellation

11.4 Block Coding

In our next endeavour in attempting to reduce the star 16-QAM BER, block coding [332] rather than convolutional coding [332] was employed by expanding the signal set to cope with the extra bits required for the FEC code. A 2/3 coding rate was considered to be appropriate, causing the number of QAM levels to increase from 16 to 64. An extension of the star constellation was selected having four amplitude rings with 16 points equispaced on each ring. Again, this was not optimal as regards to the minimum distance criterion, but simulations showed it to perform better in fading environments than constellations which were optimal for AWGN.

For block codes to perform well in the Rayleigh fading environment it is necessary to add interleaving to the system in order to randomly distribute the errors in time. The block code and the interleaving process introduce a delay, and the maximum permissible delay depends on the type of information to be transmitted. The integrity constraints of computer data transmissions are at least three orders of magnitude higher than those of digital speech transmissions. Fortunately, data channels can accept longer interleaving delays which allow effective randomisation of the bursty error statistics of the Rayleigh fading channel. This considerably increases the ability of the forward error correction (FEC) decoder to decrease the BER. For each transmission rate, propagation frequency, vehicular speed and interleaver algorithm there is a minimum interleaving depth or delay to transform the BER statistics of a fading channel into a good approximation of those encountered in a Gaussian channel. For a specific combination of system parameters, an interleaved FEC code word has to overbridge a number of channel fades.

For the propagation frequencies 1.7 to 1.9 GHz used for personal communications networks (PCN) the wavelength is about 17 cm and there is approximately one fade every 8.5 cm distance. If the data transmission rate is 64 kbit/s, yielding a signalling rate of 16 ksamples/s for the uncoded 16-level QAM, then when the mobile station is travelling at 30 mph or 13.3 m/s, there are approximately 100 QAM samples transmitted between two deep fades which are about half a wavelength apart. Diagonal interleaving over six fades randomises the bursty error statistics and has a delay of approximately 600 QAM samples corresponding to about 40 ms. This delay is typically acceptable for speech communications [361]. By increasing the interleaving depth to much higher values associated with higher delay, which is acceptable for data transmission, the BER is further improved.

We consider Bose-Chaudhuri-Hocquenghem (BCH) block codes [332] to have favourable properties for PCN transmissions. They can correct both random and bursty errors and also their error detection capability allows the BCH decoder to know when the received code word contains more errors than the correcting power of the code can cope with. Provided a systematic BCH code is used, the information part of the coded word can be separated from the parity bits without attempting error correction which would precipitate more errors due to the fact that the codec's error correction power was exceeded and hence the decoding operation would be erroneous.

A special subclass of BCH codes are the maximum minimum distance Reed-Solomon(RS) codes [332]. These codes operate on non-binary symbols and have identical error locator and symbol fields. The non-binary RS codes are optimum due to their maximum distance property, and may also be sufficiently long to overbridge channel fades without additional interleaving. However, they are more complex to implement than binary BCH codes. To explore both ends of the complexity/performance trade-off we selected an extremely long and complex RS code, the $RS(252, 168, 42)$ code over Galois field $GF(256)$ using 8-bit symbols, the moderately long and less complex $RS(44, 30, 7)$ code over $GF(64)$, as well as the low complexity short binary $BCH(63, 45, 3)$ code. All these codes have an approximately equal coding rate of $R = 0.7$. The nomenclature used here is $RS(m, n, k)$ where m is the number of encoded symbols, n is the number of information symbols and k is the number of corrected symbols in a code word. Every field element of the finite field $GF(2^l)$ is represented by an l-bit non-binary symbol in the case of RS codes.

11.5 64-level TCM

As a benchmark scheme for the block-coded arrangements a 64-level 2/3 rate convolutional coding system was devised. However, TCM as outlined in Chapter 8 cannot be used directly as it doubles the symbol set size, and here we have a quadrupling of the set size from 16 to 64 levels. Nevertheless, the principles of TCM can be retained. The 64 possible code words were divided into 16 groups of four so-called D-groups [320]. TCM only adds FEC coding to those bits that have the highest probability of error. With four input bits and six bits per symbol we can add half-rate convolutional coding to the more vulnerable two bits and leave uncoded the other two that are

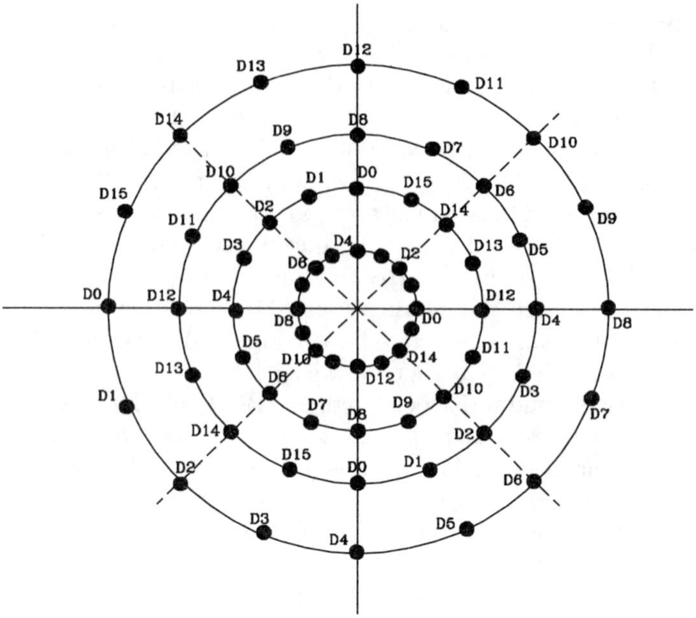

Figure 11.5: Star 64-TCM constellation

more widely spaced. The two coded bits plus the two FEC parity bits can specify one of 16 groups, and the uncoded two bits can specify one of four points within this group. The four points in each group are spaced as far apart across the constellation as possible so that there is little chance of one of these four points being mistakenly received as another in the presence of noise.

In Figure 11.5 we display the constellation of 64 points representing the 64 possible transmitted phasors.

Figure 11.5 may also be viewed as a look-up table for the differential encoding. The six bits span the decimal range 0-63 as they vary from 000000-111111. If this decimal number is in the range 0-3, 4-7, 8-11, ..., 60-63 we select subgroup D0, D1, D2, ..., D15, respectively. In each subgroup we map each of the four possible numbers onto one of the four points in that subgroup. This is done by starting on the positive x-axis and the innermost circle, and rotating anticlockwise. The first point in the subgroup that we come across is given the lowest number for that subgroup. Thus if we consider subgroup D0, we would assign the point on the positive x-axis the value 0, i.e. it is the point we map to when the incoming six bits are 000000. The point on the positive y-axis and the second circle becomes 1 corresponding to incoming bits 000001, the point on the negative x-axis becomes 2 corresponding to 000010, and the point on the negative y-axis becomes 3 corresponding to 000011. This is repeated for all subgroups. Thus the D4 point on the negative y-axis corresponds to 19 or 010011. Once we have identified the phasor for the incoming code word, we can use this to derive the differential signal which we must add onto the previous constellation point transmitted as a reference in order to derive the constellation point that we transmit. The representation is as follows.

A point on the inner ring causes us to transmit the current symbol with an identical amplitude to the previous symbol transmitted.

A point on the second ring causes us to transmit the current symbol with an amplitude on the next amplitude ring up from the previous symbol transmitted. If the previous symbol amplitude was that of the outermost ring then a change is made to the innermost ring, i.e. wrap-around occurs.

A point on the third ring causes an increase of two rings in amplitude from the previous amplitude transmitted again with wrap-around, and so on.

A point at 0 deg on the positive x-axis represents no phase change from the previous point transmitted. A point at 22.5 deg represents a phase change of +22.5 deg relative to the previous point transmitted. A point at 45 deg represents a phase change of +45 deg relative to the previous point transmitted, and so on.

As an example, if the last phasor transmitted was that corresponding to the point labelled D13 at 22.5 deg on the second ring and the input bits were 000010, we decode these input bits as the D0 point on the negative x-axis. Thus we increase the ring size by three, wrapping round from the D13 point to the innermost ring, and rotate this point by 180 deg to end up at the unlabelled D9 point between the corresponding D8 and D10 points on the innermost ring at −157.5 deg. This is the phasor transmitted.

This heuristic representation has been used as it allows set partitioning into the D-groups to be carried out in the same way as for conventional non-differential TCM.

The half-rate coder chosen was of constraint length $K = 5$ with generator polynomials

$$g1 = 1 + X^3 + X^4$$

and

$$g2 = 1 + X + X^2 + X^4$$

as recommended for the GSM system [362]. These generator polynomials are optimal for random error statistics. The encoded bits are diagonally interleaved prior to the mapping shown in Figure 11.5. This interleaving was the same as used with the block codes.

11.6 Bandwidth Efficient Coding Results

The above 64-level $R = 2/3$, $K = 5$ TCM system contrived in the previous section performed worse than the star 16-QAM at all SNRs, as shown in Figure 11.6.

When using separate block coding and absorbing the parity bits by constellation expansion at the cost of higher channel BERs and implementational complexity, but without bandwidth expansion as described in Section 11.4, the BER decreased dramatically compared to both our 64-TCM $K = 5$ system and to the uncoded star 16-QAM scheme. The $BCH(63, 45, 3)$ code and the $RS(44, 30, 7)$ code had nearly identical performances, in spite of the considerably higher block length and complexity of the RS code. The $RS(252, 168, 42)$ code offers an extra 2 dB SNR gain for a large increase in complexity. We therefore favour the BCH code which provides

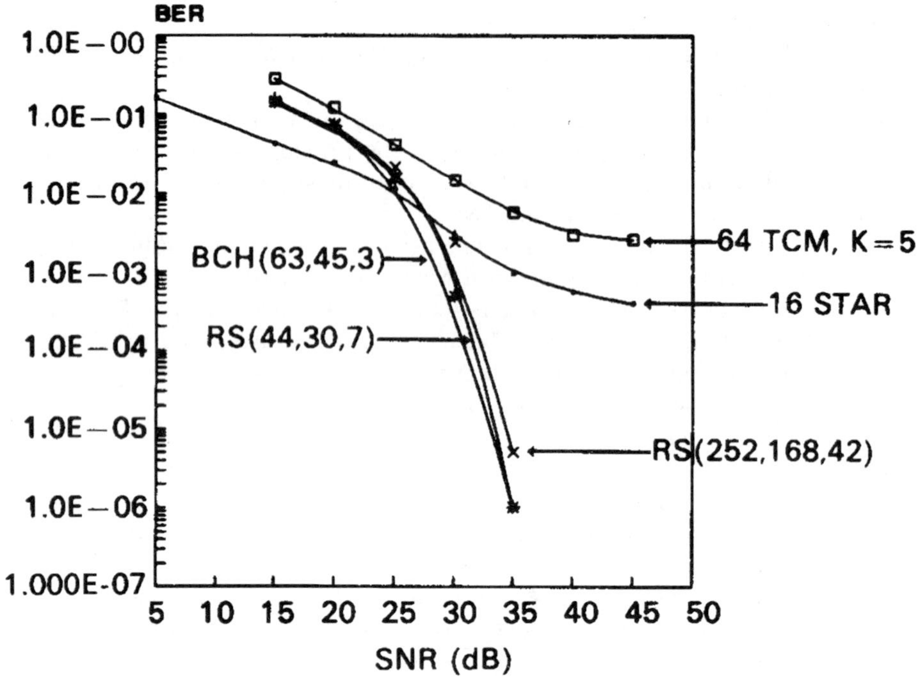

Figure 11.6: TCM and block coded performance

virtually error-free communications for channel SNR values in excess of 30-35 dB, a value that may be realisable in the small microcells to be ultimately found in a fully developed PCN. By using error correction coding and compensating for the increased bit rate by deploying higher-level modulation, we are able to provide transmissions of higher integrity, but with identical bandwidth compared to an uncoded system. The transmitted power and system complexity are also increased.

11.7 Overall Coding Strategy

We now consider 16-level and 64-level QAM schemes which are error protected, where the FEC coding now decreases the useful information transmission rate whilst increasing the integrity of the transmitted data. The previous 64-level QAM scheme having a 2/3 rate code had its code rate reduced to 1/2 whilst maintaining constant transmission rate, but decreasing the primary information rate. We selected this overall 1/2 coding rate as it is widely used in mobile radio [332,363]. A further coding rate reduction does not bring substantial coding gains and squanders channel capacity which cannot be compensated for by using more modulation levels. We also employed a 3/4 rate code in conjunction with the star 16-QAM in order to provide the same overall bit rate as the coded 64-level QAM.

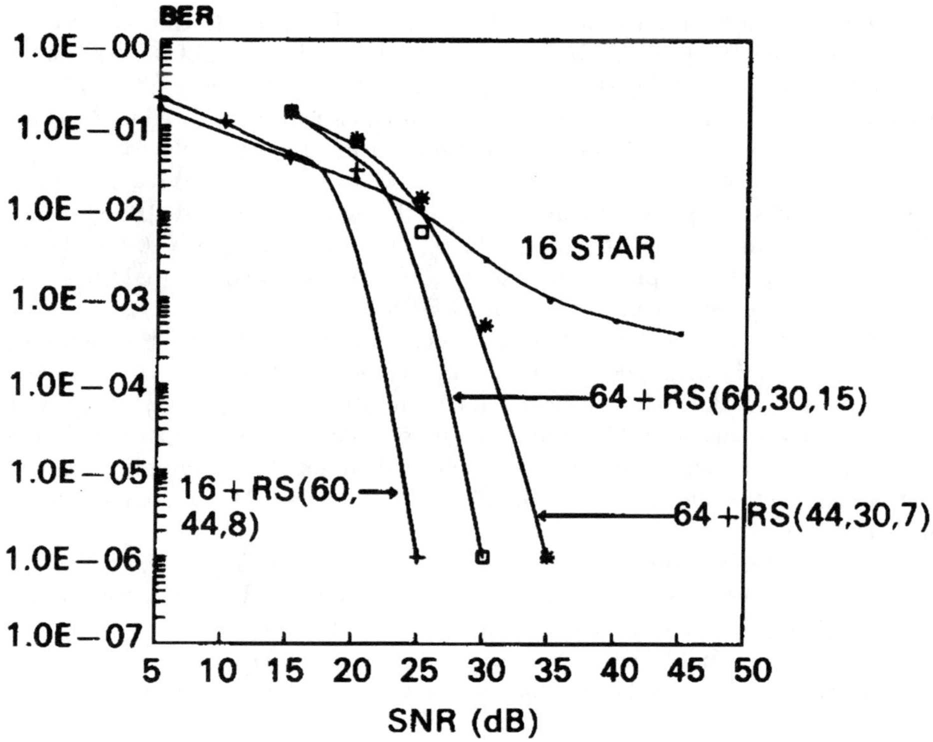

Figure 11.7: Overlaid block coding performance

Our results are depicted in Figure 11.7.

The uncoded star 16-QAM and the 64-level QAM/$RS(44,30,7)$ curves are repeated for comparison along with the 64-level QAM/$RS(60,30,15)$ and 16-level QAM/$RS(60,44,8)$ arrangements. Observe that the error correction power of the 64-QAM/$RS(60,30,15)$ scheme is about twice as strong as that of the 16-QAM/$RS(60,44,8)$ system. However, 64-QAM has typically more than twice the symbol error rate of 16-QAM and this is why the 16-QAM arrangement is likely to outperform the coded 64-QAM scheme. There is a consistent and remarkable improvement in both the 16-level QAM and 64-level QAM performance. For SNRs in excess of 25 dB there is an almost 5 dB extra SNR gain improvement due to the stronger RS code in case of the 64-level QAM scheme. By employing error correction coding, the performance of the 16-level QAM arrangement is dramatically improved and becomes superior to that of the 1/2 rate coded 64-level QAM system. Depending on the integrity required, an SNR value of 25 dB is sufficient for reliable signalling via the 16-QAM/$RS(60,44,8)$ system.

11.7.1 Square 16-QAM/PSAM/TCM scheme

The previously considered differential 32-TCM and 64-TCM star QAM schemes failed to improve the star 16-QAM performance partly due to the lack of fading compensation, and partly because of the ambitious goal of providing a 4 bit/symbol net transmission rate. Clearly, the expanded 64-TCM constellation had reduced the ability of the scheme to withstand fading compared to the 16-QAM schemes.

Ho, Cavers and Varaldi [364] embarked on a comparison of three modulation schemes with a net throughput of 2 bits/symbol. These were uncoded QPSK, rate 2/3 8-PSK/TCM as proposed by Ungerboeck [320], and rate 1/2 square 16-QAM/TCM combined with PSAM. The pilot symbol assisted fading estimates allowed the authors to improve the power efficiency over uncoded QPSK via fading channels.

The proposed square 16-QAM/TCM scheme was based on separate four-state TCM of the I and Q components. Ho *et al.* compared the BER versus channel SNR performance of their three schemes for a variety of fading rates, using a pilot periodicity of $M=7$ and a $k=11$ channel predictor. The authors also provided best-case performance estimates when using the actual complex fading sample for channel estimation. Their results are shown in Figure 11.8, where P represents a perfect channel estimate.

Observe that the best performance was achieved by 16-QAM/TCM/PSAM, since there are no unprotected bits transmitted over the channel. For BER values below 10^{-6} 16-QAM was about 5 dB more power efficient than 8-PSK, and about 25 dB more efficient than uncoded QPSK. As regards power amplifier design, it is important to compare these schemes on the basis of peak power. If all three schemes are allocated the same peak power, their corresponding peak-to-average power ratios can be used in order to derive their resulting average power. Ho *et al.* report that for a 50% excess bandwidth raised cosine pulse shaping, the corresponding peak-to-average ratios are 2.45 dB, 2.75 dB and 5.31 dB for QPSK, 8-PSK and 16-QAM, respectively. For a fair comparison the BER curves of Figure 11.8 must be shifted to the right by the above SNR values. After such a shift the 16-QAM still has a 2.44 dB peak power advantage at a BER of 10^{-6}. These figures were confirmed by other researchers [356, 365].

11.8 Distorted Constellation Star QAM

11.8.1 Introduction

In Chapters 4 and 10 we have shown that the maximum distance square 16-QAM scheme has two sub-channels with different integrities, which we have referred to as the $C1$ and $C2$ sub-channels. The BER advantage of the $C1$ channel over the $C2$ channel is not dramatic for AWGN channels, but is more significant over Rayleigh fading channels. The different sub-channel integrities can be equalised using matched FEC codes. However, in the case of coded speech it is often advantageous to exploit the unequal BERs by transmitting the more vulnerable speech data via the $C1$ channel, while the more robust speech data is sent via the $C2$ subchannel [42, 85–87]

In the 16-level star QAM systems each bit in the 4-bit symbol has an approximately equal probability of being in error. In general this is a desirable feature as data can be

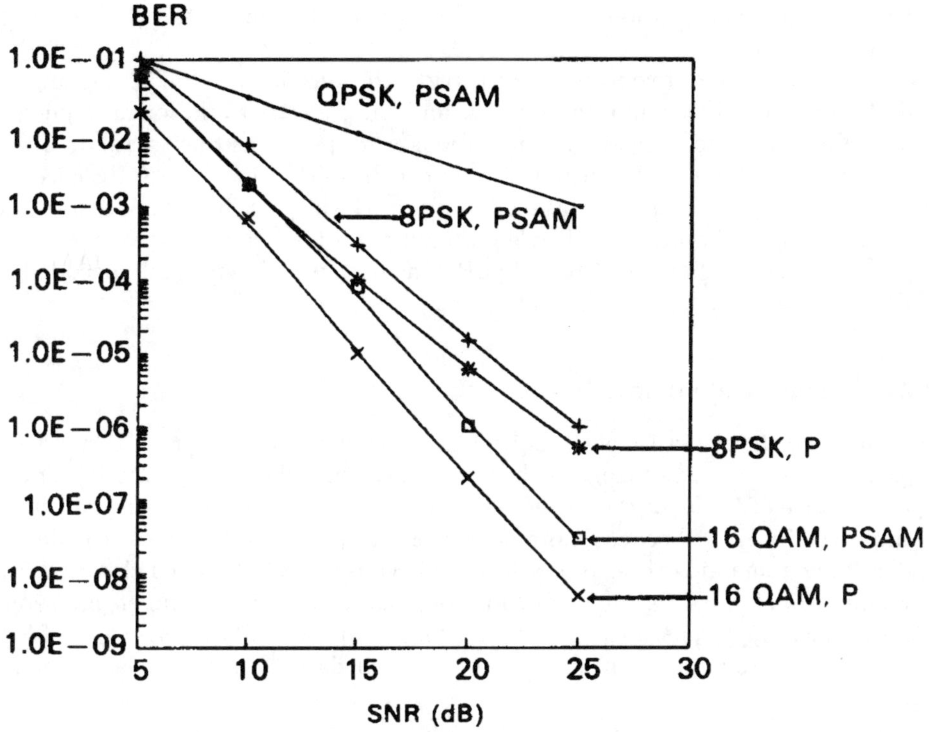

Figure 11.8: BER performance of coded 8-PSK and coded 16-QAM [364] ©IEEE, 1992, Ho *et al.*

transmitted without applying any special mapping strategy and this normally yields the best overall bit error rate (BER). There are situations, however, where equal BERs for all bits is disadvantageous. For example, when the data applied to the star QAM modem is from an n-bit PCM source, where the most significant bit corresponds to a 2^{n-1} higher voltage than the least significant bit, then the MSB needs to be received with a concomitant lower BER than the least significant bit (LSB). For information sources having a hierarchical structure in their data it is necessary to employ source-matched channel coding strategies in order to ensure that the recovered data has the requisite performance after decoding. This procedure, employing a range of channel codecs or an adaptive channel codec, is complex and expensive. We seek a simpler solution here, investigating whether distorting the star constellation may yield an acceptable performance without the need for complex source-matched channel coding.

11.8.2 Distortion of the star

As it is desirable for the average of the four individual BERs to remain as low as possible, only minor distortions to the star QAM constellation are considered, as it was found to be well suited to the Rayleigh fading environment. Accordingly the twin-level circular constellation is retained, along with the differential Gray coding of the constellation points.

Within these confines two forms of distortion are considered. One is to change the spacing between the inner and outer rings, and the other is to no longer equi-space the eight points around each ring. Simulations were performed using random data, encoded onto the star QAM constellation and transmitted on a 1.9 GHz carrier at a Baud rate of 16 ksymbols/sec corresponding to 64 kbit/s. The mobile's speed was a constant 30 mph. Appropriate Rayleigh fading and Gaussian noise was generated leading to an average channel SNR of 30 dB. The data was differentially decoded and the BER was calculated.

11.8.2.1 Amplitude distortion

Bearing in mind the structure of the star QAM constellation of Figure 11.1, increasing the spacing between the two rings of the star QAM constellation potentially improves the performance of bit b_1 at the expense of the other three bits. Bit b_1 is expected to have improved integrity because more noise is required to move it across the decision boundary. The remaining three bits will suffer because they are moved closer together on the inner ring, increasing the probability of phase error for inner-inner, inner-outer and outer-inner ring transitions, and decreasing the probability of error only for the outer-outer ring transition. In the case of random data all transitions are equally likely.

Figure 11.9 shows the simulation results for the variation of the BER of each of the four bits and the average BER as the ratio of the distance between the rings was varied. Whilst bits b_2, b_3 and b_4 generally exhibited increasing BER as would be expected, bit b_1 had a minimum BER at a ring ratio of 3, where the inner ring has a magnitude of 1 and the outer ring a magnitude of 3. The reason that the BER did not continue to fall with increasing ring ratio is that as the inner ring shrinks at the expense of the growth of the outer ring, constellation points on the inner ring become increasingly susceptible to noise, considerably distorting their amplitude and phase value. So although inner-outer and outer-inner transitions have a low probability of being decoded incorrectly, there is a relatively high probability that an inner-inner ring transition will be incorrectly decoded as either an inner-outer or outer-inner transition. Decoding this change as an outer-inner transition is possible because the receiver does not decide upon which ring the previous symbol sent was on, but looks for a significant change in amplitude between previous and current symbols. This avoids error propagation due to incorrect decisions. Simulations with the receiver attempting to make a decision as to the ring on which the previous symbol resided in order to eliminate the incorrect decoding of an inner-inner ring transition as an outer-inner ring transition showed that this scheme exhibited worse performance than the original system because of the error propagation associated with incorrect decisions.

It can be seen that there is little scope for changing the ring ratio. Variations between 1.5 and 3 would be acceptable. A ring ratio of 1.5 allows the largest variation in terms of BER between bits, with bit b_1 having a BER about 6 times greater than bits b_2 and b_4.

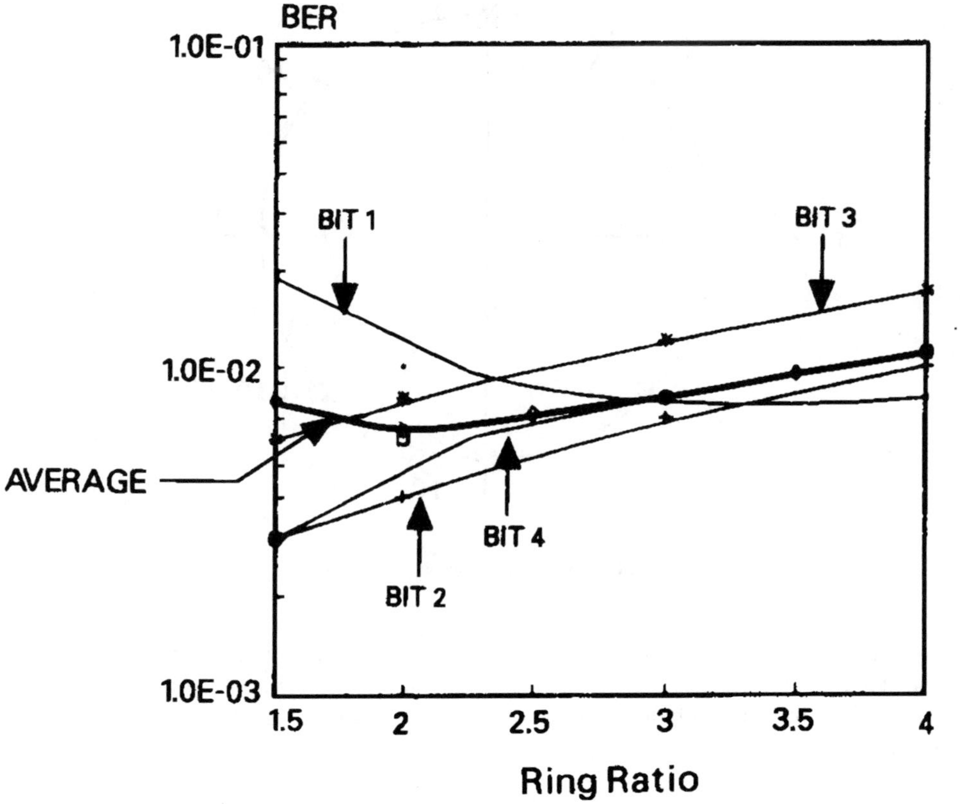

Figure 11.9: BER performance of star 16-QAM for ring ratio variations between 1.5 and 4 at 30 mph, 1.9 GHz, 16 ksymbols/s and SNR = 30 dB

The average BER exhibited a minimum at a ring ratio of 2, rising slowly as the ring ratio was increased, and rising more quickly as the ring ratio was decreased. A theoretical analysis performed by Chow *et al.* [180] for AWGN also came to similar conclusions. However, with their only being a 0.9 dB difference between ring ratios of 2 and 3, performance gains made by optimising this ring ratio will be small.

The second form of constellation distortion considered is to change the angles between each of the points on a ring. The current system has a spacing of 45 deg between each adjacent point. By increasing this angular spacing for some points and decreasing it for others, the BER of certain bits can be altered. The bitmapping onto the phasor points for one of the rings on the current system is shown in Figure 11.10.

In all cases the bit mapping on the other ring has constellation points at identical angular spacing. Only three bits are shown against each point, corresponding to bits b_2, b_3 and b_4 as b_1 is used to differentiate between the rings. It is important to remember that this diagram represents a differential mapping system. That is,

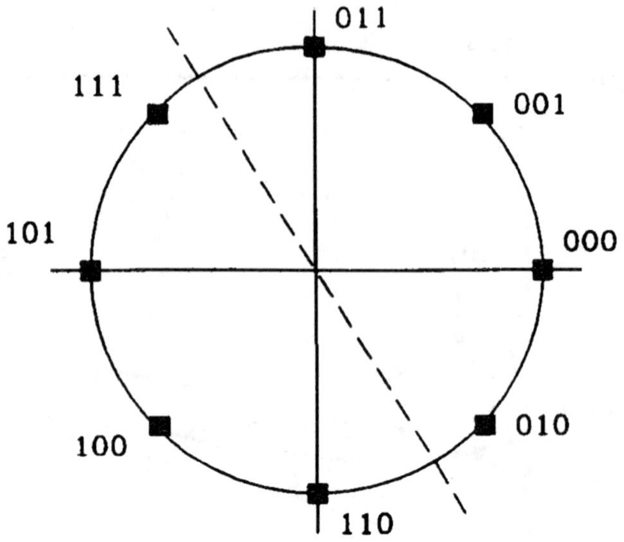

Figure 11.10: Mapping for standard star system

a phase point at an angle of 45 deg, say, implies that the current symbol should be transmitted with a phase shift of +45 deg relative to the previous symbol transmitted, interpreted as modulo 360 deg. In all subsequent simulations the ring ratio was held at a value of 3.

11.8.2.2 Phase variations

As can be seen in Figure 11.10 the 8 points on a circle can be split into two groups of 4 points each. In each of these subgroups, separated by the dashed line in the figure, the first bit, namely bit b_2 in the 4 bits mapping scheme, is the same, i.e. a logical 1 to the left of the dashed line and a logical 0 to the right of the line. The integrity of b_2 can be increased by further separating these subgroups as shown in Figure 11.11.

Here the angle θ is our variable parameter. We note from the geometry of Figure 11.11 that the angle ϕ is given by

$$\phi = 60\,\text{deg} - \frac{\theta}{3} \tag{11.8}$$

where ϕ, in common with all the angles quoted throughout, is in degrees. Hence, for example, for the undistorted case of $\theta = 45\,\text{deg}$ we have that $\phi = 45\,\text{deg}$, as expected. The variation of BER against θ is given in Figure 11.12 for our initial conditions of SNR = 30 dB, 30 mph, 1.9 GHz and 16 ksymbols/s.

Bit b_1 is not shown because it remained constant with variations in θ as would be expected. It is apparent that as θ increased above about 100 deg, the average BER rose steeply and this factor probably precludes operation in this region. Indeed, the average BER was at a minimum for $\theta = 45\,\text{deg}$, and the price paid for a different

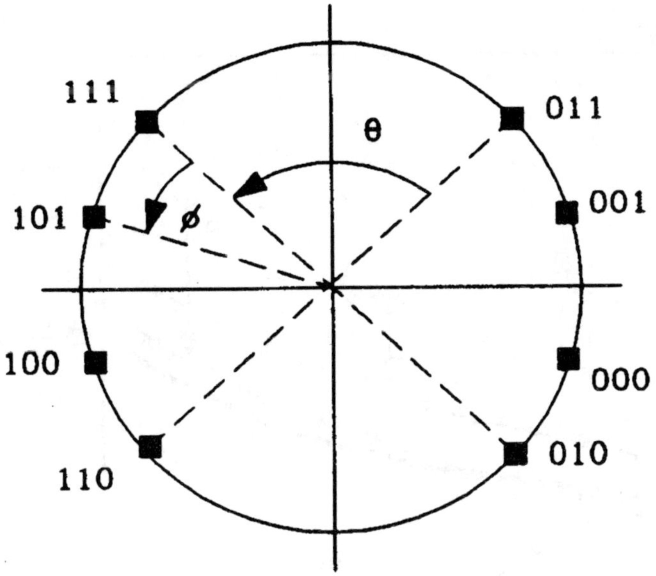

Figure 11.11: Bit b_2 distortion system

BER on each bit was a higher average BER. As desired, the BER performance of bit b_2 was improved at the expense of bits b_3 and b_4. For values of θ in excess of 80 deg there was no longer any significant improvement in the BER of bit b_2. At this point there is a ratio of about 4 between the BER of bit b_2 and bits b_3 as well as b_4. With the ring ratio held at 3, bit b_1 had a performance similar to bits b_3 and b_4.

There is a further distortion that can be applied to the constellation. This is shown in Figure 11.13 where each of the two groups of 4 points are divided into two further sub-groups.

In each of these subgroups bit b_4 has a constant value. Thus we would expect to increase the performance of bit b_4 at the expense of bit b_3 by increasing the angle α. This is because the protection distance between opposite values of b_4 is now increased, while that between opposite values of b_3 is decreased. The angles in Figure 11.13 are related by

$$\phi = 90\,\text{deg} - \frac{\theta + \alpha}{2}. \tag{11.9}$$

Again, for the undistorted case of $\theta = \alpha = 45$ deg, Equation 11.9 gives $\phi = 45$ deg. The BER performance of this system for different values of θ is shown in Figures 11.14 to 11.16.

The BER performance is displayed for only bits b_3 and b_4, as the BER for bits b_1 and b_2 remained constant. The trend in all graphs is the same, a sharply rising average BER, and only a slowly falling BER for b_4. This suggests that only small increases in

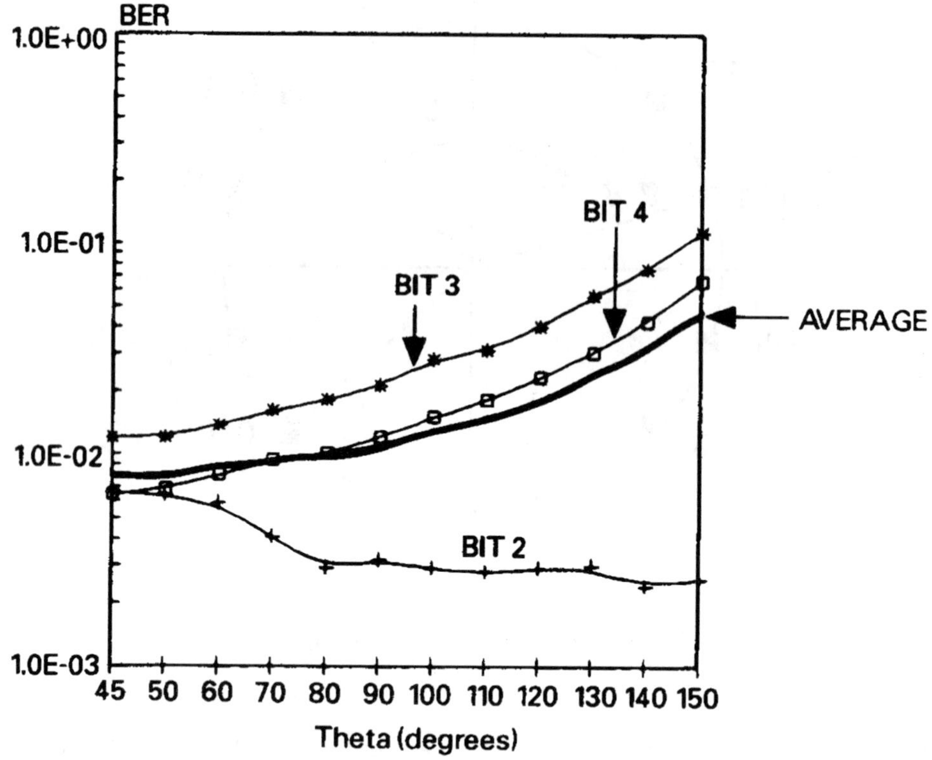

Figure 11.12: BER versus distortion angle θ for bits 2, 3 and 4 of the star 16-QAM constellation at 30 mph, 1.9 GHz, 16 ksymbols/s and SNR = 30 dB

terms of α are acceptable. For example a value of $\alpha = 50$ deg when $\theta = 90$ deg gives a ratio of 6.5 between the BER of bit b_4 and that of bits b_2 and b_3. Similarily, for the same distortion a BER ratio of 1.5 accrues between bit b_4 and bit b_1. In this instance the average BER is degraded by a factor of 1.75 which corresponds to about 2.5 dB.

11.9 Practical Considerations

11.9.1 Introduction

In previous chapters, it was assumed that no hardware imperfections were introduced. This gave an optimistic estimate of the performance. Possible imperfections include a finite number of quantisation levels, I-Q crosstalk, AM-AM and AM-PM conversion non-linearities and a finite oversampling ratio. Even if it appeared possible to implement a modem which, as far as the QAM technique employed is concerned, appears perfect, it may be sensible to reduce the performance in order to achieve a simpler

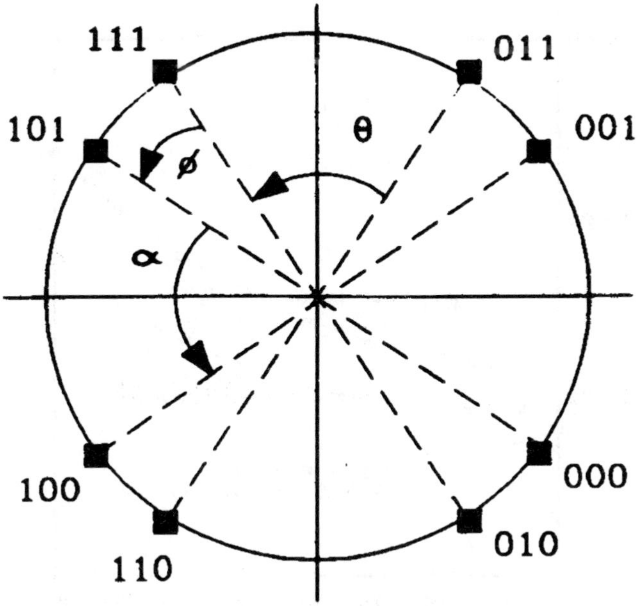

Figure 11.13: Bit b_4 distortion system

and cheaper design. In this section the effects of a number of imperfections which are likely to occur in a hardware modem are investigated, in order to establish hardware parameters.

11.9.2 Hardware imperfections

11.9.2.1 Quantisation levels

Typical QAM modems employ digital baseband circuitry, and so analogue-to-digital (A-D) conversion is required at the input of the receiver. It is important that an adequate number of bits is chosen for this convertor. If too low a number of bits are chosen then the performance will suffer badly, and if too many are chosen then the hardware complexity and cost will rise substantially whilst the maximum operating speed will fall. Initial quantisation level simulations assumed that the received signal level after pre-amplification would fall within a certain range. The quantisation levels are spread evenly throughout this dynamic range, and the incoming signal is rounded to the nearest quantisation level. Peak clipping occurs if the signal exceeds this range. Results of simulations carried out using a variety of quantisation levels are shown in Figure 11.17.

For these simulations data was transmitted at 128 ksymbols/s, at a mobile speed of 30 mph, and a propagation frequency of 1.9 GHz. Non-linear filtering (NLF) was used to shape the transmitted data and a fourth-order Butterworth lowpass

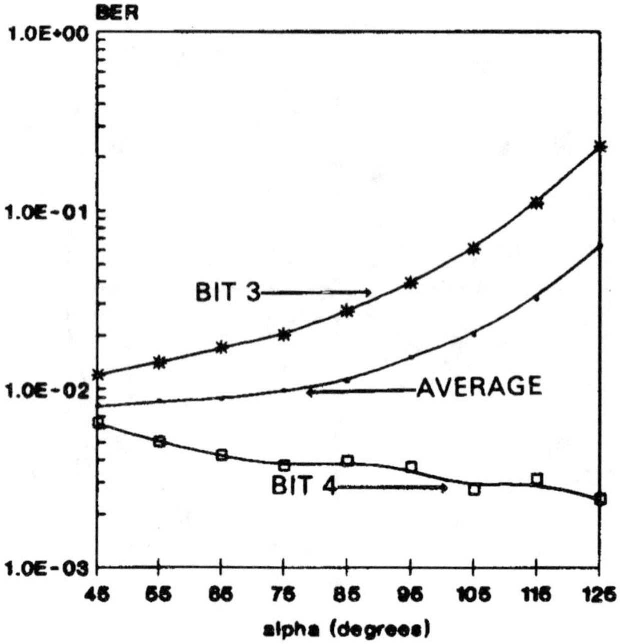

Figure 11.14: BER versus distortion angle α for bits 3 and 4 of the star 16-QAM constellation with $\theta = 45$ deg, at 30 mph, 1.9 GHz, 16 ksymbols/s and SNR = 30 dB

filter with the 3 dB cut-off point at 1.5 times the data rate (i.e. 192 kHz) was used in the receiver. Clock recovery was implemented using the modified early-late recovery system discussed in Chapter 12, and 16 times oversampling was implemented throughout the system. Unless stated otherwise, these simulation parameters can be assumed to hold true for all subsequent simulations in this chapter.

The results show that at least 14 bits will be required if the receiver is not to suffer a drop in terms of BER performance. In practice this means that 16-bit A-D convertors will be necessary, since 14-bit convertors are rare. Operation with only 12 bits would be feasible but does entail a BER penalty at high SNRs.

Consideration of the waveforms into and out of the A-D convertor showed that the full dynamic range of the convertor was not used over a short time span, but it was necessary to accommodate the peaks and troughs in the Rayleigh fading pattern. This suggested that performance could be improved by the use of a simple AGC prior to the A-D convertor. Because of the differential detection employed, the absolute signal level is unimportant. However, it is critical that the AGC must be sufficiently slow acting in order that its gain does not change significantly from one symbol to the next, as this would make the differential detection more error prone. In practice this is not a problem because there are many symbols in each fade and so the AGC can be slow acting with respect to the symbol period and still track the envelope adequately.

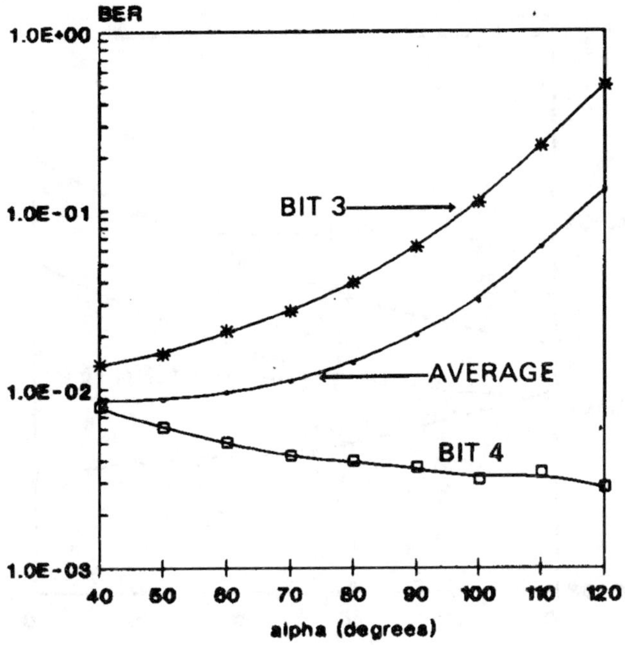

Figure 11.15: BER versus distortion angle α for bits 3 and 4 of the star 16-QAM constellation with $\theta = 60$ deg, at 30 mph, 1.9 GHz, 16 ksymbols/s and SNR = 30 dB

It is stressed that this AGC does not attempt to maintain a constant or near constant input envelope, but merely to compress the fading range in order to reduce the A-D convertor's resolution requirements. Simulations suggested that a suitable system was a slow-acting AGC, whose dynamic range could be suddenly increased if the input signal exceeded 95% of its allowed range. This tended to keep the fades within the input clipping range without unduly distorting the waveform when it was around the average value.

Results of simulations using this system are shown in Figure 11.18.

Here it can be seen that adequate performance is obtained with only 12 bit conversion, and that with only 8 bits the degradation is not unduly severe.

11.9.2.2 I-Q crosstalk

Another problem that is often encountered in the RF side of the modem is that of crosstalk between the quadrature branches. This problem can be reduced by balanced RF system design, but here an attempt is made to discover the maximum level of crosstalk which can be tolerated before unacceptable BER performance degradation occurs. No baseband counter measures are proposed, merely adequate RF system design.

Results of simulations employing 8 A-D convertor quantisation levels and the same

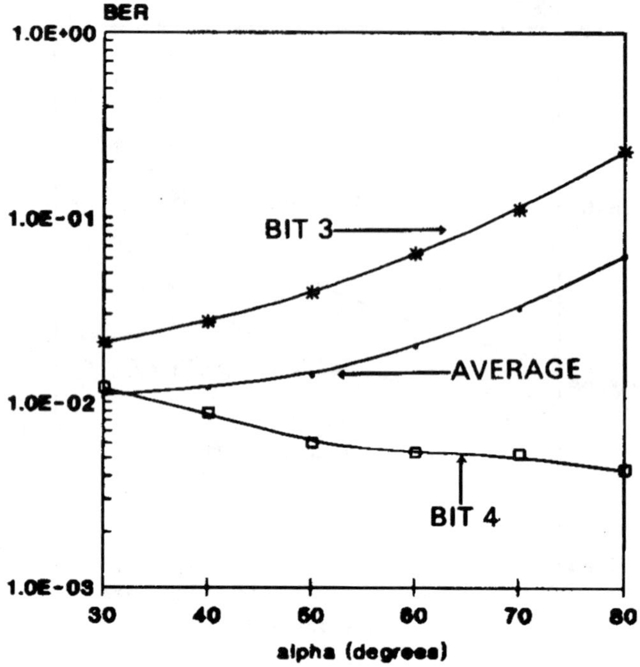

Figure 11.16: BER versus distortion angle α for bits 3 and 4 of the star 16-QAM constellation with $\theta = 90$ deg, at 30 mph, 1.9 GHz, 16 ksymbols/s and SNR = 30 dB

parameters as the previous section are shown in Figure 11.19.

Here it can be seen that levels of I-Q crosstalk up to -40 dB would seem acceptable, that is the leakage power from one quadrature branch to another should be 40 dB less than the average power in that branch.

11.9.2.3 Oversampling ratio

In order for the clock recovery system to operate satisfactorily, it is necessary to sample the received waveform a number of times during each symbol period. This allows accurate location of waveform peaks. Ideally, the modem should work with as low an oversampling ratio as possible in order to keep internal clock speeds low. In this section the complete star QAM system was simulated as the oversampling ratio was varied. For accuracy, a 16 times oversamping ratio was maintained in the simulations, but we only made a certain number of these samples available to the receiver.

The results for three different oversampling ratios are shown in Figure 11.20.

Here it can be seen that the ratio can be satisfactorily reduced to 8 samples per symbol, but a reduction to only 4 causes a significant performance degradation.

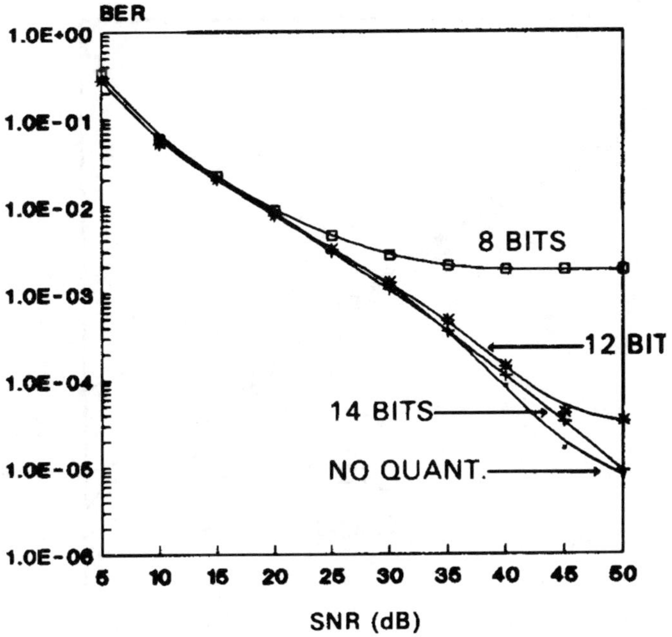

Figure 11.17: Star 16-QAM BER versus channel SNR performance using different input A-D resolutions at 128 ksymbols/s, 30 mph, 1.9 GHz

11.9.2.4 AM-AM and AM-PM distortion

We have already touched upon the problems of amplitude modulation to both amplitude modulation (AM-AM) and to phase modulation (AM-PM) conversion, when discussing the characteristics of power amplifiers in Section 5.6. A typical pair of AM-AM and AM-PM conversion curves was given in Figure 5.27. In this subsection we briefly attempt to quantify the effects of amplifier imperfections on decoded BER. This impairment is difficult to model as it is not known exactly what form the imperfections will take. In this section a number of assumptions are made about possible forms of distortion in order to assess the effect they may have on a QAM system. The AM-AM distortion is modelled by non-linear, compressed operation due to the limited dynamic range of the amplifier. This means that whilst the maximum output is maintained, the pulse shape is distorted. It was found that this had very little effect on the system, causing a slight but acceptable degradation in the clock recovery system. The degradation could be completely removed by regenerating the original pulse shape. Since the pulse shape is typically derived from a look-up table, this does not increase the complexity of the design.

The AM-PM distortion is modelled by introducing a phase shift of P degrees when the received phasor is on the outer ring. So when moving from the inner to the outer ring an undesirable differential phase shift of $\theta_{enc} + P$ is experienced where θ_{enc} is the

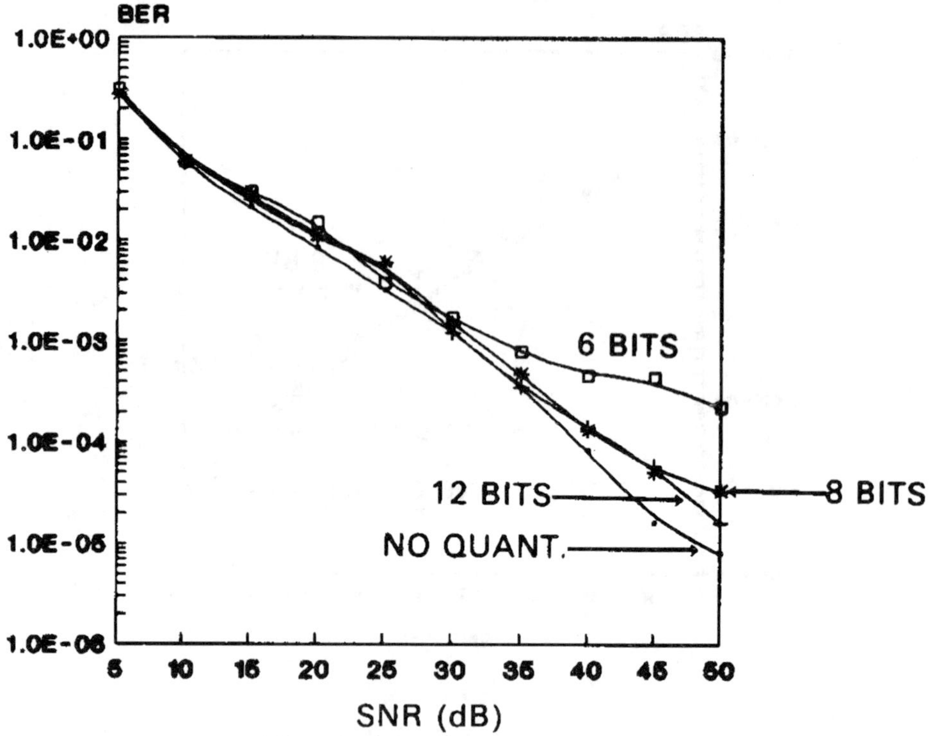

Figure 11.18: Star 16-QAM BER versus channel SNR performance using AGC and a variety of A-D convertor resolutions at 128 ksymbols/s, 30 mph, 1.9 GHz

differentially encoded transmitted angle. When moving from the outer to the inner ring a detrimental phase shift of $\theta_{enc} - P$ is experienced.

In the first set of simulations, no corrective action was attempted at the receiver, and the maximum permissible distortion which did not unduly degrade the BER performance was found. BER versus channel SNR performance curves are shown in Figure 11.21 for various AM-PM distortion values, which suggest that a maximum uncompensated phase shift of up to 10 deg could be tolerated.

However, since this phase shift is time invariant, it should be possible to take some sort of corrective action. In the differential decoder at the receiver, we can monitor the amplitude change. If there appears to have been a change from the inner to the outer ring then we subtract the known AM-PM distortion, and if there appears to have been a change from the outer to the inner ring then the known AM-PM distortion is added. This method is expected to remove most of the AM-PM distortion, but can precipitate errors if an error is made in decoding the amplitude change. BER versus channel SNR performance curves for this system are shown in Figure 11.22 for 0 deg and 40 deg conversion errors, where we can see that satisfactory performance is achieved for large values of phase distortion.

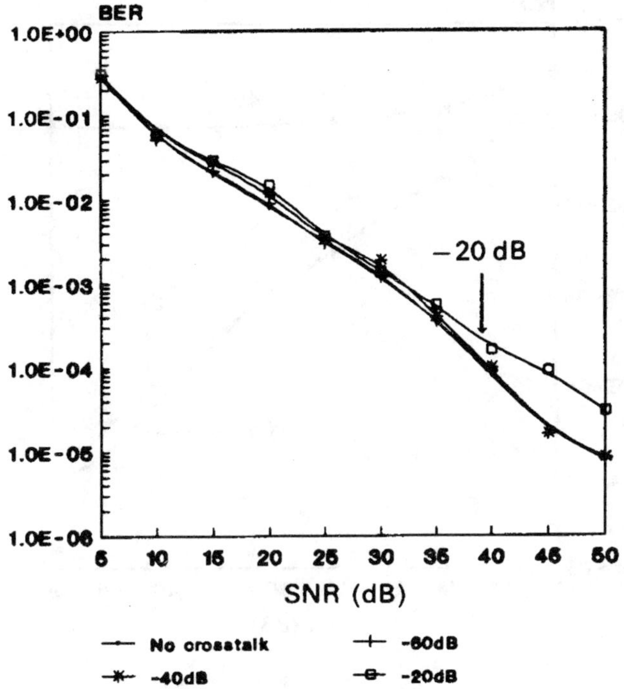

Figure 11.19: Star 16-QAM BER versus channel SNR performance using AGC and 8-bit A-D conversion when using different levels of I-Q crosstalk, at 128 ksymbols/s, 30 mph, 1.9 GHz

11.10 Summary

In this chapter we have discussed various signal processing techniques to facilitate reliable QAM transmissions in Rayleigh fading environments. The well-known 16-level square constellation was found to be unsuitable for the mobile radio environment. The sub-optimal star scheme with differential encoding and oversampling signal estimation dramatically improved the BER performance, rendering the channel appropriate for speech transmissions. The lower BERs essential for data transmissions were achieved by expanding the 16-level QAM signal set to 64 levels and using the extra channel capacity acquired for error correction coding. For SNRs in excess of 25 dB the performance of the coded 64-level QAM was superior to that of the 16-level QAM and at values of SNR in excess of 30 dB it was virtually error-free. When the overall coding rate was lowered to allow the 16-level QAM scheme to incorporate a 3/4 rate RS code and the 64-level QAM system had a 1/2 rate RS code, the 16-QAM arrangement outperformed the more complex 64-level scheme.

Unlike square QAM modems, star QAM has an approximately equal BER for each of the bits in its symbols. This is normally advantageous, but there are some instances

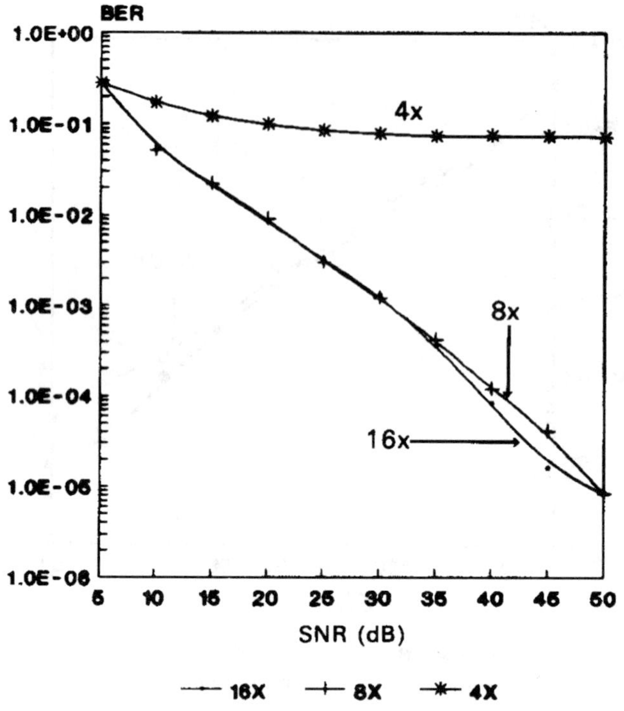

Figure 11.20: Star 16-QAM BER versus channel SNR performance using AGC, 8-bit A-D conversion, when oversampling at rates of 4, 8 and 16, 128 ksymbols/s, 30 mph, 1.9 GHz

where a differing BER for each bit can be effectively used for unequal source-matched FEC coding of speech and video sources. Ways of rendering the BER of each bit different were considered. Unfortunately, in all cases the average BER was increased, and large ratios between the BER performances of certain bits were not achieved for the scenarios considered. However, the BER ratios of approximately 6 were achieved when the average BER was degraded by a factor of 1.75.

The effects of hardware imperfections on star 16-QAM modem performance was then considered. It was established that 8-bit A-D convertors could be employed if a slow-acting AGC was deployed prior to the convertor. Mild levels of IQ crosstalk were found acceptable and the minimum oversampling ratio was found to be 8. Finally, it was determined that small AM-AM and AM-PM distortions could be tolerated, especially if correction could be applied.

In the next chapter we investigate further the use of star QAM transmission over mobile radio channels by considering the application of the clock and carrier recovery techniques introduced in Chapter 6. We show that a new clock recovery technique can be introduced which is advantageous for the particular combination of star QAM

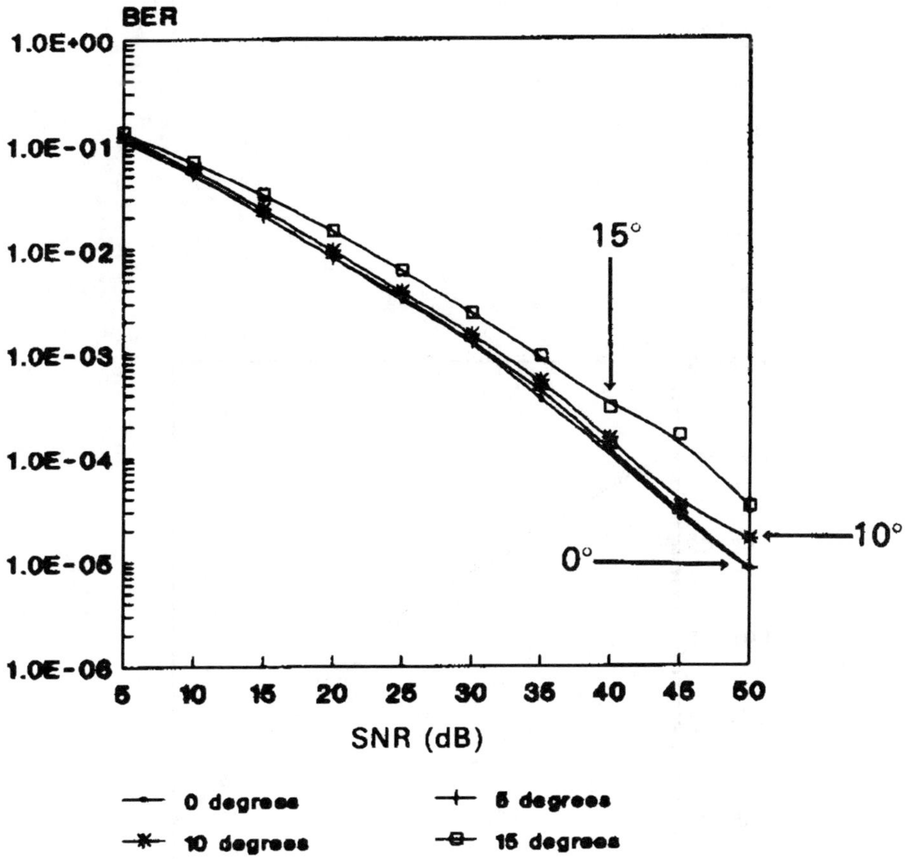

Figure 11.21: Star 16-QAM BER versus channel SNR performance at 128 ksymbols/s, 30 mph, 1.9 GHz for different values of AM-PM distortion without correction

and Rayleigh fading channels.

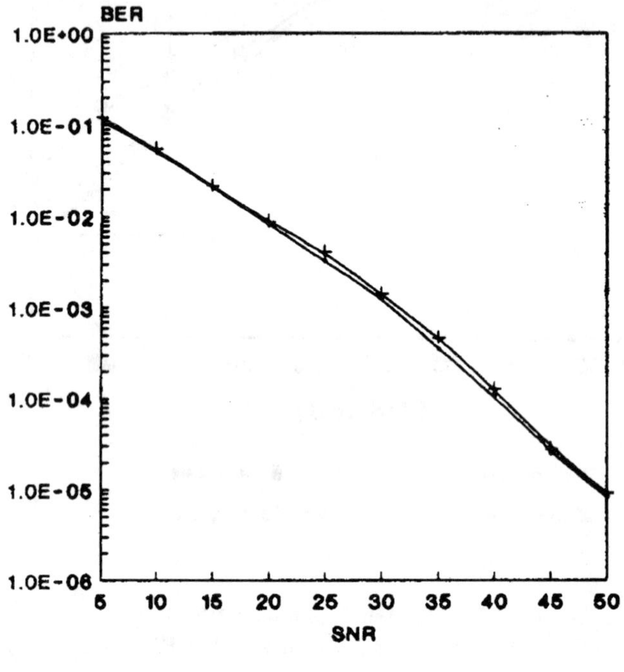

Figure 11.22: Star 16-QAM BER versus channel SNR performance at 128 ksymbols/s, 30 mph, 1.9 GHz for different values of AM-PM distortion with correction

Chapter 12

Timing Recovery for Mobile Radio

12.1 Introduction

Although the differential star QAM scheme proposed in Chapter 11 does not need carrier recovery, the characteristic 3 dB penalty associated with non-coherent detection need not be incurred if low complexity timing recovery can be implemented. Accurate clock recovery is always essential, irrespective of whether or not non-coherent detection is possible.

In this chapter clock and carrier recovery schemes for 16-level star QAM transmitted over mobile radio channels are considered, and the effect of ISI is simulated. We commence our investigations by considering the two most pertinent classic clock recovery methods, namely, the times-two and the early-late timing recovery techniques. These two techniques were discussed in Chapter 6 in the context of AWGN channels; here their performance over mobile radio channels with star QAM is considered.

12.2 Times-two Clock Recovery for QAM

In order to implement times-two clock recovery in conjunction with QAM the received signal is split into in-phase (I) and quadrature (Q) waveforms and each is processed separately, before being combined in the clock recovery system. In the case of QAM, however, the presence of high energy frequency components at half the data rate is less likely than with binary data schemes. This is because instead of changes in the baseband modulated signal about zero whenever we have a change of transmitted symbol, as with binary modulation, there can be swings between symbols on the positive side and to and from those on the negative side of the constellation resulting in fewer zero-crossings. For example, in the case of 16-level QAM and transmitted levels of ± 1 and ± 3 on the I and Q channels, a data sequence of -1, -3, -1, 1, 3, 1, -1, -3, -1 ..., -1, -3 yields a component at 1/6 of the data rate. Simulations were performed using times-two clock recovery which showed the QAM performance to be unacceptably poor. For more details on times-two clock recovery the reader is referred to Chapter 6.

12.3 Basic Early-Late Clock Recovery for Star 16-QAM

The early-late (EL) clock recovery technique described in Chapter 6 was introduced into a star QAM receiver. The simulation results showed that the EL clock recovery performed better than the times-two system. Correct locking of the clock was eventually established, and maintained moderately well during fades. The locking time, however, was unacceptably long.

The major problem with EL clock recovery for QAM transmissions is that not all QAM sequences result in time domain waveform peaks occurring every sampling period. An even more detrimental effect is that half the peaks are of the wrong polarity for the clock recovery technique. This problem is illustrated in Figure 12.1 where the polarity of the early signal minus the late signal is considered for three different received 16-QAM sequences, assuming in all cases that the sampling point is too early. Due to the squaring operation identical waveform segments having opposite polarity yield identical PLL control sequences. The early and late sampling points are shown by dashed vertical lines on the squared signal. The chosen sampling point is assumed to lie in the centre of these lines, whereas the correct sampling instant is seen to be nearer to the late sample, or the rightmost dashed line. When the sampling point is early, the early signal minus the late signal must be negative for conventional early-late schemes to function correctly, and this is represented in Figure 12.1 as "E-L Negative". In this example only the I channel is considered for simplicity. The practical implementation of this scheme could have independent EL recovery systems for both the I and Q channels, which would drive the same PLL using control logic.

In Figure 12.1(a) the QAM sequence 1, 3, 1 is shown where the first number in the sequences refers to the modulation level on the I channel transmitted in the symbol period prior to the one in which peak detection is to be performed. The second number in the sequences refers to the modulation level in the symbol period in which peak detection is attempted, while the third number refers to the modulation level in the symbol period after the one in which peak detection is attempted. Clearly not an oversampled waveform, but three independent symbols are considered. Also shown is the sequence -1, -3, -1 which leads to an identical waveform after squaring, where the squaring is an integral part of the EL process. Both of these sequences result in correct EL polarity. In Figure 12.1(b) the received sequences 3, 1, 3 and -3, -1, -3 are shown, both of which lead to an identical waveform after squaring. It can be seen that despite sampling occurring early as in (a), the EL value is of the opposite polarity to that in (a). Also problematic are sequences such as -1, 1, 3 and 1, -1, -3, both of which again lead to an identical waveform after squaring as shown in Figure 12.1(c). Observe that the monotonic waveform leads to negative early-late difference signals regardless of whether the sampling is early or late, although we only show it early.

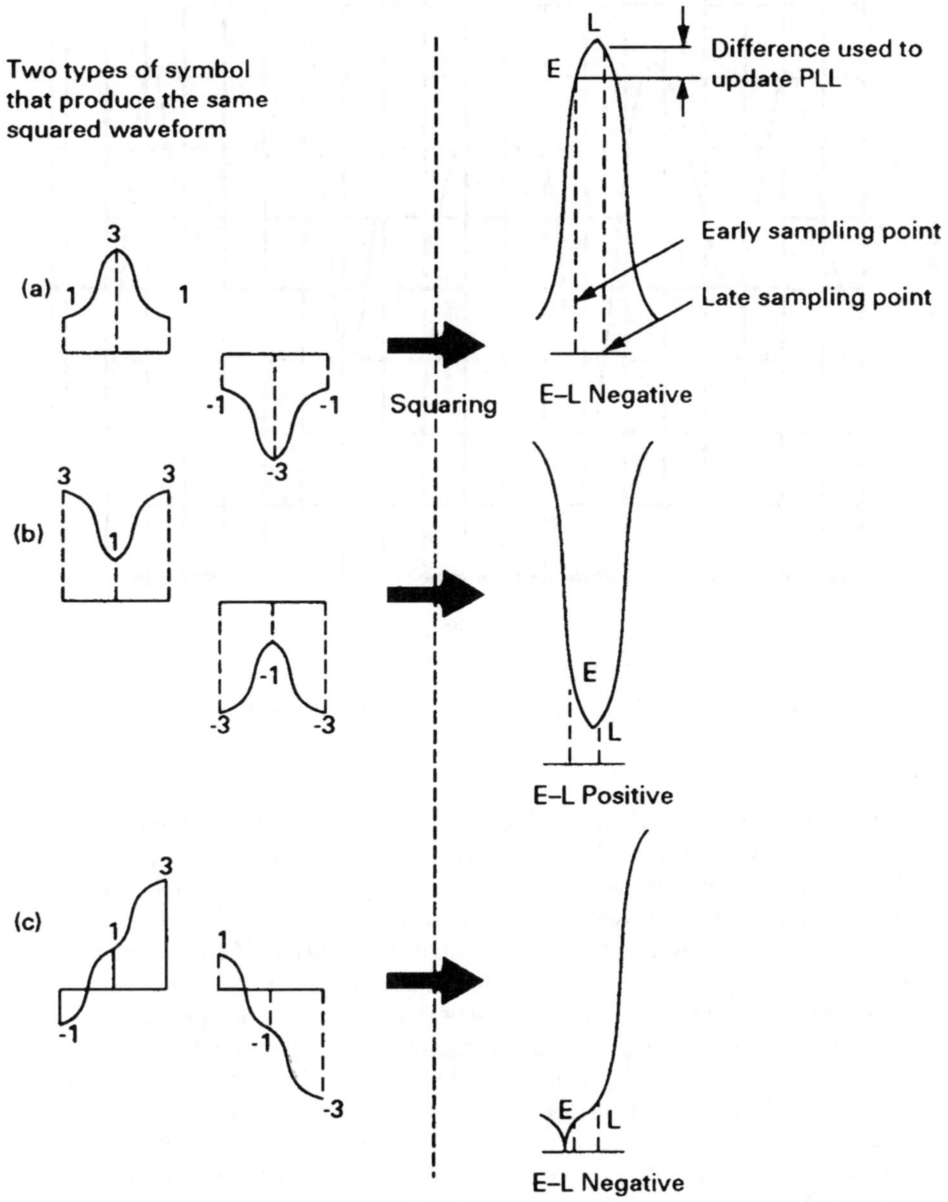

Figure 12.1: Problems with early-late clock recovery for QAM

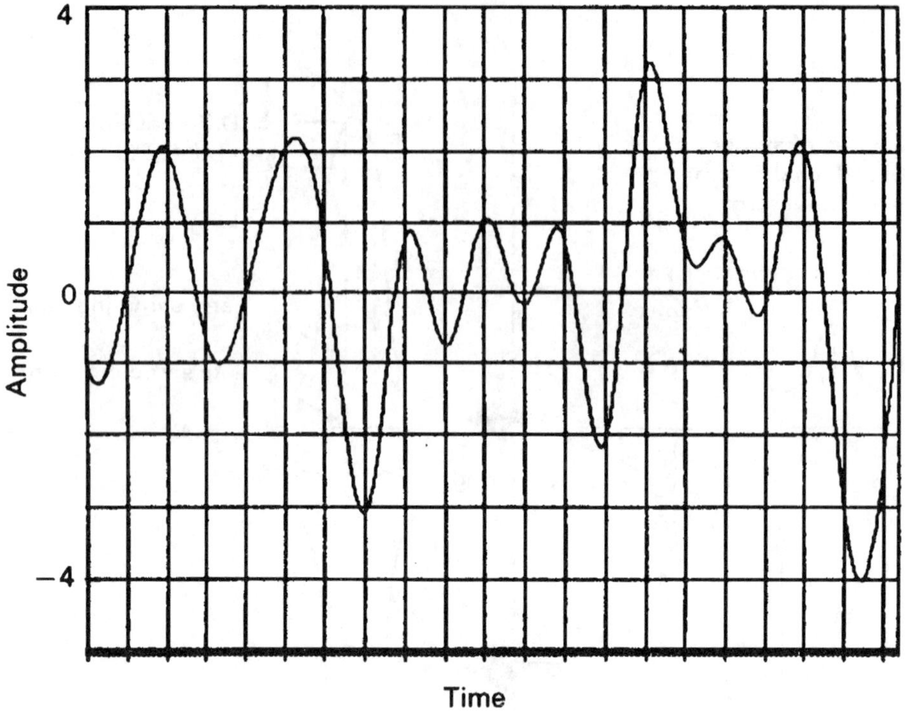

Figure 12.2: Raised cosine pulse shaping on the I channel

The consequence is that the early-late difference signal, which should be negative for all the waveforms in Figure 12.1, is correct for only a fraction of the transmitted symbols. The dips of the type in Figure 12.1(b) will cause an incorrect symbol regeneration. Since the monotonic waveforms of Figure 12.1(c) always indicate that a timing retardation is required, they will cause an equal distribution of correct and incorrect PLL control signals. These problems render the system's operation unreliable.

The pulse shaping used at the transmitter can cause further problems. If raised cosine shaping is used, as suggested by Kingsbury [182], with matched filtering at the receiver, then a typical resulting transmitted star 16-QAM waveform I component is shown in Figure 12.2. Note that for 16-level star QAM the possible magnitudes of the I channel waveform are 0, $\pm$ 0.707, ± 1 ,$\pm$ 2.12 and ± 3 for the valid symbols.

As it can be seen, the peaks do not always coincide with the equispaced sampling points shown by the vertical lines, rendering EL clock recovery inaccurate. This suggests the use of alternative pulse shaping schemes. One of the most suitable schemes would seem to be that of non-linear filtering (NLF) discussed in Chapter 4, which guarantees that the peaks are at the correct sampling point. In addition, NLF has the virtue of implementational simplicity. A typical NLF waveform is shown in Figure 12.3, where it can be seen that the peaks coincide with the optimal sampling

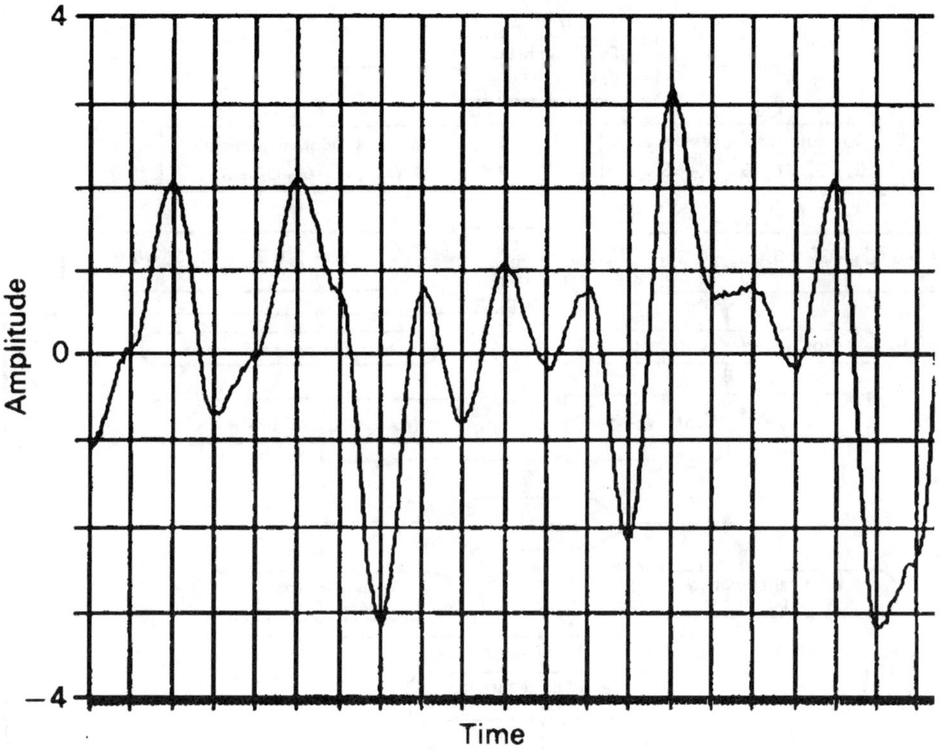

Figure 12.3: Non-linear filtered I channel waveform for star 16-QAM

points even though the zero-crossings are not equispaced. Again, the correct sampling times are given by the vertical grid lines. Based on the above arguments, it was decided to use NLF for our further investigations.

12.4 Modified Early-Late Clock Recovery for QAM

The basis of modified EL clock recovery is to search the current symbol period for a peak, and if one is found to assess its suitability for clock recovery. Only if a suitable peak is found is update information sent to the PLL. This improved system, which is based on the use of oversampling, is now described in more detail with reference to Figure 12.4.

The clock recovery system waits until all n of the oversampled observations for a symbol period have been made, where the n observations are equispaced around the current sampling point. The maximum and the minimum samples during the observation period are then identified. If the maximum is at either end of the sampling period, the search for a positive pulse is discontinued since this implies that there cannot be a valid positive peak in the current sampling period. The same applies if

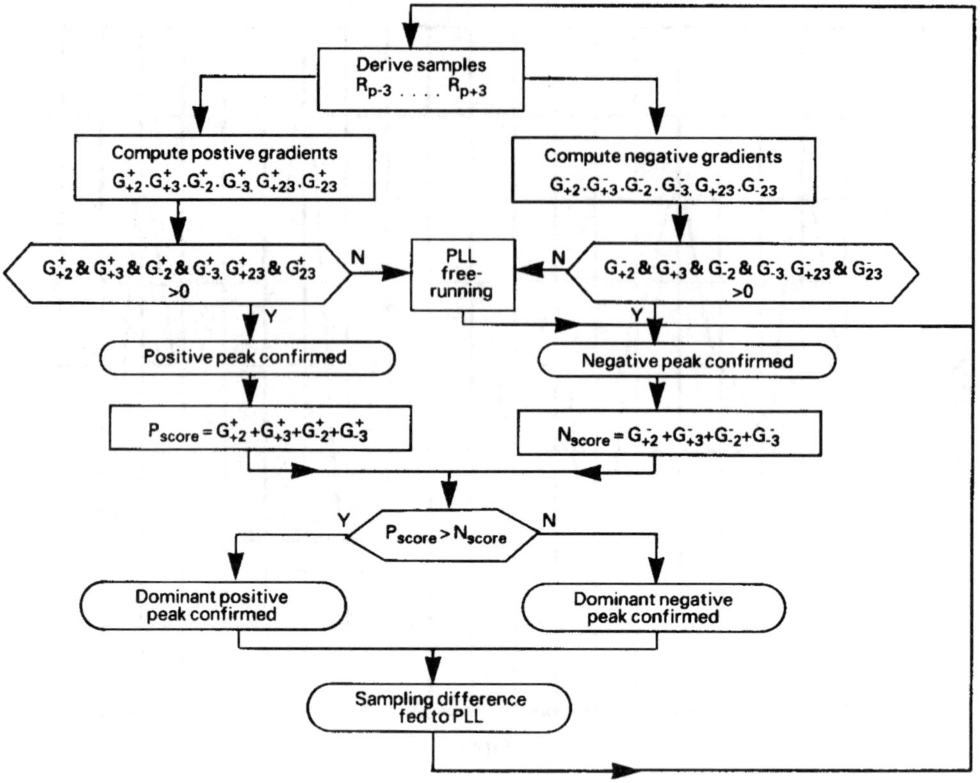

Figure 12.4: Modified early-late clock recovery flow chart

the minimum is at either end of the sampling period.

If there is a valid maximum, sample gradients on both sides of the peak are calculated and the peak is rated depending on these gradients as described below. A similar rating is calculated for the negative pulse if there is still a valid minimum. The algorithm used for rating the positive peaks is based on evaluating the following gradients:

$$\begin{aligned} G^+_{+2} &= R_p - R_{p+2} \\ G^+_{+3} &= R_p - R_{p+3} \\ G^+_{-2} &= R_p - R_{p-2} \\ G^+_{-3} &= R_p - R_{p-3} \\ G^+_{+23} &= R_{p+2} - R_{p+3} \\ G^+_{-23} &= R_{p-2} - R_{p-3} \end{aligned} \tag{12.1}$$

where R_p is the amplitude of the incoming waveform at the peak p and R_{p+k} is the amplitude at the point k oversamples offset from p, where k takes the values $\pm 2, \pm 3$. The presence of a positive peak is only confirmed if all of these gradients are positive. This implies that, moving away from the peak in both directions, the waveform falls away and that it does so with increasing steepness. The oversampled values of R_{p+1} and R_{p-1} are not used as, due to the nature of the pulse shaping, these are very similar to R_p and so the gradient being small can be easily falsified by noise. If all these quantities are positive then the peak is rated according to

$$P_{score} = G^+_{+2} + G^+_{+3} + G^+_{-2} + G^+_{-3} \tag{12.2}$$

where P_{score} is the positive peak score.

Similarly for the negative peaks

$$\begin{aligned} G^-_{+2} &= R_{p+2} - R_p \\ G^-_{+3} &= R_{p+3} - R_p \\ G^-_{-2} &= R_{p-2} - R_p \\ G^-_{-3} &= R_{p-3} - R_p \\ G^-_{+23} &= R_{p+3} - R_{p+2} \\ G^-_{-23} &= R_{p-3} - R_{p-2} \end{aligned} \tag{12.3}$$

and if all these quantities are positive then the negative peak is rated as:

$$N_{score} = G^-_{+2} + G^-_{+3} + G^-_{-2} + G^-_{-3} \tag{12.4}$$

where N_{score} is the negative peak score. If $P_{score} > N_{score}$ then the positive peak is deemed more dominant and the negative peak is discarded, otherwise the negative peak is retained and the positive peak is discarded. If a valid peak is identified, the difference between the current sampling time and the peak just identified is used to update the variable internal clock, normally a phase locked loop (PLL) system. Otherwise the PLL is allowed to continue running at its current level.

The BER performance of the star 16-QAM scheme using this clock recovery system operating without any ISI in Rayleigh fading is shown in Figure 12.5, compared to that of a system operating with perfect clock recovery.

As can be seen in the figure the performance of the two systems is almost identical over a wide range of channel SNR values, indicating correct operation of the clock recovery system.

12.5 Clock Recovery in the Presence of ISI

12.5.1 Wideband channel models

In our earlier studies we used low signalling rates, which were significantly below the channel's coherence bandwidth, and hence no ISI was introduced. When the signalling rate is increased to a few MSymbols/s, in most mobile radio environments distortion due to distant echos is introduced. In this section, unless stated, we use a vehicular speed of 30 mph, a propagation frequency of 1.9 GHz, and a signalling rate of 2 Msymbols/s. Intersymbol interference (ISI) distorts the received QAM pulse shape, complicating correct clock recovery. ISI can take many forms depending on the number of local radio scatterers and the terrain in the vicinity of the mobile station (MS). Let us consider two extreme cases.

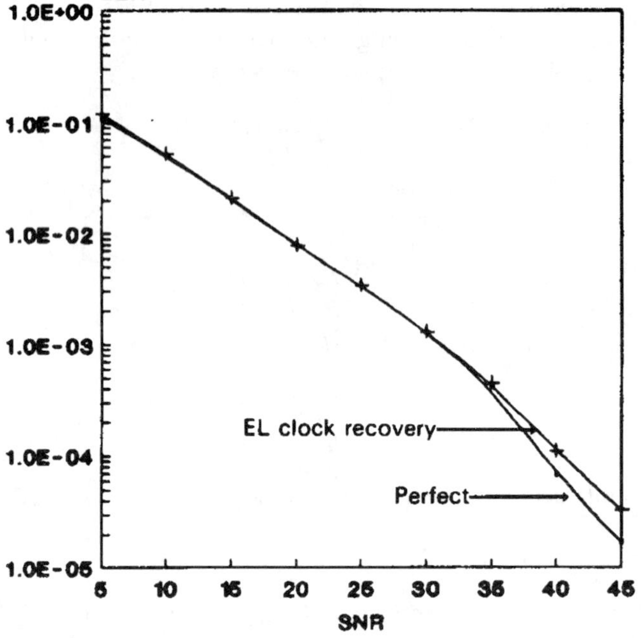

Figure 12.5: BER versus SNR performance of star 16-QAM using perfect and modified early-late clock recovery with no ISI, a mobile speed of 30 mph, and a transmission frequency of 1.9 GHz

The first extreme scenario is when the MS is in the presence of a single large reflector. Here there is a main path incident upon local scatterers leading to a flat Rayleigh fading path, and one single reflected path with a delay dependent on the position of the reflector, which is again incident upon the local scatterers, giving another independent Rayleigh distribution. This is the classic two-path dispersive channel with two delay bins. In our simulations we studied the effect of the delay τ between the direct path and the reflected path. In most practical cases τ will not be exactly equal to any multiple of the symbol period T, i.e. $\tau \neq nT$ where n represents an integer, hence the received signal peak due to the reflected path component arrives somewhere between two peaks spaced by T. There is the alternative, less likely two-path case where $\tau = nT$, that is the peak due to the reflected path component coincides with the peak of a different symbol which arrives at the receiver via the main path.

The second extreme scenario does not use the two-ray model. Instead a large number of indistinguishable incoming paths occur leading to a "peak smearing", or smoothing of the peak. We term this condition *smearing*. This situation was simulated using 16 delay bins, each one comprising 1/16 of a symbol period, and using 16-fold oversampling.

In our simulations Bultitude's wideband impulse response measurements [366]

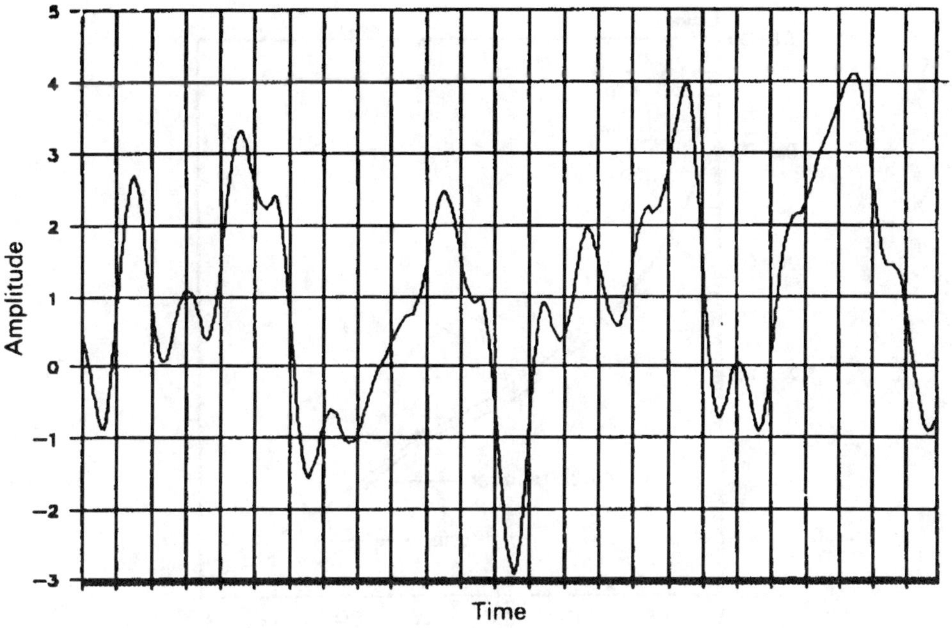

Figure 12.6: Typical star 16-QAM NLF I channel waveform with $\tau \neq nT$

made in a street microcell in Canada were used. If the results presented graphically in his paper are divided into delay bins for a transmission rate of 2 Msymbols/s with an oversample ratio of 16 then significant responses occur in bins 0, 12 and 32, each bin being 31.25 ns wide. Each path was then subjected to independent Rayleigh fading as seen in Figures 2.21 and 2.22. The approximate relative powers in these delay bins are 1, 0.28 and 0.01 respectively, i.e. if the first path is assumed to have a power of 1 then the path in delay bin 12 has a power of 0.28 times that of the first path. These results can be used directly for the $\tau \neq nT$ scenario. Furthermore, by altering the transmission rate to $16/12 \times 2$ MHz, i.e. 2.66 MHz, and ignoring the relatively insignificant path in delay bin 32, then the results can be used for the $\tau = nT$ scenario. For the smearing scenario an exponentially decaying impulse response over one symbol period was assumed. Based on measurements the smearing scenario seems less likely than the other scenarios considered here, nevertheless it is also included for the sake of completeness.

12.5.2 Clock recovery in two-path channels

12.5.2.1 Case of $\tau \neq nT$

A typical star 16-QAM NLF waveform for the ISI situation is shown in Figure 12.6.

Let us compare Figure 12.6 with the ISI-free case shown in Figure 12.3. There is considerable distortion of the received waveform due to ISI, and although the peaks

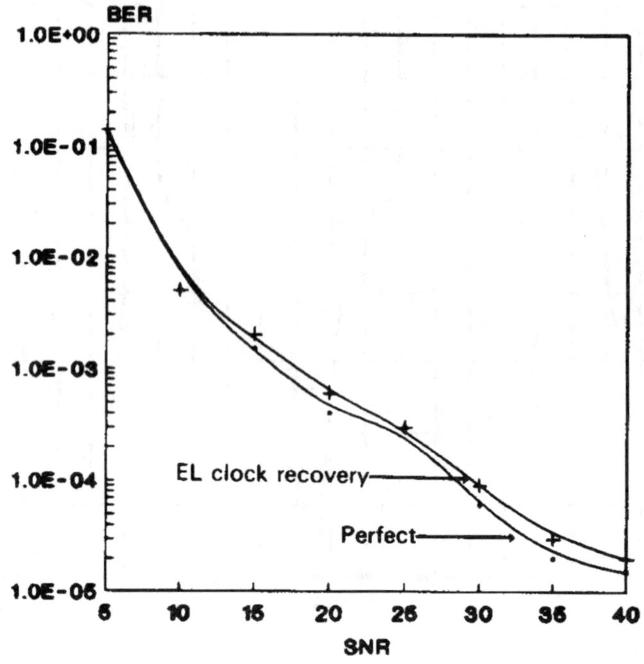

Figure 12.7: Clock recovery performance with $\tau \neq nT$, 30 mph mobile speed, 1.9 GHz carrier frequency, 2 Msymbols/s data rate

of the main pulses are still distinguishable, there are other peaks at different sampling points. Operation of the clock recovery system was closely monitored for this $\tau \neq nT$ scenario, and it was found that provided the interfering signal remained weaker than the wanted signal, the ISI did not hamper clock recovery as any additional peaks were either ignored, or given a lower peak score by the pulse rating system described previously. Once the delayed multipath signal became stronger than the main signal, which would be the case when the main signal was in a fade, the clock recovery scheme switched to the delayed signal. The delayed signal had effectively become the main signal in that situation. The BER versus channel SNR performance of the star 16-QAM modified EL clock recovery system exposed to this form of ISI is shown in Figure 12.7.

In this simulation no channel equalisation has been performed in order to keep the simulations tractable. A suitable equaliser for the star 16-QAM system is discussed in Chapter 14.

It can be seen from the figure that the performance of the system with the modified EL clock recovery scheme was almost identical to that using perfect clock recovery, showing that the clock recovery system worked adequately in this ISI-exposed scenario.

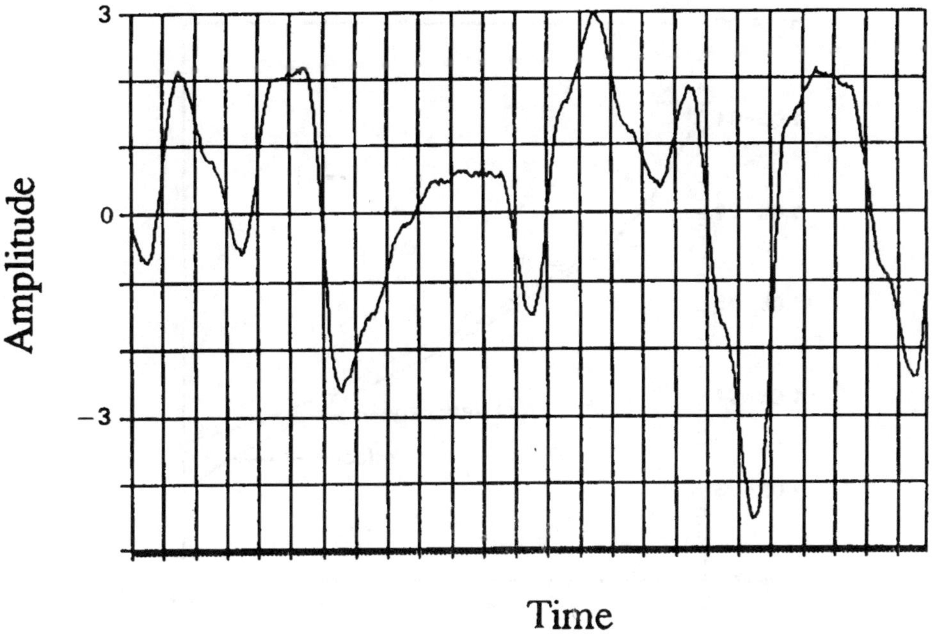

Figure 12.8: Typical star 16-QAM NLF I channel waveform with $\tau = nT$, 30 mph mobile speed, 1.9 GHz carrier frequency, 2 Msymbols/s signalling rate

12.5.2.2 Case of $\tau = nT$

A typical star 16-QAM waveform for this ISI-impaired case is shown in Figure 12.8. Again additional erroneous waveform peaks occur.

Although the waveform is considerably distorted compared to the situation seen in Figure 12.3, the clock recovery system was not unduly affected. Figure 12.9 shows the BER versus channel SNR performance curves obtained with the modified EL clock recovery system and when ideal clock recovery was used. The EL clock recovery is seen to operate almost perfectly in the presence of this form of ISI.

12.5.3 Clock recovery performance in smeared ISI

A typical waveform for the star 16-QAM NLF smeared ISI condition is shown in Figure 12.10.

Again, considerable distortion of the original waveform was experienced, but interestingly the peaks became more distinguishable, although they were often shifted from the sampling point. This shifting of peaks caused some problems to the clock recovery system, the performance of which is shown in Figure 12.11.

The performance of the system with clock recovery was about 3 dB worse in terms of channel SNR compared to ideal clock recovery. This decrease in performance may be acceptable, especially as measurements suggest that such channels are unlikely to

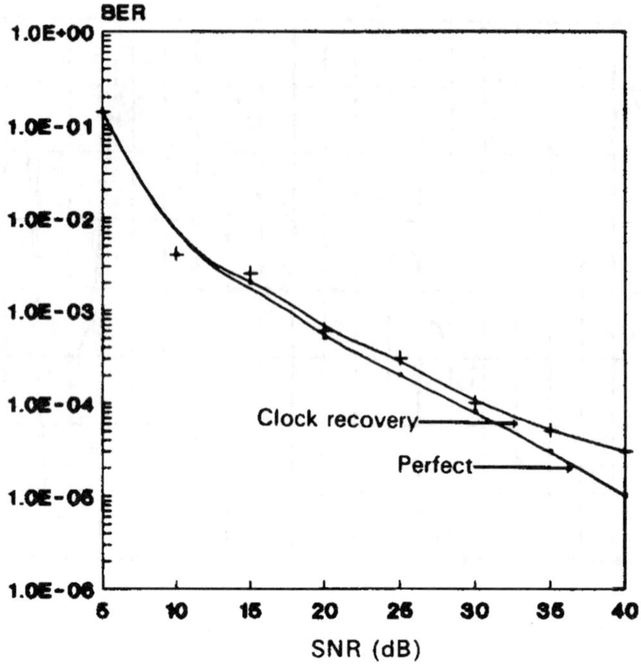

Figure 12.9: Star 16-QAM modified EL clock recovery performance with $\tau = nT$, 30 mph, 1.9 GHz, 2.66 Msymbols/s data rate

be encountered in practice. The reason for the performance of the clock recovery system diverging from that of perfect clock recovery as the SNR increases is likely to be due to random phase jitter caused by the ISI. This will tend to cause a certain number of errors regardless of the SNR, and as the BER falls, the effect of jitter-induced errors will become more significant.

12.6 Early-Late Clock Recovery Implementation Details

A possible implementation arrangement of the modified EL clock recovery system is shown in Figure 12.12.

The system is seen to consist of simple building blocks of comparators, adder/subtractors, selector switches and low capacity memory. In Figure 12.12 only the positive peak calculation system has been shown, an almost identical negative peak calculation system is also required. The received input signal passes to a comparator which compares it with the previous maximum value. If a new maximum has occurred then the value of this, along with its oversample address, is stored. The received input signal is also passed to a shift register which stores all the oversam-

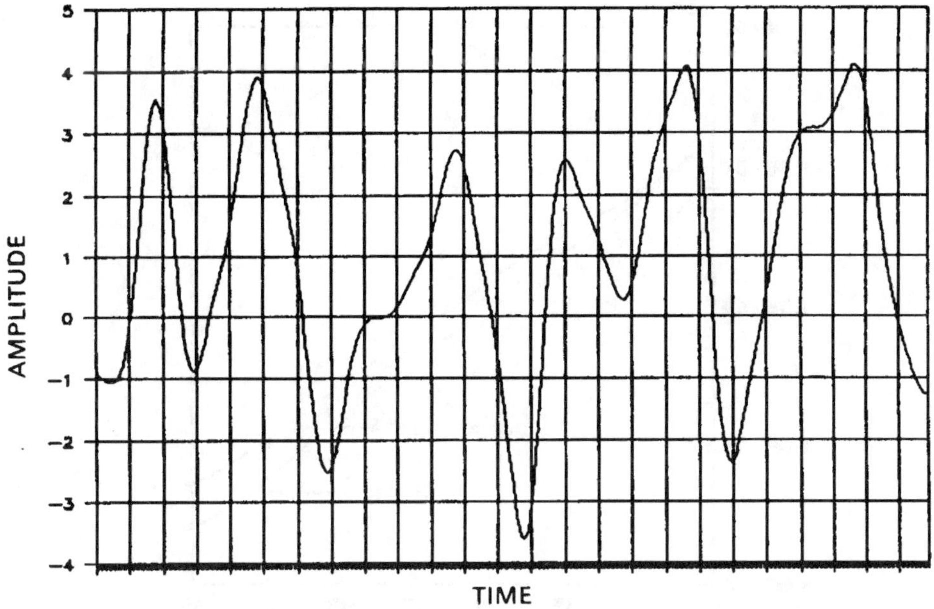

Figure 12.10: Star 16-QAM NLF I channel waveforms with smeared ISI

pled inputs occurring during the symbol period, typically eight. A bank of n-way switches is then used to latch the appropriate values into the subtractors which perform the calculations of Equation 12.1. The results are then summed according to Equation 12.2, leading to the positive peak score P_{score}. This is then compared with the negative peak score N_{score} which has been derived in an almost identical system, and the larger of the two is passed to a switch. According to whether the score exceeds the threshold and whether it was a maximum or minimum, the switch is set appropriately to select the sample point and update the PLL.

Simplifications to the implementation can be made by reducing the number of gradient calculations in Equation 12.1; this reduces performance slightly but considerably simplifies the circuit's construction.

12.7 Carrier Recovery

The problem of clock recovery is closely related to that of carrier recovery. In the case of simple modulation schemes such as BPSK, the carrier recovery operation is straightforward as there is a transmitted component at the carrier frequency for the PLL to lock onto. QAM, however, is a suppressed carrier form of modulation. This means that we either have to deliberately insert a carrier as described in the TTIB section in Chapter 10, or we employ more sophisticated recovery schemes [39]. The problem with carrier insertion methods, such as transparent tone in band (TTIB),

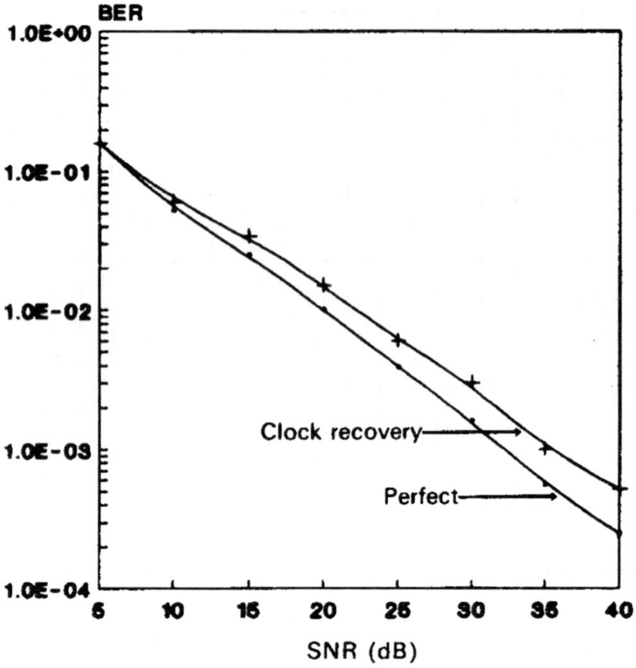

Figure 12.11: Star 16-QAM modified EL clock recovery performance with smeared ISI, 30 mph, 1.9 GHz, 2 Msymbols/s

is that they expand the required spectrum, and these defeat the underlying premise that spectral efficiency is the reason that QAM is employed. They also significantly increase the complexity. At low symbol rates, e.g. below 64 Ksymbols/s, experience has shown that it is very difficult to resolve the phase ambiguities inherent in QAM resulting in false locking, especially during fades. This is because at low signalling rates the fading envelope and phase change more dramatically between signalling instances than in the case of high symbol rates. In this respect, doubling the symbol rate is equivalent to halving the vehicular speed since they both mitigate the rapid envelope fluctuations. At these low symbol rates, differential detection is preferred and it removes the need for carrier recovery. At high symbol rates, the inherent ability of an adaptive equaliser to remove phase ambiguity renders non-differential detection viable.

Simulations with decision directed methods showed little promise. The use of the circular star constellation suggests an interesting and effective form of carrier recovery. One of the systems which has been used for QAM is *times-four carrier recovery* [39]. In this system the complex received incoming signal is raised to the fourth power. This has the effect of raising the amplitude to the fourth power and multiplying the phase angle by 4. In the case of the square QAM constellation this rotates the phase of half the constellation points (those where the magnitude of I is equal to the

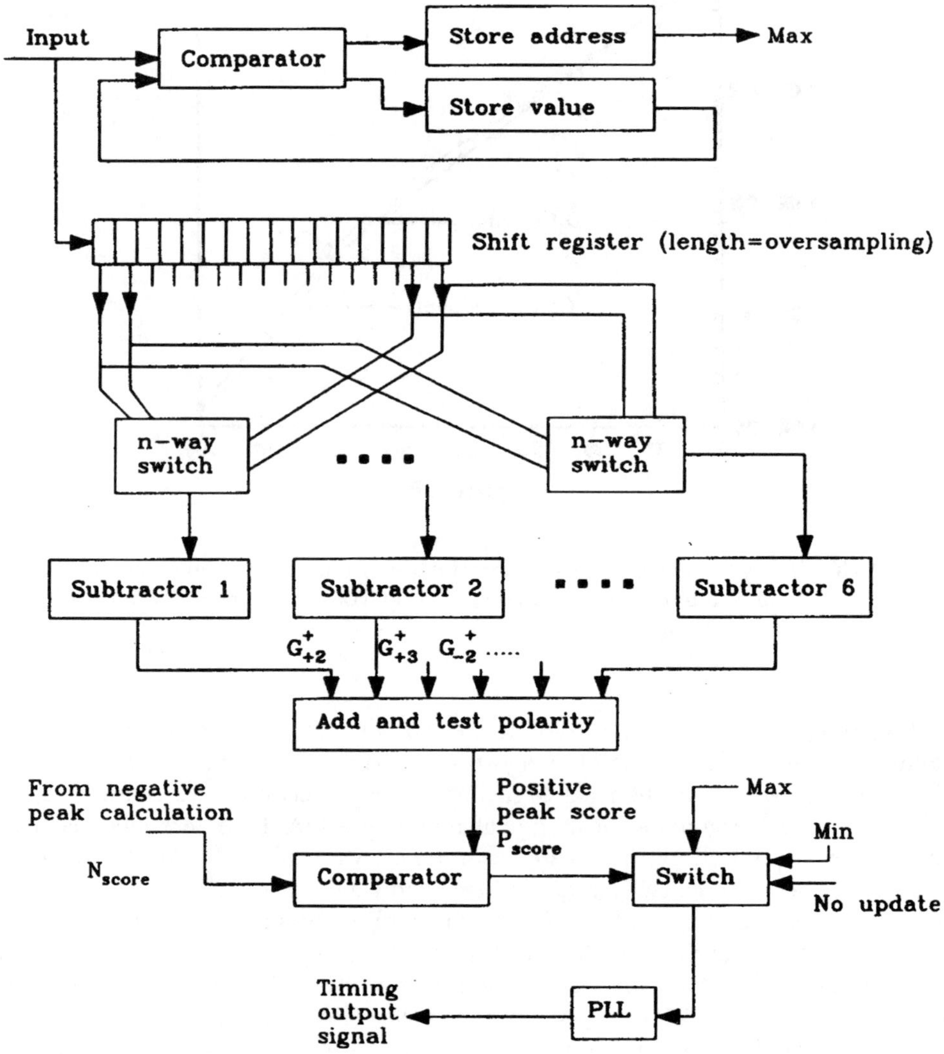

Figure 12.12: Implementation of the modified EL clock recovery system for star 16-QAM

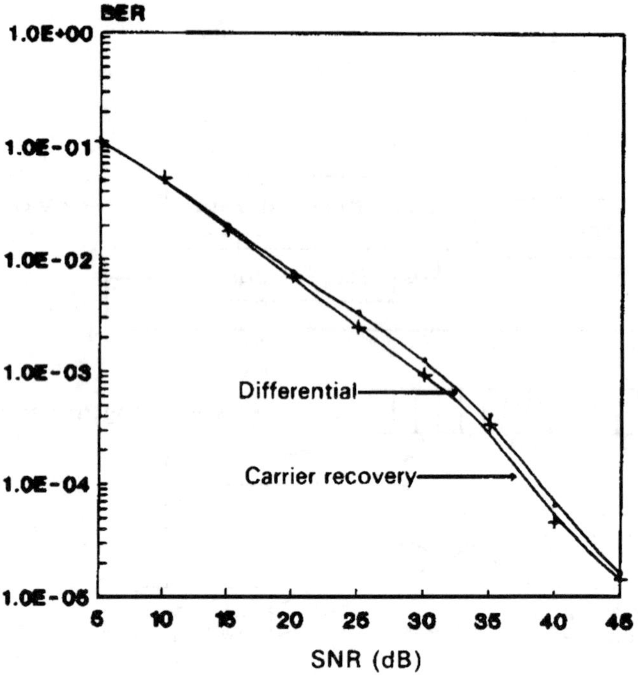

Figure 12.13: BER versus channel SNR performance of differential and coherent star 16-QAM at a mobile speed of 30 mph, 1.9 GHz carrier frequency and 64 Ksymbols/s data rate

magnitude of Q) to 180 deg, and distributes the other half of the constellation points randomly in phase. This operation generates a carrier component at four times the carrier frequency which can be used to facilitate carrier recovery. This can be easily shown using simple trigonometrical equivalences and a QAM carrier. The accuracy of this recovery is limited by the constellation points which do not map onto 180 deg. A long run of these can cause significant jitter and even loss of lock.

The circular QAM constellation offers the intriguing possibility of times-eight carrier recovery. This would have the effect of rotating *all* the constellation point to 0 deg, eliminating the jitter inherent in the previous system. This system, however, cannot resolve ambiguity between lock at any of the eight 45 deg positions and so it is necessary to periodically send a sounding sequence. The performance of this system compared to the differential system is shown in Figure 12.13. It is apparent that the system operating with carrier recovery performs slightly better than that without, but only by a small margin. The gains are lower than the expected 3 dB because of operation in Rayleigh fading rather than Gaussian noise. Here the probability of the next symbol being in error is highly correlated with the probability of the current symbol being in error. There is also the problem of catastrophic failure during deep fades. It can be concluded that the extra complexity involved in adding carrier recovery is not compensated by the gains in performance and so differential, non-coherent reception is generally preferred.

12.8 Summary

In this chapter clock and carrier recovery schemes for star 16-QAM transmissions over Rayleigh fading channels have been considered. It has been shown that a modified version of early-late clock recovery works well both in narrowband and wideband channels. Consideration of carrier recovery schemes showed that differential detection was preferable for QAM transmission.

This concludes the basic QAM schemes for mobile radio transmission. Now we will consider more advanced methods of transmission via QAM, which are generally concerned with mobile radio applications. We start the next section with an analysis of variable rate QAM whereby the number of levels is dynamically changed to reflect the changing channel.

Chapter 13

Variable Rate QAM [53]

13.1 Introduction

QAM transmissions over Rayleigh fading mobile radio channels are subjected to error bursts due to deep fades, even when the channel signal-to-noise (SNR) ratio is high, as was shown in Chapter 11. This leads to the notion of varying the number of modulation levels according to the integrity of the channel, so that when the receiver is not in a fade the number of constellation points is increased, and as the receiver enters a fade they are decreased to a value which provides an acceptable bit error rate (BER) [367]. If the required BER is specified and levels switched accordingly, a variable data throughput is obtained. Alternatively, if the throughput is held reasonably constant on average then a variable BER results. Following Steele and Webb [53], in this chapter a variety of algorithms for an adaptive QAM modems are investigated to meet different criteria.

A variable rate system can only work with duplex transmission as some method of informing the transmitter of the quality of the link as perceived by the receiver is required. The transmitter can respond by changing the number of star QAM [53] levels according to the quality criteria adopted. Successful variable rate transmission requires that the fading channel changes slowly compared to a number of symbol periods. If this condition is not met, then the frequent transmission of quality control information will significantly increase the bandwidth requirements of the system. To avoid this problem the data rate can be increased, allowing the transmission of more symbols before the channel changes significantly. The slower the mobile travels, the slower the fading rate and the lower the signalling rate required for adapting the modem to the channel.

Simulations suggest that for a mobile speed of 30 mph and propagation frequency of 1.9 GHz, the data rate should be a minimum of 512 Ksymbols/s. As the mobile speed decreased, and hence the fading rate fell, the minimum data rate reduced correspondingly, allowing a minimum of about 64 Ksymbols/s at walking speed and 1.8 GHz, and 32 Ksymbols/s at 900 MHz. For the high symbol rates associated with vehicular mobiles to occur without dispersion requires the use of microcells where the time delay spread is sufficiently small. This suggests cordless telecommunications (CT) in buildings or street microcells.

Varying the number of QAM levels in response to fading conditions results in a variable bit rate, which although nearly constant over long periods, could instantaneously vary by four times the average rate. This may be problematic for conventional speech codecs which typically generate a constant bit rate, although there do exist variable bit rate coders, but should not significantly affect computer data. Data transmissions often have the virtue that a variable throughput and relatively large delays can be tolerated, provided the BER is sufficiently low. For computer file transfer the BER must be virtually zero.

The simplest duplex arrangement for variable rate modem operation is time division duplex (TDD), where both base station (BS) and mobile station (MS) transmit over the *same* radio frequency channel, but at different times. In this case both BS and MS experience similar channel fading conditions as their transmissions are typically half a TDD frame apart. The transmission received by the MS is used to estimate the channel integrity which then dictates the number of QAM levels to be used by the MS transmitter. Similarly, the transmission received by the BS enables the number of QAM levels to be used in the subsequent BS transmission to be determined. It is critical that both the BS and MS inform the other of the number of QAM levels used by their transmitters and that this information is not corrupted by the channel, in order for the QAM demodulations to be properly performed.

Typically the data is divided into blocks, or packets that occupy a time slot, and the first few symbols in each block are reserved for signalling the side information representing the number of QAM levels. The optimum size of the block is related to the mobile speed and propagation frequency as the channel should not change significantly over the block duration. In our simulations we have used signalling blocks of 100 symbols for transmissions at 512 Ksymbols/s, a mobile speed of 30 mph and a carrier frequency of 1.9 GHz. At the start of each block a signal was transmitted representing the number of QAM levels to be used in this block. This side information was encoded onto two symbols of a 4-level QAM, i.e. QPSK system, and each of these two symbols was transmitted three times. Majority voting was performed at the receiver in order to establish the number of QAM levels to demodulate in the current block. With larger block sizes, higher integrity codes based either on FEC codes or a higher number of repetitions can be used on the information containing the number of QAM levels without significantly reducing the effective information throughput.

Figure 13.1 shows the TDD framing arrangement for adaptive QAM transmissions using one carrier per channel. If N channels per carrier are used the QAM symbols shown in the Figure are transmitted N times faster, but the time over which the fading channel must remain essentially unchanged is the same.

Frequency division duplex (FDD) operation is used by GSM, DCS 1800 and IS-54. In FDD the uplink and the downlink use different propagation frequencies that are typically spaced by some 20-40 MHz. Although the average path loss on the two channels is essentially the same, the fading on the two links may differ substantially. Both BS and MS must monitor their incoming channels and signal the required number of

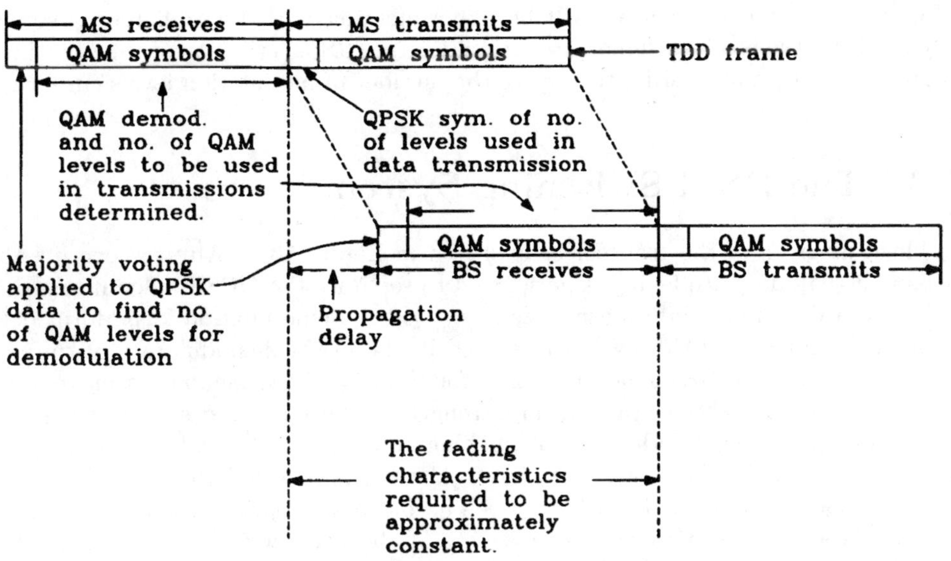

Figure 13.1: The framing structure used in the variable-level scheme

QAM levels to be used by the other in their transmissions on their outgoing channels. This results in twice the delay between estimating the number of levels required, and transmitting with this number of levels, when compared to TDD. Consequently the mobile speed in FDD must be half that in TDD for equivalent performance. All the results in this chapter are reported for TDD.

13.2 Variable QAM Constellations

As proposed in the previous section, the QAM constellation changes as the number of levels is varied. The constellation system used is based upon star QAM in conjunction with differential coding as described in Chapter 11. Constellations ranged from 1 bit/symbol corresponding to BPSK through to 6 bits/symbol star 64-QAM. The constellations for 2-level to 64-level star QAM are shown in Figure 13.2. The actual distances between the rings, and the sizes of the rings are not drawn to scale. Each constellation has the same average energy, and the radii of the rings in the 8-level and 16-level constellations are in the ratio of three-to-one.

Provided the receiver noise is sufficiently low, and implementation complexity is not an inhibiting factor, more than 6 bits/symbol can be used. As the number of bits/symbol is increased, the number of amplitude rings and the number of phase points are alternately doubled. Starting with BPSK, for 2 bits/symbol the number of phase points is doubled to get QPSK. For 3 bits/symbol the number of amplitude levels is doubled to get 2-level QPSK. For 4 bits/symbol the number of phase points is doubled to get to the 16-level star constellation, and so on until with 6 bits/sym there are 4 rings having 16 points per ring.

In the next two sections we will be experimenting with two different systems, applying either the received signal strength indicator (RSSI) criterion or the prevailing bit error rate criterion in order to control the number of modulation levels employed.

13.3 The RSSI Switching System

The block diagram of the transceiver is shown in Figure 13.3. After recovering the baseband signal, demultiplexing is performed to separate the QPSK side information and star QAM primary information signals. QPSK demodulation is performed to obtain the number of QAM levels to be used in the QAM demodulation of the primary information. QPSK demodulation is followed by QAM demodulation to yield the recovered data. The average signal strength of the baseband signal over a signalling block provides an indication of the short-term path loss and fading envelope of the radio channel. If this average is very low the mobile is either in a deep fade or at the edge of the cell. In either case it is more appropriate for it to transmit using relatively few QAM levels. Conversely, if the average is high the channel is relatively good, enabling more QAM levels to be used in its next transmission. The average over a signalling block is computed using an exponential smoothing process which gives more weight to signal levels towards the end of the block associated with more recent samples. This average signal strength is then quantised, each quantised output signifying a particular number of QAM levels to be used in the forthcoming transmission. The transmitter seen in Figure 13.3 has the inverse structure to the receiver.

The baseband signal in Figure 13.3 is related to the received signal strength indicator (RSSI) at the RF stage. This system would function identically if the RF RSSI, instead of the baseband RSSI indicator, was averaged over a signalling block. For ease of nomenclature, this method of switching the QAM levels based on the baseband signals is called baseband RSSI switching, or more simply RSSI switching.

Two criteria can be used for the switching levels generated by the quantiser in Figure 13.3. One is to select the switching levels to achieve a specified BER, resulting in a variable data throughput. In order to establish the required threshold values, simulations were conducted where graphs of BER as a function of channel SNR were plotted for the star QAM modem having 2^n, $(n = 1, 2, ..., 6)$ fixed levels. In these simulations a Gaussian channel was used, as over any short period of time when switching is performed the channel essentially exhibits a constant RSSI level plus Gaussian noise. The channel SNR value corresponding to the BER of interest identifies the switching thresholds. The SNR is therefore quantised, with the quantisation zones corresponding to different number of QAM levels in the adaptive modem. Some hysteresis should be incorporated into the switching algorithm in order to prevent continual level changing. Figure 13.4 shows an arbitrary segment of the fading envelope of the received signal applied to the QAM demodulator, and the number of bits/symbol selected during this period. The figure was produced from a simulation at a channel SNR of 30 dB with other parameters as stated previously. Because of the relatively high SNR, much of the time the modem attained its maximum 6

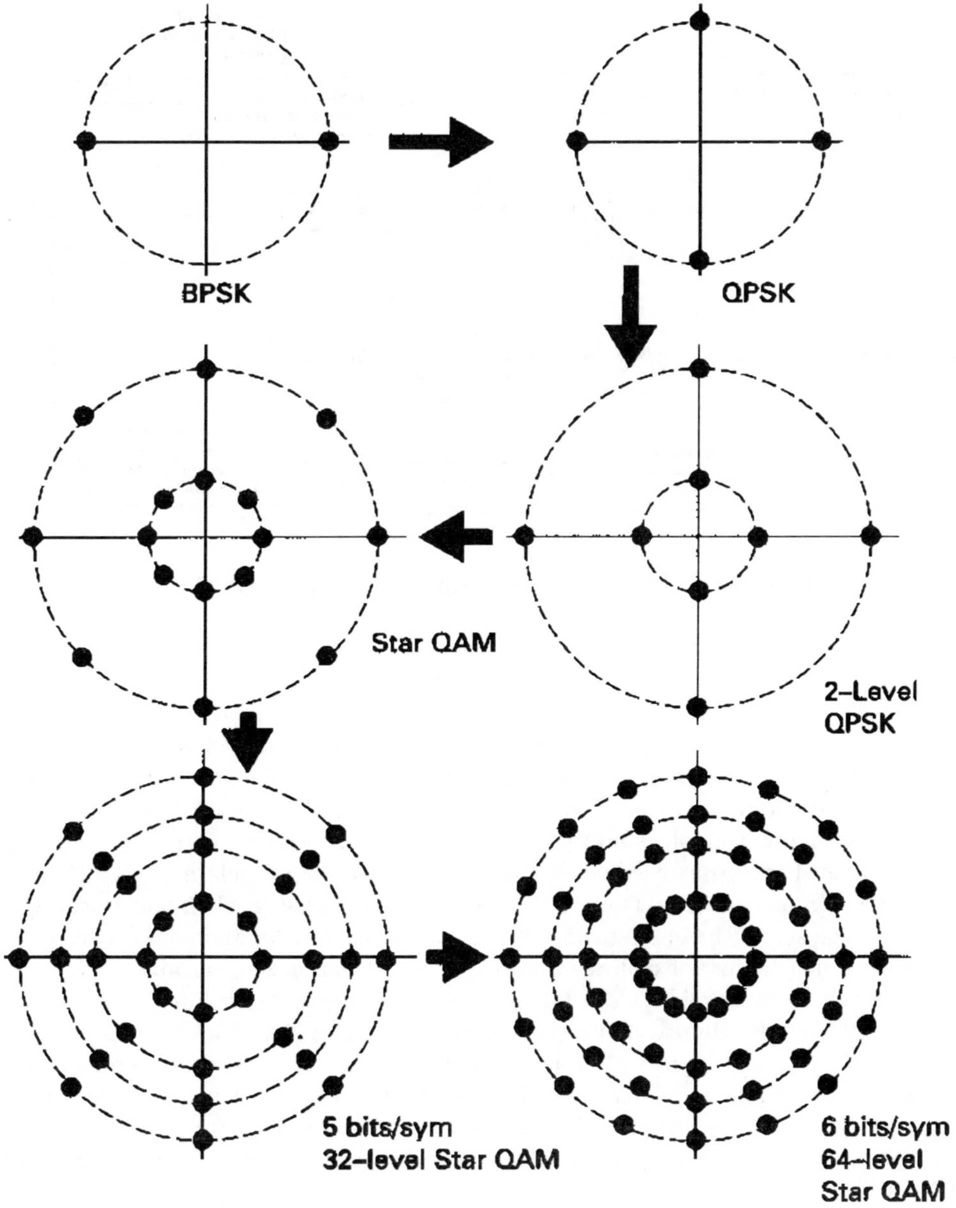

Figure 13.2: Star QAM constellations used in the variable-level scheme

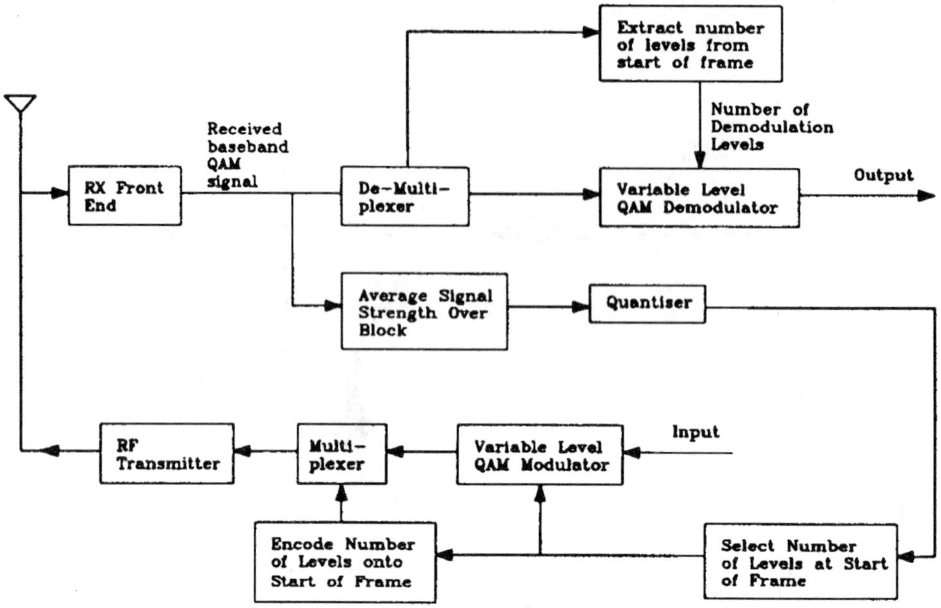

Figure 13.3: Transceiver block diagram for RSSI switching

bits/symbol.

The second criterion for choosing the switching thresholds achieves a constant average bit rate whilst accepting a variable BER. Here the thresholds derived above for the constant BER system are multiplied by a factor at the start of each block. This factor is derived from the baseband signal averaged over a number of fades. This operation merely involves adding an extra averaging circuit with the averaging window increased over that used to average over a signalling block in Figure 13.3, so that it covered many blocks. Therefore as the average signal level rises, for example when moving closer to the base station, the switching thresholds increase accordingly, maintaining a near constant average number of bits/symbol, but variable BER.

The average number of bits/symbol can be set to any level within the maximum number of bits/symbol by changing the scaling factor associated with the long-term average of the received signal. Where the average bit rate was the same for both selection criteria, they both gave identical BERs and can be viewed as equivalent systems. Figure 13.5 shows switching profiles for this constant throughput scheme over the same section of fading channel as for the previous profile. Again an SNR of 30 dB is used.

Here considerably more level changing took place in an effort to maintain an average number of bits/symbol close to the target value of 4 bits/sym. If a lower SNR had been chosen for both graphs then they would appear more similar as the former scheme would have had an average throughput closer to the 4 bits/symbol of the latter scheme.

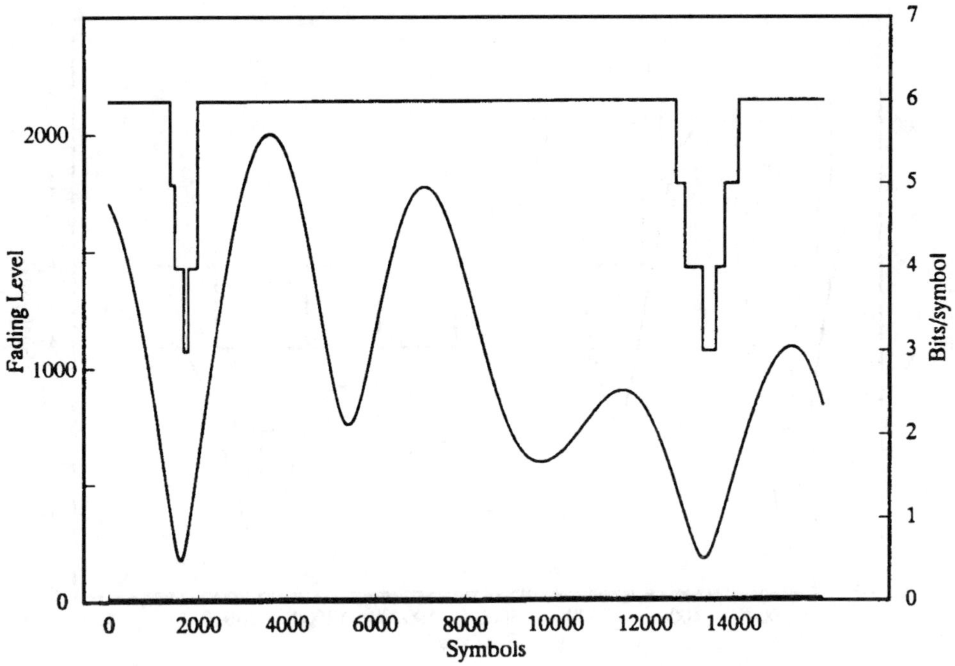

Figure 13.4: Stylised RSSI switching for variable bit rate

In an effort to further improve the system's performance, channel coding was added to both of the above systems, in our case in the form of systematic binary BCH codes. The data was encoded and block diagonally interleaved over 15,600 bits before being passed to the input of the QAM modem in Figure 13.3. At the output of the modem de-interleaving was performed in an attempt to randomise the bursty channel errors before channel decoding took place. The variable QAM schemes had the advantage over the fixed QAM scheme that errors occurred in smaller blocks because of the use of fewer QAM levels in poor channels. This allowed the channel codecs to perform more efficiently. The performance of the RSSI switching system with channel coding will be presented in contrast to FEC-coded results of another adaptive QAM modem in the next section.

13.3.1 Results

The performance of the RSSI switched variable-level star QAM system without FEC coding compared to a fixed 16-level (4 bits/symbol) star QAM system is shown in Figure 13.6.

The adaptive QAM system has had its switching thresholds adjusted so it also gives an average number of bits/symbol close to 4. The necessary signalling information of the number of QAM levels used has been taken into account in calculating this throughput. The simulation was performed for a propagation frequency of 1.9 GHz,

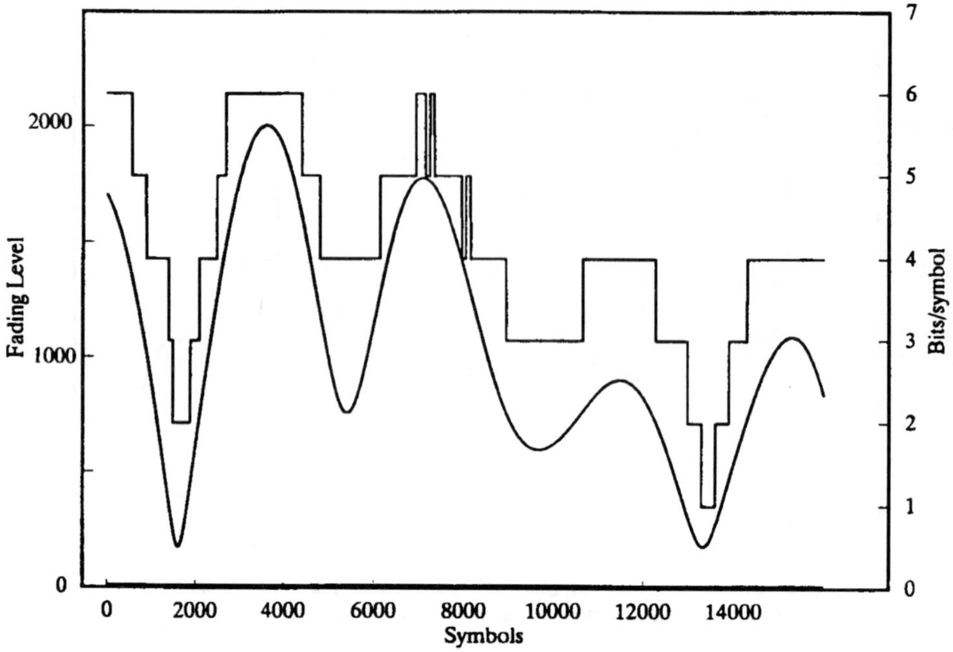

Figure 13.5: Stylised RSSI switching for constant bit rate

a vehicular speed of 30 mph, and a transmitted symbol rate of 512 KSymbol/s, providing a bit rate of 2048 Kbits/s on average. The propagation environment has been assumed to be microcellular so that ISI is insignificant. There is a significant improvement in the performance of the adaptive scheme over the fixed modem at all SNRs. This improvement is enhanced with increasing SNR. This suggests that the variable rate system has advantages in certain applications.

13.4 The Error Detector Switching System

Rather than switching the number of QAM levels in harmony with the RSSI they can be changed on the command of a channel codec, in this case a systematic binary BCH codec. Figure 13.7 shows a simplified block diagram of the BER switching transceiver. Observe that its structure is similar to that of the RSSI switching system of Figure 13.3, apart from the fact that the channel quality is now monitored by a BCH coder. In order to obtain an estimate of the channel during each received data packet, a BCH(63, 57, 1) code is overlaid onto the last 57 bits of input data in each block. The input data may already have been channel coded and interleaved using the same system as described in the previous section; the additional coding is overlaid onto whatever data was present. This codec is generally ineffective at correcting errors in the hostile fading environment as it is typically overwhelmed by

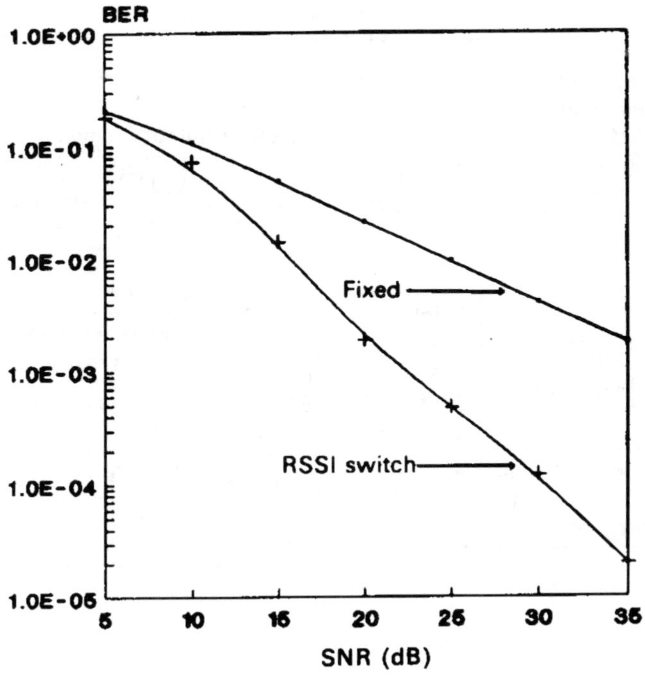

Figure 13.6: BER comparison of variable and fixed systems at 30 mph, 1.9 GHz, 512 Ksymbols/s

the non-interleaved errors in a typical error burst, but it informs the receiver that channel conditions were poor. Such a coding system involves little overhead as it is only applied to the final 57 bits in a block which contains an average of 400 bits. The coded data is applied to the adaptive QAM modulator and the modulated output is upconverted and transmitted.

The QAM demodulation is performed using the number of QAM levels extracted from the QPSK header, as described in the section on the RSSI switched QAM system. Most of the data is passed to the output where it might be subjected to de-interleaving and channel decoding. However, the final 63 bits of the recovered bit stream are passed through the BCH(63, 57, 1) codec before proceeding to the output. If no errors are detected by this BCH codec then the number of QAM levels used in the next block is doubled, otherwise it is halved. Switching profiles for the same section of fading channel envelope as used in Figures 13.4 and 13.5 are shown in Figure 13.8. Again a channel SNR of 30 dB was used. This figure shows that the BER switched system had a similar level switching profile to the variable throughput RSSI system, as might be expected. The lack of hysteresis on level switching is evident but it does not cause any additional signalling overhead or significantly alter the BER.

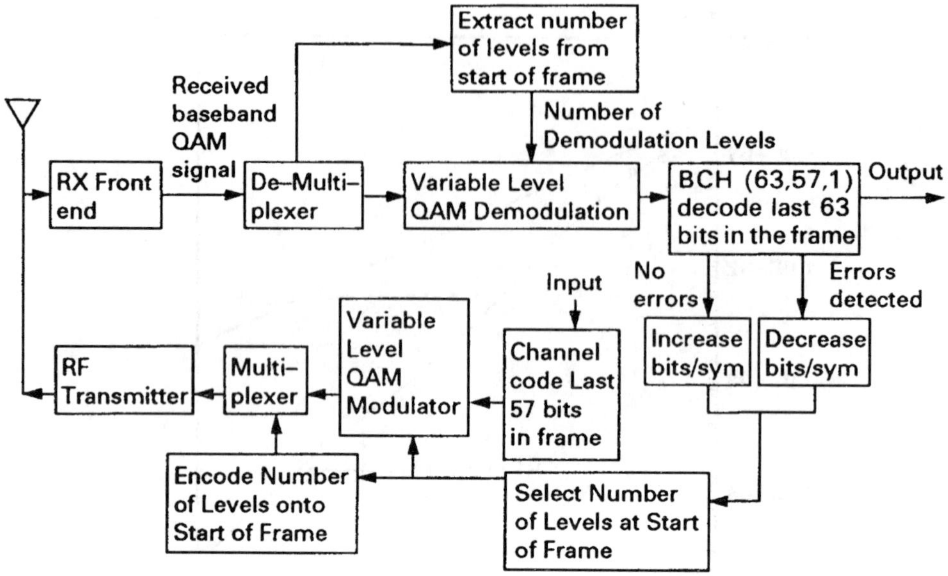

Figure 13.7: Block diagram of the adaptive BER switching star QAM receiver

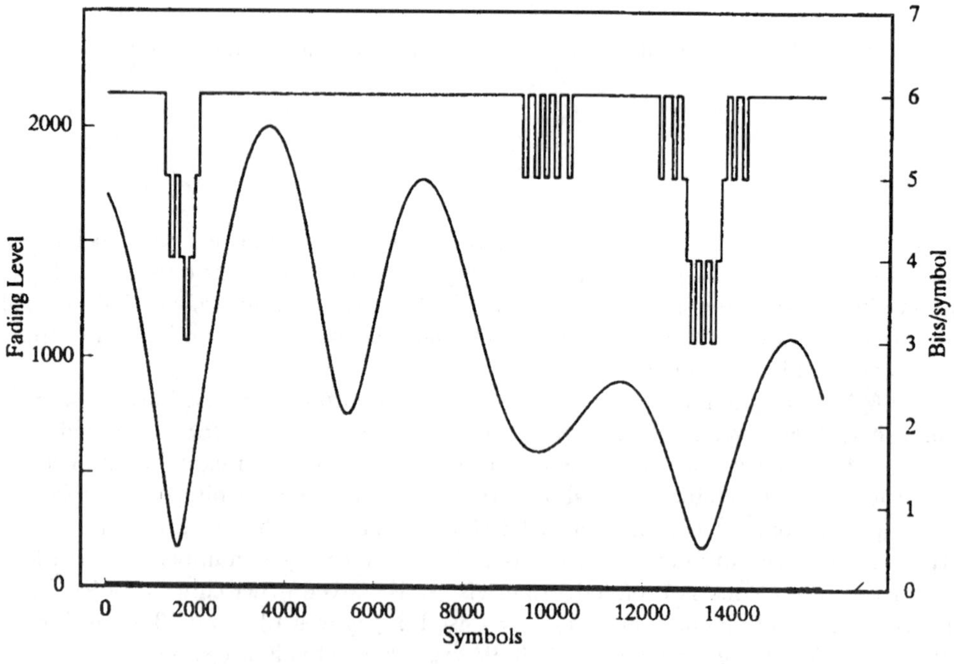

Figure 13.8: Stylised fading channel envelope and number of bits/symbol profiles for BER switching adaptive star QAM modem

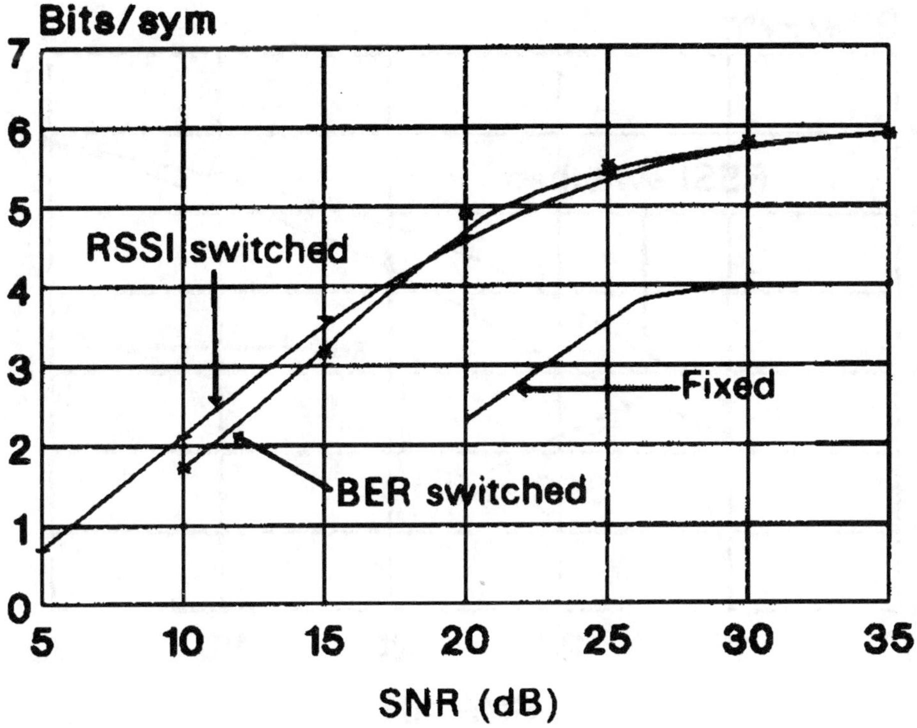

Figure 13.9: Comparison of variable and fixed star QAM systems at BER = 10^{-2}, 30 mph, 1.9 GHz, 512 Ksymbols/s

13.4.1 Results

A direct BER comparison between this adaptive system and the fixed 16-level QAM system cannot be made, since although the adaptive modem maintains a constant symbol rate, the number of bits per symbol, or bit rate, is highly variable. Instead comparison curves of the average number of bits per symbol against SNR for a variety of different BERs can be drawn. This clearly shows the trade-off between the BER and the average number of bits/symbol that can be achieved using an adaptive QAM modem. The results are shown in Figures 13.9 to 13.12.

Observe from Figure 13.9 that the number of bits/symbol of fixed star QAM was 4 when the channel SNR was high, and then exhibited the same roll-off as the adaptive modems. Both the adaptive and fixed star QAM modems decreased their number of bits/symbol with falling SNR values at a rate of approximately 2 bits/sym per 10 dB in SNR. As the SNR fell, more powerful BCH codes were progressively invoked in order to maintain the fixed BER.

For all modems the BCH codes were composed of blocks of 63 bits, the number of data bits varying between 57 and 24 as the correction power increased from 1

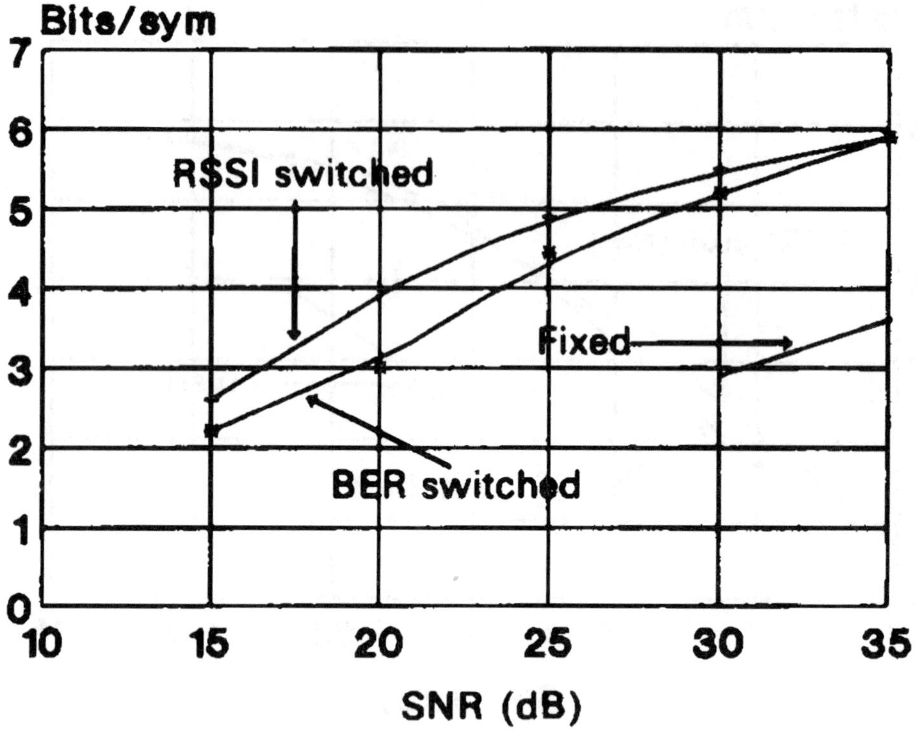

Figure 13.10: Comparison of variable and fixed star QAM systems at BER = 10^{-3}, 30 mph, 1.9 GHz, 512 Ksymbols/s

to 7 bits/block. The corresponding BCH codes are BCH(63, 57, 1), BCH(63, 51, 2), BCH(63, 45, 3), BCH(63, 39, 4), BCH(63, 36, 5), BCH(63, 30, 6) and BCH(63, 24, 7). Block diagonal interleavers of size 15,600 bits were used for all systems. Thus by having no BCH coding compared to when the maximum power BCH(63, 24, 7) code was used, the number of bits per symbol decreased from 4 to approximately $R = 24/63 \approx 1.52$ for the fixed system. For a given number of bits/symbol the adaptive modems achieved an 8-10 dB gain in SNR over the fixed system, due to their ability to adjust the number of constellation points. As the SNR increased above 25 dB, the fixed modem settled to its 4 bits/symbol while both RSSI and BER switched adaptive systems were asymptotic towards the maximum of 6 bits/sym. This was because the maximum number of bits/symbol available to the modulator was 6. In general the BER versus channel SNR curves can be expected to be asymptotic towards the maximum number of bits/symbol available in the system in use, given a sufficiently high SNR.

In all the curves, both the RSSI and BER switched adaptive star QAM systems had similar performances, with the RSSI switching system typically offering a slightly

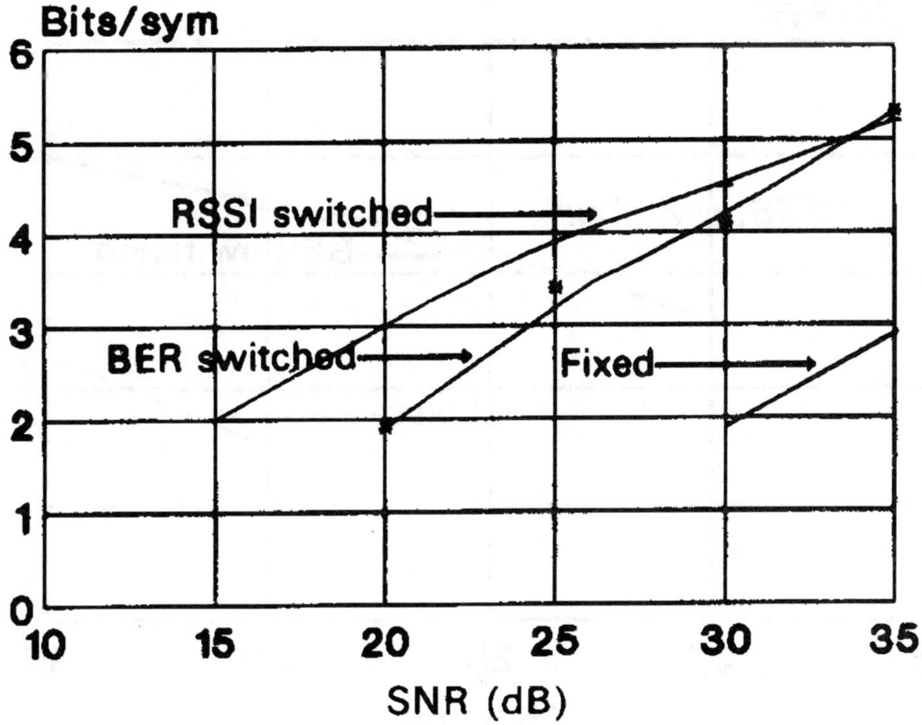

Figure 13.11: Comparison of variable and fixed star QAM systems at BER = 10^{-4}, 30 mph, 1.9 GHz, 512 Ksymbols/s

higher number of bits/symbol at low SNRs for some BERs. These small gains were due to the RSSI switching system's ability to select a lower number of levels before any errors occurred. In terms of implementational complexity the RSSI switched scheme is also slightly more attractive, since it can dispense with the additional BCH(63, 57, 1) codec.

The final comment on the curves in Figures 13.9 to 13.12 is that as the guaranteed minimum BER was increased from 10^{-5} to 10^{-3} the channel SNR required to achieve the same number of bits/symbol for the variable systems was decreased by approximately 3 dB per decade increase in BER.

13.5 Co-channel Interference

Co-channel interference is nearly always present in mobile communications. Accordingly, it is necessary to ascertain whether improvements in performance can be obtained using variable-level star QAM compared to fixed-level star QAM in the presence of a single Rayleigh-faded co-channel interferer, this being the worst case interference

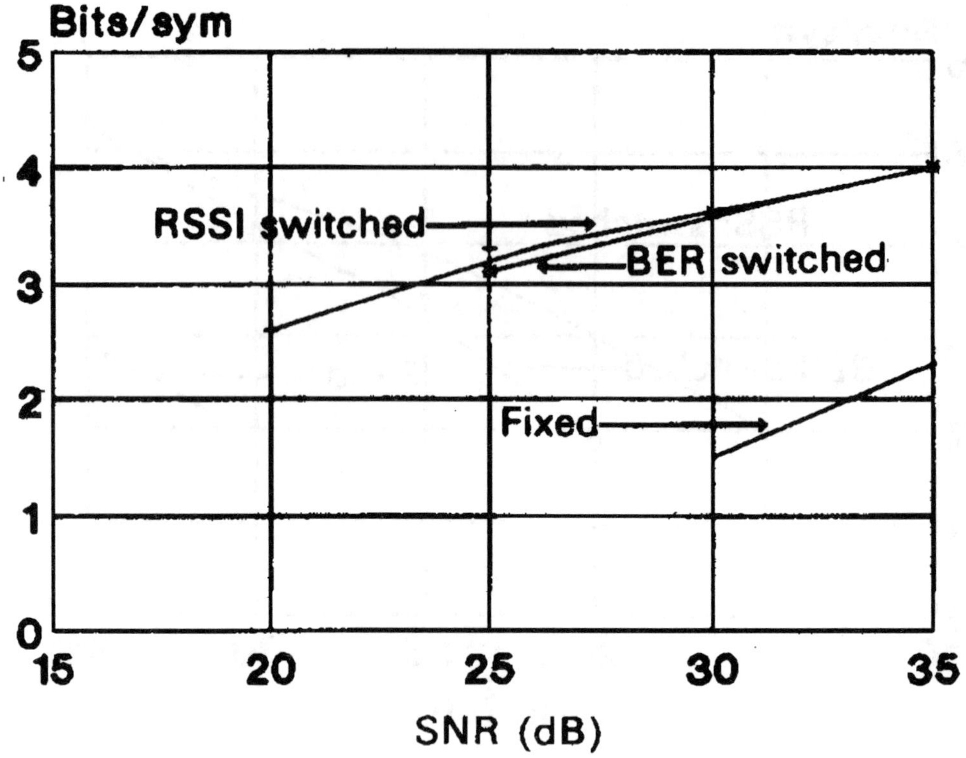

Figure 13.12: Comparison of variable and fixed star QAM systems at BER $= 10^{-5}$, 30 mph, 1.9 GHz, 512Ksymbols/s

since in the case of several co-channel interferers their effect tends towards that of AWGN of the same power due to the central limit theorem. The performance of 16-level star QAM, i.e. a non-adaptive modem is shown in Figure 13.13 for different signal-to-co-channel interference ratios (SIR). The benchmark is the curve with no co-channel interference. These curves indicate that it is necessary to operate with both SNR and SIR values at approximately 30 dB in order to achieve a BER performance better than 10^{-2}. No channel coding was used in the experiment.

When the star QAM modem with no channel coding was made adaptive using RSSI switching, this was based on the second RSSI switching criterion, where the switching thresholds were increased when encountering higher RSSI outputs, as described in the section on the RSSI switched system. This ensured that the average number of bits per symbol was 4, as in the case of the fixed modem. Recall that this RSSI switching criterion was associated with a variable BER but near constant average number of bits/symbol. Performance curves using RSSI switching are presented in Figure 13.14. For all systems in this section the simulation parameters were the same as used in previous sections, namely 30 mph, 1.9 GHz and 512 Ksymbols/s. Comparing

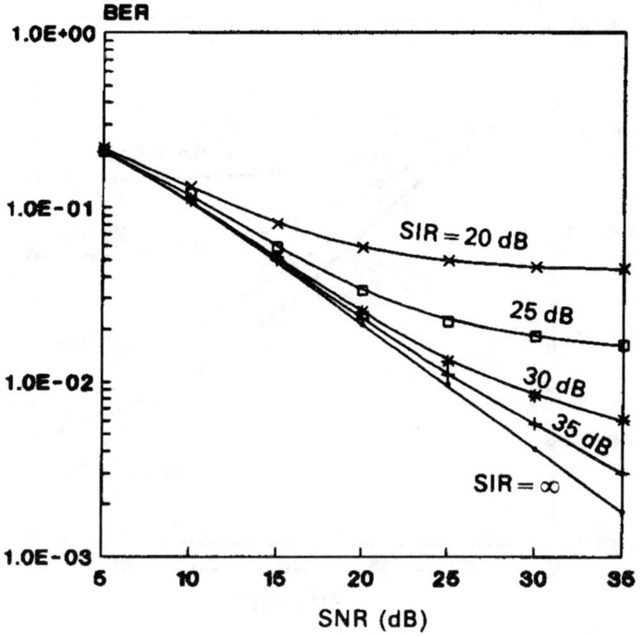

Figure 13.13: BER versus SNR performance of fixed star QAM with co-channel interference and 30 mph, 1.9 GHz, 512 ksymbols/s

Figures 13.13 and 13.14 one can see that the residual BER (measured at a SNR of 35 dB) was 2.2, 2 and 15 times lower compared to the fixed modem for SIR values of 20, 25 and 30 dB respectively.

When BER switching was used it was not possible to ensure that the average number of bits per symbol was 4. The BER and the number of bits per symbol vary in a complex manner and so BER and average number of bits/symbol are plotted as a function of SIR for a fixed channel SNR. Figures 13.15 and 13.16 show plots for SNR values of 35 and 20 dB, respectively, for both adaptive modems, and for the fixed 16-level star QAM modem. From Figure 13.15 it is evident that the non-adaptive system had the worst performance. The BER switched QAM gained in bits/symbol at the expense of poorer BER performance when compared to the RSSI switched QAM as the SIR was increased. This was to be expected because in previous sections the BER switched system was shown to operate with a higher number of bits/symbol and higher BER than the constant throughput RSSI system. Figure 13.16 shows the situation for a SNR of 20 dB, where again both adaptive modems outperformed the fixed modem.

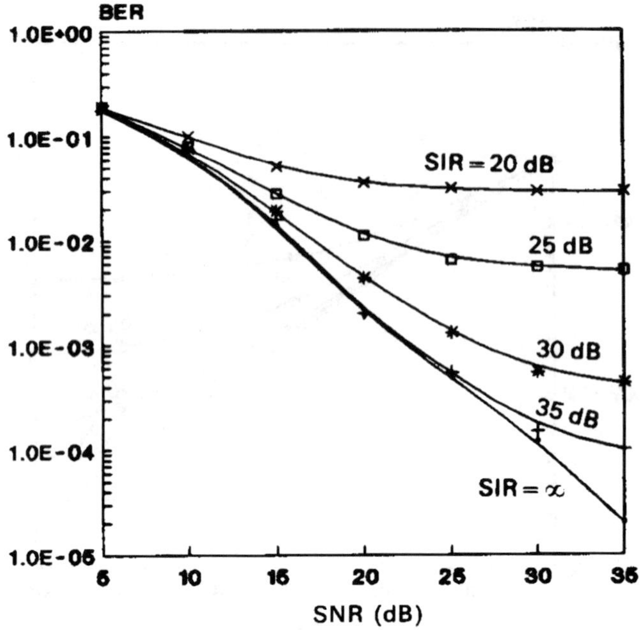

Figure 13.14: BER versus SNR performance of RSSI switched star QAM with co-channel interference and 30 mph, 1.9 GHz, 512 Ksymbols/s

13.6 Application to a DECT-Type System

The digital European cordless telecommunications (DECT) system uses TDD, supporting 12 channels per carrier. The 12 downlink and 12 uplink channels constitute a 24-slot frame lasting 10 ms. Each time slot containing a packet has a duration of 0.417 ms. The data in the packet consists of 320 bits which are transmitted at 1152 KSymbol/s. Each bit was replaced by a QAM symbol to increase system capacity. The header informed the receiver of the number of constellation points. The RSSI switching system was employed.

Figure 13.17 shows the variation of BER as a function of channel SNR for this DECT-like system for MSs travelling at different speeds. No co-channel interference was present. The DECT framing meant that the transmission and reception were separated by 12 slots or 5 ms, and consequently the channel could occasionally experience considerable changes between transmission and reception of data and therefore generate an inappropriate number of levels. Consequently all the curves for our 12-user QAM-based DECT system in Figure 13.17 demonstrate worse performance than for the single-user benchmark scheme shown for no co-channel interference in Figure 13.14, i.e. when the transmit and receive packets were concatenated. When the MS travelled faster the BER deteriorated and when the mobile's speed was 20 mph the adaptive and the fixed 4-level QAM had similar performances. However, an order

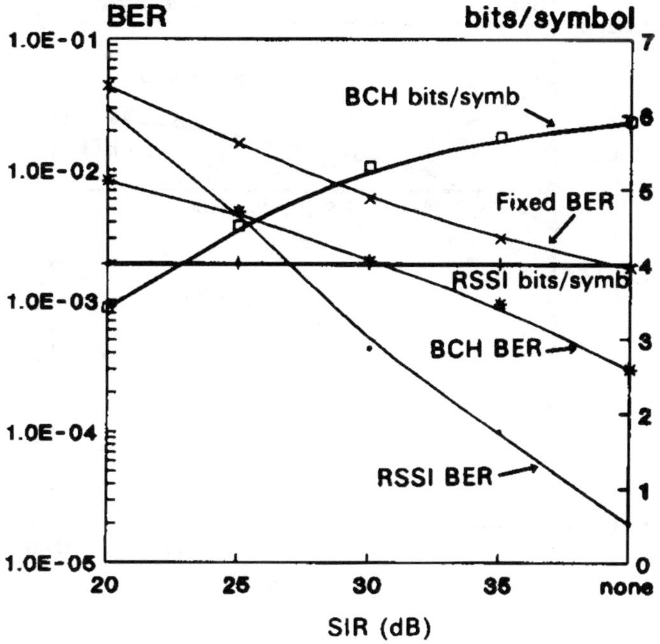

Figure 13.15: BER and bits/symbol versus SIR performance of fixed and adaptive star QAM modems at SNR=35 dB, 30 mph, 1.9 GHz, 512 Ksymbols/s

of magnitude improvement in BER was achieved for channel SNRs in excess of 20 dB when the mobile's speed was decreased from 20 to 5 mph. This was due to the slower fading rate making the channel estimation more accurate.

13.7 Summary

In this chapter we described an adaptive star QAM modem for transmission of data over Rayleigh fading channels using a variable number of modulation levels. Criteria for deciding how to vary the number of modulation levels to give a required performance characteristic were discussed. The adaptive QAM modem can be arranged to provide a nearly constant BER over a wide range of channel SNRs, although the bit rate may vary considerably. This type of performance is suitable for data services that can tolerate some delay. In general, the adaptive modem provides the flexibility to vary both the BER and the bit rate in a prescribed manner to suit a particular application. The adaptive modems have a better performance than the fixed modems both with and without co-channel interference, but require a moderate increase in complexity. A DECT-type multi-user system was suggested and it was shown that at low mobile speeds the BER of a fixed 16-level QAM system can be reduced by an order of magnitude for SNR $>$ 20 dB by employing a variable rate modem.

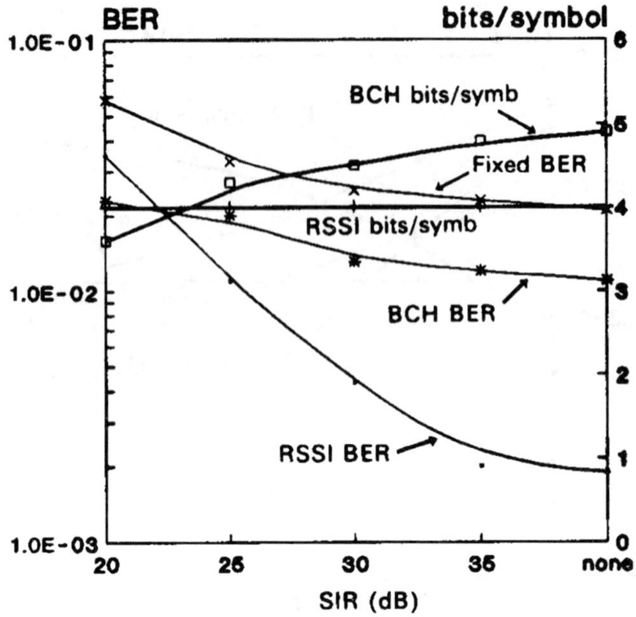

Figure 13.16: BER and bits/symbol versus SIR performance of fixed and adaptive star QAM modems at SNR=20 dB, 30 mph, 1.9 GHz, 512 Ksymbols/s

In the next chapter we investigate some of the problems that transmitting at the high bit rates suggested in this chapter can cause with mobile radio channels, and suggest some possible equaliser structures to overcome these problems.

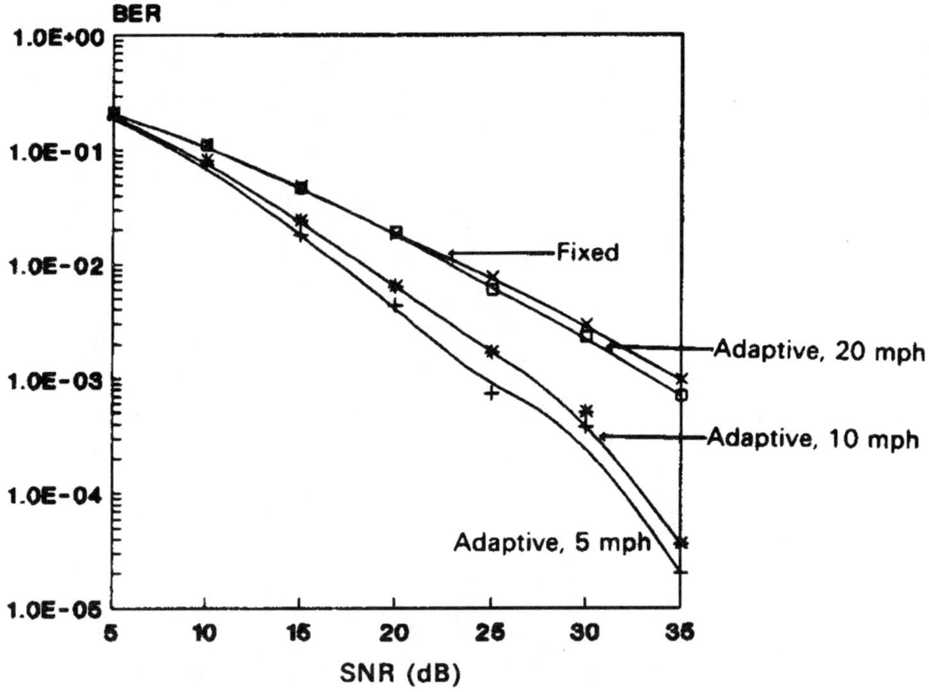

Figure 13.17: BER versus SNR performance of fixed and adaptive star QAM DECT-like modems with no co-channel interference for various vehicular speeds

Chapter 14

Wideband QAM Transmissions

14.1 Introduction

In current communications systems the majority of teletraffic is typically generated in office buildings, hence the capacity of mobile systems in these buildings must be very high. The high bit rates required to accommodate video, audio, computer data and other traffic may necessitate the symbol rate of the multilevel modulation systems being so high that the mobile radio channels exhibit dispersion. Performing equalisation of multilevel signals that are subjected to dispersion and fading is not an easy task. In this chapter methods whereby high bit rates can be conveyed via star 16-level QAM over wideband mobile radio channels are considered which we consider to be composed of three independent Rayleigh fading paths. This wideband channel response shown in Figure 14.1 is a derivative of Bultitude's indoor and outdoor measurements [366, 368] of impulse responses in microcells.

In Chapter 7 we gave a rudimentary introduction to various channel equalisers. Many of them are designed to operate with binary modulation, but can be modified to operate in conjunction with QAM with little, if any, modification. In this chapter we propose a novel channel equaliser scheme for star 16-QAM which is based on a combination of signal processing blocks, such as FEC codecs, the so-called RAKE diversity combiner to be described in the next section, and a linear equaliser (LE). We firstly detail the RAKE combiner.

14.2 The RAKE Combiner

The RAKE receiver is not an equaliser, but a combiner for diversity systems. The RAKE system inherits its name from the way it "rakes" in all the incoming pulses to form an equalised signal. A block diagram of the basic system is shown in Fig-

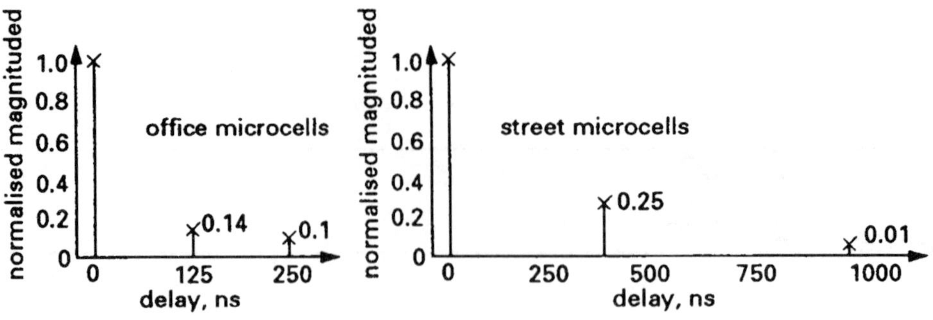

Figure 14.1: Microcellular impulse responses used in the simulations

ure 14.2. The RAKE system is identical in structure to a maximum ratio combiner. It is designed to operate in a situation where the signal to be transmitted has a narrow bandwidth, but the bandwidth available is wide, and so the transmitted signal's bandwidth can be expanded in order to use the multipath diversity inherent in the wideband system. Multipath diversity exhibits itself in the fact that in wideband channels several echoes of the transmitted signal can be observed due to far-field reflections. The channel equaliser can remove the dispersion introduced by the channel, and by locking onto the echoes appearing in the impulse response, taps within the equaliser can exploit this so-called time domain diversity effect. In order to achieve the required bandwidth expansion the baseband signal is transmitted as a series of narrow pulses with a gap between each pulse as shown in Figure 14.3. Suppose, for example, that the signalling interval duration is T and the bandwidth required when these pulses are appropriately shaped is B, but the available system bandwidth is $N \times B$. In this case an N times narrower signalling pulse of duration T/N can be used to exploit the whole bandwidth. Accordingly the RAKE receiver of Figure 14.2 must be clocked at an N times higher rate.

Because of the delay spread, the receiver perceives each transmitted pulse as a series of received pulses. In order for the RAKE combiner to operate correctly it needs to know the expected attenuation and phase change suffered by each path. Each incoming pulse is then amplified and de-rotated by these expected factors, and then the resulting partially equalised pulses are superimposed to form the input to the symbol decision section. This is shown diagrammatically in Figure 14.3. Adaptive coefficient update systems can be used instead of channel sounding methods, but these tend to be less robust due to error propagation in the coefficient update section.

14.3 The Proposed Equaliser

14.3.1 Linear equaliser

When transmitting in the Rayleigh fading environment, there are a number of counter-measures that are often employed against fading. Mobile subscribers may become

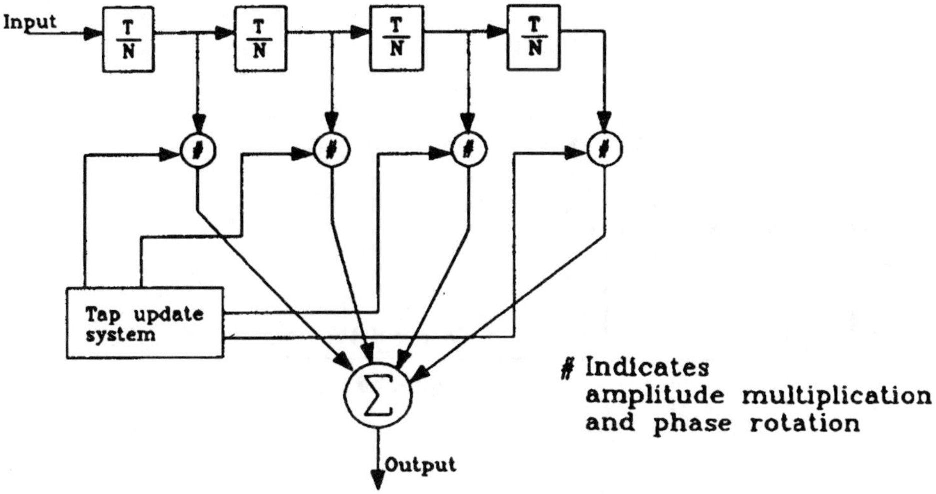

Figure 14.2: RAKE diversity combining system

stationary in a deep fade with consequent system failure. In order to combat this condition frequency hopping is often used. Accordingly, the data is divided into blocks, and at the end of each block a hop is made to a new frequency outside the coherence bandwidth of the channel. The channel is sounded at the start of the block in order to establish the impulse response of the channel at this new frequency. Sounding in the centre of the block will give better results because the impulse response of the channel cannot change so dramatically during half a block as during a whole block, and is often employed. The next block of data is then sent. The size of each block must be chosen so that the channel does not change significantly during the time the block is transmitted. However, the channel should be sounded as infrequently as possible in order to keep the data throughput high. The optimum block size is related to the mobile speed.

Simulations of adaptive linear and DFEs showed very poor performance due to catastrophic failure in fading environments. A similar problem was experienced with DFEs employing channel sounding. The linear equaliser with channel sounding offered the most promising performance.

A few slight changes are necessary as regards the channel sounding method used for binary modulation schemes in the case of QAM as the pseudo random binary sequence (PRBS) used for sounding must be mapped onto the non-binary symbol set for transmission, and both the amplitude and phase shift of each of the propagation paths must be recovered. A logical 1 from the PRBS is mapped onto the outermost phasor of the star 16-QAM constellation at0 deg and a logical 0 from the PRBS onto the outermost symbol at 180 deg. The use of the outermost symbol maximises the received SNR and by only transmitting on the I-axis, measurement of the introduced phase shift is simplified. This principle is similar to those used in the generation of

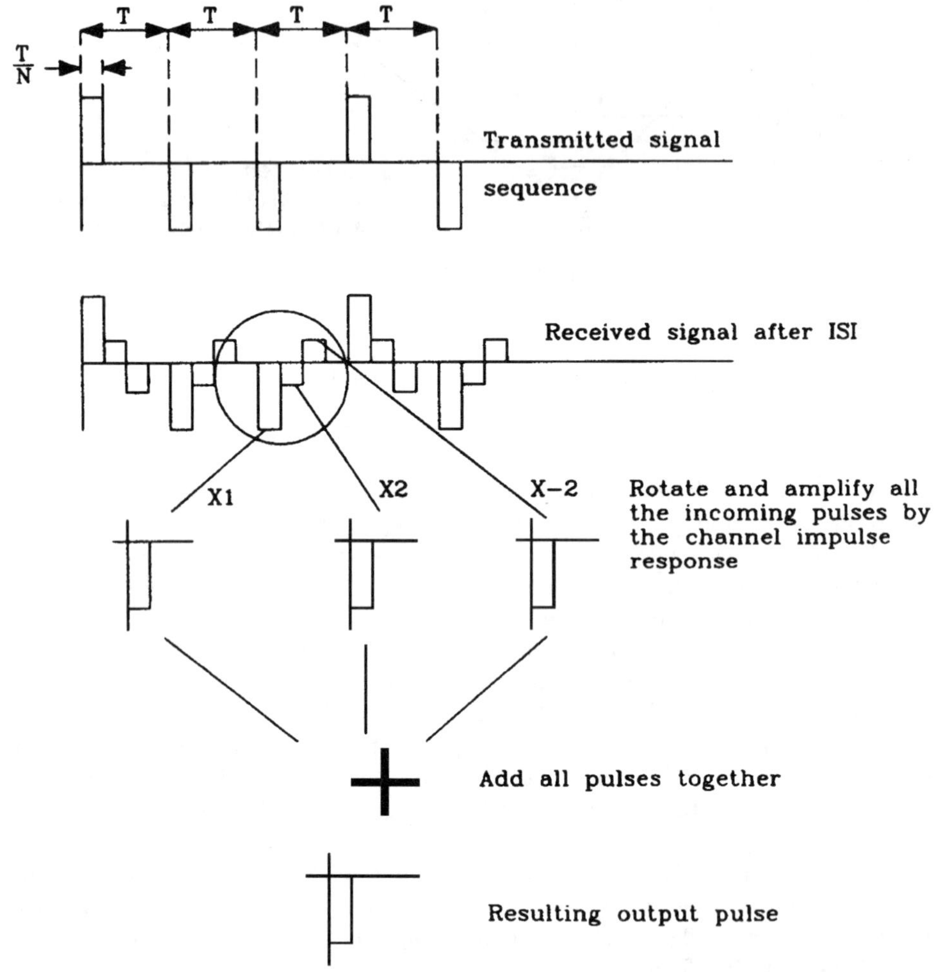

Figure 14.3: Time domain operation of the RAKE combiner

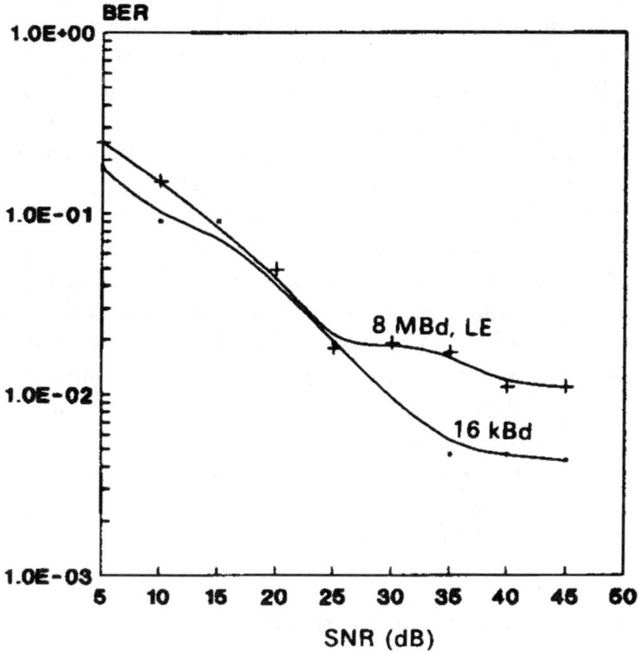

Figure 14.4: BER versus channel SNR performance of narrowband and wideband star 16-QAM at 30 mph and 1.9 GHz for 16 kBd and 8 MBd, respectively

equaliser training sequences in the CCITT V. series modems discussed in Chapter 9. Separate correlation is performed on both the I and Q axes, and the phase shift is found by taking the arctan of the results. A comparison of the BER performance of the differential 16-level star QAM modem operating at 16 kBd over a flat fading, i.e. narrowband channel, and the 8 MBd (32 Mbps) system with a LE over a frequency selective wideband channel is presented in Figure 14.4 where it can be seen that the performance of the wideband system is inferior to the narrowband system, although the difference is not substantial.

14.3.2 Iterative equaliser system

Upon studying error events using the linear equaliser it was observed in the simulations that the performance of the system would be considerably improved if it could be made more tolerant to fades which prevented the LE coefficients from decaying, as was mentioned in Chapter 7. Furthermore, the DFE is a potentially powerful equaliser if it can be prevented from failing catastrophically due to error propagation effects in the case of erroneous decisions, when incorrect information is fed back into the coefficient adjustment mechanism. A solution to this problem is to use an equaliser which exploits the advantages of both the LE and DFE but avoids their shortcomings.

The linear equaliser works satisfactorily most of the time, as previously described, except when the coefficients fail to decay. This failure to decay can be detected by comparing the values of the last few equaliser coefficients against an appropriate threshold value. If this threshold value is exceeded by the coefficient values, the decoded data from the linear equaliser will have a high error rate. A robust initial equaliser system is required which will not fail catastrophically during fades and whose output can then be further processed to reduce the BER.

In those situations where the LE failed to converge a DFE was therefore invoked which was constrained in order to prevent catastrophic failure, and a multi-stage equalisation algorithm was contrived. The first pass at equalising the channel was carried by a non-optimal equaliser based on the RAKE equaliser [164] that was described in the previous section. Although the RAKE combiner's operation is prone to errors, experience showed that these will be typically confined to neighbouring constellation points. The data regenerated by the RAKE nows has a lower BER than the received data and it can be used to keep the DFE from failing catastrophically. We refer to this non-optimal equaliser as a one symbol window equaliser (OSWE). A block diagram of the complete system proposed is shown in Figure 14.5 and its operation is discussed in depth in the following sections.

14.3.2.1 The one-symbol window equaliser

This equaliser utilises the principle that the ISI will generally exhibit itself as random noise, and calculating an appropriately weighted average of the magnitude and phase of the incoming phasors over the duration of the channel impulse response will enable it to remove this ISI. Its principle of operation is discussed below and shown diagrammatically in Figure 14.6. First the channel impulse response is estimated from the received channel sounding sequence and in Figure 14.6 three significant paths are resolved, each having an associated magnitude and phase. Next the delay bin is determined, m symbol periods after the first significant received multipath component, in which the last significant path component was received. This informs the receiver about the duration of the channel's delay spread. For each of the m delay bins a copy of the locally stored transmitted constellation is rotated by the measured rotation of the impulse response for that particular bin and held in memory as portrayed in the figure. By this means any rotation of the constellation caused by the channel can be compensated. In order to make a decision as to the rth received symbol all m received QAM phasors which include a contribution from the rth symbol are taken into account. These phasors are mapped onto the appropriately rotated constellation. Thus the QAM phasor due to the first path signal is held with the constellation rotated by the phase shift of the first path. Similarly, all other paths up to path m generate a received phasor indicated by the arrows at the bottom of Figure 14.6 for $m = 3$ consecutive signalling periods. The QAM phasor comprising the first path from the $(r + 1)$th transmitted symbol plus ISI contributions from the rth symbol would be held with the constellation rotated by the phase shift which the channel imposes on the second bin delay, and so on.

Then for each of the M possible transmitted phasors a distance summation is performed over all m rotated constellations. Specifically the distance between each

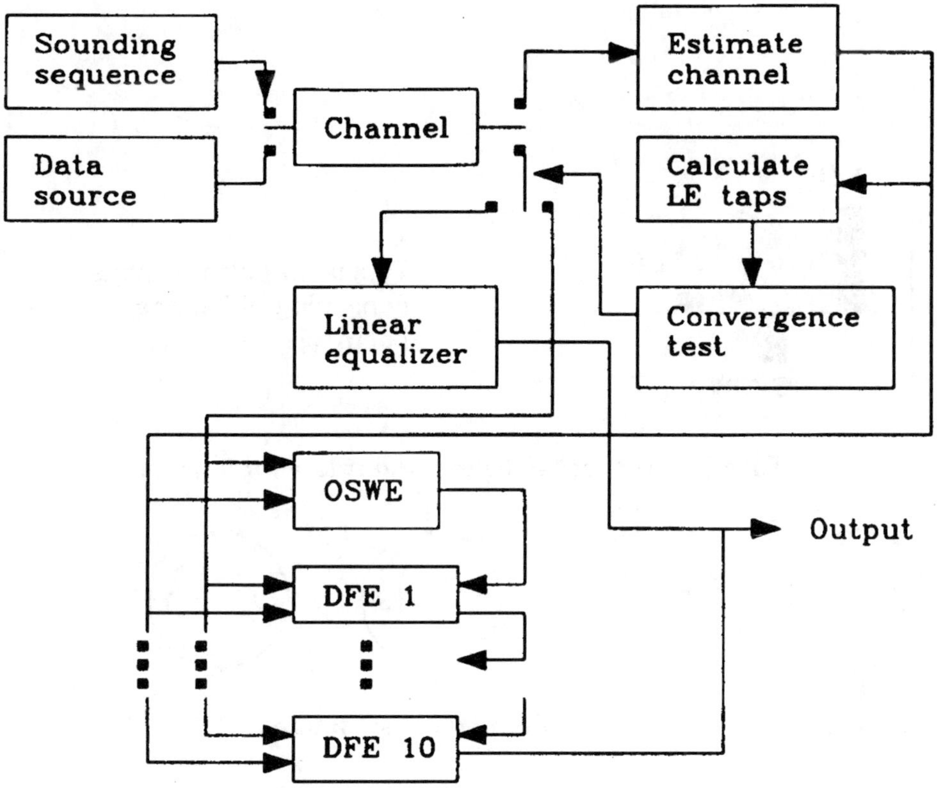

Figure 14.5: System block diagram

rotated point of the m rotated constellations to which the mth phasor has been mapped and the received symbol are found. Each distance is scaled by the ratio of the path amplitude for that delay to the path amplitude of the main tap. This scaling has the effect of reducing the importance of significantly attenuated delayed signals when regenerating the symbol and is necessary as these delayed signals will be highly corrupted by noise and ISI from other symbols.

This distance calculation is demonstrated at the foot of Figure 14.6 for $M = 4$ and $m = 3$. The Euclidean distances between the received phasors in the $m = 3$ consecutive signalling intervals, and the appropriately rotated phasor 1 are computed, giving distances of a, b and c. Weighting them with their associated path magnitude introduces the required scaling based on their significance, giving a distance of $d_1 = a + 0.5b + 0.1c$. Similarly, for phasor 2 the distance becomes $d_2 = d + 0.5e + 0.1f$, etc., up to $d_M = d_4$. Then the rotated constellation point yielding the lowest distance d is deemed to be the most likely to have been transmitted. In general terms the distance d_n, $n = 1, \ldots, M$ associated with any particular received symbol $S_n = x_t + jy_t$ at

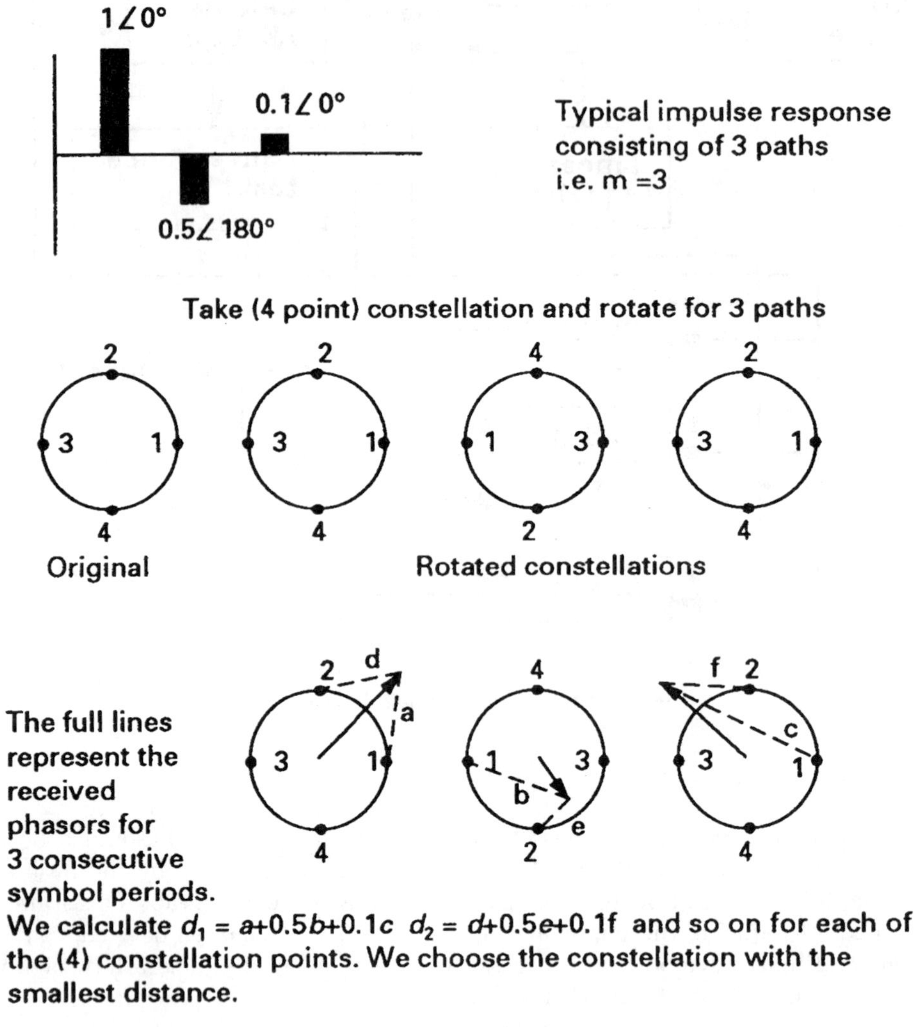

Figure 14.6: OWSE operation for $M = 4$ and $m = 3$

time t is given by

$$d_n = \sum_{i=1}^{m} \frac{a_i}{a_0} \sqrt{(x_{t+i} - u_i)^2 + (y_{t+i} - v_i)^2} \tag{14.1}$$

where m is the last significant delay bin, x_t and y_t are the Cartesian values of the received signal at time t, a_i and a_0 are the amplitudes of the measured impulse response at tap i and the main tap respectively, and u and v are the Cartesian values of the rotated constellation points and are given by

$$\begin{aligned} u_i &= r_n \cos(\theta_n + \phi_i) \\ v_i &= r_n \sin(\theta_n + \phi_i) \end{aligned} \tag{14.2}$$

r_n and θ_n are the polar amplitude and phase associated with symbol n at the transmitter, and ϕ_i is the phase shift of the ith component in the estimated channel impulse response.

Having found the total distance from between the m received phasors which include components of the rth sample and each of the M constellation points, the constellation point associated with the lowest total distance according to Equation 14.1 is selected as the most likely one to have been transmitted. This symbol along with the actual received signal levels x_t and y_t is passed to the DFE. Observe the underlying similarity between the operation of the RAKE receiver described in Section 14.2 and that of the OSWE proposed.

14.3.2.2 The limited correction DFE

The power of the DFE can be used in order to significantly reduce the error rate from the OSWE. When the LE coefficients have failed to decay, and this is detected by the convergence test block of Figure 14.5, the DFE acts on the received signal level as per normal and decides on the transmitted symbol. If this decision concurs with that of the OSWE the symbol is input to the feedback taps of the DFE and is stored in a decision register maintained for the duration of the decoding of one block. If the DFE's decision represents a neighbouring symbol in the QAM constellation to the one decoded by the OSWE then again the symbol decoded by the DFE is passed back to the feedback taps and stored as the new decoded symbol. However, should the DFE's regenerated symbol *not* be a neighbour of the OWSE-estimated symbol, the OSWE's estimated symbol is changed to one of its neighbours in such a way that it is moved nearer to the symbol position regenerated by the DFE, is input to the DFE feedback taps and stored as the new decoded symbol. This movement towards the DFE both tends to bring the DFE back towards lock and accounts for the likely inaccuracy of the OSWE during a fade.

This restriction on the symbols fed back by the DFE prevents it from failing catastrophically in all but the most severe fades. In order to improve the performance the DFE can be run iteratively a number of times using the same input signals but different feedback signals as the constraining output register changes slightly on each pass. Simulations revealed that when the LE failed completely, this iterative system typically produced a very low or zero symbol error rate. Using a knowledge of the transmitted symbol in the simulation showed that each iteration produced a lower

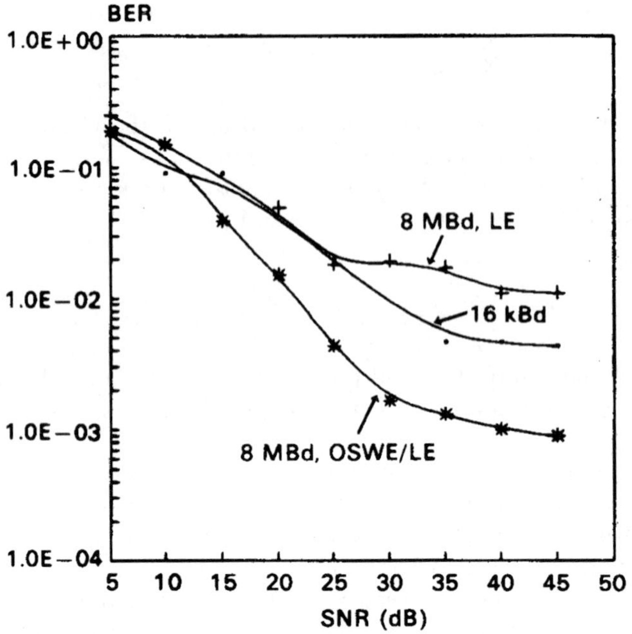

Figure 14.7: BER versus channel SNR comparison of star 16-QAM in the case of 16 kBd and 8 MBd transmission using the LE and OSWE at 30 mph and 1.9 GHz

or identical BER than the previous iteration. After approximately 10 iterations a minimum BER was reached.

The OSWE will automatically lock onto and operate on the strongest multipath component in the channel impulse response. DFEs normally have a feedforward section to allow for the strongest component not being the first path. This feedforward section is essentially a LE and the tap coefficients are set by the inverse filter process as previously explained in Chapter 7. With the combination of equalisers proposed here, it is possible to remove most of the feedforward LE section of the DFE as the OSWE has already decoded the data that would normally be decoded in the feedforward part of the DFE. The DFE operates on the untampered OSWE output, the received signal itself and the feedback data derived as previously explained. This equaliser does not have as good iterative convergence properties as when the first path is the strongest, as it has to use data from the OSWE before it can modify it, but again it shows a significant improvement in performance over the LE.

The results for this system are shown in Figure 14.7 where the performance is significantly better than that of the LE alone, and also shows an improvement over the narrowband system. This improvement over the narrowband system is because this equaliser is successfully using the inherent multipath time diversity present in the wideband channel.

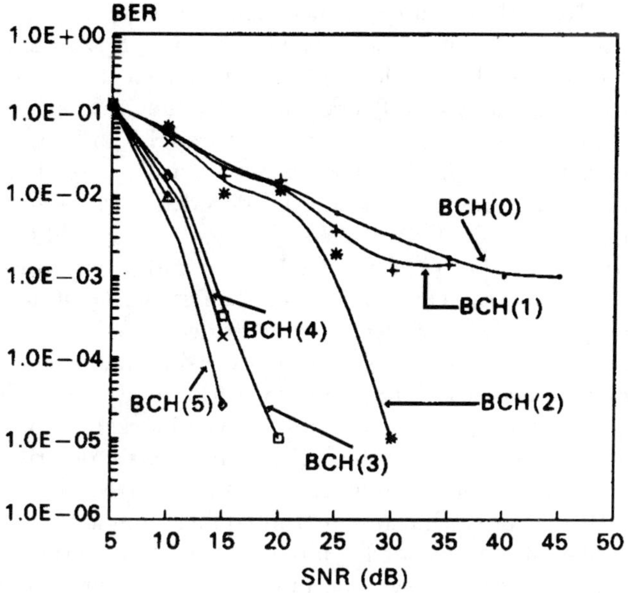

Figure 14.8: BER versus channel SNR performance of star 16-QAM at 8 MBd using the LE, the OSWE and a range of BCH FEC codes at 30 mph and 1.9 GHz

14.3.3 Use of error correction coding

The results shown in Figure 14.7 show the presence of a residual BER caused by impulse responses which cannot be equalised satisfactorily by this system. For the indoor environment, the system fails when the first component in the channel impulse response is in a fade of about 10 dB, when its magnitude becomes approximately equal to that of the other multipath components.

Error correction coding can be used to overcome the residual BER. Conventional FECs encode the bits, interleave and transmit them, perform equalisation, de-interleave the bits and then decode and correct them. The results of this strategy overlaid onto the system described in Section 14.3.2.2 are shown in Figure 14.8.

The bracketed numbers in the caption indicate the bit correction ability for a block length of 63 bits when low complexity binary BCH codecs are used. Explicitly, the codes employed are $BCH(0)$ = NoFEC,

$$BCH$$

(1) = BCH(63,57,1), BCH(2) = BCH(63,51,2), BCH(3) = BCH(63,45,3), BCH(4) = BCH(63,39,4), BCH(5) = BCH(63,36,5) and BCH(6) = BCH(63,30,6).

As it can be seen in the figure, even mild error correcting powers proved to be effective, particularly with SNRs in excess of about 20 dB. The residual BER is removed and the channel is rendered suitable for computer data transmission. Coding powers in excess of the $BCH(63, 45, 3)$ offer little extra gain for the trade-off in spectral efficiency. Note that coding reduces spectral efficiency but it has proved necessary to take this measure in most transmissions over Rayleigh fading channels.

With the system described in Section 14.3.2.2, error correction codes can be used in a more efficient manner than by merely decoding after the QAM symbol regeneration. In the same way that we use a number of passes through the DFE, we can pass the partially equalised data through the BCH decoder a number of times. If, as occurs most of the time, the LE has been used, the decoded data is stored in an array ready for de-interleaving. If the OSWE has been used then in addition to storing the decoded data in this array, both the received signal and a flag to identify that this block has been decoded by the OSWE are stored. The blocks of data, be they from the LE or the OSWE, are de-interleaved and BCH decoded.

Most de-interleaved blocks will have too many errors at this stage to allow their correction by the BCH decoder, but if any correction does occur then a marker is placed beside the corrected bits. The non-interleaved blocks which were decoded by the OSWE are then passed through the DFE as previously described but with the small modification that any bits that were marked as correct are not permitted to be altered by the DFE. Notice that differential coding cannot be used because if we have corrected a differentially decoded symbol we would know the correct difference between the previous and current symbol, but not the absolute values of these symbols. This is not a severe limitation, since differential coding is often avoided in high integrity systems because of the 3 dB SNR penalty associated with its use.

After being passed through the DFE the data is again de-interleaved and another BCH decoding pass is attempted. This process continues for as many passes as are necessary. Since a good estimate of the number of errors can be gained from the overload detector of the BCH decoder, this iterative process can be continued until either there are no errors in the array, or no more errors have been removed since the previous pass. Convergence of the DFE is improved by marking of the correct bits, helping to prevent it from making incorrect decisions. The performance of this system for the same BCH codes as before is shown in Figure 14.9. As can be seen the overall effect is very similar to the more conventional decoding method of Figure 14.8. Comparison of the two systems shows that there is a small gain of the order of 2-3 dB for most coding powers and SNRs. This is not a particularly large reward for the considerable increase in receiver complexity. This system would only be worth implementing if SNR or BER were at an absolute premium.

The reason for the gain being small is probably that although the equaliser is now forced into a better convergence, the previous system was able to remove most of the errors from an unconverged equaliser, reducing the benefit of the more complex system.

14.4 Diversity in the Wideband System

In Chapter 11 it was shown that second-order spatial diversity is beneficial in narrowband QAM. The selection or combination of the received signals from diversity receivers depends on their relative amplitudes. In the wideband system considered

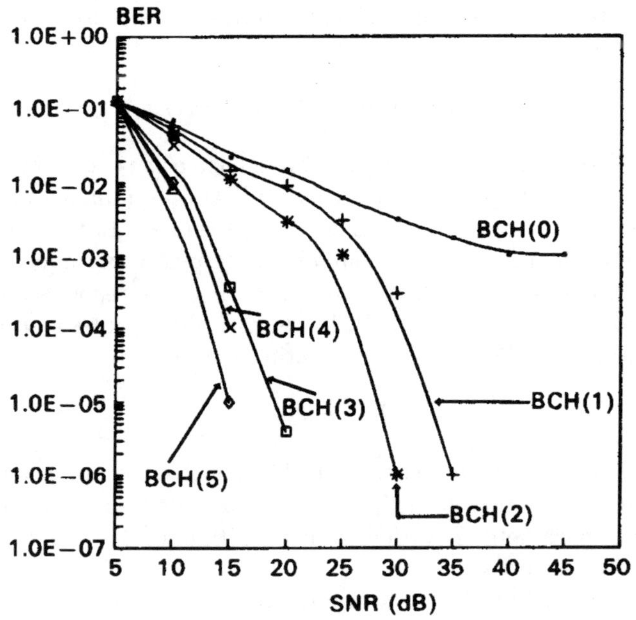

Figure 14.9: BER versus channel SNR performance of star 16-QAM at 8 MBd using the LE, the OSWE and a range of iteratively decoded BCH FEC codes at 30 mph and 1.9 GHz

here, the scenario is different. Less benefit would be expected from space diversity because the time diversity inherent in a multipath environment is used to some extent in the equalisation process. If there are two channels with different impulse responses, then to implement switched diversity reception a decision must be made as to which channel has the most favourable response. This decision is governed by the equaliser's properties.

It is necessary to produce some figure of merit that relates to the likely number of errors for a given impulse response. With switched diversity the channel with the least likelihood of errors is chosen, and with maximal ratio combining the incoming channels are weighted by the likelihood of errors. In order to identify the types of impulse responses that are more likely to yield relatively low symbol error rates, a series of simulations were performed using a channel whose impulse response has three coefficients, but in contrast to Figure 14.1 the taps weights were now varied.

The magnitude of the strongest path is designated by a_0, that of the next strongest path by a_1 and that of the weakest path by a_2. The average power of the impulse response is arranged to be a constant, and the lowest error rate was achieved when there was a dominant path a_0, and a_1 was significantly larger than a_2. For the situation when two paths were approximately equal, fewer errors were incurred when the two smaller paths were equal compared to the two larger ones being equal. The highest error rate inflicted was when all the paths had the same magnitude.

The diversity system was therefore arranged to rate an impulse response according to the system parameter

$$r = ((a_0/a_1) + (a_0/a_2)) \cdot F + (a_1/a_2) \tag{14.3}$$

where F is another system parameter which must exceed unity. The first term in r weighted by the factor F provides a measure of the dominance of the strongest path a_0, while the second unweighted term (a_1/a_2) represents the dominance of a_1 over a_2. For different channels, the one having the largest r parameter will have the most favourably shaped impulse response, and if all impulse responses have the same average power, it will give the lowest BER. In general, the impulse responses will have differing average powers and so the channel rating is modified to

$$R = a_0 r \tag{14.4}$$

as r deals in ratios and is only concerned with the shape of the impulse response. Due to multiplying r by the dominant coefficient a_0, r is scaled appropriately. The BER was examined as the parameter F varied from 1 to 10, and was found to be relatively insensitive to the value selected.

Armed with an algorithm R to rate the quality of a channel's impulse response, second-order diversity simulations can be performed. The two diversity channels had impulse responses with three coefficients each. All three coefficients had magnitudes and phases which conformed to independent Rayleigh fading statistics. The PRBS sounding sequence was transmitted over both channels for each block of data, and the channel impulse responses were estimated. The diversity branch whose impulse response had the larger R parameter was used in our switch diversity arrangements.

The variation of BER as a function of channel SNR is displayed in Figure 14.10 for this diversity system with a data rate of 8 MBd, i.e. 32 Mbps. A benchmark for the non-diversity condition is also shown. For a BER of 10^{-2} (no FEC is used here) the diversity yielded a reduction in SNR of approximately 6 dB. For comparison it shows the performance of the system which selects the incoming channel not by using R of Equation 14.4, but merely by selecting the impulse response with the strongest average power, i.e. rating the impulse responses on the basis of

$$\hat{R} = a_0^2 + a_1^2 + a_2^2 \tag{14.5}$$

The curves in Figure 14.10 show that despite the inherent diversity in the multipath signal which was exploited by the equaliser system of Figure 14.5, there were further significant gains to be made by introducing second-order spatial diversity. The method of selecting the impulse response using Equation 14.4 yielded the profile switch curve. This performed better than the conventional method of selecting the signal with the highest average power using $\hat{R}$ of Equation 14.5, the performance of which is labelled as the power switch curve in Figure 14.10.

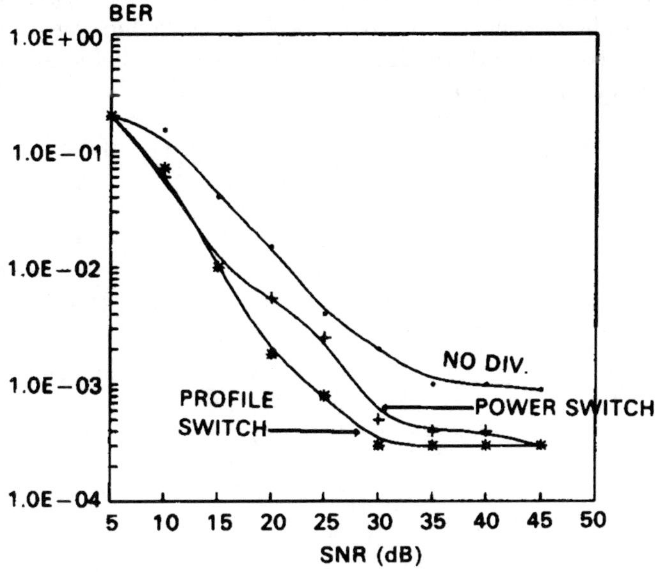

Figure 14.10: BER versus channel SNR performance of star 16-QAM at 8 MBd using the LE, the OSWE and a range of diversity combining techniques at 30 mph and 1.9 GHz

14.5 Summary

At a transmission frequency of 1.9 GHz, data at a rate of 8 Msymbols/s, corresponding to 32 Mbits/s, can be sent via star 16-level QAM over wideband dispersive channels at an integrity acceptable in many applications. In most instances a linear equaliser worked satisfactorily, but when the main component of the impulse response was in a deep fade it was advantageous to carry out a first-pass equalisation with a non-optimal equaliser and then use the partially equalised signal to prevent a DFE from failing catastrophically. When this system was used in an iterative fashion, considerable improvements in BER were achieved. The system discussed incorporates three types of equalisers in a concatenated arrangement with switching between them as conditions in the fading channel change. Improvements in BER of an order of magnitude over a linear equaliser system were obtained for SNRs in excess of 15 dB.

Second-order spatial diversity gave about 5-7 dB gain in terms of channel SNR over an equivalent uncoded system using no diversity providing that it was implemented in an intelligent way, leading to a BER of 3×10^{-4} at a SNR of 30 dB with no channel coding.

The use of channel coding provided very significant reductions in BER. With an approximately 2/3 rate BCH code, BERs as low as 1×10^{-5} were achieved for SNRs as low as 20 dB, and when an iterative channel decoding approach was adopted BERs of 1×10^{-6} were realised for the same coding power and SNR, but at the cost of a considerable increase in complexity.

In the next chapter we consider an alternative means of transmitting at high bit rates over mobile radio channels. This involves dividing the message to be transmitted into a number of streams and sending each of these streams simultaneously. It is known as orthogonal multiplexing.

Chapter 15

Quadrature-Quadrature AM

15.1 Introduction

A variety of other orthogonal signalling functions were proposed in the literature [369–371], where a number of messages can be multiplexed for transmission in the same bandwidth without interfering with each other. Orthogonal multiplexing can potentially improve the bandwidth efficiency of a communications system. One of the simplest manifestations of orthogonal multiplexing is conventional QAM, where information is transmitted using two orthogonal carriers - the in-phase and quadrature channels - within the same bandwidth. The in-phase and quadrature channels constitute two dimensions of an orthogonal space. There are a plethora of methods that can be invoked for generating *orthogonal dimensions*, which can be used for uniquely recoverable telecommunications signalling. Using a set of orthogonal modulation elements or signalling pulses $m_T(t)$ can create a further dimension for bandwidth efficient orthogonal signalling. In conventional modulation schemes, bandlimited pulses are used, but there is considerable freedom in the choice of the pulse shape. These issues will form the topic of this chapter in the context of quadrature-quadrature amplitude modulation (Q^2AM). As an introduction to of these systems, Q^2PSK [372] is examined first.

15.2 Quadrature-Quadrature Phase Shift Keying

We have seen in previous chapters that by moving from BPSK to QPSK, spectral compactness is increased. This is achieved by transmitting two bits per modulation symbol, which allows us to expand the BPSK signalling interval of T to $2T$, thereby halving the bandwidth requirement whilst maintaining the same bit rate. Alternatively, one could double the bit rate in the same bandwidth. This is because the number of signalling dimensions used in the modulation scheme is increased. BPSK can be considered to only have a single dimension - the in-phase or I channel. QPSK then has two dimensions, the in-phase and quadrature channels, or I and Q. A signal space limited in time to an interval τ and to a one-sided bandwidth W has $2\tau W$ possible dimensions [372]. For example, in QPSK the symbol duration τ is 2T, where T is the period of the bits in the incoming data stream, i.e. the signalling interval duration in BPSK. If the channel is bandlimited to $1/2T$ on each side of the carrier then the one-sided bandwidth occupancy is $W = 1/T$. The number of dimensions available within this bandwidth is $2\tau W = 4T(1/T) = 4$. Only two of these dimensions are utilised in QPSK. In theory, this bound on the number of dimensions can be reached by using the so-called prolate spheroidal wave functions but these have been shown to be unrealisable [373]. There may be other ways to approach this bound, ways which can be realised. By moving from QPSK to Q^2PSK it becomes possible to exploit all four possible dimensions.

For unshaped infinite bandwidth QPSK the input data consists of pulses $a(t)$ having period T and amplitude ± 1 which is equivalent to using the rectangular modulation elements $m_T(t) = rect(t/T)$ of the previous chapter. The incoming data stream of pulse duration T is demultiplexed into two parallel streams, $a_1(t)$ and $a_2(t)$, consisting of alternative pulses of duration $2T$. These streams are multiplied by sine and cosine carriers and summed to form the QPSK signal

$$\begin{aligned} S_{QPSK}(t) = & \frac{1}{2T}a_1(t)cos\,(2\pi f_c t) \\ & +\frac{1}{2T}a_2(t)sin\,(2\pi f_c t)\,. \end{aligned} \tag{15.1}$$

In offset QPSK (OQPSK) the bit stream $a_2(t)$ is delayed by a fraction of the symbol period $2T$, normally half, i.e. by T, in order to mitigate instantaneous changes in the carrier phase, as portrayed by Figure 15.1(a).

Although the power spectral densities of QPSK and OQPSK are identical, they react to bandlimiting and non-linear amplification differently. As opposed to QPSK, OQPSK cannot have a 180 deg phase shift, which could be caused by the coincident changes in $a_1(t)$ and $a_2(t)$. Thus OQPSK has a lower envelope fluctuation than QPSK and hence non-linear amplification will not regenerate the high frequency side lobes removed by filtering. Avoiding the above rectangular transition completely, the data streams $a_1(t)$ and $a_2(t)$ can be shaped by cosine and sine functions, respectively, which modulate the quadrature carriers, yielding:

$$\begin{aligned} S(t) = & \frac{1}{T}a_1(t)cos\left(\frac{\pi t}{2T}\right)cos(2\pi f_c t) \\ & +\frac{1}{T}a_2(t)sin\left(\frac{\pi t}{2T}\right)sin(2\pi f_c t). \end{aligned} \tag{15.2}$$

Note, however, that the modulation pulses $\cos(\frac{\pi t}{2T})$ and $\sin(\frac{\pi t}{2T})$ shown in Figure 15.1(b) are different from the more conventional raised cosine pulse introduced as non-linear filtering (NLF) in the time domain (Chapter 4).

Observe in Equation 15.3 that we have introduced two orthogonal pulse shaping

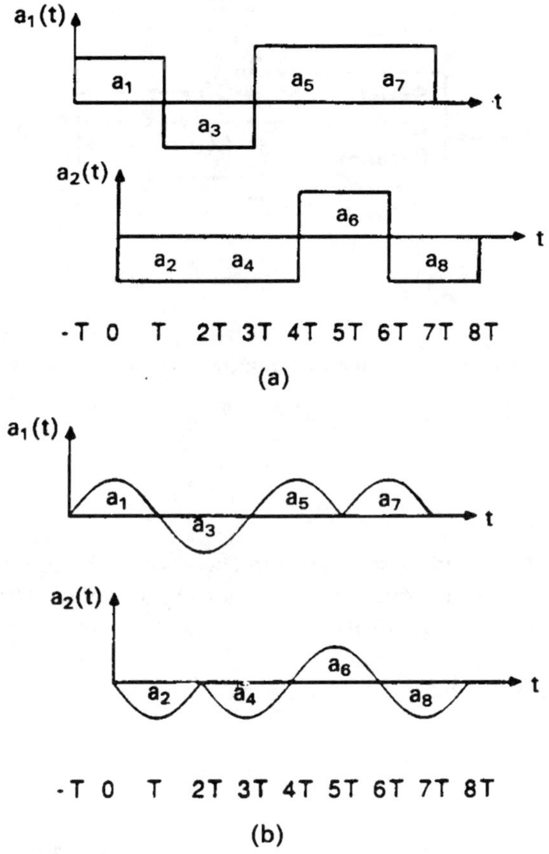

Figure 15.1: Pulse shaping: (a). rectangular, (b). sinusoidal and cosinusoidal [372] ©IEEE, 1989, Saha

functions, namely:

$$p_1(t) = sin(\pi t/2T) \tag{15.3}$$

and

$$p_2(t) = cos(\pi t/2T) \tag{15.4}$$

respectively. Armed with these functions the following set of orthogonal basis functions can be formulated:

$$\begin{aligned} s_1(t) &= p_1(t)sin2\pi f_c t \\ s_2(t) &= p_2(t)sin2\pi f_c t \\ s_3(t) &= p_1(t)cos2\pi f_c t \\ s_4(t) &= p_2(t)cos2\pi f_c t. \end{aligned} \tag{15.5}$$

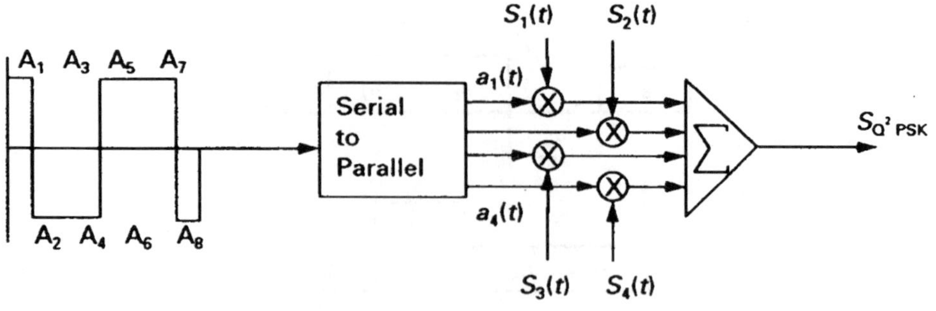

Figure 15.2: Structure of a quadrature-quadrature phase shift keying modulator

It can be seen that between any two signals in the above set there is a common factor which is either a data shaping pulse or a carrier component. The remaining factor in one is orthogonal to, or in quadrature with respect to the remaining factor in the other.

The modulated Q^2PSK signal is formulated as

$$S_{Q^2PSK} = \sum_{i=1}^{4} a_i(t)s_i(t) \tag{15.6}$$

and the block diagram of the modulator is shown in Figure 15.2.

The waveforms of the components $a_i(t)s_i(t), i = 1, 2, 3, 4$, are displayed in Figure 15.3 for the input pulses A_1, A_2, ..., A_8, where A_n is the nth binary input pulse with value -1 or 1, and whose polarities are shown in Figure 15.2.

The waveforms are displayed for two symbol periods, remembering that this modulation scheme encodes four bits per symbol. For the first symbol period the output from the serial-to-parallel convertor $a_1(t)$ is A_1. In the second symbol interval $a_1(t)$ is A_5, and in general $a_1(t) = A_{4i+1}, i = 0, 1, \ldots,$. The waveform at the top of Figure 15.3, $a_1(t)p_1(t)$, it composed of two positive sine half-cycles, corresponding to positive pulses A_1 and A_5. The waveform for $a_2(t)p_2(t)$, is composed of a negative cosine half-cycle as A_2 is -1, followed by a positive cosine half-cycle as A_6 is $+1$. Since the serial-to-parallel convertor processes four bits at a time, compared to two for QPSK, twice the amount of data per modulation interval is accommodated. Indeed, without any bandlimiting, Q^2PSK can obtain twice the throughput of QPSK at the same SNR and BER [372]. However, due to the sudden changes in the modulating signal as for example $a_4(t)p_2(t)$ in Figure 15.3, when the signal is bandlimited this statement is invalid. Bandlimiting Q^2PSK transmissions over AWGN channels requires an increase of 1.6 dB transmitted power in order to provide an identical BER performance to that of QPSK for the same channel bandwidth [372].

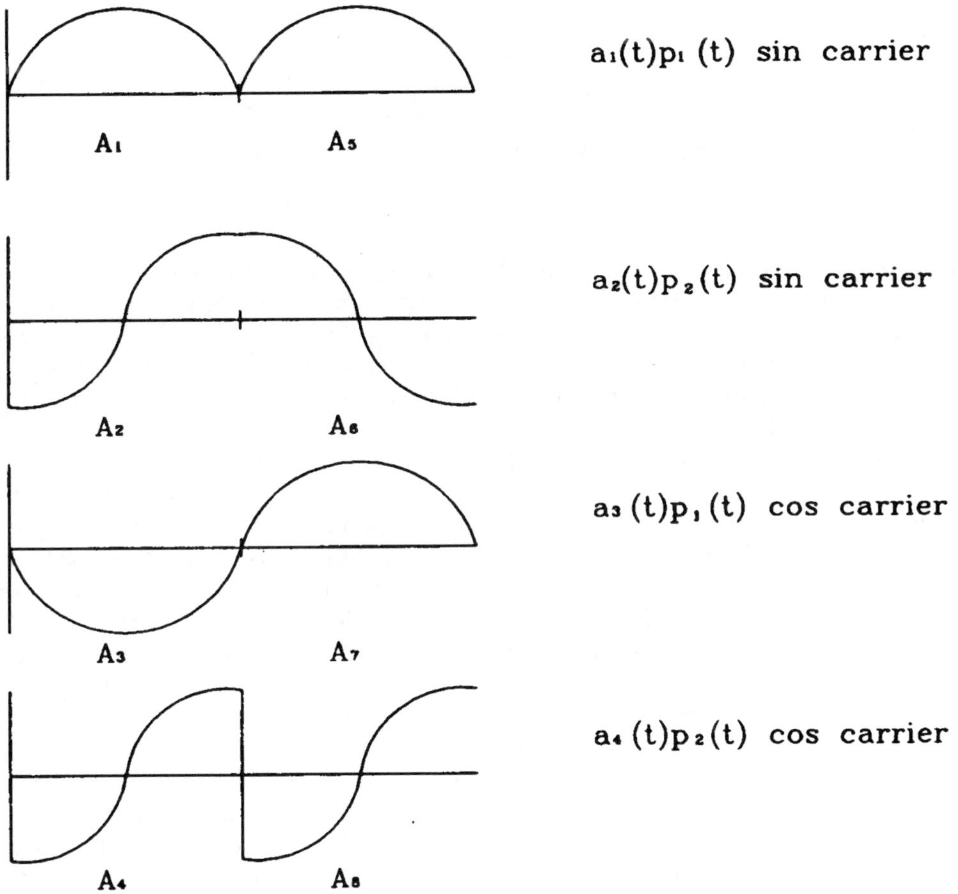

Figure 15.3: Pulse shapes for the Q^2PSK modulator

15.3 Quadrature-Quadrature Amplitude Modulation

15.3.1 Square 16-QAM

In this section the modifications necessary to include orthogonal data shaping pulses in square and star QAM are discussed.

In our approach, which is a direct expansion of Q^{2}PSK, we take four data bits for each of the I and Q channels. The structure of the modulator for this scheme is shown in Figure 15.4 and the corresponding waveforms in Figure 15.5. The four bits for the I channel, which are A_1 to A_4 for the first symbol and A_9 to A_{12} for the second symbol in Figure 15.4 are split into two groups. The first group consists of

a_n	a_{n+1}	Output
0	0	-1
0	1	-3
1	0	+1
1	1	+3

Table 15.1: Encoding scheme for Figure 15.4

$a_1(t) + a_2(t)$ where $a_1(t) = A_1$ and $a_2(t) = A_2$, representing the first two bits of an eight bit symbol. Each group of two adjacent input bits passes in parallel through a block labelled ENCODE in Figure 15.4, which takes the two binary bits and encodes them onto one of four levels, ± 1 or ± 3, corresponding to the valid amplitude levels in a square 16-QAM constellation. The mapping is Gray encoded and shown in Table 16.1 where $n = 1, 3, 5, 7$.

Extending the assignments for $s_i(t)$ in Equation 15.5 to 16-QAM by allowing not only the sign but also the magnitude of the modulation pulses to change, the amplitude derived from $a_1(t)$ and $a_2(t)$ can be encoded onto the sinusoidal pulse shape on the sinusoid, i.e. the I carrier, while that derived from $a_3(t)$ and $a_4(t)$ can be encoded onto the cosinusoidal pulse shape on the I carrier as suggested by the waveforms of Figure 15.5. The same procedure applies to the Q channel. Namely, the amplitude derived from bits $a_5(t)$ and $a_6(t)$ is encoded onto the sinusoidal pulse shape on the cosinusoid i.e., the Q carrier, and that derived from bits $a_7(t)$ and $a_8(t)$ is encoded onto the cosinusoidal pulse shape on the Q carrier.

In Figure 15.5 the same principle has been used as for Figure 15.3. Here $A_1 + A_2$ has been used to signify the amplitude derived by the mapping described previously from the bits A_1 and A_2, etc. Therefore, considering the uppermost waveform, from Figure 15.4 it can be seen that $A_1 = 1$ and $A_2 = 0$, where negative pulses are considered to have a logical value of 0. Consulting Table 15.3.1 shows that the output of the encoder has an amplitude of $+1$, consequently a positive half-cycle of a sine wave with amplitude 1 results. The next symbol period uses A_9 and A_{10}, which are both logical 0; Table 15.3.1 shows the output is -1, and accordingly a sine half-wave of amplitude -1 is observed. This allows 8 bits per symbol to be encoded onto a 16-level QAM constellation.

15.3.2 Star 16-QAM

The square constellation Q^{2}AM with 8 bits/symbol maps almost directly onto the star 16-QAM system. Instead of considering the I and Q channels separately, the system is considered to have two separate differential encoders. This is shown diagrammatically in Figure 15.6. The incoming pulse stream is split into two streams, each of which has 4 bits per symbol. Both streams are fed to identical differential encoders which both internally derive an amplitude and phase angle representing their differentially encoded data. The amplitude is designated $Z_1(t)$ and $Z_2(t)$ for differential coders 1 and 2, respectively, and the corresponding phases are $\theta_1(t)$ and $\theta_2(t)$.

The encoding scheme differentially encodes the first bit onto the amplitude and

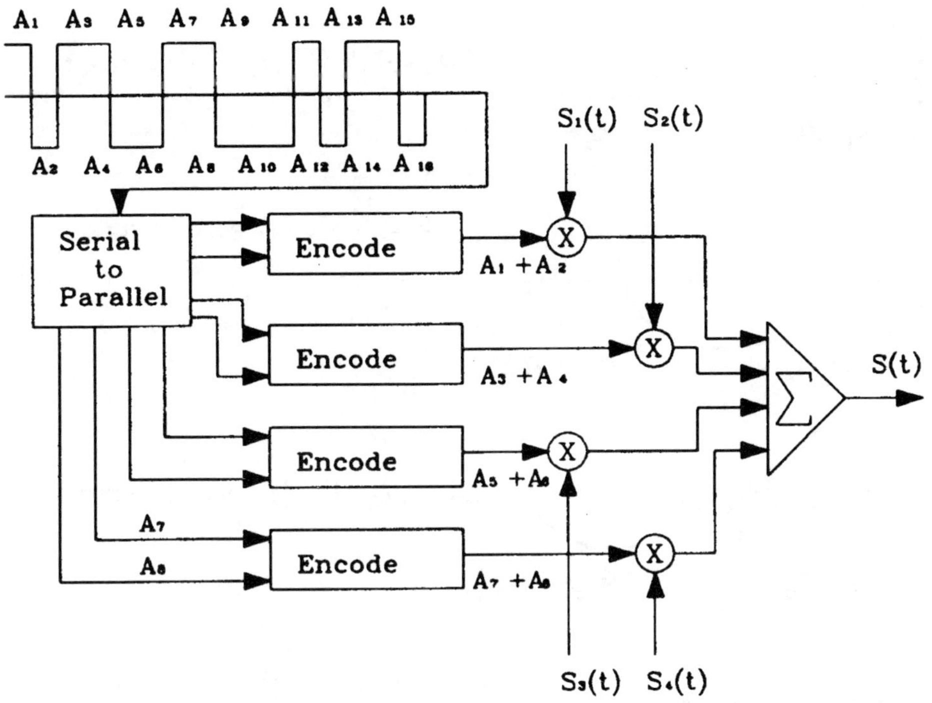

Figure 15.4: Structure of a quadrature-quadrature amplitude modulator

differentially Gray encodes the remaining three bits onto the phase. This information is then converted into two pairs of values representing I and Q magnitude for each differential coder. Thus $I_1(t) = Z_1(t)\sin\theta_1(t)$, $Q_1(t) = Z_1(t)\cos\theta_1(t)$ and so on. Note that the star 16-QAM system allows nine possible values of $I(t)$ and $Q(t)$. $I_1(t)$ is then encoded onto a sinusoidal pulse while $I_2(t)$ from differential encoder 2 is encoded onto a cosinusoidal pulse. These are then added and multiplied by $\cos w_c t$ to give the I carrier. A similar procedure is carried out for the Q channel.

At the receiver there are four matched filters, two with sinusoidal impulse responses and two with cosinusoidal impulse responses. These form two pairs used for the I and Q channels, each pair consisting of a sine and cosine filter. After the I and Q channels are separated by the quadrature mixers, each is simultaneously applied to the matched sine and cosine filter pairs for the appropriate quadrature arms. Because the encoding consisted of a sine and cosine pulse on each channel, by this matched filtering the pulses with their appropriate magnitude and polarity can be separated. The outputs of these matched filters are fed to the appropriate differential decoders which are configured so as to reconstruct the original data stream.

Simulations using this 16-Q^{2}AM system based on different orthogonal modulation pulses in an infinite bandwidth channel gave an identical BER to a 16-QAM system using a specific sinusoidal pulse shaping over one symbol alone. It should be noted

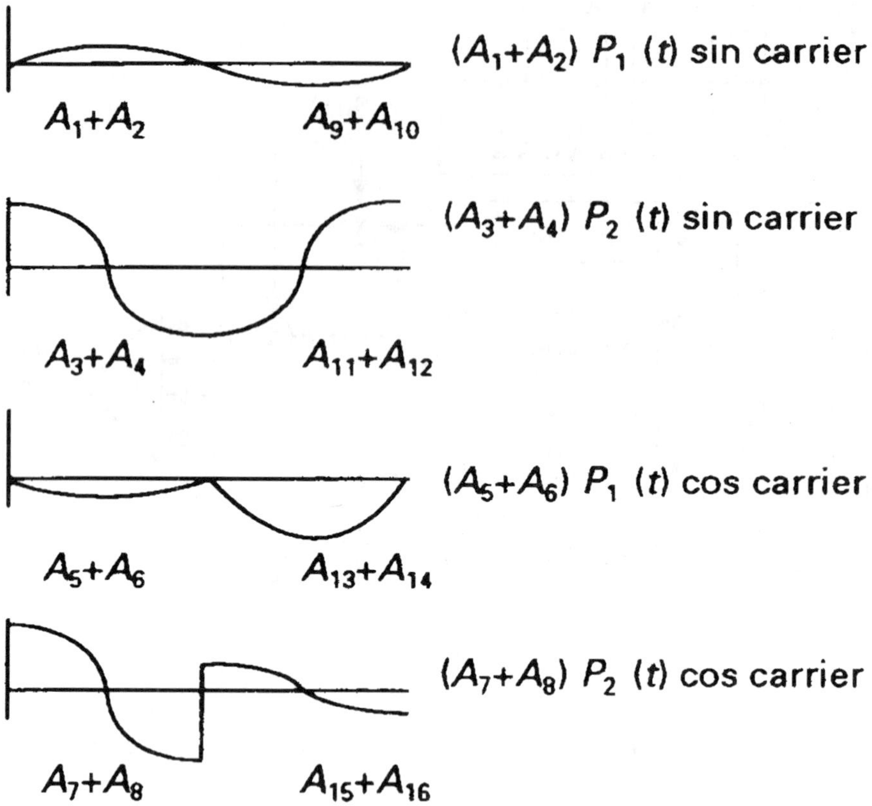

Figure 15.5: Pulse shapes for the Q^2AM modulator

that such sinusoidal pulse shaping would not normally be used; we would employ some form of raised cosine pulse shaping [182]. Conventional pulse shaping techniques have been considered in Chapter 4.

15.4 Spectral Efficiency

As expected, there is a penalty to pay for the doubling of the data throughput. The sharp changes caused mainly by the step change in the cosinusoidal pulse, as can be seen in a typical time domain waveform shown in Figure 15.7, cause considerable spectral spreading.

The spectral efficiency of Q^2AM can be improved by using different orthogonal pulse shapes, and then independently filtering these pulse shapes in order to reduce the sharp changes at modulation instants necessary in order to achieve pulse shape orthogonality. In Reference [372] a number of transmit and receiver filter pairs are suggested for this purpose. It is easy to smooth the sinusoidal pulse, but there are

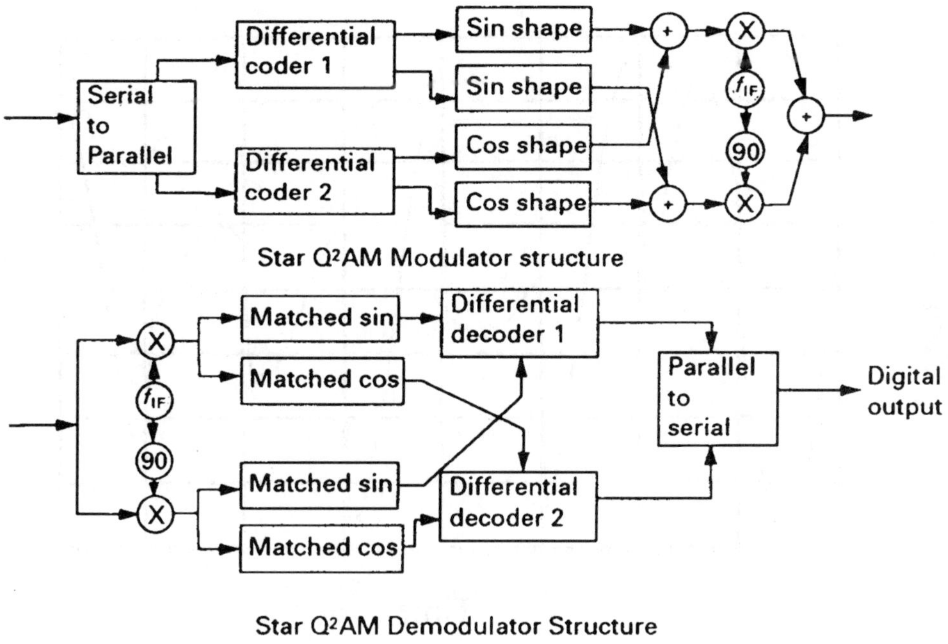

Figure 15.6: Star Q^2AM modulator and demodulator structure

severe difficulties with the cosinusoidal pulse due to the sharp changes that occur at the end of some symbol periods. When bandlimiting filtering is carried out it "smears" these changes significantly, considerably distorting the modulated signal waveform and increasing the BER. Essentially this filtering reduces the orthogonality between the data pulses. QAM is more sensitive than QPSK as more information has been encoded onto each pulse, and severe bandlimiting renders this information inaccurate.

15.5 Bandlimiting 16-Q^2AM

In order to make an accurate comparison between previous 16-QAM systems and 16-Q^2AM systems we need to compare their bandlimited performance.

Three bandlimited filtering systems are examined. The first system was the raised cosine pulse shaping of the data bits over five symbols prior to modulation as suggested in Reference [182]. This makes use of a generalised equation for the filter impulse response given as

$$g(t) = 1 - A\cos(wt) + B\cos(2wt) - C\cos(3wt)$$
$$0 < t < LT \qquad (15.7)$$

where $w = 2\pi/LT$ and LT is the duration of the impulse response. The recommended

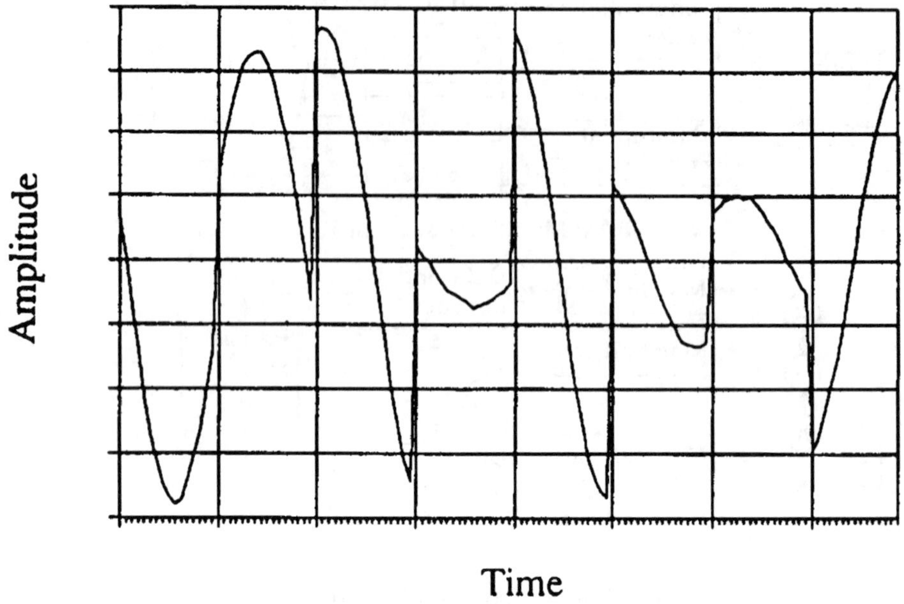

Figure 15.7: Q^2AM I channel time domain waveforms

coefficients for the transmit filter are given in Reference [182] as $A = 2.0$, $B = 1.823$, $C = 0.823$ and $L = 5$. The receiver filter is the corresponding matched filter for this response, but as the response is symmetrical about $t = 0$ it has an identical impulse response. Also considered was the non-linear filtering (NLF) system as suggested by Feher [181] and discussed in Chapter 4. This system fits a raised cosine segment between adjacent symbols if they have differing amplitudes, and a fixed level if they are of the same magnitude. Finally Q^2AM with the previously described simple unsmoothed sinusoidal and cosinusoidal pulse shapes was also included as a benchmark.

Since Q^2AM has twice the throughput of QAM, it can be considered a superior system if it can operate with the same BER as QAM but in less than *twice* the bandwidth. Indeed, it could be argued, especially at high bit rates, that Q^2AM is superior if it can operate with the same BER in just over twice the bandwidth, as the lower symbol rate will tend to reduce the effects of intersymbol interference (ISI) and thus dispense with equalisation or render it simpler and more accurate.

15.6 Performance of Bandlimited 16-QAM and 16-Q^2AM

Initial simulations using our star QAM modem were performed at a transmission rate of 2 MSymbols/s over a flat fading channel where intersymbol interference (ISI) did

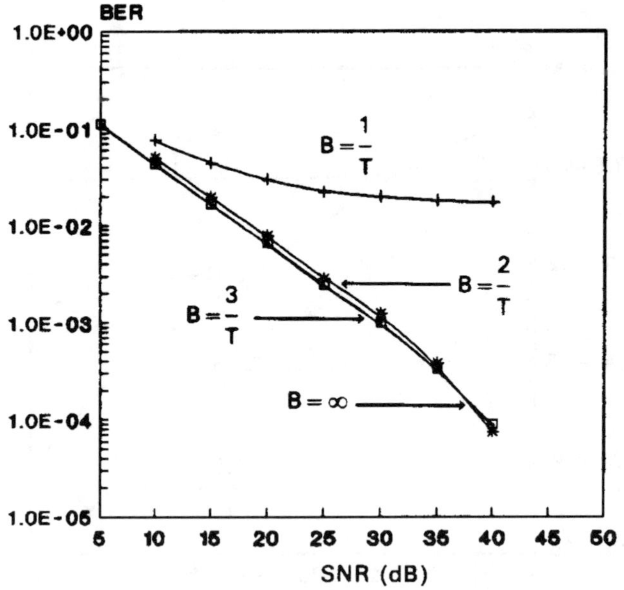

Figure 15.8: BER versus channel SNR performance of star 16-QAM at 30 mph, 1.9 GHz, 2 Msymbols/s using raised cosine filtering and various lowpass bandwidths

not occur. This situation would correspond to a microcellular or indoor environment. Lower transmission rates than 2 Msymbol/s merely had the effect of increasing the residual BER as shown in previous chapters due to the inability of the differential detector to track the rapidly changing fading envelope. The vehicular speed of the mobile station (MS) was assumed to be 30 mph. Higher MS speeds increased the residual BER, while lower speeds reduced it. The MS speed could also be directly traded off against transmission rate, so if we were to reduce our MS speed by half to 15 mph, we would be able to reduce our transmission rate by half to 1 Msymbol/s, and get identical results (assuming no ISI). The carrier frequency was 1.9 GHz.

The performance of the raised cosine filtering is shown in Figure 15.8. In Reference [182] the bandwidth of QPSK with this pulse shaping is quoted as being $1.2/T$, where T is the symbol rate. That this bandwidth is approximately true for star 16-QAM is substantiated by the results shown. When the bandwidth of the equipment filters was 3 dB down at $B = 1/T$, the performance decreased dramatically from the infinite bandwidth case. When the 3 dB point was moved to $B = 2/T$, the performance was only about 1 dB worse than the infinite bandwidth case, and by increasing the 3 dB point to $B = 3/T$, the performance of the infinite bandwidth system was essentially achieved, as shown in Figure 15.8.

Figure 15.9 shows the performance of the NLF 16-QAM system under the previous propagation conditions with various amounts of bandlimiting. Note that the BER of the non-linearly filtered but otherwise unbandlimited 16-QAM scheme was significantly higher than the case of raised cosine (RC) pulse shaping. This was be-

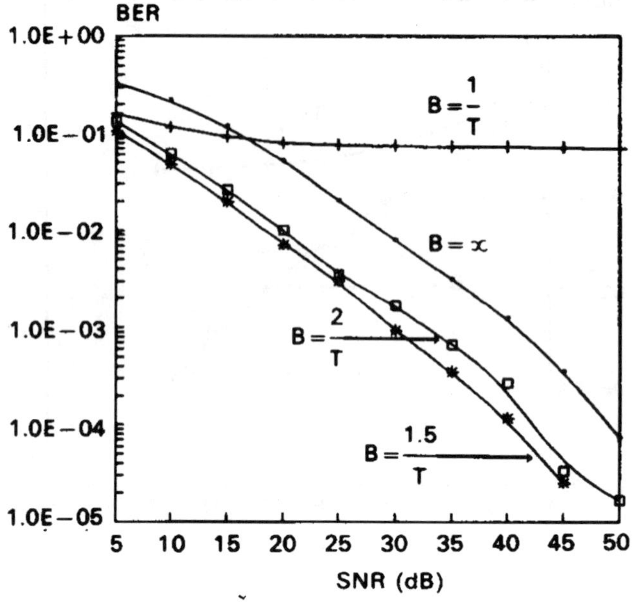

Figure 15.9: BER versus channel SNR performance of star 16-QAM at 30 mph, 1.9 GHz, 2 Msymbols/s using raised NLF filtering and various lowpass bandwidths

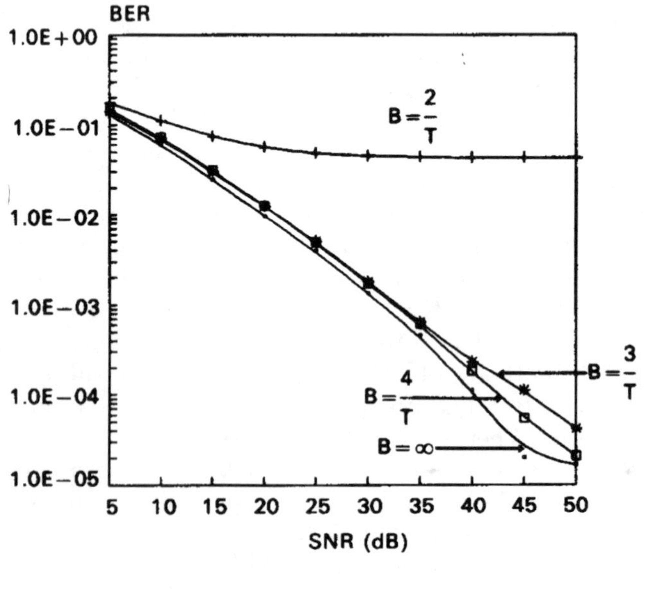

Figure 15.10: BER versus channel SNR performance of star 16-Q^{2}AM at 30 mph, 1.9 GHz, 2 Msymbols/s using various lowpass bandwidths

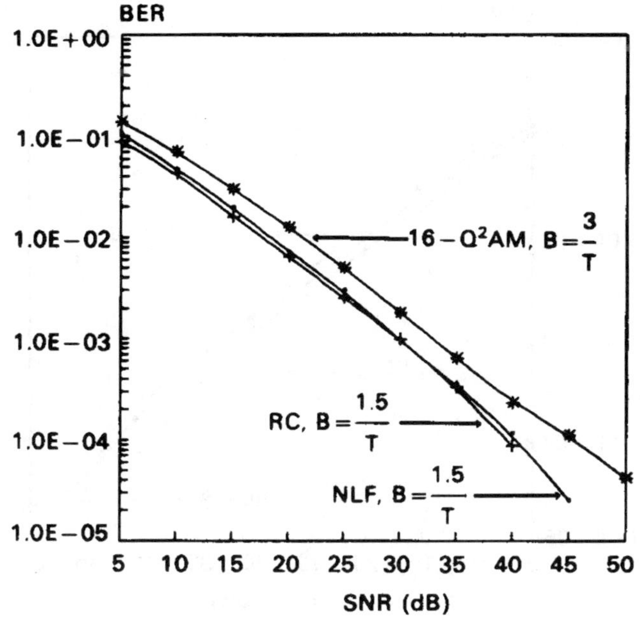

Figure 15.11: BER versus channel SNR comparison of star 16-QAM and 16-Q^2AM at 30 mph, 1.9 GHz, 2 Msymbols/s using filtering at $B = 1.5/T$ and $B = 3/T$

cause of the use of a matched filter in the raised cosine filtered receiver compared to simple sampling in the NLF 16-QAM scheme. NLF schemes cannot employ matched filtering, instead they must rely on lowpass filtering of the noise at the receiver. In our simulations we used the same receive filter as transmit filter. There was an optimum filter bandwidth for the case where the lowpass filter removes most of the noise without smearing the pulses significantly. It can be seen from the figure that this occurred at a 3 dB down cut-off frequency of about $B = 1.5/T$. When the 3 dB point was decreased to $B = 1/T$, significant performance degradation occurred.

Figure 15.10 shows the performance of the 16-Q^2AM system with various amounts of bandlimiting. The performance was approximately that of the infinite bandwidth channel for the 3 dB point above $B = 2/T$, but when this point was reduced below $B = 2/T$ the BER became unacceptable.

In Figures 15.11 and 15.12 we compare the performance of the three different filtering systems all operating at *identical bit rates.* This means that the symbol rate of Q^2AM is half that of the other two systems. Hence in Figure 15.11 we compare the performance of raised cosine and NLF with the 3 dB point at $B = 1.5/T$ to Q^2AM with the 3 dB point at $B = 3/T$. This comparison is carried out in order to show how Q^2AM performs relative to a near-optimally filtered 16-QAM modem using raised cosine pulse shaping and a practically filtered QAM modem using the NLF system. The same format is adhered to in Figure 15.12 where we consider 3 dB bandwidths

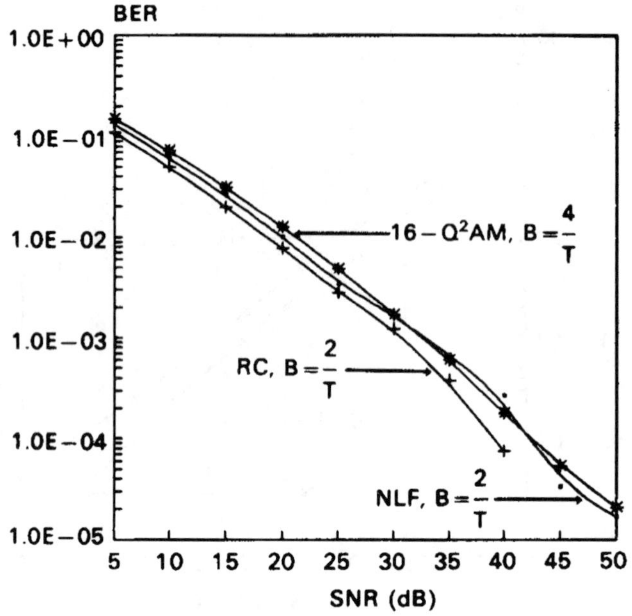

Figure 15.12: BER versus channel SNR comparison of star 16-QAM and 16-Q^2AM at 30 mph, 1.9 GHz, 2 Msymbols/s using filtering at $B = 2/T$ and $B = 4/T$

of $B = 2/T$ and $B = 4/T$ for the various systems. When bandlimited to $B = 2/T$ and $B = 4/T$ the performance of all three systems is virtually identical, although the Q^2AM scheme is slightly down on the other two. When the bandlimiting is tightened to $B = 1.5/T$ and $B = 3/T$, a level at which our preferred NLF scheme performs best, the performance of the Q^2AM scheme deteriorates to some 3 dB below that of the other two schemes. This is more significant and it suggests that Q^2AM is only worth implementing if ISI is causing serious problems.

15.7 Summary

This chapter considered how quadrature-quadrature amplitude modulation can be successfully applied for transmissions over Rayleigh fading channels. This scheme is termed Q^2AM, and for a given symbol rate Q^2AM requires approximately twice the channel bandwidth of QAM but operates at twice the bit rate. Alternatively, for the same channel bandwidth, Q^2AM operates at half the transmission rate compared to QAM, causing any ISI to be spread over fewer symbols. Decreasing the symbol rate by a factor of 2 may result in Q^2AM not needing equalisation when QAM does. In this case Q^2AM is advantageous. Otherwise it can be concluded that the necessary increase in complexity and decrease in performance is not warranted.

In the next chapter we go on to further examine the issue of spectral efficiency which was touched upon in this chapter when the need to compare different schemes within the same bandwidth was described. The next chapter considers the spectral efficiency of QAM when employed in cellular communications and compares it to other modulation schemes.

Chapter 16

Spectral Efficiency of QAM

16.1 Introduction

Within fixed communications links where co-channel interference is often not significant, the increase in efficiency, or the increase in the number of, say, speech channels available when QAM is used in place of lower efficiency modulation is often readily found. A parameter is defined known as the channel efficiency which details the number of channels that can be provided within a given bandwidth. For two systems, one using 1 bit/symbol modulation, and the other 4 bits/symbol modulation, the channel efficiency will typically be increased fourfold. It is this increased efficiency which has promoted the widespread use of QAM in fixed link systems.

However, in mobile radio applications the situation is less clear. Here we define a different efficiency which we term spectrum efficiency. We define this as the number of channels that can be provided in a cellular system with frequency reuse. For cellular systems, the system capacity is directly related to spectrum efficiency, but bears little relation to the channel efficiency. This is because changing from 1 bit/symbol to 4 bits/symbol modulation will typically require an increased co-channel interference ratio. To obtain this increased ratio frequencies must be reused less often, so the cluster size (i.e. the number of orthogonal frequency sets available within the complete system) will increase assuming a constant overall bandwidth. Increasing the cluster size will reduce the amount of frequency available within each cell. This will reduce the number of channels available in each cell. Therefore, in order to have a clear picture of the true effect of a change of modulation scheme in mobile radio applications we must consider the spectrum efficiency.

According to Lee [374] the calculation of spectrum efficiency for cellular systems proceeds as follows. Firstly an idealised cell structure with six significant interferers is postulated as shown in Figure 16.1. Regardless of the cluster size there will always be six significant interferers. These may not all be equidistant but for simplicity it is assumed they are all at their average distance from the transmitter. This leads to a small inaccuracy which will not significantly affect the result. With this assumption

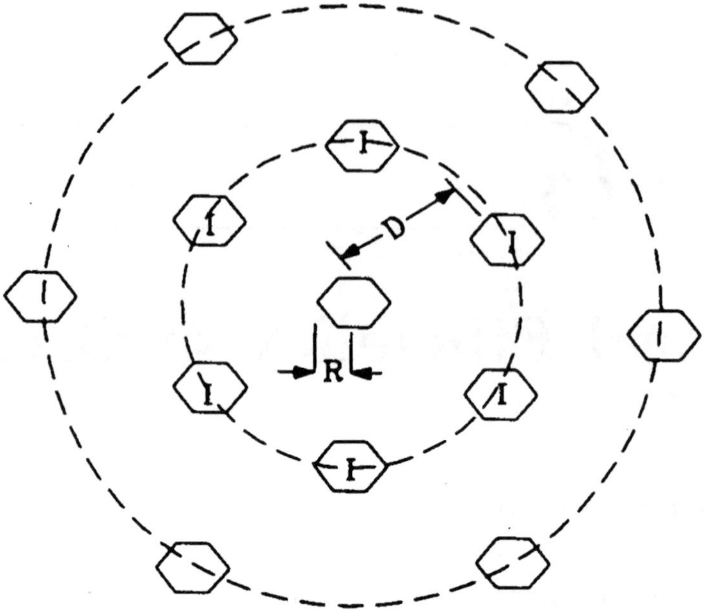

Figure 16.1: Idealised cell structure with interferers

the signal-to-interference ratio SIR is given by

$$SIR = \frac{S}{\sum_{k=1}^{6} I_k \ + \ n} \tag{16.1}$$

where S is the signal power, I_k is the interference power from the kth interfering cell, and n is the local noise. Based on a logarithmic path loss model, neglecting noise, and assuming that the signal level from all the interferers is approximately equal, Equation 16.1 may be rewritten as

$$SIR = \frac{R^{-\gamma}}{6.D^{-\gamma}} \tag{16.2}$$

where R is the radius of a cell, D the distance between interfering cells and γ the path loss law. For more details on propagation path loss models the interested reader is referred to Chapter 2. A diagrammatic representation of the interferers is shown in Figure 16.1 where the centre hexagon is the wanted cell and the significant interferers are labelled with the letter I. It then follows that

$$\frac{D}{R} = (6.SIR)^{1/\gamma}. \tag{16.3}$$

For a hexagonal cell configuration [374] we have

$$\frac{D}{R} = \sqrt{3K} \tag{16.4}$$

where K is the number of cells in a frequency reuse pattern. Equating Equations 16.3 and 16.4 yields

$$K = \frac{1}{3}(6.SIR)^{2/\gamma}. \tag{16.5}$$

The radio capacity in terms of the number of channels available is defined by

$$m = \frac{B_t}{B_c.K} \tag{16.6}$$

where B_t is the total bandwidth available and B_c is the bandwidth required per channel. Substituting for K from Equation 16.5 in terms of SIR and assuming that $\gamma = 4$, gives

$$m = \frac{B_t}{B_c\sqrt{\frac{2}{3}SIR}} \tag{16.7}$$

which is known as the radio capacity equation for a hexagonal cellular structure with a cluster size of K.

16.2 Spectrum Efficiency in Conventional Cells

We have evaluated Equation 16.7 for a wide variety of modulation schemes using a computer simulation. This simulation considered frequency shift keying (FSK), phase shift keying (PSK), and fixed-level and variable-level QAM. The QAM simulations are for differential star QAM introduced in Chapter 11, and the variable-level QAM schemes are for variable rate star QAM introduced in Chapter 13. The 8-level and 16-level star QAM constellations have two amplitude rings, whereas the 32-level and 64-level constellations have four rings.

The simulation program used includes clock and carrier recovery, co-channel and adjacent channel interferers using a range of modulation schemes, Rayleigh fading at a variety of symbol rates, and a wide range of convolutional and block coding systems. Shadow fading is not included as it does not affect the results in any way. For the simulations, near-optimal root raised cosine filtering at transmitter, receiver and co-channel transmitter was used. The interference was generated using a single interferer with the same modulation scheme and filtering as the transmitter. This was subject to a Rayleigh fading path between interferer and receiver, and arranged such that the average interference power was as selected in the simulation parameters. A carrier frequency of 1.8 GHz, a mobile speed of 15 $m/s \approx 30$ mph and a data rate of 25 kBd were assumed while the SNR was fixed at 40 dB.

Using this tool the BER against SIR performance for a range of modulation schemes, each using a range of coding techniques, was established. This led to over 60 performance curves, the most important of which are shown in Figures 16.2 and 16.3. In these figures BCH coders have been used, where the numbers in brackets refer to the number of bits corrected per 63-bit block, and the coding rate is 0.9, 0.81, 0.71, and 0.62 for correction powers of 1 to 4 respectively. Curves for square QAM have been omitted due to its poor performance, although we note that proposed schemes using pilot symbol assisted modulation (PSAM) as discussed in Chapter 10 are expected to provide similar efficiency levels to that of star QAM, albeit at a higher complexity. The curves for 16 FSK have also been omitted, again due to poor performance.

Using these curves, spectrum efficiency for a variety of BERs was established. In order to achieve this, nine BER values spanning the range 5×10^{-2} to 1×10^{-4} were selected. For each modulation scheme and each FEC coding scheme, the SIR at which these error rates were achieved, if it was possible to achieve them, was read off the performance curves drawn previously. The SIR and bandwidth required after FEC coding were then fed into Equation 16.7. The resulting number, m, is the number of channels available within an idealised cellular system, neglecting frequency-domain guard band space between adjacent channels. For any BER and any given modulation scheme there were five curves relating to the four error control codes used, and to no error control coding. Each of these led to a different channel bandwidth requirement B_c and SIR, leading to a different number of channels supported, or radio capacity m. The highest value of m was selected. This meant that at different BERs, different error control codes are optimal in spectral efficiency terms for any one particular modulation scheme. Explicit specification of the error control code favoured for every condition would be unduly complicated and unnecessary as it can be readily inferred from the graphs supplied. In general, low power error control was best for high BERs and high power error control for low BERs.

The graphs showing the radio capacity in terms of the number of channels m supported by different modems at different BER values are given in Figure 16.4. In the figures "nPSK", where n is integer, refers to n-level PSK [183], similarly for FSK, and for QAM "n fixed" refers to n-level star QAM where the number of levels are fixed over the duration of the transmission, whereas the variable-level scheme employed an average of 4 bits per symbol. From these graphs we can see that the performance of all the schemes is broadly similar. The 4PSK scheme is superior for all values of BER, closely followed by the variable rate scheme. There is a tendency for the schemes which use fewest modulation levels to give the best performance, so that 8-PSK is superior to 16-PSK and 16-QAM. Schemes which use the same number of levels, such as 16-PSK and 16-QAM, had broadly similar performances.

16.3 Spectrum Efficiency in Microcells

Whilst the above equations hold for idealised hexagonal cellular systems, they are not applicable to microcellular systems. This is because in microcells the clusters are different from those in conventional cells, and the interference is not so straightforward to calculate. This necessitates a change in Equation 16.7. It was with this in mind that an investigation was undertaken into spectrum efficiency in microcells. With this incentive a number of microcell clusters were constructed on an idealised grid pattern

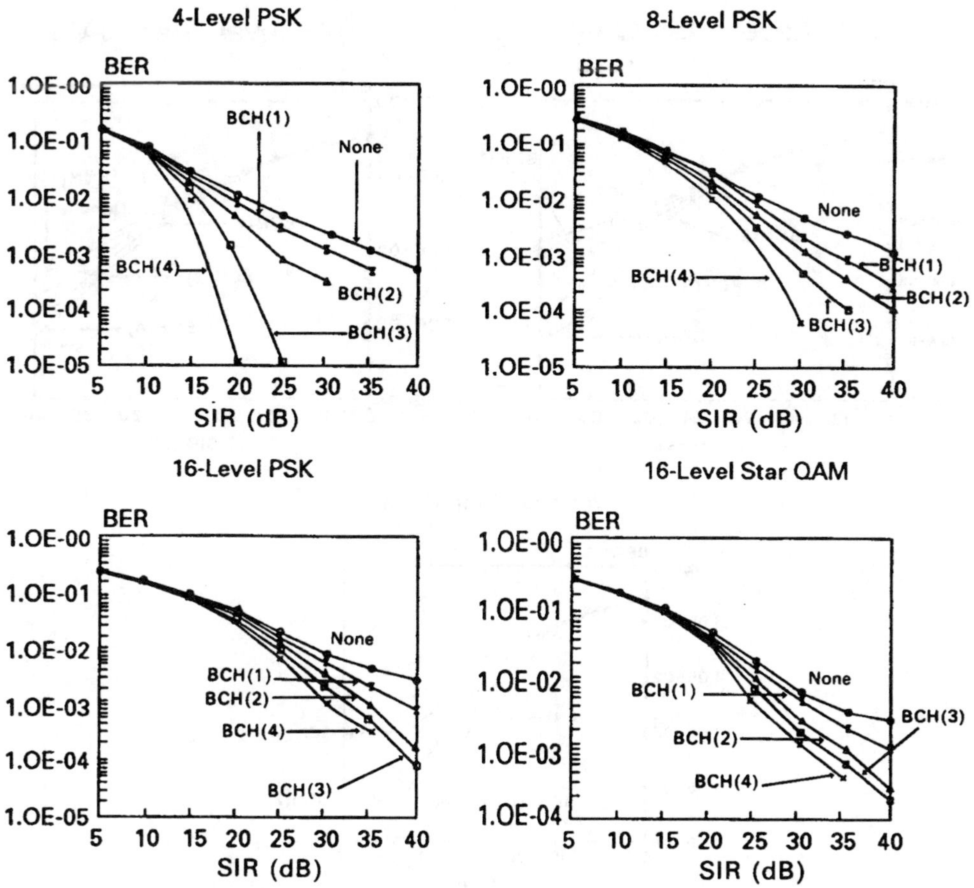

Figure 16.2: BER versus SIR performance of 4, 8, 16 PSK and 16-QAM at 30 mph, 1.8 GHz, 25 kBd, using SNR = 40 dB and various BCH codes

based on maps of New York City and the interference resulting from these clusters was established. The results were then applied to Lee's formula which allowed conclusions to be drawn as to the most appropriate modulation schemes for microcellular systems. This approach is now described in detail.

16.3.1 Microcellular clusters

In a microcell, the interference from any given co-channel interferer varies throughout the microcell, due to the small size and rapid decrease in signal level as the mobile moves away from the BS. As a result of this, it is difficult to calculate SIRs for given clusters. Instead we resorted to a microcellular prediction tool developed at Multiple Access Communications (MAC) and known as Microcellular Design System (MIDAS).

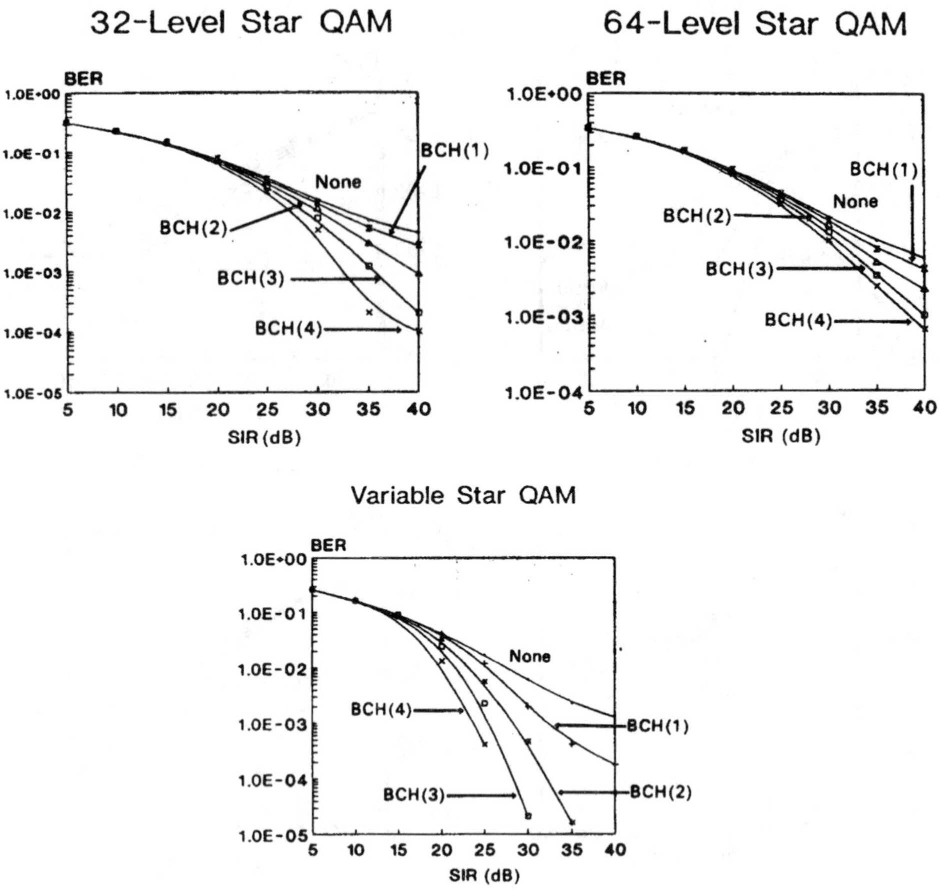

Figure 16.3: BER versus SIR performance of 32, 64 and variable rate star QAM at 30 mph, 1.8 GHz, 25 kBd, using SNR = 40 dB and various BCH codes

This tool can predict both signal and interference levels for any microcellular cluster in both practical and idealised situations and has been proven by comparison with a large database of microcellular measurement data established at MAC.

An idealised grid map based on the layout of New York City, with 100 m blocks and 20 m wide streets was loaded into MIDAS and a range of predictions were performed at both 900 MHz and 1.8 GHz. In order to establish the microcell size, the cell boundary was set based approximately on the minimum received power of the CT-2 cordless telephone system. With a 10 mW transmitter and 0 dB gain antennas, the effective radiated power was 10 dBm. Given a minimum signal level of -95 dBm at which the data can be received and a fast fading margin of 20 dB, the minimum received signal level was -75 dBm, relating to a path loss of 85 dB. This process was visualised in Figure 2.13. With microcell BSs placed midway along blocks, which is a

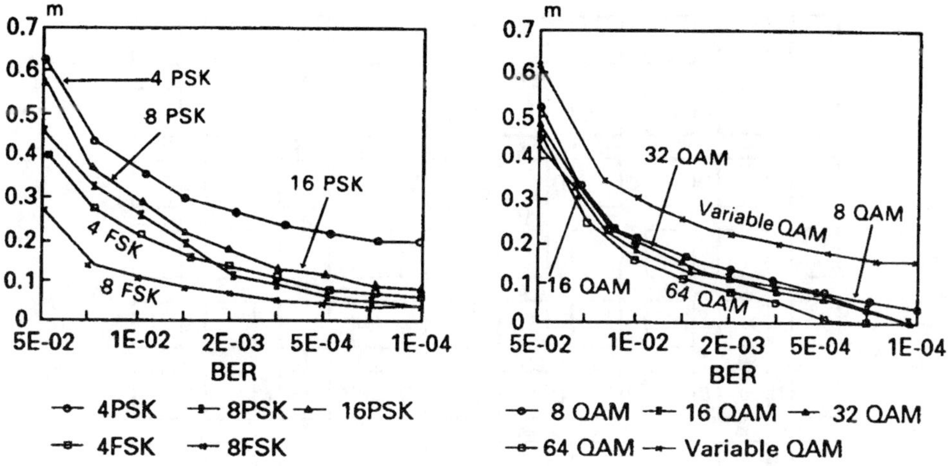

Figure 16.4: Spectrum efficiency versus BER performance of a range of modulation schemes at 30 mph, 1.8 GHz, 25 kBd, using SNR = 40 dB

suitable set of locations to produce a cluster with regular frequency assignment, this path loss of 85 dB gives a range of 180 m at 1.8 GHz, giving a microcell length of 360 m. The path loss versus distance relationship was established using the MIDAS prediction system with its six-breakpoint model rather than the less accurate $\gamma = 4$ fourth power law. This requires a microcell every third block as shown in Figure 16.5, which also illustrates the BS pattern for a four-cell cluster. In this figure, which is not drawn to scale, the square areas represent city blocks and the letters represent a BS location and its specific frequency assignment.

At 900 MHz the same path loss leads to a range of 300 m giving a microcell length of 600 m and a BS every fifth block as shown in Figure 16.6, which also shows the frequency assignment for a six-cell cluster, where the same representation as in Figure 16.5 applies.

Using these BS patterns we considered the SIRs. This was simply achieved by placing the BSs appropriately within the MIDAS modelling package, and then selecting the SIR option. The package produces a visual display of the SIR on a 1 m^2 basis which is used to determine the worst-case SIR. At 1.8 GHz using four cells per cluster only the BSs along the same street cause significant interference at a level of 25-30 dB close to the microcell edge. In contrast the use of six-cell clusters leads to there being no significant interference, since with three frequency sets available for the horizontal streets, BSs can be spaced further apart on the same street and also staggered on neighbouring streets to cause less interference. With $K = 6$ the microcell becomes SNR limited as opposed to SIR limited.

At 900 MHz the interference from BSs using the same frequency on parallel adjacent streets becomes more significant, due to the increase in reflection and diffraction around corners at this frequency. A similar method to that used to construct the cluster at 1.8 GHz was used at 900 MHz, but with the wider microcell spacing caused

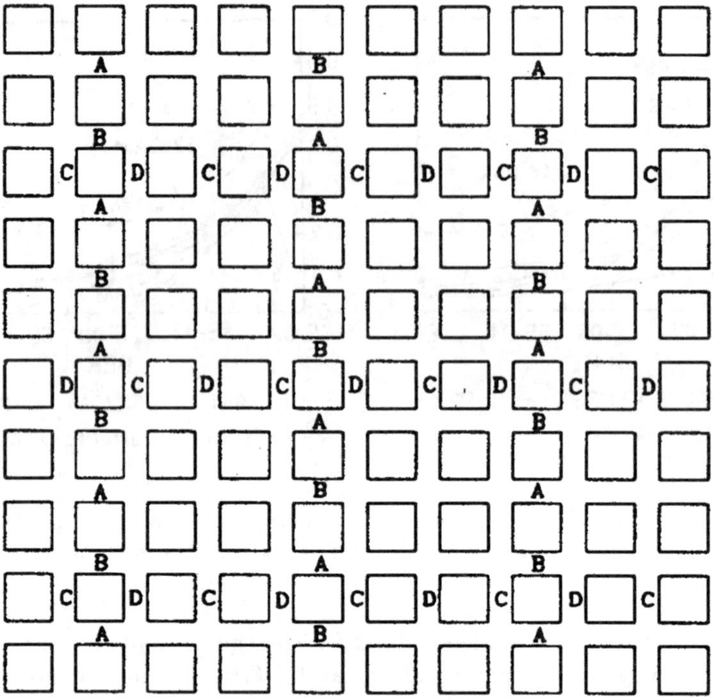

Figure 16.5: Microcellular BS pattern at 1.8 GHz for a four-cell cluster using MIDAS propagation prediction

by increased propagation distances at the lower frequency. With $K = 4$, areas where the SIR dropped to 20 dB were experienced, this SIR was lower than that for 1.8 GHz, due to the higher diffraction and reflection of energy around corners. With $K = 6$, the system again became SNR limited.

16.3.2 System design for microcells

Communications systems are generally designed for worst-case propagation. Consequently, despite the fact that microcells only experience significant interference in certain places, a communications system should nevertheless be designed to cope with this level of interference. It is possible that future designers will find ways of overcoming or reducing the interference even in four-cell clusters, by a combination of dynamic channel allocation, and switching the frequencies of mobiles experiencing significant interference to different frequencies within the same set. However, here we only consider a simple system with fixed allocation, and investigate its results. More intelligent systems will tend to reduce the cluster size.

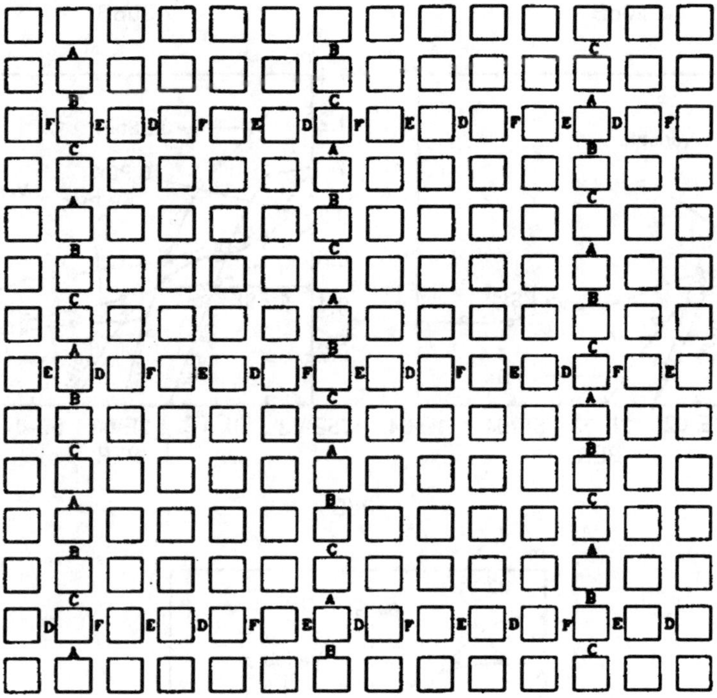

Figure 16.6: Microcellular BS pattern at 900 MHz for a six-cell cluster using MIDAS propagation prediction

16.3.3 Microcellular radio capacity

When deriving his equation for spectrum capacity, Lee used a formula for conventional cells linking SIR with the number of cells per cluster, K. We performed the above work in an effort to produce a similar equation for microcells, but because of the limited range of K that it is practical to use, we feel it inappropriate to produce a formula of this sort. Instead we substitute our results directly into Equation 16.6 for the values of K for which we have results.

Hence, for $K = 4$ we use the simulation results pertaining to 20 dB SIR, whereas for $K = 6$, where transmission is SNR limited, we use the simulation results for a range of SNRs, particularly 20, 30 and 40 dB. The BSs are maintained in the same position for each of the SNRs, and the operation of the mobiles with each SNR is examined.

Equation 16.6 is evaluated for both $K = 4$ and $K = 6$, and the cluster giving the highest number of channels is selected for each BER and each modulation scheme. In most cases a cluster size of 4 was selected. The microcellular SNR is of great importance in deciding upon the optimum modulation scheme and depends on transmitter power and receiver sensitivity for any given system. Here we give results for a range of SNRs, always using the most efficient cluster size for each SNR value to derive the

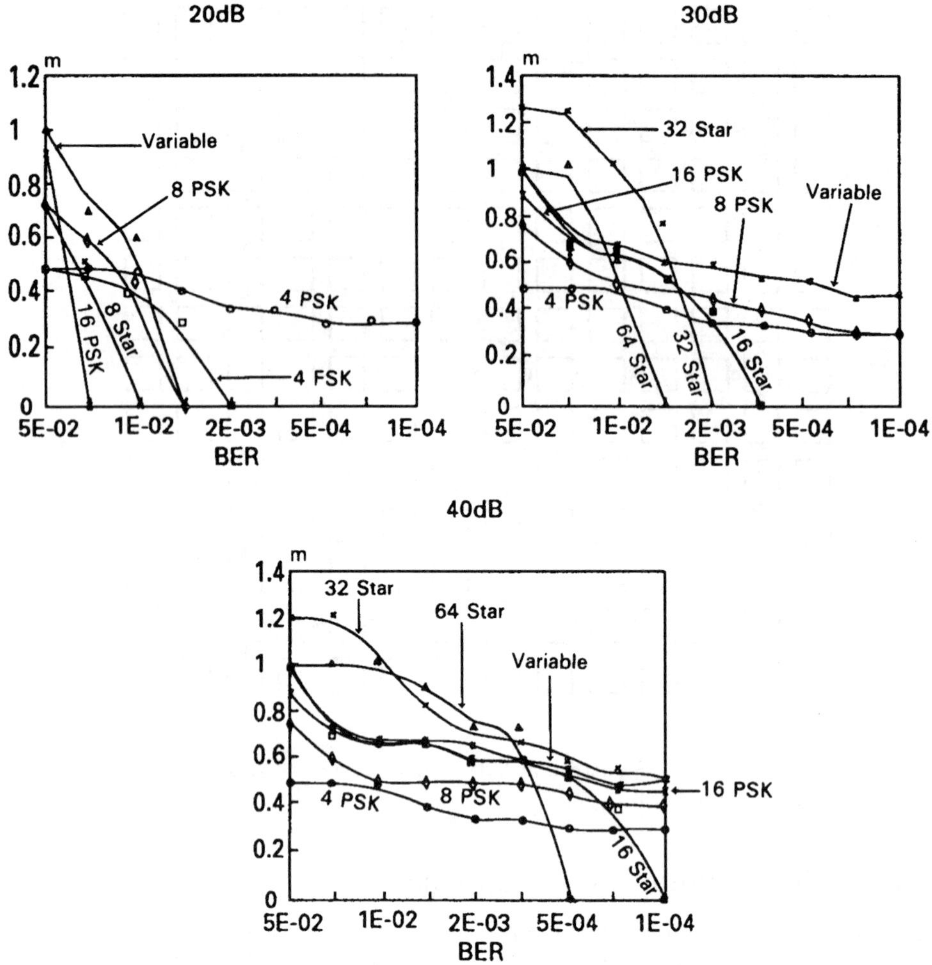

Figure 16.7: Spectrum efficiency versus BER performance in microcells of a range of modulation schemes at 30 mph, 1.8 GHz, 25 kBd, using SNR = 20 dB, 30 dB and 40 dB

number of available channels.

16.3.4 Modulation schemes for microcells

Given SNRs of 20, 30 and 40 dB and assuming a SIR of 20 dB for the four-cell cluster and infinite for the six-cell cluster, we can consider the most efficient modulation schemes to use within the microcellular environment. Graphs showing the BER against number of channels are given in Figure 16.7.

These graphs have been derived in the same way as the graphs for the conventional cells, but with appropriate SIRs assumed. The same procedure was followed as with macrocells in that for each BER along the x-axis, the SIR at which this was achieved was read off Figures 16.2 and 16.3. However, if this SIR was greater than that available within the microcell for a given cluster size the scheme was disallowed. If it was less, the scheme was allowed at the cluster size under consideration. Therefore, achieving the BER with a SIR significantly lower than that available was disadvantageous compared to achieving it with a SIR closer to that available.

This procedure was followed for $K = 4$ and $K = 6$, and the cluster size giving the best result used at each point. The QAM schemes were star QAM, using two concentric circles with four and eight amplitude points per ring for the 8-level and 16-level schemes, respectively. Curves with very poor performance have been omitted for clarity. In all cases a Rayleigh fading channel is assumed with transmission at 1.8 GHz, mobile speed of 15 m/s and data rate of 25 kBd. The results at 900 MHz are similar because almost the same SIR and SNR values apply to both frequencies. Differential transmission and reception have been used along with near-optimal root raised cosine filtering, modified early-late clock recovery and with the interferer using the same modulation scheme as the transmitter. The same nomenclature has been used as in Figure 16.4.

For the 20 dB SNR case, variable QAM is most effective for high BERs, with 4-PSK giving the best performance for BERs below 1×10^{-2}. When the SNR is 30 dB the variable-level QAM is superior for all but the highest BERs. If only fixed-level schemes are considered then 8-PSK proves slightly better than 4-PSK and 4-FSK for low BERs. For the 40 dB SNR case for high BERs the higher-level modulation schemes, namely 32-QAM and 64-QAM prove best. At BERs below 2×10^{-4} the variable rate QAM schemes are in line with the 32-level QAM, with the 4-level and 8-level schemes giving substantially worse performance due to their inability to make the best use of the SNR and SIR available. We note that if variable rate QAM giving an average of 5 bits/symbol had been employed, we would have expected it to outperform 32-level QAM in all scenarios.

16.4 Summary

In this chapter the interference levels that might be expected in a range of conventional cellular and microcellular clusters have been considered. Simulation work has suggested that for conventional cell sizes the lower-level modulation schemes such as 4-PSK are best, with variable rate QAM schemes only incurring a slight penalty. For microcells with communication at both 900 MHz and 1.8 GHz, four-cell or six-cell clusters are advocated depending on the SNR expected. Based on the expected SNRs and interference levels we then surmised that variable rate QAM schemes are often superior to the other modulation schemes considered here. For low BERs and SNR = 20 dB, 4-PSK provides the best performance; whereas for high BERs, particularly when the SNR is as high as 30 or 40 dB, then 32-level and 64-level star QAM are most suitable.

Having considered the performance of the modulation scheme in isolation, in the next chapter we proceed to investigate QAM and other modulation schemes when used for the transmission of speech, and investigate means whereby the different bit error rates of different bits on the same QAM constellation can be linked advantageously with the properties of modern speech codecs.

Chapter 17

QAM-based Speech Systems and Bench-Mark Schemes

17.1 Introduction

In the previous chapter we portrayed a variety of 1-6 bits/symbol modems and studied their true spectral efficiency in macro- and microcellular environments with a range of frequency reuse patterns. Our only concern was in maximising the capacity of the cellular system when expressed in terms of speech channels supported in both SNR and SIR limited environments.

In this closing Chapter we propose a variety of QAM-based mobile radio speech communications systems including speech and FEC codecs, 1-4 bits/symbol modems, and a dynamic multiple access scheme in order to explore the range of trade-offs and to illustrate ways in which QAM schemes might be employed [375–378]. We consider systems both with and without error correction coding using the 32 kbps CCITT G721 adaptive differential pulse code modulation (ADPCM) standard and the 13 kbps regular pulse excited (RPE) speech codecs used in the pan-European digital mobile radio system GSM. Simulation results are presented giving BER, bandwidth occupancy and an estimate of complexity for 4 bit/symbol 16-Star QAM modems, 2 bit/symbol $\frac{\pi}{4}$-shifted differential quadrature phase shift keying ($\frac{\pi}{4}$-DQPSK) modems and binary Gaussian minimum shift keying (GMSK) modems.

The chapter is organised as follows. After a short system overview in the next section, in Sections 17.2 and 17.3 we briefly consider radio modems and speech codecs. Speech quality measures and discontinuous transmissions are discussed in Sections 17.4 and 17.5, while Section 17.6 is devoted to bit protection issues and channel codecs. Section 17.7 is dedicated to the description of a menu of candidate speech systems and Section 17.8 to a dynamic multiple access scheme known as packet reservation multiple access (PRMA).

17.2 Modem Schemes

The choice of modem is influenced by the constraints imposed by equipment complexity, power consumption, spectral efficiency, robustness against channel errors, cochannel and adjacent channel interference and the propagation phenomena, which depends on the cell size [158]. Most of these issues have been addressed in Chapter 17. Equally important are the associated issues of linear or non-linear amplification and filtering addressed in Chapter 4, the applicability of non-coherent differential detection (Chapter 11), soft-decision detection, equalisation (Chapters 7 and 14) and so forth. Here we briefly describe two schemes which we will use for comparison with our QAM system. These are Gaussian minimum shift keying (GMSK) modulation, which is used in the European second generation digital mobile radio systems and $\frac{\pi}{4}$-shifted differential quadrature phase shift keying (DQPSK) modulation favoured by the Americans in their IS-54 D-AMPS network as well as by the Japanese digital cellular system.

17.2.1 GMSK Modulation

Gaussian Minimum Shift Keying (GMSK) was introduced in Section 7.6 in the context of Viterbi Equalisation of partial response modems, where the phase transitions between adjacent phase states are given by the convolution of a Gaussian pulse shaping function with the symbol sequence [158]. If the length of the Gaussian pulse exceeds the symbol period, the response of the modulator to a symbol is spread over several symbol periods introducing controlled intersymbol interference (CISI). We have seen the effect of signalling pulse width extension in the context of the OMPX modem of Chapter 15, although there no partial-response type overlapping signalling was allowed. Permitting the modulation pulses to overlap has the advantage of reducing the bandwidth of the modulated signal but requires a more complex receiver. The length of the Gaussian response is controlled by the normalised bandwidth, $B_N = B_b T$, where B_b is the Gaussian lowpass filter bandwidth, and T is the symbol period. If the phase transitions are unfiltered, $B_N = \infty$ and we have a Minimum Shift Keying (MSK) modulator. As B_N is reduced, more and more CISI occurs, but the out-of-band spectral spillage is reduced correspondingly. The Pan-European system, GSM uses a GMSK modulator with $B_N = 0.3$ having an impulse response extending over approximately three signalling periods, which requires sophisticated equalisation to achieve good performance. The digital European cordless telecommunications (DECT) system's modulator uses $B_N = 0.5$, which eases clock recovery. The typical spectral occupancy in terms of bits per Hz of a GSM-like GMSK modem is 1.35.

As B_N is reduced, the spectrum becomes more compressed as the phase transitions follow a smoother curve, but more CISI is introduced, rendering bit-by-bit detection using a frequency discriminator more prone to channel impairments. Full-response continuous phase modulation (CPM) schemes where no CISI is introduced, such as MSK can be demodulated by a frequency discriminator, by simply measuring the change of phase over a symbol period. When the phase increases, its derivative, the frequency is positive and a logical '1' is detected, and vice versa, when the phase decreases. Maximum Likelihood Sequence Estimation (MLSE) using the Viterbi algo-

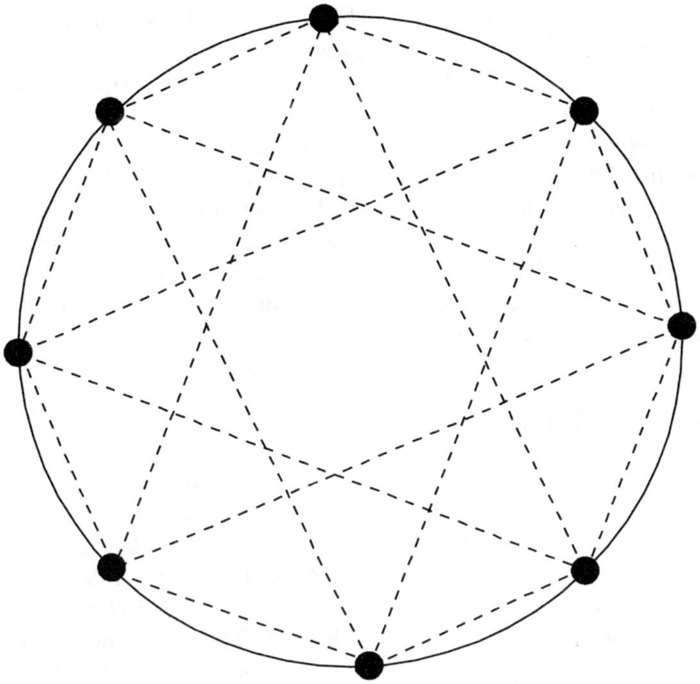

Figure 17.1: $\frac{\pi}{4}$-DQPSK phasor constellation

rithm [158] is generally employed to remove both the CISI and the ISI introduced by the channel. Furthermore, with decreasing B_N the eye-pattern becomes more complicated and the reduced eye-opening makes clock and carrier recovery more difficult to achieve.

17.2.2 $\frac{\pi}{4}$–DQPSK Modulation

A particular form of differential quadrature phase shift keying, known as $\frac{\pi}{4}$–DQPSK, transmits 2 bits/symbol, allowing the bandwidth to be reduced compared to binary schemes at the cost of increased complexity and degraded SNR and signal-to-interference ratio (SIR) performance. Careful filtering and linear amplification of the modulated signal as described for QAM in Chapter 4 is required in order to maintain low spectral side-lobes and hence achieve good adjacent channel performance. Because there are 2 bits/symbol, the bitrate can be higher than with GMSK before it is necessary to use a channel equaliser.

In this linear modulation scheme the amplitude variations are reduced compared to conventional DQPSK due to the rotation of the constellation by $\frac{\pi}{4}$ every symbol period, ensuring that no phasor transition path passes through the zero amplitude point as demonstrated by the phasor constellation of Figure 17.1.

The data stream is differentially encoded to allow low-complexity differential detection to be employed. Alternatively, MLSE Viterbi detection [158] can be implemented to achieve better performance at higher complexity.

In the scheme we simulated a 129th order Blackman-windowed root raised cosine finite impulse response (FIR) Nyquist transmit and receive filters with a roll-off factor of $\alpha = 0.35$ are employed and an equaliser is only required when operating in a wide-band environment. The $\frac{\pi}{4}$-DQPSK spectral occupancy is 1.62 bit/Hz.

This differentially encoded scheme is readily demodulated by a non-coherent differential detector which evaluates the phasor changes between adjacent sampling instants. Since coherent carrier recovery is not necessary, reduced complexity is achieved and carrier recovery false locking problems are eliminated. However, at low SNR and SIR values some performance penalty is incurred.

17.3 Speech Codecs

In recent years speech coders have experienced dramatic increases in complexity resulting in codecs with bit rates as low as 4.8 kbps [158][Chapter 3]. The choice of an appropriate speech codec is based on the requirements of speech quality, computational complexity, power consumption, bit rate, delay, and robustness against channel errors. The two speech codecs used in our experiments are the CCITT G 721. 32 kbps adaptive differential pulse code modulation (ADPCM) codec proposed for the digital European cordless telephone (DECT) system portrayed in Section 13.6, and the 13 kbps Regular Pulse Excited (RPE), Long Term Predictor (LTP) assisted codec used in the GSM system.

17.3.1 Adaptive Differential Pulse Code Modulation

The 32 kbit/s transmission rate ADPCM codec is specified in the CCITT G.721 Recommendation. The encoder/decoder pair is shown in Figure 17.2.

The A-law or μ-law companded PCM signal is first converted into linear PCM format. Since this is a predictive codec, each linear PCM-coded sample is estimated by an adaptive predictor on the basis of the locally reconstructed past samples. This estimate is subtracted from the PCM sample to be encoded in order to form the prediction error or difference signal. This reduced variance difference signal can be quantised with approximately the same noise variance as the original linear PCM signal, using a 4-bit adaptive quantiser. Hence an 8-bit PCM sample is represented by a 4-bit ADPCM sample, giving a transmission rate of 32 kbit/s at a sampling rate of 8 kHz. This ADPCM stream is transmitted to the decoder. Also it is locally decoded, using the inverse adaptive quantiser, to deliver the quantised difference signal, which is added to the previous signal estimate to yield the locally reconstructed signal. Based on the quantised difference signal and the locally reconstructed signal the adaptive predictor derives the subsequent signal estimate. This local decoder is identical to the ADPCM decoder of the receiver, and hence both decoders generate the same signal estimates. These are then added to the difference signal in order to generate the reconstructed signal.

The ADPCM decoder is constituted by the local decoder part of the encoder, and additionally it comprises the linear PCM to A-law or μ-law converter. The

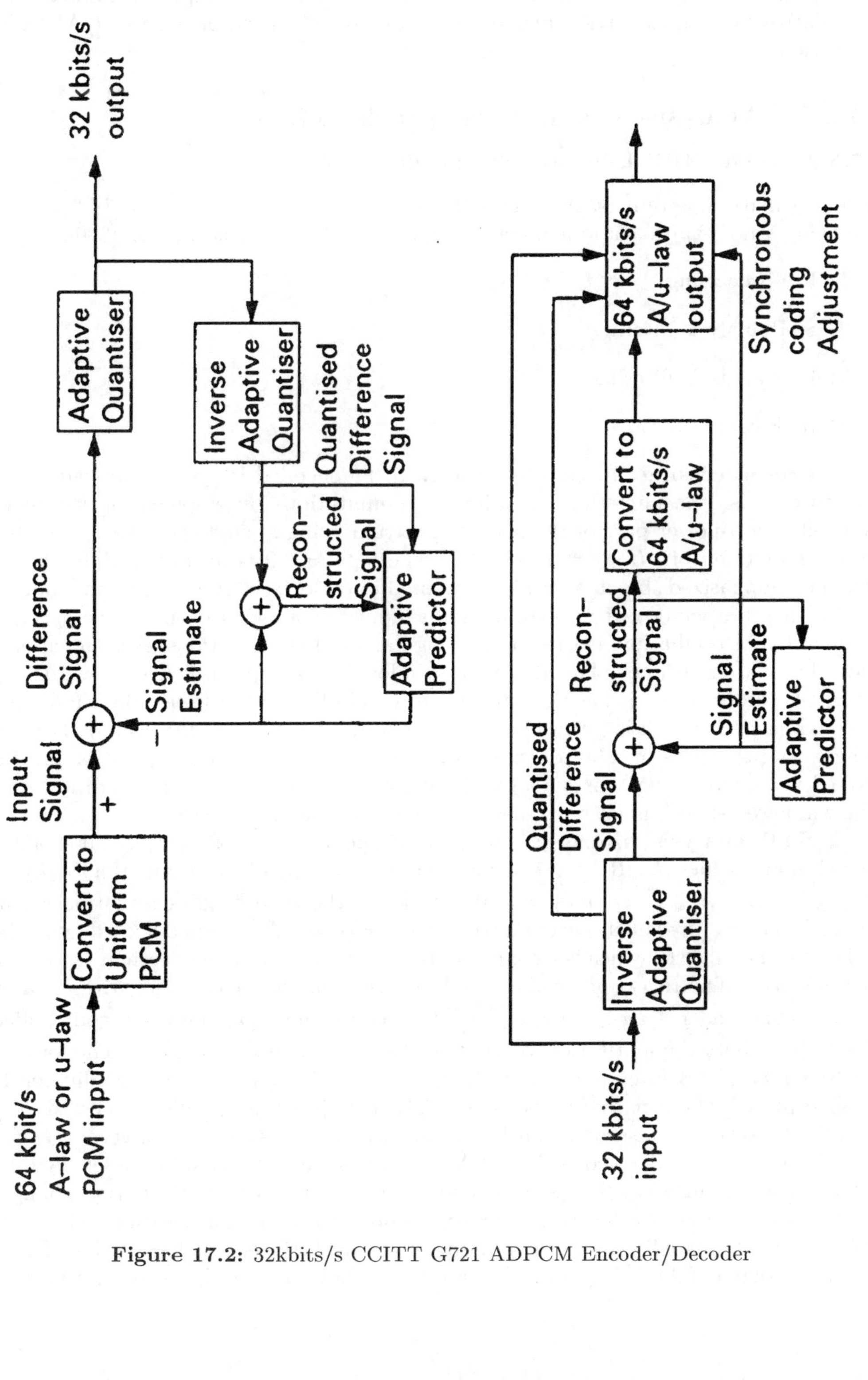

Figure 17.2: 32kbits/s CCITT G721 ADPCM Encoder/Decoder

synchronous coding adjustment block shown in Figure 17.2 attempts to eliminate the cumulative tandem distortion occurring in subsequent synchronous PCM/ADPCM operations.

17.3.2 Analysis-by-synthesis speech coding

17.3.2.1 The RPE-LTP Speech Encoder

The schematic diagram of the RPE-LTP encoder is shown in Figure 17.3, where following Vary's approach four functional parts can be recognised [379], [380]:

1) Pre-processing

2) STP analysis filtering

3) LTP analysis filtering

4) RPE computation

1 Pre-processing: Initially the speech to be encoded is pre-emphasized in order to increase the numerical precision in computations by emphasizing the high-frequency, low-power part of the speech spectrum using a single-pole filter with the transfer function of: $H(z) = 1 - c_1 z^{-1}$, where $c_1 \approx 0.9$ is an appropriate choice. The pre-emphasized speech $s_P(n)$ is segmented into blocks of 160 samples in a buffer, where they are windowed by a Hamming-window in order to mitigate the spectral domain Gibbs oscillation, caused by rectangular windowing of the speech signal outside the analysis frame. The Hamming-window has a tapering effect towards the edges of a block, while it has no influence in its middle ranges. Hence the windowed, pre-emphasized speech is given by: $s_{PSW}(n) = s_{PS}(n) \cdot c_2 \cdot (0.54 - 0.46 \cos 2\pi n/L)$, where $s_{PS}(n)$ represents the pre-emphasized, segmented speech, $s_{PSW}(n)$ is its windowed version and the constant $c_2 = 1.5863$ is determined from the condition that the windowed speech must have the same power as the input speech.

2 STP analysis filtering: For each segment of L=160 samples nine autocorrelation coefficients R(k) are computed from $s_{PSW}(n)$ by evaluating $R(k) = \sum_{n=0}^{L-1-k} s_{PSW}(k) s_{PSW}(n+k), k = 0 \ldots 8$. From the speech autocorrelation coefficients R(k) eight so-called reflection coefficients $K(i)$ are determined according to the Schur-recursion [381], which is a computationally efficient recursive method for inverting matrices. The reflection coefficients $K(i)$ are then converted to logarithmic area ratios defined as $LAR(i) = \log_{10}(\frac{1+K(i)}{1-K(i)})$, because the logarithmically companded LARs have better quantisation properties than the coefficients $K(i)$. The various LAR(i) $i = 1 \ldots 8$ filter parameters have different dynamic ranges and differently shaped probability density functions (PDFs). This justifies the allocation of 6, 5, 4 and 3 bits to the first, second, third and fourth pairs of LARs, respectively.

The quantised LAR(i) coefficients LAR'(i) are locally decoded into the set LAR"(i) as well as transmitted to the speech decoder. In order to mitigate the abrupt changes in the nature of the speech spectral envelope around the STP analysis frame edges, the LAR parameters are linearly interpolated, and towards the edges of an analysis frame the interpolated LAR"'(i) parameters are used. Now the locally decoded reflection

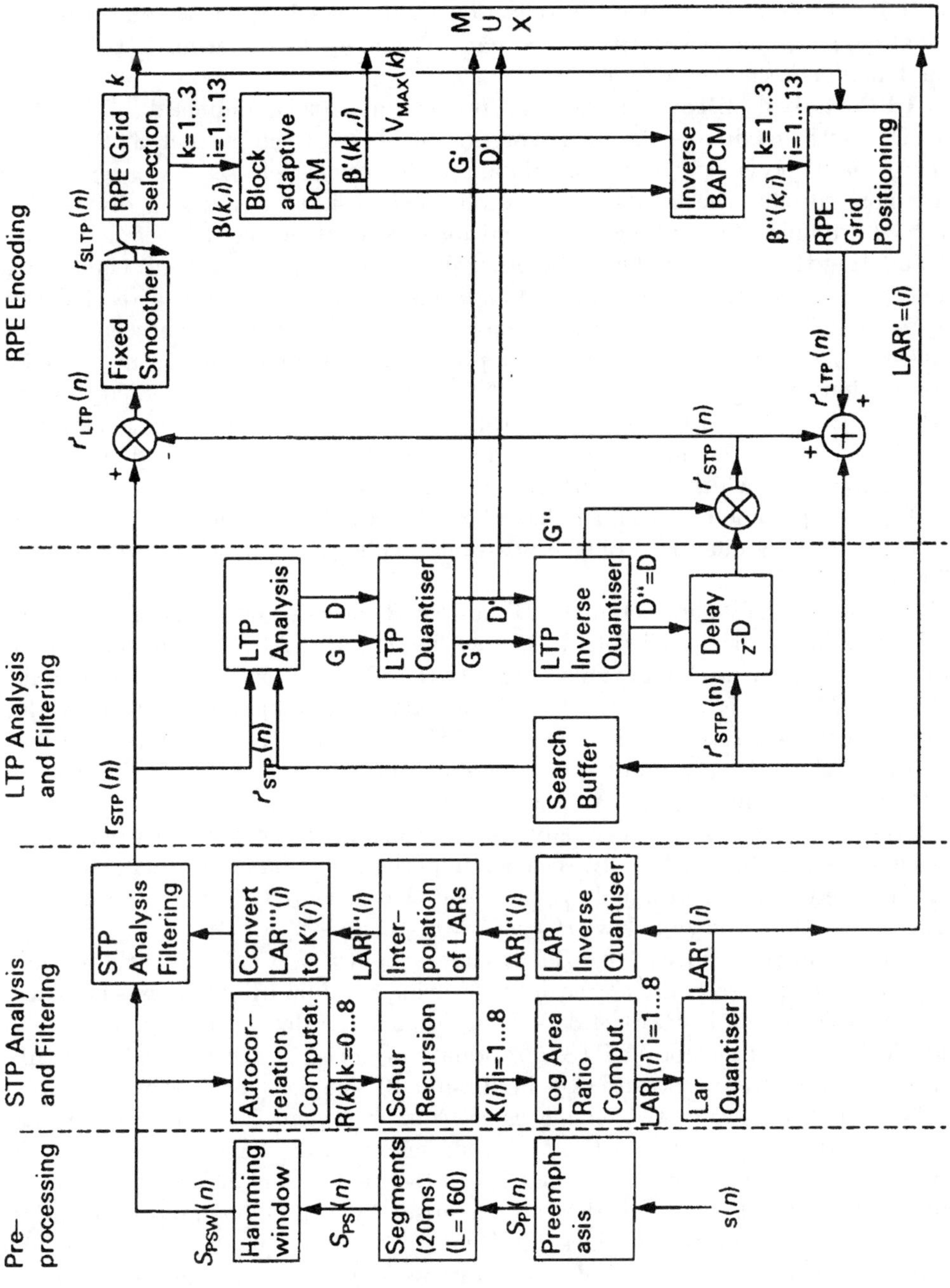

Figure 17.3: Blockdiagram of the RPE-LTP Encoder

coefficients K'(i) are computed by converting LAR"'(i) back into K'(i), which are used to compute the STP residual $r_{STP}(n)$ in a so-called PARCOR structure. The PARCOR scheme directly uses the reflection coefficients K(i) to compute the STP residual $r_{STP}(n)$, and it constitutes the natural analogy to the acoustic tube model of the human speech production.

3 LTP analysis filtering: The LTP prediction error is minimised by that LTP delay D, which maximises the cross correlation between the current STP residual $r_{STP}(n)$ and its previously received and buffered history at delay D, i.e., $r_{STP}(n-D)$. To be more specific, the L=160 samples long STP residual $r_{STP}(n)$ is divided into four N=40 samples long subsegments, and for each of them one LTP is determined by computing the cross correlation between the presently processed subsegment and a continuously sliding N=40 samples long segment of the previously received 128 samples long STP residual segment $r_{STP}(n)$. The maximum of the correlation is found at a delay D, where the currently processed subsegment is the most similar to its previous history. This is most probably true at the pitch periodicity or at a multiple of the pitch periodicity. Hence the most redundancy can be extracted from the STP residual, if this highly correlated segment is subtracted from it, multiplied by a gain factor G, which is the normalised cross correlation found at delay D. Once the LTP filter parameters G and D have been found, they are quantised to give G' and D', where G is quantised only by two bits, while to quantise D' seven bits are sufficient.

The quantised LTP parameters (G',D') are locally decoded into the pair (G",D") in order to produce the locally decoded STP residual $r'_{STP}(n)$ for use in the forthcoming subsegments to provide the previous history of the STP residual for the search buffer, as shown in Figure 17.3. With the LTP parameters just computed the LTP residual $r_{LTP}(n)$ is calculated as the difference of the STP residual $r_{STP}(n)$ and its estimate $r''_{STP}(n)$, which has been computed by the help of the locally decoded LTP parameters (G",D) as shown below: $r_{LTP}(n) = r_{STP}(n) - r''_{STP}(n), r''_{STP}(n) = G''r'_{STP}(n-D)$. Here $r'_{STP}(n-D)$ represents an already known segment of the past history of $r'_{STP}(n)$, stored in the search buffer. Finally, the content of the search buffer is updated by using the locally decoded LTP residual $r'_{LTP}(n)$ and the estimated STP residual $r''_{STP}(n)$ to form $r'_{STP}(n)$, as shown below: $r'_{STP}(n) = r'_{LTP}(n) + r''_{STP}(n)$.

4 RPE computation: The LTP residual $r_{LTP}(n)$ is weighted with the fixed smoother, which is essentially a gracefully decaying band limiting low-pass filter with a cut-off frequency of 4 kHz/3=1.33 kHz according to a decimation by three about to be deployed. The smoothed LTP residual $r_{SLTP}(n)$ is decomposed into three excitation candidates, constituted by 14, 13 and 13 samples. Then the energies E1, E2, E3 of the three decimated sequences are computed, and the candidate with the highest energy is chosen to be the best representation of the LTP residual. The excitation pulses are afterwards normalised to the highest amplitude $v_{max}(k)$ in the sequence of the 13 samples, and they are quantised by a three bit uniform quantiser, whereas the logarithm of the block maximum $v_{max}(k)$ is quantised with six bits. According to three possible initial grid positions k, two bits are needed to encode the initial offset of the grid for each subsegment. The pulse amplitudes $\beta(k,i)$, the grid positions k and the block maxima $v_{max}(k)$ are locally decoded to give the LTP residual $r'_{LTP}(n)$, where the 'missing pulses' in the sequence are filled with zeros.

17.3.2.2 The RPE-LTP Speech Decoder

The block diagram of Vary's RPE-LTP decoder [379], [380] is shown in Figure 17.4, which exhibits an inverse structure, constituted by the functional parts of:

1) RPE decoding

2) LTP synthesis filtering

3) STP synthesis filtering

4) Post-processing

1 RPE decoding: In the decoder the grid position k, the subsegment excitation maxima $v_{max}(k)$ and the excitation pulse amplitudes $\beta'(k,i)$ are inverse quantised, and the de-normalised pulse amplitudes are computed by multiplying the decoded amplitudes with their corresponding block maxima. The LTP residual model $r'_{LTP}(n)$ is recovered by appropriately positioning the pulse amplitudes $\beta(k,i)$ according to the initial offset k.

2 LTP synthesis filtering: Firstly the LTP filter parameters (G',D') are inverse quantised in order to derive the LTP synthesis filter. Then the recovered LTP excitation model $r'_{LTP}(n)$ is invoked to excite this LTP synthesis filter (G',D') to recover a new subsegment of length N=40 of the estimated STP residual $r'_{STP}(n)$. To do so, the past history of the recovered STP residual $r'_{STP}(n)$ is used, appropriately delayed by D' samples and multiplied by G' to deliver the estimated STP residual $r''_{STP}(n)$, according to: $r''_{STP}(n) = G'.r'_{STP}(n - D')$, and then $r''_{STP}(n)$ is used to compute the most recent subsegment of the recovered STP residual, as given by: $r'_{STP}(n) = r''_{STP}(n) + r'_{LTP}(n)$.

3 STP synthesis filtering: To compute the synthesized speech $\hat{s}(n)$ the PARCOR synthesis is used, where similarly to the STP analysis filtering the reflection coefficients k(i) i=1...8 are required. The LAR'(i) parameters are decoded by using the LAR inverse quantiser to give LAR"(i), which are again linearly interpolated towards the analysis frame edges between parameters of the adjacent frames to prevent abrupt changes in the character of the speech spectral envelope. Finally, the interpolated parameter set is transformed back into reflection coefficients, where filter stability is guaranteed, if recovered reflection coefficients, which fell outside the unit circle are reflected back into it, by taking their reciprocal values. The inverse formula to convert LAR(i) back into k(i) is given by: $k(i) = \frac{10^{LAR(i)}-1}{10^{LAR(i)}+1}$.

4 Post-processing: The post-processing is constituted by the de-emphasis, using the inverse filter given below: $H(z) = 1 + c_1 z^{-1}$.

In our implementation of the RPE-LTP codec we used ten Line Spectral Frequencies (LSFs) [158], rather than eight LARs for the short-term predictor coefficients and encoded them using again a total of 36 bits. Due to the LSFs' ordering property channel errors can be detected and the appropriate ordering can be restored at the receiver. This improves the codec's robustness. A further variation in our codec was that we used four bits to quantise the long-term predictor's gain (LTPG) instead of the standardised two bits. The summarised RPE-LTP bit allocation scheme is tabulated in Table 17.1 for a period of 20 ms, which is equivalent to the encoding of L=160 samples.

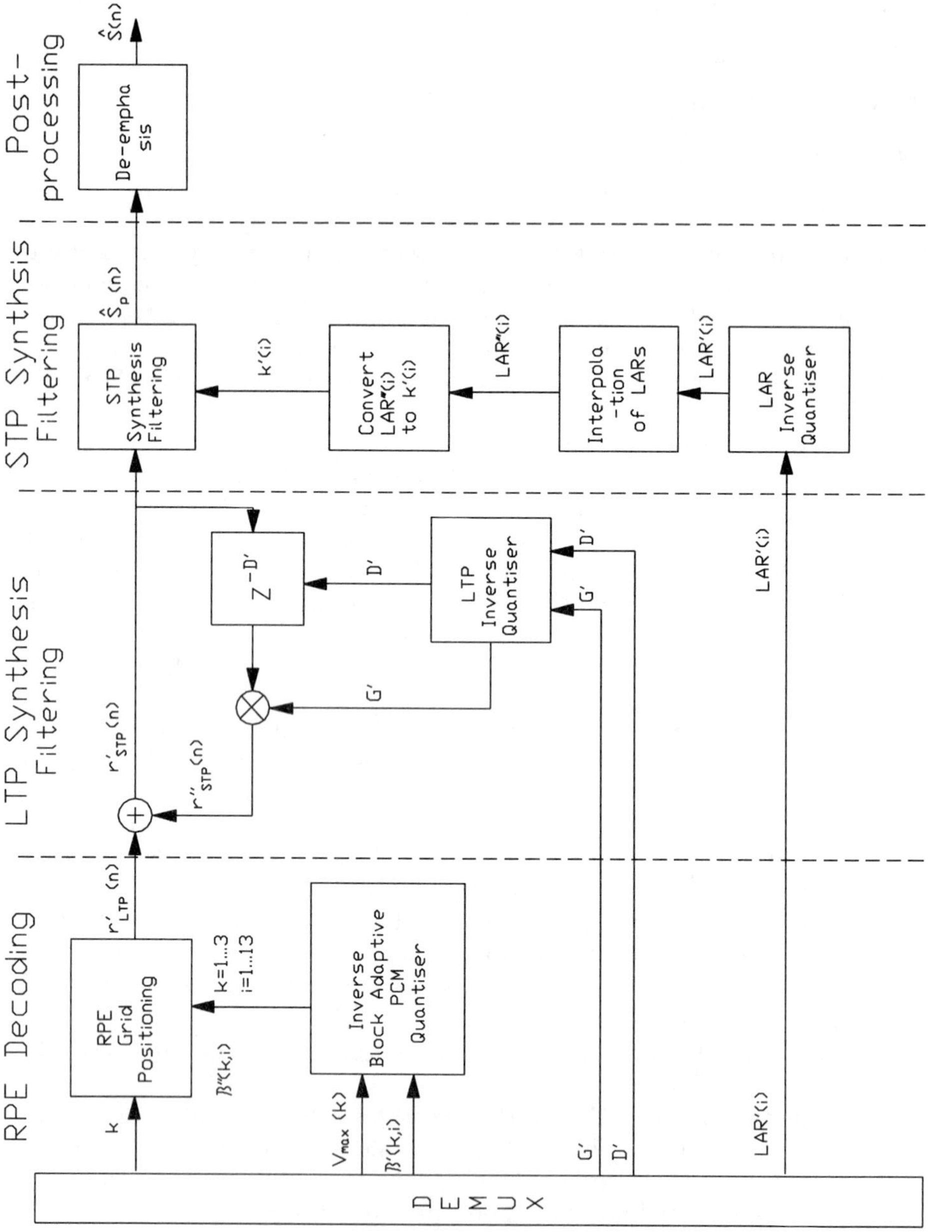

Figure 17.4: Blockdiagram of the RPE-LTP Decoder

Parameter to be encoded	No. of bits	Bitpos.
10 LSF coefficients	36	1-36
4 RPE Grid-positions	4 x 2 = 8	37,38, etc.
4 RPE Block maxima	4 x 6 = 24	39-44, etc.
4x13=52 Pulse amp.	52 x 3 = 156	45-83, etc.
4 LTP Gain (LTPG)	4 x 4 = 16	84-90, etc.
4 LTP Delay (LTPD)	4 x 7 = 28	91-94, etc.
Total no. of bits/20ms	268	
Transmission bitrate	13.4 kbit/s	

Table 17.1: Summary of the 13.4 kbits/s RPE-LTP bit-allocation scheme

17.4 Speech Quality Measures

In general the speech quality of communications systems is difficult to assess and quantify. The most reliable quality evaluation methods are subjective, such as the so-called mean opinion score (MOS), which uses a five-point scale between one and five. MOS-tests use evaluation of speech by untrained listeners, but their results depend on the test conditions. Specifically, the selection and ordering of the test material, the language, and listener expectations all influence their outcome. A variety of other subjective measures is discussed in references [382], [383], [384], but subjective measures are tedious to derive and difficult to quantify during system development.

Objective speech quality measures do not provide results that could be easily converted into MOS values, but they facilitate quick comparative measurements during research and development. Most objective speech quality measures quantify the distortion between the speech communications system's input and output either in time or in frequency domain. The conventional SNR can be defined as

$$SNR = \frac{\sigma_{in}^2}{\sigma_e^2} = \frac{\sum_n s_{in}^2(n)}{\sum_n [s_{out}(n) - s_{in}(n)]^2}, \tag{17.1}$$

where $s_{in}(n)$ and $s_{out}(n)$ are the sequences of input and output speech samples, while σ_{in}^2 and σ_e^2 are the variances of the input speech and that of the error signal, respectively. A major drawback of the conventional SNR is its inability to give equal weighting to high- and low-energy speech segments, because its computation will be dominated by the higher-energy voiced speech segments. Therefore the reconstruction fidelity of voiced speech is given higher priority than that of low-energy unvoiced sounds, when computing the arithmetic mean of the SNR. Hence a system optimised for maximum SNR usually is suboptimum in terms of subjective speech quality.

Some of the problems of SNR computation can be overcome by using the segmental SNR (SEGSNR)

$$SEGSNR^{dB} = \frac{1}{M} \sum_{m=1}^{M} 10 \log_{10} \frac{\sum_{n=1}^{N} s_{in}^2(n)}{\sum_{n=1}^{N} [s_{out}(n) - s_{in}(n)]^2}, \tag{17.2}$$

where N is the number of speech samples within a segment of typically 15–25 ms, while M is the number of 15–25 ms segments, over which $SEGSNR^{dB}$ is evaluated. The advantage of using $SEGSNR^{dB}$ over conventional SNR is that it averages the SNR^{dB} values of 15–20 ms segments, giving a better weighting to low-energy unvoiced segments by effectively computing the geometric mean of the SNR values due to averaging in the logarithmic domain instead of the arithmetic mean. Hence the SEGSNR values correlate better with subjective speech quality measures, such as the MOS.

For linear predictive hybrid speech codecs, such as the 13 kbps RPE GSM codec, spectral domain measures typically have better correlation with perceptual assessments than time domain measures. The so-called cepstral distance measure physically represents the logarithmic spectral envelope distortion and is computed as [382,383] :

$$CD = \sqrt{\left[C_0^{in} - C_0^{out}\right]^2 + 2\sum_{j=1}^{3p}\left[C_j^{in} - C_j^{out}\right]^2}, \tag{17.3}$$

where C_j^{in} and C_j^{out}, $j = 0 \ldots 3p$ are the cepstral coefficients of the input and output speech, respectively, and p is the order of the short term predictor filter which is typically 8–10. These cepstral coefficients can be readily computed from the coefficients of the short-term predictor using the results of Reference [383].

17.5 Discontinuous Transmission

The idea of discontinuous transmission (DTX) has long been known in bandwidth and power limited satellite systems, where spectral efficiency is improved using Digital Speech Interpolation (DSI). In mobile radio communications however, the Pan-European system [158] (Chapter 8), GSM is the first digital system to use voice activity detection (VAD) and DTX to further reduce the MS's power consumption and increase spectral efficiency through reducing interference during silent periods. Assuming an average speech activity of 50% and a high number of interferers combined with frequency hopping to randomise the interference load, significant spectral efficiency gains can be achieved. The fundamental problem with DTX is differentiating between speech and noise, while keeping false noise triggering and speech spurt clipping as low as possible. In vehicle-mounted MSs the speech/noise recognition problem is aggravated by the vehicle noise. This problem can be resolved by deploying a combination of threshold comparisons [89,385], and spectral domain techniques [386], [387]. Another important associated problem is the annoying effect of noiseless inactive segments, which is mitigated by introducing comfort noise in these segments at the receiver. This problem is also addressed in references [386], [387].

Another important application of VADs is found in packet reservation multiple access (PRMA) schemes. PRMA is a statistical multiplexing arrangement contrived to improve the efficiency of conventional TDMA systems which will be discussed later in Section 17.8.

17.6 Channel Coding and Bit-mapping

In mobile communications both convolutional and block codes having different strengths and weaknesses are deployed [158][Chapter 4].

In the higher complexity, more robust, lower bit rate speech transmission systems based on the RPE-LTP codec [388] we favour binary Bose-Chaudhuri-Hocquenghem (BCH) block codes. They combine low computational complexity with high error correcting power. They also have reliable error detection properties, which can be used in speech post-enhancement schemes [42], when the received speech is corrupted due to an FEC decoding failure.

For robustness against channel errors some speech bits must be better protected than others. In order to identify the effect of bit errors on speech quality we systematically corrupted each of the 268 RPE-LTP bits and performed a bit sensitivity analysis based on a combination of Segmental SNR (SEGSNR) and Cepstral Distance (CD) degradation [389], [388] as defined in Section 17.4. Specifically, the SEGSNR and CD degradations are shown in Figure 17.5, respectively, where the first 94 bits of a 268-bit frame are portrayed. These bits represent the 36 line spectral frequency(LSF) bits as well as the 58 excitation bits of the first subsegment. Impairments caused by the second, third and fourth subsegments for the bits in positions 95-268 in the frame are identical to those of the first one, i.e. to bits 37-94, whence they are not depicted here. The LSFs determine the spectral envelope, while the LTP parameters determine the fine spectral structure. However, the SEGSNR sensitivities seem to emphasise the low-frequency LSFs and the LTP parameters, while the CD suggests that almost all LSFs are equally important, but underestimates the role of the LTP parameters. This is intuitively likely as the CD excels in measuring the envelope distortion, but not in estimating waveform fidelity. This is why the excitation bits hardly inflict CD degradation, although their role is clearly recognised down to the least (LSB), medium, and most significant bit (MSB) in terms of SEGSNRs in Figure 17.5. Nonetheless, CD objective measures are regarded to have the highest correlation with subjective listening tests.

Our basic observation is that corruption of the bits at beginning and end of both of the SEGSNR and CD figures results in more serious degradations than that of the central bits. The reason for using degradations, rather than the absolute values of SEGSNR and CD is that their absolute values are speaker-dependent, while we are only interested in their degradations due to specific bit errors when compared to their values for unimpaired transmissions. In order to form a balanced view of objective bit-sensitivities and take proper account of subjective quality degradations we decided to amalgamate the SEGSNR and CD measures into a combined sensitivity figure S. Therefore the bits were assigned a sensitivity rank in terms of both CD and SEGSNR, where the most sensitive bit inflicting the highest SEGSNR and CD degradation was allocated the sensitivity figure 1, while the least sensitive one was assigned the sensitivity figure of 94. Then for each bit between 1 and 94 the SEGSNR and CD sensitivity indices deduced from Figure 17.5 were added, in order to form the dimensionless combined sensitivity figure S. Following this the bits were sorted in ascending order of the combined sensitivity figure S, Armed with the sensitivity figures S, an arbitrarily complex embedded FEC scheme can be implemented. For

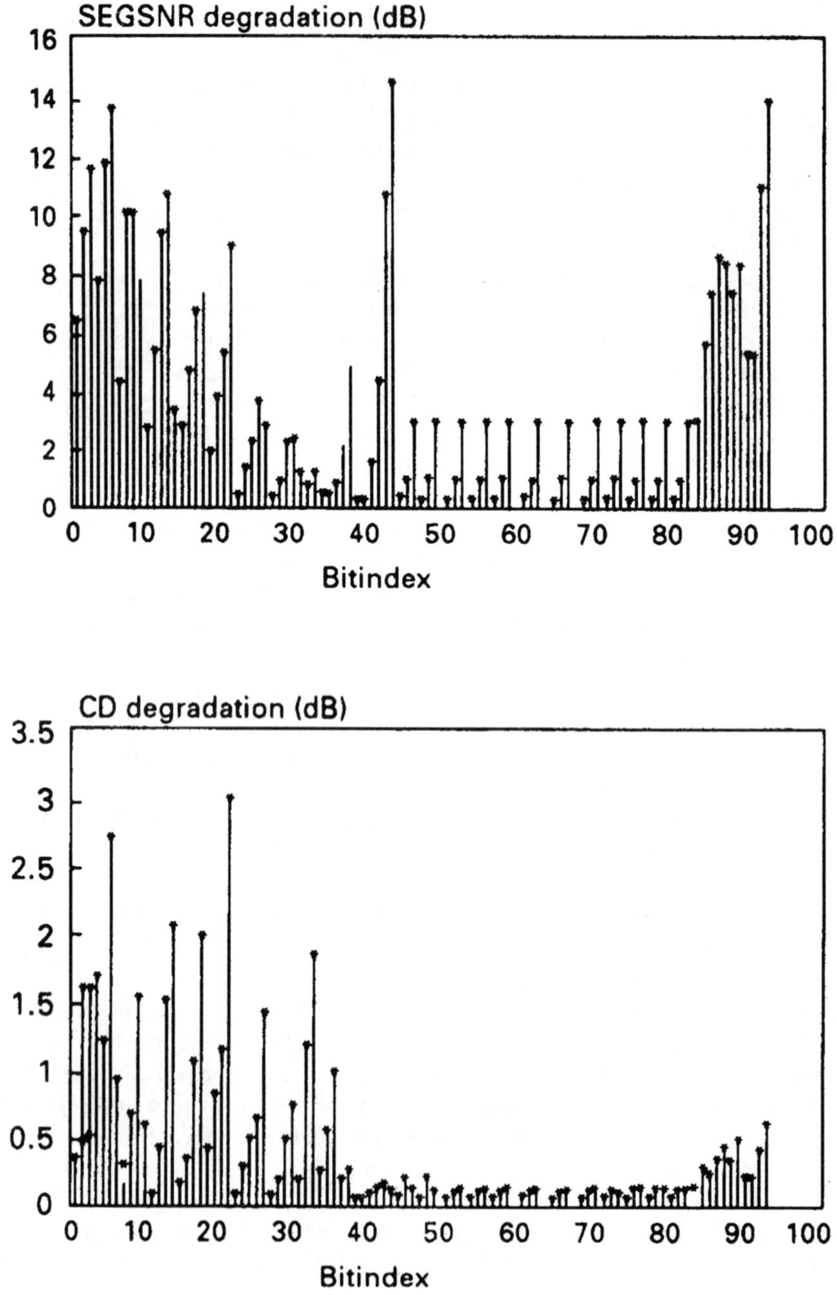

Figure 17.5: Bit-Sensitivities in SEGSNR and CD Degradation for the 13.4 kbps RPE Codec

practical reasons here we opted for two protection classes. Note that assigning any subsegment bit in the RPE frame with an index $95 > X > 36$ to a specific category implies also the inclusion of the corresponding bits $(X + 58)$, $(X + 116)$ and $(X + 174)$ of the other subsegments in the same class, provided that his class could accommodate more bits.

After careful consideration we accommodated 116 bits in the high protection class and assigned a shortened binary BCH(62,29,6) code, while the remaining 152 less significant speech bits are protected by a less powerful shortened BCH(62,38,4) code. The codes are appropriately shortened to embrace the 116 and 152 class one and class two speech bits by four rectangularly interleaved BCH codewords in each category. Then in case of serious speech impairments the reliable error detection capability of the high protection BCH code is invoked to activate the correlative speech post-processing algorithm proposed by D.J. Goodman *et al.* [390], portrayed in [42]. The bit rate of the RPE/BCH system is computed as follows. The 116 highly protected bits are coded into four consecutive BCH(62,29,6) codewords giving a total of 248 bits. Similarly, the 152 low-protection bits are encoded by four BCH(62,38,4) codewords yielding equally 248 bits. Then a total of 496 bits per 20 ms LPC frame are transmitted.

17.7 Speech Transmission Systems

In our simulations we used the GMSK, $\frac{\pi}{4}$-DQPSK and 16-Star QAM modems combined with both the unprotected low-complexity 32 kbps ADPCM codec (as in DECT and CT2) and the 13 kbps RPE-LTP codec with its twin-class FEC. Each modem had the option of either a low or a high complexity demodulator. The high complexity demodulator for the GMSK modem was a maximum likelihood sequence estimator based on the Viterbi algorithm [158], while the low complexity one was a frequency discriminator. For the two multilevel modems either low complexity non-coherent differential detection or a maximum likelihood correlation receiver (MLH-CR) was invoked. Synchronous transmissions and perfect channel estimation were used in evaluating the relative performances of the systems listed in Table 17.7. Our results represent performance upper bounds, allowing relative performance comparisons under identical circumstances.

The system performances apply to microcellular conditions. The carrier frequency was 2GHz, the data rate 400 kBd, and the mobile speed 15m/s. At 400 kBd in microcells the fading is flat and usually Rician. The best and worst Rician channels are the Gaussian and Rayleigh fading channels, respectively, and we performed our simulations for these channels to obtain upper and lower bound performances. Our conditions of 2GHz, 400 kBd and 15m/s is arbitrary. They correspond to a fading pattern that can be obtained for a variety of different conditions, for example, at 900 MHz, 271 kBd and 23m/s. We compare the performances of the systems defined in Table 17.7 when operating according to our standard conditions.

Returning to Table 17.7, the first column shows the system classification letter, the next the modulation used, the third the demodulation scheme employed, the fourth the FEC scheme and the fifth the speech codec deployed. The sixth column gives the estimated relative order of the complexity of the schemes, where the most complex one having a complexity parameter of 12 is the 16-Star QAM, MLH-CR, BCH, RPE-LTP arrangement. All the BCH-coded RPE-LTP schemes have complexity parameters larger than six, while the unprotected ADPCM systems are characterised by values of one to six, depending on the complexity of the modem used. The speech Baud rate and the TDMA user bandwidth are given next. A signalling rate of 400 kBd was chosen for all our experiments, irrespective of the number of modulation levels, to provide a fair comparison for all the systems, under identical propagation conditions. The 400 kBd systems have a total bandwidth of $400/1.35 = 296$ kHz, $2{\cdot}400/1.62 = 494$ kHz and $4 \cdot 400/2.4 = 667$ kHz, respectively. When computing the user bandwidth requirements we will take account of the different bandwidth constraints of GMSK, $\frac{\pi}{4}$-DQPSK and 16-QAM, assuming an identical Baud rate.

In order to establish the speech performance of systems A-L we evaluated the SEGSNR versus channel SNR, and CD versus channel SNR characteristics of these schemes. These experiments yielded 24 curves for AWGN, and 24 curves for Rayleigh fading channels, constituting the best and worst case channels, respectively. Then for the twelve different systems and two different channels we derived the minimum required channel SNR value for near-unimpaired speech quality in terms of both CD and SEGSNR. These values are listed in columns 13 and 14 of Table 17.7.

17.8 Packet Reservation Multiple Access

Packet reservation multiple access (PRMA) is a statistical multiplexing method for conveying speech signals via time division multiple access (TDMA) systems. A full treatment of PRMA was given in references [391], [392] by Goodman *et al.*, while various PRMA-assisted CT systems were proposed in [89, 375, 393]. The operation of PRMA is based on the voice activity detector (VAD) being able to reliably detect inactive speech segments. In our systems a VAD similar to that of the GSM system was used, which is described in Chapter 8 of Reference [158]. Inactive users' TDMA time slots are allocated to other users, who become active. The users who are just becoming active have to contend for the available time slots with a certain permission probability, which is an important PRMA parameter. Previously colliding users contend for the next available time-slot with a less than unity permission probability in order to prevent them from consistently colliding in their further attempts to attain reservation. If more than one user is contending for a free slot, neither of them will be granted it. If, however, only one user requires the time slot, he can reserve it for future use until he becomes inactive. Under heavily loaded network conditions, when many users are contending for a reservation, a speech packet might have to contend for a number of consecutive slots. When the contention delay exceeds 32 ms, the speech packet of 20 ms duration must be dropped. The probability of packet dropping must be kept below 1%, a value inflicting minimal degradation in perceivable speech quality.

Further important PRMA parameters are listed in Table 17.3. The transmitted Baud rate was fixed to 400 kBd, while the source rate was variable, depending on the system used. The source Baud rates for systems A-L are listed in column 7 of Table 17.7. The PRMA frame duration was 20 ms and the number of PRMA slots

1 Syst.	2 Modulator	3 Detector	4 FEC	5 Speech Codec	6 Comp-lexity Order	7 Baud Rate (KBd)	8 TDMA User Bandw. (kHz)	9 No of TDMA Users per Carrier	10 No of PRMA Users per Carrier	11 No of PRMA Users per slot	12 PRMA User Bandw. (kHz)	13 Min SNR (dB) AWGN	14 Min SNR (dB) Rayleigh
A	GMSK	Viterbi	No	ADPCM	2	32	23.7	11	18	1.64	14.5	7	∞
B	GMSK	Freq. Discr.	No	ADPCM	1	32	23.7	11	18	1.64	14.5	21	31
C	$\frac{\pi}{4}$-DQPSK	MLH-CR	No	ADPCM	4	16	19.8	22	42	1.91	10.4	10	28
D	$\frac{\pi}{4}$-DQPSK	Differential	No	ADPCM	3	16	19.8	22	42	1.91	10.4	10	28
E	16-StQAM	MLH-CR	No	ADPCM	6	8	13.3	44	87	1.98	6.7	20	∞
F	16-StQAM	Differential	No	ADPCM	5	8	13.3	44	87	1.98	6.7	21	31
G	GMSK	Viterbi	BCH	RPE-LTP	8	24.8	18.4	12	22	1.83	10.1	1	15
H	GMSK	Freq. Discr.	BCH	RPE-LTP	7	24.8	18.4	12	22	1.83	10.1	8	18
I	$\frac{\pi}{4}$-DQPSK	MLH-CR	BCH	RPE-LTP	10	12.4	15.3	24	46	1.92	8	5	20
J	$\frac{\pi}{4}$-DQPSK	Differential	BCH	RPE-LTP	9	12.4	15.3	24	46	1.92	8	6	18
K	16-StQAM	MLH-CR	BCH	RPE-LTP	12	6.2	10.3	48	96	2.18	4.7	13	25
L	16-StQAM	Differential	BCH	RPE-LTP	11	6.2	10.3	48	96	2.18	4.7	16	24

GMSK: Gaussian Minimum Shift Keying
$\frac{\pi}{4}$-DQPSK: Differential Phase Shift Keying
16-StQAM: 16-level Quadrature Amplitude Modulation
MLH-CR: Maximum Likelihood Correlation Receiver
BCH: Bose Chaudhuri Hocquenghem FEC coding
ADPCM: Adaptive Differential Pulse Code Modulation
RPE-LTP: Regular Pulse Excited speech codec with Long Term Prediction
TDMA: Time Division Multiple Access
PRMA: Packet Reservation Multiple Access

Table 17.2: System parameters [375], ©IEEE, 1994, Williams, Hanzo *et al.*

PRMA parameter	
Channel rate	400 kBd
Source rate	variable (Table 17.7, col. 7)
Frame duration	20 ms
No. of slots (users/carrier)	variable (Table 17.7, col. 9)
Slot duration	variable (20 ms/No. of slots)
Header length	64 bits
Maximum speech delay	32 ms
Permission probability	variable

Table 17.3: Summary of PRMA parameters

was identical to the number of TDMA slots per frame or TDMA users per carrier.

The number of TDMA and PRMA slots per frame is computed from the frame duration as follows. During a PRMA frame of 20 ms the 32 kbps ADPCM speech encoder generates 640 bits of speech information. A header length of 64 bits was assumed to serve signalling and control purposes and so the 704 bits/20 ms = 35.2 kbps source information rate allowed us to support the integer of 400/35.2, namely 11 time slots, i.e., 11 TDMA users per 400 kBd bearer in case of unprotected binary transmissions, such as in systems A and B. When using the 2 and 4 bit/Hz $\frac{\pi}{4}$-DQPSK and 16-Star QAM modems, the packet length was decreased by factors of 2 and 4, respectively, causing the number of slots per frame to increase by similar factors. Thus the use of multilevel modulation increases the number of PRMA slots.

For the FEC-protected scenarios using the 13.4 kbits/s RPE-LTP codec a $2 \times 29 =$ 58 bits long header was used. Each 29 bits block in the header was BCH(62,29,6) coded to give an FEC-coded header of 124 bits. Recall from Section 17.6 that the 268 RPE-LTP bits were FEC coded to 496 bits and hence after adding the signalling header the 20 ms PRMA frame was constituted by 496 + 124 = 620 bits, yielding a bitrate of 31 kbps. The number of PRMA slots supported in case of the 1, 2 and 4 bits/symbol scenarios was 12, 24 and 48, respectively, as seen in column 9 of Table 17.7.

17.8.1 PRMA performance

PRMA efficiency was evaluated in terms of packet dropping probability (P_{drop}) versus number of users for systems A-L, using the number of slots listed in column 9 of Table 17.7. Our results are shown in Figure 17.6 for both the unprotected and FEC-coded scenarios represented by systems A-F and G-L, respectively.

The number of users supported by the schemes providing 11, 22 and 44 time slots at a 1% packet dropping probability is seen to be 18, 42 and 87, respectively. Their values are listed in column 10 of Table 17.7. Then the relative number of PRMA users per time slot is computed as 18/11=1.64, 37/20=1.85 and 67/35=1.91 users/slot, respectively. As seen from columns 9 and 10 of Table 17.7, the corresponding figures for the 1, 2, and 4 bits/symbol modem, 32 kbps speech codec combinations are 18/11=1.64, 42/22=1.91 and 87/44=1.98, respectively, emphasizing the higher efficiency of the systems with multi-level modems. This fact is further accentuated by their less steep dropping probability curves, yielding a more benign overloading characteristic and higher grade of service. As seen from Figure 17.6, similar tendencies are valid for our FEC-protected RPE-LTP schemes as well. The number of users supported in case of the 1, 2, and 4 bit/symbol systems G-L is 22, 46 and 96, respectively. The associated PRMA efficiencies are expressed using columns 9 and 10 of Table 17.7 as 22/12=1.83, 46/24=1.92 and 2.18 users/slot, respectively. The relative number of PRMA users/slot values are listed in column 11 of Table 17.7. The normalised PRMA user bandwidth figures listed in column 12 of Table 17.7 can be derived by dividing the TDMA user bandwidth values in column 8 by the relative number of PRMA users/slot given in column 11 of Table 17.7. The normalised PRMA bandwidth figures vary between the extreme values of 14.5 and 4.7, emphasizing again the superiority of the multi-level schemes.

We note, however, that these gains are reduced in SIR-limited scenarios as we have seen in Chapter 17, due to the less dense frequency reuse of multilevel modems. Nevertheless, multilevel modulation schemes result in higher PRMA gains than their lower level counterparts.

17.9 Summary

In this Chapter we compared the performance of a QAM system against binary and quartenary modulation schemes, showing the advantages which can be gained in certain applications, in this case that of PRMA. This highlights the complex range of choices which face a mobile radio system designer and shows that the modulation scheme chosen can have far-reaching consequences as regards to system performance. A variety of further QAM-based speech, wideband audio and image communication schemes can be found in References [42, 42, 85, 87, 88, 88, 89, 89–95, 375, 376, 388, 389, 393, 394].

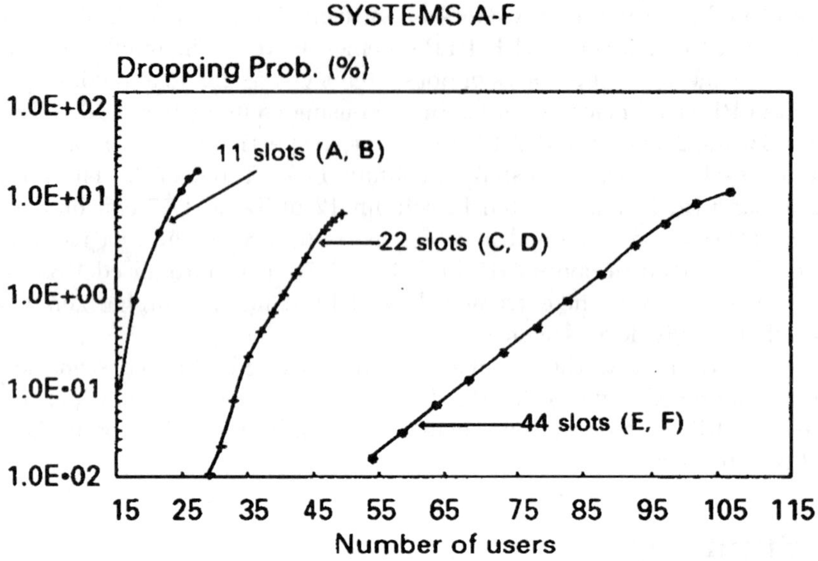

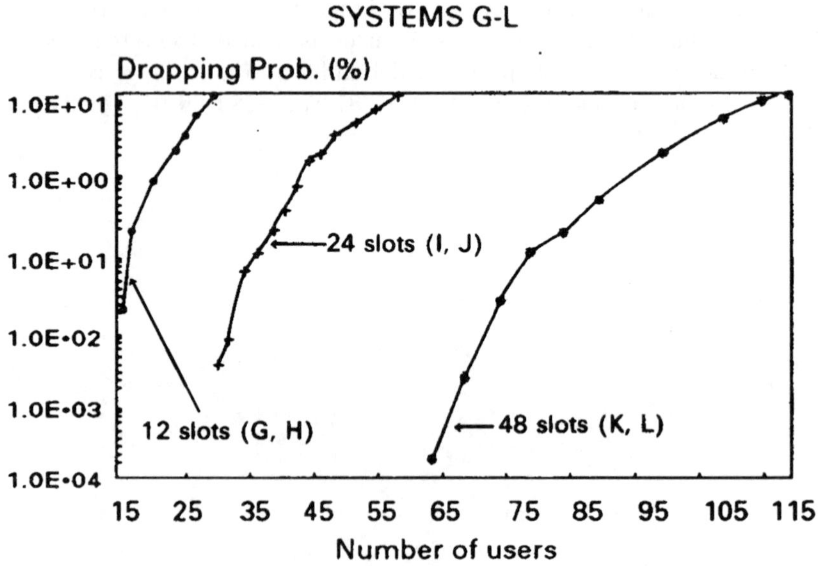

Figure 17.6: Packet dropping versus number of users performance of systems A-L with and without channel coding

Chapter 18

Introduction to Orthogonal Frequency Division Multiplexing

18.1 Introduction

In this chapter we examine orthogonal frequency division multiplexing (OFDM) as a means of dealing with the channel-induced linear distortion problems encountered when transmitting over a dispersive radio channel. The fundamental principle of orthogonal multiplexing originates from Chang [71], and over the years a number of researchers have investigated this technique [72–74,77–82,97–99]. Despite its conceptual elegance, until recently its deployment has been mostly limited to military applications due to implementational difficulties. However, it has recently been adopted as the new European digital audio broadcasting (DAB) standard, and this consumer electronics application underlines its significance as a broadcasting technique [75,76,83,137,138].

In the OFDM scheme of Figure 18.1 the serial data stream of a traffic channel is passed through a serial-to-parallel convertor which splits the data into a number of parallel channels. The data in each channel is applied to a modulator, such that for N channels there are N modulators whose carrier frequencies are f_0, f_1, ..., f_{N-1}. The difference between adjacent channels is Δf and the overall bandwidth W of the N modulated carriers is $N\Delta f$, as shown in Figure 18.2.

These N modulated carriers are then combined to give an OFDM signal. We may view the serial-to-parallel convertor as applying every Nth symbol to a modulator. This has the effect of interleaving the symbols into each modulator, e.g. symbols S_0, S_N, S_{2N}, ... are applied to the modulator whose carrier frequency is f_1. At the receiver the received OFDM signal is demultiplexed into N frequency bands, and the N modulated signals are demodulated. The baseband signals are then recombined using a parallel-to-serial convertor.

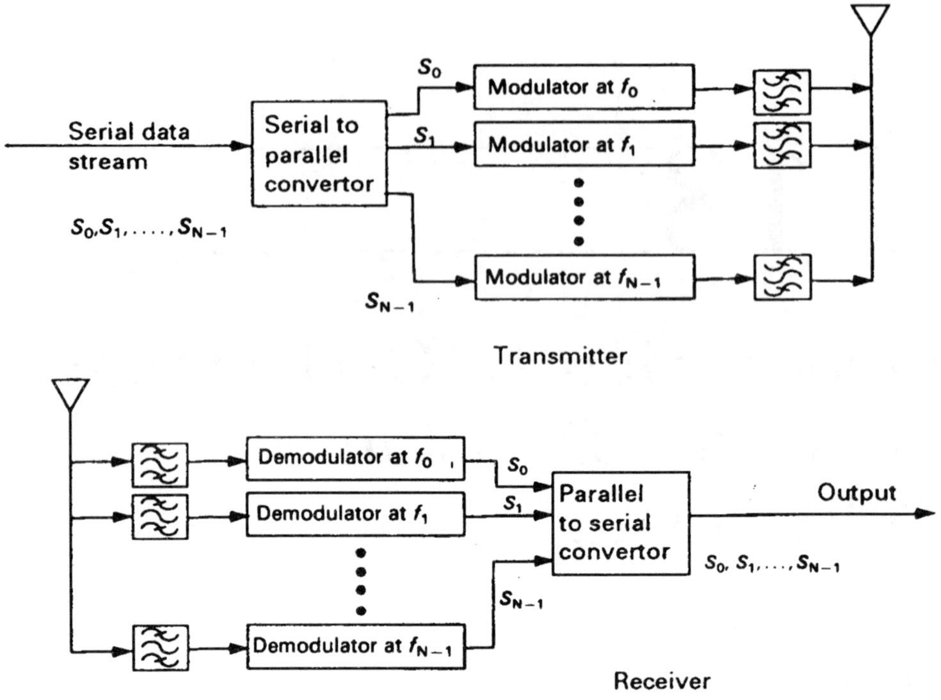

Figure 18.1: Basic OFDM system

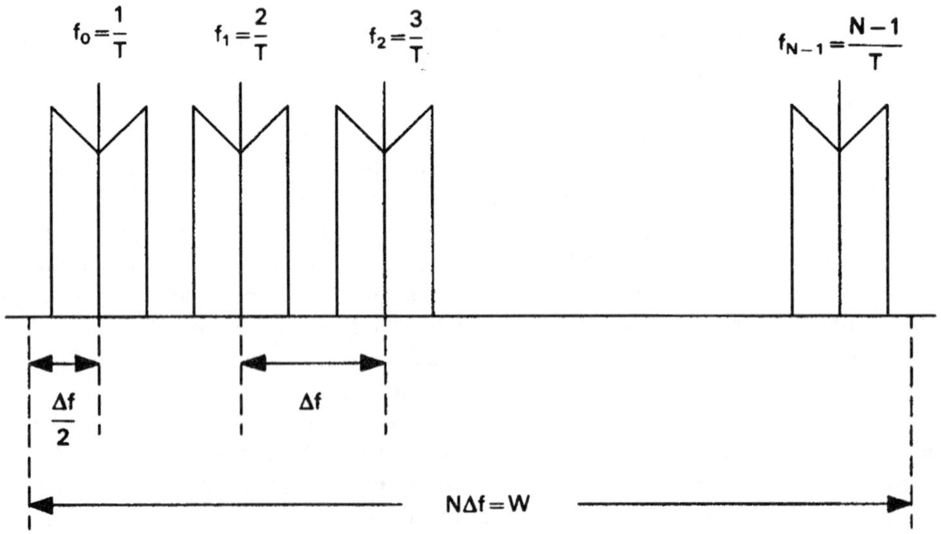

Figure 18.2: FDM frequency assignment

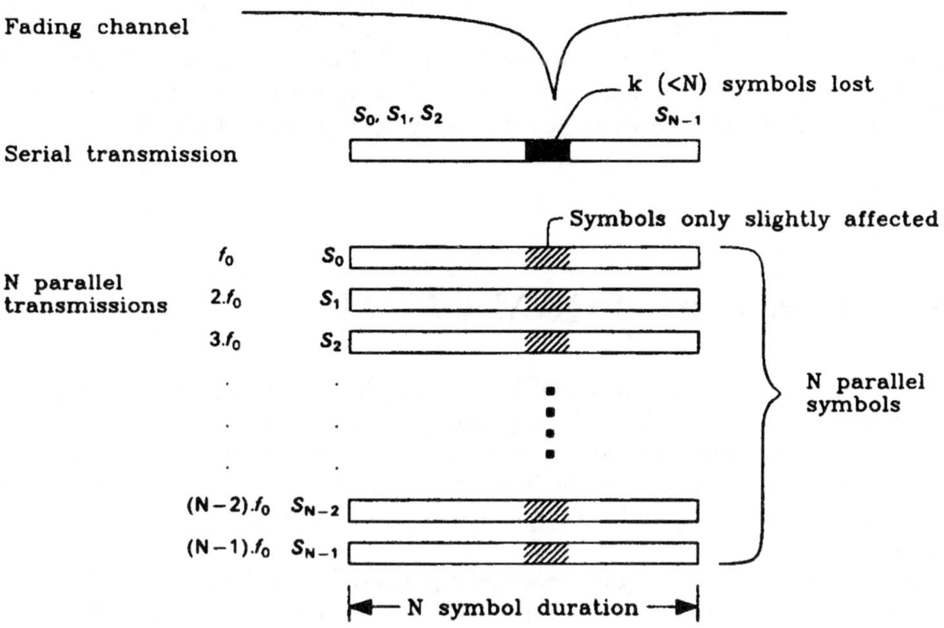

Figure 18.3: Effect of a fade on serial and parallel systems

In the more conventional approach the traffic data is applied directly to the modulator operating with a carrier frequency at the centre of the transmission band $f_0, \ldots, f_{N-1}$, ie at $(f_{N-1} + f_0)/2$, and the modulated signal occupies the entire bandwidth W. When the data is applied sequentially the effect of a deep fade in a mobile channel is to cause burst errors. Figure 18.3 shows the serial transmission of symbols S_0, S_1, ..., S_{N-1}, while the solid shaded block indicates the position of the error burst which affects $k < N$ symbols. By contrast, during the N-symbol period of the conventional serial system, each OFDM modulator carries only one symbol, and the error burst causes severe signal degradation for the duration of k serial symbols. This degradation is shown cross-hatched. However, if the error burst is only a small fraction of the symbol period then each of the OFDM symbols may only be slightly affected by the fade and they can still be correctly demodulated. Thus while the serial system exhibits an error burst, no errors or few errors may occur using the OFDM approach.

A further advantage of OFDM is that because the symbol period has been increased, the channel delay spread is a significantly shorter fraction of a symbol period than in the serial system, potentially rendering the system less sensitive to ISI than the conventional serial system.

A disadvantage of the OFDM approach shown in Figure 18.1 is the increased complexity over the conventional system caused by employing N modulators and filters at the transmitter and N demodulators and filters at the receiver. This complexity can be reduced by the use of the discrete Fourier transform (DFT), typically implemented as a fast Fourier transform (FFT), as will be shown in Section 18.3. The simple conceptual reason for this is that – as it will be argued in the context of Figure 18.4 – the OFDM sub-channel modulators use a set of harmonically related sinusoidal and cosinusoidal sub-channel carriers, just like the basis functions of the DFT.

18.2 Principles of QAM-OFDM

The simplest version of the basic OFDM system has N sub-bands, each separated from its neighbour by a sufficiently large guard band in order to prevent interference between signals in adjacent bands. However, the available spectrum can be used much more efficiently if the spectra of the individual sub-bands are allowed to overlap. By using coherent detection and orthogonal sub-band tones the original data can be correctly recovered.

In the system shown in Figure 18.4, the input serial data stream is rearranged into a sequence $\{d_n\}$ of N QAM symbols at baseband. Each serial QAM symbol is spaced by $\Delta t = 1/f_s$ where f_s is the serial signalling or symbol rate. At the nth symbol instant, the QAM symbol $d(n) = a(n)+jb(n)$ is represented by an in-phase component $a(n)$ and a quadrature component $b(n)$. A block of N QAM symbols are applied to a serial-to-parallel convertor and the resulting in-phase symbols $a(0)$, $a(1)$, ..., $a(N-1)$, and quadrature symbols $b(0)$, $b(1)$, ..., $b(N-1)$ are applied to N pairs of balanced modulators. The quadrature components $a(n)$ and $b(n)$, $n = 0, 1, ..., N-1$, modulate the quadrature carriers $\cos w_n t$ and $\sin w_n t$, respectively. Notice that the signalling interval of the sub-bands in the parallel system is N times longer than that of the serial system giving $T = N\Delta t$, which corresponds to an N-times lower signalling rate. The sub-band carrier frequencies $w_n = 2\pi n f_0$ are spaced apart by $w_0 = 2\pi/T$.

The modulated carriers $a(n)\cos w_n t$ and $b(n)\sin w_n t$ when added together constitute a QAM signal, and as $n = 0, 1, ..., N-1$ we have N QAM symbols at RF, where the nth QAM signal is given by:

$$\begin{aligned} X_n(t) &= a(n)\cos w_n t + b(n)\sin w_n t \\ &= \gamma(n)\cos(w_n t + \psi_n) \end{aligned} \tag{18.1}$$

where

$$\gamma(n) = \sqrt{a^2(n) + b^2(n)} \tag{18.2}$$

and

$$\psi_n = \tan^{-1}\left(\frac{b(n)}{a(n)}\right) \tag{18.3}$$
$$n = 0, 1, ..., N-1.$$

By adding the outputs of each sub-channel signal $X_n(t)$ whose carriers are offset by

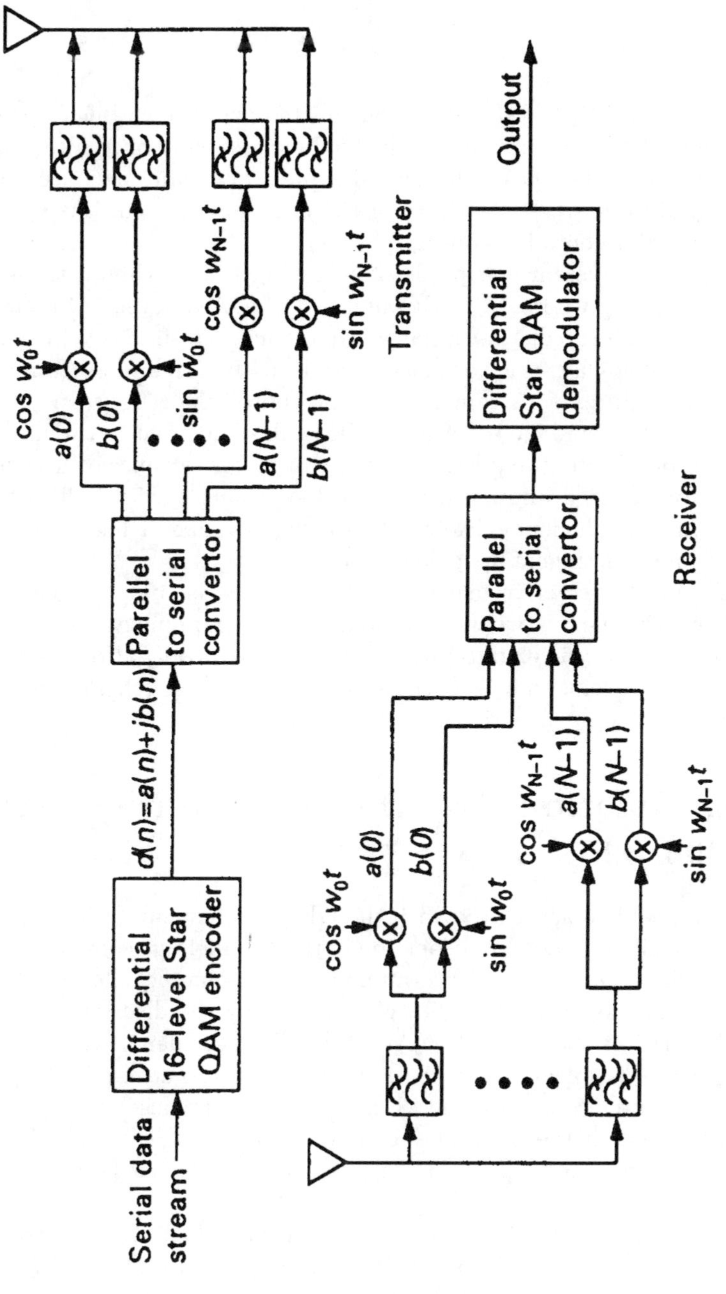

Figure 18.4: Detailed OFDM system

$\omega_0 = 2\pi/T$ we obtain the FDM/QAM signal

$$D(t) = \sum_{n=0}^{N-1} X_n(t). \tag{18.4}$$

This set of N FDM/QAM signals is transmitted over the mobile radio channel. At the receiver this OFDM signal is demultiplexed using a bank of N filters to regenerate the N QAM signals. The QAM baseband signals $a(n)$ and $b(n)$ are recovered and turned into serial form $\{d_n\}$. Recovery of the data ensues using the QAM baseband demodulator and differential decoding.

In theory, such a system is capable of achieving the maximum transmission rate of $log_2\ Q\ bits/s/Hz$, where Q is the number of QAM levels. In practice, there is some spectral spillage due to adjacent frequency sub-bands which reduces this efficiency. Spectral spillage due to the sub-bands at the top and bottom of the overall frequency band requires a certain amount of guard space between adjacent users. Furthermore, spectral spillage between OFDM sub-bands due to the imperfections of each of the sub-band filters requires that the sub-bands be spaced further apart than the theoretically required minimum amount, decreasing spectral efficiency. In order to obtain the highest efficiency, the block size should be kept high and the sub-band filters made to meet stringent specifications.

One of the most attractive features of this scheme is that the bandwidth of the sub-channels is very narrow when compared to the communications channel's coherence bandwidth. Therefore, flat-fading narrowband propagation conditions apply. The sub-channel modems can use almost any modulation scheme, and QAM is an attractive choice in some situations.

18.3 Modulation by Discrete Fourier Transform [395, 396]

A fundamental problem associated with the OFDM scheme described is that in order to achieve high resilience against dispersion in the channels we consider, the block size, N, has to be high, requiring a large number of sub-channel modems. Fortunately, it can be shown mathematically [395, 396] that taking the discrete Fourier transform (DFT) of the original block of N QAM symbols and then transmitting the DFT coefficients serially is exactly equivalent to the operations required by the OFDM transmitter of Figure 18.4. Substantial hardware simplifications can be achieved for OFDM transmissions if the bank of sub-channel modulators/demodulators is implemented using the computationally efficient pair of inverse fast Fourier transform and fast Fourier transform (IFFT/FFT).

Observe that the stream of complex baseband symbols $X(k)$ to be transmitted can be described both in terms of in-phase $I(k)$ and quadrature phase $Q(k)$ components, as well as by magnitude and phase. In case of a square-shaped constellation which was used in Chapter 10, $X(k) = I(k) + jQ(k)$ might be a more convenient formalism; for the star QAM constellation introduced in Chapter 11, $X(k) = |X(k)|e^{j\Phi(k)}$ appears to be more attractive.

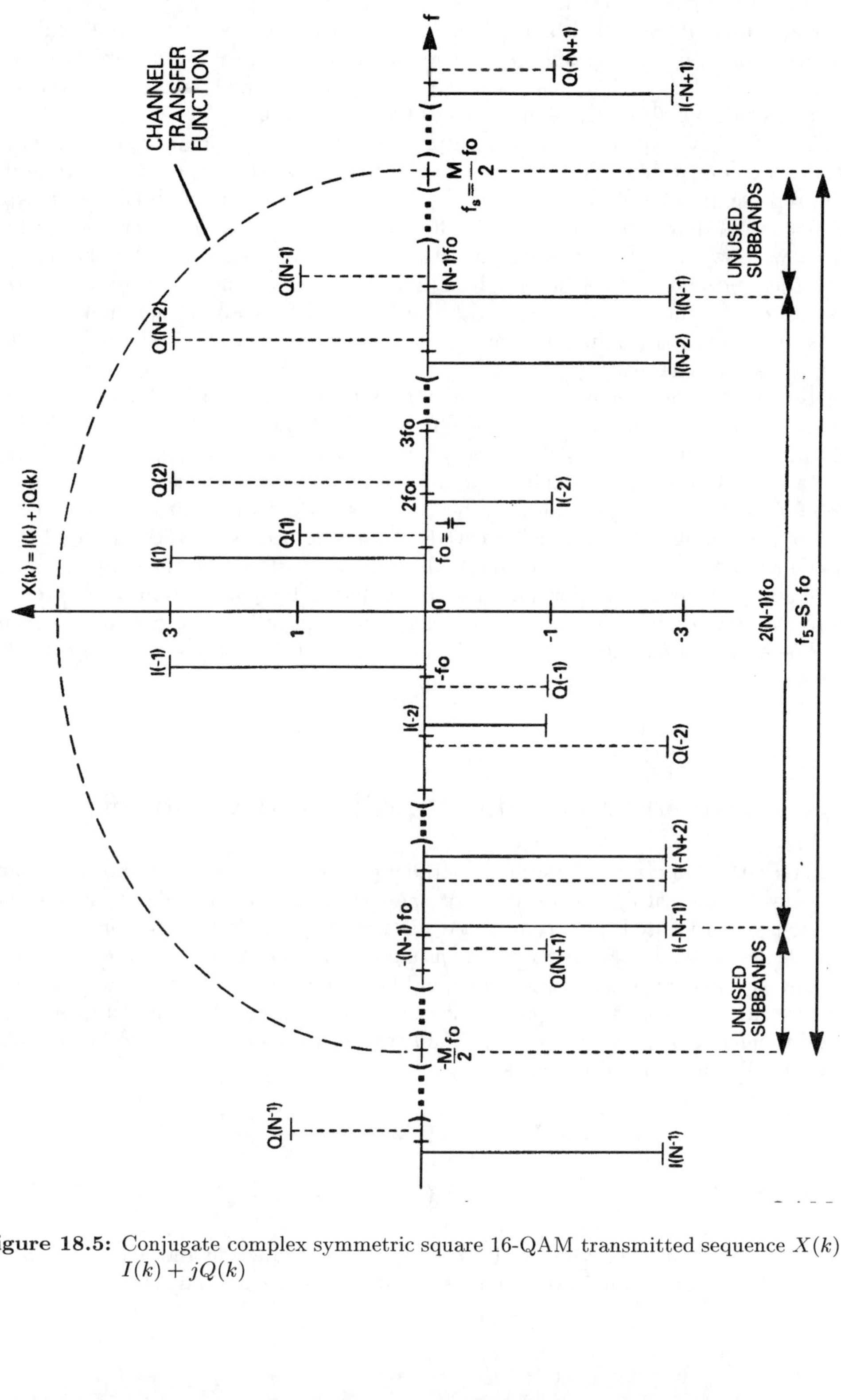

Figure 18.5: Conjugate complex symmetric square 16-QAM transmitted sequence $X(k) = I(k) + jQ(k)$

Observe in Figure 18.5 that the frequency region $(N-1)f_0 < f_n \le \frac{M}{2}f_0$ represents the typical unused transition band of the communications channel, where it exhibits significant amplitude and group delay distortion. However, in the narrow sub-bands the originally wide band frequency selective fading channel can be rendered frequency-flat upon using a sufficiently high number of sub-channels.

The set of QAM symbols can be interpreted as a spectral domain sequence, which due to its conjugate complex symmetry will have a real IDFT pair in the time domain, representing the real modulated signal. The rectangular modulation elements $m_T(t) = rect\frac{t}{T}$ have an infinite bandwidth requirement, since in effect they rectangularly window the set of orthogonal basis functions of constituted by the carriers $rect\frac{t}{T} \cdot \cos w_{0l}t$ and $rect\frac{t}{T} \cdot \sin w_{0l}t$. It is possible to use a time-domain raised cosine pulse instead of the $rect\frac{t}{T}$ function [73], but this will impose further slight impairments on the time-domain modulated signal which is also exposed to the hostile communications channel.

Observe that the representation of $m_0(t)$ by its $\Delta t = 1/f_s$-spaced samples is only correct if $m_0(t)$ is assumed to be periodic and bandlimited to $2(N-1)f_0$. This is equivalent to saying that $m_0(t)$ can only have a bandlimited frequency domain representation if in the time domain it expands from $-\infty$ to ∞. Due to sampling at a rate of $f_s = 1/\Delta t$ the spectral lobes become periodic at the multiples of f_s, but if the Nyquist sampling theorem is observed, no aliasing occurs. In order to fulfil these requirements, the modulating signal $m_0(t)$ derived by IFFT from the conjugate complex symmetric baseband information signal X_l has to be quasi-periodically extended before transmission via the channel, at least for the duration of the channel's memory. The effects of bandlimited transmission media will be discussed in the following section.

18.4 Transmission via Bandlimited Channels

The DFT/IDFT operations assume that the input signal is periodic in both time and frequency domain with a periodicity of M samples. If the modulated sample sequence is periodically repeated and transmitted via the lowpass filter (LPF) preceding the channel, the channel is excited with a continuous, periodic signal. However, in order not to waste precious transmission time and hence channel capacity we would like to transmit only one period of $m_0(t)$ constituted by M samples. Assuming a LPF with a cut-off frequency of $f_c = 1/2\Delta t = f_s/2$ and transmitting only one period of $m_0(i\Delta t)$ the channel's input signal becomes:

$$\begin{aligned} m_{0,LPF}(t) &= m_0(i\Delta t) * \frac{1}{\Delta t}\frac{\sin(\pi t/\Delta t)}{\pi t/\Delta t} \\ &= m_0(i\Delta t) * \frac{1}{\Delta t} sinc\frac{\pi t}{\Delta t}, \end{aligned} \tag{18.5}$$

where the LPF's impulse response is given by the *sinc* function and hence $m_{0,LPF}(t)$ has an infinite time domain duration. One period of the periodic modulated signal

$m_{0,p}(i\Delta t)$, is given by

$$m_0(i\Delta t) = m_{0,p}(i\Delta t) rect\frac{t}{T}. \tag{18.6}$$

The convolution in Equation 18.5 can be written as:

$$m_{0,LPF}(t) = \sum_{i=0}^{M-1} m_0(i\Delta t)\frac{1}{\Delta t} sinc\frac{\pi(t - i\Delta t)}{\Delta t}. \tag{18.7}$$

In the spectral domain this is equivalent to writing

$$M_{0,LPF}(f) = M_0(f) rect\frac{f}{f_c}, \tag{18.8}$$

where $M_0(f) = FFT\{m_0(i\Delta t)\}$ and $H_{LPF}(f) = rect\frac{f}{f_c}$ is the LPF's frequency domain transfer function. Transforming Equation 18.6 into the frequency domain yields:

$$\begin{aligned} M_0(f) &= FFT\{m_{0,p}(i\Delta t) rect\frac{t}{T}\} \\ &= M_{0,p}(f) * \frac{1}{f_0} sinc\frac{\pi f}{f_0}, \end{aligned} \tag{18.9}$$

where $M_{0,p}(f)$ is the frequency domain representation of $m_{0,p}(i\Delta t)$, which is convolved with the Fourier transform of the $rect\frac{t}{T}$ function. Now $M_0(f)$ of Equation 18.9 is lowpass filtered according to Equation 18.8, giving:

$$\begin{aligned} M_{0,LPF}(f) &= M_0(f) rect\frac{f}{f_c} \\ &= \left[M_{0,p}(f) * \frac{1}{f_0} sinc\frac{\pi f}{f_0}\right] rect\frac{f}{f_c}. \end{aligned} \tag{18.10}$$

So the effect of the time domain truncation of the periodic modulated signal $m_{0,p}(i\Delta t)$ to a single period as in Equation 18.6 manifests itself in the frequency domain as the convolution of Equation 18.9, generating the infinite bandwidth signal $M_0(f)$. When $M_0(f)$ is lowpass filtered according to Equation 18.10, it becomes bandlimited to f_c, and its Fourier transform pair $m_{0,LPF}(t)$ in Equation 18.5 has an infinite time domain support due to the convolution with $sinc(\pi t/\Delta t)$. This phenomenon results in interference due to time domain overlapping between consecutive transmission blocks, which can be mitigated by quasi-periodically extending $m_0(i\Delta t)$ for the duration of the memory of the channel before transmission. At the receiver only the un impaired central section is used for signal detection.

In order to portray a practical OFDM scheme and to aid the exposition here and in the previous section, in Figure 18.6 we plotted a few characteristic signals.

Figure 18.6(a) shows the transmitted spectrum using $M = 128$ and rectangular 16-QAM having $I, Q = \pm 1, \pm 3$.

In contrast to Figure 18.5, where the I and Q components were portrayed next to each other in order to emphasise that they belong to the same frequency, here the I and Q components associated with the same frequency are plotted with the same spacing

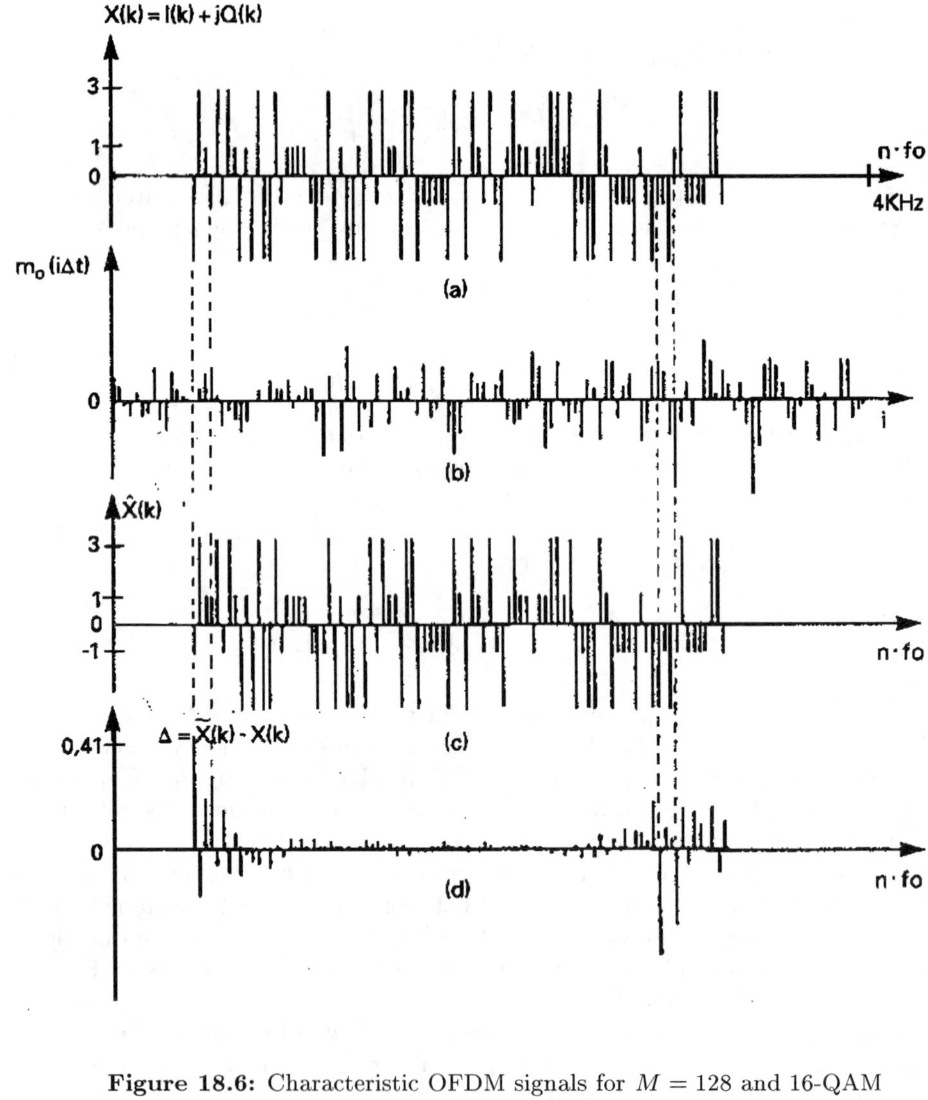

Figure 18.6: Characteristic OFDM signals for $M = 128$ and 16-QAM

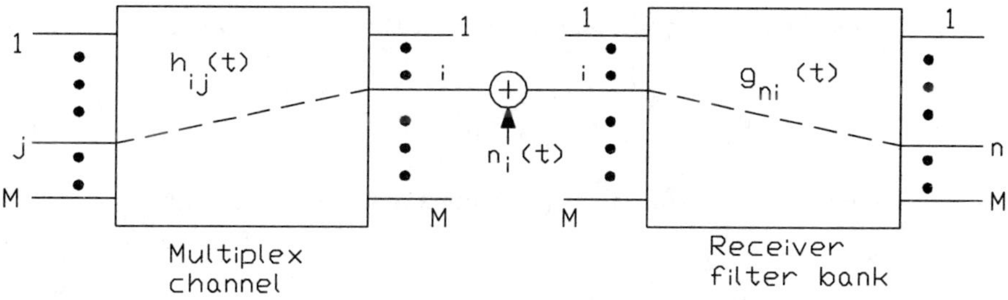

Figure 18.7: Multidimensional interference model [398] ©IEEE, 1975, Van Etten

as adjacent sub-bands. In Figure 18.6(a) there are 64 legitimate frequencies between 0 and 4 kHz, corresponding to 128 lines. Observe, however, that for frequencies of 0-300 Hz and 3.4-4.0 kHz we have allocated no QAM symbols. This is because this signal was transmitted after conjugate complex extension and IFFT over the M1020 CCITT telephone channel simulator, which has a high attenuation in these frequency slots. The real modulated signal after IFFT is plotted in Figure 18.6(b), which is constituted by 128 real samples. At the receiver this signal is demodulated by FFT in order to derive the received signal $\tilde{X}(k)$, which is then subjected to hard decision delivering the sequence $\hat{X}(k)$ seen in Figure 18.6(c). Lastly, Figure 18.6(d) portrays the error signal $\Delta(k) = X(k) - \tilde{X}(k)$, where large errors can be observed towards the transmission band edges due to the M1020 channel's attenuation. The interpretation of these characteristic signals will be further augmented in Section 18.6, as further details of the OFDM system are unravelled.

18.5 Generalised Nyquist Criterion

Using frequency division or orthogonal multiplexing (OFDM) two different sources of interference can be identified [397–401]. Intersymbol interference (ISI) is defined as the crosstalk between signals within the same sub-channel of consecutive FFT frames, which are separated in time by the signalling interval T. Inter channel interference (ICI) is the cross talk between adjacent sub-channels or frequency slots of the same FFT frame. Since the effects of these interference sources and their mitigation methods are similar, it is convenient to introduce the term multidimensional interference (MDI) [399].

In order to describe MDI we assume the linear system model depicted in Figure 18.7, which is the concatenation of the M-sub-channel multiplex transmission medium to the M-sub-channel receiver filter bank. The channel's impulse response between its input j and output i is denoted by $h_{ij}(t)$, while that of the receiver filter bank is $g_{ni}(t)$. With the assumption of this linear system model the output signal of the receiver filter is the superposition of the system responses due to all input signals 1, ..., M. Our objective is to optimise the linear receiver in order to minimise the effects of AWGN and MDI, which will reduce the BER of the receiver.

Assuming a sub-channel noise spectral density of N_i and receiver filter transfer functions of $G_{ni}(f)$ between the input i and output n, the noise variance at output n becomes:

$$\sigma_n{}^2 = \sum_{i=1}^{M} N_i \int_0^\infty G_{ni}^2(f)df, \tag{18.11}$$

which can also be expressed in terms of $g_{ni}(t)$ using Parseval's theorem:

$$\sigma_n{}^2 = \sum_{i=1}^{M} N_i \int_0^\infty g_{ni}^2(\tau)d\tau. \tag{18.12}$$

Following the approaches in [399–401] the noise variance can be minimised as a function of the receiver filter impulse response $g_{ni}(t)$. Supposing that the sub-channel input signals a_j $j = 1, \ldots, M$ can take any value from the mutually independent, equiprobable set of L-ary alphabet, then the channel's input vector in the kth signalling interval is given by:

$$\mathbf{u^T(k)} = \left[a_1^k, a_2^k, \ldots, a_M^k\right]. \tag{18.13}$$

Observe that the number of possible input vectors is L^M.

From Figure 18.7 we can see that the system's response at the receiver output can be computed through convolution of the channel and receiver filter impulse responses. The input signal of sub-channel j generates an output at every receiver filter input $i = 1, \ldots, M$, each of which gives rise to a response at the nth receiver filter output. In order to derive the nth output, the contributions must be summed first fixing the jth sub-channel input over the receiver filter input index i. Then, since the input signal vector $\mathbf{u^T(k)}$ excites all sub-channel inputs $j = 1, \ldots, M$, summing over the nth index j gives the output signal of the receiver filter:

$$s_n{}^k(t) = \sum_{j=1}^{M} a_j{}^k \sum_{i=1}^{M} \int_0^\infty g_{ni}(\tau)h_{ij}(t-\tau)d\tau. \tag{18.14}$$

At the received sampling instants $(t_s + lT)$ we have

$$s_n{}^k(t_s + lT) = \sum_{j=1}^{M} a_j{}^k \sum_{i=1}^{M} \int_0^\infty g_{ni}(\tau)h_{ij}(t_s + lT - \tau)d\tau. \tag{18.15}$$

The optimisation problem now is to find the system transfer functions, which minimise the noise variance at the output of the nth receiver filter under the constraint of constant $s_n{}^k(t_s + lT)$, for all l and k. Using the technique of variational calculus, the following functional must be minimised [398]:

$$F_n = \sigma_n^2 - 2\sum_{k=1}^{L^M}\sum_{l} \lambda_{nkl} s_n{}^k(t_s + lT), \tag{18.16}$$

where λ_{nkl} is the so-called Lagrange multiplier. On substituting Equations 18.12

and 18.15 into Equation 18.16 we have:

$$\begin{aligned} F_n &= \sum_{i=1}^{M} N_i \int_0^{\infty} g_{ni}{}^2(\tau)d\tau \\ &- 2\sum_{k=1}^{L^M} \sum_l \lambda_{nkl} \sum_{j=1}^{M} a_j{}^k \sum_{i=1}^{M} \int_0^{\infty} g_{ni}(\tau)h_{ij}(t_s + lT - \tau)d\tau. \end{aligned} \tag{18.17}$$

The minimisation [398] of F_n in Equation 18.17 yields:

$$g_{ni}(t) = \frac{1}{N_i} \sum_{k=1}^{L^M} \sum_l \lambda_{nkl} \sum_{j=1}^{M} a_j{}^k h_{ij}(t_s + lT - t). \tag{18.18}$$

Assuming identical noise spectral density in all sub-channels, i.e. that $N_i = N$ for all i and introducing the shorthand

$$c_{njl} = \frac{1}{N} \sum_{k=1}^{L^M} a_j{}^k \lambda_{nkl}, \tag{18.19}$$

then Equation 18.18 can be simplified to:

$$g_{ni}(t) = \sum_{j=1}^{M} \sum_l c_{njl} h_{ij}(t_s + lT - t). \tag{18.20}$$

As seen from Equation 18.20, the set of optimum receiver filters $g_{ni}(t)$ depends on the sub-channel impulse responses $h_{ij}(t)$. The receiver filter impulse responses $g_{ni}(t)$ from the ith filter input to all filter outputs have to be matched to the responses $h_{ij}(t)$ from all sub-channel inputs $j = 1, \ldots, M$ to the ith subchannel output or received filter input, as detailed in section 4.5.2.

In order to derive the matched receiver filter response $g_{ni}(t)$ according to Equation 18.20, we have to superimpose the responses of all receiver filters that are matched to the OFDM sub-channel outputs due to the jth OFDM sub-channel input. The summation over the index l weighted by the coefficients c_{njl} in Equation 18.20 represents the ISI due to the T-spaced h_{ij} contributions from previous sampling instants.

Using our general MDI model in Figure 18.7, the overall impulse response at the sampling instant (lT) between the jth OFDM sub-channel input and the nth receiver filter output can be denoted by $f_{nj}(lT)$, $n, j = 1, \ldots, M$ or in matrix form:

$$\mathbf{F_l} = \begin{bmatrix} f_{11}(lT) & f_{12}(lT) & \ldots & f_{1M}(lT) \\ f_{21}(lT) & f_{22}(lT) & \ldots & f_{2M}(lT) \\ \vdots & & & \\ f_{M1}(lT) & f_{M2}(lT) & \ldots & f_{MM}(lT) \end{bmatrix}. \tag{18.21}$$

The total accumulated MDI at the nth receiver filter output due to both ISI and ICI can be defined [398] as:

$$MDI_n = \frac{|f_{nn}(0)| - \sum_l \sum_{i=1}^{M} |f_{ni}(l)|}{f_{nn}(0)}. \tag{18.22}$$

It is plausible that the condition of MDI-free transmission is met if there is no cross-talk amongst the sub-channels, i.e. the matrix of Equation 18.21 is the diagonal identity matrix for any arbitrary sampling instant (lT). In formal terms this means that for $n, j = 1, 2, \ldots, M$, $l = 0, \pm 1, \pm 2, \ldots, \pm\infty$ the generalised Nyquist criterion proposed by Schnidman [397] must be satisfied:

$$\sum_l f_{nj}(lT) = \begin{cases} 1 & \text{if } n = j \\ 0 & \text{otherwise.} \end{cases} \tag{18.23}$$

If Equation 18.23 is satisfied, the MDI term in Equation 18.22 becomes zero and the generalised Nyquist criterion can be fulfilled by the appropriate choice of the coefficients c_{njl}.

Quite clearly, the generalised Nyquist criterion in Equation 18.23 requires not only that the conventional Nyquist criterion shall be met, i.e. $f_{nn}(lT) = \delta_l$, $l = 0, 1, \ldots, \infty$, where δ_l is the Kronecker delta, in order to render the ISI from other signalling intervals zero. It also requires that all the other $M(M-1)$ sub-channels with their matched receiver filters shall have a zero-crossing in their impulse responses, i.e. for $n, j = 1, 2, \ldots, M$ and $l = 0, \pm 1, \ldots, \pm\infty$ we have that

$$F_{nj}(lT) = \delta_{nj}\delta_l \tag{18.24}$$

shall be satisfied, where

$$\delta_{nj} = \begin{cases} 1 & \text{if } n = j \\ 0 & \text{otherwise.} \end{cases} \tag{18.25}$$

Therefore, the generalised Nyquist criterion requires the equivalent folded-in baseband transfer characteristic introduced in Chapter 4 to be an ideal lowpass characteristic for $n = j$ and zero otherwise.

18.6 Basic OFDM Modem Implementations

A historic OFDM modem implementation was proposed by Hirosaki in Reference [78] and is shown in Figure 18.8. This implementation is based on staggered QAM (SQAM) or offset QAM (OQAM), where the quadrature components are delayed by half a signalling interval with respect to each other in order to reduce the signal envelope fluctuation (see Chapter 4). The system is constituted by M synchronised baseband channels operating with a Baud rate of $f_0 = 1/T$. The baseband modulating signals of sub-channels i and $(i + M/2)$ are amplitude modulated onto the carriers $f_i = [f_0 + (i-1)f_0]$, $i = 1, \ldots, M/2$, with the carrier suppressed. This implies that the sub-channels are spaced according to the Baud rate of $f_0 = 1/T$. Then the sum of the sub-channel signals i and $(i + M/2)$, $i = 1, \ldots, M/2$ form the ith SQAM sub-channel, where the in-phase and quadrature phase modulating signals are shifted by $T/2$ with respect to each other. The transmit and receive filters $G(\omega)$ are identical square-root raised-cosine Nyquist filters and an equaliser implementation is also proposed in [78].

However, Hirosaki's conceptually simple OFDM implementation may become im-

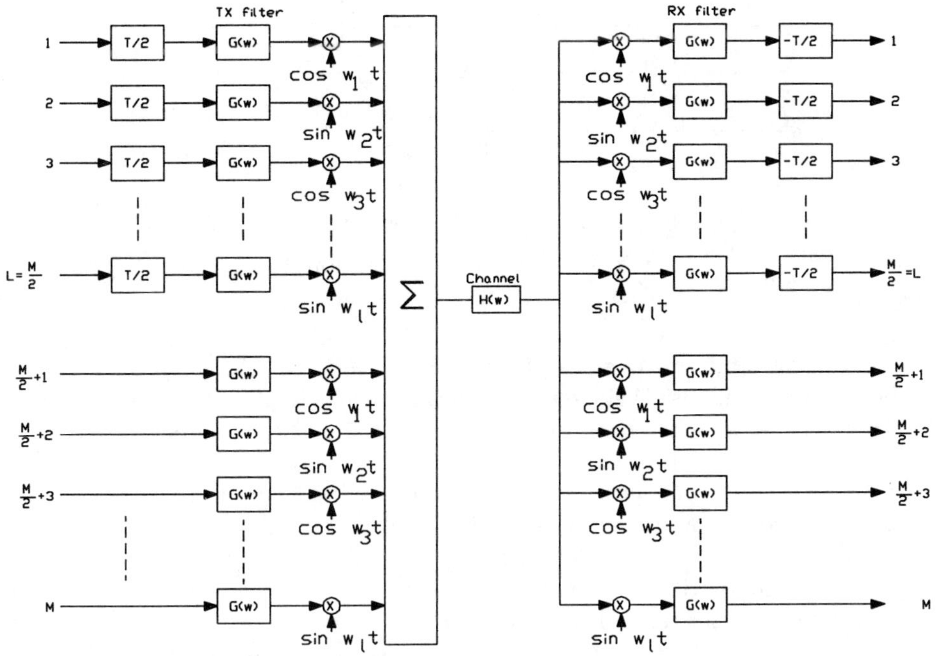

Figure 18.8: Filter-bank implementation of the FDM modem [78] ©IEEE, 1981, Hirosaki

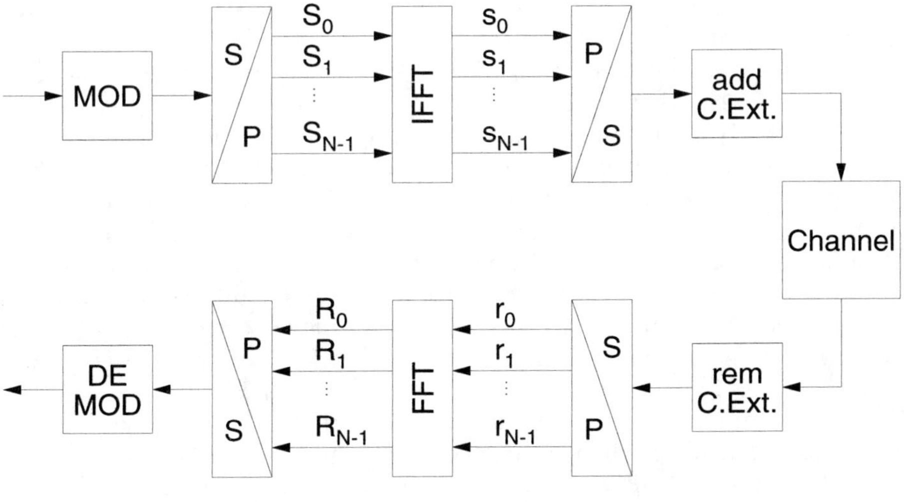

Figure 18.9: Schematic of N-subcarrier OFDM transmission system

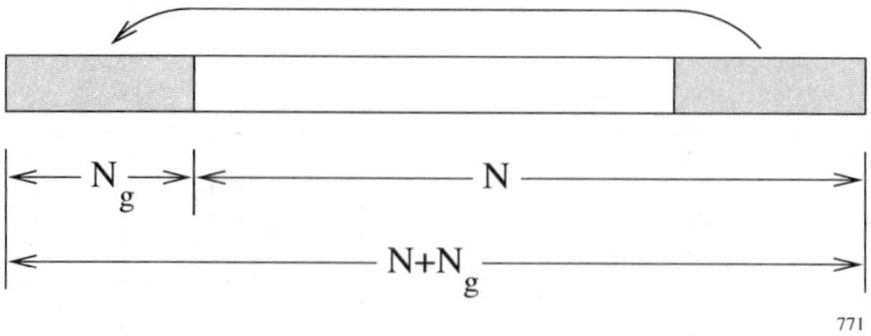

Figure 18.10: Stylised plot of N-subcarrier OFDM time domain signal with a cyclic extension of N_g samples

plementationally prohibitive in terms of complexity and cost [78], especially for a high number of sub-channels. Weinstein [73] suggested the digital implementation of OFDM subcarrier modulators/demodulators based on the discrete Fourier transform (DFT).

The DFT and its more efficient implementation, the fast Fourier transform (FFT) are typically employed in practice in the baseband OFDM modulation/demodulation process, as can be seen in the schematic of Figure 18.9. The serial data stream is mapped to data symbols with a symbol rate of $1/T_s$, employing a general phase and amplitude modulation scheme, and the resulting symbol stream is demultiplexed into a vector of N data symbols S_0 to S_{N-1}. The parallel data symbol rate is $1/N \cdot T_s$, ie. the parallel symbol duration is N times longer than the serial symbol duration T_s. The inverse FFT (IFFT) of the data symbol vector is computed and the coefficients s_0 to s_{N-1} constitute an OFDM symbol. The s_n are the time domain samples of the OFDM symbol and are transmitted sequentially over the channel at a symbol rate of $1/T_s$. At the receiver, a spectral decomposition of the received time domain samples r_n is computed employing an N-tap FFT, and the recovered data symbols R_n are restored in serial order and demultiplexed.

If we assume that the bandwidth of the OFDM spectrum is finite, then simple Fourier theory dictates that the corresponding time domain signal has an infinite duration. The underlying assumption upon invoking the IFFT for modulation is that although N frequency domain samples produce N time domain samples, the time domain signal is assumed to be periodically repeated, theoretically, for an infinite duration. In practice, however, it is sufficient to repeat the time domain signal periodically for the duration of the channel's memory, i.e. for a duration that is comparable to the length of the CIR. Hence, for transmission over time dispersive channels, each time domain OFDM symbol is extended by the so-called cyclic extension (C. Ext. in Figure 18.9) or a guard interval of N_g samples duration, in order to overcome the inter-OFDM symbol interference due to the channel's memory.

The samples of the cyclic extension are copied from the end of the time domain OFDM symbol, generating the transmitted time domain signal ($s_{N-N_g-1}, \ldots, s_{N-1}, s_0, \ldots, s_{N-1}$) depicted in Figure 18.10. At the receiver, the samples of the cyclic extension are discarded. Clearly, the need for a cyclic extension in time dispersive environments reduces the efficiency of OFDM transmissions by a factor of $N/(N+N_g)$. Since the duration N_g of the necessary cyclic extension depends only on the channel's

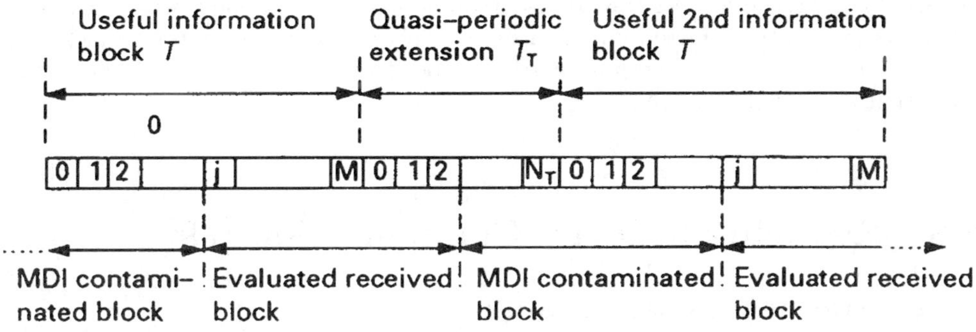

Figure 18.11: Reducing MDI by quasi-periodic block extension

memory, OFDM transmissions employing a high number of carriers N are desirable for efficient operation. Typically a guard interval length of not more than 10% of the OFDM symbol's duration is employed.

18.7 Cyclic OFDM Symbol Extension

As we have seen in Section 18.4, the response of the low pass filtered OFDM channels to a block of modulated signal is theoretically infinite. In Section 18.5 we introduced the generalised Nyquist criterion, which allowed infinite-length system responses, as long as their periodic zero-crossings yielded no MDI. If, however, the zero-crossings or the sampling instants fluctuate, the MDI soon becomes prohibitively high.

An alternative method of combating the previously characterised multi-dimensional interference (MDI) is to transmit quasi-periodically extended time domain blocks, i.e. OFDM symbols, after modulation by IFFT [79]. The length of the added quasi-periodic extension depends on the memory length of the channel, in other words, on the length of the transient response of the channel to the quasi-periodic excitation constituted by the modulated signal. The method is best explained with reference to Figure 18.11. Every block of length-T modulated signal segment is quasi-periodically extended by a length of T_T transient duration simply repeating N_T samples of the useful information block. Then the total sequence length becomes $(M + N_T)$ samples, corresponding to a duration of $(T + T_T)$. Trailing and leading samples of this extended block are corrupted by the channel's transient response, hence the receiver is instructed to ignore the first j number of samples of the received block and also disregard $(M + N_T - j)$ trailing samples. Only the central M number of samples are demodulated by FFT at the receiver, which are essentially unaffected by the channel's transient response, as seen in Figure 18.11.

The number of extension samples N_T required depends on the length of the channel's transient response and the number of modulation levels. If the number of modulation levels is high, the maximum acceptable MDI due to channel transients must be kept low in order to maintain sufficient noise margins before the data is corrupted. This then requires a longer quasi-periodic extension, i.e. N_T must be also higher. It must also be appreciated that this MDI-corrupted extension actually wastes channel capacity as well as transmitted power. However, if the useful information blocks are long, i.e. $M \geq 128$, the extension length can be kept as low as 10% of the useful information block length.

18.8 Reducing MDI by Compensation [79]

18.8.1 Transient system analysis

Following Kolb's [79] and Schüssler's [80] approach, we can describe the OFDM transmission system by using input and output vectors $\mathbf{u}(k)$ and $\mathbf{y}(k)$, respectively, that are related by

$$\mathbf{y}(k) = \mathbf{S}\mathbf{u}(k), \tag{18.26}$$

where the convolution matrix $\mathbf{S}$ [402] is given by:

$$\mathbf{S} = \begin{bmatrix} h_0 & 0 & \dots & \dots & 0 \\ h_1 & h_0 & \dots & \dots & 0 \\ h_2 & h_1 & h_0 & \dots & 0 \\ \vdots & \vdots & \vdots & & \vdots \end{bmatrix}. \tag{18.27}$$

Due to causality the vectors $\mathbf{u}(k)$ and $\mathbf{y}(k)$ are of infinite length for positive sample indices only. Let us assume furthermore that the input vector $\mathbf{u}(k)$ represents $L \geq 2$ periods of an M-sample periodic sequence and that the channel's impulse response is shorter than M samples. Then, for example, for $L = 3$ periods and a channel impulse response length of $n = M - 1$ samples the convolution in Equation 18.26 can be written as:

$$\begin{bmatrix} \mathbf{y}_1 \\ \mathbf{y}_2 \\ \mathbf{y}_3 \\ \mathbf{y}_4 \end{bmatrix} = \begin{bmatrix} \mathbf{S}_0 & 0 & 0 \\ \mathbf{S}_1 & \mathbf{S}_0 & 0 \\ 0 & \mathbf{S}_1 & \mathbf{S}_0 \\ 0 & 0 & \mathbf{S}_1 \end{bmatrix} \begin{bmatrix} \mathbf{u} \\ \mathbf{u} \\ \mathbf{u} \end{bmatrix}, \tag{18.28}$$

where

$$\mathbf{S}_0 = \begin{bmatrix} h_0 & 0 & \dots & \dots & 0 \\ h_1 & h_0 & \dots & \dots & 0 \\ h_2 & h_1 & h_0 & \dots & 0 \\ \vdots & \vdots & \vdots & & \vdots \\ h_n & h_{n-1} & h_{n-2} & \dots & h_0 \end{bmatrix} \tag{18.29}$$

and

$$\mathbf{S}_1 = \begin{bmatrix} 0 & h_n & h_{n-1} & \dots & h_1 \\ 0 & 0 & h_n & \dots & h_2 \\ \vdots & \vdots & \vdots & & \vdots \\ 0 & 0 & 0 & \dots & h_n \\ 0 & 0 & 0 & \dots & 0 \end{bmatrix}. \tag{18.30}$$

Upon exciting our transmission channel with a periodic signal, its response is constituted by a transient response $\mathbf{y}_{in}(k)$ plus a stationary periodic response $\mathbf{y}_{st}(k)$. The stationary response has the periodicity M of the excitation, while the channel's transient response dies down after $n = M - 1$ samples.

The system's response during the first period of M samples is then given by:

$$\begin{aligned} \mathbf{y}_1 &= \mathbf{y}_{in} + \mathbf{y}_{st} = \mathbf{S}_0\mathbf{u} \\ &= [h_0u_0; (h_1u_0 + h_0u_1); (h_2u_0 + h_1u_1 + h_0u_2); \\ &\quad \ldots; (h_nu_0 + h_{n-1}u_1 + h_{n-2}u_2 + \ldots + h_0u_n)]^T, \end{aligned} \tag{18.31}$$

where we also exploited Equation 18.28 and $[\,]^T$ means the transpose of $[\,]$. During the second cycle of the excitation the transient does not affect the system's response since $u < M$, and hence by the help of Equation 18.28 we get:

$$\begin{aligned} \mathbf{y}_2 &= \mathbf{y}_{st} = [\mathbf{S}_1\mathbf{S}_0][\mathbf{uu}]^T \\ &= [\mathbf{S}_1 + \mathbf{S}_0]\mathbf{u} = \mathbf{H}\mathbf{u}, \end{aligned} \tag{18.32}$$

where the matrix $\mathbf{H} = [\mathbf{S}_1 + \mathbf{S}_0]$ is computed from Equations 18.29 and 18.30 as follows:

$$\mathbf{H} = \mathbf{S}_1 + \mathbf{S}_0 = \begin{bmatrix} h_0 & h_n & h_{n-1} & \ldots & h_1 \\ h_1 & h_0 & h_n & \ldots & h_2 \\ \vdots & & & & \\ h_n & h_{n-1} & h_{n-2} & \ldots & h_0 \end{bmatrix}. \tag{18.33}$$

Then the system's response for the second cycle from Equation 15.46 can be written as:

$$\mathbf{y}_2 = \begin{bmatrix} h_0u_0 + h_nu_1 + \ldots + h_1u_n \\ h_1u_0 + h_0u_1 + \ldots + h_2u_n \\ \vdots \\ h_nu_0 + h_{n-1}u_1 + \ldots + h_0u_n \end{bmatrix}. \tag{18.34}$$

Similarly, the system's response during the third period of the excitation is unaffected by transients, and hence given by:

$$\mathbf{y}_3 = \mathbf{H}\mathbf{u}. \tag{18.35}$$

The switch-off transient $\mathbf{y}_4 = \mathbf{y}_{out}$ of the system describes the output signal after the excitation died down. From Equation 18.28 we get:

$$\begin{aligned} \mathbf{y}_{out} = \mathbf{y}_4 &= \mathbf{S}_1\mathbf{u} = \mathbf{S}_1\mathbf{u} + \mathbf{S}_0\mathbf{u} - \mathbf{S}_0\mathbf{u} \\ &= (\mathbf{S}_1 + \mathbf{S}_0)\mathbf{u} - \mathbf{S}_0\mathbf{u}. \end{aligned} \tag{18.36}$$

Combining Equations 18.32 and 18.33 gives:

$$\mathbf{y}_{out} = \mathbf{y}_4 = \mathbf{y}_{st} - (\mathbf{y}_{in} + \mathbf{y}_{st}) = -\mathbf{y}_{in}, \tag{18.37}$$

which implies that the trailing transient is the "inverse" of the leading transient.

Based on the previous analysis our aim is to transmit each information block only once ($L = 1$) and remove the effects of MDI by cancellation, rather than reduce the system's effective transmission rate by the factor $M/(M + N_T)$, where N_T is the number of samples used for the quasi-periodic transmission block expansion. The kth received signal vector $\mathbf{y}^k$ is constituted by an initial transient due to the trailing transient $\mathbf{y}_{out}^{k-1}$ of the $(k-1)$th transmitted vector, a stationary response $\mathbf{y}_{st}^k$ due to the kth transmitted vector plus the initial transient $\mathbf{y}_{in}^{k-1}$ due to the kth transmitted vector:

$$\mathbf{y}^k = \mathbf{y}_{out}^{k-1} + \mathbf{y}_{st}^k + \mathbf{y}_{in}^k. \tag{18.38}$$

Equation 18.37, for the trailing transient from block $(k-1)$, gives us:

$$\mathbf{y}_{out}^{k-1} = -\mathbf{y}_{in}^{k-1}, \tag{18.39}$$

which can be substituted in Equation 18.38 to yield:

$$\mathbf{y}^k = \mathbf{y}_{in}^k + \mathbf{y}_{st}^k - \mathbf{y}_{in}^{k-1}. \tag{18.40}$$

18.8.2 Recursive MDI compensation

Our objective is now to determine the transmitted vector $\mathbf{u}^k$, which can be inferred if $\mathbf{y}^k$ is known. Hence the compensation of MDI ensues as follows:

(1) Initially a known preamble sequence $\mathbf{u}_0$ is transmitted, sending $L \geq 2$ repetitions to determine the channel's impulse response, i.e. the matrix $\mathbf{H}$. This is possible using Equations 18.32 and 18.33, where the system's response is in its stationary phase during the second cycle of the periodic preamble. The system's impulse response can be conveniently measured by transmitting a specific preamble $\mathbf{u}_0$, which is constituted by a "white" spectrum with the real part set to one and the imaginary part set to zero. This input signal modulates the sub-channel modulators, where the modulated signal after IFFT becomes the Kronecker delta. When this preamble signal excites the transmission system, its response is the impulse response itself. Once the impulse response is known, the matrices $\mathbf{H}$, $\mathbf{S}_0$ and $\mathbf{S}_1$ are also known.

(2) Since the system responses $\mathbf{y}_1^0$ and $\mathbf{y}_2^0$ due to the preamble $\mathbf{u}_0$ during the first two periods are now known, the system's initial transient can be computed from Equations 15.45 and 15.46 using:

$$\mathbf{y}_1^0 = \mathbf{y}_{in}^0 + \mathbf{y}_{st}^0 = \mathbf{S}_0\mathbf{u} \tag{18.41}$$

and

$$\mathbf{y}_2^0 = \mathbf{y}_{st}^0 = [\mathbf{S}_1 + \mathbf{S}_0]\mathbf{u} \tag{18.42}$$

as follows:

$$\mathbf{y}_{in}^0 = \mathbf{y}_1^0 - \mathbf{y}_{st}^0 = \mathbf{y}_1^0 - \mathbf{y}_2^0 = -\mathbf{S}_1\mathbf{u}. \tag{18.43}$$

With the knowledge of this initial transient response the trailing effects corrupting the consecutive blocks can be recursively compensated.

(3) The preamble $\mathbf{u}_0$ is followed by the useful information blocks $\mathbf{y}^k$, $k = 1, 2, \ldots, \infty$, transmitted only once ($L = 1$). Upon receiving the system's response

$\mathbf{y}^k$ due to $\mathbf{u}^k$, $k = 1, 2, \ldots, \infty$, the trailing transient of the previous block can be subtracted as follows:

$$\mathbf{y}^k - \mathbf{y}_{out}^{k-1} = \mathbf{y}^k + \mathbf{y}_{in}^{k-1}. \tag{18.44}$$

When considering Equations 18.38 and 18.32 we get:

$$\mathbf{y}^k - \mathbf{y}_{out}^{k-1} = \mathbf{y}_{st}^k + \mathbf{y}_{in}^k = \mathbf{S}_0\tilde{\mathbf{u}}^k, \tag{18.45}$$

where we used $\tilde{\mathbf{u}}^k$ which is different from $\mathbf{u}^k$ due to channel effects. The estimated transmitted signal $\tilde{\mathbf{u}}^k$ can be recovered using Equations 18.44 and 18.45:

$$\tilde{\mathbf{u}}^k = \mathbf{S}_0^{-1}(\mathbf{y}^k - \mathbf{y}_{out}^{k-1}) = \mathbf{S}_0^{-1}(\mathbf{y}^k + \mathbf{y}_{in}^{k-1}). \tag{18.46}$$

From Equation 18.36 we have $\mathbf{y}_{out}^{k-1} = \mathbf{S}_1\mathbf{u}^{k-1}$ that can be substituted in Equation 18.46, giving

$$\tilde{\mathbf{u}}^k = \mathbf{S}_0^{-1}(\mathbf{y}^k - \mathbf{S}_1\mathbf{u}^{k-1}), \tag{18.47}$$

which is an explicit formula for the estimated transmitted vector $\tilde{\mathbf{u}}^k$ in terms of the received vector $\mathbf{y}^k$, the previously recovered transmitted vector $\mathbf{u}^{k-1}$ and the matrices $\mathbf{S}_0$ and $\mathbf{S}_1$, which depend only on the system's impulse response.

(4) Now the compensated vector $\tilde{\mathbf{u}}^k$ in Equation 18.47 can be demodulated by FFT and the demodulated signal $\tilde{\mathbf{U}}^k$ is subjected to hard decisions, which are represented as $D\{\ \}$, giving the recovered information signal in the following form:

$$\mathbf{U}^k = D\{\tilde{\mathbf{u}}^k\} = D\{FFT[\tilde{\mathbf{u}}^k]\}. \tag{18.48}$$

(5) Since $\mathbf{u}^k$ is needed in Equation 18.47 for the next recursive compensation step,

$$\mathbf{u}^k = IFFT[\mathbf{U}^k] \tag{18.49}$$

is computed to conclude the compensation process.

This method bears a strong resemblance to the conventional partial response technique, allowing signals belonging to adjacent signalling intervals to overlap. The controlled ISI can then be recursively compensated, if the channel can be considered slowly changing.

18.9 Adaptive Channel Equalisation

In this section it will be assumed that the MDI was removed by compensation, quasi-periodic block extension or by obeying the generalised Nyquist criterion. The linear distortions introduced by the unequalised channel transfer function $H(f)$ can be removed by estimating $H(if_0)$ and then dividing the received signal spectrum by $H(if_0)$, before hard decision decoding is performed using the function $D\{\ \}$.

In the previous section we mentioned how $H(if_0)$ can be measured using a preamble having a real-valued "white" spectrum. By setting all real spectral lines to unity and all imaginary lines to zero in the preamble data frame, after modulation by the IFFT the transmitted signal is the Kronecker delta. Therefore the received signal is the channel's impulse response, which after demodulation by FFT gives the channel's frequency domain transfer function $H(if_0)$. The received signal's spectrum after demodulation by FFT becomes:

$$\tilde{X}_i = H(if_0)X_i, \; i = 1, \ldots, \; M, \tag{18.50}$$

where X_i is the ith transmitted spectral line at frequency if_0. After equalisation by dividing the received spectral line $\tilde{X}_i$ with the estimated channel transfer function $H(if_0)$ and taking hard decisions, we obtain the recovered sequence:

$$\hat{X}_i = D\left\{\frac{\tilde{X}_i}{H(if_0)}\right\}, \; i = 1, \ldots, \; M. \tag{18.51}$$

If the recovered sequence $\hat{X}_i$ is error-free, it lends itself to the recursive recomputation of the channel's frequency response in order to cope with slowly time-varying transmission media [79]. The up-dated transfer function is given by:

$$H_a(if_0) = \frac{\tilde{X}_i}{\hat{X}_i} = \frac{\tilde{X}_i}{D\left\{\frac{\tilde{X}_i}{H(if_0)}\right\}}, \; i = 1, \ldots, \; M. \tag{18.52}$$

In order to retain robustness for transmissions over channels having high bit error rate a leaky algorithm can be introduced to generate a weighted average of the previous and current transfer function using the leakage factor β in the following fashion:

$$H_{\text{adaptive}}(if_0) = \beta H_a^k(if_0) + (1 - \beta)H_a^{k-1}(if_0). \tag{18.53}$$

The leakage factor β is a parameter determined by the prevailing channel bit error rate.

An interesting aspect of our OFDM scheme is that if the channel varies slowly, then using differential coding between corresponding sub-channels of consecutive OFDM transmission frames removes the requirement for a channel equaliser as long as the difference is computed before hard decision takes place [73]. This is due to the fact that spectral lines of the same sub-channel or same frequency will suffer the same attenuation due to the channel's linear distortion. Hence the effect of channel attenuation and phase shift drops out before hard decision takes place. Similar arguments can be exploited between adjacent lines of the same OFDM transmission frame as well, if the channel's transfer characteristic is sufficiently smooth.

If the frequency domain transfer function $H(if_0)$ is more erratic as a function of frequency or time, a number of known pilot tones can be included in the transmitted spectrum. These pilots facilitate the more accurate estimation and equalisation of the channel transfer function. This technique was proposed by Cimini for wide-band, frequency-selective multipath mobile channels in reference [74] and it will be widely used throughout the forthcoming chapters. In case of narrowband Rayleigh-fading mobile radio channels the fading is "frequency flat fading", hence each transmitted frequency component suffers the same attenuation and phase shift. In this case time domain pilot symbols employed as in the PSAM schemes of Chapter 10 are useful in supporting the tracking of the Rayleigh fading envelope [12].

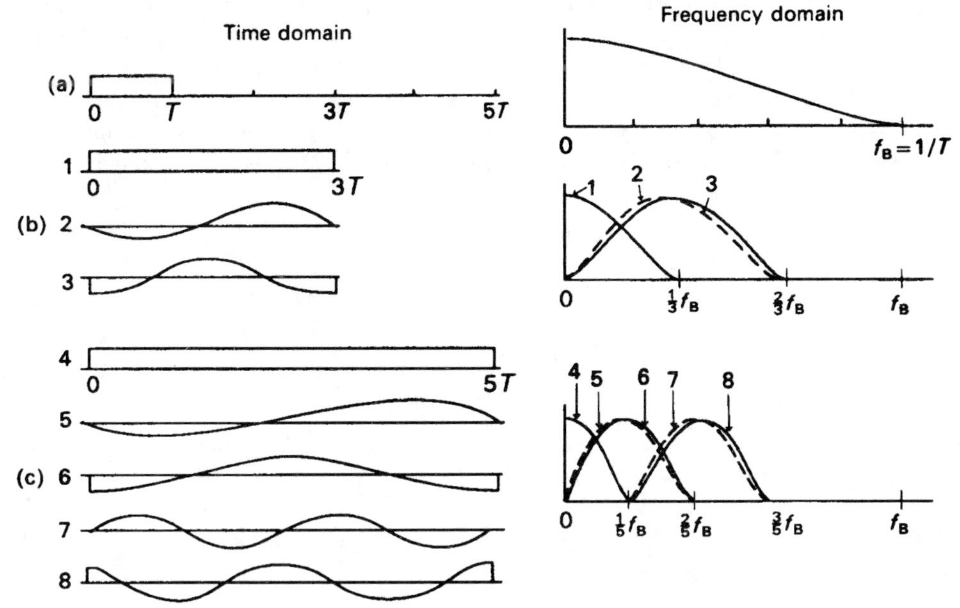

Figure 18.12: OFDM carriers, their stylised spectra and bandwidth requirement: (a) one-carrier system, (b) three-carrier system, (c) five-carrier system [369] ©Springer, 1969, Harmuth

18.10 OFDM Bandwidth Efficiency

In the OFDM system each symbol to be transmitted modulates an assigned carrier of a set of wide-sense orthogonal basis functions and these modulated sub-channel signals are superimposed for transmission via the communications channel. The received signal can be demodulated for example by the correlation receiver described in the previous section or by FFT.

A set of suitable wide-sense orthogonal basis functions of gradually increasing length T, $3T$ and $5T$, similar to those used in our OFDM schemes, is depicted in Figure 18.12 for a one-carrier, three-carrier and five-carrier system, respectively [369], Harmuth2.

Both their time domain waveforms and stylised spectra are shown in Figure 18.12. In a simplistic approach here we assume that the signal spectra can be band limited to the bandwidth of its main spectral lobe, as suggested by the figure.

Using an essentially serial system with one carrier, as seen in Figure 18.12(a), the minimum bandwidth required is $f_B = 1/T$ and the bandwidth efficiency is $\eta =$ 1 Bd/Hz because the spectrum of this pulse is represented by the sinc function whose first zero is at $f_B = 1/T$. The three-carrier system of Figure 18.12(b) expands the length of the basis functions to $3T$, thereby reducing the bandwidth requirement to $B = \frac{2}{3}f_B$, giving $\eta = 1.5\,\text{Bd/Hz}$. This is because the rectangularly windowed sin and cos spectra are represented by the convolution of a tonal spectral line and a frequency domain sinc function describing the spectrum of the rectangular time domain window. The five-carrier scheme using basis functions of $5T$ length further reduces the bandwidth to $B = \frac{3}{5} f_B$ and increases the spectral efficiency to 1.67 Bd/Hz.

Similarly, the approximate bandwidth of a $(2M + 1)$-carrier system using an impulse as well as M sine and M cosine carriers of length $(2M + 1)T$ becomes

$$B = \frac{M+1}{2M+1} \frac{1}{T}, \tag{18.54}$$

yielding a bandwidth efficiency of

$$\eta = \frac{2M+1}{M+1} \text{ Bd/Hz}. \tag{18.55}$$

When $M \to \infty$, we have $\lim_{M\to\infty} \eta = 2\,\text{Bd/Hz}$, which for a typical value of $M = 64$ gives $\eta = 129/65 = 1.98\,\text{Bd/Hz}$. The interferences caused by the above band limitation are given in closed form in Reference [370] for a variety of carriers, but for the more attractive scenarios using a higher number of carriers it can only be estimated by simulation studies.

18.11 Summary

In this chapter we have investigated orthogonal multiplexing as an alternative means of transmitting high-rate information over highly dispersive mobile radio channels. The essential premise of orthogonal multiplexing is to divide the serial input data stream into a number of parallel streams and to transmit these low-rate parallel streams simultaneously. This offered two main advantages. Firstly, by increasing the time – due to the low sub-channel signalling rate – over which a symbol was transmitted, the probability of it being completely destroyed in a fade was reduced. Secondly, by decreasing the bandwidth of each channel, the need for a channel equaliser was diminished. Modulation on each of these orthogonal channels could be either binary or multilevel, and both were considered.

However, orthogonal multiplexing also has a number of disadvantages. It requires an increase in the complexity of the transmitting and receiving equipment, and may also engender an increased transmission delay. It will typically also require pilot symbols or some other similar redundancy such that the receiver can be informed of how each individual channel has been modified by the mobile radio channel.

A number of sections were concerned with the issue of multidimensional interference, whereby interference might occur not only between consecutive symbols of the same sub-channel, but also from one sub-channel to the adjacent sub-channels, and methods to ameliorate this were suggested.

Following this rudimentary introduction to OFDM in the forthcoming chapters we will delve into OFDM more deeply, considering a range of associated research and implementational aspects.

Chapter 19

OFDM Transmission over Gaussian Channels

High data rate communications over additive white Gaussian noise (AWGN) channels are limited not only by noise, but especially with increasing symbol rates, often more significantly by the intersymbol interference (ISI) due to the memory of the dispersive wireless communications channel. Explicitly, this channel memory is caused by the dispersive channel impulse response (CIR) due to the different length propagation paths between the transmitting and the receiving antennae. This dispersion effect could theoretically be measured by transmitting an infinitely short impulse and "receiving" the channel impulse response itself. On this basis, several measures of the effective duration of the impulse response can be calculated, one being the delay spread. The multipath propagation of the channel manifests itself by different echos of possibly different transmitted symbols overlapping at the receiver, which leads to error rate degradation.

This effect occurs not only in wireless communications, but also over all types of electrical and optical waveguides, although for these media the relative time differences are comparatively small, mostly due to multimode transmission or incorrect electrical or optical adaption at interfaces.

In wireless communications systems the duration and the shape of the channel impulse response depend heavily on the propagation environment of the communications system in question. While indoor wireless networks typically exhibit only short relative delays, outdoor networks, like the Global System for Mobile (GSM) [403] can face maximum delays spreads in the order of 15 μs.

As a general rule, the effects of ISI on the transmission error statistics are negligible, as long as the delay spread is significantly shorter than the duration of one transmitted symbol. This implies that the symbol rate of communications systems is practically limited by the channel's memory. For higher symbol rates, there is typically significant deterioration of the system's error rate performance.

If symbol rates exceeding this limit are to be transmitted over the channel, mechanisms must be implemented in order to combat the effects of ISI. Channel equalisation techniques can be used to suppress the echos caused by the channel. To perform this operation, the channel impulse response must be estimated. Significant research efforts were invested into the development of such channel equalisers, and most wireless systems in operation use equalisers to combat ISI.

There is, however, an alternative approach towards transmitting data over a multipath channel. Instead of attempting to cancel the effects of the channel's echoes, orthogonal frequency division multiplexing (OFDM) [403] modems employ a set of harmonically related carriers in order to transmit information symbols in parallel over the channel. Since the system's data throughput is the sum of all the parallel channels' throughputs, the data rate per sub-channel is only a fraction of the data rate of a conventional single-carrier system having the same throughput. This allows us to design a system supporting high data rates, while maintaining symbol durations much longer than the channel's memory, thus circumventing the need for channel equalisation.

19.1 Orthogonal Frequency Division Multiplexing

19.1.1 History

Frequency division multiplexing (FDM) or multi-tone systems have been employed in military applications since the 1960s, for example by Bello [96], Zimmerman [72], Powers and Zimmerman [97], and others. orthogonal frequency division multiplexing (OFDM), which employs multiple carriers overlapping in the frequency domain, was pioneered by Chang [71, 98]. Saltzberg [99] studied a multicarrier system employing orthogonal time-staggered quadrature amplitude modulation (O-QAM) on the carriers.

The use of the discrete Fourier transform (DFT) to replace the banks of sinusoidal generators and the demodulators was suggested by Weinstein and Ebert [73] in 1971, which significantly reduces the implementation complexity of OFDM modems. In 1980, Hirosaki [84] suggested an equalisation algorithm in order to suppress both intersymbol and intersubcarrier interference caused by the channel impulse response or timing and frequency errors. Simplified OFDM modem implementations were studied by Peled [77] in 1980, while Hirosaki [78] introduced the DFT based implementation of Saltzberg's O-QAM OFDM system. From Erlangen University, Kolb [79], Schüßler [80], Preuss [81] and Rückriem [82] conducted further research into the application of OFDM. Cimini [74] and Kalet [83] published analytical and early seminal experimental results on the performance of OFDM modems in mobile communications channels.

More recent advances in OFDM transmission are presented in the impressive state-of-the-art collection of works edited by Fazel and Fettweis [100], including the research by Fettweis *et al.* at Dresden University, Rohling *et al.* at Braunschweig University, Vandendorp at Loeven University, Huber *et al.* at Erlangen University, Lindner *et al.* at Ulm University, Kammeyer *et al.* at Bremen University and Meyr *et al.* [101, 102] at Aachen University, but the individual contributions are too numerous to mention.

While OFDM transmission over mobile communications channels can alleviate the problem of multipath propagation, recent research efforts have focused on solving a set of inherent difficulties regarding OFDM, namely the peak-to-meanpower ratio, time and frequency synchronisation, and on mitigating the effects of the frequency selective fading channel. These issues are addressed below in more depth.

19.1.1.1 Peak-to-mean power ratio

It is plausibe that the OFDM signal – which is the superposition of a high number of modulated sub-channel signals – may exhibit a high instantaneous signal peak with respect to the average signal level. Furthermore, large signal amplitude swings are encountered, when the time domain signal traverses from a low instantaneous power waveform to a high power waveform, which may results in a high out-of-band (OOB) harmonic distortion power, unless the transmitter's power amplifier exhibits an extremely high linearity across the entire signal level range, as discussed in Section 4.5.1. This then potentially contaminates the adjacent channels with adjacent channel interference. Practical amplifiers exhibit a finite amplitude range, in which they can be considered near-linear. In order to prevent severe clipping of the high OFDM signal peaks, which is the main source of OOB emissions, the power amplifier must not be driven into saturation and hence they are typically operated with a certain so-called back-off, creating a certain "headroom" for the signal peaks, which reduces the risk of amplifier saturation and OOB emmission. Two different families of solutions have been suggested in the literature, in order to mitigate these problems, either reducing the peak-to-mean power ratio, or improving the amplification stage of the transmitter.

More explicitly, Shepherd [103], Jones [104], and Wulich [105] suggested different coding techniques which aim to minimise the peak power of the OFDM signal by employing different data encoding schemes before modulation, with the philosophy of choosing block codes whose legitimate code words exhibit low so-called crest factors or peak-to-mean power envelope fluctuation. Müller [106], Pauli [107], May [108] and Wulich [109] suggested different algorithms for post-processing the time domain OFDM signal prior to amplification, while Schmidt and Kammeyer [110] employed adaptive subcarrier allocation in order to reduce the crest factor. Dinis and Gusmão [111–113] researched the use of two-branch amplifiers, while the clustered OFDM technique introduced by Daneshrad, Cimini and Carloni [114] operates with a set of parallel partial FFT processors with associated transmitting chains. OFDM systems with increased robustness to non-linear distortion have been proposed by Okada, Nishijima and Komaki [115] as well as by Dinis and Gusmão [116].

19.1.1.2 Synchronisation

Time and frequency synchronisation between the transmitter and receiver are of crucial importance as regards to the performance of an OFDM link [117, 118]. A wide variety of techniques have been proposed for estimating and correcting both timing and carrier frequency offsets at the OFDM receiver. Rough timing and frequency acquisition algorithms relying on known pilot symbols or pilot tones embedded into the OFDM symbols have been suggested by Claßen [101], Warner [119], Sari [120], Moose [121], as well as Brüninghaus and Rohling [122]. Fine frequency and timing tracking algorithms exploiting the OFDM signal's cyclic extension were published by

Moose [121], Daffara [123] and Sandell [124].

19.1.1.3 OFDM/CDMA

Combining OFDM transmissions with code division multiple access (CDMA) allows us to exploit the wideband channel's inherent frequency diversity by spreading each symbol across multiple subcarriers. This technique has been pioneered by Yee, Linnartz and Fettweis [125], by Chouly, Brajal and Jourdan [126], as well as by Fettweis, Bahai and Anvari [127]. Fazel and Papke [128] investigated convolutional coding in conjunction with OFDM/CDMA. Prasad and Hara [129] compared various methods of combining the two techniques, identifying three different structures, namely multicarrier CDMA (MC-CDMA), multicarrier direct sequence CDMA (MC-DS-CDMA) and multi-tone CDMA (MT-CDMA). Like non-spread OFDM transmission, OFDM/CDMA methods suffer from high peak-to-mean power ratios, which are dependent on the frequency domain spreading scheme, as investigated by Choi, Kuan and Hanzo [130].

19.1.1.4 Adaptive antennas

Combining adaptive antenna techniques with OFDM transmissions was shown to be advantageous in suppressing co-channel interference in cellular communications systems. Li, Cimini and Sollenberger [131–133], Kim, Choi and Cho [134], Lin, Cimini and Chuang [135] as well as Münster *et al.* [136] have investigated algorithms for multi-user channel estimation and interference suppression.

19.1.1.5 OFDM applications

Due to their implementational complexity, OFDM applications have been scarce until quite recently. Recently, however, OFDM has been adopted as the new European digital audio broadcasting (DAB) standard [75, 76, 137–139] as well as for the terrestrial digital video broadcasting (DVB) system [120, 140].

For fixed-wire applications, OFDM is employed in the asynchronous digital subscriber line (ADSL) and high-bit-rate digital subscriber line (HDSL) systems [141–144] and it has also been suggested for power line communication systems [145, 146] due to its resilience to time dispersive channels and narrowband interferers.

More recently, OFDM applications were studied within the European *4th* Framework Advanced Communications Technologies and Services (ACTS) programme [147]. The MEDIAN project investigated a 155 Mbps wireless asynchronous transfer mode (WATM) network [148–151], while the Magic WAND group [152, 153] developed a wireless local area network (LAN). Hallmann and Rohling [154] presented a range of different OFDM-based system proposals that were applicable to the European Telecommunications Standardisation Institute's (ETSI) recent personal communications oriented air interface concept [155].

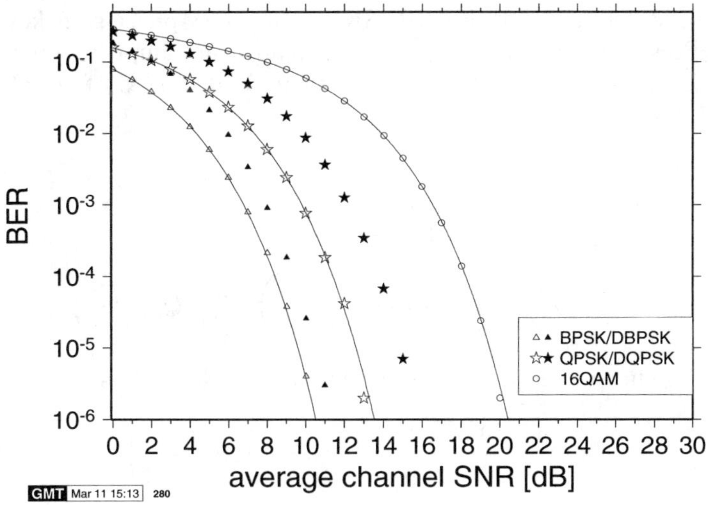

Figure 19.1: BER versus SNR curves for the OFDM modem in AWGN channel using BPSK, DBPSK, QPSK, DQPSK, and 16-QAM. The lines indicate the theoretical performance of the coherently detected modulation schemes in a serial modem over AWGN channels

19.2 The Frequency Domain Modulation

Modulation of the OFDM subcarriers is analogous to the modulation in conventional serial systems. The modulation schemes of the subcarriers are generally quadrature amplitude modulation (QAM) or phase shift keying (PSK) [403] in conjunction with both coherent and non-coherent detection. Differential coded star-QAM (DSQAM) [403] can also be employed. If coherently detected modulation schemes are employed, then the reference phase of the OFDM symbol must be known, which can be acquired with the aid of pilot tones [12] embedded in the spectrum of the OFDM symbol, as will be discussed in Chapter 20. For differential detection the knowledge of the absolute subcarrier phase is not necessary, and differentially coded signalling can be invoked either between neighbouring subcarriers or between the same subcarriers of consecutive OFDM symbols.

19.3 OFDM System Performance over AWGN Channels

As the additive white Gaussian noise (AWGN) in the time domain channel corresponds to AWGN of the same average power in the frequency domain, an OFDM modem's performance in an AWGN channel is identical to that of a serial modem. Analogously to a serial system, the bit error rate (BER) versus signal-to-noise rate (SNR) characteristics are determined by the frequency domain modulation scheme used. In Figure 19.3 the simulated BER versus SNR curves for binary phase shift keying (BPSK), differential BPSK (DBPSK), quaternary phase shift keying (QPSK), differential QPSK (DQPSK) and coherent 16-quadrature amplitude modulation (16-QAM) are shown, together with the theoretical BER curves of serial modems, as derived in [257]:

$$p_{e,BPSK}(\gamma) = Q(\sqrt{2\gamma}), \tag{19.1}$$

$$p_{e,QPSK}(\gamma) = Q(\sqrt{\gamma}), \tag{19.2}$$

$$p_{e,16QAM}(\gamma) = 0.75 \cdot Q\left(\sqrt{\frac{\gamma}{5}}\right) 0.25 \cdot Q\left(\sqrt{\frac{\gamma}{5}}\right), \tag{19.3}$$

where γ is the SNR and the Gaussian $Q()$-function is defined as

$$Q(y) = \frac{1}{\sqrt{2\pi}} \int_y^{\infty} e^{-x^2/2} dx = \text{erfc}\left(\frac{y}{\sqrt{2}}\right).$$

It can be seen from the figure that the experimental BER performance of the OFDM modem is in very good accordance with the theoretical BER curves of conventional serial modems in AWGN channels.

19.4 Clipping Amplification

19.4.1 OFDM signal amplitude statistics

The time domain OFDM signal is constituted by the sum of complex exponential functions, whose amplitudes and phases are determined by the data symbols transmitted over the different carriers. Assuming random data symbols, the resulting time domain signal exhibits an amplitude probability density function (PDF) approaching the two-dimensional or complex Gaussian distribution for a high number of subcarriers. Figure 19.2(a) explicitly shows that the measured amplitude histogram of the in-phase (I) component of a 256-subcarrier OFDM signal obeys a Gaussian distribution with a standard deviation of $\sigma = 1/\sqrt{2}$. The mean power of the signal is $2\sigma^2 = 1$. As the amplitude distribution in both the in-phase and the quadrature phase (Q) component is Gaussian, the two-dimensional amplitude histogram follows a complex Gaussian, or in other words, a Rayleigh distribution [257] with the same standard deviation. The observed amplitude histogram of the 256-subcarrier OFDM signal and the corresponding Rayleigh probability density function are depicted in Figure 19.2(b). The vertical grid lines in the figure indicate the relative amplitude above the mean value in decibels. It can be seen that unlike full response serial modulation schemes – which have a more limited range of possible output amplitudes – the parallel modem's output signal exhibits strong amplitude fluctuations. Note that the standard deviation σ of the probability density functions depicted in Figure 19.2(a) is independent of the number of subcarriers employed, since the mean power of the signal is normalised to 1.

The probability of the instantaneous signal amplitude being above a given threshold, the cumulative density function (CDF), is depicted in Figure 19.3(a). It can be

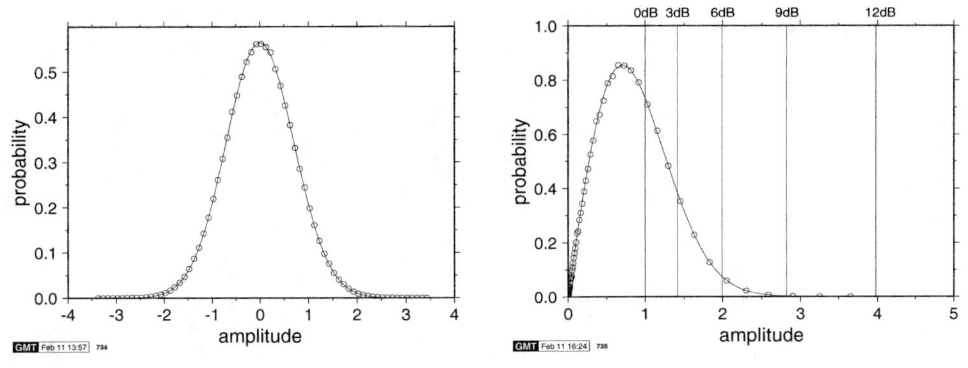

(a) I/Q PDF

(b) amplitude PDF

Figure 19.2: Statistics of OFDM time domain signal: (a) amplitude histogram of the I component of a 256-subcarrier OFDM signal (markers) obeying a Gaussian PDF (continuous line) with $\sigma = 1/\sqrt{2}$; (b) two-dimensional amplitude histogram (markers) following a Rayleigh PDF (continuous line) with $\sigma = 1/\sqrt{2}$. The vertical lines in (b) indicate the corresponding powers above mean in decibels

noted that the probability of the signal amplitude exceeding the 6 dB mark is about 1.1%, while the 9dB mark is exceeded with a probability of about $3.5 \cdot 10^{-4}$. The signal amplitude is higher than the average by 10.6 dB with a probability of 10^{-5}.

19.4.2 Clipping amplifier simulations

In order to evaluate the effects of a non-linear amplifier on the performance of an OFDM system, simulations have been conducted employing a simple clipping amplifier model. This clipping amplifier limits the amplitude of the transmitted signal to a given level, without perturbing the phase information. This amplitude limitation of the time domain signal affects both the received symbols on all subcarriers in the OFDM symbol, as well as the frequency domain out-of-band emissions, and therefore increases the interference inflicted on adjacent carriers.

In Figure 19.3(a) the probability of the instantaneous signal amplitude exceeding a given level was shown. As amplitude limitation in the time domain affects all the subcarriers in the OFDM symbol, the probability of the time domain peak amplitude per OFDM symbol period being clipped by the amplifier is the probability of at least one of the N time domain samples exceeding a given amplitude limit. This clipping probability for a given maximal amplifier amplitude a is displayed in Figure 19.3(b). It can be observed that the clipping probability per OFDM symbol is dependent on the number of subcarriers employed.

Given the information in Figure 19.3(b), the necessary amplifier back-off for an OFDM transmitter can be determined. If the acceptable clipping probability per OFDM symbol is 10^{-5}, then the necessary amplifier back-off values would be 12.1 dB,

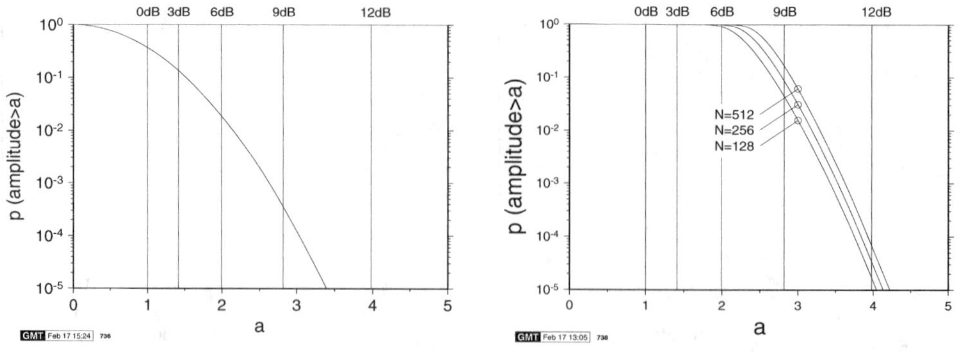

(a) signal CDF

(b) peak CDF

Figure 19.3: OFDM time domain signal: probability of (a) the instantaneous signal amplitude and (b) the peak amplitude per OFDM symbol being above a given threshold value. The lines indicate the corresponding power levels above the mean. The peak value CDF depends on the number of subcarriers, and curves are shown for 128, 256 and 512 carriers

12.3 dB, and 12.5 dB for a 128, 256, and 512 subcarrier OFDM modem, respectively. If a clipping probability of 10^{-4} is acceptable, then these back-off values can be reduced by about 0.6 dB.

19.4.2.1 Peak-power reduction techniques

Two main types of peak-to-mean power ratio reduction techniques have been investigated in the literature, which rely on either introducing redundancy in the data stream or on post-processing the time domain OFDM signal before amplification, respectively.

Shepherd [103], Jones [104], and Wulich [105] suggest different coding techniques which aim to ensure that only low peak power OFDM symbols are chosen for transmission, whilst excluding the transmission of the specific combinations of modulating bits that are known to result in the highest peak factors. These schemes can be viewed as simple k out of n block codes. Depending on the tolerable peak to mean power ratio, the set of acceptable OFDM symbols is computed and the data sequences are mapped onto OFDM symbols from this set only. These techniques reduce the data throughput of the system, but mitigate the severity of clipping amplification.

Müller [106] proposes techniques based on generating a set of OFDM signals by multiplying the modulating data vector in the frequency domain with a set of different phase vectors known to both the transmitter and the receiver, before applying the IDFT. The transmitter will then choose the resulting OFDM symbol exhibiting the lowest peak factor and transmits this together with the chosen phase vector's identification. This technique reduces the average peak factors and requires only a low signalling overhead, but cannot guarantee a given maximum peak factor.

Another group of peak amplitude limiting techniques is based on modifying the time domain signal, where its amplitude exceeds the given peak amplitude limit. Straightforward clipping falls into this category, but induces strong frequency domain out-of-band emissions. In order to avoid the spectral domain spillage of hard clipping, both multiplicative and additive time domain modifications of the OFDM signal have been investigated. Pauli [107] proposes a multiplicative correction of the peak values and their adjacent samples using smooth Gaussian functions for limiting out-of-band emissions. However, multiplying the time domain signal with the time-varying amplitude limiting function introduces intersubcarrier interference.

A similar technique, using additive instead of multiplicative amplitude limiting, is described by May [108]. A smoothly varying time domain signal is added to the time domain signal, which has been optimised for low out-of-band emissions. As the DFT is a linear operation, adding a correction signal in the time domain will overlay the frequency domain data symbols with the DFT of the correction signal, introducing additional noise.

As an alternative solution, Wulich [109] suggests measuring the peak amplitude in each OFDM time domain symbol and scaling the amplitude for the whole OFDM symbol, so that the peak value becomes constant from symbol to symbol. At the receiver, an amplitude correction similar to pilot-assisted scaling over fading channels has to be performed. The advantage of this technique is that the system's throughput is not reduced by introducing redundant information, nor is the orthogonality of the subcarriers impaired. As the OFDM symbol energy is fluctuating with the peak power exhibited by the different OFDM symbols, however, the BER performance of the system is impaired.

A further approach to reducing the amplifier linearity requirements, and which relies on the findings of Figure 19.3(b), is to split an OFDM symbol into groups of subcarriers, which are processed by IFFT, amplified and transmitted over separate transmit chains. This system, referred to as clustered OFDM, has been proposed by Daneshrad and Cimini [114], and combines the advantages of reduced peak power in each transmitter chain with additional spatial diversity from the use of multiple transmit antennas, which can be exploited by coding or OFDM-CDMA techniques.

19.4.2.2 BER performance using clipping amplifiers

If lower back-off values are chosen and no peak power reduction techniques are employed in the OFDM transmission system, then the clipping amplifier will influence both the performance of the OFDM link as well as the out-of-band emissions. The effects of the amplifier back-off value on the BER performance in an AWGN channel for coherently detected 512-subcarrier BPSK and QPSK OFDM transmission are depicted in Figure 19.4.

The investigated amplifier back-off values range from 9 dB down to 0 dB. According to Figure 19.3(b), an amplifier back-off of 9 dB results in a clipping probability per OFDM symbol of about 17%, while an amplifier back-off of 0 dB results in nearly certain clipping of every transmitted OFDM symbol.

It can be observed in Figure 19.4 for both BPSK and QPSK that for an amplifier back-off value of 9 dB the BER performance is indistinguishable from the non-limited

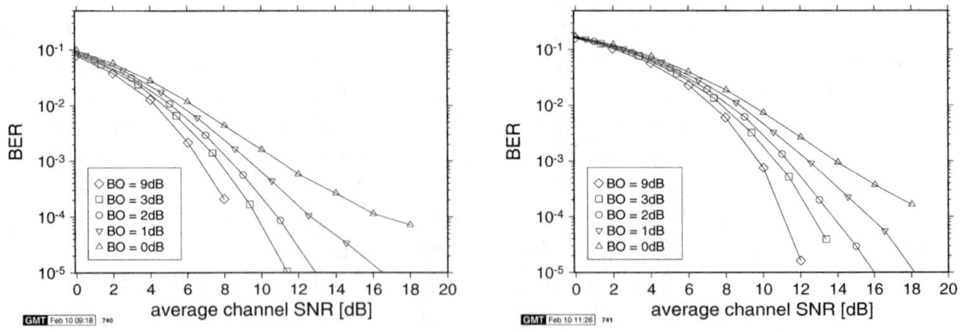

(a) BPSK **(b)** QPSK

Figure 19.4: Influence of amplitude clipping on BER performance of an OFDM modem

case, which was portrayed in Figure 19.3. For smaller back-off values, however, the effects of clipping on the BER performance are more severe. A 3 dB amplifier back-off results in a SNR penalty of about 1.3 dB at a BER of 10^{-4}, and a back-off of 0 dB results in more than 8 dB SNR penalty at the same BER.

The BER performance of the OFDM modem can be improved slightly if the receiver's decision boundaries are adjusted according to the amplitude loss in each subcarrier due to the overall signal power loss, which was observed by O'Neill and Lopes [404]. In fading channels this amplitude adjustment would be performed in conjunction with the channel estimation.

19.4.2.3 Signal spectrum with clipping amplifier

The simulated spectrum of a 512-subcarrier OFDM signal is shown in Figure 19.5. Figure 19.5(a) depicts the spectrum of the transmitted signal, when all 512 subcarriers are used for data transmission. A raised cosine Nyquist filter with an excess bandwidth of $\alpha = 0.35$ was used for pulse shaping, resulting in a relatively slow fall-off of the power spectral density outside the FFT bandwidth of $\pm 256\Delta f$. In order to avoid adjacent channel interference due to spectral overlapping, an adjacent carrier must be placed at a frequency distance considerably higher than the data symbol rate. If a tighter spectral packing of adjacent carriers is important, then pulse shaping filters with a steeper frequency transfer function have to be employed, which are more complex to implement. Since in this scenario all the subcarriers are used for data transmission, the time domain samples are uncorrelated and therefore the spectrum in Figure 19.5(a) is equivalent to that of a serial modem transmitting at the same sample rate and employing the same pulse shaping filter.

Figure 19.5(b) depicts the signal spectrum of a 512-subcarrier modem employing 200 virtual subcarriers at the edges of the bandwidth, which are effectively disabled and carry no energy. Although the same pulse shaping filter was employed as in Figure 19.5(a), the resulting spectral shape is nearly rectangular, with a drop of

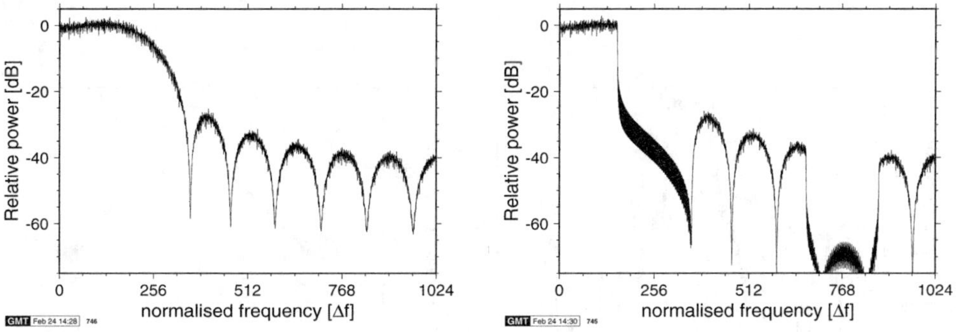

(a) no virtual carriers

(b) 200 virtual carriers

Figure 19.5: Spectrum of a 512-subcarrier OFDM signal using raised-cosine Nyquist pulse shaping filter with an excess bandwidth of $\alpha = 0.35$: (a) no virtual subcarriers, (b) 2×100 subcarriers at the edges of the spectrum. The characteristic sinc-signal shaped curve of this figure is a consequence of rectangular windowing used in the spectral estimation. This effect can be mitigated by more smooth windowing.

approximately 20 dB within one subcarrier distance Δf outside the signal bandwidth. Comparing this spectrum with Figure 19.5(a) it can be observed that the overall shape of the spectrum is determined by the pulse shaping filter, while the employment of disabled virtual subcarriers cuts out parts of the spectrum. It can be seen from the spectrum that by employing virtual subcarriers a very tight power spectrum can be obtained with a simple pulse shaping filter. This rectangular signal spectrum allows tight packing of adjacent carriers, therefore enhancing the spectral efficiency of an OFDM system.

The effects of a clipping amplifier on the OFDM signal spectrum are demonstrated in Figure 19.6. Specifically, Figure 19.6(a) shows the spectrum corresponding to the system employed in the context of Figure 19.5(b), but with the amplifier back-off set to 9 dB. The frequency axis is normalised to the signal bandwidth of $B = 312 \cdot \Delta f$ due to the disabled carriers and the power spectral density is normalised to the average signal power in the signal band. Comparison of Figure 19.6(a) and 19.5(b) shows no discernible change in the spectral domain.

Reducing the amplifier back-off results in increased out-of-band emissions. Figure 19.6(b) shows the averaged power spectral density of the same system with an amplifier back-off of 0 dB. In this case, the spectral spillage is apparent directly outside the signal band.

Since the amplitude limiting of the transmitted time domain signal is due to the power amplifier, no bandlimiting filtering is performed after the amplitude distortion. Li and Cimini [405] proposed to employ a hard limiter and a lowpass filter before the non-linear amplification stage. In this case, the out-of-band emissions can be reduced, while the BER penalty due to hard limiting the time domain signal still applies.

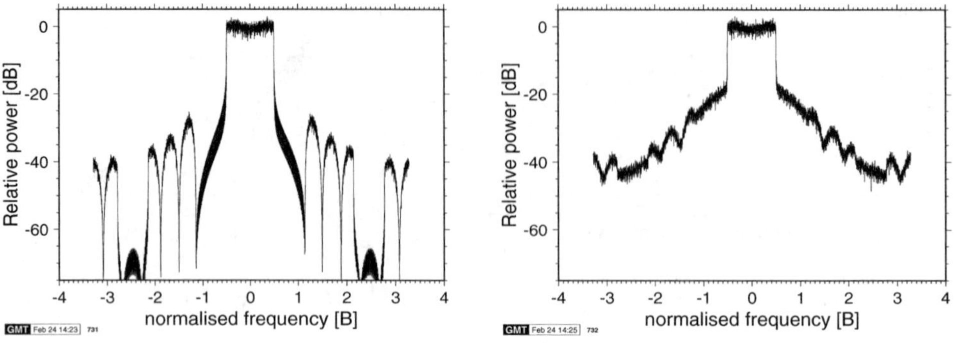

(a) BO = 9dB

(b) BO = 0dB

Figure 19.6: Power spectra of a 512-subcarrier OFDM modem employing 200 virtual carriers and Nyquist filtering with an excess bandwidth of 0.35. The frequencies are normalised to the signal bandwidth of $B = 312 \cdot \Delta f$: (a) 9 dB amplifier back-off (BO), (b) 0 dB amplifier back-off

(a) interference power (dB) vs. carrier separation and BO

(b) iso-interference plot (dB)

Figure 19.7: Integrated adjacent channel interference power (dB) in a signal bandwidth of B for different values of normalised carrier separation and amplifier back-off

In order to evaluate the amount of adjacent channel interference caused by non-linear amplification of the OFDM signals, the total interference power in a frequency domain window of width B having a variable frequency offset was integrated and normalised by the in-band power for varying values of amplifier back-off. Figure 19.7(b) shows the results of this integration for amplifier back-off values up to 9 dB and adjacent carrier separations from B to $2B$. A carrier separation of B corresponds to directly adjacent signal bands, and such a set-up would approach the maximal possible system throughput of 2 symbols/s/Hz or 2 Baud/Hz.

It can be observed from the figure that the integrated interference levels are indistinguishable for amplifier back-off values of 6 dB and 9 dB. The integrated interference level in this case is below -30 dB with directly adjacent carriers and below -33 dB for all carrier separation values above $1.03B$. The drop in interference for carrier separations above $1.7B$ is due to the frequency domain image of the virtual carriers that can be seen in Figure 19.6(a) at frequencies between $2.2B$ and $3.8B$. For a back-off value of 3 dB the interference is approximately 1 dB worse than for the higher back-off values for carrier separations above $1.5B$, and this difference is up to 5 dB for smaller carrier separations.

For a back-off value of 0 dB, which corresponds to the spectrum shown in Figure 19.6(b), the integrated interference is below -22.5 dB for all carrier separations. The figure also shows two graphs for negative amplifier back-off values for comparison, which exhibit considerably higher interference values. Since negative back-off values imply amplifier clipping below the nominal mean power of the signal, this would not be a likely case at the transmitter's power amplifier.

19.4.3 Clipping amplification - summary

We have seen that for OFDM transmission a considerable amplifier back-off is necessary, in order to maintain the BER and spectral characteristics of the ideal OFDM system. This complicates the transmitter design and leads to inefficient power amplifier applications.

Although a 512-subcarrier OFDM modem is capable of producing a peak-to-average power ratio of up to 27 dB, we have seen that much lower amplifier back-off values result in acceptable system performance. A back-off of 12.5 dB results in only one of 10^5 OFDM symbols suffering clipping at all, and we have seen for a 6 dB back-off that both the BER performance as well as the averaged spectrum are indistinguishable from the ideally amplified scenario.

19.5 Analogue-to-Digital Conversion

We have seen in Figure 19.2 that the time domain OFDM signal follows a two-dimensional or complex Gaussian distribution. The distribution of signal amplitudes determines the necessary parameters of the analogue-to-digital convertor (ADC) at the receiver. We have investigated the effects of the receiver input ADC on the BER performance of an OFDM system in an AWGN channel environment.

Since an ADC transforms the continuous analogue input signal to discrete-valued digital information, the original input signal is not preserved perfectly. As the dy-

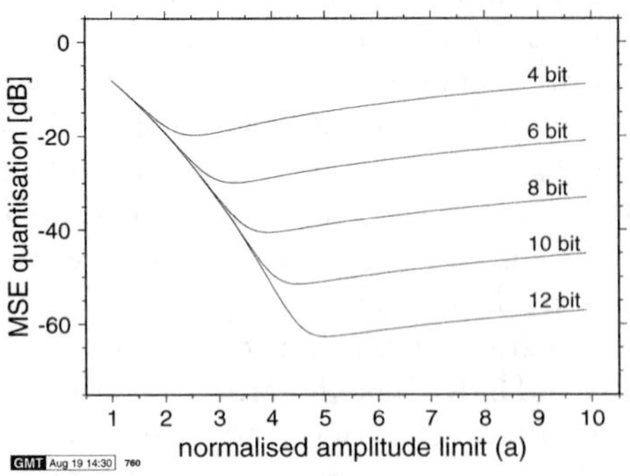

(a)

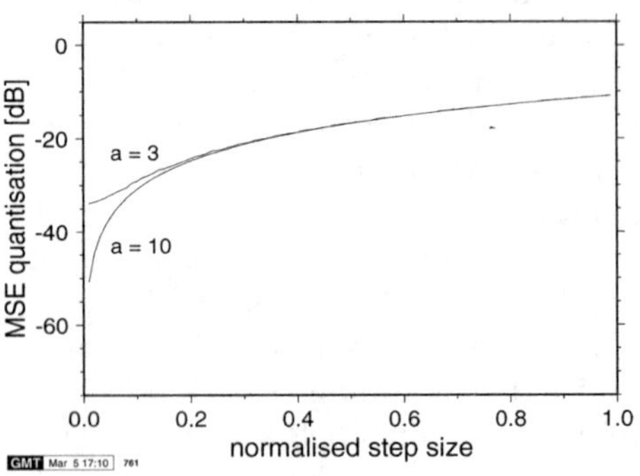

(b)

Figure 19.8: Mean square quantisation error for linearly spaced ADC at OFDM receiver: (a) quantisation noise for different amplitude limits normalised to the average input signal amplitude σ versus ADC resolution; (b) quantisation noise for amplitude limits of $a = 3$ and $a = 10$ versus the linear step size

namic range of an ADC is limited, there are amplitude clipping effects if the input signal exceeds the ADC's maximal input amplitude.

A linear analogue-to-digital convertor quantises the input signal into a set of n equidistant output values, which span an interval with clipping values of $-a$ and a. In this case, the ADC error stems from quantisation error due to the finite step size of $2a/n$ for input values within the dynamic range $[-a, a]$, as well as from amplitude clipping for input values outside this range.

Taking into account the Gaussian nature of the I and Q components of the OFDM signal, the mean square ADC error can be determined for different ADC set-ups. Figure 19.8(a) shows the mean square ADC error normalised to the average input signal power versus clipping amplitude for different quantisation resolution values. For this and the following figures, the input signal is modelled by a white Gaussian noise function with a standard deviation σ. The mean square quantisation error is relative to the mean input power σ^2, and the amplitude limit a is in units of σ.

The mean square ADC error value can be interpreted as the mean power of an additional random noise superimposed on the quantised signal. This ADC noise is not Gaussian, hence estimating the BER performance degradation with the Gaussian noise approximation is inadequate.

It can be seen from the figure that a small dynamic range of $2a$ results in high ADC noise irrespective of the number of ADC bits, which is due to predominant clipping. For $a = 1$, the normalised ADC noise power is about -8 dB, again independently of the resolution of the ADC. For higher values of a, the total ADC noise is dominated by the effects of increasing step size. Depending on the AGC's given resolution, there is an optimal value of a, for which the ADC noise, constituted by clipping and granular noise, is minimal.

In order to separate the effects of the amplitude limitation and the quantisation

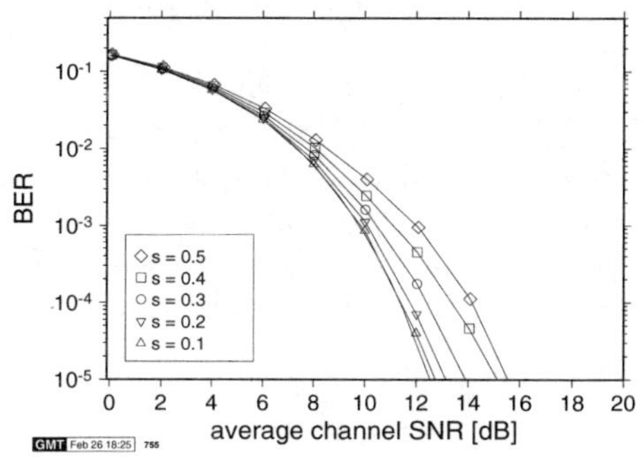

(a) coherent QPSK

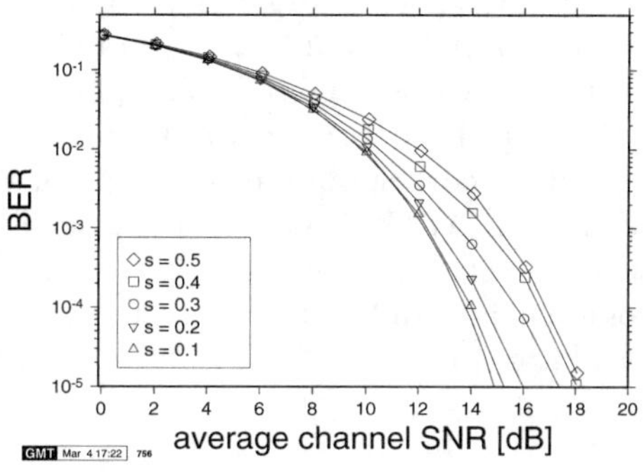

(b) differential QPSK

Figure 19.9: BER versus channel SNR for a QPSK modem with coherent and differential detection using different normalised receiver ADC resolution values in an AWGN channel. The maximal normalised amplitude is $a = 3$

step size, BER simulations have been conducted with a fixed amplitude limiter and a range of ADC step sizes. The linear ADC step size is given for a range of ADC resolutions b and two amplitude limits a in Table 19.1. The effects of clipping on the OFDM performance have been discussed in the context of non-linear amplification, and Figure 19.4 showed that an amplifier back-off of 9 dB results in no BER performance degradation due to amplifier clipping in an AWGN channel for a 512-subcarrier OFDM modem. Setting the amplitude limits of the simulated ADC to $a = 3$ corresponds to a 9.5 dB dynamic range, hence the amplitude clipping contribution to the ADC noise does not significantly impair the modem's BER performance.

$a \backslash b$	4	5	6	7	8	9	10	11	12
3	0.375	0.188	0.094	0.0469	0.0234	0.0117	0.00586	0.00293	0.00146
10	0.625	0.313	0.156	0.0781	0.039	0.0195	0.00977	0.00488	0.00244

Table 19.1: Linear ADC step size for amplitude limits of $a = 3$ and $a = 10$ and ADC resolutions from $b = 4$ bits/sample to $b = 12$ bits/sample

The average ADC noise power for the normalised amplitude limits of $a = 3$ and $a = 10$ are depicted in Figure 19.8(b). The amplitude limit of $a = 3$ results in a residual ADC noise of about -34 dB due to amplitude clipping, while the amplitude clipping noise contribution for a limit of $a = 10$ is below -50 dB. For step sizes above 0.2, which corresponds to ADC resolutions of 5 bits or less, the gap between the two curves is closed, and the noise power is dominated by the quantisation noise.

The BER performance of a 512-subcarrier OFDM modem employing coherently as well as differentially detected QPSK is depicted in Figure 19.9. These experiments have been conducted for a maximal amplitude of $a = 3$ with step sizes from 0.1 to 0.5, which roughly corresponds to ADC resolutions between 6 and 3 bits. In both graphs the curve without markers represents the modem's BER performance in the absence of ADC quantisation and clipping noise.

It can be observed in both figures that the SNR penalty at a BER of 10^{-4} with an ADC step size of 0.1 is 0.15 dB and 0.2 dB for QPSK and DQPSK, respectively. This resolution corresponds to 60 steps in the range from -3 to 3, and therefore would require an ADC of 6-bit resolution. For higher step sizes the SNR penalty is more severe; a step size of 0.5 results in a SNR penalty of about 2.8 dB and 3.2 dB for coherent and differential detection, respectively. Let us now consider the effects of oscillator noise.

19.6 Phase Noise

The presence of phase noise is an important limiting factor for an OFDM system's performance [117, 118, 406], and depends on the quality and the operating conditions of the system's RF hardware. In conventional mobile radio systems around a carrier frequency of 2GHz the phase noise constitutes typically no severe limitation, however in the 60GHz carrier frequency, 225MHz bandwidth system considered in Section 20.1.1 its effects were less negligible and hence had to be investigated in more depth. Oscillator noise stems from oscillator inaccuracies in both the transmitter and receiver and manifests itself in the baseband as additional phase and amplitude modulation of the received samples [407]. The oscillator noise influence on the signal depends on the noise characteristics of the oscillators in the system and on the signal bandwidth. It is generally split in amplitude noise $A(t)$ and phase noise $\Phi(t)$, and the influence of the amplitude noise $A(t)$ on the data samples is often neglected. The time domain functions $A(t)$ and $\Phi(t)$ have Gaussian histograms, and their time domain correlation is determined by their respective long-term power spectra through the Wiener-Khintchine theorem.

If the amplitude noise is neglected, imperfect oscillators are characterised by the long-term power spectral density (PSD) $N_p(f')$ of the oscillator output signal's phase noise, which is also referred to as the phase noise mask. The variable f' represents the frequency distance from the oscillator's nominal carrier frequency in a bandpass model, or equivalently, the absolute frequency in the baseband. An example of this phase noise mask for a practical oscillator is given in Figure 19.10(a). If the phase noise PSD $N_p(f')$ of a specific oscillator is known, then the variance of the phase error $\Phi(t)$ for noise components in a frequency band $[f_1, f_2]$ is the integral of the phase noise spectral density over this frequency band as in [407]:

$$\bar{\Phi}^2 = \int_{f_1}^{f_2} \left(\frac{2N_p(f')}{C} \right) df', \tag{19.4}$$

where C is the carrier power and the factor 2 represents the double-sided spectrum of the phase noise. The phase noise variance $\bar{\Phi}^2$ is also referred to as the integrated phase jitter, which is depicted in Figure 19.10(b).

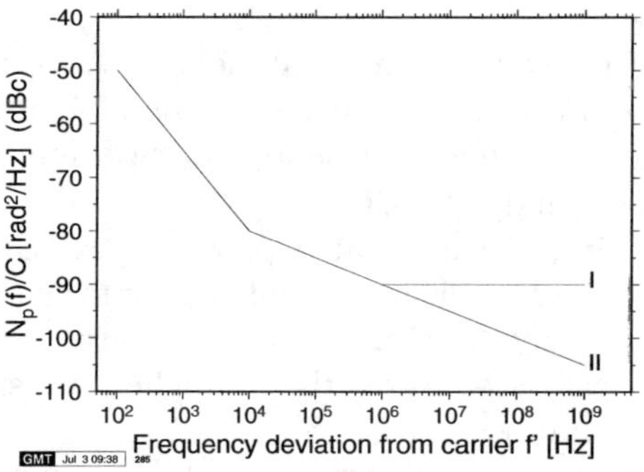

(a) $N_p(f')/C$

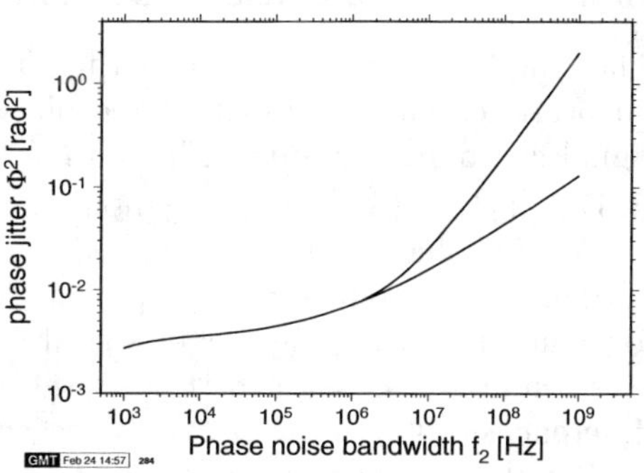

(b) Φ^2

Figure 19.10: Phase noise characterisation: (a) spectral phase noise density (phase noise mask), (b) integrated phase jitter for two different phase noise masks

19.6.1 Effects of phase noise

The phase noise contribution of both the transmitter and receiver can be viewed as an additional multiplicative effect of the radio channel, like fast and slow fading. In serial modulation schemes phase noise manifests itself as random phase errors of the received samples. The effects of this additional random phase component on the received samples depend on the modulation scheme employed, and as expected, they are more pronounced in differential detection schemes than in coherently detected arrangements invoking carrier recovery. The carrier recovery, however, is affected by the phase noise, which in turn degrades the performance of a coherently detected scheme.

For OFDM schemes, multiplication of the received time domain signal with a time-varying channel transfer function is equivalent to convolving the frequency domain spectrum of the OFDM signal with the frequency domain channel transfer function. Since the phase noise spectrum's bandwidth is wider than the subcarrier spacing, this results in energy spillage into other sub-channels and therefore in intersubcarrier interference, an effect which will be quantified below.

19.6.2 Phase noise simulations

In our studies two different models of the phase noise in an OFDM communications system have been investigated: the simple bandlimited white phase noise model, which is solely based on the value of the integrated phase jitter of Equation 19.4, and a coloured phase noise model, which also takes into account the phase noise mask of Figure 19.10(a). Both of these models will be described below.

19.6.2.1 White phase noise model

The simplest way of modelling the effect of phase noise in simulations is to assume uncorrelated Gaussian distributed phase errors with a standard deviation of $\bar{\Phi}$ at the signal sampling instants. This corresponds to the phase noise exhibiting a uniform PSD $N_p(f') = (\bar{\Phi})^2 / B$ throughout the signal bandwidth B.

As the phase noise is assumed to be white, the integrated phase jitter, or the variance of the phase noise given in Equation 19.4 constitute sufficient information for generating the phase error signal $\Phi(t)$. Clearly, since no correlation between the noise samples is assumed, this is the worst-case scenario with the highest possible differences of $\Phi(t)$ between two consecutive samples.

Simulations have been performed for both a 512-subcarrier OFDM modem, as well as for a serial modem for comparison. The serial modem's BER performance curves are depicted in Figure 19.12 for BPSK and QPSK employing both coherent and differential detection, for integrated phase jitter values of $\bar{\Phi}^2 = 0.05\ rad^2$ to $\bar{\Phi}^2 = 0.25\ rad^2$. For coherently detected BPSK, as shown in Figure 19.12(a), white phase noise of $\bar{\Phi}^2 = 0.05\ rad^2$ already causes a measurable SNR penalty of about 0.6 dB at a BER of 10^{-4}, while a phase jitter of 0.1 rad^2 results in a SNR penalty of 1.8 dB at this BER. For phase jitter values above 0.1 rad^2, residual bit error rates of $6 \cdot 10^{-5}$, $5 \cdot 10^{-4}$ and $2 \cdot 10^{-3}$ were observed at phase jitters of 0.15, 0.2 and 0.25 rad^2, respectively.

19.6.2.1.1 Serial modem The serial QPSK modem's performance is more vulnerable to the effects of phase noise than that of BPSK, as is shown in Figure 19.12(a) and 19.12(b). For all simulated values of $\bar{\Phi}^2$, the bit error rate exceeds 10^{-4}. For a phase jitter of 0.25 rad^2, the observed residual bit error rate is 6%. The curves for differential detection of BPSK and QPSK are shown in Figure 19.12(c) and 19.12(d), respectively. Since for differentially detected PSK schemes both the received modulated symbol as well as the reference phase are corrupted by phase noise, the phase noise effects are severe. The DBPSK modem exhibits residual bit errors of more than 10^{-4} for all simulated phase jitter values above 0.05 rad^2, while the DQPSK modem's BER performance is worse than 0.9% for all simulated values of $\bar{\Phi}^2$.

19.6.2.1.2 OFDM modem While for serial modems the phase noise results in phase errors of the received samples, the FFT modem's performance is affected by the inter-subcarrier interference caused by the convolution of the phase noise spectrum with the OFDM spectrum. The average amount of power spillage from one subcarrier into the rest of the OFDM spectrum observed for different values of integrated phase jitter is depicted in Figure 19.11(a). It can be seen that the white phase noise results in uniform spreading of inter-subcarrier interference over all subcarriers, and a loss of signal power in the desired subcarrier number 256. It can be noted that doubling the variance of the phase noise results approximately in doubling the interference power in the other subcarriers, as demonstrated by the 3 dB differences seen in Figure 19.11(a). The received signal power in the central subcarrier drops accordingly.

The total interference power received in each subcarrier is the sum of all active subcarriers' contributions, which is shown in Figure 19.11(b) for a 512-subcarrier OFDM

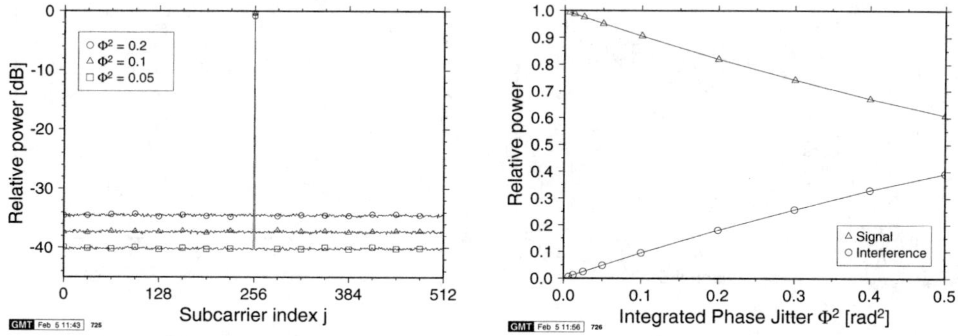

(a) interference from single subcarrier

(b) signal and interference level

Figure 19.11: The effect of white phase noise on OFDM: (a) power spillage due to white phase noise, (b) average signal and interference levels for one subcarrier

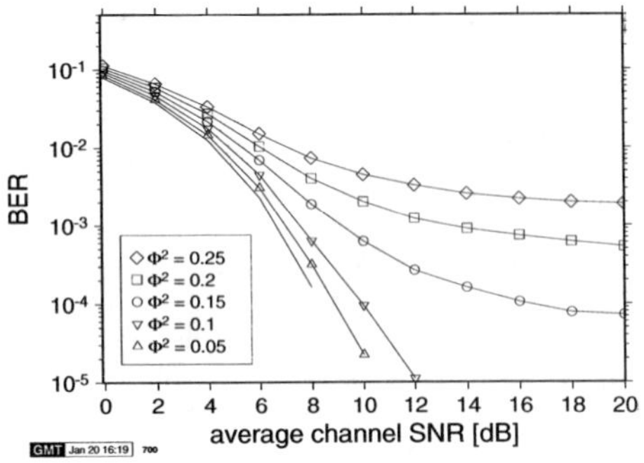

(a) coherent BPSK

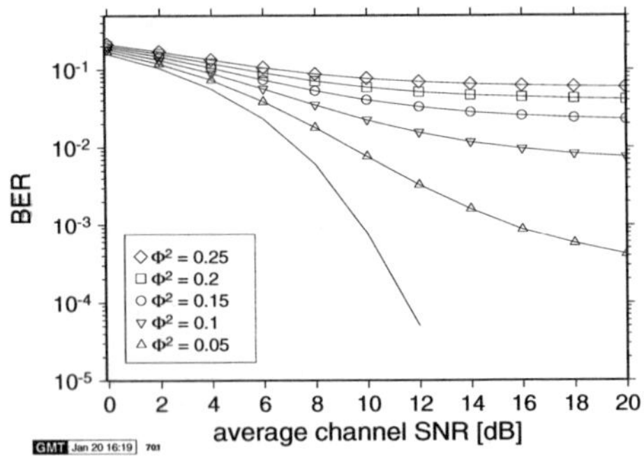

(b) coherent QPSK

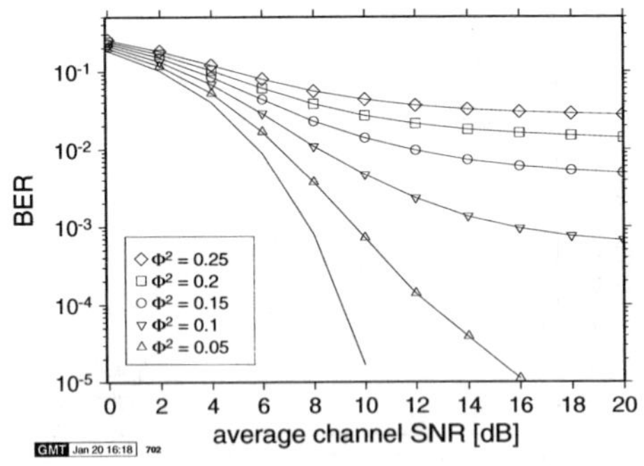

(c) differential BPSK

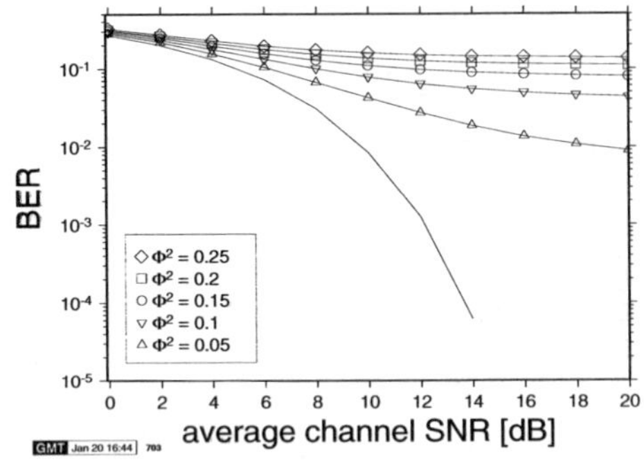

(d) differential QPSK

Figure 19.12: Bit error rate versus channel SNR for a serial modem in an AWGN channel and exposed to white phase noise for integrated phase jitter values of 0.05 rad^2 to 0.25 rad^2: (a) BPSK, (b) QPSK, (c) DBPSK, (d) DQPSK

modem. It can be seen that the signal energy drops while the interference energy rises for rising values of the integrated phase jitter, with the signal-to-interference ratio (SIR) dropping to about 1.5, which corresponds to 2 dB for $\bar{\Phi}^2 = 0.5$. For small values of phase jitter up to approximately 0.1, the interference power rises linearly with the phase jitter, and more slowly for higher values of phase jitter. The interchannel-interference (ICI) experienced in each subcarrier for the white phase noise channel depending on the phase jitter is shown in Figure 19.13(a). The SIR is above 10 dB for phase jitter values below approximately 0.1, and 20 dB for values phase jitter values of 0.01. This is in accordance to the linear relationship of interference power and phase jitter for small values observed above.

Figure 19.14 shows the BER performance results observed for a 512-subcarrier OFDM modem in the white phase noise channels. For coherently detected OFDM BPSK the BER performance is generally worse than for the serial modem, characterised in Figure 19.12(a). For differential detection, however, the OFDM modem's

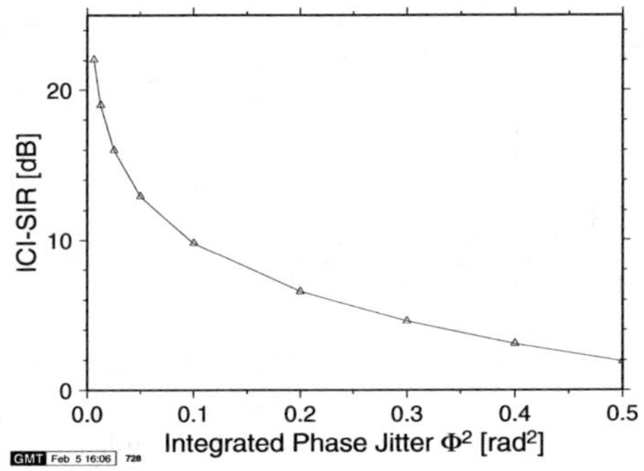

(a) ICI-SIR for OFDM modem in white phase noise

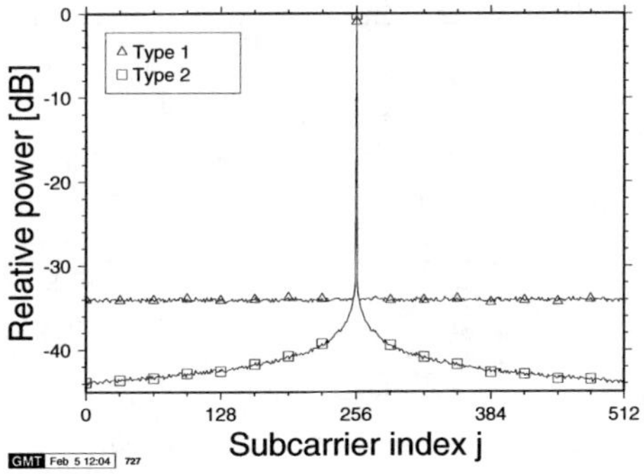

(b) interference from single subcarrier, coloured phase noise

Figure 19.13: OFDM in phase noise channel: (a) ICI-SIR for OFDM modem in white phase noise channel; (b) average signal and interference levels for one subcarrier in the coloured phase noise channels corresponding to the phase noise masks shown in Figure 19.10(a)

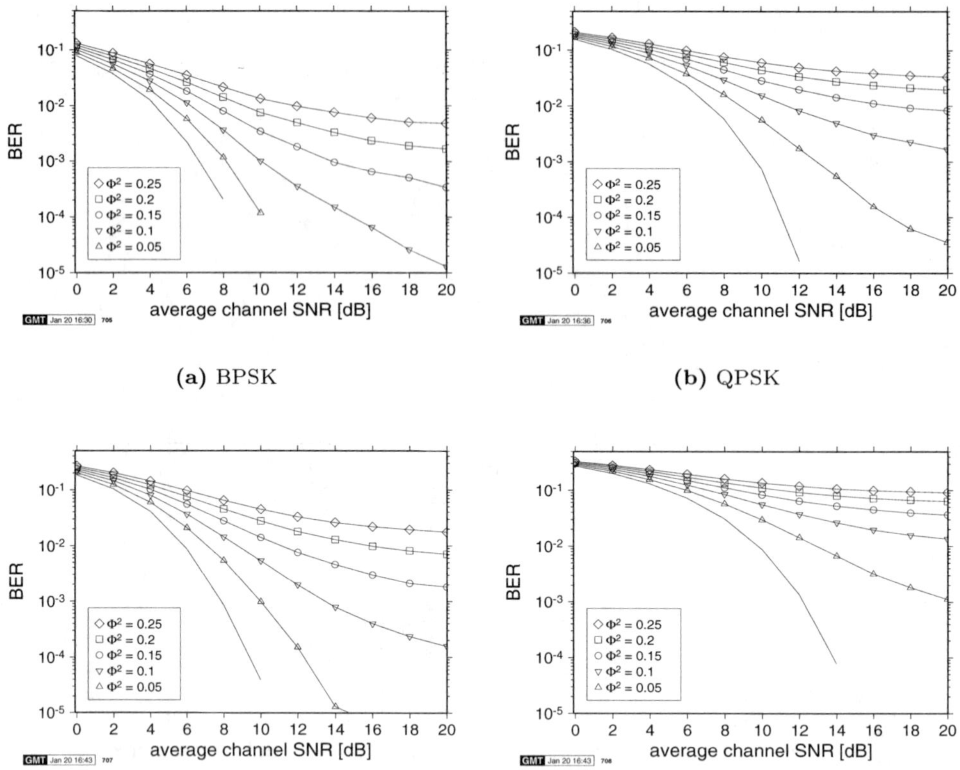

(a) BPSK

(b) QPSK

(c) differential BPSK

(d) differential QPSK

Figure 19.14: Bit error rate versus channel SNR for a 512 subcarrier OFDM modem in an AWGN channel and white phase noise for different values of integrated phase jitter of 0.05 rad^2 to 0.25 rad^2: (a) BPSK, (b) QPSK, (c) DBPSK, (d) DQPSK (compared with Figure 19.12)

BER performance is better than the serial modem's depicted in Figure 19.12(c). This different behaviour can be explained by the different effects of the phase noise on the different systems. It has been observed in Figure 19.11(b) that for the investigated values of the integrated phase jitter the mean interference power in each subcarrier of an OFDM modem increases approximately linearly with the mean square of the phase error in the serial modem. For the same values of noise variance, coherent serial BPSK is more robust to the phase errors in the received serial symbols than DBPSK, while the noise-like interference observed in the OFDM system is less detrimental to the differentially detected BPSK than to coherent BPSK.

For both differential and coherent QPSK the OFDM modem performs better than the serial modem, as it can be seen comparing Figure 19.14(b) with 19.12(b) and 19.14(d) with 19.12(d). There is, however, considerable degradation of the BER performance in both cases. Having considered the worst-case scenario of uncorrelated white phase noise, let us now focus our attention on the more practical case of coloured phase noise sources.

19.6.2.2 Coloured phase noise model

The integral $\bar{\Phi}^2$ of Equation 19.4 characterises the long-term statistical properties of the oscillator's phase and frequency errors due to phase noise. In order to create a time domain function satisfying the standard deviation $\bar{\Phi}^2$, a white Gaussian noise spectrum was filtered with the phase noise mask $N_p(f')$ depicted in Figure 19.10(a), which was transformed into the time domain. A frequency resolution of about 50 Hz was assumed in order to model the shape of the phase noise mask at low frequencies, which led to a FFT transform length of $2^{22} = 4194304$ samples for the frequency range of Figure 19.10(a).

The resulting time domain phase noise channel data is a stream of phase error samples, which were used to distort the incoming signal at the receiver. The double-sided phase noise mask used for the simulations is given in Table 19.2. Between the

f'[Hz]	100	1k	10k	100k	1M
N_p/C[dB]	-50	-65	-80	-85	-90

Table 19.2: Two-sided phase noise mask used for simulations. f' = frequency distance from carrier, N_p/C = normalised phase noise density

points given in Table 19.2, a log-linear interpolation is assumed, as shown in Figure 19.10(a). As the commercial oscillator's phase noise mask used in our investigations was not specified for frequencies beyond 1 MHz, two different cases were considered for frequencies beyond 1 MHz: (I) a phase noise floor at -90 dB, and (II) an $f^{-1/2}$ law. Both of these extended phase noise masks are shown in Figure 19.10(a). The integrated phase jitter has been calculated using Equation 19.4 for both scenarios, and the value of the integral for different noise bandwidths is depicted in Figure 19.10(b).

For the investigated 155 Mbits/s wireless ATM (WATM) system's [148–151] double-sided bandwidth of 225 MHz, the integration of the phase noise masks results in phase jitter values of $\bar{\Phi}^2 = 0.2303\ rad^2$ and $\bar{\Phi}^2 = 0.04533\ rad^2$ for the phase noise mask with and without noise floor, respectively.

The simulated BER performance of a 512-subcarrier OFDM system with a subcarrier distance $\Delta f = 440\ kHz$ over the two different phase noise channels is depicted in Figure 19.15 for coherently detected BPSK and QPSK. In addition to the BER graphs corresponding to the coloured phase noise channels described above, graphs of the modems' BER performance over white phase noise channels with the equivalent integrated phase jitter values was also plotted in the figures. It can be observed that the BER performance for both modulation schemes and for both phase noise masks is very similar for the coloured and the white phase noise models.

The interference caused by one OFDM subcarrier in the two coloured phase noise channels is depicted in Figure 19.13(b). It can be seen that the Type 1 channel exhibits a virtually white interference spectrum, very similar to the white phase noise channels depicted in Figure 19.11(a). Only the subcarriers directly adjacent to the

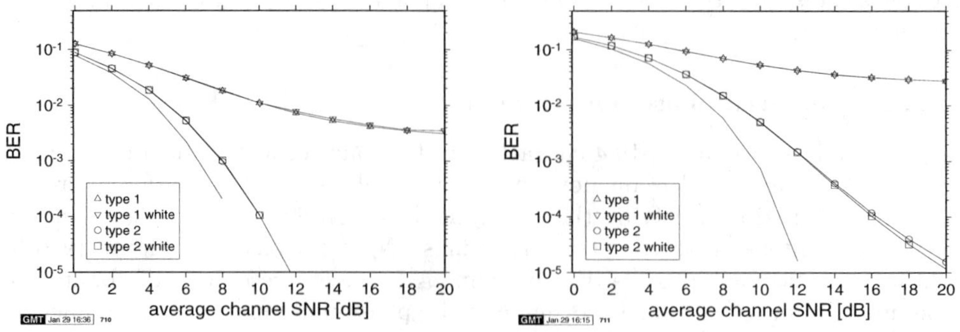

(a) BPSK

(b) QPSK

Figure 19.15: Bit error rate versus channel SNR for a 512-subcarrier OFDM modem in the presence of phase noise. Type 1 represents the coloured phase noise channel with the phase noise mask depicted in Figure 19.10(a) assuming a noise floor of 90 rad^2/Hz, while Type 2 is the channel without phase noise floor. The curves designated "white" are the corresponding white phase noise results. The lines without markers give the corresponding results in the absence of phase noise

signal bearer show higher interference influence than in the white phase noise channel. This is due to the Type 1 phase noise density given in the phase noise mask; it is non-flat only for frequencies below 1 MHz, which corresponds to approximately 2.3 subcarrier distances at a separation of 440 kHz. It can be observed in Figure 19.13(b) that higher interference was measured in the two subcarriers adjacent to the signal carrier. If the interference caused by all subcarriers is combined, then all but the very closest subcarriers have equal contributions to the interference signal, resulting in interference like Gaussian noise.

The Type 2 channel, which exhibits no noise floor in the phase noise mask, results in interference that is dependent on the distance from the interfering subcarrier. The summation of the interference is dominated by the interference contribution of the carriers in close vicinity, hence the resulting interference is less Gaussian than for the Type 1 channel.

The simulated BER results shown in Figure 19.15 show virtually indistinguishable performance for the modems in both the coloured and the white phase noise channels. Only a slight difference can be observed for QPSK between the Type 1 and Type 2 phase noise masks, where the corresponding white phase noise results in a better performance than the coloured noise. This difference can be explained with the interference being caused by fewer interferers compared to the white phase noise scenario, resulting in a non-Gaussian error histogram.

19.6.3 Phase noise - Summary

Phase noise, like all time-varying channel conditions experienced by the time-domain signal, results in intersubcarrier interference in OFDM transmissions. If the bandwidth of the phase noise is high compared to the OFDM subcarrier spacing, then this interference is caused by a high number of contributions from different subcarriers, resulting in a Gaussian noise interference. Besides this noise inflicted upon the received symbols, the signal level in the subcarriers drops by the amount of energy spread over the adjacent subcarriers.

The integral over the phase noise mask, termed phase jitter, is a measure of the signal-to-interference ratio that can be expected in the received subcarriers, if the phase noise has a wide bandwidth and is predominantly white. The relationship between the phase jitter and the SIR is shown in Figure 19.13(a). For narrowband phase noise this estimation is pessimistic.

Chapter 20

OFDM Transmission over Wideband Channels

Orthogonal frequency division multiplexing modems have been originally conceived in order to transmit data reliably in time dispersive or frequency selective channels without the need for a complex time domain channel equaliser. In this chapter, the techniques employed for the transmission of quadrature amplitude modulated (QAM) OFDM signals over a time dispersive channel are discussed and channel estimation methods are investigated.

20.1 The Channel Model

The channel model assumed in this chapter is that of a finite impulse response (FIR) filter with time-varying tap values. Every propagation path i is characterised by a fixed delay τ_i and a time-varying amplitude $A_i(t) = a_i \cdot g_i(t)$, which is the product of a complex amplitude a_i and a Rayleigh fading process $g_i(t)$. The Rayleigh processes g_i are independent from each other, but they are all exhibiting the same normalised Doppler frequency f'_d, depending on the parameters of the simulated channel.

The ensemble of the p propagation paths constitutes the impulse response

$$h(t,\tau) = \sum_{i=1}^{p} A_i(t) \cdot \delta(\tau - \tau_i) = \sum_{i=1}^{p} a_i \cdot g_i(t) \cdot \delta(\tau - \tau_i), \tag{20.1}$$

which is convolved with the transmitted signal.

All the investigations carried out in this chapter were based on one of the system and channel models characterised below. Each of the models represents a framework for a class of similar systems, grouped in three categories: wireless asynchronous transfer mode (WATM), wireless local area networks (WLAN) and a time division multiple access (TDMA) OFDM form of a personal communications type scheme, in a similar framework to the universal mobile telecommunication system (UMTS).

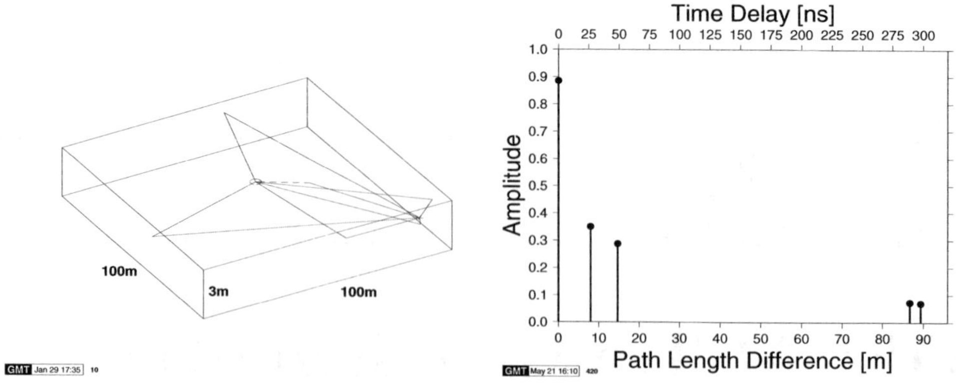

(a) paths **(b)** impulse response

Figure 20.1: The 60GHz, 225MHz bandwidth WATM channel: five-path model and resulting impulse response

20.1.1 The wireless asynchronous transfer mode system

The wireless asynchronous transfer mode (WATM) system parameters used for our investigations follow closely the specifications of the Advanced Communications Technologies and Services (ACTS) Median system, which is a proposed wireless extension to fixed-wire ATM-type networks. In the Median system, the OFDM FFT length is 512, and each symbol is padded with a cyclic prefix of length 64. The sampling rate of the Median system is 225 Msamples/s, and the carrier frequency is 60 GHz. The uncoded target data rate of the Median system is 155 MBps.

20.1.1.1 The WATM channel

The WATM channel employed here is a pessimistic model of the operating environment of an indoor wireless ATM network similar to that of the ACTS Median system. For our simulations, we assumed a vehicular velocity of about 50 km/h or 13.9 m/s, resulting in a normalised Doppler frequency of $f'_d = 1.235 \cdot 10^{-5}$. The impulse response was determined by simple ray tracing in a warehouse-type environment of $100m \times 100m \times 3$ m^3, which is shown schematically in Figure 20.1(a). The resulting impulse response, shown in Figure 20.1(b) exhibits a maximum path delay of 300 ns, which corresponds to 67 sampling intervals. The five taps of the impulse response were derived assuming free space propagation, using the inverse second power law.

The Fourier transform of this impulse response leads to the static frequency domain channel transfer function depicted in Figure 20.2(a). Note that the central frequency of the bandwidth is in FFT bin 0, hence the spectrum appears wrapped around in this graph. Throughout this work the subcarrier index 0 contains the central frequency, with the subcarriers 1 to $N/2$ spanning the positive relative frequencies, and the subcarriers $N/2-1$ to $N-1$ containing the negative relative frequency range.

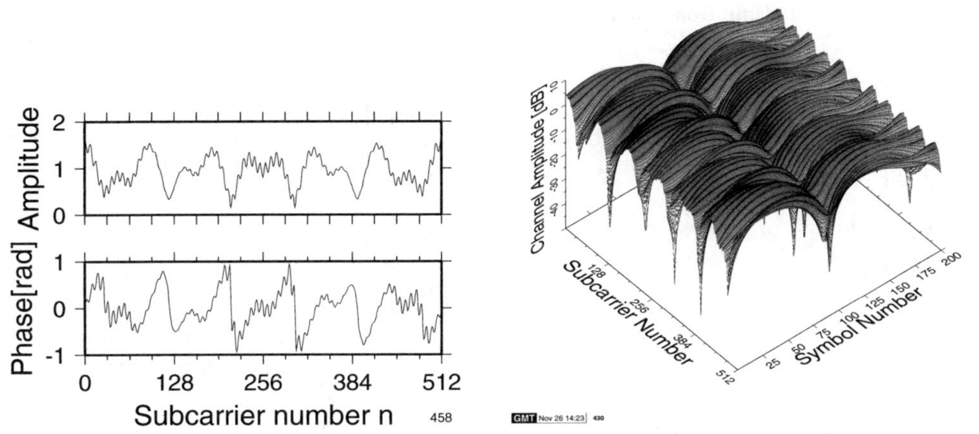

(a) channel frequency response **(b)** time varying amplitude

Figure 20.2: WATM channel: (a) unfaded frequency domain channel transfer function $H(n)$; (b) time-varying channel amplitude for 200 consecutive OFDM symbols

Since the assumed impulse response is real, the resulting channel transfer function is conjugate complex symmetric around the central frequency bin.

If each of the impulses in Figure 20.1(b) is faded according to a Rayleigh fading process with the normalised Doppler frequency f'_d, then the resulting time-varying impulse response will lead to a time-varying frequency domain channel transfer function, which is depicted in Figure 20.2(b). In this figure, the amplitude of the channel transfer function has been plotted over the bandwidth of the OFDM system for 200 consecutive OFDM symbol time slots.

Figure 20.2(b) reveals a relatively high correlation of the channel transfer function amplitude both along the time and the frequency axes. The correlation in the frequency domain is dependent on the impulse response; the longer the tap delays, the lower the correlation in the transfer function. The similarity of the channel transfer functions along the time axis stems from the slowly changing nature, i.e. the low normalised Doppler frequency, of the assumed narrowband channels. The impulse response changes only little between consecutive OFDM symbol intervals; the correlation between neighbouring channel transfer functions is therefore affected by the Doppler frequency of the constituting narrowband fading channels relative to the duration of one OFDM symbol. For this graph, the impulse response was kept constant for the duration of each OFDM symbol. We will show in Section 20.2.2 that the notion of defining a frequency domain channel transfer function is problematic in environments exhibiting rapid changes of the impulse response.

As the channel impulse response depicted in Figure 20.1(b) is of the line-of-sight (LOS) type with one dominant path, fading of this LOS path results in an amplitude variation across the whole signal bandwidth. This os seen in Figure 20.2(b) at around symbol number 100, where a substantial amplitude fade across all subcarriers occurs.

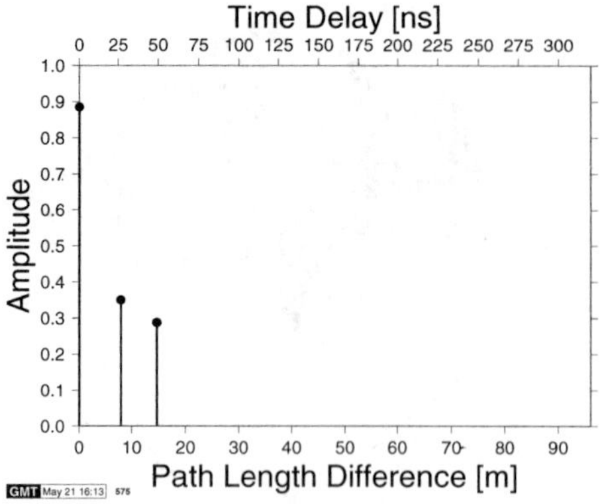

(a) channel impulse response

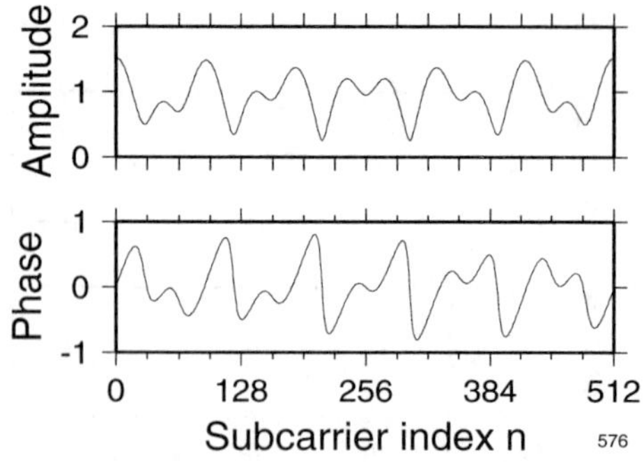

(b) channel frequency response

Figure 20.3: Short WATM channel: (a) impulse response, (b) unfaded frequency domain channel transfer function $H(n)$

20.1.1.2 The shortened WATM channel

Since the WATM channel discussed above is the worst-case scenario for an indoor wireless high data rate network, we have introduced a truncated version of the original WATM channel impulse response depicted in Figure 20.1(b), by only retaining the first three impulses. This reduces the total length of the impulse response, with the last path arriving at a delay of 48.9 ns due to the reflection with an excess path length of about 15 m with respect to the line-of-sight path, which corresponds to 11 sample periods. Omitting the two long delay pulses in the impulse response does not affect the total impulse response energy significantly, lowering the power by only 0.045 dB, hence no renormalisation of the impulse response was necessary. The resulting impulse response exhibits a root mean squared (RMS) delay spread of $1.5276 \cdot 10^{-8}$s, and it is shown in Figure 20.3(a). The resulting frequency domain transfer function for the unfaded short WATM impulse response is given in Figure 20.3(b), and comparison with Figure 20.2(a) reveals a close correspondence with the original WATM channel transfer function. As the first three low delay paths are the same for both channel models, the general shape of the channel transfer function is very similar; the last two paths in the WATM impulse response result in fast ripple on top of the dominant lower frequency components of the channel transfer function of Figure 20.2(a), which is absent in the shortened channel of Figure 20.3(b).

20.1.2 The wireless local area network system

The main wireless local area network (WLAN) system parameters we opted for were loosely based on the high performance local area network (HIPERLAN) system [408], leading to a sampling rate of 20 MHz and a carrier frequency of 17 GHz. The assumed data symbols utilise a 1024-point FFT and a cyclic extension of 168 samples.

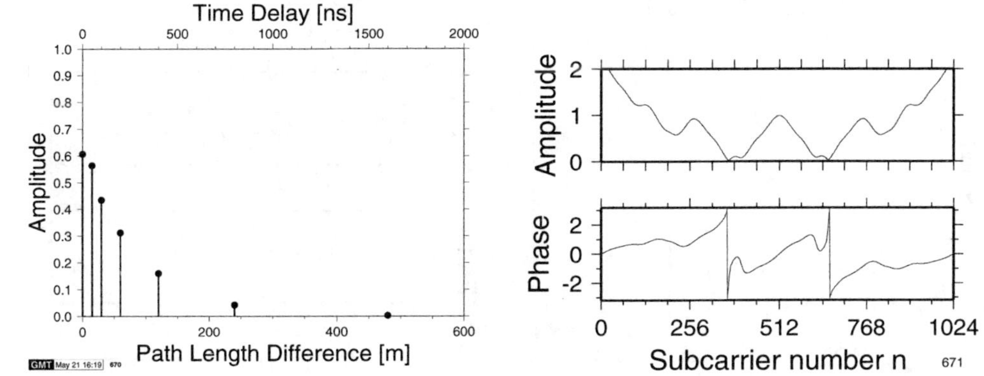

(a) modified HIPERLAN CM1 150

(b) channel frequency response

Figure 20.4: WLAN channel: (a) modified HIPERLAN CM1 150 ns impulse response, (b) unfaded frequency domain channel transfer function $H(n)$

20.1.2.1 The WLAN channel

The WLAN channel's impulse response is based on the HIPERLAN CM1 impulse response with a RMS delay spread of 150 ns, as described by Tellambura *et al.* [408]. The second and third path of the original impulse response were combined to achieve a symbol-spaced impulse response pattern. The resulting impulse response is shown in Figure 20.4(a). The corresponding frequency domain channel transfer function is displayed in Figure 20.4(b). Assuming a worst-case vehicular velocity of 50km/h, fading channel simulations were conducted employing Rayleigh fading narrowband channels with a normalised Doppler frequency of $f'_d = 3.94 \cdot 10^{-5}$.

20.1.3 The UMTS system

The set of parameters employed for the UMTS-type investigations was based on a version of the ACTS FRAMES Mode 1 proposal [409], resulting in a sampling rate of 2.17 MHz, a carrier frequency of 2 GHz, and a bandwidth of 1.6 MHz. A 1024-point FFT OFDM symbol with 410 virtual subcarriers is employed, in order to comply with the spectral constraints. The data segment of the OFDM symbol is padded with a 168-sample cyclic extension.

20.1.3.1 The UMTS type channel

A discretised COST 207 bad urban (BU) impulse response, as described in the COST 207 final report [410], was chosen for the UMTS-type channel. The symbol-spaced discretised impulse response employed for our investigations is depicted in Figure 20.5(a). This impulse response results in a strongly frequency selective channel, as shown in Figure 20.5(b). The area shaded in grey shows the location of the virtual subcarriers in the OFDM spectrum. Since no signal is transmitted in this

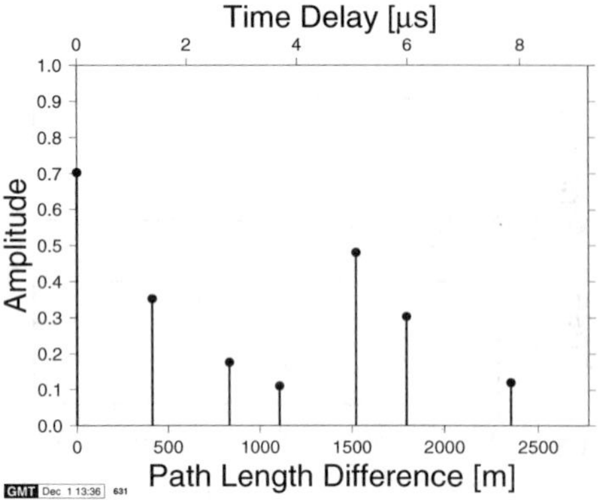

(a) COST 207 BU

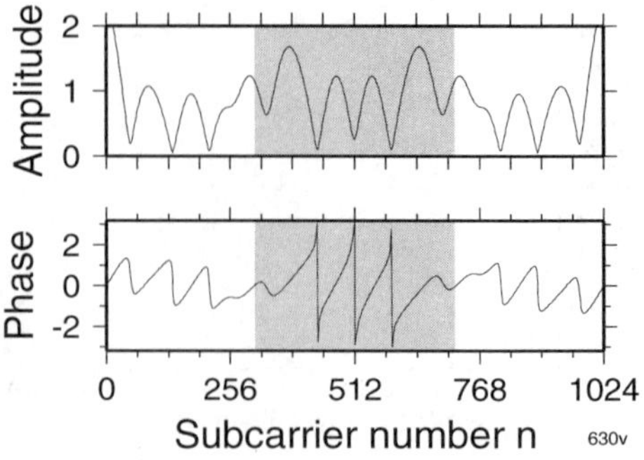

(b) channel frequency response

Figure 20.5: UMTS channel: (a) COST 207 BU impulse response, (b) unfaded frequency domain channel transfer function $H(n)$

range of frequencies, the channel transfer function for the virtual subcarriers does not affect the modem's performance. For simulations under Rayleigh fading conditions, a carrier frequency of 1.9 GHz and a vehicular velocity of 50 km/h was assumed, leading to a normalised Doppler frequency of $f'_d = 4.0534 \cdot 10^{-5}$.

	f_c	$1/T_s$	$f_{d,max}$	$f'_{d,max}$	τ_{max}	RMS(τ)
UMTS	2 GHz	2.17 MHz	87.9 Hz	$4.05 \cdot 10^{-5}$	7.83 μs	3.28 μs
WLAN	17 GHz	20 MHz	787 Hz	$3.94 \cdot 10^{-5}$	1.6 μs	0.109 μs
WATM	60 GHz	225 MHz	2278 Hz	$2.345 \cdot 10^{-5}$	300 ns	34.3 ns
WATM - short	-"-	-"-	-"-	-"-	48.9 ns	16.9 ns

Table 20.1: Carrier frequency f_c, sample rate $1/T_s$, maximal Doppler frequency $f_{d,max}$, normalised maximal Doppler frequency $f'_{d,max}$, maximal path delay τ_{max}, and channel RMS delay spread RMS(τ) of the various system frameworks

20.2 Effects of Time Dispersive Channels on OFDM

The effects of the time variant and time dispersive channels on the data symbols transmitted in an OFDM symbol's subcarriers are diverse. Firstly, if the impulse response of the channel is longer than the duration of the OFDM guard interval, then energy will spill over between consecutive OFDM symbols, leading to inter OFDM symbol interference. We will not investigate this effect here, as the length of the guard interval is generally chosen to be longer than the longest anticipated channel impulse response.

If the channel is changing only slowly compared to the duration of an OFDM symbol, then a time invariant impulse response can be associated with each transmitted OFDM symbol, which may change between different OFDM symbols. In this case, the frequency selective effects of the channel result in a frequency-dependent multiplicative distortion of the received symbols, very similar to the effects of a time domain fading channel envelope in a serial modem. If the channel impulse response duration is shorter than the OFDM guard interval, then no intersubcarrier interference (ISI) is experienced, a case corresponding to a narrowband fading channel in the case of a serial modem. This will be investigated in Section 20.2.1.

A non-stationary channel, however, will introduce inter subcarrier interference due to the time variant impulse response. The effects of this ISI will be studied in Section 20.2.2.

20.2.1 Effects of the stationary time dispersive channel

Here a channel is referred to as stationary if the impulse response does not vary significantly over the duration of one OFDM symbol, but it is time variant over longer periods of time. In this case, the time domain convolution of the transmitted time domain signal with the channel impulse response corresponds simply to the multiplication of the spectrum of the signal with the channel frequency transfer function $H(f)$:

$$s(t) * h(t) \longleftrightarrow S(f) \cdot H(f), \tag{20.2}$$

where the channel's frequency domain transfer function $H(f)$ is the Fourier transform of the impulse response $h(t)$:

$$h(t) \longleftrightarrow H(f). \tag{20.3}$$

Since the information symbols $S(n)$ are encoded into the amplitude of the transmitted spectrum at the subcarrier frequencies f_n, the received symbols $r(n)$ are the product of the transmitted symbol with the channel's frequency domain transfer function $H(n)$ plus the additive complex Gaussian noise samples $n(n)$:

$$r(n) = S(n) \cdot H(n) + n(n). \tag{20.4}$$

20.2.2 Non-stationary channel

A channel is classified here as non-stationary if the impulse response changes significantly over the duration of an OFDM symbol. In this case, the frequency domain transfer function is time variant during the transmission of an OFDM symbol and this time-varying frequency domain transfer function leads to the loss of orthogonality between the OFDM symbol's subcarriers. The amount of this intersubcarrier interference depends on the rate of change in the impulse response.

The simplest environment to study the effects of non-stationary channels is the narrowband channel, whose impulse response consists of only one fading path. If the amplitude of this path is varying in time, then the received OFDM symbol's spectrum will be the original OFDM spectrum convolved with the spectrum of the channel variation during the transmission of the OFDM symbol. As this short-term

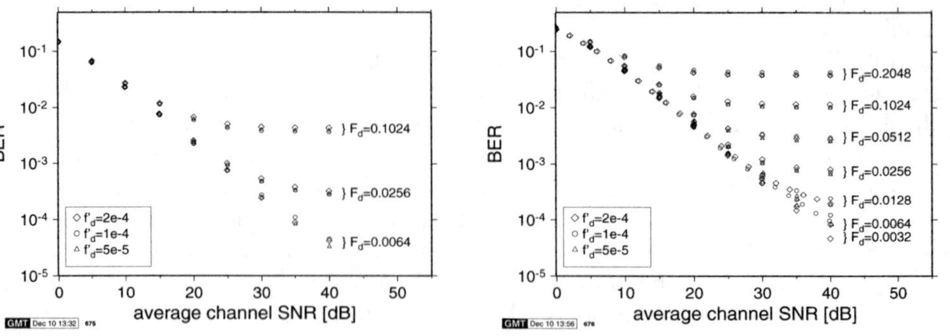

(a) perfect phase recovery BPSK

(b) differential BPSK

Figure 20.6: BPSK OFDM modem performance in a fading narrowband channel for normalised Doppler frequencies of $f'_d = 5 \cdot 10^{-5}$, $1 \cdot 10^{-4}$ and $2 \cdot 10^{-4}$ and FFT lengths between 16 and 4096. The FFT length for a given normalised Doppler frequency f'_d and an OFDM symbol normalised Doppler frequency F_d can be obtained from Table 20.2

channel spectrum is varying between different transmission bursts, we will investigate the effects of the time-varying narrowband channel averaged over a high number of transmission bursts.

Since the interference is caused by the variation of the channel impulse response during the transmission of each OFDM symbol, we introduce the "OFDM symbol normalised" Doppler frequency F_d:

$$F_d = f_d \cdot NT_s = f'_d \cdot N, \tag{20.5}$$

where N is the FFT length, $1/T_s$ is the sampling rate, f_d is the Doppler frequency characterising the fading channel and $f'_d = f_d \cdot T_s$ is the conventional normalised Doppler frequency.

f'_d / N	16	32	64	128	256	512	1024	2048	4096
5×10^{-5}	-	-	0.0032	0.0064	0.0128	0.0256	0.0512	0.1024	0.2048
1×10^{-4}	-	0.0032	0.0064	0.0128	0.0256	0.0512	0.1024	0.2048	-
2×10^{-4}	0.0032	0.0064	0.0128	0.0256	0.0512	0.1024	0.2048	-	-

Table 20.2: OFDM symbol normalised Doppler frequency $F_d \doteq f_d N T_s = f'_d N$ for FFT lengths between 16 and 4096 and the set of system parameters employed for non-stationary channel experiments

The BER performance for the OFDM modem configurations shown in Table 20.2 was determined by simulation and the simulation results for BPSK are given in Figure 20.6. Figure 20.6(a) depicts the BER performance of an OFDM modem employing BPSK with perfect narrowband fading channel estimation, where it can be observed

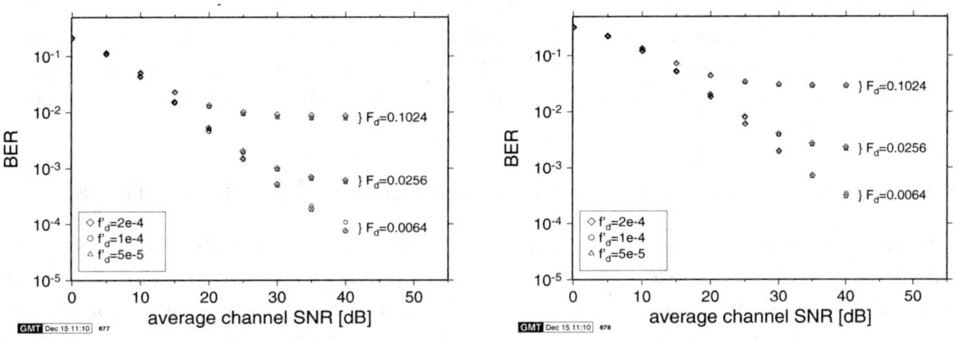

(a) QPSK

(b) 16QAM

Figure 20.7: OFDM modem performance assuming perfect phase recovery in a fading narrowband channel with normalised Doppler frequencies of $f_d' = 5 \cdot 10^{-5}$, $1 \cdot 10^{-4}$ and $2 \cdot 10^{-4}$ and FFT lengths between 16 and 2048 symbols. The FFT length for a given normalised Doppler frequency f_d' and an OFDM symbol normalised Doppler frequency F_d can be obtained from Table 20.2

that for any given value of F_d the different FFT lengths and channels behave similarly. For an F_d value of 0.0256, a residual bit error rate of about $2.8 \cdot 10^{-4}$ is observed, while for $F_d = 0.1024$ the residual BER is about 0.37%.

Figure 20.6(b) shows the performance of an OFDM system without the assumption of perfect phase recovery. Instead, differential detection is employed at the receiver, and, as in Figure 20.6(a), there is good correspondence between the BER curves for the same value of F_d. It can be seen that for F_d values from 0.0128 on, there is already a visible performance penalty at a SNR value of 40 dB. The highest observed residual BER is 4.2% for $f_d = 0.2048$.

Higher-order modulation schemes, such as QPSK and 16-QAM, are more affected by the intersubcarrier interference caused by the time-varying narrowband fading channel characteristics, as shown in Figure 20.7. Again, the correspondence between different set-ups resulting in the same value of F_d is very good. Since the BER performance is limited by the intersubcarrier interference instead of noise, the BER curves of Figures 20.6(a), 20.7(a) and 20.7(b) do not exhibit the expected fixed SNR shift of 3 dB between BPSK and QPSK and 9.5 dB between BPSK and 16-QAM, as experienced for noise-only limited systems. From the BER curves and residual BER values of Figures 20.6 and 20.7, it can be deduced that the effects of the intersubcarrier interference for $F_d = 0.1024$ and $F_d = 0.0256$ are equivalent to the effects of noise at SNR values of $\gamma = 17$ dB and $\gamma = 28$ dB, respectively. These values are long-term averages and appear to be valid for all the modulation schemes used.

20.2.2.1 Summary of time-variant channels

Time variant channels are of major influence on the BER performance of OFDM systems. The intersubcarrier interference caused by the time-varying nature of the transmission channel limits the attainable bit error rates for the UMTS and the WLAN scenario, where normalised Doppler frequencies of around $4 \cdot 10^{-5}$ and FFT lengths of 1024 are assumed, leading to $F_d \approx 0.04$. Therefore, an error floor of about 10^{-3} has to be faced by these systems, even if perfect channel estimation is assumed.

20.2.3 Signalling over time dispersive OFDM channels

Analogously to the case of serial modems in narrowband fading channels, the amplitude and phase variations inflicted by the channel's frequency domain transfer function $H(n)$ upon the received symbols will severely affect the bit error probabilities, where different modulation schemes suffer to different extents from the effects of the channel transfer function. Coherent modulation schemes rely on the knowledge of the symbols' reference phase, which will be distorted by the phase of $H(n)$. If such a modulation scheme is to be employed, then this phase distortion has to be estimated and corrected. For multilevel modulation schemes, where also the magnitude of the received symbol bears information, the magnitude of $H(n)$ will affect the demodulation. Clearly, the performance of such a system depends on the quality of the channel estimation.

A simpler approach to signalling over fading channels is to employ differential modulation, where the information is encoded in the difference between consecutive symbols. Differential phase shift keying (DPSK) employs the phase of the previously received symbol as phase reference, encoding information in the phase difference between symbols. DPSK is thus only affected by the differential channel phase distortion between two consecutive symbols, rather than by the channel phase distortion's absolute value.

20.3 Channel Estimation

Frequency domain channel estimation algorithms generate the channel transfer function estimates $\hat{H}(n)$ for subsequent correction of the received symbols prior to demodulation. The accuracy of the algorithm influences the total system performance to a great extent, especially for systems employing multilevel modulation and coherent detection. We have investigated several different wideband channel estimation techniques, which can be split into two groups: those operating on the spectrum of the received OFDM symbol and those employing time domain correlation algorithms.

20.3.1 Frequency domain channel estimation

Frequency domain channel estimation algorithms operate on the basis of the received samples in the frequency domain, that is after the receiver's FFT processing stage. At this stage, the frequency domain channel transfer function $H(n)$ can be estimated by using known frequency domain pilot symbols embedded in the OFDM symbol's spectrum and exploiting the strong correlation between consecutive samples of the frequency domain channel transfer function $H(n)$.

20.3.1.1 Pilot symbol assisted schemes

Pilot symbol assisted modulation (PSAM) schemes obtain a channel transfer function estimate on the basis of known frequency domain pilot symbols that are interspersed with the transmitted data symbols [12]. Conventionally, PSAM schemes are utilised in narrowband fading environments for serial modems. For each received pilot symbol the instantaneous channel fading value is estimated as the quotient of the received and the expected symbol, and the channel fading estimation for the data symbols is derived from these pilot fading estimates by means of interpolation. A range of different interpolation techniques were comparatively studied by Torrance *et al.* [411] for serial modems in Rayleigh fading narrowband channels, ultimately favouring linear interpolation due to its low complexity and good performance. We will address the issues of interpolation schemes in the context of OFDM transmission in fading wideband channels below.

In parallel modems, PSAM can be utilised for estimating the frequency domain channel transfer function $H(n)$ in a time dispersive environment, if no inter-subcarrier interference is assumed. Accordingly, n_p pilot symbols P_i are transmitted in the subcarriers with indices $p_i, i = 1, \ldots, n_p$ within the total OFDM symbol bandwidth of N subcarriers. At the receiver, the channel transfer function $\hat{H}(p_i)$ at the pilot subcarriers is estimated from the received samples $r(p_i)$:

$$\hat{H}(p_i) = r(p_i)/P_i. \tag{20.6}$$

In a second step, the values of the channel transfer function are estimated for the unknown data symbols by interpolation using the $\hat{H}(p_i)$ values of Equation 20.6. Clearly, the placement of the pilots and the interpolation technique will influence the quality of the channel estimation. In this chapter we will investigate two different interpolation techniques, linear and lowpass interpolation, both with a varying number of pilot subcarriers.

20.3.1.1.1 Linear interpolation for PSAM The simplest interpolation technique that can be used to estimate the channel transfer function for the subcarriers between two neighbouring channel estimation samples $\hat{H}(p_i)$ and $\hat{H}(p_{i+1})$ is a linear function:

$$\hat{H}(n) = \hat{H}(p_i) + \frac{\hat{H}(p_{i+1}) - \hat{H}(p_i)}{p_{i+1} - p(i)} \cdot (n - p_i) \quad \text{for} \quad p_i \leq n \leq p_{i+1}. \tag{20.7}$$

Figure 20.8 shows the amplitude and the phase of the stationary WATM channel's frequency domain transfer function $H(n)$ for different numbers of equidistant pilot symbols in the frequency domain in an ideal noiseless environment. It can be seen that the channel estimation accuracy improves with increasing number of pilot tones in the spectrum. The number of pilot tones in the OFDM spectrum necessary to sample the channel transfer function can be determined on the basis of the sampling theorem as follows.

The frequency domain channel's transfer function $H(f)$ is the Fourier transform of the channel impulse response $h(t)$. Each of the impulses in the impulse response

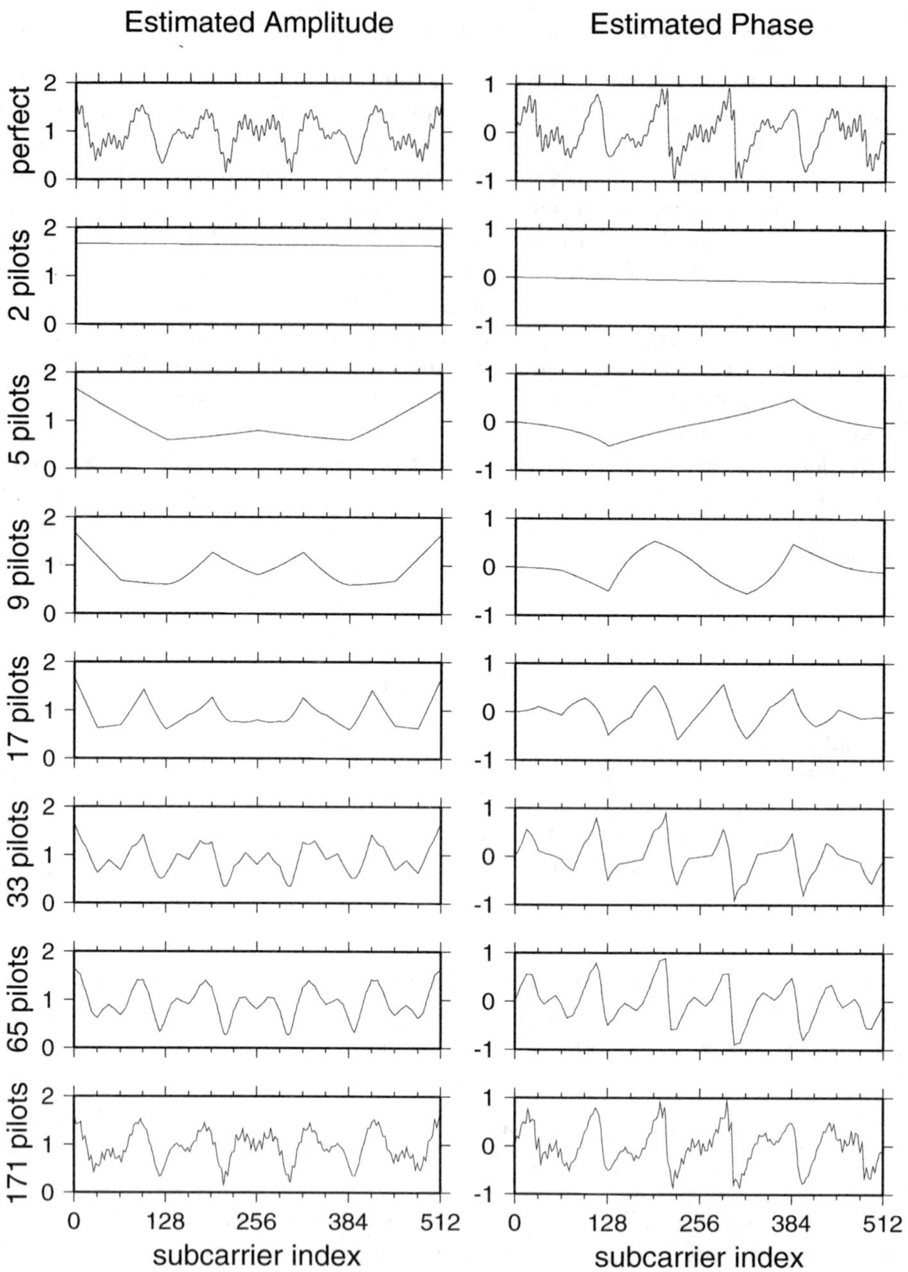

Figure 20.8: Estimation of the stationary WATM channel of Figure 20.2, a noiseless environment using linear interpolation between pilots employing 2...171 pilots per 512-subcarrier OFDM symbol

will result in a complex exponential function $e^{-(j2\pi\tau/T_s)f}$ in the frequency domain, depending on its time delay τ. In order to sample this contribution to $H(f)$ according to the sampling theorem, the maximum pilot spacing Δp in the OFDM symbol is:

$$\Delta p \leq \frac{N}{2\tau/T_s}\Delta f. \tag{20.8}$$

In the case of the WATM channel impulse response with a maximum delay of $\tau/T_s = 67$, the resulting pilot spacing in the OFDM symbol is three subcarriers, requiring 171 pilot carriers per 512-subcarrier OFDM symbol. If the channel estimation is to resolve only the effects of the first three dominant paths with a maximum delay of $\tau/T_s = 11$ in Figure 20.1(b), then the minimal pilot spacing can be increased from 3 to 23 subcarriers.

Inspection of Figure 20.8 underlines these points: in order to resolve the full detail of the original transfer function, as depicted at the top of the figure, 171 pilots must be used. Furthermore, even using 171 pilots in the OFDM symbol, the estimated curve is not a very close match of the original. This is due to the linear interpolation algorithm used.

Nonetheless, the estimations performed with 33 and 65 pilots, respectively, show very similar levels of detail. Both schemes can appropriately sample the effects of the first three paths, but they are undersampled for capturing information concerning the two long-delay paths. If the pilot spacing is higher than the calculated number of 23, then the accuracy of the estimation declines further.

The shortened WATM channel, as characterised in Figure 20.3(a), can be sampled adequately utilising any pilot spacing of less than 23 subcarriers, which for our 512-subcarrier system corresponds to a minimum number of 23 pilots per OFDM symbol. Figure 20.9 shows the linearly interpolated channel estimation values for PSAM channel estimation using 2, ..., 171 pilots in the 512-subcarrier OFDM symbol for the noiseless shortened WATM channel. Again, it can be seen that the accuracy of the channel transfer function estimation improves with the number of pilots, but that a total of 171 pilots is necessary to yield a channel estimation that is indistinguishable from the perfect channel estimation curve in this scale. As 33 and 65 pilot subcarriers already exceed the 23 pilots required to sample the frequency domain channel transfer function according to the Nyquist sampling theorem, it becomes clear that the estimation errors are due to the linear interpolation algorithm and that – for optimal channel estimation accuracy – substantial oversampling of the channel has to be employed, thus deteriorating the system's bandwidth efficiency.

20.3.1.1.2 Ideal lowpass interpolation for PSAM The perfect interpolator for the measured channel transfer function samples $\hat{H}(p_i)$ is the ideal lowpass filter with a cut-off frequency of $1/\Delta p$, where Δp is the spectral distance between consecutive pilot subcarriers. Unlike the linear interpolation, this requires equidistant pilot placement in the OFDM symbol.

While an ideal rectangular filter transfer function is impossible to implement perfectly in the time domain due to its infinitely long impulse response, we employed an ideal rectangular frequency domain filter based on a FFT/IFFT operation. A perfect FFT-based lowpass filter can be implemented if its bandwidth coincides with a FFT bin of the chosen Fourier transform, which limits the set of possible pilot numbers. In our case, a 512-point FFT/IFFT operation was employed, therefore allowing for

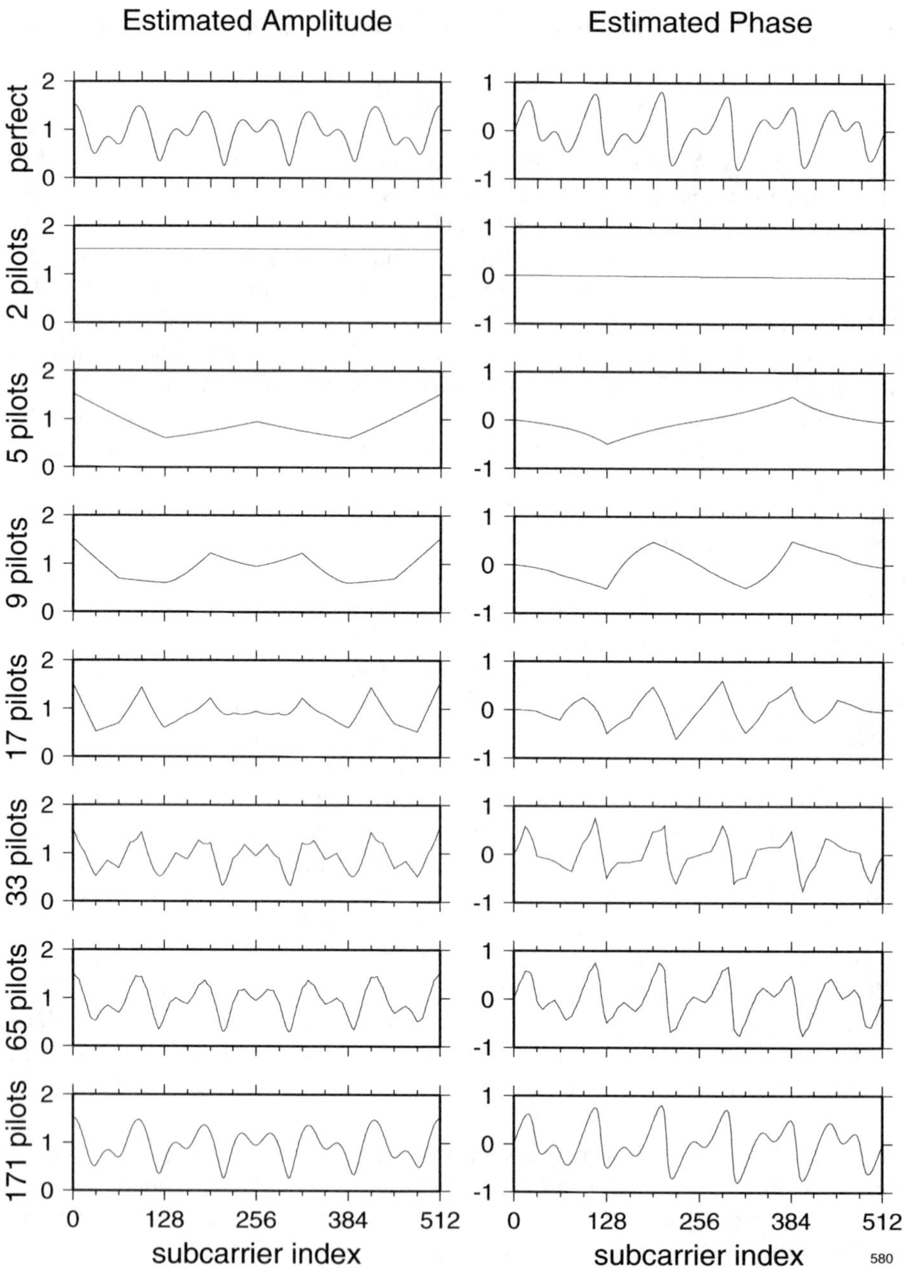

Figure 20.9: Estimation of the stationary shortened WATM channel of Figure 20.3, a noiseless environment using linear interpolation between pilots employing 2 ..., 171 pilots per 512-subcarrier OFDM symbol

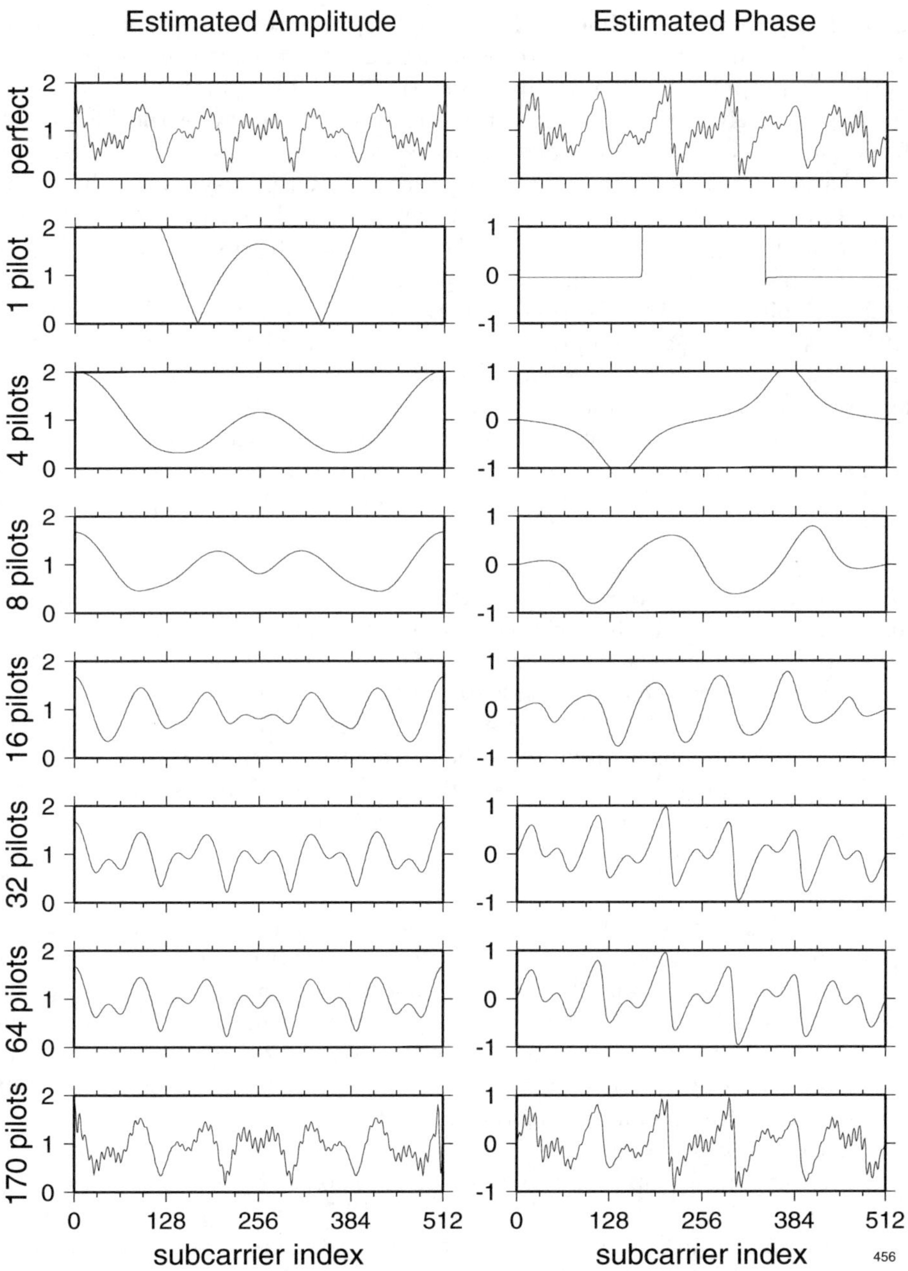

Figure 20.10: Estimation of the stationary WATM channel of Figure 20.2, a noiseless environment using ideal lowpass interpolation between pilots employing 2, ..., 171 pilots per 512-subcarrier OFDM symbol

sets of 2^n with $1 \leq n \leq 9$ pilots.

For comparison with the linear interpolation, the same set of pilot distances Δp has been employed for the experiments. Since the lowpass interpolator does not require a pilot in the last subcarrier, the corresponding number of pilots is reduced by one for the lowpass interpolator, when compared to the linear interpolator.

Figure 20.10 gives an overview of the estimation accuracy of the PSAM scheme with ideal lowpass interpolation in a noiseless environment. Again, the resolution of the estimation depends on the number of pilots per OFDM symbol. The estimated transfer functions for 32 and for 64 pilots are identical and exactly correspond to the effect of the first three impulses in $h(t)$. In order to resolve $H(f)$ more finely, at least 170 pilot subcarriers must be employed. Closer inspection of the estimated $\hat{H}(n)$ for 170 pilots reveals relatively high errors in the estimation. This is due to non-optimal lowpass filtering; the optimum bandwidth of the lowpass filter corresponding to $\Delta p = 3\Delta f$ does not line up with the filter's FFT bins, hence the passband is too wide. This leads to considerable estimation errors, especially at the highest subcarrier indices, since there is no pilot symbol at the last subcarrier. If a perfect lowpass interpolation scheme was to be used over the WATM channel, then a pilot distance of $\Delta p = 2\Delta f$ would have to be used, resulting in 256 pilot subcarriers per 512-subcarrier OFDM symbol.

The equivalent estimated channel transfer functions for the case of the shortened WATM channel are depicted in Figure 20.11. It can be seen from the figure that the channel estimation is indistinguishable from the perfect case for the 32-pilot scenario and the 64-pilot scenario. Since the number of pilots necessary for the perfect sampling of this channel is 23, which corresponds to a pilot spacing of 23, the channel estimation utilising less than 23 pilots yields inaccurate results. If 170 pilot subcarriers are employed in each OFDM symbol, then the same effect as observed for the full-length WATM channel applies; because the frequency resolution of the FFT employed for the lowpass filter implementation is insufficient, the filter passband does not correspond exactly to the pilot frequency.

20.3.1.1.3 Summary The Nyquist sampling theorem applies to the channel estimation in the frequency domain utilising pilot subcarriers in the OFDM symbol, but delivers only a lower bound for the number of pilots necessary, which is valid for perfect lowpass interpolation. If linear interpolation is employed, then oversampling beyond the Nyquist rate is necessary, in order to achieve high estimation accuracy. The number of pilot subcarriers that has to be used is dependent on the modulation scheme employed for signalling over the subcarriers.

The ideal lowpass interpolator yields better estimation accuracy than the linear interpolator for a given number of pilots, but the choice of pilot sets is limited by the implementation of the lowpass filter. If a FFT/IFFT based filter implementation is chosen, then only a limited set of possible pilot distances is available. This, in the WATM channel, leads to the need for a substantially higher number of pilots than would be necessary for the more flexible linear interpolator.

20.3.2 Time domain channel estimation

An alternative approach to channel estimation in OFDM transmission systems is to estimate the channel directly from the time domain received signal. Analogous to

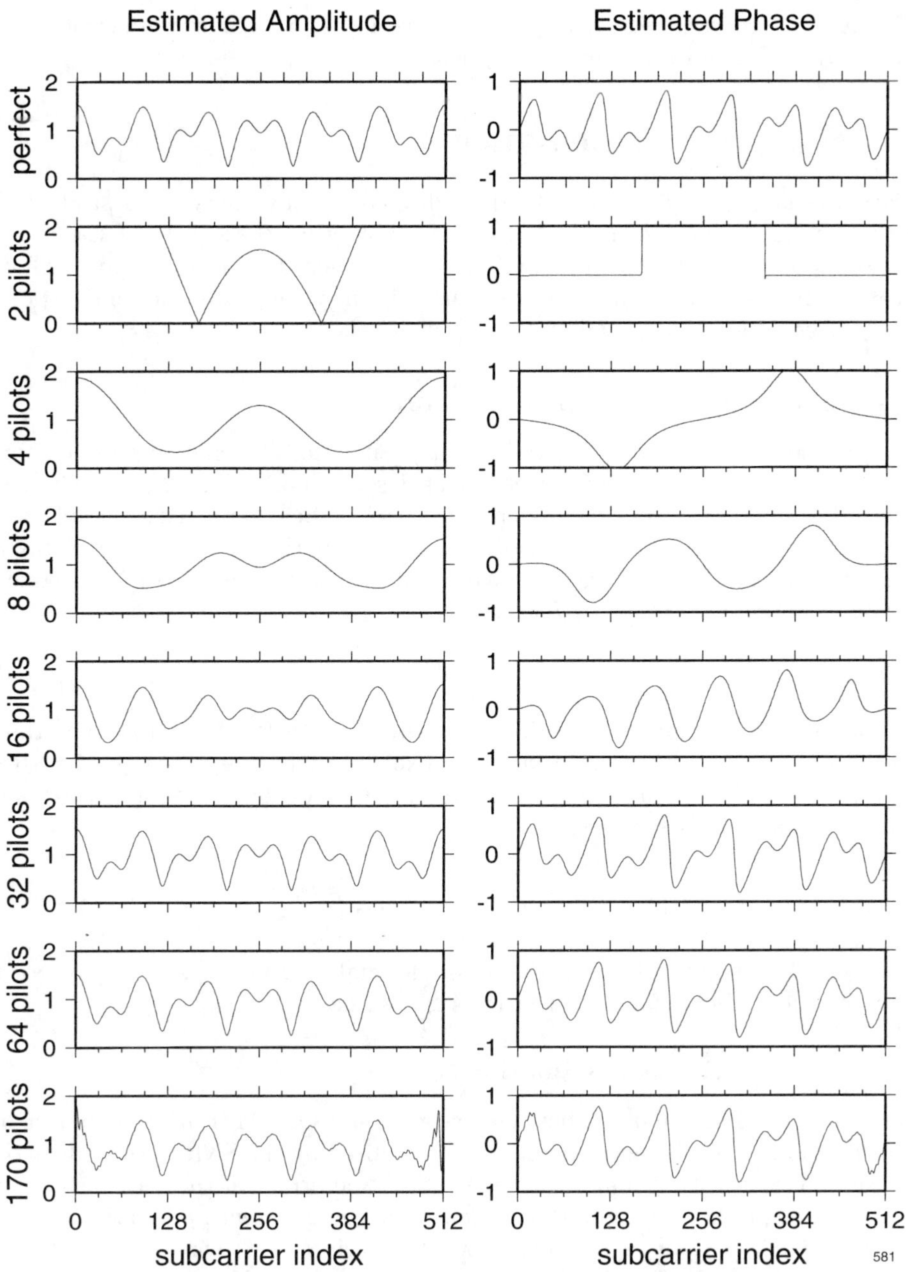

Figure 20.11: Estimation of the stationary shortened WATM channel of Figure 20.3, a noiseless environment using ideal lowpass interpolation between pilots employing 2, ..., 171 pilots per 512-subcarrier OFDM symbol

wideband estimation in serial modems, a training sequence in the transmitted data stream can be employed in order to perform a correlation-based impulse response estimation. The corresponding frequency domain channel transfer function can be computed by FFT from this impulse response.

20.4 System Performance

The various channel estimation techniques discussed above have been studied in a perfectly noiseless environment and without considering the effects of the channel estimation upon the system's bit error rate performance. In this section, we will consider the attainable performance of data transmission systems over the time dispersive channel, for both non-fading and stationary channels.

20.4.1 Static time dispersive channel

The static time dispersive channel exhibits a time invariant channel impulse response, which inflicts no intersubcarrier interference as described in Section 20.2.2. Therefore, the OFDM subcarriers are independent and each subcarrier n corresponds to an AWGN channel with a signal-to-noise ratio of γ_n. If the average channel SNR is γ, then the sub-channel SNR values γ_n depend on the magnitude of the frequency domain channel transfer function $H(n)$:

$$\gamma_n = \gamma \cdot |H(n)|^2 . \tag{20.9}$$

If we assume perfect channel estimation, corresponding to $\hat{H}(n) = H(n)$, then the bit error rate $p_{e,n}$ for each subcarrier n depends exclusively on the sub-channel SNR γ_n, which allows the calculation of the overall system bit error rate p_e by simply averaging over the sub-channel SNR values:

$$p_e = \frac{1}{N_u}\sum_n p_{e,n} = \frac{1}{N_u}\sum_n p_e(\gamma_n), \tag{20.10}$$

where N_u is the number of subcarriers used for data signalling and the use of the same modulation scheme for all subcarriers is assumed.

20.4.1.1 Perfect channel estimation

Perfect channel estimation is the best-case scenario for OFDM transmission over time dispersive channels, as the performance is only limited by the SNR of the subcarriers, not by the reference phase and amplitude estimation errors in the receiver. Simulations have been performed for both coherent and non-coherent modulation schemes, where the frequency domain channel transfer function $H(n)$ was calculated at the receiver by applying the Fourier transform to the perfectly known noiseless impulse response of the channel, as depicted in Figure 20.1(b). The resulting channel transfer function is shown in Figure 20.2(a).

Figure 20.12(a) shows the measured BER per subcarrier for different levels of average OFDM SNR for the simulated 512-subcarrier modem over the WATM channel employing coherent binary phase shift keying (BPSK) in the subcarriers. It is apparent that the bit error probability varies significantly between different subcarriers,

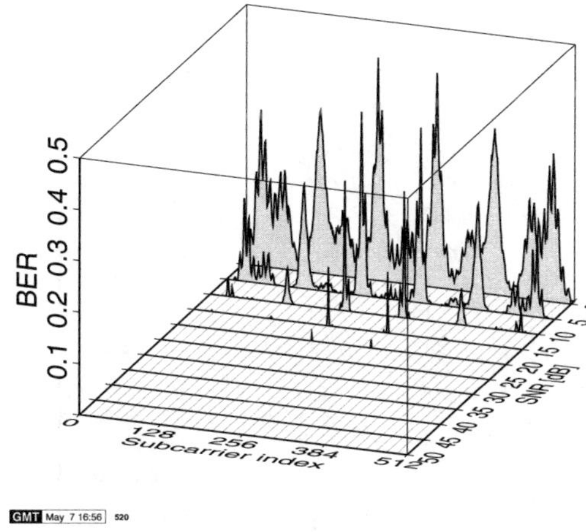

(a) BPSK

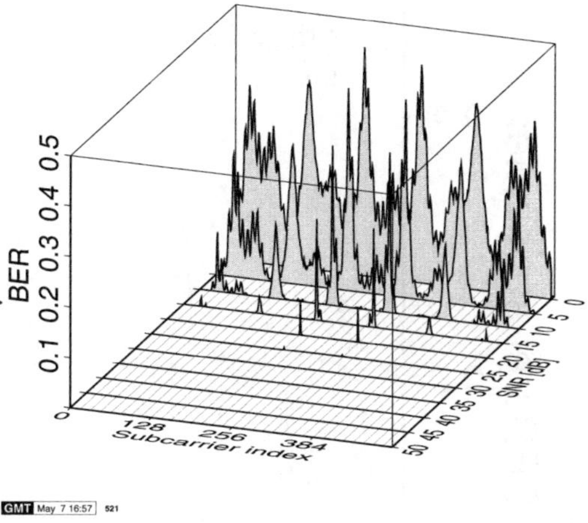

(b) QPSK

Figure 20.12: Simulated BER per subcarrier over the static WATM channel of Figure 20.1(b) with perfect channel estimation for (a) coherent BPSK transmission and (b) coherent QPSK transmission. The symmetry of the BER curves indicates that the unfaded CIR was real.

from about 2% in good sub-channel conditions, up to 40% in the deep fades of the channel transfer function $H(f)$ at an average OFDM SNR of 0 dB. At an average SNR of 5 dB, groups of virtually error-free subcarriers can be observed, interspersed with bundles of carriers exhibiting high bit error probabilities. At 15 dB, only a very small number of carriers experience measurable transmission error rates at all.

The recorded bit error rates in the simulation correspond very closely to the theoretical results obtained by calculating the sub-channel SNR γ_n for each subcarrier index n following Equation 20.9 and evaluating the appropriate bit error probability function for the chosen modulation scheme. Specifically, the BER over the AWGN channel was calculated employing Equations 19.1 to 19.3 For coherent modulation, the bit error probabilities $p_e(\gamma)$ for BPSK, QPSK and 16-QAM are depicted in Figure 19.3, and Figure 20.13 shows good correspondence between the obtained theoretical bit error rates and the simulation results.

The corresponding graph for QPSK, as shown in Figure 20.12(b), is similar to the BPSK case, only exhibiting higher bit error probabilities. Again, the bit error rate per subcarrier varies strongly with the channel transfer function, which is due to the variation in sub-channel SNR.

The OFDM bit error probability averaged over all subcarriers versus the channel SNR for BPSK, QPSK and 16-QAM transmission over the non-fading WATM channel is shown in Figure 20.14(a). The QPSK curve is shifted to the right by 3 dB compared to the BPSK performance curve, which is in accordance with the performance in a narrowband channel, as demonstrated in Section 19.3. The same observation can be made for the performance difference between 16-QAM and the other signalling

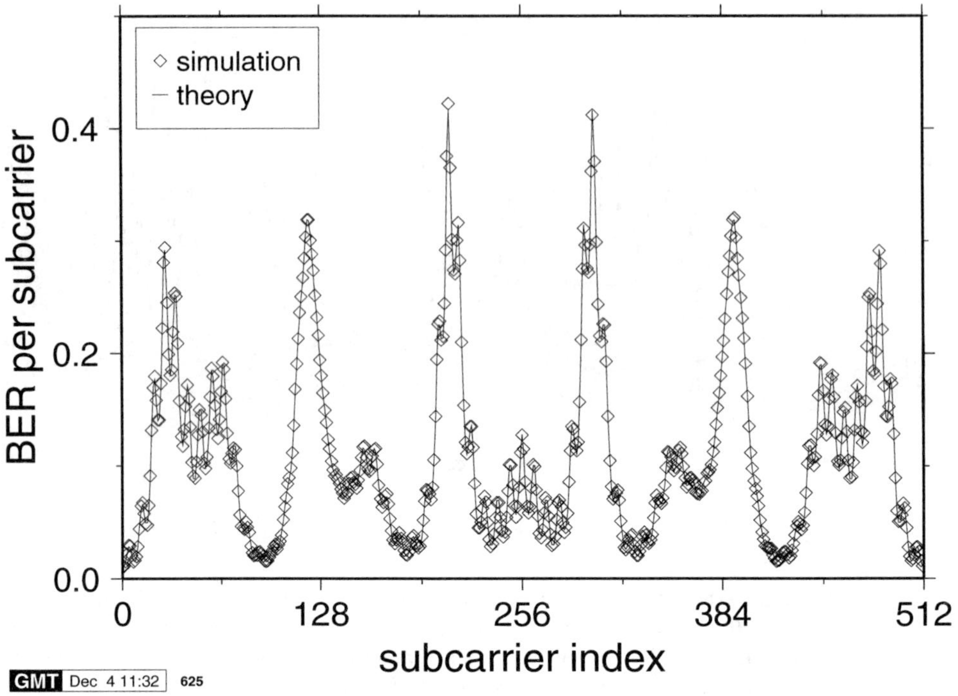

Figure 20.13: Simulated and theoretical BER per subcarrier for coherently detected BPSK transmission with perfect channel estimation over the static WATM channel of Figure 20.1(b) at an average OFDM SNR of 0 dB

schemes.

Although the relative BER performance relationship between the different modulation schemes corresponds to the BER results found in the narrowband AWGN channel, the absolute bit error rate values depend on the channel transfer function. Figure 20.14(b) depicts the corresponding BER versus channel SNR graph for the shortened WATM channel of Figure 20.3. The relative BER performance relationship between the different modulation schemes is the same as over the other channels, but the shape of the BER curve is different from that of the WATM channel. Comparing the frequency domain channel transfer function of the shortened WATM channel of Figure 20.3(b) and that of the original WATM channel of Figure 20.2(a) it can be seen that the fading is less severe in the shortened WATM channel, leading to better BER performance at high SNR values.

The UMTS channel depicted in 20.5(b) exhibits even stronger fading in the frequency domain than the two WATM channels, and consequently the BER performance is worse than that of both the WATM systems. At a BER of 10^{-3} the required SNR for the corresponding modulation schemes is about 9 dB higher for the UMTS system than for the WATM system.

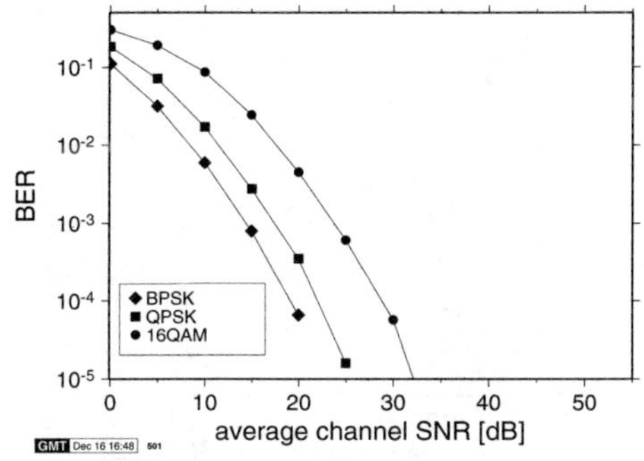

(a) WATM channel

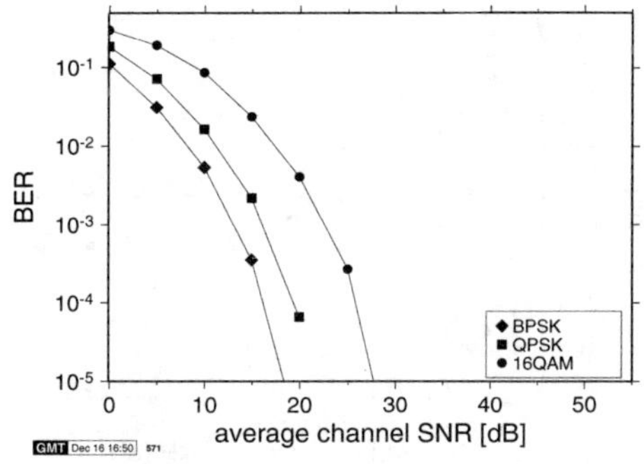

(b) short WATM channel

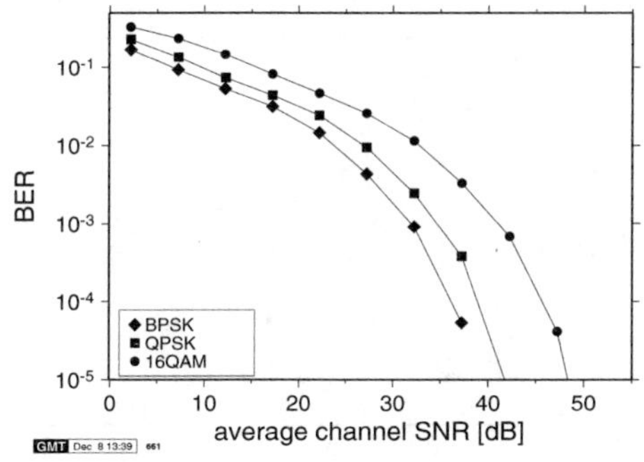

(c) WLAN channel

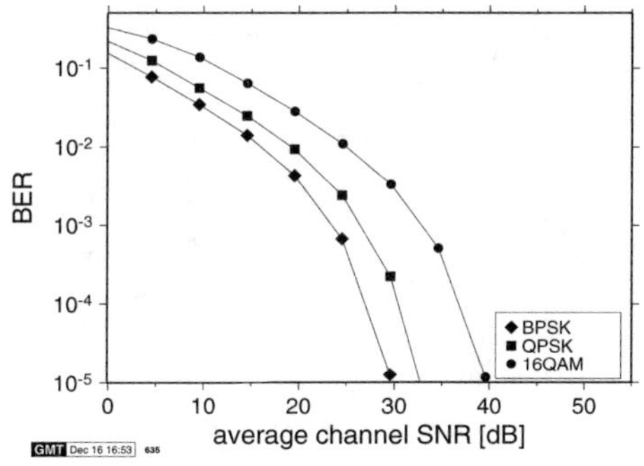

(d) UMTS channel

Figure 20.14: Bit error rate versus channel SNR for static (a) WATM channel (Figure 20.1(b)), (b) shortened WATM channel (Figure 20.3), (c) WLAN channel (Figure 20.4), (d) UMTS channel (Figure 20.5(a)); perfect channel estimation and coherent detection

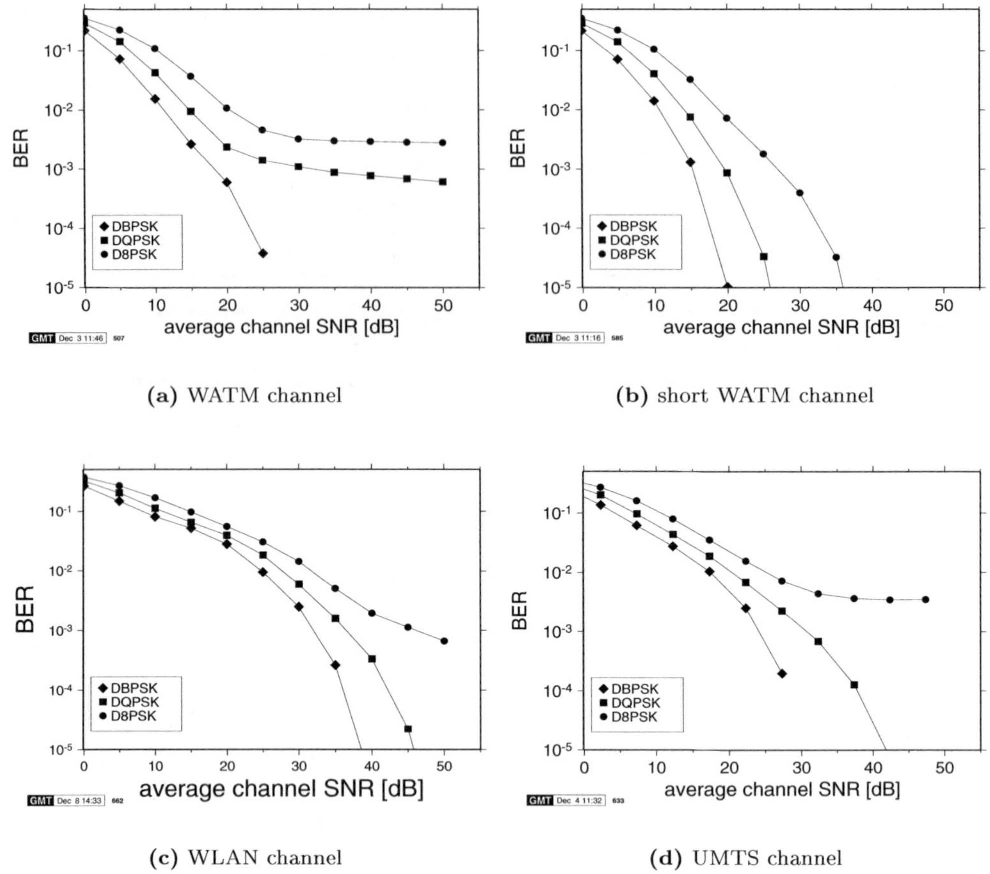

Figure 20.15: Bit error rate versus channel SNR for static (a) WATM channel (Figure 20.1(b)), (b) shortened WATM channel (Figure 20.3), (c) WLAN channel (Figure 20.4), (d) UMTS channel (Figure 20.5(a)), using differential detection

20.4.1.2 Differentially coded modulation

Differentially coded modulation enables the receiver to detect the data symbols without knowledge of the reference phase, therefore no channel estimation is required, if differential phase shift keying (DPSK) is used for modulating the subcarriers. Since the transmitted information is encoded in the phase difference between two consecutive subcarriers, there needs to be one phase reference subcarrier per OFDM symbol, typically in the the first subcarrier.

While the absolute phase of the frequency domain channel transfer function $H(n)$ at subcarrier n does not affect the reception, the phase change of $H(n)$ between two consecutive subcarriers does influence the reception of the symbols. Note that the BER performance of both the WATM system in Figure 20.15(a) as well as that of

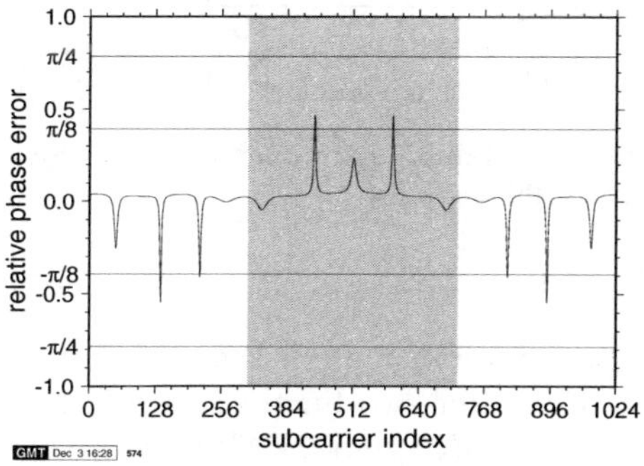

(a) differential phase error

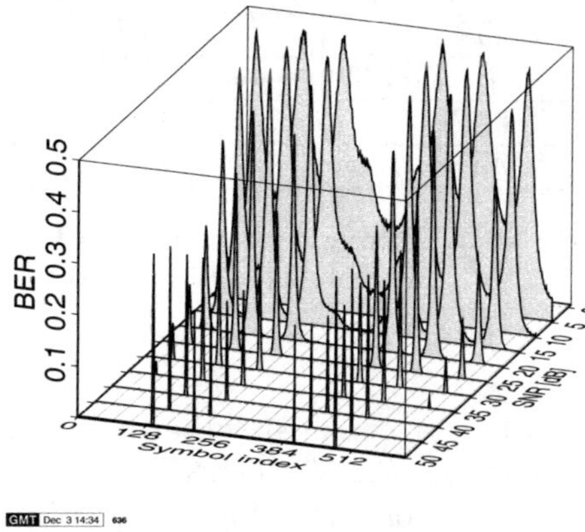

(b) BER per subcarrier

Figure 20.16: Differential modulation over the static UMTS channel of Figure 20.5. (a) Phase difference between consecutive frequency domain samples of the frequency domain channel transfer function $H(n)$; the grey area designates the virtual subcarriers, which are not used for data transmission. (b) Simulated BER per subcarrier in static UMTS channel for D8PSK without channel estimation

the UMTS system in Figure 20.15(d) exhibit a residual bit error rate for high SNR levels for certain modulation schemes. These noise-independent errors are caused by channel effects and are investigated for the UMTS system below.

In differential phase shift keying, the information to be transmitted is mapped to the phase difference between two consecutively received symbols. As all the symbols transmitted over the wideband OFDM channel suffer different phase rotations from the channel's frequency domain transfer function $H(n)$, the channel phase rotation difference $(\measuredangle H(n) - \measuredangle H(n-1))$ between subcarrier $(n-1)$ and subcarrier n offsets the receiver's decision boundaries for detected signals. Figure 20.16(a) shows the phase difference $(\measuredangle H(n) - \measuredangle H(n-1))$ between adjacent subcarriers for the UMTS channel of Figure 20.5. The grey area in the graph marks the positions of the virtual subcarriers, which are not used for transmission, and therefore do not have an impact on the system's performance. The horizontal lines at $\pm\pi/8$ and $\pm\pi/4$ mark the receiver's decision boundaries for DQPSK and D8PSK, respectively. If the channel transfer function's differential phase shift crosses the appropriate decision boundary for a given subcarrier index, then the data symbol transmitted over this subcarrier suffers from residual bit errors. In our example of Figure 20.16(a), there are four visible peaks comprising actually 6 out of 612 data-bearing subcarriers crossing the $\pi/8$ decision boundary for D8PSK. As the data is grey mapped on the PSK symbols, exactly one residual bit error will occur on each of the six subcarriers, resulting in a residual bit error rate of $6/1836 = 3.268 \cdot 10^{-3}$. Figure 20.16(b) gives an overview of the resulting BER per subcarrier for the data-bearing subcarriers for SNR values from 0 dB up to 50 dB. It can be seen that the residual bit errors are concentrated

System	FFT length	Max. delay $[T_s]$	max. Δp $[\Delta f]$	min. N_p
WATM	512	67	3	171
short WATM	512	11	23	22
WLAN	1024	32	16	64
UMTS	1024	17	30	34

Table 20.3: Theoretical maximum pilot distance and minimal number of pilots for the WATM, short WATM, WLAN and UMTS systems

in four bursts, comprising the six symbols with subcarrier indices 136, 137, 312, 401, 477, and 478. At 50 dB SNR the symbols in the outer two bursts have a bit error rate of 1/3, corresponding to one bit error per D8PSK symbol. The bit error rate of the inner two error bursts is lower, at about 23%, but is rising with higher SNR values to 1/3. This rising BER with higher SNR values for the two inner error bursts is explained by the channel transfer function shown in Figure 20.16(a). The channel induces exactly one bit error for the symbols in each of the corresponding subcarriers in the absence of noise. Since the phase error due to the channel's phase rotation is very close to the decision 8-DPSK boundary, the presence of noise can influence the reception towards the correct decision, therefore reducing the long-term bit error rate.

In addition to residual bit errors, the decision boundary offset caused by the differential phase shift of the channel transfer function as depicted in Figure 20.16(a) will also compromise the noise sensitivity of the subcarriers that are not directly affected by residual errors.

20.4.1.3 Pilot symbol assisted modulation

As discussed in Section 20.3.1.1, pilot symbol assisted modulation (PSAM) schemes acquire a channel transfer function estimate $\hat{H}(n)$ by sampling $H(n)$ with the aid of known pilot symbols embedded in the transmission burst and interpolating between the received pilots. We have seen that the minimal number of pilots needed to sample the channel transfer function follows the sampling theorem and that the maximal pilot distance in the OFDM symbol is given by Equation 20.8. Furthermore, we have seen that this minimal number of pilots only applies to the ideal lowpass filter interpolation algorithm, and more pilot symbols have to be employed, if linear rather than lowpass interpolation is to be used, therefore trading receiver complexity for transmission overhead.

Here we will characterise the performance of the four different OFDM systems in their respective non-fading channel environments. Each system was investigated using a range of pilot numbers. Using Equation 20.8, we can predict the number of pilot symbols necessary for each scenario, if ideal interpolation is assumed, and these numbers are given in Table 20.3. The performance evaluation for all these scenarios was carried out for 8, 16, 32 and 64 pilots with ideal lowpass interpolation and with the equivalent 9, 17, 33 and 65 pilot sets for linear interpolation. We will concentrate on our most robust and least robust modulation schemes, namely BPSK and 16-QAM.

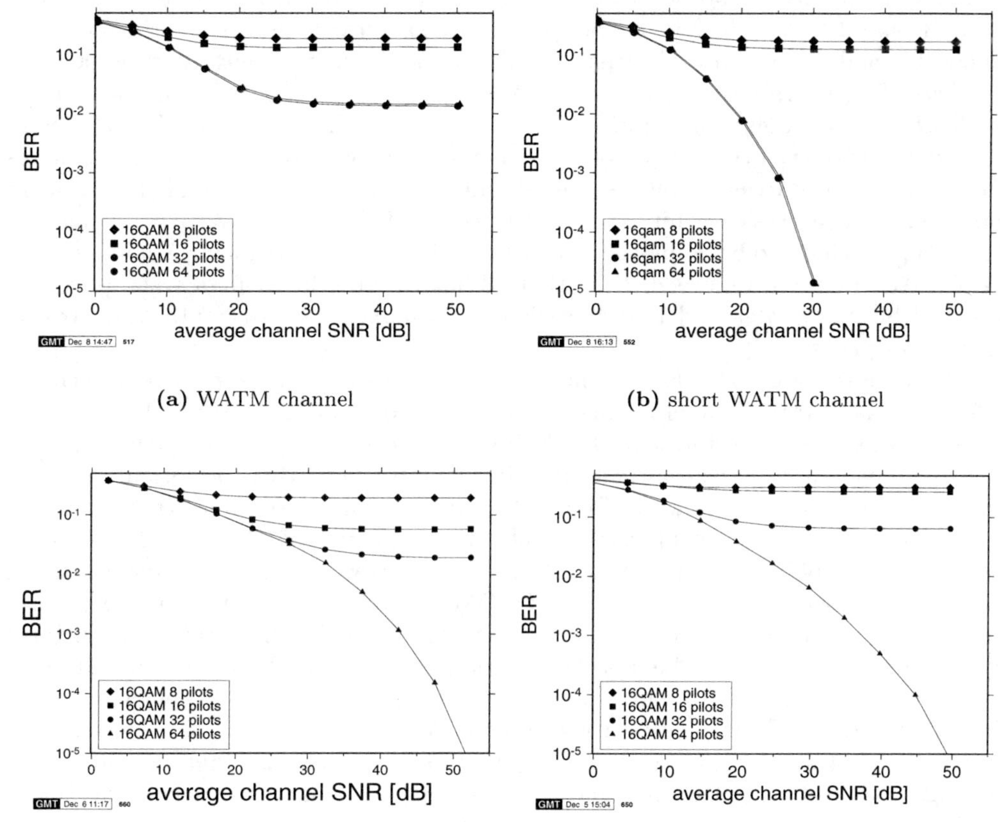

(a) WATM channel

(b) short WATM channel

(c) WLAN channel

(d) UMTS channel

Figure 20.17: Bit error rate versus channel SNR for static (a) WATM channel (Figure 20.1(b)), (b) shortened WATM channel (Figure 20.3), (c) WLAN channel (Figure 20.4), (d) UMTS channel (Figure 20.5(a)), PSAM-16QAM with ideal lowpass interpolation

Figure 20.17 shows the BER performance for all the discussed systems in their respective non-fading channels employing PSAM-16-QAM with ideal lowpass interpolation as modulation scheme. It can be seen in Figure 20.17(a) that the WATM system faces high residual bit error rates for all sets of pilot symbols investigated. This is expected, because the number of pilot subcarriers employed in these simulations was lower than the minimum number of 171 stated in Table 20.3. Therefore, the channel transfer function is not adequately sampled in the frequency domain and the resulting channel estimation is of insufficient accuracy for coherently detecting 16-QAM. Note that the residual bit error rate drops with increasing number of pilots employed for the channel estimation, but that the curves for 32-pilot and 64-pilot PSAM exhibit virtually equal performance. Figure 20.17(a) reveals that the estimated channel transfer function for these two cases is the same, owing to the twin-burst structure of the WATM impulse response of Figure 20.1(b). Therefore, increasing the number of pilots from 32 to 64 will not increase the system's performance, since this does not allow the modem to resolve the channel effects due to the two paths around 90 m.

According to Table 20.3, the short WATM channel can be adequately sampled with 22 pilot symbols equidistantly interspersed with the data symbols. This is in accordance with the system's performance curves in Figure 20.17(b). The BER curves for 32 and 64 subcarriers exhibit essentially identical performance, while for 8 and 16 pilot symbols the residual bit error rate is above 10%. For the WLAN channel, at least 64 pilots have to be employed in order to sample all the frequency components of $H(f)$. We can see in Figure 20.17(c) that all but the 60 pilot symbol systems exhibit residual bit error rates and that the BER performance increases with an increased number of pilots.

If the more robust BPSK modulation scheme is used in the subcarriers, then the effects of inaccurate channel estimation are less pronounced. Figure 20.18 gives an overview of the BER performance for BPSK modems in the four operational environments, employing a perfect lowpass filter interpolator. The BER performance curves for the WATM system, shown in Figure 20.18(a), exhibit no residual errors. Like the 16-QAM curves, the performance is identical, irrespective of whether 32 or 64 pilot symbols are employed. Unlike in the 16-QAM case, however, the performance is best for 16-pilot subcarriers employed in the OFDM symbol. This oddity can be explained comparing the channel phase estimation errors for 16 and for 32 subcarriers, which are depicted in Figure 20.19. The phase estimation error for the WATM channel transfer function estimated with 16 pilot symbols is shown in Figure 20.19(a), which exhibits high fluctuations of the phase error of up to 0.7 radians. The phase estimation error for 32 pilot symbols, as shown in Figure 20.19(b), while substantially smaller on average, exhibits a peak error value of about 1.0 rad, which is closer to the BPSK decision boundary of $\pm\pi/2$ than the maximum error of the 16-pilot estimation. For higher SNR values the overall BER will be dominated by the bit errors occurring in the subcarriers with the highest phase estimation errors, therefore leading to the 16-pilot system performing better than the overall more accurate 32 and 64 pilot symbol estimation systems.

The WATM system's performance over the short WATM channel of Figure 20.3 is depicted in Figure 20.18(b). No residual bit error rates are observed and the BER performance with 32 and 64 pilots is identical. The 16-pilot BER curve is close to the 32- and 64-pilot results, and the modem performs about 18 dB worse with 8 pilot symbols. Both the WLAN as well as the UMTS channels exhibit residual bit errors for 8 and 16 pilot symbols and show an improving BER performance with increasing numbers of pilots. The WLAN system, whose performance is shown in Figure 20.18(c), performs without residual errors only when using 64 pilot symbols. For the UMTS channel, however, 32-pilot channel estimation results in a BER performance comparable to that of the 64-pilot estimation.

If the linear interpolation algorithm is employed instead of the ideal lowpass filter, then the BER performance of all the systems investigated is degraded. This is in accordance with the channel estimation experiments, as evidenced for example by the WATM channel estimation plots in Figures 20.8 and 20.10. The BER performance curves for 16-QAM transmission employing linearly interpolated PSAM are given in

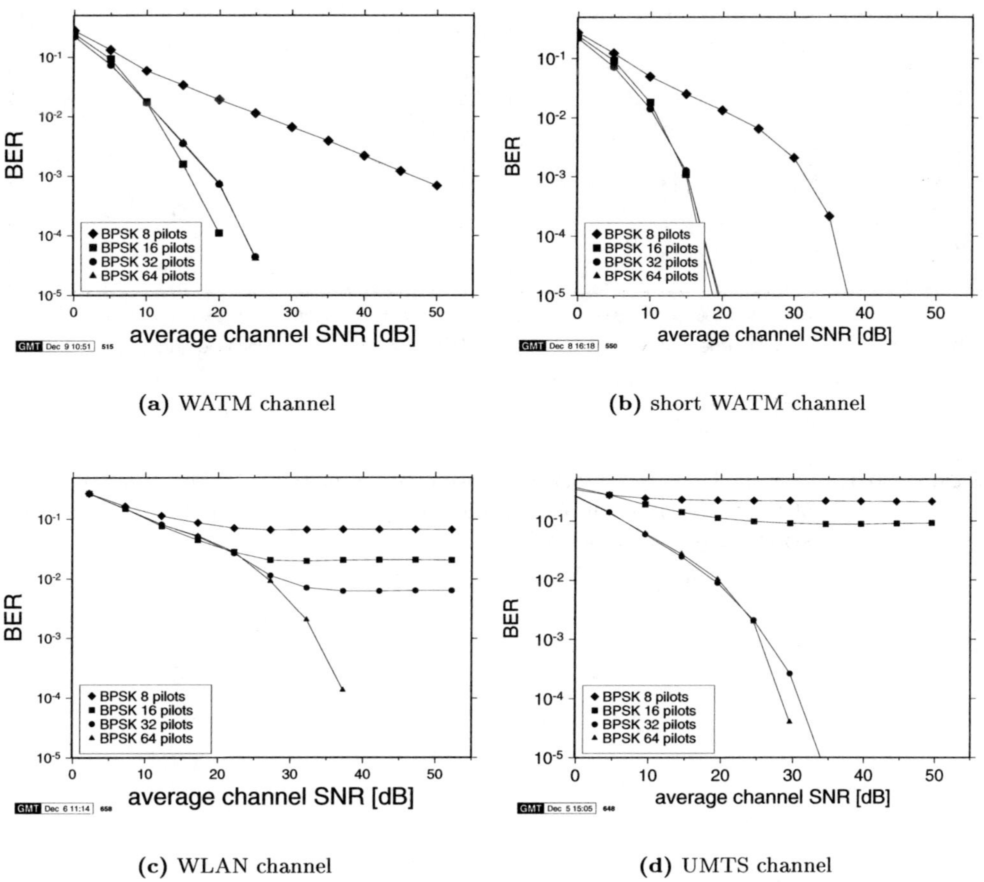

Figure 20.18: Bit error rate versus channel SNR for static (a) WATM channel (Figure 20.1(b)), (b) shortened WATM channel (Figure 20.3), (c) WLAN channel (Figure 20.4), (d) UMTS channel (Figure 20.5(a)), PSAM-BPSK with ideal lowpass interpolation

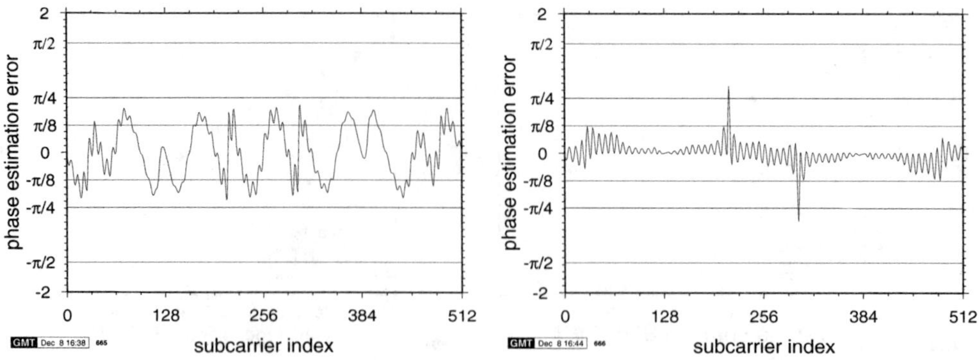

(a) 16 pilots

(b) 32 pilots

Figure 20.19: Channel transfer function phase estimation error for the WATM channel of Figure 20.1(b) employing ideal lowpass filter interpolator: (a) 16 pilot symbols, (b) 32 pilot symbols

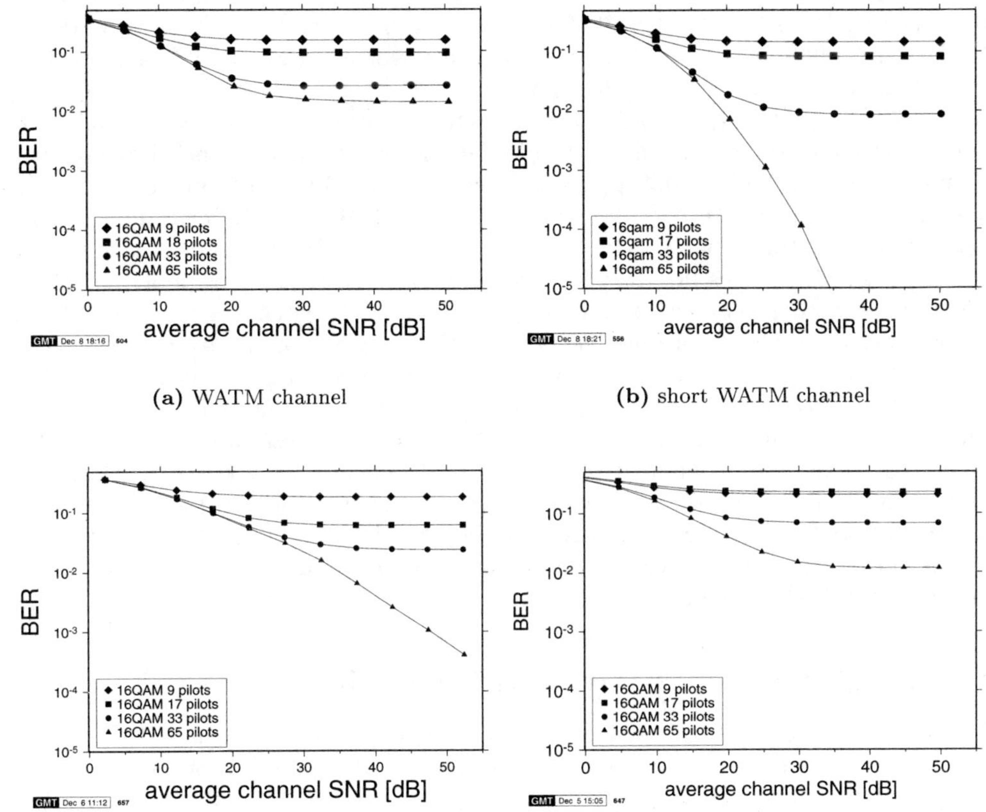

(a) WATM channel

(b) short WATM channel

(c) WLAN channel

(d) UMTS channel

Figure 20.20: Bit error rate versus channel SNR for static (a) WATM channel (Figure 20.1(b)), (b) shortened WATM channel (Figure 20.3), (c) WLAN channel (Figure 20.4), (d) UMTS channel (Figure 20.5(a)), using PSAM 16-QAM with linear interpolation

Figure 20.20, and it can be seen that for all modems, the performance is worse than that associated with the lowpass interpolation curves shown in Figure 20.17. The least differences between the two interpolation algorithms are visible for the WATM modem. In both cases, 64 pilot symbols are insufficient for combating residual bit errors and the residual bit error rates for both schemes are similar. In the short UMTS channel, the differences are more obvious. While the performance of the 32- and 64-pilot estimation systems was the same for lowpass interpolation, the linear interpolation algorithm shows residual errors for 32 pilot symbols. Compared to the lowpass interpolation, the BER results for the linear interpolator using 64 pilots are about 3 dB worse.

The BER performance for the WLAN system, as depicted in Figures 20.20(c) and 20.17(c), shows relatively minor differences between lowpass and linear interpolation.

The residual bit error rates for 8, 16 and 32 pilots are similar, while for 64 pilots there is a SNR loss of about 5 dB for the linearly interpolated system at a BER of 10^{-3}. The UMTS, whose performance curves are plotted in Figure 20.20(d), exhibits a residual BER of about 1% for 64 pilots.

For BPSK-PSAM transmission over the OFDM systems, the difference in performance between lowpass and linear interpolation algorithms are much less apparent than for 16-QAM. Figure 20.21 gives an overview of the BER performance of the simulated systems in their respective environments using BPSK with linear interpolation based channel estimation. The BER performance of the WATM system, as depicted in Figure 20.21(a), does not vary significantly with the number of pilots employed. This differs from the curves observed for the lowpass interpolator in Figure 20.18(a), where much higher bit error rates were observed, if only 8 pilots were employed for the channel estimation. Both the linear as well as the lowpass interpolator results show best performance for 17 and 16 pilot symbols, respectively. Over the short WATM channel, systems employing either interpolator exhibit an essentially identical performance for all but 8 and 9 pilots, respectively. In the latter case, there is a performance difference of 13-dB SNR for a BER of 10^{-3} in favour of the linear interpolation. For the WLAN and the UMTS, the BER performance with lowpass and linear interpolation is fairly similar, the difference being slightly lower residual bit error rates for the linearly interpolated channel estimation.

20.4.2 Slowly varying time-dispersive channel

The slowly varying time-dispersive channel is characterised by an impulse response that is time-varying, but which is assumed constant for the duration of one OFDM symbol. Hence we also refer to this channel model as an OFDM symbol time-invariant channel. This stationary property prevents intersubcarrier interference and renders the wideband channel estimation possible. In this case, the channel can be viewed as a succession of static impulse responses generated upon fading the constituting paths. The rate of change for each of the paths' fading is described by the Doppler frequency of the fading channel.

As we have seen in Section 20.2.2, the time-varying nature of the channel transfer function will introduce intersubcarrier interference depending on the symbol normalised Doppler frequency $F_d = f_d \cdot T_s \cdot N$. Performance degradations can be observed even for small values of f_d. We will investigate the validity of the assumption of the stationary channel for our four scenarios in the case of perfect channel estimation.

20.4.2.1 Perfect channel estimation

Perfect channel estimation is the best-case scenario for transmission over wideband channels and the system performance in these circumstances constitutes an upper bound to the achievable performance with realistic channel estimation algorithms. As the results for perfect channel estimation are not influenced by sub-optimal channel estimation, the effects of the time-varying channel on the transmitted data symbols can be studied and the validity of the static channel assumption for the four operating frameworks, namely WATM, short WATM, WLAN and UMTS, can be verified.

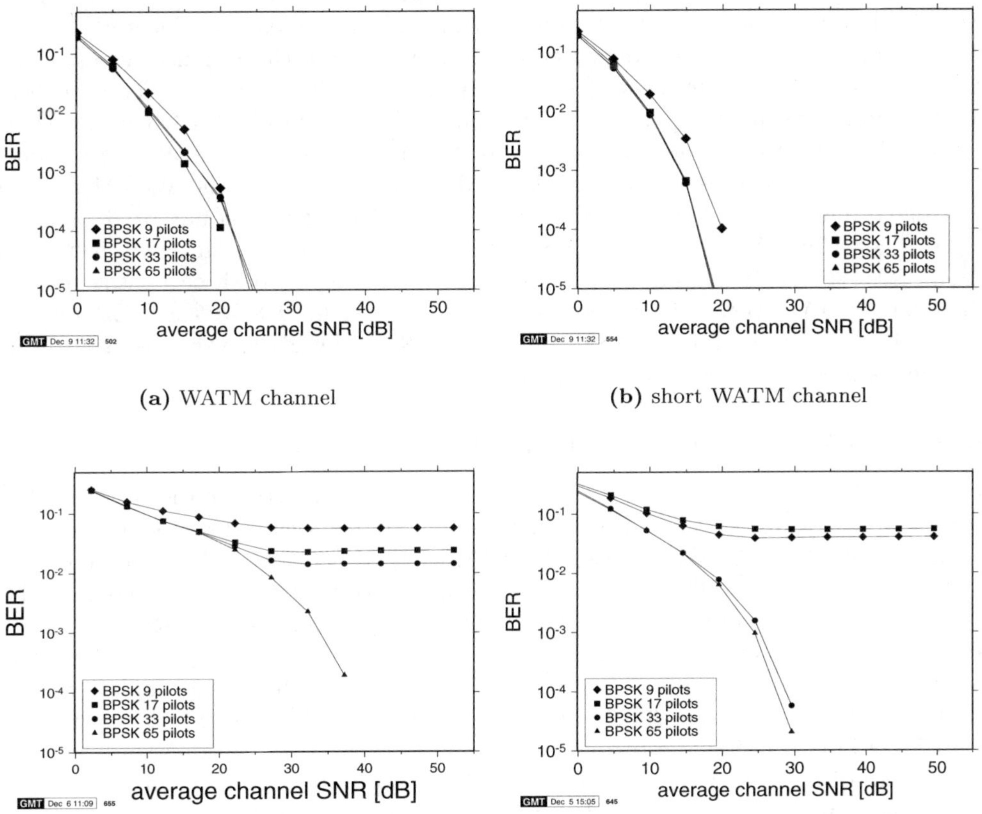

(a) WATM channel

(b) short WATM channel

(c) WLAN channel

(d) UMTS channel

Figure 20.21: Bit error rate versus channel SNR for static (a) WATM channel (Figure 20.1(b)), (b) shortened WATM channel (Figure 20.3), (c) WLAN channel (Figure 20.4), (d) UMTS channel (Figure 20.5(a)), using PSAM-BPSK with linear interpolation

Perfect channel estimation of the time-varying wideband channel was achieved by taking a snapshot of the channel impulse response in the centre of the OFDM symbol and calculating the corresponding frequency domain channel transfer function from this impulse response by the help of the FFT. This transfer function does not fully characterise the channel, unlike in the static channel case, since it does not give any information concerning the intersubcarrier interference and only takes into account the time-varying nature of the impulse response. Nonetheless, the achieved channel estimation constitutes a best-case benchmark for realistic channel estimation algorithms.

In order to isolate the influence of the time-domain variations of the channel during the transmission of an OFDM symbol, experiments have been conducted for all four operating frameworks of Table 20.1 with two different models of the fading channel.

Firstly, the channel impulse response was updated after every transmitted sample, resulting in a realistic model for a time-varying wideband channel. The second set of experiments employed a stationary channel model, where the impulse response was kept constant for the duration of each transmitted OFDM symbol. The resulting BER results from these experiments are shown in Figure 20.22. A BER performance degradation due to the time-varying channel can be observed for all the systems. The BER degradation is low for the WATM system over both the WATM and the shortened WATM channel with $F_d = 0.0063$, exhibiting a residual BER of under $2 \cdot 10^{-4}$ for 16-QAM. For the WLAN system and the UMTS, whose F_d values are about 0.04, the BER degradation is much more evident. In both cases, the residual BER for 16-QAM transmission is about $6 \cdot 10^{-3}$ and for BPSK the BER residual is $7.4 \cdot 10^{-4}$. Clearly, for the UMTS and WLAN at the given vehicular speed, the channel cannot be assumed stationary.

20.4.2.2 Pilot symbol assisted modulation

In order to investigate the effects of imperfect channel estimation on the system's BER performance in wideband fading channels separately from the effects of time-domain variations of the impulse response, the PSAM experiments have been conducted in the stationary channel. Therefore, the observed performance degradation compared to the perfect channel estimation results depicted in Figure 20.22 are caused by incorrect amplitude and phase estimations. The experimental results for 16-QAM transmission employing PSAM in conjunction with ideal lowpass and linear interpolation techniques are given in Figures 20.23 and 20.24, respectively. The estimation accuracy of the PSAM interpolation algorithms depends on each faded impulse response actually encountered by the modem, with some impulse amplitude combinations resulting in very poor estimation quality for the linear interpolation algorithm.

20.5 Conclusion

The performance of OFDM transmission over time-varying and time-dispersive channels is mainly limited by two factors, long-term shape and the time variability of the channel's impulse response. The impulse response determines the frequency domain channel transfer function, which manifests itself as frequency domain fading in the bandwidth of the OFDM symbol. The effects of this fading can be combated by the same methods as employed in serial modems, namely by pilot symbol assisted modulation (PSAM) or differential detection of the received data symbols. Analogous to time domain transmission systems, PSAM requires the transmission of additional pilot symbols and hence increases the system's overhead, while differential detection typically shows a 3 dB SNR performance loss compared to coherent schemes. For both differential detection and PSAM the structure of the channel's impulse response determines the performance. Long channel delays result in rapid fading in the frequency domain, requiring more pilots per data symbol for PSAM or decreasing the BER performance for differential detection.

Variations of the channel's impulse response, caused by fast fading of the constituting paths, result in interference between the OFDM symbol's subcarriers. This

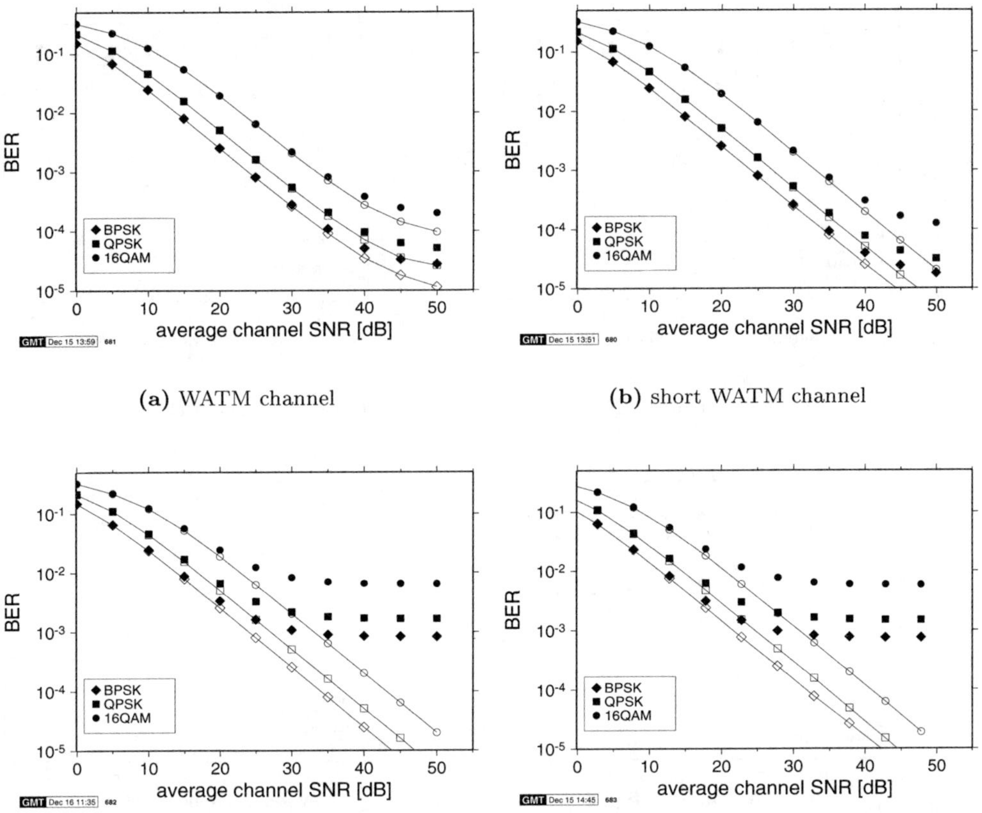

(a) WATM channel

(b) short WATM channel

(c) WLAN channel

(d) UMTS channel

Figure 20.22: Bit error rate versus channel SNR for fading (a) WATM channel (Figure 20.1(b)), (b) shortened WATM channel (Figure 20.3), (c) WLAN channel (Figure 20.4), (d) UMTS channel (Figure 20.5(a)), using coherent detection and perfect channel estimation. The open symbols correspond to the stationary channel model and the filled symbols correspond to the continuously fading channel model

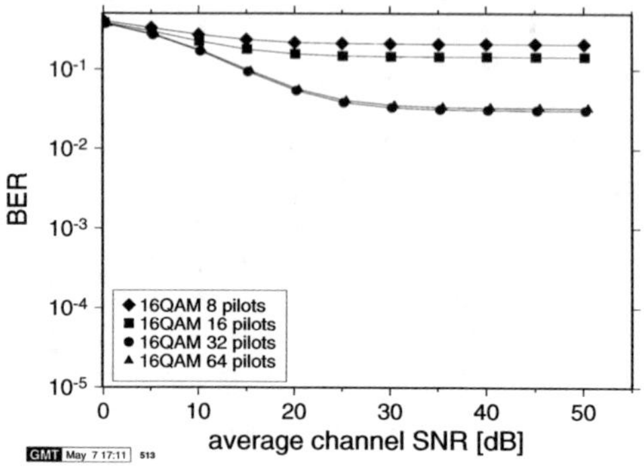

(a) WATM channel

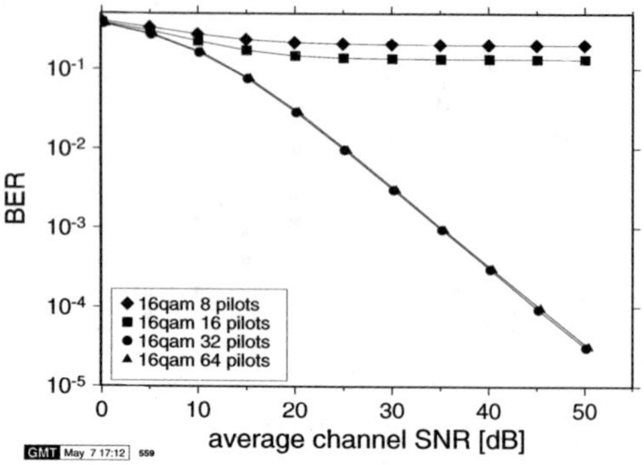

(b) short WATM channel

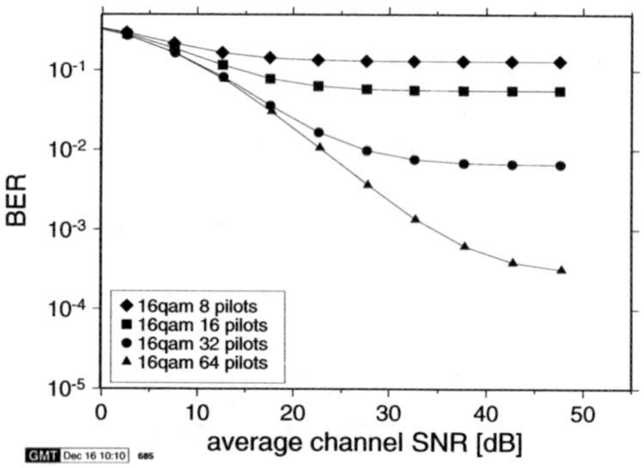

(c) WLAN channel

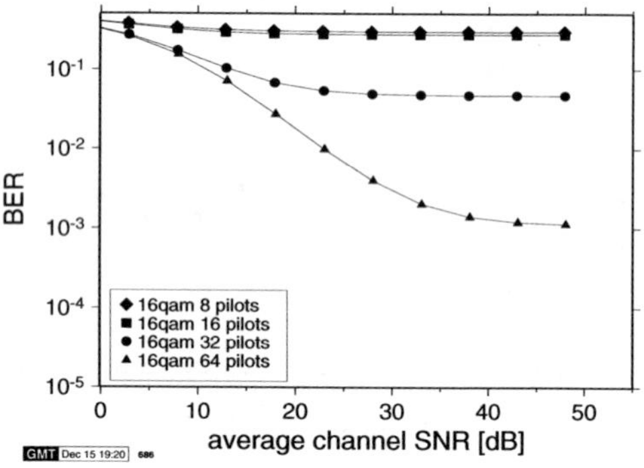

(d) UMTS channel

Figure 20.23: Bit error rate versus channel SNR for fading (a) WATM channel (Figure 20.1(b)), (b) shortened WATM channel (Figure 20.3), (c) WLAN channel (Figure 20.4), (d) UMTS channel (Figure 20.5(a)), using coherent detection and pilot symbol assisted channel estimation with ideal lowpass filter interpolation for 16-QAM

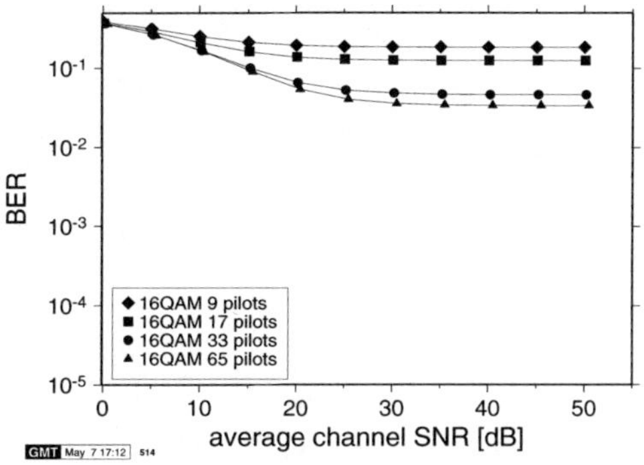

(a) WATM channel

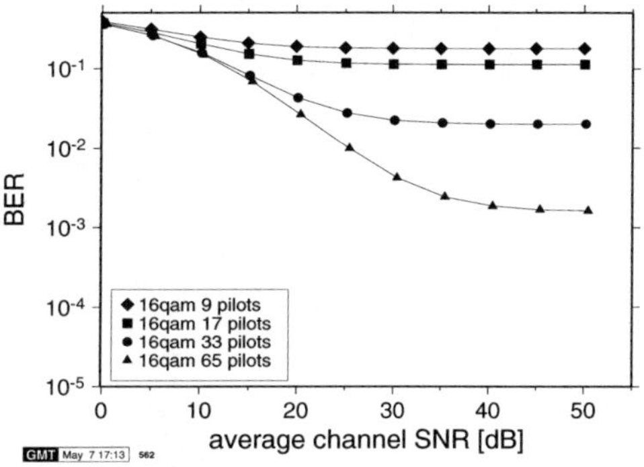

(b) short WATM channel

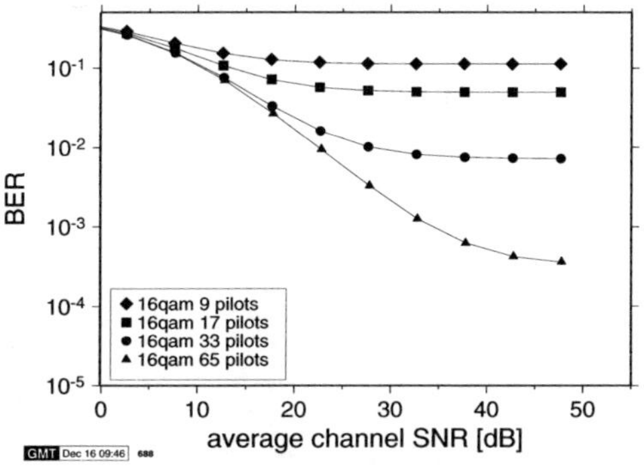

(c) WLAN channel

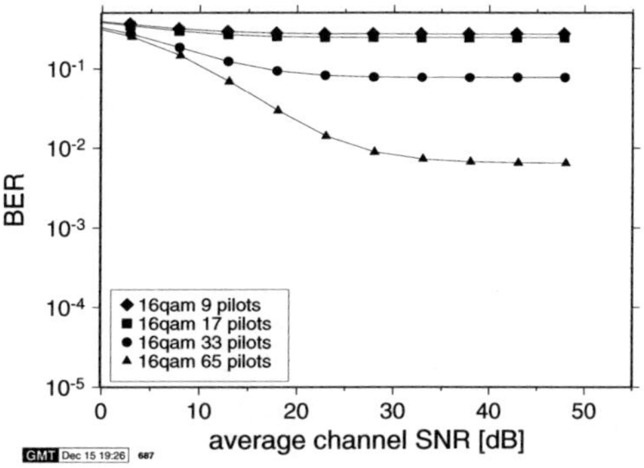

(d) UMTS channel

Figure 20.24: Bit error rate versus channel SNR for fading (a) WATM channel (Figure 20.1(b)), (b) shortened WATM channel (Figure 20.3), (c) WLAN channel (Figure 20.4), (d) UMTS channel (Figure 20.5(a)), using coherent detection and pilot symbol assisted channel estimation with linear interpolation for 16-QAM intersubcarrier interference is low, if the channel varies only insignificantly during one OFDM symbol, but causes severe performance degradation, if the rate of the channel variation is not much slower than the OFDM symbol rate. The effect of intersubcarrier interference therefore limits the maximum number of subcarriers employed in an OFDM system, depending on the channel's Doppler frequency and the robustness of the modulation scheme employed.

Chapter 21

OFDM Time and Frequency Domain Synchronisation

In this chapter we will investigate the effects of time and frequency domain synchronisation errors on the performance of an OFDM system, and two different synchronisation algorithms will be presented for time-domain burst-based OFDM communications systems.

21.1 System Performance with Frequency and Timing Errors

The performance of the synchronisation subsystem, in particular the accuracy of the frequency and timing error estimations, is of major influence on the overall OFDM system performance. In order to investigate the effects of carrier frequency and time domain FFT window alignment errors, a series of investigations has been performed over different channels.

21.1.1 Frequency shift

Carrier frequency errors result in a shift of the received signal's spectrum in the frequency domain. If the frequency error is an integer multiple n of the subcarrier spacing Δf, then the received frequency domain subcarriers are shifted by $n \cdot \Delta f$. The subcarriers are still mutually orthogonal, but the received data symbols, which were mapped to the OFDM spectrum, are in the wrong position in the demodulated spectrum, resulting in a bit error rate of 0.5.

If the carrier frequency error is not an integer multiple of the subcarrier spacing, then energy is spilling over between the subcarriers, resulting in loss of their mutual orthogonality. In other words, interference is observed between the subcarriers, which deteriorates the bit error rate of the system. The amount of this intersubcarrier

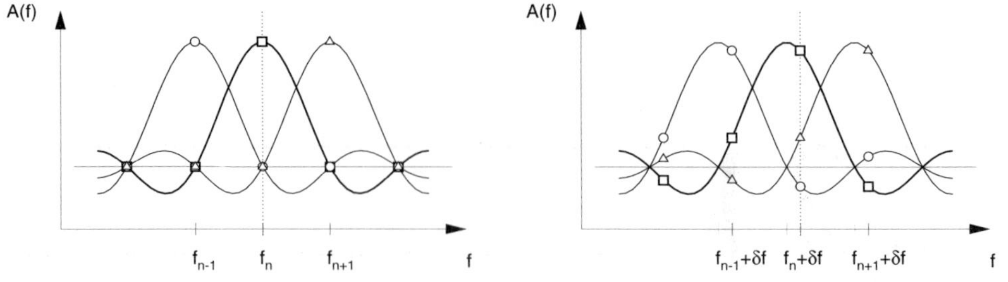

(a) perfect sampling **(b)** frequency error

Figure 21.1: Stylised plot of OFDM symbol spectrum with sampling points for three subcarriers. The symbols on the curves signify the contributions of the three subcarriers to the sum at the sampling point: (a) no frequency offset between transmitter and receiver, (b) frequency error δf present

interference can be evaluated by investigating the spectrum of the OFDM symbol.

21.1.1.1 The spectrum of the OFDM signal

is derived from its time domain representation transmitted over the channel. A single OFDM symbol in the time domain can be described as:

$$u(t) = \left[\sum_{n=0}^{N-1} a_n e^{j\omega_n \cdot t}\right] \times \text{rect}\left(\frac{t}{N \cdot T_s}\right), \tag{21.1}$$

which is the sum of N subcarriers $e^{\omega_n \cdot t}$, each modulated by a QAM symbol a_n and windowed by a rectangular window of the OFDM symbol duration T_s. The Fourier transform of this rectangular window is a frequency domain sinc function, which is convolved with the Dirac delta subcarriers, determining the spectrum of each of the windowed complex exponential functions, leading to the spectrum of the nth single subcarrier in the form of

$$A_n(\omega) = \frac{\sin(N \cdot T_s \cdot \omega/2)}{N \cdot T_s \cdot \omega/2} * \delta(\omega - \omega_n).$$

Replacing the angular velocities ω by frequencies and using the relationship $N \cdot T_s = 1/\Delta f$, the spectrum of a subcarrier can be expressed as:

$$A_n(f) = \frac{\sin(\pi \frac{f-f_n}{\Delta f})}{\pi \frac{f-f_n}{\Delta f}} = \text{sinc}\left(\frac{f-f_n}{\Delta f}\right).$$

The OFDM receiver samples the received time domain signal, demodulates it by invoking the FFT and in case of a carrier frequency shift it generates the sub-channel signals in the frequency domain at the sampling points $f_n + \delta f$, which are spaced from each other by the subcarrier spacing Δf and misaligned by the frequency error δf. This scenario is shown in Figure 21.1. Figure 21.1(a) shows the sampling of the subcarrier at frequency f_n at the optimum frequency raster, resulting in a maximum signal amplitude and no intersubcarrier interference. If the frequency reference of the receiver is offset with respect to that of the transmitter by a frequency error of δf, then the received symbols suffer from intersubcarrier interference, as depicted in Figure 21.1(b).

The total amount of intersubcarrier interference experienced by subcarrier n is the sum of the interference amplitude contributions of all the other subcarriers in the OFDM symbol:

$$I_n = \sum_{j, j \neq n} a_j \cdot A_j(f_n + \delta f).$$

Since the QAM symbols a_j are random variables, the interference amplitude in subcarrier n, I_n, is also a random variable, which cannot be calculated directly. If the number of interferers is high, however, then the power spectral density of I_n can be approximated with that of a Gaussian process, according to the central limit theorem. Therefore, the effects of the intersubcarrier interference can be modelled by additional white Gaussian noise superimposed on the frequency domain data symbols.

The variance of this Gaussian process σ_{ISI_n} is the sum of the variances of the interference contributions,

$$\sigma^2_{ISI_n} = \sum_{j, j \neq n} \sigma^2_{a_j} \cdot |A_j(f_n + \delta f)|^2 .$$

The quantities $\sigma^2_{a_j}$ are the variances of the data symbols, which are the same for all j in a system that is not varying the average symbol power across different subcarriers. Additionally, because of the constant subcarrier spacing of Δf, the interference amplitude contributions can be expressed more conveniently as:

$$A_j(f_n + \delta f) = A_j n(\delta f) = sinc((n - j) + \frac{\delta f}{\Delta f}).$$

The sum of the interferer powers leads to the intersubcarrier interference variance expression:

$$\sigma^2_{ISI} = \sigma^2_a \cdot \sum_{i=-N/2-1}^{N/2} \left| sinc(i + \frac{\delta f}{\Delta f}) \right| .$$

The value of the intersubcarrier interference (ISI) variance for FFT lengths of $N = 64$, 512 and 4096 and for a range of frequency errors δf is shown in Figure 21.2. It can be seen that the number of subcarriers does not influence the ISI noise variance for OFDM symbol lengths of more than 64 subcarriers. This is due the to rapid decrease of the interference amplitude with increasing frequency separation, so that only the interference from close subcarriers contributes significantly to the interference load on the subcarriers.

In order to investigate the accuracy of the Gaussian approximation, simulations were conducted and histograms of the measured interference amplitude were pro-

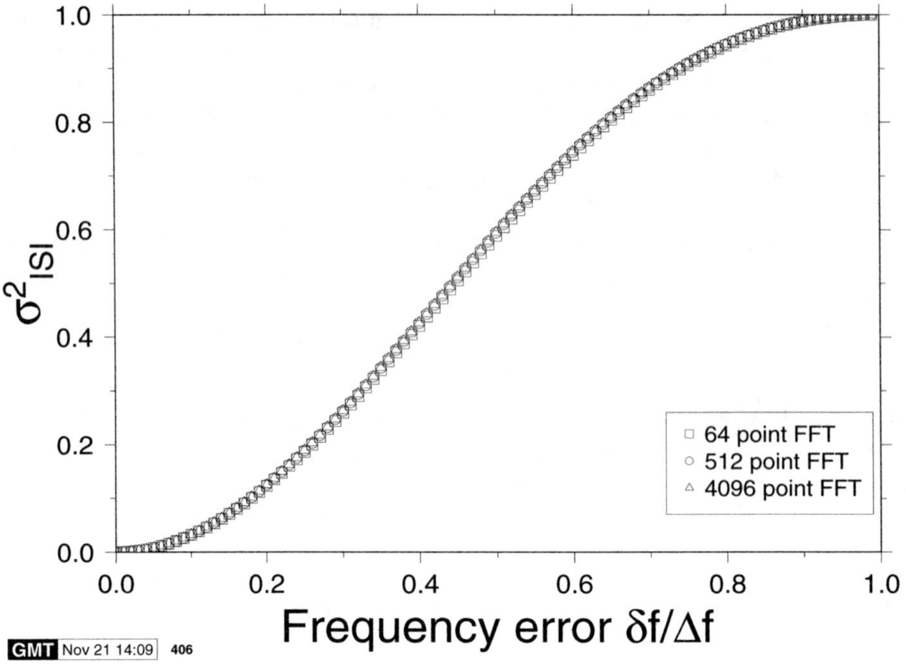

Figure 21.2: Intersubcarrier interference variance due to a frequency shift δf FFT lengths of $N = 64$, 512 and 4096 for normalised frequency errors $\delta f/\Delta f$ between 0 and 1

duced for QPSK and 16-QAM modulation of the subcarriers. The triangles in Figure 21.3 depict the histograms of ISI noise magnitudes recorded for a 512-subcarrier OFDM modem employing QPSK and 16-QAM in a system having a frequency error of $\delta f = 0.3\Delta f$. The continuous line drawn in the same graph is the corresponding approximation of the histogram by a Gaussian probability density function (PDF) of the variance calculated using Equation 21.1.1.1. It can be observed that the Gaussian curve is a reasonable approximation for both histograms in the central region, but that for the tails of the distributions the Gaussian function exhibits high relative errors. The histogram of the interference caused by the 16-QAM signal is, however, closer to the Gaussian curve than the QPSK interference histogram.

However, the frequency mismatch between the transmitter and receiver of an OFDM system not only results in intersubcarrier interference, but it also reduces the useful signal amplitude at the frequency domain sampling point by a factor of $f(\delta f) = \text{sinc}(\delta f/\Delta f)$. Using this and σ^2_{ISI}, the theoretical influence of the intersubcarrier interference, approximated by a Gaussian process, can be calculated for a given modulation scheme in an AWGN channel. In the case of coherently detected QPSK, the closed-form expression for the BER $P_e(\gamma)$ at a channel SNR γ is given [257] by:

$$P_e(\gamma) = Q(\sqrt{\gamma}),$$

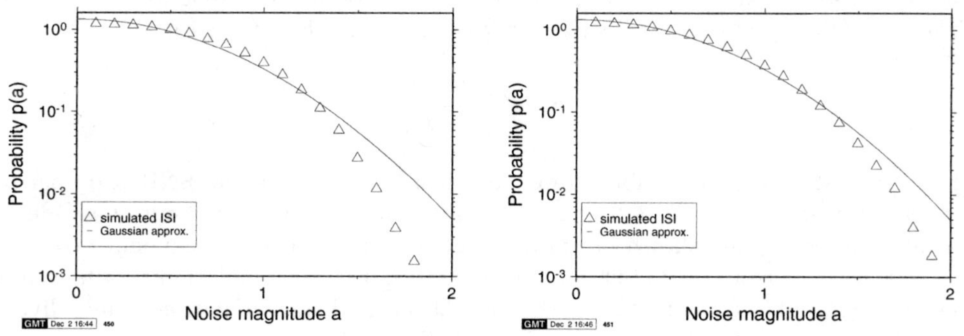

(a) QPSK (b) 16QAM

Figure 21.3: Histogram of the ISI magnitude for a simulated 512-subcarrier OFDM modem using QPSK or 16-QAM for $\delta f = 0.3\Delta f$; the line represents the Gaussian approximation having the same variance

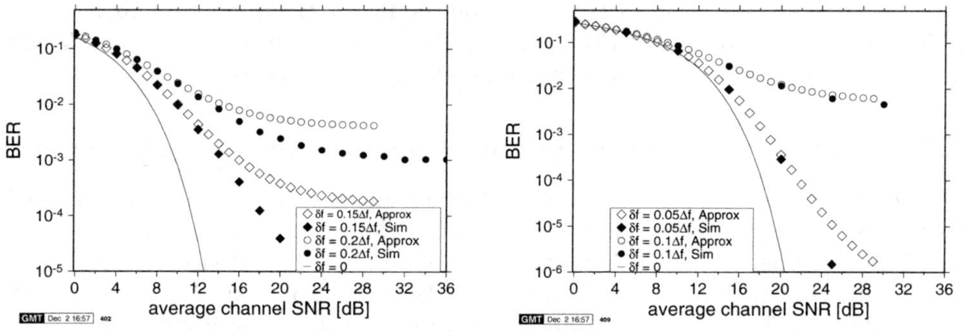

(a) QPSK (b) 16QAM

Figure 21.4: The effect of intersubcarrier interference due to frequency synchronisation error on the BER over AWGN channels. (a) Bit error probability versus channel SNR for frequency errors of $0.15\Delta f$ and $0.2\Delta f$ for a QPSK modem. (b) BER versus channel SNR for frequency errors of $0.05\Delta f$ and $0.1\Delta f$ for a 16-QAM modem. In both graphs, the filled symbols are simulated BER results, while the open symbols are the predicted BER curves using the Gaussian intersubcarrier interference model

where the Gaussian $Q()$-function is defined as

$$Q(y) = \frac{1}{\sqrt{2\pi}} \int_y^\infty e^{-x^2/2} dx = \text{erfc}\left(\frac{y}{\sqrt{2}}\right).$$

Assuming that the effects of the frequency error can be approximated by white

Gaussian noise of variance σ_{ISI}^2 and taking into account the attenuated signal magnitude $f(\delta f) = sinc(\pi\delta f/\Delta f)$, we can adjust the equivalent SNR to:

$$\gamma' = \frac{f(\delta f) \cdot \sigma_a^2}{\sigma_{ISI}^2 + \sigma_a^2/\gamma},$$

where σ_a^2 is the average symbol power and γ is the real channel SNR. Comparison between the theoretical BER calculated using γ' and simulation results for different frequency errors δf are shown in Figure 21.4(a). While for both frequency errors the theoretical BER using the Gaussian approximation fits the simulation results well for channel SNR values of up to 12 dB, the predictions and the simulation results diverge for higher values of SNR. The pessimistic BER prediction is due to the pronounced discrepancy between the histogram and the Gaussian curve in Figure 21.3 at the tail ends of the amplitude histograms, since for high noise amplitudes the Gaussian model is a poor approximation for the intersubcarrier interference.

The equivalent experiment, conducted for coherently detected 16-QAM, results in the simulated and predicted bit error rates depicted in Figure 21.4(b). For 16-QAM transmission, the noise resilience is much lower than for QPSK, hence for our experiments smaller values of δf have been chosen. It can be observed that the Gaussian noise approximation is a much better fit for the simulated BER in a 16-QAM system than for a 4-QAM modem. This is in accordance with Figure 21.3, where the histograms of the interference magnitudes were depicted.

21.1.1.2 Effects of frequency mismatch on different modulation schemes

In order to investigate the effects of frequency mismatch on different modulation schemes, a series of simulations were conducted employing both coherently and differentially detected, as well as pilot symbol assisted QPSK systems. Figures 21.5(a), 21.6(a) and 21.7(a) show the performance of the QPSK, pilot symbol assisted QPSK (PSA-QPSK) and differential QPSK (DQPSK) OFDM schemes, respectively. As a benchmark, the BER performance of the equivalent serial modulation scheme is plotted [257] as a line on all the graphs, which also represents the performance that is achieved by the parallel modem, when $\delta f = 0$.

21.1.1.2.1 Coherent modulation Figure 21.5(a) reveals the BER performance degradation due to increasing the carrier frequency offset δf. The adjacent subchannel interference effects are considerable, even for small frequency errors. It is clear that for a carrier frequency offset of $\delta f = 0.2 \cdot \Delta f$, the BER reaches a residual value of about 10^{-3} at approximately 26 dB of SNR. The minimal possible BER for different values of carrier frequency mismatch are plotted in Figure 21.5(b). For relative frequency errors above $0.18 \cdot \Delta f$, the attainable bit error rate is worse than 10^{-4}.

21.1.1.2.2 PSAM Figure 21.6(a) shows the bit error rate performance of the pilot-assisted QPSK system over an AWGN channel, which is consistently worse than that of an equivalent coherently demodulated system without pilot assistance. This is

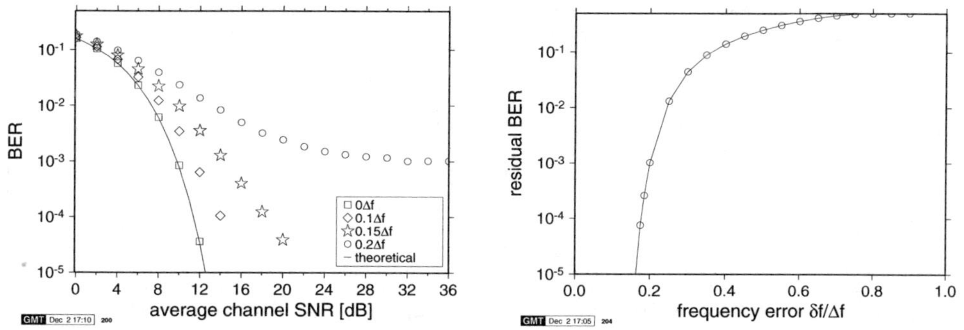

(a) BER versus SNR

(b) Residual BER

Figure 21.5: Bit error rate versus channel SNR performance for a non-pilot-assisted QPSK OFDM modem in an AWGN channel: (a) bit error rate versus signal-to-noise ratio plot for different constant frequency errors, (b) plot of residual bit error rate versus the frequency error

in accordance with the performance of PSAM schemes over AWGN channels, where PSAM systems are generally handicapped by errors in the channel estimation, which is unnecessary for non-fading channels. Since the frequency error mismatch results in additional noise in the received frequency domain pilots, the quality of the channel estimation deteriorates. This reduces the BER performance of a PSAM system compared to a coherently demodulated system refraining from using pilots.

The bit error rate curves given in the figure were computed for three pilot subcarriers per 512-subcarrier OFDM symbol, which were invoked to mitigate the effects of the channel's fading envelope in a time or frequency fading environment. The corresponding relationship between the residual bit error rate and the frequency error is given in Figure 21.6(b). In the simulations presented here linear interpolation was employed between the pilots. Clearly, the BER performance of the system depends on the number of pilots employed as well as on the interpolation method, and these effects are discussed in Section 20.3.1.1 in more depth.

21.1.1.2.3 Differential modulation The corresponding simulation results for differentially encoded QPSK are shown in Figure 21.7. Again, the impact of the intersubcarrier interference is severe even for small relative frequency errors. Figure 21.7(b) shows that a frequency error of only $0.12\Delta f$ results in a BER residual of 10^{-4}.

21.1.1.2.4 Frequency error - summary A frequency error in an OFDM system results in a shift of the received frequency domain symbols relative to the receiver's raster. This leads to intersubcarrier interference, whose nature is noise-like, owing to the great number of contributing interfering subcarriers. Because of the frequency domain aliasing, the variance of this interference is constant for all subcarriers in the

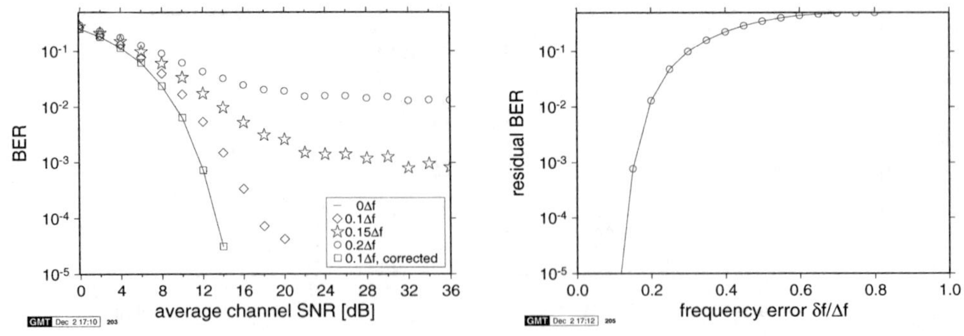

(a) BER versus SNR **(b)** Residual BER

Figure 21.6: Bit error rate versus channel SNR performance for a 3-pilot-assisted QPSK OFDM modem in an AWGN channel: (a) bit error rate versus signal-to-noise ratio plot for different fixed frequency errors, (b) plot of residual bit error rate versus the frequency error

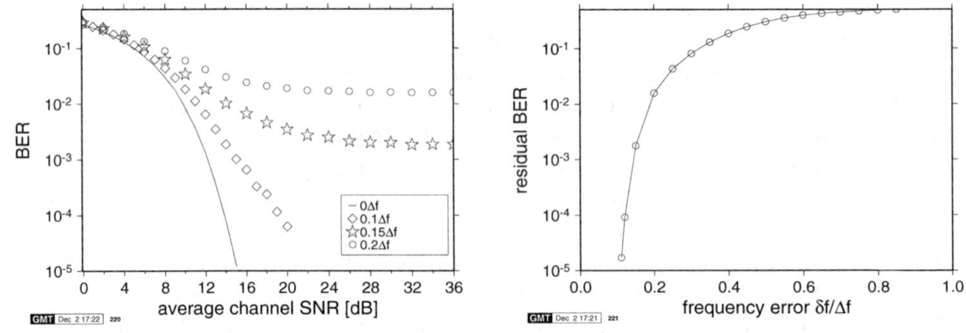

(a) BER versus SNR **(b)** Residual BER

Figure 21.7: Bit error rate versus channel SNR performance for differential QPSK modulation scheme in AWGN channel. The OFDM FFT length is 512. (a) Bit error rate versus signal-to-noise ratio plot for different fixed frequency errors, (b) plot of residual bit error rate versus the frequency error

OFDM symbol, if all subcarriers carry the same average power. The variance of the interference can be computed and used to model the interference effects by additional white noise superimposed on the data symbols. Depending on the modulation scheme employed, this model typically yields a good estimation of the actual BER.

Different modulation schemes are affected differently by the presence of frequency errors in the system, analogously to their performance in purely AWGN environments. Coherent detection suffers the least penalty, followed by differential detection and

PSAM schemes. Let us now consider the effects of time domain synchronisation errors.

21.1.2 Time domain synchronisation errors

Unlike frequency mismatch, as discussed above, time synchronisation errors do not result in intersubcarrier interference. Instead, if the receiver's FFT window spans samples from two consecutive OFDM symbols, inter OFDM symbol interference occurs.

Additionally, even small misalignments of the FFT window result in an evolving phase shift in the frequency domain symbols, leading to BER degradation. Initially, we will concentrate on these phase errors.

If the receiver's FFT window is shifted with respect to that of the transmitter, then the time shift property of the Fourier transform, formulated as:

$$\begin{aligned} f(t) &\longleftrightarrow F(\omega) \\ f(t-\tau) &\longleftrightarrow e^{-j\omega\tau}F(\omega) \end{aligned}$$

describes its effects on the received symbols. Any misalignment τ of the receiver's FFT window will introduce a phase error of $2\pi\Delta f\tau/T_s$ between two adjacent subcarriers. If the time shift is an integer multiple m of the sampling time T_s, then the phase shift introduced between two consecutive subcarriers is $\delta\phi = 2\pi m/N$, where N is the FFT length employed. This evolving phase error has a considerable influence on the BER performance of the OFDM system, clearly depending on the modulation scheme used.

21.1.2.1 Coherent demodulation

Coherent modulation schemes suffer the most from FFT window misalignments, since the reference phase evolves by 2π throughout the frequency range for every sampling time misalignment. Clearly, this results in a total loss of the reference phase, and hence coherent modulation cannot be employed without phase correction mechanisms, if imperfect time synchronisation has to be expected.

21.1.2.2 Pilot symbol assisted modulation

Pilot symbol assisted modulation (PSAM) schemes can be employed in order to mitigate the effects of spectral attenuation and the phase rotation throughout the FFT bandwidth. Pilots are interspersed with the data symbols in the frequency domain and the receiver can estimate the evolving phase error from the received pilots' phases.

This operation is performed with the aid of the wideband channel estimation discussed in Section 20.3.1.1, and the number of pilot subcarriers necessary for correctly estimating the channel transfer function depends on the maximum anticipated time shift τ. Following the notion of the frequency domain channel transfer function $H(n)$ introduced in Chapter 20, the effects of phase errors can be written as:

$$H(f) = e^{-j2\pi f\tau}. \tag{21.2}$$

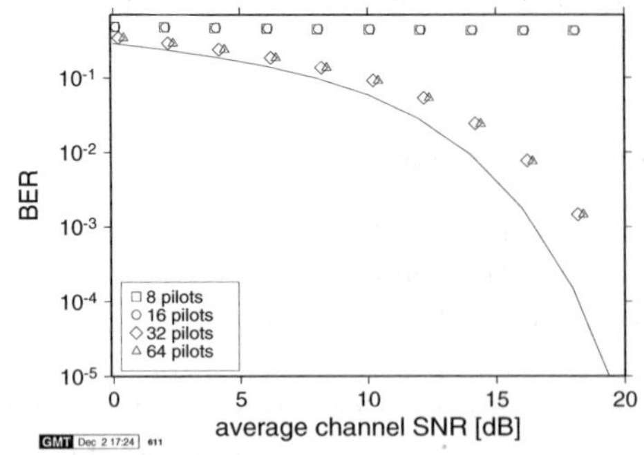

(a) perfect lowpass interpolation

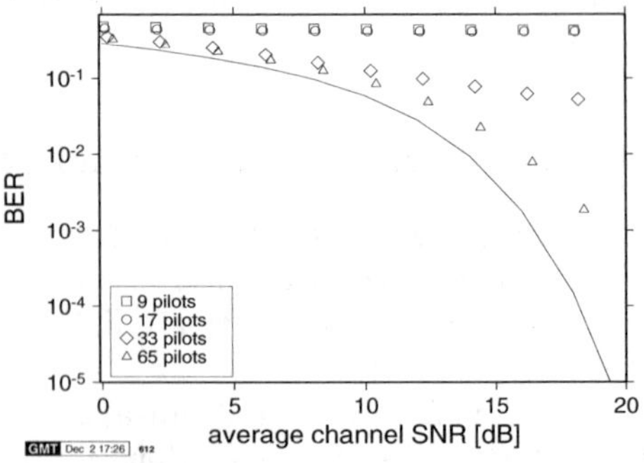

(b) linear interpolation

Figure 21.8: Bit error rate versus channel SNR performance for 16-level PSA-QAM in an AWGN channel for different pilot subcarrier spacings in the presence of a fixed FFT window misalignment of $\tau = 10T_s$. The OFDM FFT length is 512. (a) PSAM interpolation using ideal lowpass interpolator, (b) PSAM using linear interpolator. In both graphs, the line marks the coherently detected 16-QAM performance in the absence of both FFT window misalignment and PSAM

Replacing the frequency variable f by the subcarrier index n, where $f = n\Delta f = n/(NT_s)$ and normalising the time misalignment τ to the sampling time T_s, so that $\tau = m \cdot T_s$, the frequency domain channel transfer function can be expressed as:

$$H(n) = e^{-j2\pi \frac{nm}{N}}. \tag{21.3}$$

The number of pilots necessary for correctly estimating this frequency domain channel transfer function $H(n)$ is dependent on the normalised time delay m. Following the Nyquist sampling theorem, the distance Δp between two pilot tones in the OFDM spectrum must be less than or equal to half the period of $H(n)$, so that

$$\Delta p \leq \frac{N}{2m}. \tag{21.4}$$

The simulated performance of a 512-subcarrier 16-QAM PSAM modem in the presence of a constant timing error of $\tau = 10T_s$ in an AWGN channel is depicted in Figure 21.8 for both of the PSAM interpolation algorithms investigated in Section 20.3.1.1. Following Equation 21.4, the maximum acceptable pilot subcarrier distance required for resolving a normalised FFT window misalignment of $m = \tau/T_s = 10$ is $\Delta p = N/20 = 512/20 = 25.6$, requiring at least 20 pilot subcarriers equidistantly spaced in the OFDM symbol. We can see in both graphs of Figure 21.8 that the bit error rate is 0.5 for both schemes if less than 20 pilot subcarriers are employed in the OFDM symbol. For pilot numbers above the required minimum of 20, however, the performance of the ideal lowpass interpolated PSAM scheme does not vary with the number of pilots employed, while the linearly interpolated PSAM scheme needs higher numbers of pilot subcarriers for achieving a similar performance to the lowpass interpolator scheme. The continuous lines in the graphs show the BER curve for a coherently detected 16-QAM OFDM modem in the absence of timing errors, while utilising no PSAM. The BER penalty of PSAM in a narrowband AWGN channel as well as the performance differences between the two PSAM schemes are in accordance with the results found in Section 20.3.1.1.

21.1.2.3 Differential modulation

Differential encoding of OFDM symbols can be implemented both between corresponding subcarriers of consecutive OFDM symbols or between adjacent subcarriers of the same OFDM symbol. We found the latter more advantageous in TDMA environments and hence this principle is employed here. The BER performance of differentially encoded modulation schemes is affected by the phase shift $\delta\phi$ between adjacent subcarriers introduced by timing errors and this influence can be evaluated for example for DPSK systems, as will be shown below.

The two-dimensional probability density function of a noisy phasor in polar coordinates is calculated in Appendix 21.6 and is given by Equation 21.48, if we assume the transmitted phase to be zero. Integration of this function over the magnitude r gives the phase probability function $p_\phi(\phi)$:

$$p_\phi(\phi) = \int_0^\infty \frac{r}{2\pi\sigma^2} \cdot e^{-(r^2+\mathcal{A}^2-2r\mathcal{A}\cos\phi)/2\sigma^2} dr, \tag{21.5}$$

where $\mathcal{A}$ is the amplitude of the noiseless phasor. For differential phase modulation, the error of the difference between the phases of two consecutive symbols, $\Phi = \phi_k - \phi_{k-1}$, determines the symbol error rate (SER). The PDF $p_\Phi(\Phi)$ of this difference can be expressed by a variable transform, which results in the following integral:

$$p_\Phi(\Phi) = \int_{-\pi}^{\pi} p_\phi(\psi - \Phi) \cdot p_\phi(\psi) d\psi, \tag{21.6}$$

where ψ is an auxiliary variable. The received symbol will be demodulated correctly, if the difference error $\Delta\phi$ is within the decision boundaries of the PSK constellation. For M-DPSK, the symbol error rate (SER) is given by:

$$SER = 1 - \int_{-\pi/M}^{\pi/M} p_\Phi(\Phi) d\Phi. \tag{21.7}$$

If there is an FFT window misalignment induced phase shift $\delta\phi$ between consecutive symbols, then the integration limits in Equation 21.7 are biased by this shift:

$$SER = 1 - \int_{-\pi/M+\delta\phi}^{\pi/M+\delta\phi} p_\Phi(\Phi) d\Phi. \tag{21.8}$$

Simulations have been performed for a 512-subcarrier OFDM system, employing DBPSK and DQPSK for different FFT window misalignment values. The BER

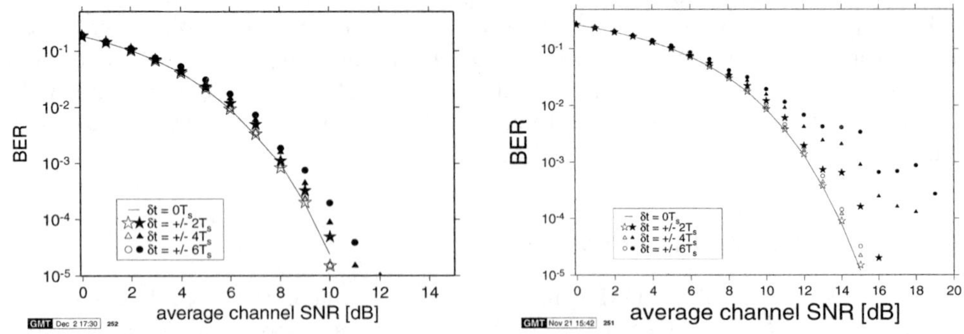

(a) DBPSK

(b) DQPSK

Figure 21.9: Bit error rate versus SNR over AWGN channels for a 512-subcarrier OFDM modem employing DBPSK and DQPSK, respectively. Positive time shifts imply time-advanced FFT window or delayed received data

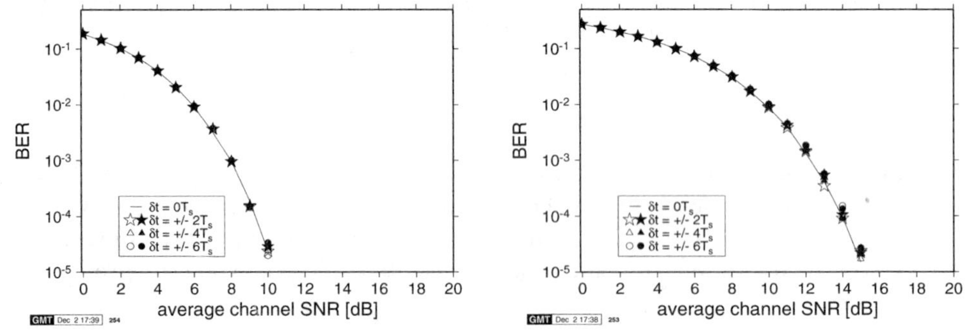

(a) DBPSK

(b) DQPSK

Figure 21.10: Bit error rate versus SNR over AWGN channels for a 512-subcarrier OFDM modem employing a postamble of 10 symbols for DBPSK and DQPSK, respectively. Positive time shifts correspond to time-advanced FFT window or delayed received data

performance curves for timing errors up to six sampling intervals are displayed in Figure 21.9. Note that one sample interval misalignment represents a phase error of $2\pi/512$ between two consecutive samples, which explains why the BER effects of the simulated positive timing misalignments marked by the hollow symbols are negligible for DBPSK. Specifically, a maximum SNR degradation of 0.5 dB was observed for DQPSK.

Positive FFT window time shifts correspond to a delayed received data stream and hence all samples in the receiver's FFT window belong to the same quasi-periodically extended OFDM symbol. In the case of negative time shifts, however, the effects on the bit error rate are much more severe due to inter OFDM symbol interference. Since the data is received prematurely, the receiver's FFT window contains samples of the forthcoming OFDM symbol, not from the cyclic extension of the wanted symbol.

This non-symmetrical behaviour of the OFDM receiver with respect to positive and negative relative timing errors can be mitigated by adding a short postamble, consisting of copies of the OFDM symbol's first samples. Figure 21.10 shows the BER versus SNR curves for the same offsets, while using a 10-sample postamble. Now, the behaviour for positive and negative timing errors becomes symmetrical. Clearly, the required length of this postamble depends on the largest anticipated timing error, which adds further redundancy to the system. This postamble can be usefully employed, however, to make an OFDM system more robust to time misalignments and thus to simplify the task of the timedomain FFT window synchronisation system.

21.1.2.3.1 Time-domain synchronisation errors - summary Misalignment of the receiver's FFT window relative to the received sample stream leads to possible inter OFDM symbol interference as well as to an evolving shift of the reference phase throughout the received frequency domain OFDM symbol. While the effects of inter OFDM symbol interference can be mitigated for moderate misalignments by appending a cyclic postamble to the OFDM symbol, the phase errors in the frequency domain make it impossible to use coherently detected modulation schemes without phase recovery methods. Instead, differentially detected schemes can be employed, which nonetheless suffer from performance degradation due to the phase errors. Alternatively, pilot symbol assisted channel estimation schemes can be employed in conjunction with coherent detection.

21.2 Synchronisation Algorithms

The results of Section 21.1 indicate that the accuracy of a modem's time and frequency domain synchronisation system dramatically influences the overall BER performance. We have seen that carrier frequency differences between the transmitter and the receiver of an OFDM system will introduce additional impairments in the frequency domain caused by intersubcarrier interference, while FFT window misalignments in the time domain will lead to phase errors between the subcarriers. Both of these effects will degrade the system's performance and have to be kept to a minimum by the synchronisation system.

In a TDMA-based OFDM system, the frame synchronisation between a master station – in cellular systems generally the base station – and the portable stations has also to be maintained. For these systems, a reference symbol marking the beginning of a new time frame is commonly used. This added redundancy can be exploited for both frequency synchronisation and FFT window alignment, if the reference symbol is correctly chosen.

In order to achieve synchronisation with a minimal amount of computational effort at the receiver, while also minimising the amount of redundant information added to the data signal, the synchronisation process is normally split into an acquisition phase and a tracking phase, if the characteristics of the random frequency and timing errors are known. In the acquisition phase, an initial estimate of the errors is acquired, using more complex algorithms and possibly a higher amount of synchronisation information in the data signal, whereas later the tracking algorithms only have to correct for small short-term deviations.

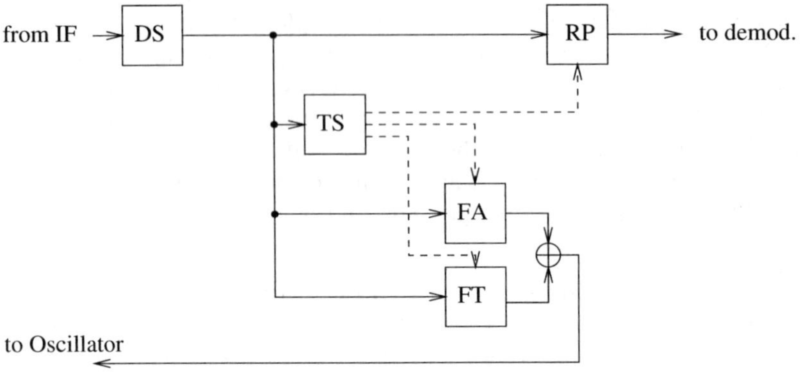

Figure 21.11: Block diagram of the synchronisation system: **DS**-downsampling and clock recovery, **TS**-time synchronisation, **FA**-frequency acquisition, **FT**-frequency tracking, **RP**-remove cyclic extension prefix

The block diagram of a possible synchronisation system is shown in Figure 21.11. The down sampling and clock recovery module DS has to determine the optimum sampling instant. The time synchronisation TS controls the frequency acquisition FA, the frequency tracking FT as well as the timedomain alignment of the FFT window, which is carried out by the "remove prefix" or RP block. These operations will be detailed during our further discourse.

At the beginning of the synchronisation process neither the frequency error nor the timing misalignment are known, hence synchronisation algorithms must be found that are sufficiently robust to initial timing and frequency errors.

21.2.1 Coarse frame and OFDM symbol synchronisation

Coarse frame and symbol synchronisation algorithms presented in the literature all rely on additional redundancy inserted in the transmitted data stream. The pan-European digital video broadcasting (DVB) system uses a so-called null symbol as the first OFDM symbol in the time frame. No energy is transmitted [412], during the null symbol and it is detected by monitoring the received baseband power in the time domain, without invoking FFT processing. Claßen [101] proposed an OFDM synchronisation burst of at least three OFDM symbols per time frame. Two of the OFDM symbols in the burst would contain synchronisation subcarriers bearing known symbols along with normal data transmission carriers, but one of the OFDM symbols would be the exact copy of one of the other two, thus resulting in more than one OFDM symbol synchronisation overhead per synchronisation burst. For the ALOHA environment, Warner [119] proposed the employment of a power detector and subsequent correlation-based detection of a set of received synchronisation subcarriers embedded in the data symbols. The received synchronisation tones are extracted from the received time domain signal using an iterative algorithm for updating the synchronisation tone values once per sampling interval. For a more detailed discussion on these techniques the interested reader is referred to the literature [101, 119], while for a variety of further treatises to the contributions by Mammela and his team [413–415].

21.2.2 Fine symbol tracking

Fine symbol tracking algorithms are generally based on correlation operations either in the time or in the frequency domain. Warner [119] and Bingham [416] employed frequency domain correlation of the received synchronisation pilot tones with known synchronisation sequences, while de Couasnon [417] utilised the redundancy of the cyclic prefix by integrating over the magnitude of the difference between the data and the cyclic extension samples. Sandell [124] proposed exploiting the autocorrelation properties of the received time domain samples imposed by the cyclic extension for fine time domain tracking.

21.2.3 Frequency acquisition

The frequency acquisition algorithm has to provide an initial frequency error estimate, which is sufficiently accurate for the subsequent frequency tracking algorithm to operate reliably. Generally the initial estimate must be accurate to half a subcarrier spacing. Sari [120] proposed the use of a pilot tone embedded into the data symbol, surrounded by zero-valued virtual subcarriers, so that the frequency-shifted pilot can be located easily by the receiver. Moose [121] suggested a shortened repeated OFDM symbol pair, analogous to his frequency tracking algorithm to be highlighted in the next section. By using a shorter DFT for this reference symbol pair, the subcarrier distance is increased and thus the frequency error estimation range is extended. Claßen [101,102] proposed to use binary pseudo-noise (PN) or so-called CAZAC training sequences carried by synchronisation subcarriers, which are also employed for the frequency tracking. The frequency acquisition, however, is performed by a search for the training sequence in the frequency domain. This is achieved by means of frequency domain correlation of the received symbol with the training sequence.

21.2.4 Frequency tracking

Frequency tracking generally relies on an already established coarse frequency estimation having a frequency error of less than half a subcarrier spacing. Moose [121] suggested the use of the phase difference between subcarriers of repeated OFDM symbols in order to estimate frequency deviations of up to one-half of the subcarrier spacing, while Claßen [101] employed frequency domain synchronisation subcarriers embedded into the data symbols, for which the phase shift between consecutive OFDM symbols can be measured. Daffara [123] and Sandell [124] used the phase of the received signal's autocorrelation function, which represents a phase shift between the received data samples and their repeated copies in the cyclic extension of the OFDM symbols.

Following the above brief literature survey, we will investigate two different synchronisation algorithms, both making use of a reference symbol marking the beginning of a new time frame. This limits the use of both algorithms to systems whose channel access scheme is based on time division multiple access (TDMA) frames.

21.2.5 Time- and Frequency-Domain Synchronisation Based on Autocorrelation

Both the frequency and the time domain synchronisation control signals can be derived from the received signal samples' cyclic nature upon exploiting the OFDM symbols' cyclic time domain extension by means of correlation techniques. A range of symbol timing and fine frequency tracking algorithms were proposed by Mandarini and Falaschi [418]. Originally, Moose [121] proposed a synchronisation algorithm using repeated data symbols, and methods for the frequency error estimation using the cyclic extension of OFDM symbols were presented by Daffara *et al.* [123] and Sandell *et al.* [124]. The frequency acquisition and TDMA frame synchronisation proposed here are based on similar principles, employing a dedicated reference symbol exploited in the time domain [419].

No added redundancy in the data symbols and no a priori knowledge of the synchronisation sequences constituting the reference symbol are required, since only the repetitive properties of the OFDM symbols and those of the reference symbol (REF) in the proposed time division duplex (TDD) frame structure seen in Figure 21.13 are exploited. All the processing is carried out in the time domain, hence no FFT-based demodulation of the reference symbol is necessary.

21.2.6 Multiple Access Frame Structure

The proposed 64-slot TDMA/TDD frame structure is depicted at the top of Figure 21.12, which is constituted by a null symbol, a reference symbol and 62 data symbols. Let us initially consider the role of the reference symbol.

21.2.6.1 The reference symbol

The reference symbol shown in Figure 21.13 was designed to assist in the operation of the synchronisation scheme and consists of repetitive copies of a synchronisation pattern SP of N_s pseudorandom complex samples. The synchronisation algorithm at the receiver needs no knowledge of the employed synchronisation pattern, hence this sequence could be used for channel-sounding training sequences or for base station identification signals. Note therefore that there are three hierarchical periodic time domain structures in the proposed framing scheme: the short-term intrinsic periodicity in the reference symbol of Figure 21.13, the medium-term periodicity associated with the quasi-periodic extension of the OFDM symbols and the long-term periodicity of the OFDM TDMA/TDD frame structure, repeating the reference symbol every 64 OFDM symbols, as portrayed in Figure 21.12. The long-term reference symbol periodicity is exploited in order to maintain OFDM frame synchronisation, while the medium-term synchronism of the cyclic extension assists in the process of OFDM

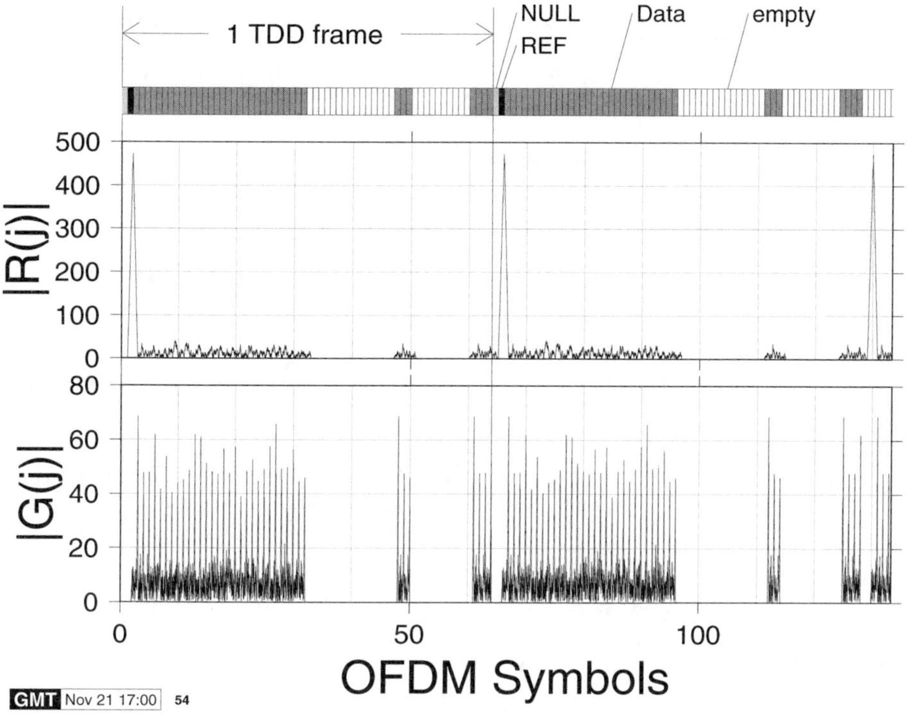

Figure 21.12: Time synchronisation: plots of the correlation terms $R(j)$ and $G(j)$ from Equation 21.9 and 21.10 for two consecutive 64-slot TDD frames under perfect channel conditions. The peaks indicate the correct TDD frame and OFDM symbol synchronisation instants, respectively

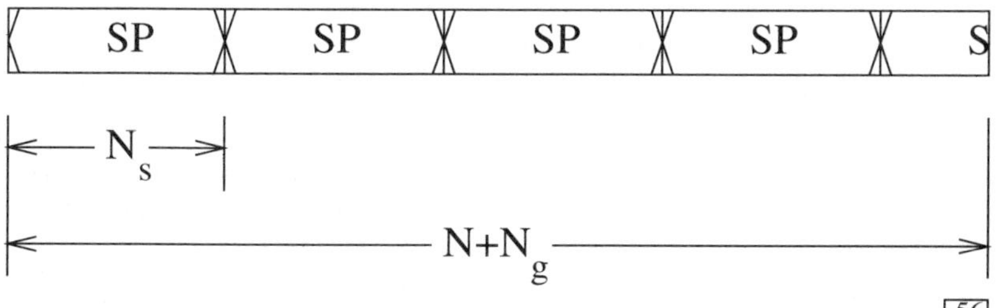

Figure 21.13: Reference symbol consisting of consecutive copies of a synchronisation pattern (SP) in the time domain

symbol synchronisation. A detailed discussion of this figure will be provided during our further discourse. Let us begin with the macroscopic structure.

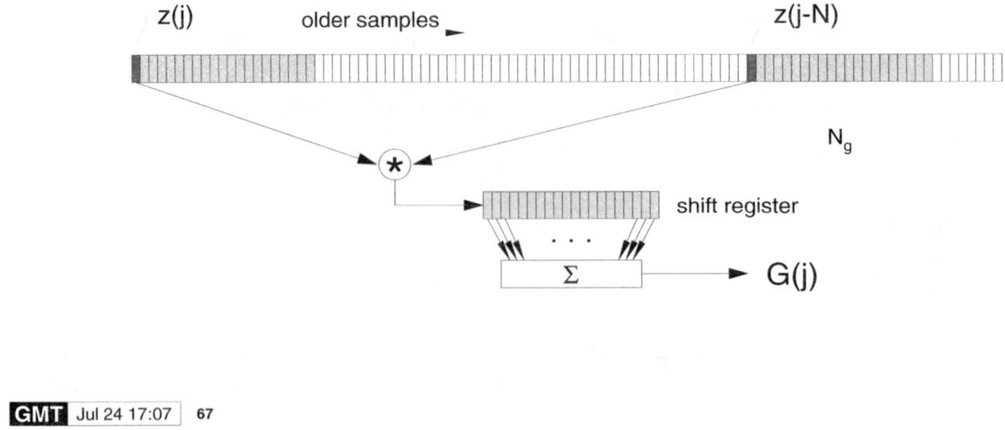

Figure 21.14: Schematic plot of the computation of the correlation function $G(j)$. The grey area represents the memory of the shift register

21.2.6.2 The correlation functions

The synchronisation algorithms rely on the evaluation of the following correlation functions $G(j)$ and $R(j)$, where j is the index of the most recent input sample:

$$G(j) = \sum_{m=0}^{N_g-1} z(j-m) \cdot z(j-m-N)^* \tag{21.9}$$

$$R(j) = \sum_{m=0}^{N+N_g-N_s-1} z(j-m) \cdot z(j-m-N_s)^*, \tag{21.10}$$

where $z(j)$ are the received complex signal samples, N is the number of subcarriers per OFDM symbol, N_g is the length of the cyclic extension and N_s is the periodicity within the reference symbol, as seen in Figure 21.13. The asterisk * denotes the conjugate of a complex value.

$G(j)$ is used for both frequency tracking and OFDM symbol synchronisation, expressing the correlation between two sequences of N_g samples length spaced by N in the received sample stream, as shown in Figure 21.14. The second function, $R(j)$, is the corresponding expression for the reference symbol, where the period of the repetitive synchronisation pattern is N_s, as seen in Equation 21.10 and Figure 21.13. In this case, $(N_g + N - N_s)$ samples are taken into account for the correlation computation, and they are spaced by a distance of N_s samples. Having defined the necessary correlation functions for quantifying the time and frequency synchronisation error, let us now concentrate on how the synchronisation algorithms rely on their evaluation.

21.2.7 Frequency tracking and OFDM symbol synchronisation

In this section we consider details of the frequency tracking and OFDM symbol synchronisation algorithms, which make use of $G(j)$, as defined by Equation 21.9.

21.2.7.1 OFDM symbol synchronisation

The magnitude of $G(j_{max})$ is maximum, if $z(j_{max})$ is the last sample of the current OFDM symbol, since then the guard samples constituting the cyclic extension and their copies in the current OFDM symbol are perfectly aligned in the summation windows. Figure 21.12 shows the simulated magnitude plots of $G(j)$ and $R(j)$ for two consecutive WATM frames, with $N = 512$ and $N_g = N_s = 50$, under perfect channel conditions. The observed correlation peaks of $|G(j)|$ can be easily identified at the last sample of an OFDM symbol. The amplitudes of the correlation peaks fluctuate, since the transmitted OFDM data symbols differ. The correlation peak magnitude is equal to the energy contained in the N_g samples of the cyclic extension and averages 50 for our system with $N_g = 50$ and an average sample power of unity.

The simulated accuracy of the OFDM symbol synchronisation in an AWGN channel is characterised in Figure 21.15. Observe in the figure that for SNRs in excess of about 7 dB the histogram is tightly concentrated around the perfect estimate, typically resulting in OFDM symbol timing estimation errors below $\pm 20T_s$. However, as even slightly misaligned time domain FFT windows cause phase errors in the frequency domain, this estimation accuracy is insufficient. In order to improve the OFDM symbol timing synchronisation, the estimates must be lowpass filtered. Let us now concentrate on the issues of fine frequency tracking.

21.2.7.2 Frequency tracking

A carrier frequency error of δf results in an evolving phase error $\Psi(j)$ of the received samples $z(j)$:

$$\begin{aligned} \Psi(\delta f, j) &= 2\pi\delta f \cdot j \cdot T_s && (21.11) \\ &= 2\pi \frac{j\delta f}{N\Delta f}. && (21.12) \end{aligned}$$

Clearly, the phase error difference between two received time domain samples $z(j_1)$ and $z(j_2)$ is a function of the frequency error and their time delay, and is given by $\Psi(\delta f, j_2) - \Psi(\delta f, j_1) = \Psi(\delta f, |j_2 - j_1|)$. If the original phase difference between the two received symbols $z(j_1)$ and $z(j_2)$ is known, and all other phase distortion is absent, then the phase difference error can be used to determine the frequency error δf.

Since the time domain samples of the cyclic extension or the guard interval are known to be a copy of the last N_g data samples of the OFDM symbol, the frequency error can be estimated using each of these N_g pairs of identical samples. In order to improve the estimation accuracy when exposed to noise and other channel impairments, averaging can be carried out over the N_g estimates.

The phase of $G(j)$ at $j = j_{max}$ equals the averaged phase shift between the guard time samples and the corresponding data samples of the current OFDM symbol. Since

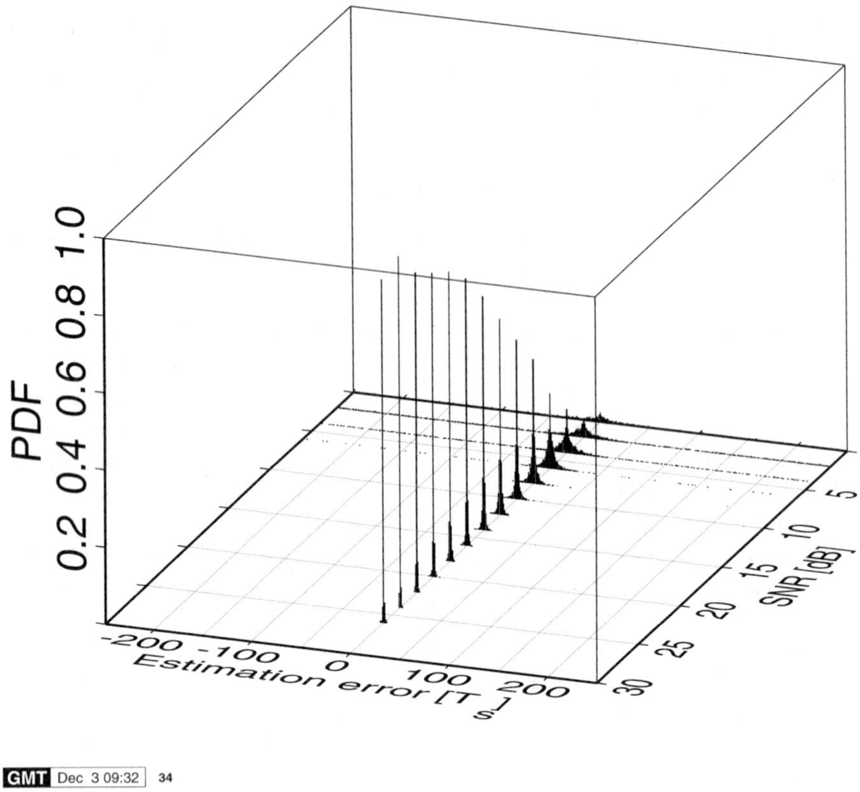

Figure 21.15: Histogram of the symbol timing estimation errors normalised to the sample interval T_s in an AWGN channel for $N = 512$ and $N_g = 50$ with no lowpass filtering of the estimates

the corresponding sample pairs are spaced by N samples, rearranging Equation 21.12 leads to the fine frequency error estimation δf_t given by:

$$\delta f_t = \frac{\Delta f}{2\pi} \cdot \measuredangle G(j_{max}). \tag{21.13}$$

Because of the 2π ambiguity of the phase, the frequency error must be smaller than $\Delta f/2$. Therefore, the initial frequency acquisition must ensure a rough frequency error estimate with an accuracy of better than $\Delta f/2$, if the proposed fine frequency tracking is used.

Assuming perfect estimation of the position j_{max} of the correlation peak $G(j_{max})$, as we have seen in Figure 21.12, the performance of the fine frequency error estimation in an AWGN environment is shown in Figure 21.17(a). Observe that for AWGN SNR values above 10 dB, the estimation error histogram is concentrated to errors below about $0.02\Delta f$, where Δf is the subcarrier spacing.

Having resolved the issues of OFDM symbol synchronisation and fine frequency tracking, let us now focus our attention on the aspects of frequency acquisition and frame synchronisation.

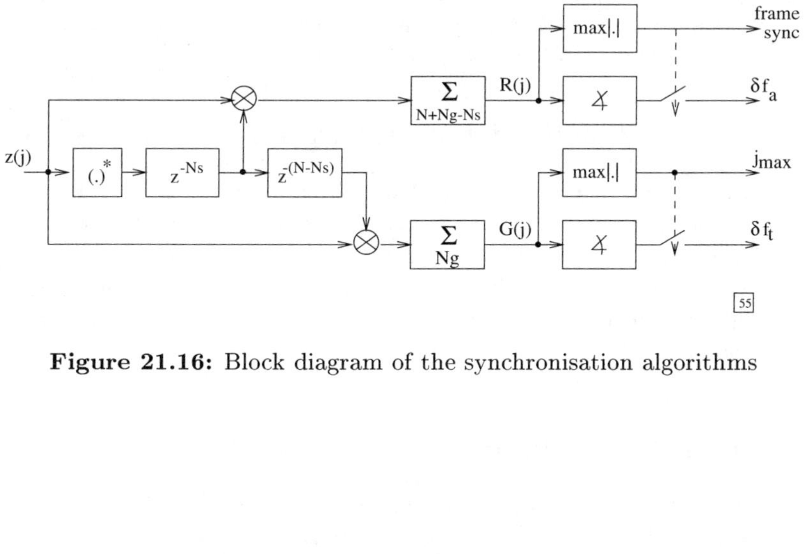

Figure 21.16: Block diagram of the synchronisation algorithms

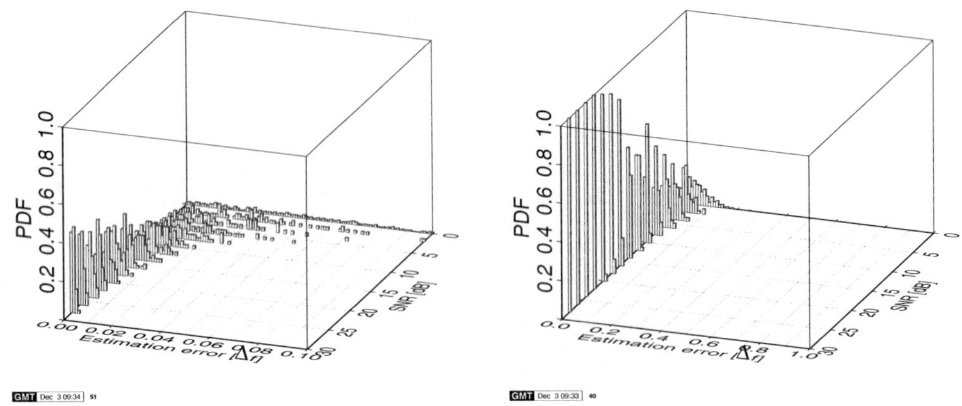

(a) frequency tracking **(b)** frequency acquisition

Figure 21.17: Histogram of the simulated frequency estimation error for the frequency acquisition and frequency tracking algorithms using $N = 512$ and $N_s = N_g = 50$ over AWGN channels

21.2.8 Frequency acquisition and frame synchronisation

Our proposed frequency acquisition and frame synchronisation techniques are based on the same algorithms as the fine frequency and OFDM symbol synchronisation. However, instead of using the medium-term periodicity of the cyclic extension of the OFDM data symbols, the dedicated reference symbol with shorter cyclic period $N_s < N$ is exploited in order to improve the frequency capture range, as it will be highlighted below.

21.2.8.1 Frame synchronisation

Similarly to the OFDM symbol synchronisation, the magnitude of $R(j)$ in Equation 21.10 and Figure 21.12 is maximum when the periodic synchronisation segments SP of length N_s of the reference symbol shown in Figure 21.13 perfectly overlap. Again, the magnitude of $R(j)$ for two simulated TDD frames is shown at the top of Figure 21.12. The OFDM frame timing is synchronised with the peak of $R(j)$, which can additionally be taken into account for the OFDM symbol synchronisation. The peak height is constant under perfect channel conditions, owing to the fixed reference symbol.

21.2.8.2 Frequency acquisition

The frequency acquisition algorithm uses the same principle as the frequency tracking scheme of Section 21.2.7.2. Specifically, the phase of $R(j)$ at the last sample of the reference symbol j_{max} contains information on the frequency error:

$$\measuredangle R(j_{max}) = 2\pi \cdot \delta f_a \cdot N_s \cdot T_s = 2\pi \cdot \delta f_a \cdot \frac{N_s}{N \cdot \Delta f} \tag{21.14}$$

leading to

$$\delta f_a = \frac{N}{N_s}\frac{\Delta f}{2\pi} \cdot \measuredangle R(j_{max}). \tag{21.15}$$

Since the spacing between the sample pairs used in the computation of $R(j)$ is smaller than in the case of $G(j)$, which was used for the frequency tracking ($N_s < N$), the maximum detectable frequency error is now increased from $\Delta f/2$ to $N/N_s \cdot \Delta f/2$, where Δf is the subcarrier spacing of the OFDM symbols. The simulated performance of the frequency acquisition algorithm for $N = 512$, $N_s = 50$ and perfect time synchronisation is shown in Figure 21.17(b) for transmissions over an AWGN channel. As seen in the figure, the scheme maintains an acquisition error below $\Delta f/2$ even for SNR values down to 0 dB, which exceeds the system's operational specifications.

21.2.8.3 Block diagram of the synchronisation algorithms

To summarise our previous elaborations, Figure 21.16 shows the detailed block diagram of the synchronisation algorithms. The received samples are multiplied with the complex conjugate of the delayed input sequences and summed up over $(N+N_g-N_s)$ and N_g samples, respectively. The magnitude maxima of the two sequences $G(j)$ and $R(j)$ are detected and they trigger the sampling of the phase estimates $\measuredangle G$ and $\measuredangle R$ in order to derive the two frequency error estimates δf_a and δf_t. Let us now describe an alternative frequency synchronisation technique, which can be used as a benchmark in order to assess the potential of our previously described technique.

21.2.9 Frequency acquisition using frequency-domain pilots

The algorithm described in this section has been proposed by Mandarini and Falaschi [418]. In contrast to the reference symbol of Figure 21.13, the technique advocated by Mandarini and Falaschi relies on the reference symbol shown in Figure 21.18 containing a set of pilot tones, which is transmitted once per TDD frame. The reference symbol is processed in the frequency domain, after demodulation by FFT.

The idea of the pilot tone algorithm is that a frequency mismatch between the transmitter and the receiver will result in a shift of the pilot tones in the received spectrum. This shift is measured at the receiver by searching for the FFT frequency bin with the maximum amplitude, resulting in an estimation accuracy of half a subcarrier spacing. This estimation is improved further in a second stage by considering the relative amplitudes of the adjacent FFT bins.

21.2.9.1 The reference symbol

The reference symbol is transmitted once at the beginning of every TDD frame, as depicted in Figure 21.12. This reference symbol consists of a set of M pilot tones spanning the OFDM signal's bandwidth, spaced from each other by ΔN subcarriers or by a frequency gap of $\Delta F = \Delta N \cdot \Delta f$. Accordingly, the maximum detectable frequency error is $\Delta F/2 = \Delta N/2 \cdot \Delta f$ and hence the higher the number of pilots, the lower the frequency capture range. A stylised plot of the reference symbol in the frequency domain is shown in Figure 21.18.

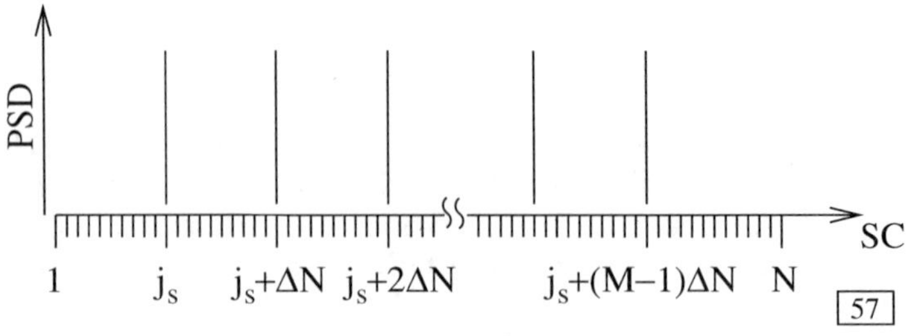

Figure 21.18: Stylised plot of the power spectral density (PSD) of the pilot-tone-based reference symbol using M pilot tones in the N subcarrier (SC) spectrum, spaced by $(\Delta N - 1)$ blank subcarriers

In order to keep the overall OFDM symbol energy constant, the total energy of the reference symbol is set equal to the average OFDM symbol energy and split equally between the M pilot tones.

21.2.9.2 Frequency acquisition

The frequency acquisition algorithm estimates the frequency error by searching for the position of the pilot tones in the spectrum of the received reference symbol. In order to accomplish this, the received time domain signal of the reference symbol is demodulated by FFT. In order to minimise the influence of noise, the sum of the received power spectral amplitudes over all the M frequency ranges depicted in Figure 21.18 is then calculated as follows:

$$V(j) = \sum_{m=0}^{M-1} |Y(j + j_s + m\Delta N)|^2 \quad \text{for} \quad -\frac{\Delta N - 1}{2} \leq j \leq \frac{\Delta N - 1}{2}, \tag{21.16}$$

where $Y(j)$ represents the frequency domain samples of the received demodulated reference symbol. Observe that for the sample position identified by j $V(j)$ corresponds to the superposition of the powers of the ΔN-spaced $Y(j)$ samples, one from each of the M frequency ranges. In Figure 21.19 simulated values for both $|Y(j)|$ and $V(j)$ are given for SNR values of 0 dB and 50 dB, respectively, for a 10-pilot reference symbol in a 512-subcarrier OFDM system with $\delta f = 0.3\Delta f$ frequency error. It can be seen that the positions of the pilot tones in the received spectrum are easily determined even for SNR values of 0 dB, thanks to the high power of the pilot tones relative to the average data symbol power. The total energy of the reference is 512, leading to a pilot tone amplitude of $\sqrt{512/10} = 7.15$. Because of the subcarrier energy spillover caused by the frequency error, the amplitude of the peaks of $Y(j)$ in the figure is lowered to about 6.2. Accordingly, the peak value of $V(j)$ is about 390 instead of 512, with the rest of the received energy located in the adjacent FFT bins.

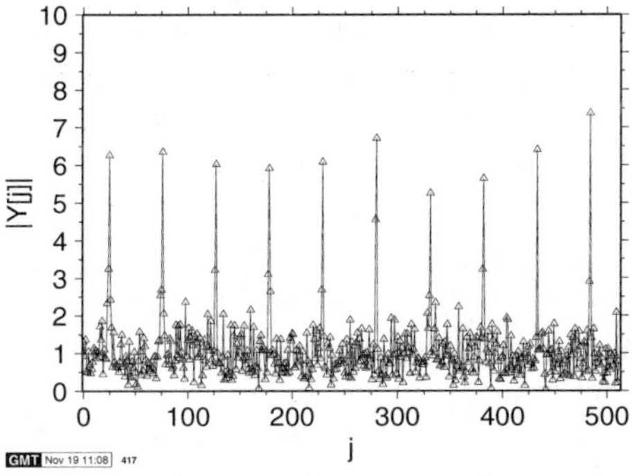

(a) $|Y(j)|$ for $\gamma = 0$dB

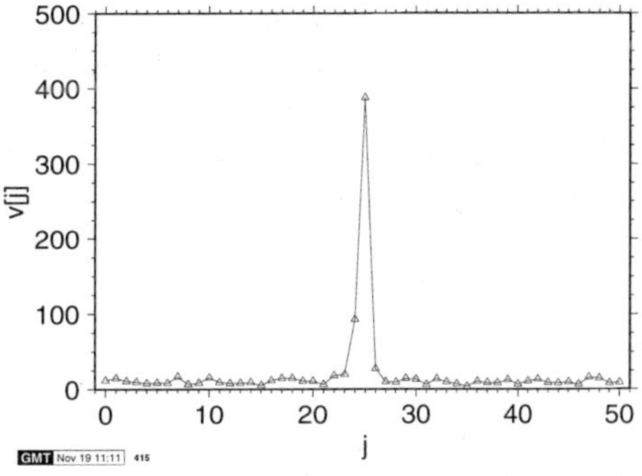

(b) $V(j)$ for $\gamma = 0$dB

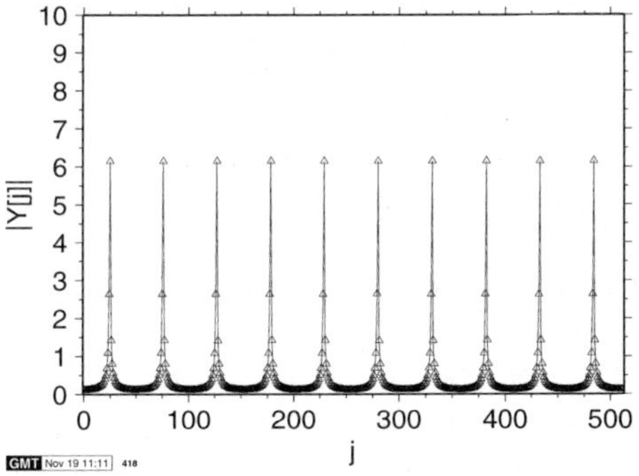

(c) $|Y(j)|$ for $\gamma = 50$dB

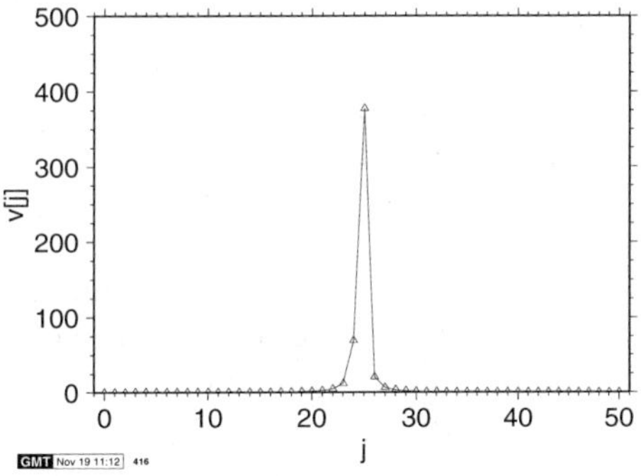

(d) $V(j)$ for $\gamma = 50$dB

Figure 21.19: Magnitude of the received frequency domain samples $|Y(j)|$ and averaged samples $V(j)$ for SNR values of $\gamma = 0$ dB and $\gamma = 50$ dB for a 512-subcarrier OFDM modem employing 10 pilot tone subcarriers in the reference symbol. The simulated frequency mismatch between the transmitter and the receiver was $\delta f = 0.3\Delta f$

The frequency error is determined by estimating the position of the highest peak in $V(j)$ over the range of j given in Equation 21.16. First, the index j_{max} of the maximum value of $V(j)$ is found. This gives a rough estimate of the frequency error with a frequency resolution of the subcarrier spacing Δf. As can be observed in Figure 21.19, this rough frequency error estimation is very noise resilient, resulting in a reliable detection of the frequency error. The accuracy of this estimation, however, is insufficient for low BER modem operation. If a subsequent frequency error tracking algorithm like the proposed time domain correlation-based approach is employed in the system, then the $0.5\Delta f$ estimation accuracy can be an adequate starting point for the tracking algorithm. If a better estimate is needed, however, then the amplitudes in the neighbouring bins around the peak value in $V(j)$ can be exploited to refine the first estimate, as follows.

Since the frequency error is not generally an integer multiple of the resolution Δf of the FFT, a better frequency error estimate can be derived by determining the position of the peak value of $V(j)$ around j_{max}. The proposed algorithm exploits the amplitude of $V(j)$ at $j = j_{max}$ and the two adjacent values $V(j_{max} \pm 1)$ in order to derive the following quantities:

$$\rho_1 = \frac{\sqrt{V(j_{max}+1)}}{\sqrt{V(j_{max})}} \tag{21.17}$$

$$\rho_{-1} = \frac{\sqrt{V(j_{max}-1)}}{\sqrt{V(j_{max})}}, \tag{21.18}$$

which represent the normalised frequency domain pilot values at the positions adjacent to the $V(j)$ peak at $j = j_{max}$. Their difference $(\rho_1 - \rho_{-1})$ is computed and the rough frequency error estimate of $j_{max} \cdot \Delta f$ is corrected by the help of the estimated deviation d as follows:

$$d = \frac{\rho_1 - \rho_{-1}}{2} \tag{21.19}$$

$$\delta f = \Delta f \cdot (j_{max} + \text{sgn}\{d\} \cdot \sqrt{|d|}), \tag{21.20}$$

where the value of $\sqrt{|d|}$ is the estimation of the true peak position between the peak $V(j_{max})$ and the higher of the two adjacent values of $V(j_{max} \pm 1)$.

Clearly, the number of pilot tones included in the reference symbol, M, determines the maximal frequency deviation that can be detected using this algorithm, which is given by $\delta f_{max} = \Delta f \cdot \lfloor N/2M \rfloor$, where $\lfloor x \rfloor$ means the highest integer number smaller or equal to x. Not only the frequency capture range, but also the peak-to-average power ratio or crest factor (CF) of the reference symbol in the time domain depends on the number of pilot tones used. If the phase of all the M pilot tones is equal, then the upper bound for the resulting crest factor equals M, which would be the result of all M sinusoids adding constructively to a peak value. Experimental results in Figure 21.21(b) show that the actually measured CF for pilot symbols with a varying number of pilot tones is very close to the stated upper bound. For comparison, the histogram of the simulated CF values for pseudorandom OFDM data symbols is shown in Figure 21.21(b). This PDF shows that the average CF value for a 512-subcarrier 4-QAM OFDM system is about 8.5 dB, although peak values up to 12 dB have been occasionally observed. For small numbers of pilot tones, the crest factor of the reference symbol is therefore not worse than that of the average data symbols.

21.2.9.3 Performance of the pilot based frequency acquisition in AWGN Channels

A series of simulations was performed in order to characterise the performance of the pilot tone based frequency acquisition algorithm in an AWGN environment. For different fixed frequency errors and noise levels, the frequency error estimation was invoked and the difference between the estimated and the actual error was recorded.

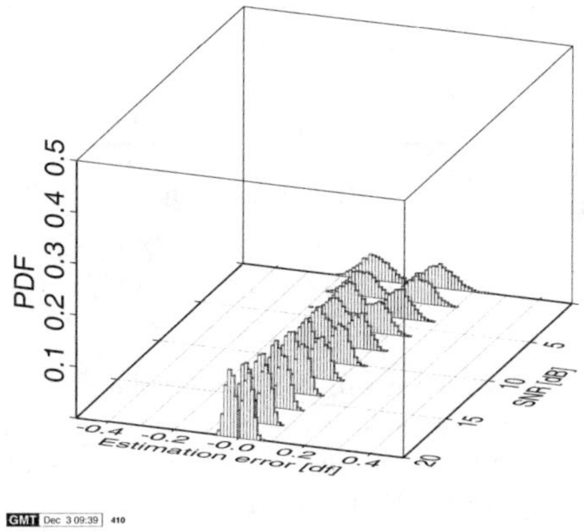

(a) 10 pilots

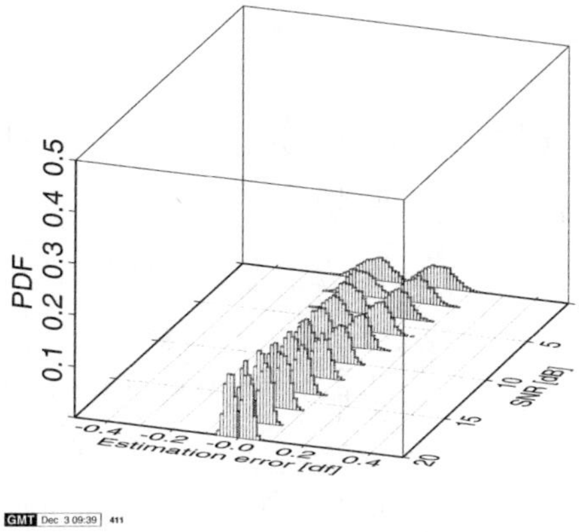

(b) 20 pilots

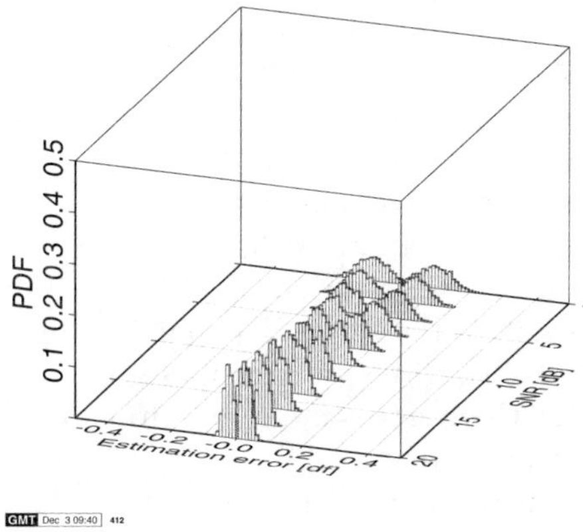

(c) 30 pilots

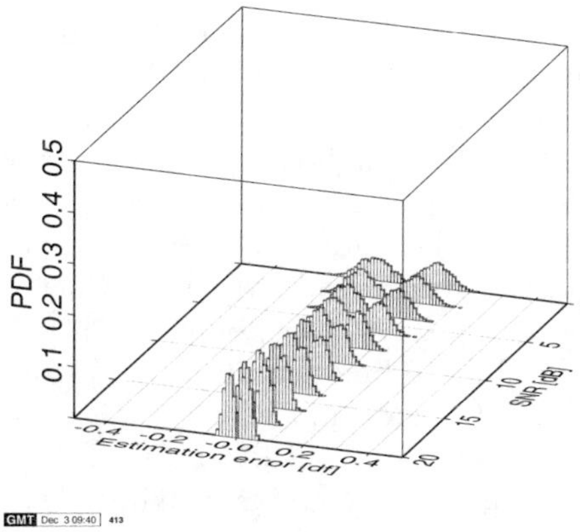

(d) 40 pilots

Figure 21.20: Histograms of simulated frequency mismatch estimation error for a 512-subcarrier OFDM modem in a narrowband AWGN channel for 10, 20, 30 and 40 pilot tones. The simulated frequency error was $\delta f = 0.0$

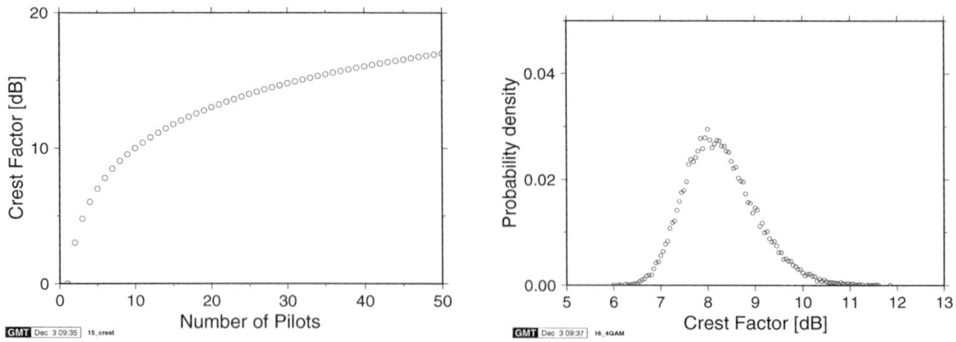

(a) reference symbol

(b) data symbols

Figure 21.21: Characterisation of the coarse frequency synchronisation using the reference symbol of Figure 21.18: (a) crest factor versus number of pilot tones for non-optimised reference symbols, (b) histogram of simulated data symbol crest factors for a 512-subcarrier, 4-QAM OFDM system

Figure 21.20 shows the histograms of the frequency mismatch estimation errors for 10, 20, 30 and 40 pilots and for a range of SNR values in the case of an actual frequency error of $\delta f = 0.0$.

Inspection of the figure reveals that the estimation accuracy is not sensitive to the number of pilot tones used in the reference symbol; in fact, the differences in the simulated histograms are hardly visible. The estimation accuracy is significantly better than $0.5\Delta f$ throughout the simulated SNR range and hence meets the requirements of the subsequent fine frequency synchronisation for all simulated signal-to-noise ratios. It can be observed, however, that there is a bifurcation of the estimation error histogram which leads to two distinct probability maxima on both sides of the true value, with a low probability for a perfect estimation. This behaviour can be explained by the fine peak location estimation, which exhibits a high noise sensitivity for low δf values and hence small differences in the noise samples on either side of the correlation peak lead to considerable etimation errors.

The gap between the maxima closes with increasing SNR, but even at 20 dB there are frequency estimation errors of up to 8% of the subcarrier spacing, which underlines the need for subsequent fine frequency synchronisation. Alternatively, averaging the estimated frequency errors for several estimations would improve the estimation accuracy, but it would slow down the system's response time to frequency deviations. The histogram is symmetric around the perfect estimation value, hence a longer-term averaging would yield the correct frequency error estimate.

The estimation accuracy of the pilot tone frequency acquisition algorithm depends on the actual frequency error. In order to illustrate this, frequency estimation error histograms for a given frequency mismatch of $\delta f = 0.3\Delta f$ are shown in Figure 21.22, and they exhibit rather different characteristics from the previous case with $\delta f = 0$. The histograms are offset with respect to the perfect estimation value, hence averaging over multiple estimations will result in a residual estimation error. Subsequent fine

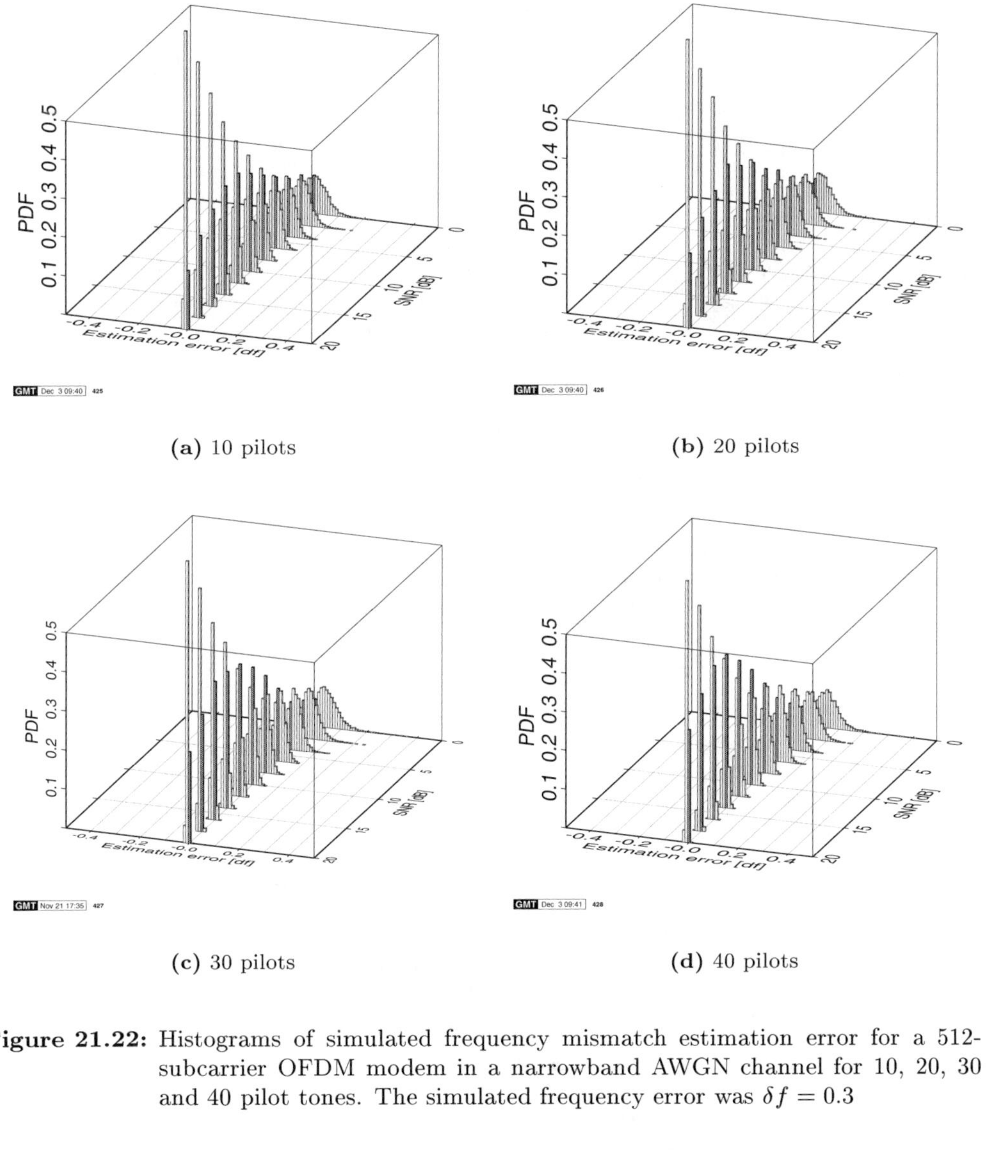

(a) 10 pilots

(b) 20 pilots

(c) 30 pilots

(d) 40 pilots

Figure 21.22: Histograms of simulated frequency mismatch estimation error for a 512-subcarrier OFDM modem in a narrowband AWGN channel for 10, 20, 30 and 40 pilot tones. The simulated frequency error was $\delta f = 0.3$

frequency sychronisation is therefore necessary for low SNR values . For low noise levels, however, the estimation accuracy is greatly improved with respect to the earlier case. For SNRs in excess of 10 dB, the estimation errors do not exceed 5% of the subcarrier spacing, and will therefore not affect the system performance to a great extent. It can be observed that the asymmetry of the estimation errors is dependent on the number of pilots employed.

The frequency deviation estimation error histogram for $\delta f = 0.5\Delta f$ is shown in Figure 21.23, where it can be observed that in this case the histogram is split into two distinct parts, and that each of the groups is centred at 8% of the subcarrier distance

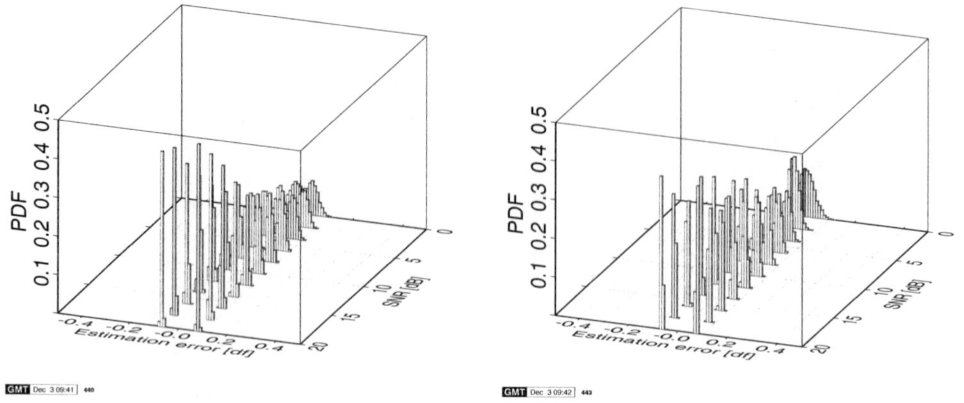

(a) $\delta f = 0.5\Delta f$, 10 pilots (b) $\delta f = 0.5\Delta f$, 40 pilots

Figure 21.23: Histograms of simulated frequency mismatch estimation error for a 512-subcarrier OFDM modem in a narrowband AWGN channel for 10 and 40 pilot tones. The simulated frequency error was $\delta f = 0.5\Delta f$

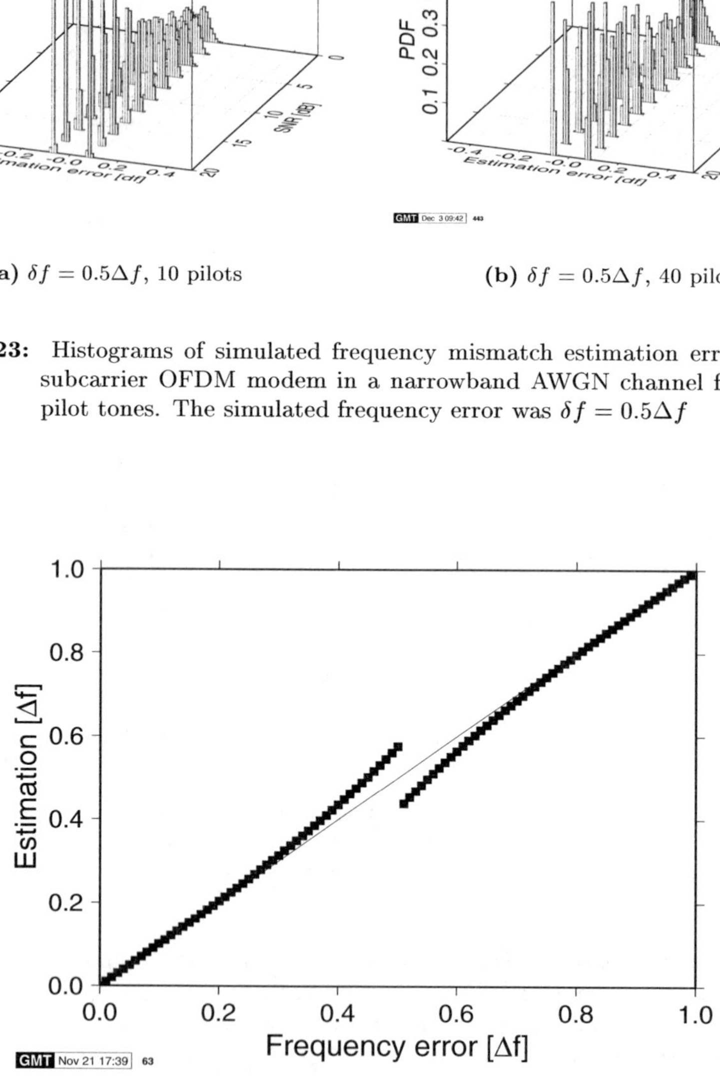

Figure 21.24: Estimated versus actual frequency error for the pilot-tone-based coarse frequency synchronisation algorithm in a perfect noiseless channel

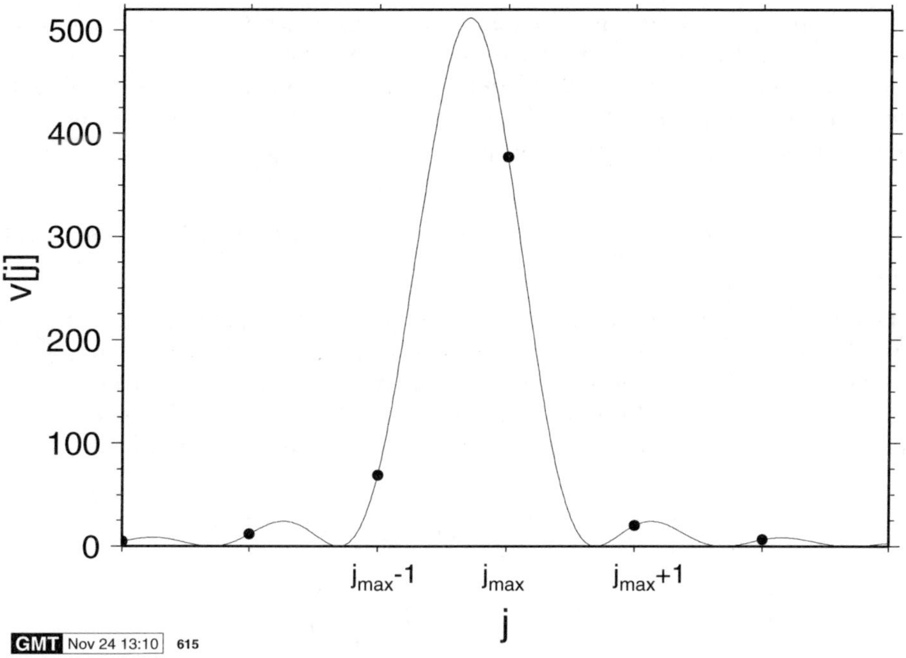

Figure 21.25: Detail from Figure 21.19(d): $V(j)$ for an SNR value of $\gamma = 50$ dB for a 512-subcarrier OFDM modem employing 10 pilot tones in the reference symbol. The simulated frequency mismatch is $0.3\Delta f$. The symbols represent the simulated $V(j)$ values, while the continuous line corresponds to the sinc^2 approximation of Equation 21.22

above and below the correct value. Again, fine frequency synchronisation would be necessary to ensure optimal system performance.

In order to evaluate the residual estimation errors, Figure 21.24 depicts the estimated versus the actual frequency error for a perfect, noiseless channel. For frequency errors close to an integer multiple of Δf the estimation accuracy is very good, but values close to $(n + 1/2)\Delta f$ are estimated with errors of up to 8% of the subcarrier spacing. This observation is in accordance with the results of the above experiments, which showed a residual estimation error of about 1% of the carrier separation at $\delta f = 0.3\Delta f$ and of $\pm 8\%$ at $\delta f = 0.5\Delta f$.

21.2.9.4 Alternative frequency error estimation for frequency-domain pilot tones

In order to improve the frequency error dependent performance of the frequency domain pilot-based frequency synchronisation algorithm outlined above, an alternative peak position estimation has been investigated in order to enhance the algorithm based on Equation 21.20. As we have seen earlier, the continuous spectrum of each OFDM subcarrier follows a sinc function centred around the subcarrier frequency. In the presence of a frequency error between the transmitter and receiver, the receiver's sampling raster in the frequency domain is not aligned with the received spectrum's maximum and nulls, but shifted by the frequency error δf.

The spectrum of the received reference symbol – which contains M pilot tones at the frequencies $(j_s + m\Delta N) \cdot \Delta f$ with $0 \le m \le (M-1)$ – in the presence of a frequency error δf in a noiseless environment can be expressed as:

$$Y(j) = \sqrt{\frac{N}{M}} \cdot \sum_{m=0}^{M-1} \operatorname{sinc}\left(j - j_s - m\Delta N + \frac{\delta f}{\Delta f}\right). \tag{21.21}$$

The factor $\sqrt{N/M}$ is the amplitude of each pilot tone, ensuring that the overall energy of the reference symbol is equal to the average OFDM symbol energy. If the frequency distance ΔN between two consecutive pilot tones is sufficiently large so that the received spectra of the different pilot tones do not significantly overlap, then the vector $V(j)$, as defined by Equation 21.16, can be approximated as:

$$V(j) = \sum_{m-0}^{M-1} |Y(j + j_s + m\Delta N)|^2 \tag{21.22}$$

$$\approx N \cdot \operatorname{sinc}^2\left(j + \frac{\delta f}{\Delta f}\right). \tag{21.23}$$

Figure 21.25 shows the simulated values of $V(j)$ from Figure 21.19(d) in greater detail around the peak, along with the continuous line corresponding to the sinc approximation of Equation 21.22. It can be seen that there is a very good correspondence between the simulated and the approximated values for the pilot spacing of $M = 50$ employed in this case.

Using the sinc^2 approximation of Equation 21.22, the values for $V(j_{max})$, $V(j_{max}+1)$ and $V(j_{max}-1)$ can be expressed in terms of the normalised fine frequency error estimation $\nu = \delta f/\Delta f - j_{max}$ as follows:

$$V(j_{max}) \approx N\operatorname{sinc}^2(\nu), \tag{21.24}$$

$$V(j_{max}+1) \approx N\operatorname{sinc}^2(1+\nu), \quad \text{and} \tag{21.25}$$

$$V(j_{max}-1) \approx N\operatorname{sinc}^2(-1+\nu). \tag{21.26}$$

Then the following normalised terms can be defined:

$$\rho_1 = \frac{\sqrt{V(j_{max}+1)}}{\sqrt{V(j_{max})}} \tag{21.27}$$

$$\approx \frac{\sqrt{N}\,|\sin(\pi(1+\nu))|}{|\pi(1+\nu)|} \cdot \frac{|\pi\nu|}{\sqrt{N}\,|\sin(\pi\nu)|} \tag{21.28}$$

$$= \frac{|\nu|}{|1+\nu|} \quad \text{and} \tag{21.29}$$

$$\rho_{-1} = \frac{\sqrt{V(j_{max}-1)}}{\sqrt{V(j_{max})}} \tag{21.30}$$

$$\approx \frac{\sqrt{N}\,|\sin(\pi(-1+\nu))|}{|\pi(-1+\nu)|} \cdot \frac{|\pi\nu|}{\sqrt{N}\,|\sin(\pi\nu)|} \tag{21.31}$$

$$= \frac{|\nu|}{|-1+\nu|}. \tag{21.32}$$

The value d, defined as half the difference between ρ_1 and ρ_{-1} in Equation 21.19, can therefore be approximated as:

$$d = \frac{\rho_1 - \rho{-1}}{2} \approx \frac{1}{2}\left(\frac{|\nu|}{|1+\nu|} - \frac{|\nu|}{|-1+\nu|}\right). \tag{21.33}$$

Solving Equation 21.33 for ν values smaller than one subcarrier distance yields:

$$\nu = \begin{cases} -\sqrt{\frac{d}{d+1}} & \text{for} \quad -1 < \nu < 0 \quad (\text{or} \quad (d > 0) \\ \sqrt{\frac{d}{d-1}} & \text{for} \quad 0 \leq \nu < 1 \quad (\text{or} \quad d \leq 0) \end{cases}. \tag{21.34}$$

An alternative pilot-based frequency synchronisation algorithm can therefore be contrived by calculating the peak position estimate ν from d, as defined in Equation 21.19, using Equation 21.34 and replacing the δf estimate in Equation 21.20 by:

$$\delta f = \Delta f \cdot (j_{max} + \nu). \tag{21.35}$$

A series of simulations were conducted in order to investigate the performance of this modified peak position estimation algorithm in noisy conditions. All the investigations have been performed for a 512-subcarrier system employing a 10-pilot reference symbol in a narrowband Gaussian white noise channel.

Histograms of the estimation errors for fixed frequency errors of $\delta f = 0$, $\delta f = 0.3\Delta f$ as well as for $\delta f = 0.5\Delta f$ are given in Figure 21.26. In all cases, the estimation accuracy was better than $0.5\Delta f$, therefore allowing the subsequent use of the OFDM data symbol based tracking algorithm of Section 21.2.7. Comparison of Figure 21.26(a) with the corresponding results for the original peak position estimation algorithm in Figure 21.20(a) reveals a similar performance for both algorithms. This was expected, since the modified estimation terms $\sqrt{d/(d-1)}$ and $-\sqrt{d/(d+1)}$ are close to the term $\sqrt{d}$ in Equation 21.20 of the original algorithm for small values of d.

For a frequency error of $\delta f = 0.3\Delta f$, as depicted in Figure 21.26(b), the estimation accuracy was dramatically improved, when compared to the case of $\delta f = 0$. This was in accordance with the results achieved using the original peak position estimation, as shown in Figure 21.22(a). Although both schemes achieved accurate estimation results, the modified algorithm of Equation 21.35 exhibited a lower estimation bias and it was slightly more accurate than the original algorithm of Equation 21.20.

Imposing the third simulated frequency error of $\delta f = 0.5\Delta f$ revealed the different behaviour of the two investigated algorithms. While the original version of the algorithm exhibited a split histogram with systematic estimation errors of about 8% of the subcarrier distance, as can be seen in Figure 21.23(a), the modified algorithm delivered a significantly more accurate frequency error estimate. The estimation accuracy of the modified algorithm in a noiseless environment is demonstrated in Figure

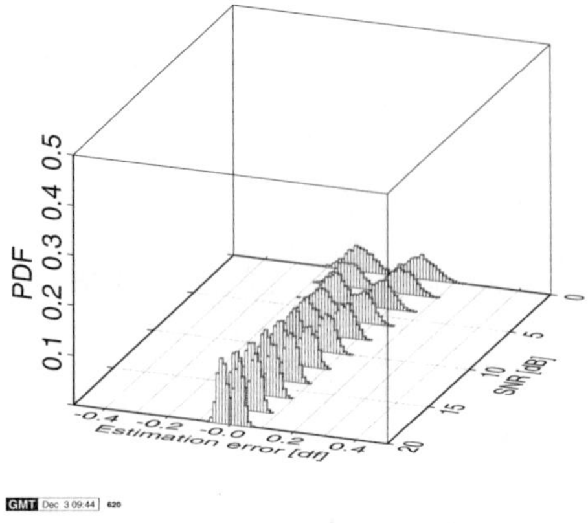

(a) $\delta f = 0$

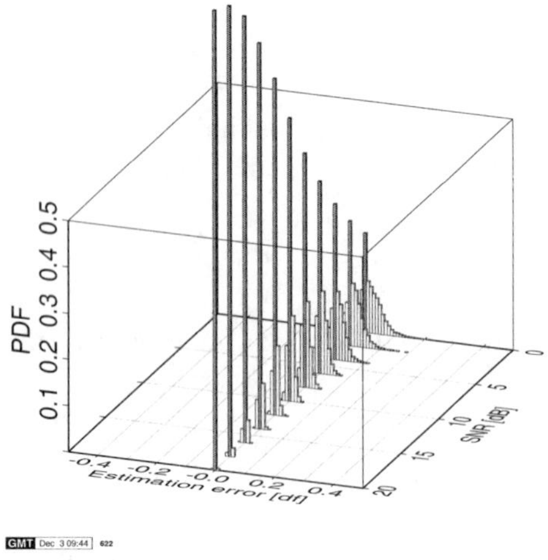

(b) $\delta f = 0.3\Delta f$

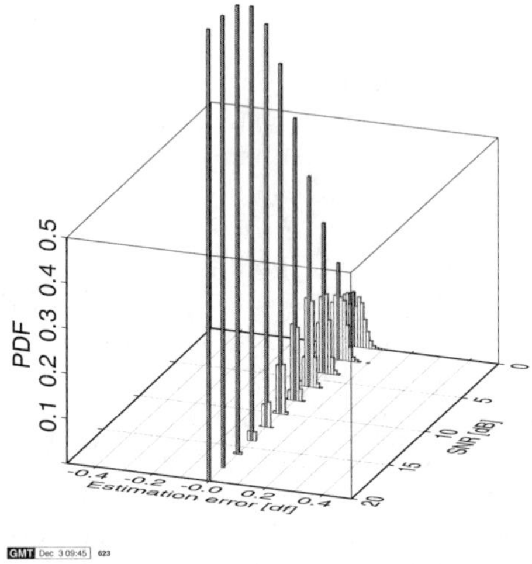

(c) $\delta f = 0.5\Delta f$

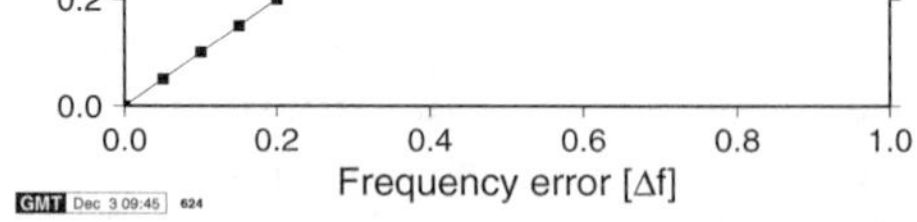

(d) residual accuracy

Figure 21.26: (a, b, c): Histograms of the simulated frequency mismatch estimation error for a 512-subcarrier OFDM modem in a narrowband AWGN channel for 10 pilot tones employing the alternative peak position estimation algorithm. The simulated frequency error was $\delta f = 0.3$. (d) Estimated versus actual frequency error for pilot-tone-based frequency synchronisation employing the alternative peak position estimation algorithm

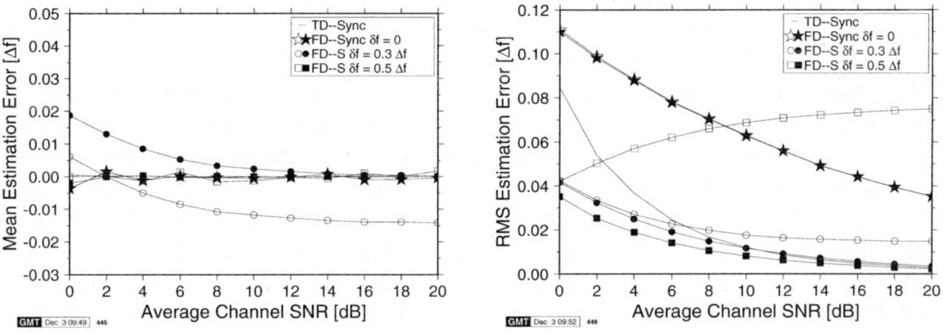

(a) mean error　　　　**(b)** RMS

Figure 21.27: Comparison of the frequency acquisition algorithms over AWGN channels: (a) mean estimation error, (b) RMS estimation error. TD = time domain estimation algorithm, FD = frequency domain estimation algorithm employing a 10-pilot reference symbol. The open symbols show the performance associated with the original peak position estimation of Equation 21.20, while the closed symbols signify the alternative peak position estimation algorithm of Equation 21.35 for given frequency errors δf of 0, 0.3 and 0.5 times the subcarrier distance Δf

21.26(d). There are no systematic estimation errors for the whole frequency error range that the peak position estimation is expected to handle.

21.3 Comparison of the Frequency Acquisition Algorithms

Both the frequency acquisition algorithms examined achieve the minimum requirements of an estimation accuracy better than half a subcarrier spacing for AWGN channel SNR levels down to 0 dB and hence ensure reliable operation of the subsequent fine frequency synchronisation algorithm.

There is, however, a difference in their absolute accuracy. While the pilot tone algoruthm of Section 21.2.9 suffers from residual estimation errors for certain ranges of frequency errors, the correlation-based algorithm of Section 21.2.5 delivers the same estimation accuracy throughout its frequency capture range. Values for the long-term mean and RMS estimation error are shown in Figure 21.27. The solid lines without markers in the two graphs are the mean and RMS error curves for the time domain correlation algorithm, which are valid for all frequency errors within the frequency capture range of the algorithm ($\pm 5.12\Delta f$ for $N = 512$ and $N_s = 50$). The curves with markers correspond to the pilot tone synchronisation system employing a 10-pilot reference symbol, for frequency errors of 0, 0.3, and 0.5 times the subcarrier distance Δf. For these frequency error values, curves are drawn for the original as well as for the modified peak position estimation algorithm. The performance of both the pilot tone algorithms is clearly dependent on the frequency error, both in terms of the mean as well as the RMS errors.

The mean estimation error curves in Figure 21.27(b) show the long-term averaged estimation errors for the different schemes. While none of the algorithms exhibits mean errors above 0.5% of the subcarrier distance for frequency deviations of 0 or 0.5% Δf over the SNR range investigated, frequency errors of $0.3\Delta f$ cause both the pilot tone estimation algorithms to deliver biased estimates. The amount of this mean estimation error depends on the SNR, and the noise level dependency of the estimation is the same for both the original as well as for the alternative peak search algorithm. The residual estimation error in the absence of noise is different for the two peak position estimators, however. While the alternative estimation algorithm delivers a non-biased estimation for high SNR values, the original algorithm exhibits an estimation error of about 1.5% of the subcarrier distance. This is in accordance with Figure 21.24, which shows the residual estimation errors for the original peak position estimation. It is interesting to note that the original peak search algorithm does not result in a biased non-zero mean error estimate for frequency errors of $\delta f = 0.5\Delta f$, although the instantaneous estimation errors at this point are at a maximum, as can be seen from Figures 21.24 and 21.23(a). This is because of the symmetry of the histogram, which will result in a non-biased mean estimation value. For all other frequency errors between 0 and $0.5\Delta f$, however, the histogram of estimation errors will be asymmetrical and the estimation will therefore be biased, as seen for example in Figure 21.22.

The modified peak search algorithm of Section 21.2.9.4, too, delivers biased estimations for frequency errors between 0 and $0.5\Delta f$, as seen in Figure 21.27. Since these are due to the noise affecting the estimation, there is no residual estimation error in noiseless environments. For SNR values in excess of 8 dB, the mean estimation error is smaller than 0.5% of the subcarrier distance for $\delta f = 0.3\Delta f$. The time domain frequency synchronisation algorithm of Section 21.2.5 does not exhibit any significant bias in the estimation error for the range of SNR values investigated. Therefore, long-term averaging of the time domain algorithm's estimates will result in correct frequency error estimation at the cost of reduced agility.

Figure 21.27(b) depicts the RMS estimation error curves for the investigated synchronisation algorithms for the same set of frequency errors. It can be seen that for both frequency domain pilot-based error estimation algorithms the estimation quality varies greatly with the frequency error to be estimated. The original peak position estimation algorithm exhibits residual RMS values of $0.015\Delta f$ and about $0.8\Delta f$ for frequency errors δf of $0.3\Delta f$ and $0.5\Delta f$, respectively. For $\delta f = 0.3\Delta f$ this corresponds to the bias of 1.5% of the subcarrier distance, which has been observed in Figure 21.27(a). For $\delta f = 0.5\Delta f$, this residual estimation error can be observed in the histogram, as shown in Figure 21.23(a). Although the long-term mean of the estimation is the correct value, each single estimation exhibits an error of about 8% of the subcarrier distance. This value corresponds to the residual level of RMS estimation error for $\delta f = 0.5\Delta f$ in Figure 21.27(b).

The frequency domain pilot-based algorithm employing the alternative peak position estimation of Section 21.2.9.4 does not exhibit residual RMS estimation errors.

The estimation accuracy does, however, vary with the frequency error to be estimated. For $\delta f = 0$, its performance is virtually the same as that of the original peak position estimation algorithm, with high estimation errors for the whole AWGN SNR range. At $\delta f = 0.3\Delta f$ and $0.5\Delta f$, the estimation accuracy is much better, at RMS estimation errors of around 1% of the subcarrier distance for SNR values in excess of 12 dB and 10 dB, respectively.

The time domain correlation-based algorithm's accuracy does not depend on the frequency error to be estimated. Its performance is comparable to that of the frequency domain algorithm employing the alternative peak search algorithm of Section 21.2.9.4 for frequency errors δf of $0.3\Delta f$ or $0.5\Delta f$ for SNR values above 10 dB, but it performs significantly better at $\delta f = 0$.

Comparing frequency error the estimation accuracy of the different algorithms it becomes clear that all the algorithms would satisfy the estimation accuracy of $0.5\Delta f$, necessary if subsequent fine frequency tracking mechanisms are used data symbol by data symbol. If no fine frequency tracking is to be employed, however, then the constraints on the estimation accuracy are significantly tighter. We have seen in Section 21.1.1.2 that, depending on the modulation scheme used, frequency errors of 5–10% of the subcarrier spacing lead to degradation of the system's SNR performance. In this case – and if no lowpass filtering of the estimated value is to be employed – the time domain correlation-based synchronisation algorithm of Section 21.2.5 would be the only applicable algorithm, since its accuracy is consistently high and its estimation is unbiased throughout the frequency error range.

If the pilot tone reference symbol based frequency synchronisation algorithms are to be employed without subsequent frequency tracking, then the estimation accuracy has to be improved using lowpass filtering techniques of the estimates. Let us now consider the system's BER performance when subjected to frequency synchronisation errors.

21.4 BER Performance with Frequency Synchronisation

In order to investigate the effects of the proposed frequency synchronisation algorithm on an OFDM modem, a series of experiments were conducted. The synchronisation algorithm chosen was the time domain correlation algorithm of Section 21.2.5, because of its frequency error independent estimation accuracy. We modelled a system employing one reference symbol and one data symbol per 64-slot TDMA frame. The frequency error estimates were not lowpass filtered, instead every estimate was used directly to correct the frequency error of the subsequent data symbol. Coherently detected BPSK, QPSK and 16-QAM were employed as modulation schemes in the subcarriers and the BER performance of the modem was investigated in both narrowband AWGN as well as in fading and non-fading wideband channels. In all the simulations, a 512-subcarrier OFDM scheme employing no virtual carriers and perfect wideband channel estimation was used.

The BER performance of the simulated modems in all the channels is given in Figure 21.28. Figure 21.28(a) depicts the BER versus channel SNR for BPSK, QPSK and

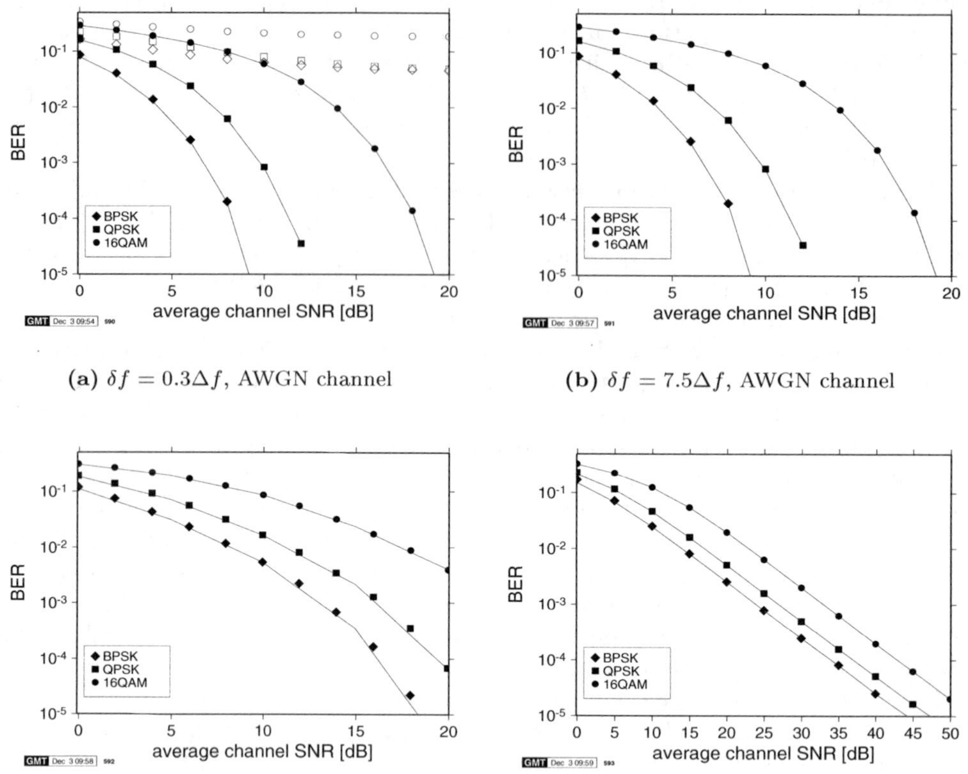

(a) $\delta f = 0.3\Delta f$, AWGN channel

(b) $\delta f = 7.5\Delta f$, AWGN channel

(c) $\delta f = 0.3\Delta f$, short WATM channel, non fading

(d) $\delta f = 0.3\Delta f$, short WATM channel, fading

Figure 21.28: BER versus channel SNR performance curves for the 512-subcarrier OFDM system in the presence of fixed frequency errors. The lines indicate the performance for perfectly corrected frequency error, and the open symbols show the performance for uncorrected frequency errors. The filled symbols indicate simulations with our proposed frequency error estimation using the time domain correlation technique. The short WATM channel impulse response is shown in Figure 20.3(a).

16-QAM in an AWGN channel with a frequency error of $\delta f = 0.3\Delta f$. The white symbols in the graph portray the BER performance of an OFDM modem employing no frequency synchronisation. It can be seen that the not frequency error results in heavy intersubcarrier interference, which manifests itself by a high residual bit error rate of about 5% for BPSK and QPSK and about 20% for 16-QAM. The lines in the graph characterise the performance of the modem in the absence of frequency errors. The black markers correspond to the BER recorded with the frequency synchronisation algorithm in operation. It can be seen that the performance of the modem employing the proposed frequency synchronisation algorithm of Section 21.2.5 is nearly indistinguishable from the perfectly synchronised case. In Figure 21.28(b), the modem's BER curves for an AWGN channel at a frequency error of $7.5\Delta f$ are depicted. Since the synchronisation algorithm's accuracy does not vary with varying frequency errors, the modem's BER performance employing the proposed synchronisation algorithm at $\delta f = 7.5\Delta f$ is the same as at $\delta f = 0.3\Delta f$. The BER for the non-synchronised modem is, however, 50% and the corresponding markers are off the graph.

The synchronised modem's BER performance in wideband channels is given in Figure 21.28(c) and 21.28(d). The impulse response used in these investigations was the short WATM impulse response, which is depicted in Figure 20.3(a). Perfect knowledge of the channel impulse response was assumed for perfect phase and amplitude correction of the data symbols with coherent detection. Again, the BER curves for both the non-fading and the fading channels show a remarkable correspondence between the ideal performance lines and the performance of the synchronised modems. In all the investigated environments, the modem's performance was unaffected by the estimation accuracy of the time domain reference symbol synchronisation algorithm.

21.5 Conclusion

In this chapter, the effects of frequency and timing errors in OFDM transmissions have been characterised. While frequency errors result in frequency domain inter-subcarrier interference, timing errors lead to time domain inter OFDM symbol interference and to frequency domain phase rotations.

In order to overcome the effects of moderate timing errors, a cyclic postamble and the use of pilot symbol assisted modulation or differential detection was proposed. Different frequency and timing error estimation algorithms were portrayed, and their performance was investigated. A new combined frequency and timing synchronisation algorithm based on a dedicated reference symbol exploited in the time domain was proposed and the system's performance employing this algorithm was investigated in AWGN and fading channels. The solution of Section 21.2.5 was finally advocated for implementation in a 34Mbit/s real-time demonstration testbed.

21.6 Appendix: Theoretical Performance of OFDM Synchronisation Algorithms

21.6.1 Frequency synchronisation in an AWGN channel

The correlation operation used in the frequency synchronisation algorithm is based on the conjugate complex multiplication of the noisy input sample with its noisy and phase–shifted copy. To derive the theoretical performance of this algorithm, the influence of the noise on the phase of this product must be known. Throughout the calculations, we will assume no frequency– and time invariant phase errors, in order to simplify the notation.

21.6.1.1 One phasor in AWGN environment

21.6.1.1.1 Cartesian coordinates Every received sample is a superposition of the transmitted phasor and two statistically independent quadrature noise samples; if the received phasor is real and of magnitude $\mathcal{A}$, the noisy signal is described by the complex stochastic variable Z:

$$\begin{aligned} Z &= \mathcal{A} + N && (21.36)\\ &= X + jY && (21.37)\end{aligned}$$

As the two quadrature noise processes are statistically independent, the joint probability density function (PDF) of the two quadrature components x and y is given by the product of two one–dimensional Gaussian PDFs:

$$\begin{aligned} p_{x,y}(x,y) &= p(x)\cdot p(y) && (21.38)\\ &= \left(\frac{1}{\sqrt{2\pi}\sigma}\right)^2 \cdot e^{-(x-\mathcal{A})^2/2\sigma^2} \cdot e^{-y^2/2\sigma^2} && (21.39)\\ &= \frac{1}{2\pi\sigma^2}\cdot e^{-((x-\mathcal{A})^2+y^2)/2\sigma^2} && (21.40)\end{aligned}$$

21.6.1.1.2 Polar coordinates The stochastic variable Z can be expressed in polar coordinates:

$$\begin{aligned} Z &= R\cdot e^{j\Phi} && (21.41)\\ withR &= \sqrt{X^2+Y^2} && (21.42)\\ \Phi &= \tan^{-1}\tfrac{Y}{X} && (21.43)\end{aligned}$$

To derive the PDF in polar coordinates $p_{r,\phi}(r,\phi)$ from the expression in Cartesian coordinates $p_{x,y}(x,y)$, given in Equation 21.40, the variable transform $(x,y) \Rightarrow (r,\phi)$ is performed:

$$\begin{aligned} x(r,\phi) &= r\cdot\cos\phi && (21.44)\\ y(r,\phi) &= r\cdot\sin\phi && (21.45)\end{aligned}$$

The determinant J of the corresponding Jacobean matrix is:

$$J = \begin{vmatrix} \frac{\partial x(r,\phi)}{\partial r} & \frac{\partial y(r,\phi)}{\partial r} \\ \frac{\partial x(r,\phi)}{\partial \phi} & \frac{\partial y(r,\phi)}{\partial \phi} \end{vmatrix} = \begin{vmatrix} \cos\phi & \sin\phi \\ -r\cdot\sin\phi & r\cdot\cos\phi \end{vmatrix} = r \qquad (21.46)$$

This leads to the probability density function of one noisy phasor in polar coordinates:

$$\begin{aligned} p_{r,\phi}(r,\phi) &= |J|\cdot p_{x,y}(x(r,\phi),y(r,\phi)) && (21.47)\\ &= \tfrac{r}{2\pi\sigma^2}\cdot e^{-(r^2+\mathcal{A}^2-2r\mathcal{A}\cos\phi)/2\sigma^2} && (21.48)\end{aligned}$$

21.6.1.2 Product of two noisy phasors

21.6.1.2.1 Joint probability density To derive the probability density function of the product of two noisy phasors, we will assume that both of the phasors are stochastic variables with a PDF given in Equation 21.48, with the same signal amplitude $\mathcal{A}$ and zero signal phase. The new stochastic variable Π represents the product of the two noisy phasors Z_1 and Z_2:

$$\Pi = \Xi * e^{j\Psi} = Z_1 \cdot Z_2 \tag{21.49}$$

The joint probability density function $p_\Pi(\Xi = \xi, \Psi = \psi)$ is derived from Equation 21.48 using complex multiplication:

$$\Xi = R_1 \cdot R_2 \tag{21.50}$$

$$\Psi = \Phi_1 + \Phi_2 \tag{21.51}$$

Two auxiliary variables must be introduced in order to solve the system. We choose $\Lambda = R_2$ and $\Omega = \Phi_2$. The resulting transformation functions and their inverse functions are:

$$\left.\begin{array}{lcl} \xi &=& r_1 \cdot r_2 \\ \psi &=& \phi_1 + \phi_2 \\ \lambda &=& r_2 \\ \omega &=& \phi_2 \end{array}\right\} \Leftrightarrow \left\{\begin{array}{lcl} r_1 &=& \xi/\lambda \\ \phi_1 &=& \psi - \omega \\ r_2 &=& \lambda \\ \phi_2 &=& \omega \end{array}\right. \tag{21.52}$$

The determinant J of the corresponding Jacobean matrix is:

$$J = \begin{vmatrix} \frac{\partial r_1}{\partial \xi} & \frac{\partial r_2}{\partial \xi} & \frac{\partial \phi_1}{\partial \xi} & \frac{\partial \phi_2}{\partial \xi} \\ \frac{\partial r_1}{\partial \psi} & \frac{\partial r_2}{\partial \psi} & \frac{\partial \phi_1}{\partial \psi} & \frac{\partial \phi_2}{\partial \psi} \\ \frac{\partial r_1}{\partial \lambda} & \frac{\partial r_2}{\partial \lambda} & \frac{\partial \phi_1}{\partial \lambda} & \frac{\partial \phi_2}{\partial \lambda} \\ \frac{\partial r_1}{\partial \omega} & \frac{\partial r_2}{\partial \omega} & \frac{\partial \phi_1}{\partial \omega} & \frac{\partial \phi_2}{\partial \omega} \end{vmatrix} = \begin{vmatrix} 1/\lambda & 0 & 0 & 0 \\ 0 & 1 & 0 & 0 \\ \xi \ln \lambda & 0 & 1 & 0 \\ 0 & 0 & 0 & 1 \end{vmatrix} = 1/\lambda \tag{21.53}$$

The resulting probability density function depending on the four new variables is:

$$p_\Pi(\xi, \psi, \lambda, \omega) = |J| \cdot p_Z(r_1 = \xi/\lambda, \phi_1 = \psi - \omega, r_2 = \lambda, \phi_2 = \omega) \tag{21.54}$$

The two probability density functions $p_Z(r1, \phi_1)$ and $p_Z(r2, \phi_2)$ are statistically independent, therefore

$$p_Z(r_1, \phi_1, r_2, \phi_2) = p_Z(r1, \phi_1) \cdot p_Z(r2, \phi_2) \tag{21.55}$$

Using the Equations 21.48 and 21.55 in 21.54, we find the expression for the PDF of the product of two noisy phasors as a function of the four transformed variables ξ, ψ, λ and ω:

$$p_\Pi(\xi, \psi, \lambda, \omega) = \frac{\xi}{\psi} \cdot \frac{1}{4\pi^2\sigma^4} \cdot e^{-\frac{\xi^2/\lambda^2 + \lambda^2 + 2\mathcal{A}^2 - 2\mathcal{A}(\lambda \cos \omega + \xi/\lambda \cos(\psi - \omega)))}{2\sigma^2}} \tag{21.56}$$

Eliminating the auxiliary variables λ and ω by integration of 21.56 yields:

$$p_{\Pi}(\Xi = \xi, \Psi = \psi) = \int_0^{\infty} \int_{-\pi}^{\pi} p_{\Pi}(\xi, \psi, \lambda, \omega) d\omega d\lambda \tag{21.57}$$

21.6.1.2.2 Phase distribution The distribution of the phase of Π is obtained by integrating equation 21.57 over the amplitude ξ:

$$p_{\Pi}(\Psi = \psi) = \int_0^{\infty} p_{\Pi}(\xi, \psi) d\xi \tag{21.58}$$

21.6.1.2.3 Numerical integration The integrals in Equations 21.57 and 21.58 can not be solved analytically, therefore numerical integration has to be employed to determine the probability density functions $p_{\Pi}(\xi, \psi)$ and $p_{\Pi}(\psi)$. This numerical integral has been evaluated for a range of different signal-to-noise (SNR) values γ:

$$\gamma = \frac{\mathcal{A}}{2\sigma^2} \tag{21.59}$$

Simulations have been performed for the same SNR values to verify the expressions 21.57 and 21.58. The simulated histograms have been employed to perform a χ^2–test with the numerical results. The χ^2 value gives a distance measure between an expected probability distribution and a N–binned set of experimental data:

$$\chi^2 = \sum_{i=0}^{N-1} \frac{(p(i) - h(i))^2}{p(i)} \tag{21.60}$$

where $h(i)$ is the value of the ith data bin, and $p(i)$ its expected value. As an example, Figure 21.6.1.2.3 illustrates both the numerical evaluation of Equation 21.57 and the histogram of the simulation for a SNR value of 6dB and a phasor amplitude of $\mathcal{A} = 2$. The figures reflect a high degree of correspondence between the numerical and the simulation results. This is confirmed by a quantitative measure, the χ^2 test, which yields a confidence–level in excess of 99% when testing the hypothesis that the simulated results are of the same distribution as given in equation 21.57. Table 21.1 displays the values of χ^2 and the resulting confidence level $\Gamma(2100, \chi^2)$ for SNR values between 0 and 10dB. The numerically integrated phase probability

SNR	χ^2	Confidence level
0	3.959337e-1	> 99%
2	3.344832e-1	–"–
4	3.012187e-1	–"–
6	2.110520e-1	–"–
8	1.835500e-1	–"–
10	1.718824e-1	–"–

Table 21.1: χ^2 values and confidence levels for $p_{\Pi}(\xi, \psi)$ for different SNR values

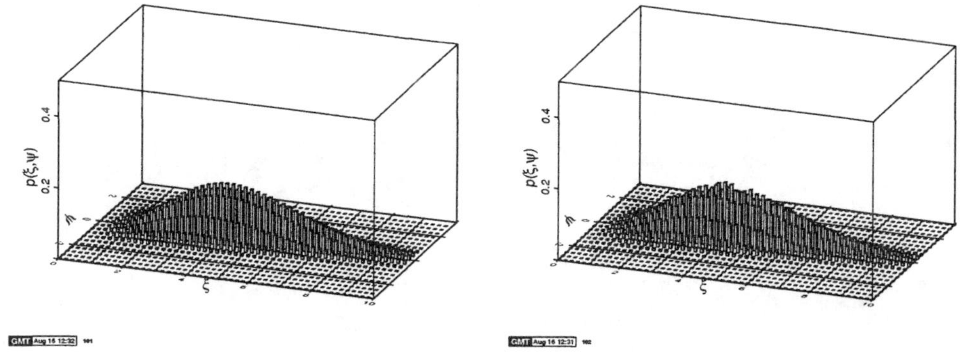

(a) $p_{\Pi}(\xi, \chi)$ (b) simulated histogram

Figure 21.29: Product of two noisy phasors with $\mathcal{A} = 2$ and γ = 6dB: (a): numerical solution of Equation 21.57, (b): histogram of 100000 simulated products

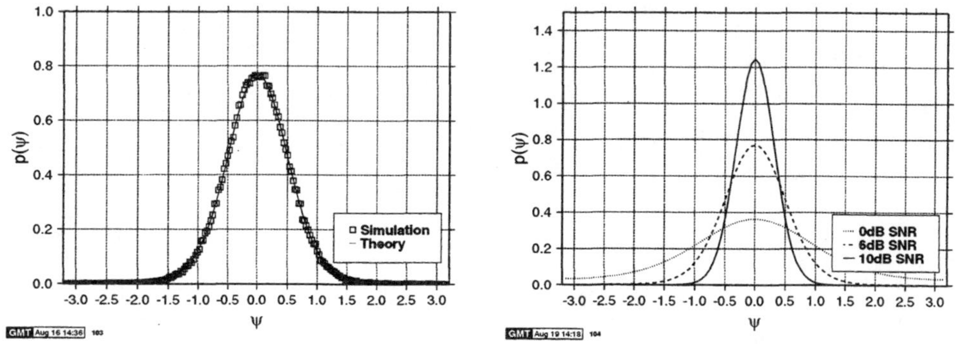

(a) phase PDF, theory and simulation (b) phase PDFs for different SNR values

Figure 21.30: Product of two noisy phasors: phase probability density functions. (a): theoretical PDF and simulated histogram of the phase distribution for SNR=6dB (b): theoretical PDFs for different SNR values

density function Equation 21.58 and the histogram of the simulated values are shown in Figure 21.30(a). Figure 21.30(b) depicts the phase PDF for different SNR values.

Chapter 22

Adaptive Single- and Multi-user OFDM Techniques

22.1 Introduction

Steele and Webb [53] proposed adaptive modulation for exploiting the time-variant Shannonian channel capacity of fading narrowband channels, and further research was conducted at Osaka University by Sampei *et al.* [420], at the University of Stanford by Goldsmith *et al.* [421], by Pearce, Burr and Tozer at the University of York [422], Lau and McLeod at the University of Cambridge [423], and at Southampton University [424,425]. The associated principles can also be invoked in the context of parallel modems, as it has been demonstrated by Kalet [83], Czylwik *et al.* [426] as well as by Chow, Cioffi and Bingham [427].

22.1.1 Motivation

We have seen in Figure 20.12 that the bit error probability of different OFDM subcarriers transmitted in time dispersive channels depends on the frequency domain channel transfer function. The occurrence of bit errors is normally concentrated in a set of severely faded subcarriers, while in the rest of the OFDM spectrum often no bit errors are observed. If the subcarriers that will exhibit high bit error probabilities in the OFDM symbol to be transmitted can be identified and excluded from data transmission, the overall BER can be improved in exchange for a slight loss of system throughput. As the frequency domain fading deteriorates the SNR of certain subcarriers, but improves others' above the average SNR value, the potential loss of throughput due to the exclusion of faded subcarriers can be mitigated by employing higher-order modulation modes on the subcarriers exhibiting high SNR values.

In addition to excluding sets of faded subcarriers and varying the modulation modes employed, other parameters such as the coding rate of error correction coding schemes can be adapted at the transmitter according to the perceived channel transfer function. This issue will be addressed in Chapter 23.

Adaptation of the transmission parameters is based on the transmitter's perception of the channel conditions in the forthcoming time slot. Clearly, this estimation of future channel parameters can only be obtained by extrapolation of previous channel estimations, which are acquired upon detecting each received OFDM symbol. The channel characteristics therefore have to be varying sufficiently slowly compared to the estimation interval.

Adapting the transmission technique to the channel conditions on a time slot by time slot basis for serial modems in narrowband fading channels has been shown to considerably improve the BER performance [428] for time division duplex (TDD) systems assuming duplex reciprocal channels. However, the Doppler fading rate of the narrowband channel has a strong effect on the achievable system performance: if the fading is rapid, then the prediction of the channel conditions for the next transmit time slot is inaccurate, and therefore the wrong set of transmission parameters may be chosen. If, however, the channel varies slowly, then the data throughput of the system is varying dramatically over time, and large data buffers are required at the transmitters in order to smoothen the bit rate fluctuation. For time critical applications, such as interactive speech transmission, the potential delays can become problematic. A given single-carrier adaptive system in narrowband channels will therefore operate efficiently only in a limited range of channel conditions.

Adaptive OFDM modem channels can ease the problem of slowly time-varying channels, since the variation of the signal quality can be exploited in the time domain and the frequency domain. The channel conditions still have to be monitored based on the received OFDM symbols, and relatively slowly varying channels have to be assumed, since we have seen in Section 20.2.2 that OFDM transmissions are not well suited to rapidly varying channel conditions.

22.1.2 Adaptive techniques

Adaptive modulation is only suitable for duplex communication between two stations, since the transmission parameters have to be adapted using some form of two-way transmission in order to allow channel measurements and signalling to take place.

Transmission parameter adaptation is a response of the transmitter to time-varying channel conditions. In order to efficiently react to the changes in channel quality, the following steps have to be taken:

(1) *Channel quality estimation:* In order to appropriately select the transmission parameters to be employed for the next transmission, a reliable estimation of the channel transfer function during the next active transmit time slot is necessary.

(2) *Choice of the appropriate parameters for the next transmission:* Based on the prediction of the channel conditions for the next time slot, the transmitter has to select the appropriate modulation modes for the subcarriers.

(3) *Signalling or blind detection of the employed parameters:* The receiver has to be informed, as to which demodulator parameters to employ for the received packet. This information can either be conveyed within the OFDM symbol

itself, with the loss of effective data throughput, or the receiver can attempt to estimate the parameters employed by the remote transmitter by means of blind detection mechanisms.

22.1.2.1 Channel quality estimation

The transmitter requires an estimate of the expected channel conditions for the time when the next OFDM symbol is to be transmitted. Since this knowledge can only be gained by prediction from past channel quality estimations, the adaptive system can only operate efficiently in an environment exhibiting relatively slowly varying channel conditions.

The channel quality estimation can be acquired from a range of different sources. If the communication between the two stations is bidirectional and the channel can be considered reciprocal, then each station can estimate the channel quality on the basis of the received OFDM symbols, and adapt the parameters of the local transmitter to this estimation. We will refer to such a regime as *open-loop adaptation*, since there is no feedback between the receiver of a given OFDM symbol and the choice of the modulation parameters. A time division duplex (TDD) system in absence of interference is an example of such a system, and hence a TDD regime is assumed for generating the performance results below. Channel reciprocity issues were addressed for example in [429, 430].

If the channel is not reciprocal, as in a frequency division duplex (FDD) system, then the stations cannot determine the parameters for the next OFDM symbol's transmission from the received symbols. In this case, the receiver has to estimate the channel quality and explicitly signal this perceived channel quality information to the transmitter in the reverse link. Since in this case the receiver explicitly instructs the remote transmitter as to which modem modes to invoke, this regime is referred to as *closed-loop adaptation*. With the aid of this technique the adaptation algorithms cas take into account effects such as interference as well as non-reciprocal channels. If the communication between the stations is essentially unidirectional, then a low-rate signalling channel must be implemented from the receiver to the transmitter. If such a channel exists, then the same technique as for non-reciprocal channels can be employed.

Different techniques can be employed to estimate the channel quality. For OFDM modems, the bit error probability in each subcarrier is determined by the fluctuations of the channel's instantaneous frequency domain channel transfer function H_n, if no interference is present. The estimate of the channel transfer function $\hat{H}_n$ can be acquired by means of pilot tone based channel estimation, as demonstrated in Section 20.3.1.1. More accurate measures of the channel transfer function can be gained by means of decision directed or time domain training sequence based techniques. The estimate of the channel transfer function $\hat{H}_n$ does not take into account effects, such as co-channel or intersubcarrier interference. Alternative channel quality measures that include interference effects can be devised on the basis of the error correction decoder's soft output information or by means of decision feedback local SNR estimations.

The delay between the channel quality estimate and the actual transmission of the OFDM symbol in relation to the maximal Doppler frequency of the channel is crucial to the adaptive system's performance. If the channel estimate is obsolete at the time of transmission, then poor system performance will result. For a closed-loop adaptive system the delays between channel estimation and transmission of the packet are generally longer than for an open-loop adaptive system, and therefore the Doppler frequency of the channel is a more critical parameter for the system's performance than in the context of open-loop adaptive systems.

22.1.2.2 Parameter adaptation

Different transmission parameters can be adapted to the anticipated channel conditions, such as the modulation and coding modes. Adapting the number of modulation levels in response to the anticipated local SNR encountered in each subcarrier can be employed, in order to achieve a wide range of different trade-offs between the received data integrity and throughput. Corrupted subcarriers can be excluded from data transmission and left blank or perhaps used for crest factor reduction.

The adaptive channel coding parameters entail code rate, adaptive interleaving and puncturing for convolutional and turbo codes, or varying block lengths for block codes [403]. These techniques can be combined with adaptive modulation mode selection and will be discussed in Chapter 23.

Based on the estimated frequency domain channel transfer function, spectral predistortion at the transmitter of one or both communicating stations can be invoked, in order to partially or fully counteract the frequency selective fading of the time dispersive channel. Unlike frequency domain equalisation at the receiver – which corrects for the amplitude and phase errors inflicted upon the subcarriers by the channel but cannot improve the signal-to-noise ratio in poor quality channels – spectral predistortion at the OFDM transmitter can deliver near-constant signal-to-noise levels for all subcarriers and can be thought of as power control on a subcarrier-by-subcarrier basis.

In addition to improving the system's BER performance in time dispersive channels, spectral predistortion can be employed in order to perform all channel estimation and equalisation functions at only one of the two communicating duplex stations. Low cost, low power consumption mobile stations can communicate with a base station that performs the channel estimation and frequency domain equalisation of the uplink, and uses the estimated channel transfer function for predistorting the down-link OFDM symbol. This set-up would lead to different overall channel quality on the uplink and the downlink, and the superior downlink channel quality could be exploited by using a computationally less complex channel decoder having weaker error correction capabilities in the mobile station than in the base station.

If the channel's frequency domain transfer function is to be fully counteracted by the spectral predistortion upon adapting the subcarrier power to the inverse of the channel transfer function, then the output power of the transmitter can become excessive, if heavily faded subcarriers are present in the system's frequency range. In order to limit the transmitter's maximal output power, hybrid channel predistortion and adaptive modulation schemes can be devised, which would deactivate transmission in deeply faded sub-channels, while retaining the benefits of predistortion in the

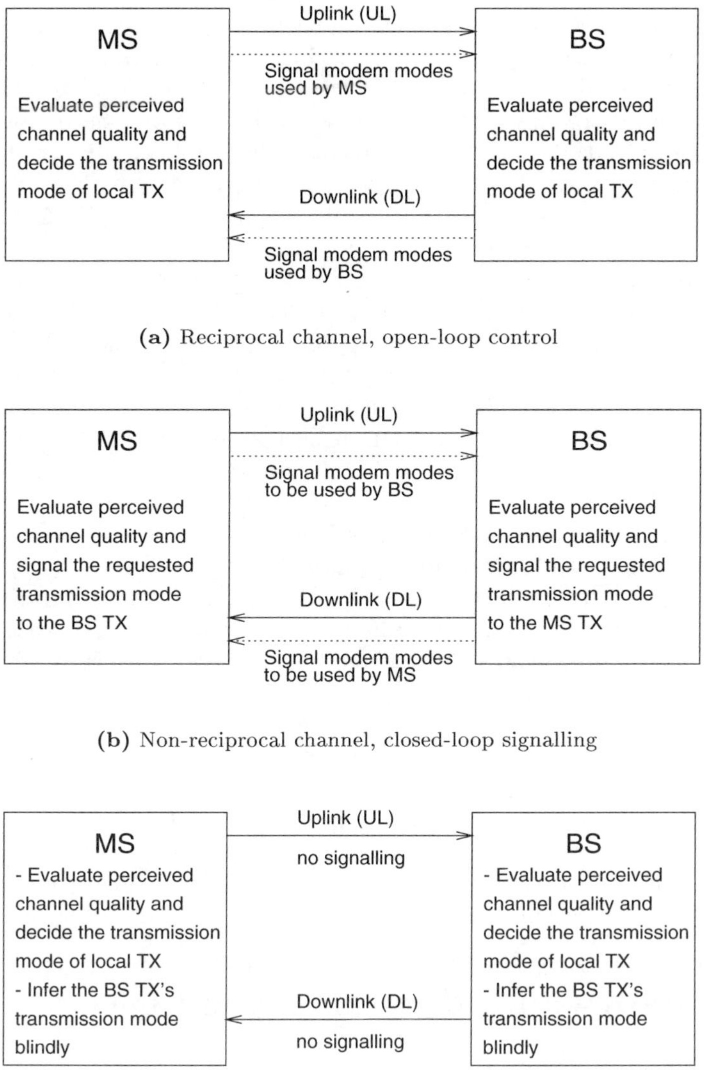

(a) Reciprocal channel, open-loop control

(b) Non-reciprocal channel, closed-loop signalling

(c) Reciprocal channel, blind modem-mode detection

Figure 22.1: Signalling scenarios in adaptive modems

remaining subcarriers.

22.1.2.3 Signalling the parameters

Signalling plays an important role in adaptive systems and the range of signalling options is summarised in Figure 22.1 for both open-loop and closed-loop signalling, as well as for blind detection. If the channel quality estimation and parameter adaptation have been performed at the transmitter of a particular link, based on open-loop adaptation, then the resulting set of parameters has to be communicated to the receiver in order to successfully demodulate and decode the OFDM symbol. If the receiver itself determines the requested parameter set to be used by the remote transmitter, the closed-loop scenario, then the same amount of information has to be transported to the remote transmitter in the reverse link. If this signalling information is corrupted, then the receiver is generally unable to correctly decode the OFDM symbol corresponding to the incorrect signalling information.

Unlike adaptive serial systems, which employ the same set of parameters for all data symbols in a transmission packet [424, 425], adaptive OFDM systems have to react to the frequency selective nature of the channel, by adapting the modem parameters across the subcarriers. The resulting signalling overhead may become significantly higher than that for serial modems, and can be prohibitive for subcarrier-by-subcarrier modulation mode adaptation. In order to overcome these limitations, efficient and reliable signalling techniques have to be employed for practical implementation of adaptive OFDM modems.

If some flexibility in choosing the transmission parameters is sacrificed in an adaptation scheme, like in the sub-band adaptive OFDM schemes described below, then the amount of signalling can be reduced. Alternatively, blind parameter detection schemes can be devised, which require little or no signalling information, respectively. Two simple blind modulation scheme detection algorithms are investigated in Section 22.2.6.2 [431].

22.1.3 System aspects

The effects of transmission parameter adaptation for OFDM systems on the overall communication system have to be investigated in at least the following areas: data buffering and latency due to varying data throughput, the effects of co-channel interference and bandwidth efficiency.

22.2 Adaptive Modulation for OFDM

22.2.1 System model

The system model of the N-subcarrier orthogonal frequency division multiplexing (OFDM) modem is shown in Figure 22.2 [71]. At the transmitter, the modulator generates N data symbols S_n, $0 \leq n \leq N-1$, which are multiplexed to the N subcarriers. The time domain samples s_n transmitted during one OFDM symbol are generated by the inverse fast Fourier transform (IFFT) and transmitted over the channel after the cyclic extension (C. Ext.) has been inserted. The channel is modelled by its time-variant impulse response $h(\tau, t)$ and AWGN. At the receiver, the cyclic extension is removed from the received time domain samples, and the data samples r_n are fast Fourier transformed, in order to yield the received frequency-domain data symbols R_n.

The channel's impulse response is assumed to be time invariant for the duration of

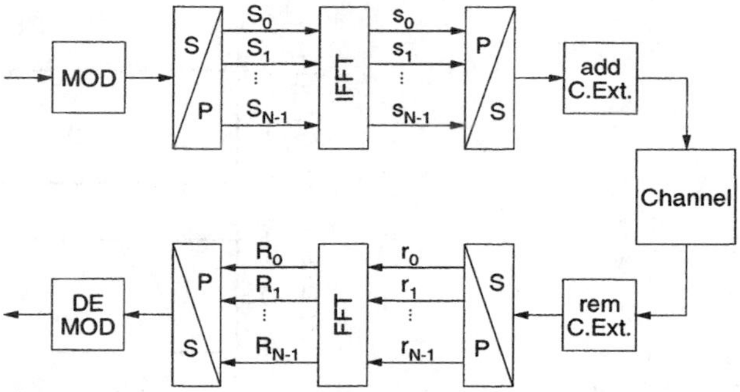

Figure 22.2: Schematic model of the OFDM system

one OFDM symbol, therefore it can be characterised for each OFDM symbol period by the N-point Fourier transform of the impulse response, which is referred to as the frequency domain channel transfer function H_n. The received data symbols R_n can be expressed as:

$$R_n = S_n \cdot H_n + n_n,$$

where n_n is an AWGN sample. Coherent detection is assumed for the system, therefore the received data symbols R_n have to be defaded in the frequency domain with the aid of an estimate of the channel transfer function H_n. This estimate $\hat{H}_n$ can be obtained by the use of pilot subcarriers in the OFDM symbol, or by employing time domain channel sounding training sequences embedded in the transmitted signal. Since the noise energy in each subcarrier is independent of the channel's frequency domain transfer function H_n, the "local" signal-to-noise ratio SNR in subcarrier n can be expressed as

$$\gamma_n = |H_n|^2 \cdot \gamma,$$

where γ is the overall SNR. If no signal degradation due to intersubcarrier interference (ISI) or interference from other sources appears, then the value of γ_n determines the bit error probability for the transmission of data symbols over the subcarrier n.

The goal of adaptive modulation is to choose the appropriate modulation mode for transmission in each subcarrier, given the local SNR γ_n, in order to achieve a good trade-off between throughput and overall BER. The acceptable overall BER varies depending on other systems parameters, such as the correction capability of the error correction coding and the nature of the service supported by this particular link.

The adaptive system has to fulfill these requirements:

(1) Channel quality estimation

(2) Choice of the appropriate modulation modes

(3) Signalling or blind detection of the modulation modes

We will examine these three points with reference to Figure 22.1 in the following sections for the example of a 512-subcarrier OFDM modem in the shortened WATM channel of Section 20.1.1.2.

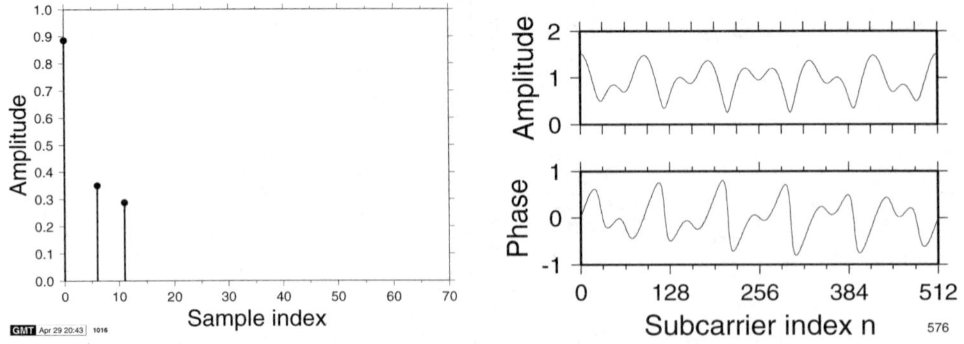

(a) $h(n)$ (b) H_n

Figure 22.3: WATM wideband channel: (a) unfaded symbol spaced impulse response, (b) the corresponding frequency domain channel transfer function

22.2.2 Channel model

The impulse response $h(\tau, t)$ used in our experiments was generated on the basis of the symbol-spaced impulse response shown in Figure 22.3(a) by fading each of the impulses obeying a Rayleigh distribution of a normalised maximal Doppler frequency of $f'_d = 1.235 \cdot 10^{-5}$, which corresponds to the WLAN channel experienced by a modem transmitting at a carrier frequency of 60 GHz with a sample rate of 225 MHz and a vehicular velocity of 50 km/h. The complex frequency domain channel transfer function H_n corresponding to the unfaded impulse response is shown in Figure 22.3(b).

22.2.3 Channel estimation

The most convenient setting for an adaptive OFDM (AOFDM) system is a time division duplex (TDD) system in a slowly varying reciprocal channel, allowing open-loop adaptation. Both stations transmit an OFDM symbol in turn, and at each station the most recent received symbol is used for the channel estimation employed for the modulation mode adaptation for the next transmitted OFDM symbol. The channel estimation on the basis of the received symbol can be performed by PSAM (see Section 20.3.1.1), or upon invoking more sophisticated methods, such as decision directed channel estimation. Initially, we will assume perfect knowledge of the channel transfer function during the received time slot.

22.2.4 Choice of the modulation modes

The two communicating stations use the open-loop predicted channel transfer function acquired from the most recent received OFDM symbol, in order to allocate the appropriate modulation modes to the subcarriers. The modulation modes were chosen from the set of binary phase shift keying (BPSK), quadrature phase shift keying

(QPSK), 16-quadrature amplitude modulation (16-QAM), as well as "no transmission", for which no signal was transmitted. These modulation modes are denoted by M_m, where $m \in (0, 1, 2, 4)$ is the number of data bits associated with one data symbol of each mode.

In order to keep the system complexity low, the modulation mode is not varied on a subcarrier-by-subcarrier basis, but instead the total OFDM bandwidth of 512 subcarriers is split into blocks of adjacent subcarriers, referred to as sub-bands, and the same modulation scheme is employed for all subcarriers of the same sub-band. This substantially simplifies the task of signalling the modem mode and renders the employment of alternative blind detection mechanisms feasible, which will be discussed in Section 22.2.6.

Three modulation mode allocation algorithms were investigated in the sub-bands: a fixed threshold controlled algorithm, an upper bound BER estimator and a fixed throughput adaptation algorithm.

22.2.4.1 Fixed threshold adaptation algorithm

The fixed threshold algorithm was derived from the adaptation algorithm proposed by Torrance for serial modems [428]. In the case of a serial modem, the channel quality is assumed to be constant for all symbols in the time slot, and hence the channel has to be slowly varying in order to allow accurate channel quality prediction. Under these circumstances, all data symbols in the transmit time slot employ the same modulation mode, chosen according to the predicted SNR. The SNR thresholds for a given long-term target BER were determined by Powell optimisation [432]. Torrance assumed two uncoded target bit error rates: 1% for a high data rate "speech" system, and 10^{-4} for a higher integrity, lower data rate "data" system. The resulting SNR thresholds l_n for activating a given modulation mode M_n in a slowly Rayleigh fading narrowband channel for both systems are given in Table 22.1. Specifically, the modulation mode

	l_0	l_1	l_2	l_4
speech system	$-\infty$	3.31	6.48	11.61
data system	$-\infty$	7.98	10.42	16.76

Table 22.1: Optimised switching levels for adaptive modulation over Rayleigh fading channels for the "speech" and "data" systems, shown in instantaneous channel SNR [dB] (from [432])

M_n is selected if the instantaneous channel SNR exceeds the switching level l_n.

This adaptation algorithm originally assumed a constant instantaneous SNR over all of the block's symbols, but in the case of an OFDM system in a frequency selective channel the channel quality varies across the different subcarriers. For sub-band adaptive OFDM transmission, this implies that if the the sub-band width is wider than the channel's coherence bandwidth [403], then the original switching algorithm cannot be employed. For our investigations, we have therefore employed the lowest quality subcarrier in the sub-band for the adaptation algorithm based on the thresholds given in Table 22.1. The performance of the 16 sub-band adaptive system over the shortened

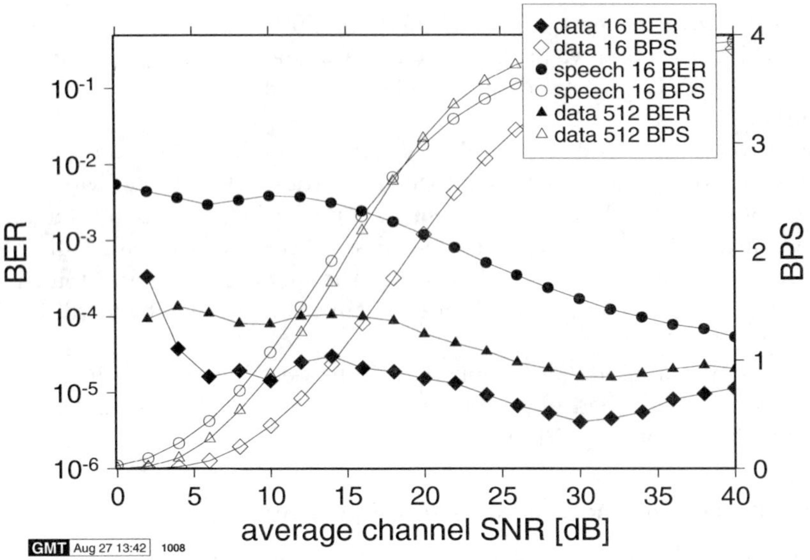

Figure 22.4: BER and BPS throughput performance of the 16 sub-band, 512 subcarrier switching level adaptive OFDM modem employing BPSK, QPSK, 16-QAM and "no transmission" over the Rayleigh fading time dispersive channel of Figure 20.3 using the switching thresholds of Table 22.1

WATM Rayleigh fading channel of Figure 20.3 is shown in Figure 22.4.

Adjacent or consecutive time slots have been used for the uplink and downlink slots in these simulations, so that the delay between channel estimation and transmission was rendered as short as possible. Figure 22.4 shows the long-term average BER and throughput of the studied modem for the "speech" and "data" switching levels of Table 22.1 as well as for a subcarrier-by-subcarrier adaptive modem employing the "data" switching levels. The results show the typical behaviour of a variable throughput AOFDM system, which constitutes a trade-off between the best BER and best throughput performance. For low SNR values, the system achieves a low BER by transmitting very few bits and only when the channel conditions allow. With increasing long-term SNR, the throughput increases without significant change in the BER. For high SNR values the BER drops as the throughput approaches its maximum of 4 bits per symbol, since the highest-order constellation was 16-QAM.

It can be seen from the figure that the adaptive system performs better than its target bit error rates of 10^{-2} and 10^{-4} for the "speech" and "data" systems, respectively, resulting in measured bit error rates lower than the targets. This can be explained by the adaptation regime, which was based on the conservative principle of using the lowest quality subcarrier in each sub-band for channel quality estimation, leading to a pessimistic channel quality estimate for the entire sub-band. For low values of SNR, the throughput in bits per data symbol is low and exceeds the fixed BPSK throughput of 1 bit/symbol only for SNR values in excess of 9.5 dB and 14 dB

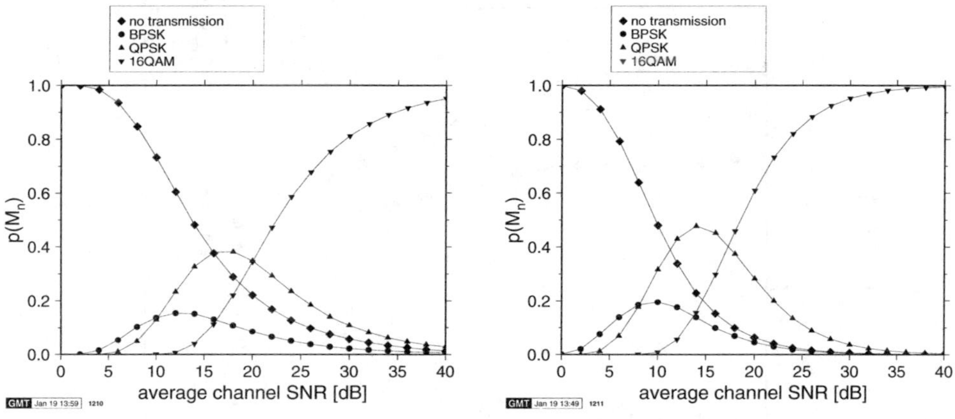

(a) 16 sub-bands **(b)** subcarrier-by-subcarrier

Figure 22.5: Histograms of modulation modes versus channel SNR for the "data" switching level adaptive 512 subcarrier, 16 sub-band OFDM modem over the Rayleigh fading time dispersive channel of Figure 20.3 using the switching thresholds of Table 22.1

for the "speech" and "data" systems, respectively.

The upper bound performance of the system with subcarrier-by-subcarrier adaptation is also portrayed in the figure, shown as 512 independent sub-bands, for the "data" optimised set of threshold values. It can be seen that in this case the target BER of 10^{-4} is closely met over a wide range of SNR values from about 2 dB to 20 dB, and that the throughput is considerably higher than in the case of the 16 sub-band modem. This is the result of more accurate subcarrier-by-subcarrier channel quality estimation and fine-grained adaptation, leading to better exploitation of the available channel capacity.

Figure 22.5 shows the long-term modulation mode histograms for a range of channel SNR values for the "data" switching levels in both the 16 sub-band and the subcarrier-by-subcarrier adaptive modems using the switching thresholds of Table 22.1. Comparison of the graphs shows that higher-order modulation modes are used more frequently by the subcarrier-by-subcarrier adaptation algorithm, which is in accordance with the overall throughput performance of the two modems in Figure 22.4.

The throughput penalty of employing sub-band adaptation depends on the frequency domain variation of the channel transfer function. If the sub-band bandwidth is lower than the channel's coherence bandwidth, then the assumption of constant channel quality per sub-band is closely met, and the system performance is equivalent to that of a subcarrier-by-subcarrier adaptive scheme.

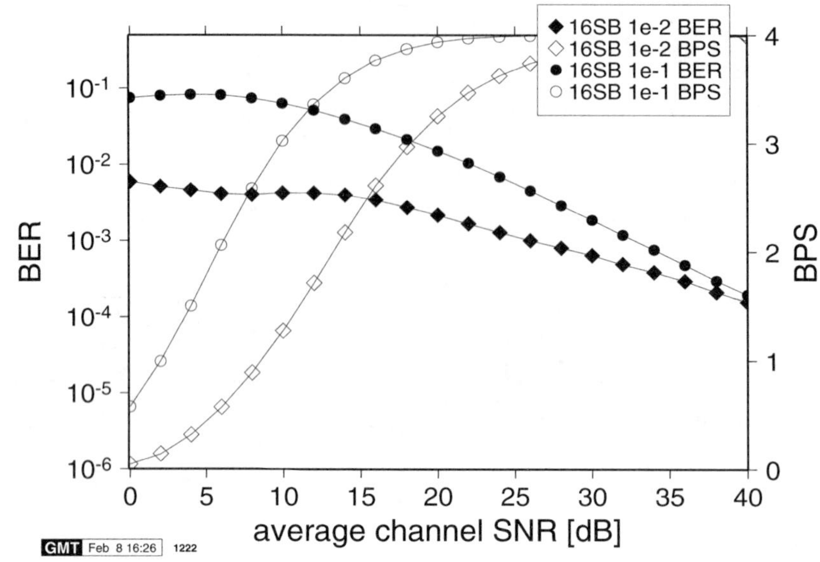

Figure 22.6: BER and BPS throughput performance of the 16 sub-band, 512 subcarrier BER estimator adaptive OFDM modem employing BPSK, QPSK, 16-QAM and "no transmission" over the Rayleigh fading time dispersive channel of Figure 20.3

22.2.4.2 Sub-band BER estimator adaptation algorithm

We have seen above that the fixed switching level based algorithm leads to a throughput performance penalty, if used in a sub-band adaptive OFDM modem, when the channel quality is not constant throughout each sub-band. This is due to the conservative adaptation based on the subcarrier experiencing the most hostile channel in each sub-band.

An alternative scheme taking into account the non-constant SNR values γ_j across the N_s subcarriers in the jth sub-band can be devised by calculating the expected overall bit error probability for all available modulation modes M_n in each sub-band, which is denoted by $\bar{p}_e(n) = 1/N_s \sum_j p_e(\gamma_j, M_n)$. For each sub-band, the mode having the highest throughput, whose estimated BER is lower than a given threshold, is then chosen. While the adaptation granularity is still limited to the sub-band width, the channel quality estimation includes not only the lowest quality subcarrier, which leads to an improved throughput.

Figure 22.6 shows the BER and throughput performance for the 16 sub-band adaptive OFDM modem employing the BER estimator adaptation algorithm in the Rayleigh fading time dispersive channel of Figure 20.3. The two sets of curves in the figure correspond to target bit error rates of 10^{-2} and 10^{-1}, respectively. Comparing the modem's performance for a target BER of 10^{-2} with that of the "speech" modem in Figure 22.4, it can be seen that the BER estimator algorithm results in significantly higher throughput while meeting the BER requirements. The BER estimator algo-

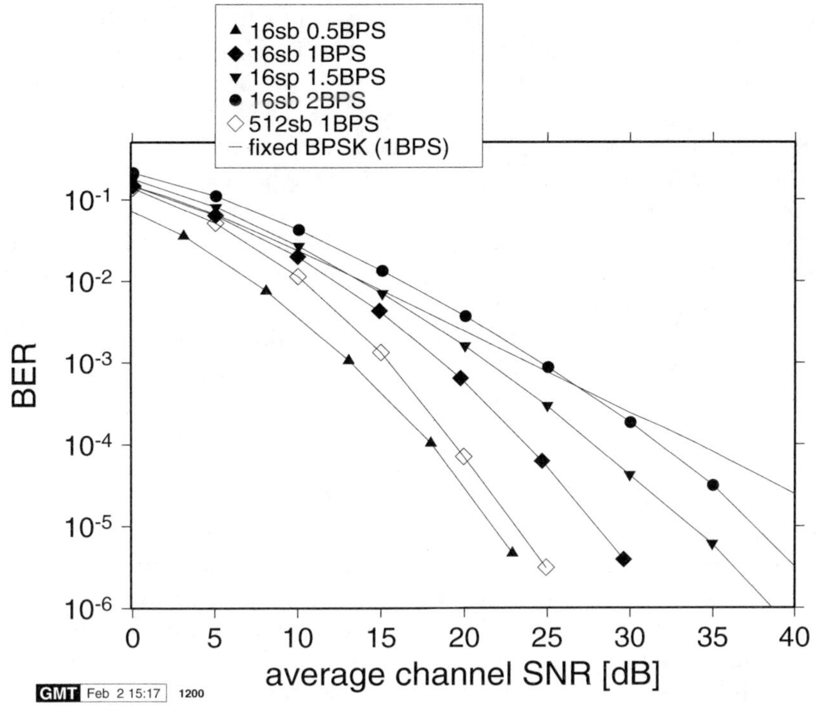

Figure 22.7: BER performance versus SNR for the 512 subcarrier, 16 sub-band constant throughput adaptive OFDM modem employing BPSK, QPSK, 16-QAM and "no transmission" in the Rayleigh fading time dispersive channel of Figure 20.3 for 0.5, 1, 1.5 and 2 bits per symbol (BPS) target throughput

rithm is readily adjustable to different target bit error rates, which is demonstrated in the figure for a target BER of 10^{-1}. Such adjustability is beneficial when combining adaptive modulation with channel coding, as will be discussed in Section 22.2.7.

22.2.5 Constant throughput adaptive OFDM

The time-varying data throughput of an adaptive OFDM modem operating with either of the two adaptation algorithms discussed above makes it difficult to employ such a scheme in a variety of constant rate applications. Torrance [428] studied the system implications of variable-throughput adaptive modems in the context of narrowband channels, stressing the importance of data buffering at the transmitter, in order to accommodate the variable data rate. The required length of the buffer is related to the Doppler frequency of the channel, and a slowly varying channel – as required for adaptive modulation – results in slowly varying data throughput and therefore the need for a high buffer capacity. Real-time interactive audio or video transmission is sensitive to delays, and therefore different modem mode adaptation algorithms are needed for such applications.

The constant throughput AOFDM scheme proposed here exploits the frequency selectivity of the channel, while offering a constant bit rate. Again, sub-band adaptivity is assumed, in order to simplify the signalling or the associated blind detection of the modem schemes.

The modulation mode allocation of the sub-bands is performed on the basis of a cost function to be introduced below, based on the expected number of bit errors in each sub-band. The expected number of bit errors $e_{n,s}$, for each sub-band n and for each possible modulation mode index s, is calculated on the basis of the estimated channel transfer function $\hat{H}$, taking into account the number of bits transmitted per sub-band and per modulation mode, $b_{n,s}$.

Each sub-band is assigned a state variable s_n holding the index of a modulation mode. Each state variable is initialised to the lowest-order modulation mode, which in our case is 0 for "no transmission". A set of cost values $c_{n,s}$ is calculated for each sub-band n and state s:

$$c_{n,s} = \frac{e_{n,s+1} - e_{n,s}}{b_{n,s+1} - b_{n,s}} \tag{22.1}$$

for all but the highest modulation mode index s. This cost value is related to the expected increase in the number of bit errors, divided by the increase of throughput, if the modulation mode having the next higher index is used instead of index s in sub-band n. In other words, Equation 22.1 quantifies the expected incremental bit error rate of the state transition $s \rightarrow s + 1$ in sub-band n.

The modulation mode adaptation is performed by repeatedly searching for the block n having the lowest value of c_{n,s_n}, and incrementing its state s_n. This is repeated until the total number of bits in the OFDM symbol reaches the target number of bits. Because of the granularity in bit numbers introduced by the sub-bands, the total number of bits may exceed the target. In this case, the data is padded with dummy bits for transmission.

Figure 22.7 gives an overview of the BER performance of the fixed throughput 512-subcarrier OFDM modem over the time dispersive channel of Figure 20.3 for a range of target bit numbers. The curve without shapes represents the performance of a fixed BPSK OFDM modem over the same channel and where the modem transmits, i.e. 1 bit over each data subcarrier per OFDM symbol. The diamond shapes give the performance of the equivalent throughput adaptive scheme for the 16 sub-band arrangement filled shapes and for the subcarrier-by-subcarrier adaptive scheme open shapes. It can be seen that the 16 sub-band adaptive scheme yields a significant improvement in BER terms for SNR values above 10 dB. The SNR gain for a bit error rate of 10^{-4} is 8 dB compared to the non-adaptive case. Subcarrier-by-subcarrier adaptivity increases this gain by a further 4 dB. The modem can readily be adapted to the system requirements by adjusting the target bit rate, as shown in Figure 22.7. Halving the throughput to 0.5 BPS, the required SNR is reduced by 6 dB for a BER of 10^{-4}, while increasing the throughput to 2 BPS deteriorates the noise resilience by 8 dB at the same BER.

22.2.6 Signalling and blind detection

The adaptive OFDM receiver has to be informed of the modulation modes used for the different sub-bands. This information can either be conveyed using signalling subcarriers in the OFDM symbol itself, or the receiver can employ blind detection techniques in order to estimate the transmitted symbols' modulation modes, as seen in Figure 22.1.

22.2.6.1 Signalling

The simplest way of signalling the modulation mode employed in a sub-band is to replace one data symbol by an M-PSK symbol, where M is the number of possible modulation modes. In this case, reception of each of the constellation points directly signals a particular modulation mode in the current sub-band. In our case, for four modulation modes and assuming perfect phase recovery, the probability of a signalling error $p_s(\gamma)$, when employing one signalling symbol, is the symbol error probability of QPSK. Then the correct sub-band mode signalling probability is:

$$(1 - p_s(\gamma)) = (1 - p_{b,QPSK}(\gamma))^2 ,$$

where $p_{b,QPSK}$ is the bit error probability for QPSK:

$$p_{b,QPSK}(\gamma) = Q(\sqrt{\gamma}) = \frac{1}{2} \cdot \mathrm{erfc}\left(\sqrt{\frac{\gamma}{2}}\right) ,$$

which leads to the expression for the modulation mode signalling error probability of

$$p_s(\gamma) = 1 - \left(1 - \frac{1}{2} \cdot \mathrm{erfc}\left(\sqrt{\frac{\gamma}{2}}\right)\right)^2 .$$

The modem mode signalling error probability can be reduced by employing multiple signalling symbols and maximum ratio combining of the received signalling symbols $R_{s,n}$, in order to generate the decision variable R'_s prior to decision:

$$R'_s = \sum_{n=1}^{N_s} R_{s,n} \cdot \hat{H}^*_{s,n},$$

where N_s is the number of signalling symbols per sub-band, the quantities $R_{s,n}$ are the received symbols in the signalling subcarriers, and $\hat{H}_{s,n}$ represents the estimated values of the frequency domain channel transfer function at the signalling subcarriers. Assuming perfect channel estimation and constant values of the channel transfer function across the group of signalling subcarriers, the signalling error probability for N_s signalling symbols can be expressed as:

$$p'_s(\gamma, N_s) = 1 - \left(1 - \frac{1}{2} \cdot \mathrm{erfc}\left(\sqrt{\frac{N_s\gamma}{2}}\right)\right)^2 . \tag{22.2}$$

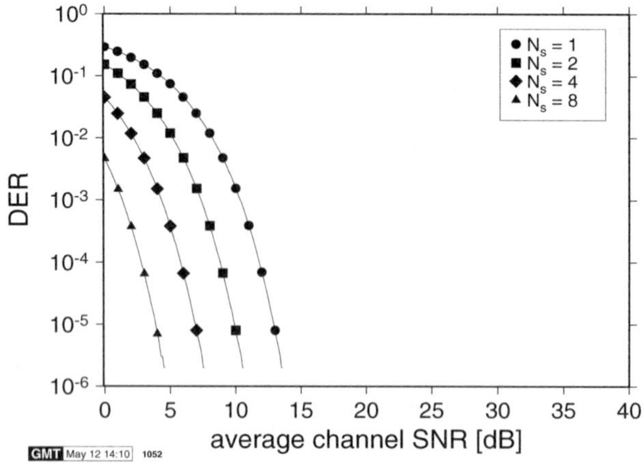

Figure 22.8: Modulation mode detection error ratio (DER) if signalling with maximum ratio combining is employed for QPSK symbols in an AWGN channel for 1, 2, 4 and 8 signalling symbols per sub-band, evaluated from Equation 22.2

Figure 22.8 shows the signalling error rate in an AWGN channel for 1, 2, 4 and 8 signalling symbols per sub-band, respectively. It can be seen that doubling the number of signalling subcarriers improves the performance by 3 dB. Modem mode detection error ratios (DER) below 10^{-5} can be achieved at 10 dB SNR over AWGN channels if two signalling symbols are used. The signalling symbols for a given sub-band can be interleaved across the entire OFDM symbol bandwidth, in order to benefit from frequency diversity in fading wideband channels.

As seen in Figure 22.1, blind detection algorithms aim to estimate the employed modulation mode directly from the received data symbols, therefore avoiding the loss of data capacity due to signalling subcarriers. Two algorithms have been investigated, one based on geometrical SNR estimation and another incorporating error correction coding.

22.2.6.2 Blind detection by SNR estimation

The receiver has no a-priori knowledge of the modulation mode employed in a particular received sub-band and estimates this parameter by quantising the defaded received data symbols $R_n/\hat{H}_n$ in the sub-band to the closest symbol $\hat{R}_{n,m}$ for all possible modulation modes M_m for each subcarrier index n in the current sub-band. The decision directed error energy e_m for each modulation mode is calculated:

$$e_m = \sum_n \left(R_n/\hat{H}_n - \hat{R}_{n,m}\right)^2$$

and the modulation mode M_m which minimises e_m is chosen for the demodulation of the sub-band.

The DER of the blind modulation mode detection algorithm described in this sec-

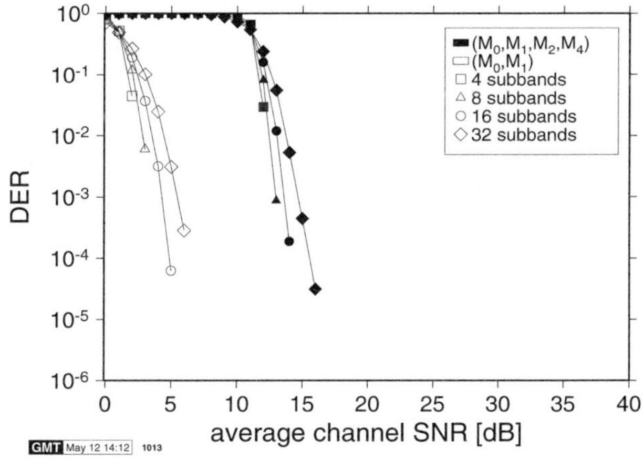

Figure 22.9: Blind modulation mode detection error ratio (DER) for 512-subcarrier OFDM systems employing (M_0, M_1) as well as for (M_0, M_1, M_2, M_4) for different numbers of sub-bands in an AWGN channel

tion for a 512-subcarrier OFDM modem in an AWGN channel is depicted in Figure 22.9. It can be seen that the detection performance depends on the number of symbols per sub-band, with fewer sub-bands and therefore longer symbol sequences per sub-band leading to a better detection performance. It is apparent, however, that the number of available modulation modes has a more significant effect on the detection reliability than the block length. If all four legitimate modem modes are employed, then reliable detection of the modulation mode is only guaranteed for AWGN SNR values of more than 15-18 dB, depending on the number of sub-bands per OFDM symbol. If only M_0 and M_1 are employed, however, the estimation accuracy is dramatically improved. In this case, AWGN SNR values above 5-7 dB are sufficient to ensure reliable detection. The estimation accuracy could be improved by using the estimate of the channel quality, in order to predict the modulation mode, which is likely to have been employed at the transmitter. For example, at an estimated channel SNR of 5 dB it is unlikely that 16-QAM was employed as the modem mode and hence this a priori knowledge can be exploited, in order increase our confidence in the corresponding decision.

Figure 22.10 shows the BER performance of the fixed threshold "data" 16 sub-band adaptive system in the fading wideband channel of Figure 20.3 for both sets of modulation modes, namely for (M_0, M_1) and (M_0, M_1, M_2, M_4) with blind modulation mode detection. Erroneous modulation mode decisions were assumed to yield a BER of 50% in the received block. This is optimistic, since in a realistic scenario the receiver would have no knowledge of the number of bits actually transmitted, leading to loss of synchronisation in the data stream. This problem is faced by all systems having a variable throughput and not employing an ideal reliable signalling channel. This impediment must be mitigated by data synchronisation measures.

It can be seen from Figure 22.10 that while blind modulation mode detection yields

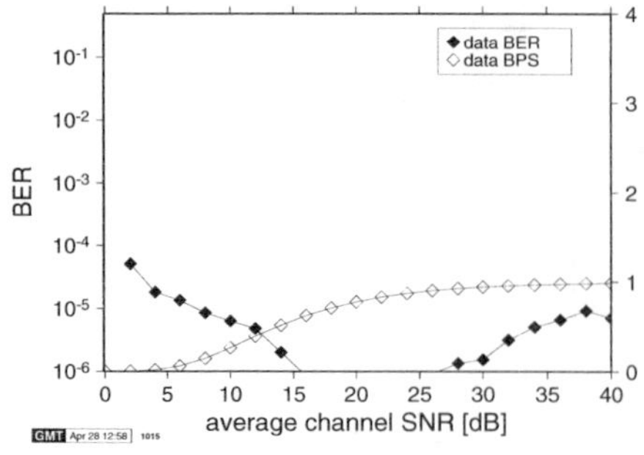

(a) (M_0, M_1)

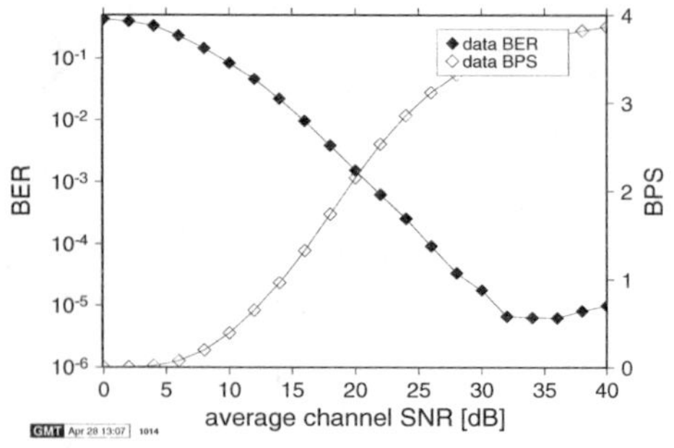

(b) (M_0, M_1, M_2, M_4)

Figure 22.10: BER and BPS throughput performance of a 16 sub-band, 512 subcarrier adaptive OFDM modem employing (a) no transmission (M_0) and BPSK (M_1), or (b) (M_0, M_1, M_2, M_4), both using the data-type switching levels of Table 22.1 and the SNR-based blind modulation mode detection of Section 22.2.6.2 over the Rayleigh fading time dispersive channel of Figure 20.3

poor performance for the quadruple-mode adaptive scheme, the twin-mode scheme exhibits BER results consistently better than 10^{-4}.

22.2.6.3 Blind detection by multi-mode trellis decoder

If error correction coding is invoked in the system, then the channel decoder can be employed to estimate the most likely modulation mode per sub-band. Since the number of bits per OFDM symbol is varying in this adaptive scheme, and the channel encoder's block length is therefore not constant, for the sake of implementational convenience we have chosen a convolutional encoder at the transmitter. Once the modulation modes to be used are decided upon at the transmitter, the convolutional encoder is employed to generate a zero-terminated code word having the length of the OFDM symbol's capacity. This code word is modulated on the subcarriers according to the different modulation modes for the different sub-bands, and the OFDM symbol is transmitted over the channel.

At the receiver, each received data subcarrier is demodulated by all possible demodulators, and the resulting hard decision bits are fed into parallel trellises for Viterbi decoding. Figure 22.11 shows a schematic sketch of the resulting parallel trellis if 16-QAM (M_4), QPSK (M_2), BPSK (M_1), and "no transmission" (M_0) are employed, for a convolutional code having four states. Each sub-band in the adaptive scheme corresponds to a set of four parallel trellises, whose inputs are generated independently by the four demodulators of the legitimate modulation modes. The number of transitions in each of the trellises depends on the number of output bits received from the different demodulators, so that the 16-QAM (M_4) trellis contains four times as many transitions as the BPSK and "no transmission" trellises. Since in

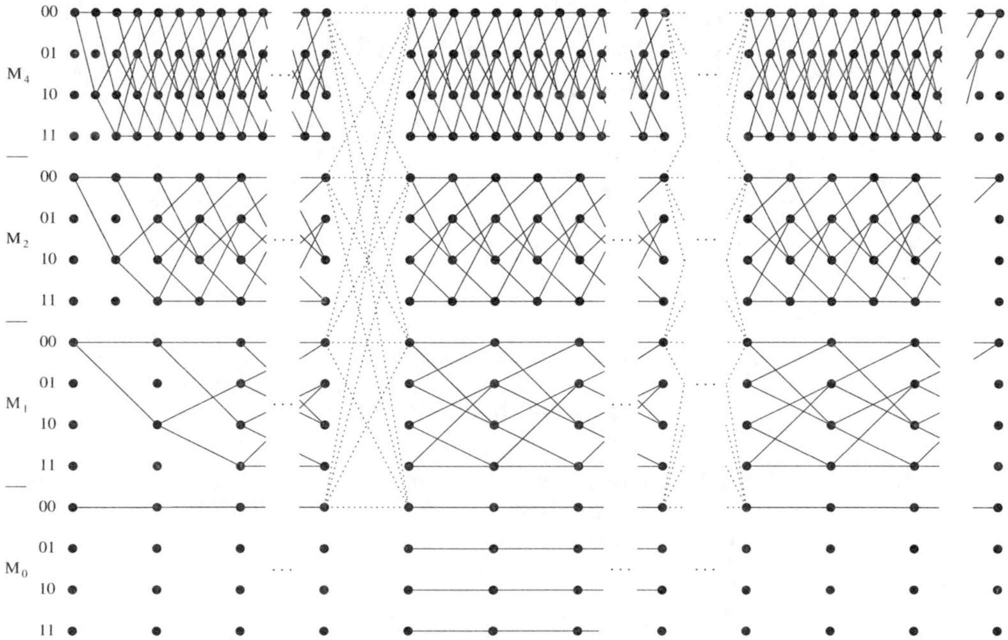

Figure 22.11: Schematic plot of the parallel trellises for blind modulation mode detection employing convolutional coding. In this example, a four-state 00-terminated convolutional encoder was assumed. The dotted lines indicate the intersub-band transitions for the 00 state, and are omitted for the other three states

the case of "no transmission" no coded bits are transmitted, the state of the encoder does not change. Therefore, legitimate transitions for this case are only horizontal ones.

At sub-band boundaries, transitions are allowed between the same state of all the parallel trellises associated with the different modulation modes. This is not a transition due to a received bit, and therefore preserves the metric of the originating state. Note that in the figure only the possible allowed transitions for the state 00 are drawn; all other states originate the equivalent set of transitions. The initial state of the first sub-band is 00 for all modulation modes, and since the code is terminated in 00, the last sub-band's final states are 00.

The receiver's Viterbi decoder calculates the metrics for the transitions in the parallel trellises, and once all the data symbols have been processed, it traces back through the parallel trellis on the surviving path. This backtracing commences at the most likely 00 state at the end of the last sub-band. If no termination was used at the decoder, then the backtracing would start at the most likely of all the final states of the last block.

Figure 22.12 shows the modulation mode detection error ratio (DER) for the parallel trellis decoder in an AWGN channel for 16 and 8 sub-bands, if a convolutional code of constraint length 7 is used. Comparison with Figure 22.9 shows consider-

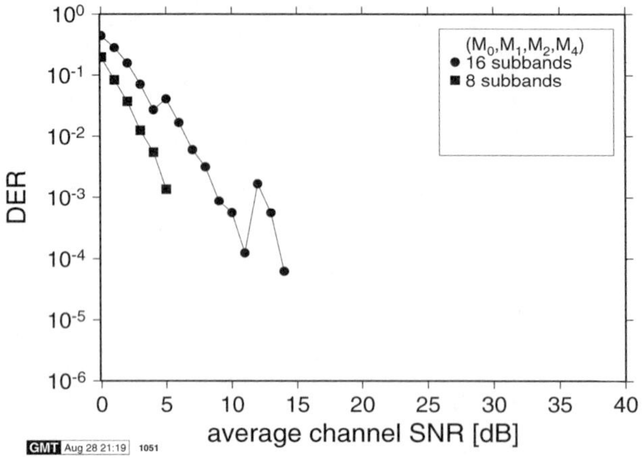

Figure 22.12: Blind modulation mode detection error ratio (DER) using the parallel trellis algorithm of Section 22.2.6.3 with a $K = 7$ convolutional code in an AWGN channel for a 512-subcarrier OFDM modem

able improvements relative to the BER estimation based blind detection scheme of Section 22.2.6.2, both for 16 and 8 sub-bands. Higher sub-band lengths improve the estimation accuracy by a greater degree than has been observed for the BER estimation algorithm of Figure 22.9. A DER of less than 10^{-5} was observed for AWGN SNR values of 6 dB and 15 dB in the 8 and 16 sub-band scenarios, respectively. The use of stronger codes could further improve the estimation accuracy, at the cost of higher complexity.

22.2.7 Sub-band adaptive OFDM and channel coding

Adaptive modulation can reduce the BER to a level where channel decoders can perform well. Figure 22.13 shows both the uncoded and coded BER performance of a 512-subcarrier OFDM modem in the fading wideband channel of Figure 20.3, assuming perfect channel estimation. The channel coding employed in this set of experiments was a turbo coder [433] with a data block length of 1000 bits, employing a random interleaver and 8 decoder iterations. The log-MAP decoding algorithm was used [434]. The constituent half-rate convolutional encoders were of constraint length 3, with octally represented generator polynomials of $(7, 5)$ [403]. It can be seen that the turbo decoder provides a considerable coding gain for the different fixed modulation schemes, with a BER of 10^{-4} for SNR values of 13.8 dB, 17.3 dB and 23.2 dB for BPSK, QPSK and 16-QAM transmission, respectively.

Figure 22.14 depicts the BER and throughput performance of the same decoder employed in conjunction with the adaptive OFDM modem for different adaptation algorithms. Figure 22.14(a) shows the performance for the "speech" system employing the switching levels listed in Table 22.1. As expected, the half-rate channel coding results in a halved throughput compared to the uncoded case, but offers low BER transmission over the channel of Figure 20.3 for SNR values of down to 0 dB, main-

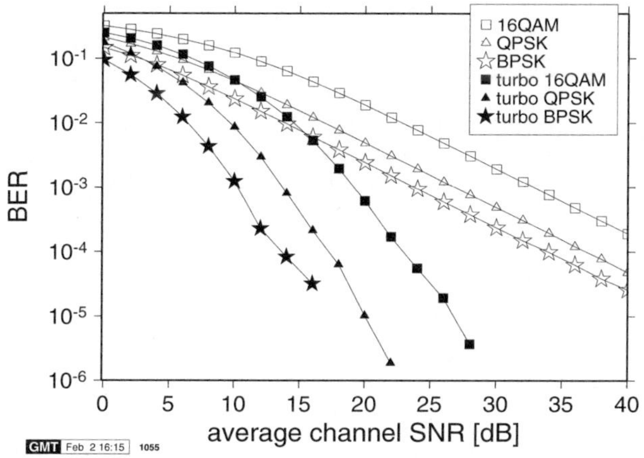

Figure 22.13: BER performance of the 512-subcarrier OFDM modem in the fading time dispersive channel of Figure 20.3 for both uncoded and half-rate turbo-coded transmission, using 8-iteration log-MAP turbo decoding, 1000-bit random interleaver, and a constraint length of 3

taining a BER below 10^{-6}.

Further tuning of the adaptation parameters can ensure a better average throughput, while retaining error-free data transmission. The switching level based adaptation algorithm of Table 22.1 is difficult to control for arbitrary bit error rates, since the set of switching levels was determined by an optimisation process for uncoded transmission. Since the turbo codec has a non-linear BER versus SNR characteristic, direct switching level optimisation is an arduous task. The sub-band BER predictor of Section 22.2.4.2 is easier to adapt to a channel codec, and Figure 22.14(b) shows the performance for the same decoder, with the adaptation algorithm employing the BER prediction method having an upper BER bound of 1%. It can be seen that the less stringent uncoded BER constraints when compared to Figure 22.14(a) lead to a significantly higher throughput for low SNR values. The turbo-decoded data bits are error-free, hence a further increase in throughput is possible while maintaining a high degree of coded data integrity.

The second set of curves in Figure 22.14(b) show the system's performance, if an uncoded target BER of 10% is assumed. In this case, the turbo decoder's output BER is below 10^{-5} for all the SNR values plotted, and shows a slow decrease for increasing values of SNR. The throughput of the system, however, exceeds 0.5 data bits per symbol for SNR values of more than 2 dB.

22.2.8 The effect of channel Doppler frequency

Since the adaptive OFDM modem employs the most recently received OFDM symbol in order to predict the frequency domain transfer function of the reverse channel for the next transmission, the quality of this prediction suffers from the time variance

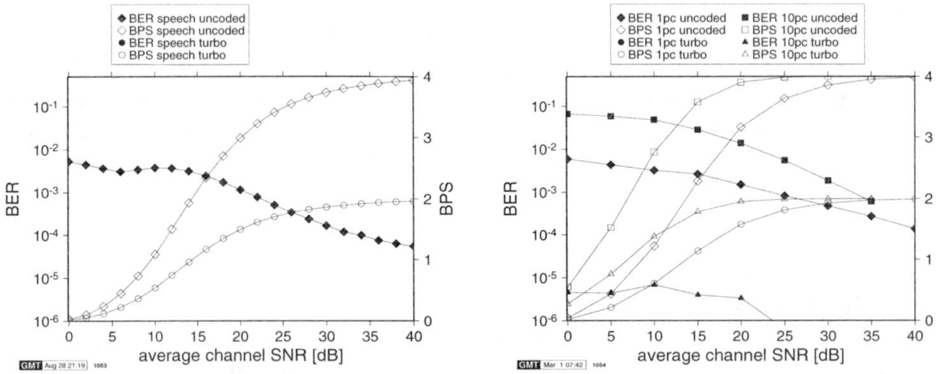

(**a**) speech system

(**b**) maximal BER 1% and 10%

Figure 22.14: BER and BPS throughput performance of 16 sub-band, 512 subcarrier adaptive turbo coded and uncoded OFDM modem employing (M_0, M_1, M_2, M_4) for (a) speech-type switching levels of Table 22.1 and (b) a maximal estimated sub-band BER of 1% and 10% over the channel of Figure 20.3. The turbo-coded transmission over the speech system and the 1% maximal BER system are error-free for all examined SNR values and therefore the corresponding BER curves are omitted from the graphs, hence the lack of black circles on (a) and (b)

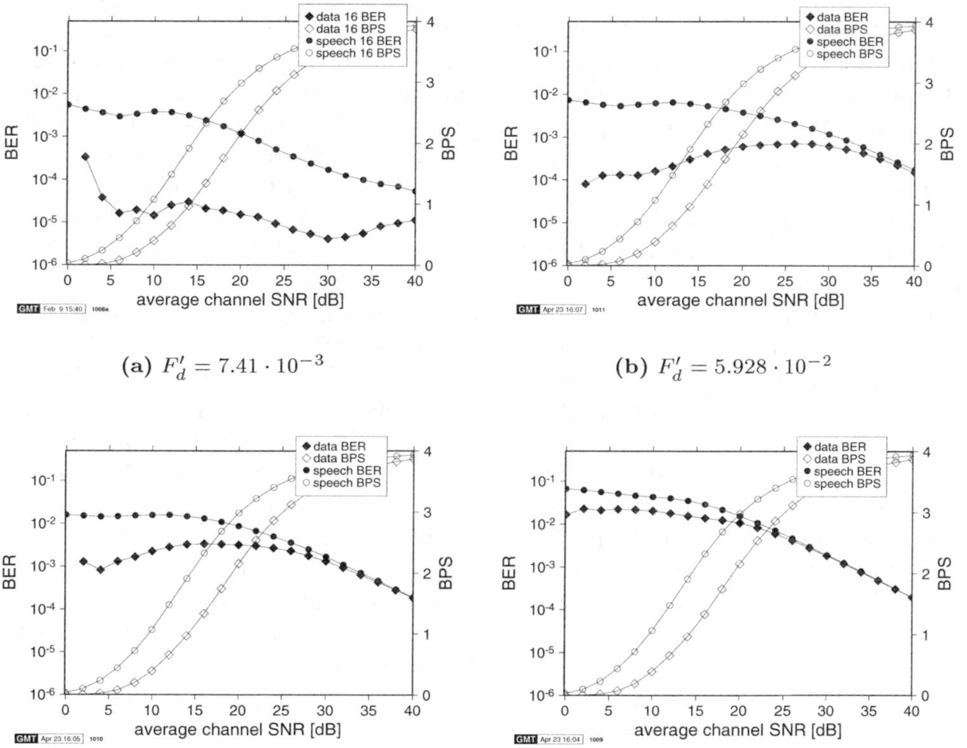

(a) $F'_d = 7.41 \cdot 10^{-3}$

(b) $F'_d = 5.928 \cdot 10^{-2}$

(c) $F'_d = 1.1856 \cdot 10^{-1}$

(d) $F'_d = 2.3712 \cdot 10^{-1}$

Figure 22.15: BER and BPS throughput performance of 16 sub-band, 512 subcarrier adaptive OFDM modem employing (M_0, M_1, M_2, M_4) for both data-type and speech-type switching levels with perfect modulation mode detection and different frame normalised Doppler frequencies F'_d over the channel of Figure 20.3. The triangular markers in (a) show the performance of a subcarrier-by-subcarrier adaptive modem using the data-type switching levels of Table 22.1 for comparison

of the channel transfer function between the uplink and downlink time slots. We assume that the time delay between the uplink and downlink slots is the same as the delay between the downlink and uplink slots, and we refer to this time as the frame duration T_f. We normalise the maximal Doppler frequency f_d of the channel to the frame duration T_f, and define the *frame normalised Doppler frequency* F'_d as $F'_d = f_d \cdot T_f$. Figure 22.15 depicts the fixed switching level (see Table 22.1) modem's BER and throughput performance in bits per symbol (BPS) for values of F'_d between $7.41 \cdot 10^{-3}$ and $2.3712 \cdot 10^{-1}$. These values stem from the studied WATM system with a time slot duration of 2.67 μs and up/downlink delays of 1, 8, 16 and 32 time slots at a channel Doppler frequency of 2.78 kHz. As mentioned in Section 20.1.1.1, this corresponds to a system employing a carrier frequency of 60 GHz, a sampling rate of 225 Msamples/s and a vehicular velocity of 50 km/h or $13.\bar{8}$ m/s.

Figure 22.15(a) shows the BER and BPS throughput of the studied modems in a framework with consecutive uplink and downlink time slots. This corresponds to $F_d' = 7.41 \cdot 10^{-3}$, while the target bit error rates for the speech and data system are met for all SNR values above 4 dB, and the BER performance is generally better than the target error rates. This was explained above with the conservative choice of modulation modes based on the most corrupted subcarrier in each sub-band, resulting in lower throughput and lower bit error rates for the switching level based sub-band adaptive modem.

Comparing Figure 22.15(a) with the other performance curves, it can be seen that the bit error rate performance for both the speech and the data system suffer from increasing decorrelation of the predicted and actual channel transfer function for increasing values of F_d'. In Figure 22.15(b) an 8 time slot delay was assumed between uplink and downlink time slots, which corresponds to $F_d' = 5.928 \cdot 10^{-2}$, and therefore the BER performance of the modem was significantly deteriorated. The "speech" system still maintains its target BER, but the "data" system delivers a BER of up to 10^{-3} for SNR values between 25 and 30 dB. It is interesting to observe that the delayed channel prediction mainly affects the higher-order modulation modes, which are employed more frequently at high SNR values. This explains the shape of the BER curve for the "data" system, which is rising from below 10^{-4} at 2 dB SNR up to 10^{-3} at 26 dB SNR. The average throughput of the modem is mainly determined by the statistics of the estimated channel transfer function at the receiver, and this is therefore not affected by the delay between the channel estimation and the packet transmission.

22.2.9 Channel estimation

All the adaptive modems above rely on the estimate of the frequency domain channel transfer function, both for equalisation of the received symbols at the receiver, as well as for the modem mode adaptation of the next transmitted OFDM symbol. Figure 22.16 shows the BER versus SNR curves for the 1% target BER modem, as it was presented above, if pilot symbol assisted channel estimation [12] is employed instead of the previously used delayed, but otherwise perfect, channel estimation.

Comparing the curves for perfect channel estimation and for the 64-pilot lowpass interpolation algorithm, it can be seen that the modem falls short of the target bit error rate of 1% for channel SNR values of up to 20 dB. More noise resilient channel estimation algorithms can improve the modem's performance. If the passband width of the interpolation lowpass filter (see Section 20.3.1.1) is halved, which is indicated in Figure 22.16 as the reduced bandwidth (red. bw.) scenario, then the BER gap between the perfect and the pilot symbol assisted channel estimation narrows, and a BER of 1% is achieved at a SNR of 15 dB. Additionally, employing pairs of pilots with the above bandwidth-limited interpolation scheme further improves the modem's performance, which results in BER values below 1% for SNR values above 5 dB. The averaging of the pilot pairs improves the noise resilience of the channel estimation, but introduces estimation errors for high SNR values. This can be observed in the

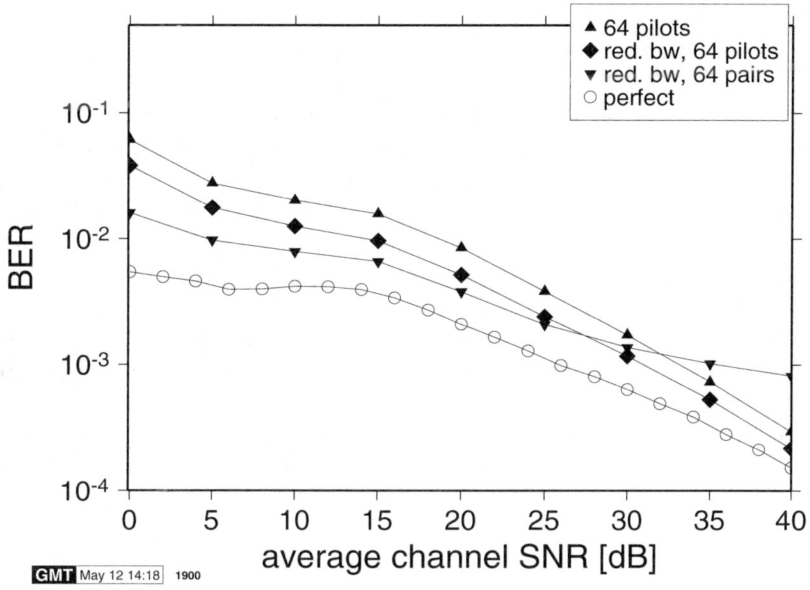

Figure 22.16: BER versus channel SNR performance for the 1% target BER adaptive 16 sub-band, 512 subcarrier OFDM modem employing pilot symbol assisted channel transfer function estimation over the channel of Figure 20.3

residual BER in the figure.

Having studied a range of different AOFDM modems, let us now embark on a system design study in the context of an adaptive interactive speech system.

22.3 Adaptive OFDM Speech System[1]

22.3.1 Introduction

In this section we introduce a bidirectional high-quality audio communications system, which will be used to highlight the systems aspects of adaptive OFDM transmissions over time dispersive channels. Specifically, the channel-coded adaptive transmission characteristics and a potential application for joint adaptation of modulation, channel coding and source coding are studied.

The basic principle of adaptive modulation is to react to the anticipated channel capacity for the next OFDM symbol transmission burst, by employing modulation modes of different robustness to channel impairments and of different data throughput. The trade-off between data throughput and integrity can be adapted to different system environments. For data transmission systems, which are not dependent on

[1]T. Keller, M. Münster, L. Hanzo: A Turbo-coded Burst-by-burst Adaptive Wideband Speech Transceiver, submitted to ©IEEE JSAC, 2000 [435]. Details concerning the source codec and its performance are discussed in the above publication.

a fixed data rate and do not require low transmission delays, variable-throughput adaptive schemes can be devised that operate efficiently with powerful error correction coders, such as long block length turbo codes [436]. Real-time audio or video communications employing source codecs, which allow variable bit rates, can also be used in conjunction with variable-rate adaptive schemes, but in this case block-based error correction coders cannot be readily employed.

Fixed-rate adaptive OFDM systems, which sacrifice a guaranteed BER performance for the sake of a fixed data throughput, are more readily integrated into interactive communications systems, and can coexist with long block based channel coders in real-time applications.

For these investigations, we propose a hybrid adaptive OFDM scheme, based on a multimode constant throughput algorithm, consisting of two adaptation loops: an inner constant throughput algorithm, having a bit rate consistent with the source and channel coders, and an outer mode switching control loop, which selects the target bit rate of the whole system from a set of distinct operating modes. These issues will become more explicit during our further discourse.

22.3.2 System overview

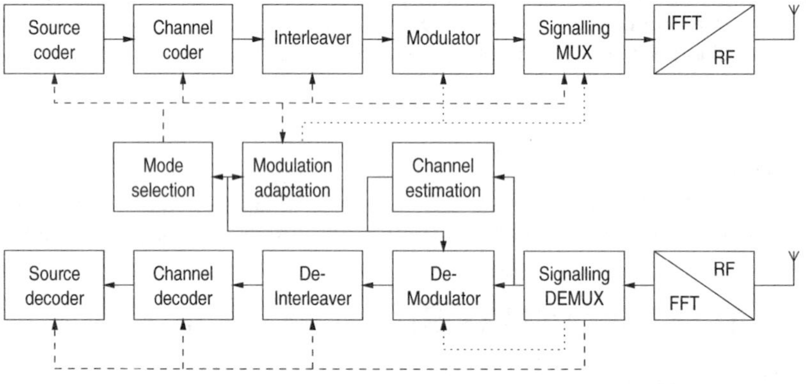

Figure 22.17: Schematic model of the multimode adaptive OFDM system

The structure of the studied adaptive OFDM modem is depicted schematically in Figure 22.17. The top half of the diagram is the transmitter chain, which consists of the source and channel coders, a channel interleaver decorrelating the channel's frequency domain fading, an adaptive OFDM modulator, a multiplexer adding signalling information to the transmit data, and an IFFT/RF OFDM block. The receiver, at the lower half of the schematic, consists of a RF/FFT OFDM receiver, a demultiplexer extracting the signalling information, an adaptive demodulator, a de-interleaver/channel decoder and the source decoder. The parameter adaptation linking the receiver and transmitter chain consists of a channel estimator, and the throughput mode selection as well as the modulation adaptation blocks.

The open-loop control structure of the adaptation algorithms can be seen in the figure: the receiver's operation is controlled by the signalling information that is contained in the received OFDM symbol, while the channel quality information generated by the receiver is employed to determine the remote transmitter's matching parameter set by the modulation adaptation algorithms. The two distinct adaptation loops distinguished by the dotted and dashed lines are the inner and outer adaptation loops, respectively. The outer adaptation loop controls the overall throughput of the system, which is chosen from a finite set of predefined modes, so that a fixed delay decoding of the received OFDM data packets becomes possible. This outer loop controls the block length of the channel encoder and interleaver, and the target throughput of the inner adaptation loop. The operation of the adaptive modulator, controlled by the inner loop, is transparent to the rest of the system. The operation of the adaptation loops is described in more detail below.

22.3.2.1 System parameters

The transmission parameters have been adopted from the TDD mode of the UMTS system of Section 20.1.3, with a carrier frequency of 1.9 GHz, a time frame and time slot duration of 4.615 ms and 122 μs, respectively. The sampling rate is assumed to be 3.78 MHz, leading to a 1024-subcarrier OFDM symbol with a cyclic extension of 64 samples in each time slot. For spectral shaping of the OFDM signal, there are a total of 206 virtual subcarriers at the bandwidth boundaries.

The 7 kHz bandwidth PictureTel audio codec[2] has been chosen for this system because of its good audio quality, robustness to packet dropping and adjustable bit rate. The channel encoder/interleaver combination is constituted by a convolutional turbo codec [433] employing block turbo interleavers in conjunction with a subsequent pseudorandom channel interleaver. The constituent half-rate recursive systematic convolutional (RSC) encoders are of constraint length 3, with octal generator polynomials of $(7, 5)$ [403]. At the decoder, 8 iterations are performed, utilising the so-called maximum a posteriori (MAP) [434] algorithm and log-likelihood ratio soft inputs from the demodulator.

The channel model consists of a four-path COST 207 typical urban impulse response [410], where each impulse is subjected to independent Rayleigh fading with a normalised Doppler frequency of $2.25 \cdot 10^{-6}$, corresponding to a pedestrian scenario with a walking speed of 3 mph.

The unfaded impulse response and the time- and frequency-varying amplitude of the channel transfer function are depicted in Figure 22.18.

22.3.3 Constant throughput adaptive modulation

The constant throughput adaptive algorithm attempts to allocate a given number of bits for transmission in subcarriers exhibiting a low BER, while the use of high BER subcarriers is minimised. We employ the open-loop adaptive regime of Figure 22.1, basing the decision concerning the next transmitted OFDM symbol's modulation scheme allocation on the channel estimation gained at the reception of the most recently received OFDM symbol by the local station. Sub-band adaptive modulation

[2]see http://www.picturetel.com

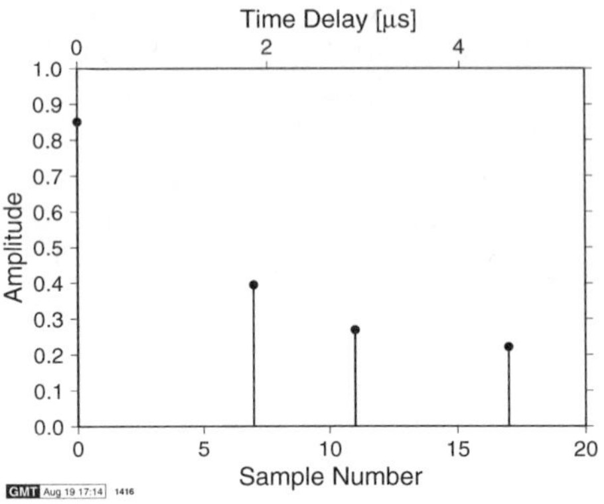

(a) unfaded impulse response

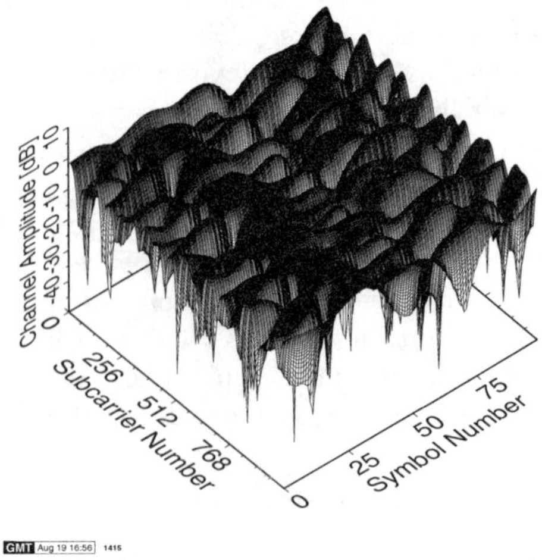

(b) time varying amplitude

Figure 22.18: Channel for PictureTel experiments: (a) unfaded channel impulse response (b) time-varying channel amplitude for 100 OFDM symbols

[437] – where the modulation scheme is adapted not on a subcarrier-by-subcarrier basis, but for sub-bands of adjacent subcarriers – is employed in order to simplify the signalling requirements. The adaptation algorithm was highlighted in Section 22.2.5. For these investigations we employed 32 sub-bands of 32 subcarriers in each OFDM symbol. Perfect channel estimation and signalling were used.

22.3.3.1 Constant-rate BER performance

Figure 22.19 characterises the fixed-throughput adaptive modulation scheme's performance under the channel conditions characterised above, for a block length of 578 coded bits. As a comparison, the BER curve of a fixed BPSK modem transmitting the same number of bits in the same channel, employing 578 out of 1024 subcarriers, is also depicted. The number of useful audio bits per OFDM symbol was based on a 200-bit target data throughput, which corresponds to a 10 kbps data rate, padded with 89 bits, which can contain a checksum for error detection and high-level signalling information. Furthermore, half-rate channel coding was used.

The BER plotted in the figure is the hard decision based bit error rate at the receiver before channel decoding. It can be seen that the adaptive modulation scheme yields a significantly improved performance, which is also reflected in the frame error rate (FER). This FER approximates the probability of a decoded block containing errors, in which case it is unusable for the audio source decoder and hence it is dropped. This error event can be detected by using the checksum in the OFDM data symbol.

As an example, the modulation mode allocation for the 578 data bit adaptive

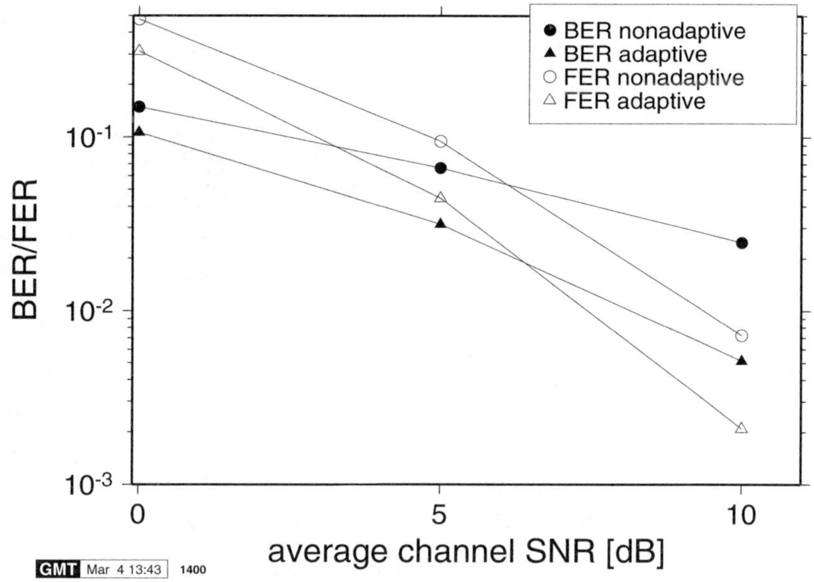

Figure 22.19: BER and FER performance for the fixed-throughput adaptive and non-adaptive OFDM modems in the fading time dispersive channel of Section 20.1.3 for a block length of 578 coded bits

modem at an average channel SNR of 5 dB is given in Figure 22.20(a) for 100 consecutive OFDM symbols. The unused sub-bands with indexes 15 and 16 contain the virtual carriers, and therefore do not transmit any data. It can be seen that the constant throughput adaptation algorithm of Section 22.2.5 allocates data to the higher quality subcarriers on a symbol-by-symbol basis, while keeping the total number of bits per OFDM symbol constant. As a comparison, Figure 22.20(b) shows the equivalent overview of the modulation modes employed for a fixed bit rate of 1458 bits per OFDM symbol. It can be seen that in order to meet this increased throughput target, hardly any sub-bands are in "no transmission" mode, and overall higher-order modulation schemes have to be employed.

22.3.4 Multimode adaptation

While the fixed-throughput adaptive algorithm described above copes with the frequency domain fading of the channel, there is a medium-term variation of the overall channel capacity due to time domain channel quality fluctuations as indicated in Figure 22.18(b). While it is not straightforward to employ powerful block-based channel coding schemes, such as turbo coding, in variable-throughput adaptive OFDM schemes for real-time applications like voice or video telephony, a multimode adaptive system can be designed that allows us to switch between a set of different source and channel codecs as well as transmission parameters, depending on the overall channel quality. We have investigated the use of the estimated overall BER at the output

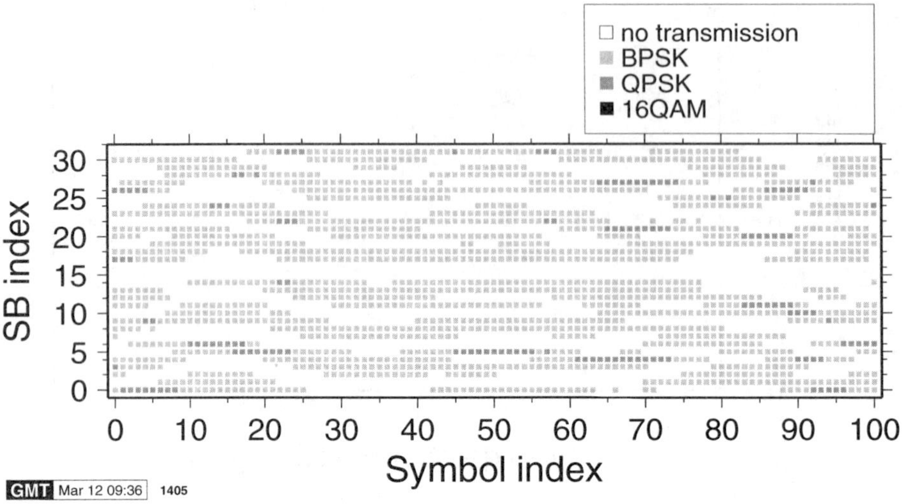

(a) 578 data bits per OFDM symbol

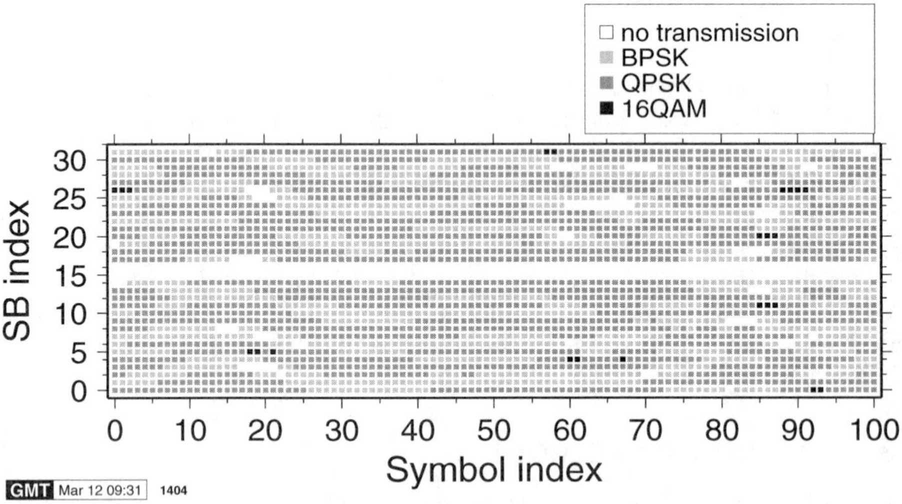

(b) 1458 data bits per OFDM symbol

Figure 22.20: Overview of modulation mode allocation for fixed-throughput adaptive modems over the fading time dispersive channel of Figure 22.18(b) at 5 dB average channel SNR of the receiver, which is the sum of all the $e(j, s_j)$ quantities of Equation 22.1 after adaptation. On the basis of this expected bit error rate at the input of the channel decoder, the probability of a frame error (FER) must be estimated and compared with the estimated FER of the other modem modes. Then the mode having the highest throughput exhibiting an estimated FER of less than 10^{-6} – or alternatively the mode exhibiting the lowest FER – is selected and the source encoder, the channel encoder and the adaptive modem are set up accordingly.

We have defined four different operating modes, which correspond to unprotected audio data rates of 10, 16, 24, and 32 kbps at the source codec's interface. With half-rate channel coding and allowing for checksum and signalling overhead, the number of transmitted coded bits per OFDM symbol was 578, 722, 1058 and 1458 for the four modes, respectively.

22.3.4.1 Mode switching

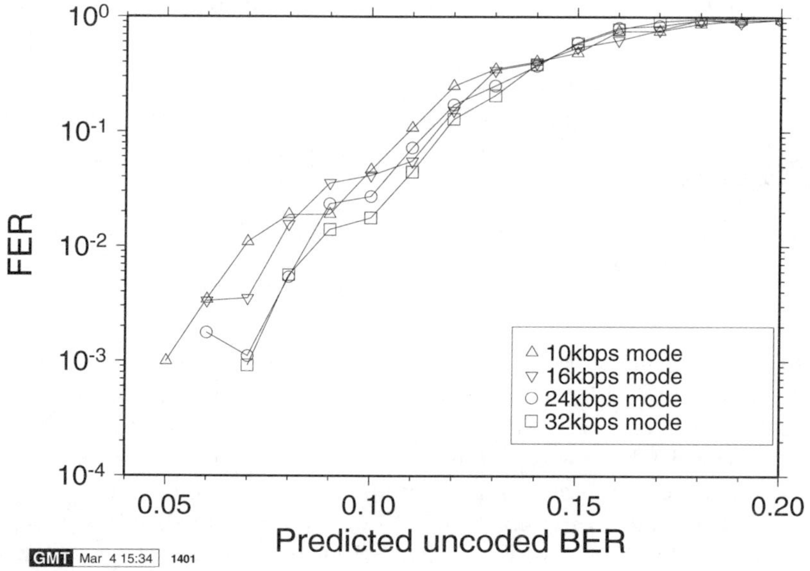

Figure 22.21: Frame error rate versus the predicted unprotected BER for 10 kbps, 16 kbps, 24 kbps and 32 kbps modes

Figure 22.21 shows the *observed* FER for all four modes versus the unprotected BER that was *predicted* at the transmitter. The predicted unprotected BER was discretised into intervals of 1%, and the channel-coded FER was evaluated over these BER intervals. It can be seen from the figure that for estimated protected BER values below 5% no frame errors were observed for any of the modes. For higher estimated unprotected BER values, the higher throughput modes exhibited a lower FER than the lower throughput modes, which was consistent with the turbo coder's performance increase for longer block lengths. A FER of 1% was observed for a 7%

predicted unprotected error rate for the 10 kbps mode, and BER values of 8% to 9% were allowed for the longer blocks, whilst maintaining a FER of less than 1%

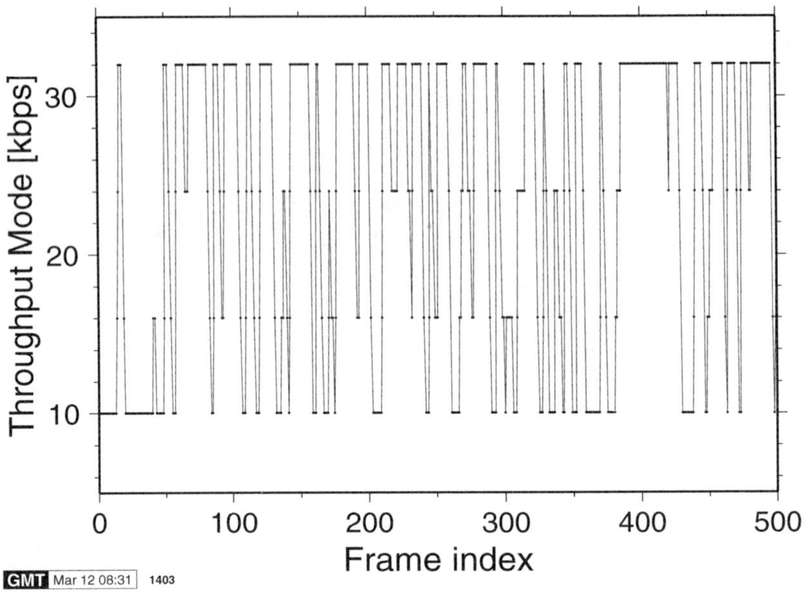

Figure 22.22: Mode switching pattern at 5 dB channel SNR over the channel of Figure 22.18(b)

For this experiment, we assumed the best-case scenario of using the actual measured FER statistics of Figure 22.21 for the mode switching algorithm rather than estimating the FER on the basis of the estimated uncoded BER. In this case, the previously observed FER corresponding to the predicted overall BER values for the different modes were compared, and the mode having the lowest FER was chosen for transmission. The mode switching sequence for the first 500 OFDM symbols at 5 dB channel SNR over the channel of Figure 22.18(b) is depicted in Figure 22.22. It can be seen that in this segment of the sequence 32 kbps transmission is the most frequently employed mode, followed by the 10 kbps mode. The intermediate modes are mostly transitory, as the improving or deteriorating channel conditions makes it necessary to switch between the 10 kbps and 32 kbps modes. This behaviour is consistent with Table 22.2, for the Switch-I scheme.

22.3.5 Simulation results

The comparison between the different adaptive schemes will be based on a channel SNR of 5 dB over the channel of Figure 22.18(b), since the audio codec's performance is unacceptable for SNR values around 0 dB, and as the adaptive modulation is most effective for channel SNR values below 10 dB.

Scheme	FER [%]	Rate-10kbps [%]	Rate-16kbps [%]	Rate-24kbps [%]	Rate-32kbps [%]
Fixed-10kbps	4.45	95.55	0.0	0.0	0.0
Fixed-16kbps	5.58	0.0	94.42	0.0	0.0
Fixed-24kbps	10.28	0.0	0.0	89.72	0.0
Fixed-32kbps	18.65	0.0	0.0	0.0	81.35
Switch-I	4.44	21.87	13.90	11.59	48.20
Switch-II	5.58	0.0	34.63	11.59	48.20

Table 22.2: FER and relative usage of different bit rates in the fixed bit rate schemes and the variable schemes Switch-I and Switch-II (successfully transmitted frames) for a channel SNR of 5 dB over the channel of Figure 22.18(b)

22.3.5.1 Frame error results

The audio experiments [435] have shown that the audio quality is acceptable for frame dropping rates of about 5%, and that the perceived audio quality increases with increasing throughput. Table 22.2 gives an overview of the frame error rates and mode switching statistics of the system for a channel SNR of 5 dB over the channel of Figure 22.18(b). It can be seen that for the fixed modes the FER increases with the throughput, from 4.45% in the 10 kbps mode up to 18.65% for the 32 kbps mode. This is because the turbo codec performance improves for longer interleavers, the OFDM symbol had to be loaded with more bits, hence there was a higher unprotected BER. The time-variant bit rate mode-switching schemes, referred to as Switch-I and Switch-II for the four- and three-mode switching regimes used, deliver frame dropping rates of 4.44% and 5.58%, respectively. Both these FER values are acceptable for the audio transmission. It can be seen that upon incorporating the 10 kbps mode in the switching regime Switch-I of Table 22.2 the overall FER is lowered only by an insignificant amount, while the associated average throughput is reduced considerably.

22.3.5.2 Audio segmental SNR

Figure 22.23 displays the cumulative density function (CDF) of the segmental SNR (SEGSNR) [403] obtained from the reconstructed signal of an audio test sound for all the modes of Table 22.2 discussed above at a channel SNR of 5 dB over the channel of Figure 22.18(b).

Focusing our attention on the figure, we can draw a whole range of interesting conclusions. As expected, for any given SEGSNR it is desirable to maintain as low a proportion as possible of the audio frames' SEGSNRs below a given abscissa value. Hence we conclude that the best SEGSNR CDF was attributable to the Switch-II scheme, while the worst performance was observed for the fixed 10 kbps scheme. Above a SEGSNR of 15 dB the CDFs of the fixed 16, 24 and 32 kbps modes follow our expectations. Viewing matters from a different perspective, the Switch-II scheme exhibits a SEGSNR of less than 20 dB with a probability of 0.8, compared to 0.95 for the fixed 10 kbps scheme.

Before concluding we also note that the CDFs do not have a smoothly tapered tail, since for the erroneous audio frames a SEGSNR of 0 dB was registered. This results in the step-function-like behaviour associated with the discontinuities corresponding to the FER values in the FER column of Table 22.2.

22.4 Pre-equalisation

We have seen above how the receiver's estimate of the channel transfer function can be employed by the transmitter in order to dramatically improve the performance of an OFDM system by adapting the subcarrier modulation modes to the channel conditions. For sub-channels exhibiting a low signal-to-noise ratio, robust modulation modes were used, while for subcarriers having a high SNR, high throughput multi-level modulation modes can be employed. An alternative approach to combating the frequency selective channel behaviour is to apply pre-equalisation to the OFDM symbol prior to transmission on the basis of the anticipated channel transfer function. The optimum scheme for the power allocation to the subcarriers is Shannon's water-pouring solution [438]. We will investigate a range of related topics in this section.

22.4.1 Motivation

As discussed above, the received data symbol R_n of subcarrier n over a stationary time dispersive channel can be characterised by:

$$R_n = S_n \cdot H_n + n_n,$$

where S_n is the transmitted data symbol, H_n is the channel transfer function of subcarrier n, and n_n is a noise sample.

The frequency domain equalisation at the receiver, which is necessary for non-differential detection of the data symbols, corrects the phase and amplitude of the received data symbols using the estimate of the channel transfer function $\hat{H}_n$ as follows:

$$R'_n = R_n/\hat{H}_n = S_n \cdot H_n/\hat{H}_n + n_n/\hat{H}_n.$$

If the estimate $\hat{H}_n$ is accurate, this operation defades the constellation points before decision. However, upon defading, the noise sample n_n is amplified by the same amount as the signal, therefore preserving the SNR of the received sample.

Pre-equalisation for the OFDM modem operates by scaling the data symbol of subcarrier n, S_n, by a predistortion function E_n, computed from the inverse of the anticipated channel transfer function, prior to transmission. At the receiver, no equalisation is performed, hence the received symbols can be expressed as:

$$R_n = S_n \cdot E_n \cdot H_n + n_n.$$

Since no equalisation is performed, there is no noise amplification at the receiver. Similarly to the adaptive modulation techniques illustrated above, pre-equalisation is only applicable to a duplex link, since the transmitted signal is adapted to the specific channel conditions perceived by the receiver. As in other adaptive schemes, the transmitter needs an estimate of the current frequency domain channel transfer

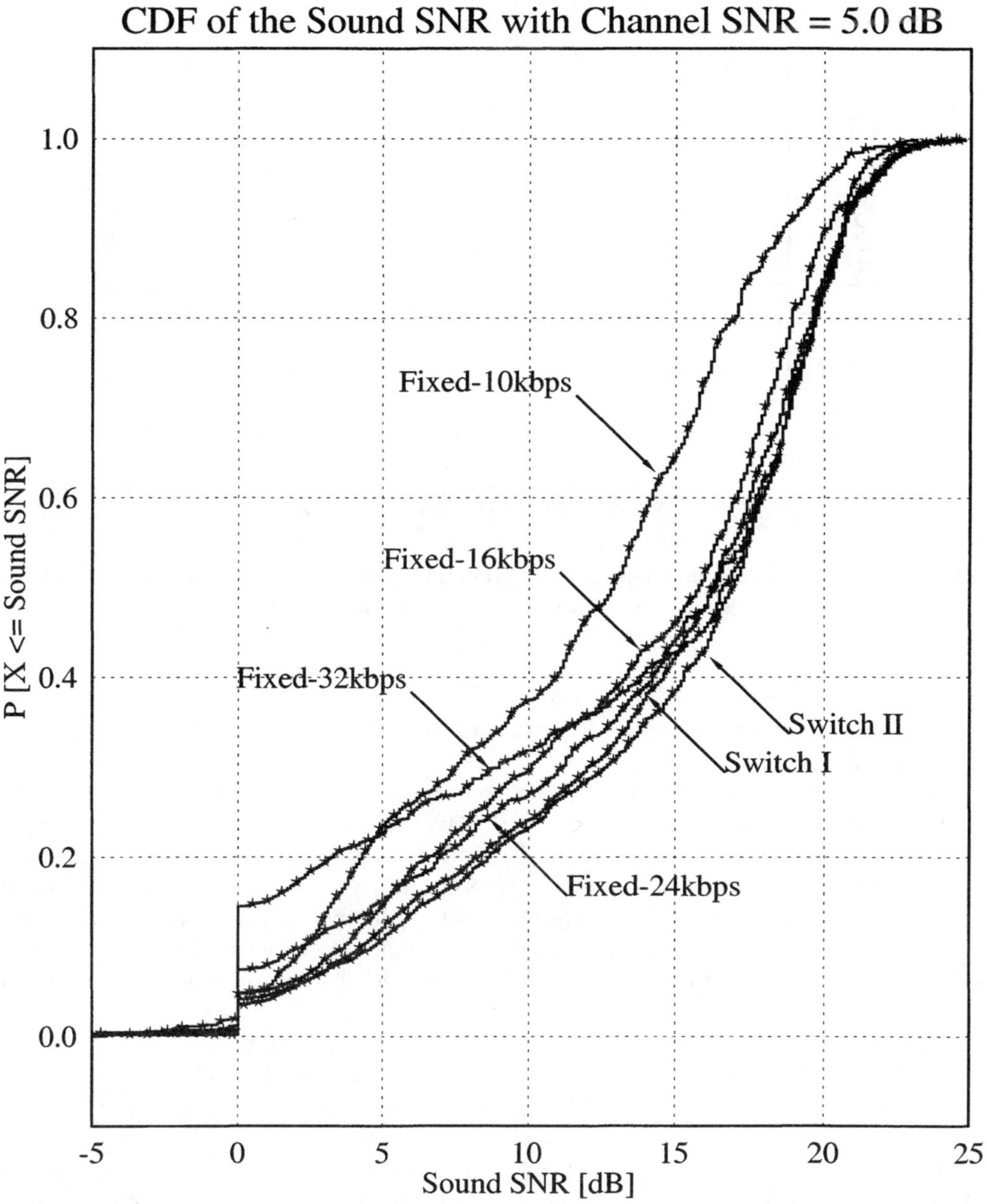

Figure 22.23: Typical CDF of the segmental SNR of a reconstructed audio signal transmitted over the fading time dispersive channel of Section 20.1.3 at a channel SNR of 5 dB (from [435])

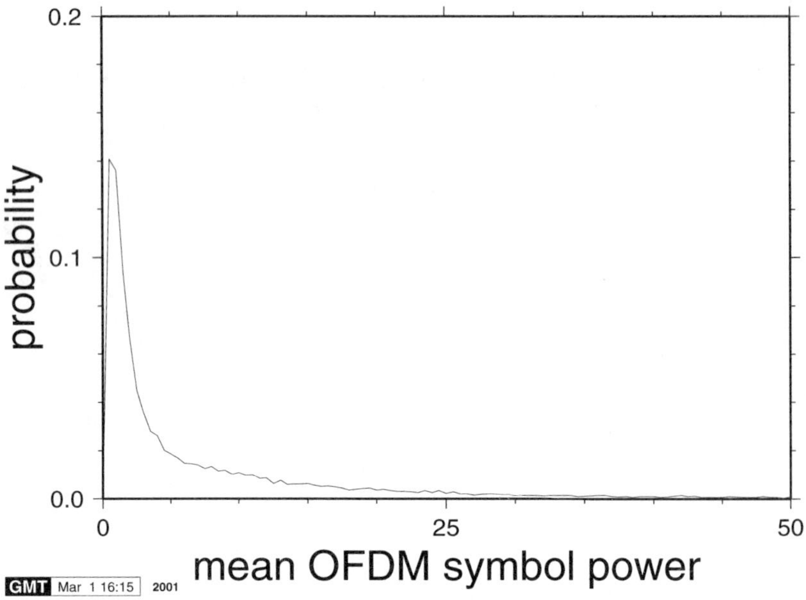

Figure 22.24: OFDM symbol energy histogram for 512-subcarrier 16-QAM with full channel inversion over the short WATM channel of Figure 20.3

function, which can be obtained from the received signal in the reverse link, as seen in Figure 22.1.

The simplest choice of the pre-equalisation transfer function E_n is the inverse of the estimated frequency domain channel transfer function, $E_n = 1/\hat{H}_n$. If the estimation of the channel transfer function is accurate, then perfect channel inversion would result in an AWGN-like channel perceived at the receiver, since the anticipated time and frequency dependent behaviour of the channel is precompensated at the transmitter. The BER performance of such a system, accordingly, is identical to that of the equivalent modem in an AWGN channel with respect to the received signal power.

Since the pre-equalisation algorithm amplifies the power in each subcarrier by the corresponding estimate of the channel transfer function, the transmitter's output power fluctuates in an inverse fashion with respect to time variant channel. The fades in the frequency domain channel transfer function can be deep, hence the transmit power in the corresponding subcarriers may be high. Figure 22.24 shows the histogram of the total OFDM symbol energy at the transmitter's output for the short WLAN channel of Figure 20.3 in conjunction with "full channel inversion", normalised to the fixed average output energy. It can be seen that the OFDM symbol energy fluctuates widely, with observed peak values in excess of 55. The long-term mean symbol energy was measured to be 22.9, which corresponds to an average output power increase of 13.6 dB. This naturally would impose unacceptable constraints on the required perfectly linear dynamic range of the power amplifier. Hence in practice only limited

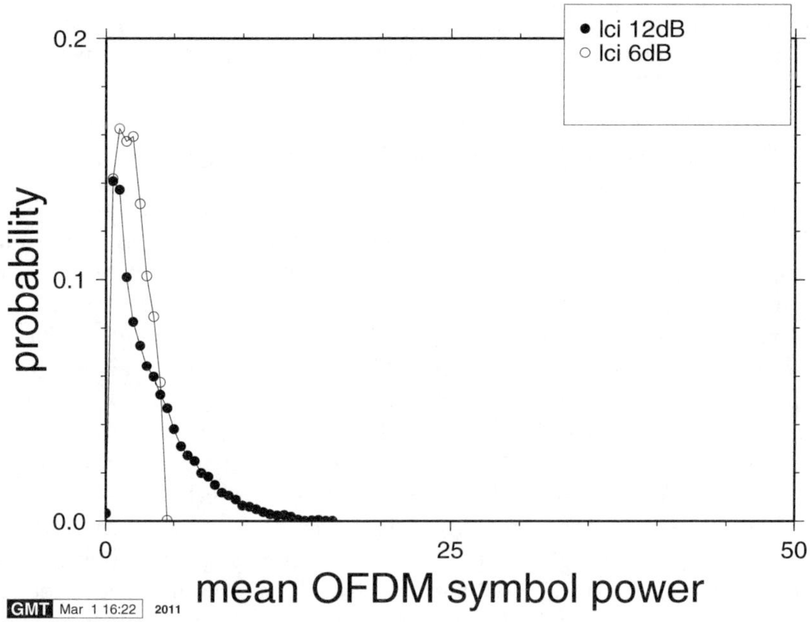

Figure 22.25: OFDM symbol energy histogram for 512-subcarrier 16-QAM transmission with limited dynamic range channel inversion (lci) over the WATM channel of Figure 20.3.

dynamic scenarios can be considered.

Explicitly, in order to limit the associated transmit power fluctuations, the dynamic range of the pre-equalisation algorithm can be limited to a value l, so that the following relations apply:

$$E_n = a_n \cdot e^{-j\phi_n}, \text{with} \tag{22.3}$$

$$\phi_n = \angle \hat{H}_n \quad \text{and} \tag{22.4}$$

$$a_n = \begin{cases} \left|\hat{H}_n\right| & \text{for } \left|\hat{H}_n\right| \leq l \\ l & \text{otherwise.} \end{cases} \tag{22.5}$$

Limiting the values of E_n to the value of l does not affect the phase of the channel pre-equalisation. Depending on the modulation mode employed for transmission, reception of the symbols affected by the amplitude limitation is still possible, perhaps for phase shift keying. Multilevel modulation modes exploiting the received symbol's amplitude will be affected by the imperfect pre-equalisation. The associated mean OFDM symbol power histogram is shown in Figure 22.25. Given - for example - a maximum allowed amplification factor of 6 and 12 dB, the normalised transmitted OFDM symbol power in the figure should be limited to 4 and 16, respectively. However, in practice even higher values may be observed, which is due to the OFDM symbol's energy fluctuation as a function of the specific data sequence, if multilevel modulation modes are used. In order to circumvent the above peak-to-mean envelope fluctuation problems, it is in practical terms more attractive to combine pre-equalisation with sub-band blocking, which is the topic of the next section.

22.4.2 Pre-equalisation with sub-band blocking

Limiting the maximal amplification of the subcarriers leads to a reduced BER performance compared to the full channel inversion, but the system performance can be improved by identifying the subcarriers that cannot be fully pre-equalised and by disabling subsequent transmission in these subcarriers. This "blocking" of the transmission in certain subcarriers can be seen as adaptive modulation with two modulation modes, which introduces the problem of modulation mode signalling. As has been discussed in the context of Figure 22.1 for the adaptive modulation modems above, this signalling task can be solved in different ways, namely by blind detection of blocked subcarriers, or by transmitting explicit signalling information contained in the data block. We have seen above that employing sub-band adaptivity - rather than subcarrier-by-subcarrier adaptivity - simplifies both the modem mode detection as well as the mode signalling, at the expense of a lower system throughput. In order to keep the system's complexity low and to allow for simple modem mode signalling or blind detection, we will assume a 16 subband adaptive scheme here.

Analogously to the adaptive modulation schemes above, the transmitter decides for all subcarriers in each sub-band, whether to transmit data or not. If pre-equalisation is possible under the power constraints, then the subcarriers are modulated with the pre-equalised data symbols. The information on whether or not a sub-band is used for transmission, this is signalled to the receiver.

Since no attempt is made to transmit in the sub-bands that cannot be power-efficiently pre-equalised, the power conserved in the blank subcarriers can be used for "boosting" the data-bearing sub-bands. This scheme allows for a more flexible pre-equalisation algorithm than the fixed threshold method described above; here is a summary:

(1) Calculate the necessary transmit power p_n for each sub-band n, assuming perfect pre-equalisation.

(2) Sort sub-bands according to their required transmit power p_n.

(3) Select sub-band n with the lowest power p_n, and add p_n to the total transmit power. Repeat this procedure with the next lowest power, until no further sub-bands can be added without the total power $\sum p_j$ exceeding power limit l.

Figure 22.26 depicts the 16-QAM BER performance over the short WATM channel of Figure 20.3. The BER floor stems from the channel's time-variant nature, since there is a delay between the channel estimation instant and the instant of transmission. The average throughput figures for the 6 dB and 12 dB symbol energy limits are 3.54 and 3.92 bits per data symbol, respectively. It can be noted that the BER floor is lower for $l = 6$ dB than for $l = 12$ dB. This is because the effects of the channel variation due to the delay between the instants of channel estimation and reception in the faded subcarriers on the equalisation function are more dramatic, than in the higher

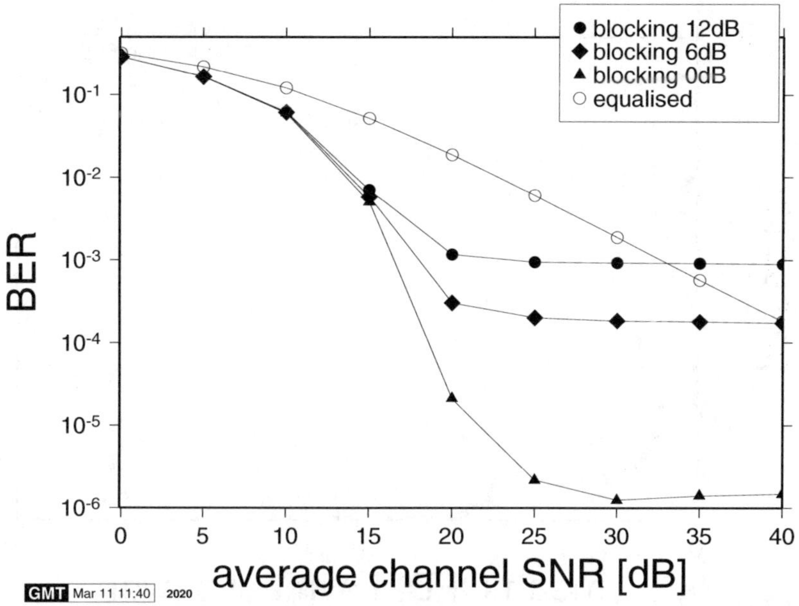

Figure 22.26: BER performance of the 512-subcarrier 16-QAM OFDM modem over the fading short WATM channel of Figure 20.3 employing 16 sub-band pre-equalisation with blocking and a delay of 1 time slot between the instants of perfect channel estimation and reception. Note that the transmit power is not shown in this figure

quality subcarriers. The lower the total symbol energy limit l, the smaller the number of low quality subcarriers used for transmission. If the symbol energy is limited to 0 dB, then the BER floor drops to $1.5 \cdot 10^{-6}$ at the expense of the throughput, which attains 2.5 BPS. Figure 22.27 depicts the mean OFDM symbol energy histogram for this scenario. It can be seen that, compared with the limited channel inversion scheme of Figure 22.25, the allowable symbol energy is more efficiently allocated, with a higher probability of high energy OFDM symbols. This is the result of the flexible reallocation of energy from blocked sub-bands, instead of limiting the output power on a subcarrier-by-subcarrier basis.

22.4.3 Adaptive modulation with spectral predistortion

The pre-equalisation algorithms discussed above invert the channel's anticipated transfer function, in order to transform the resulting channel into a Gaussian-like non-fading channel, whose SNR is dependent only on the path loss. Sub-band blocking has been introduced above, in order to limit the transmitter's output power, while maintaining the near-constant SNR across the used subcarriers. The pre-equalisation algorithms discussed above do not cancel out the channel's path loss, but rely on the receiver's gain control algorithm to automatically account for the channel's average

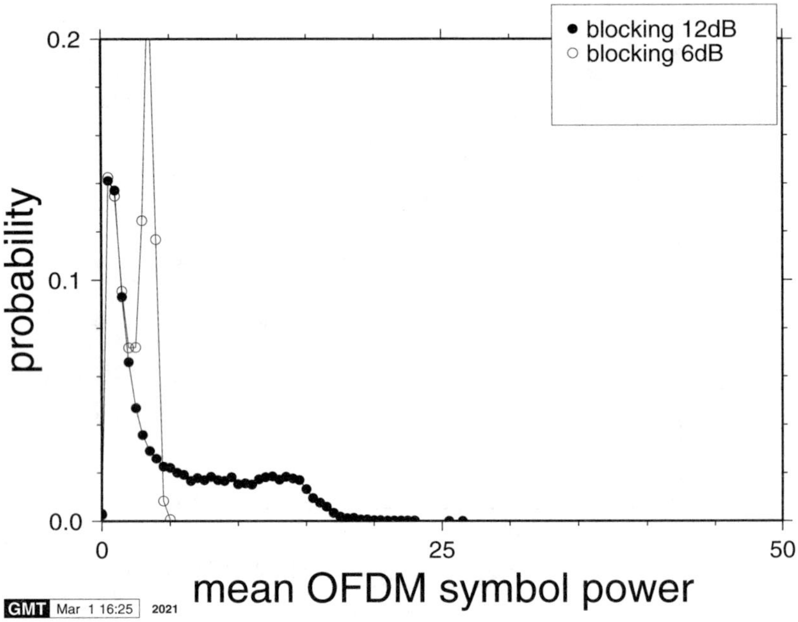

Figure 22.27: OFDM symbol energy histogram for 512 subcarrier, 16 sub-band pre-equalisation with blocking over the short WATM channel of Figure 20.3 using 16-QAM. The corresponding BER curves are given in Figure 22.26

path loss.

We have already that maintaining Gaussian channel characteristics is not the most efficient way of exploiting the channel's time variant capacity. If maintaining a constant data throughput is not required by the rest of the communications system, then a fixed BER scheme in conjunction with error correction coding can assist in maximising the system's throughput. The results presented for the target BER adaptive modulation scheme in Figure 22.14(b) showed that, for the particular turbo coding scheme used, an uncoded BER of 1% resulted in error-free channel coded data transmission, and that for an uncoded target BER of 10% the turbo decoded data BER was below 10^{-5}. We have seen that it is impossible to exactly reach the anticipated uncoded target BER with the adaptive modulation algorithm, since the adaptation algorithm operates in discrete steps between modulation modes.

Combining the target BER adaptive modulation scheme and spectral predistortion allows the transmitter to react to the channel's time and frequency variant nature, in order to fine-tune the behaviour of the adaptive modem in fading channels. It also allows the transmitter to invest the energy that is not used in "no transmission" sub-bands into the other sub-bands without affecting the equalisation at the receiver.

The combined algorithm for adaptive modulation with spectral predistortion described here does not intend to invert the channel's transfer function across the OFDM

target BER	10^{-4}	1%	10%
SNR(BPSK)[dB]	8.4	4.33	−0.85
SNR(QPSK)[dB]	11.42	7.34	2.16
SNR(16QAM)[dB]	18.23	13.91	7.91

Table 22.3: Required target SNR levels for 1% and 10% target BER for the different modulation schemes over an AWGN channel

symbol's range of subcarriers, it is therefore not a pure pre-equalisation algorithm. Instead, the aim is to transmit a sub-band's data symbols at a power level which ensures a given target SNR at the receiver, that is constant for all subcarriers in the sub-band, which in turn results in the required BER. Clearly, the receiver has to anticipate the different relative power levels for the different modulation modes, so that error-free demodulation of the multilevel modulation modes employed can be ensured.

The joint adaptation algorithm requires the estimates of the noise floor level at the receiver as well as the channel transfer function, which includes the path loss. On the basis of these values, the necessary amplitude of E_n required to transmit a data symbol over the subcarrier n for a given received SNR of γ_n can be calculated as follows:

$$|E_n| = \frac{\sqrt{N_0 \cdot \gamma_n}}{\left|\hat{H}_n\right|},$$

where N_0 is the noise floor at the receiver. The phase of E_n is used for the pre-equalisation, and hence:

$$\angle E_n = -\angle \hat{H}_n.$$

The target SNR, of subcarrier n, γ_n, is dependent on the modulation mode that is signalled over the subcarrier, and determines the system's target BER. We have identified three sets of target SNR values for the modulation modes, with uncoded target BER values of 1% and 10% for use in conjunction with channel coders, as well as 10^{-4} for transmission without channel coding. Table 22.3 gives an overview of these levels, which have been read from the BER performance curves of the different modulation modes in a Gaussian channel.

Figure 22.28 shows the performance of the joint predistortion and adaptive modulation algorithm over the fading time-dispersive short WATM channel of Figure 20.3 for the set of different target BER values of Table 22.3, as well as the comparison curves of the perfectly equalised 16-QAM modem under the same channel conditions. It can be seen that the BER achieved by the system is close to the BER targets. Specifically, for a target BER of 10%, no perceptible deviation from the target has been recorded, while for the lower BER targets the deviations increase for higher channel SNRs. For a target BER of 1%, the highest measured deviation is at the SNR of 40 dB, where the recorded BER is 1.36%. For the target BER of 10^{-4}, the BER deviation is small at 0 dB SNR, but at an SNR of 40 dB the experimental BER is $2.2 \cdot 10^{-3}$. This increase of the BER with increasing SNR is due to the rapid channel

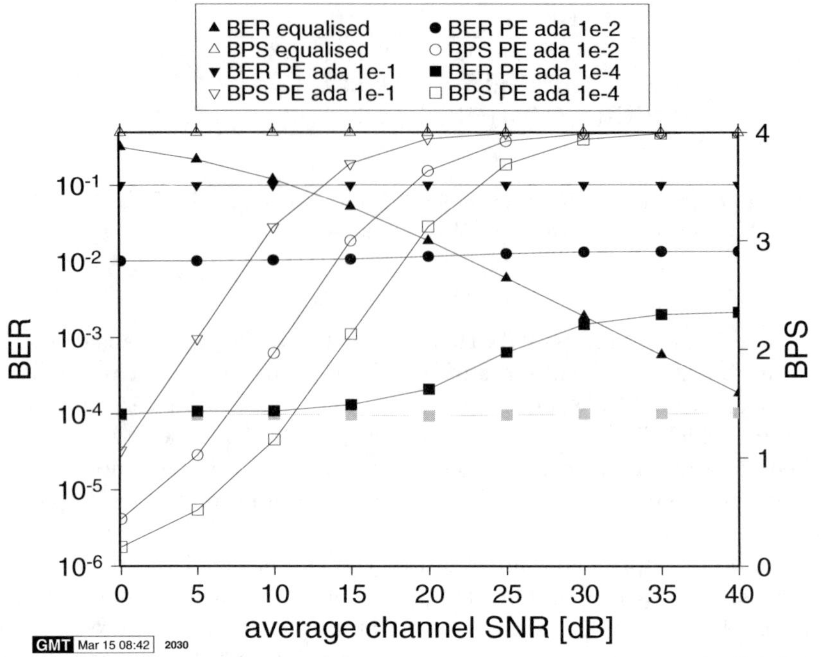

Figure 22.28: BER performance and BPS throughput of the 512 subcarrier, 16 sub-band adaptive OFDM modem with spectral predistortion over the Rayleigh fading time dispersive short WATM channel of Figure 20.3, and that of the perfectly equalised 16-QAM modem. The light grey BER curve gives the performance of the adaptive modem for a target BER of 10^{-4} with no delay between channel estimation and transmission, while the other results assume 1 time slot delay between uplink and downlink.

variations in the deeply faded subcarriers, which are increasingly used at higher SNR values. The light grey curve in the figure denotes the system's performance if no delay is present between the channel estimation and the transmission. In this case, the simulated BER shows only very little deviation from the target BER value. This is consistent with the behaviour of the full channel inversion pre-equalising modem.

22.5 Comparison of the Adaptive Techniques

Figure 22.29 compares the different adaptive modulation schemes discussed in this chapter. The comparison graph is split into two sets of curves, depicting the achievable data throughput for a data BER of 10^{-4} highlighted for the fixed throughput systems in Figure 22.29(a), and for the time-variant -throughput systems in Figure 22.29(b).

The fixed throughput systems, highlighted in black in Figure 22.29(a), comprise the non-adaptive BPSK, QPSK and 16-QAM modems, as well as the fixed-throughput adaptive scheme, both for coded and uncoded applications. The non-

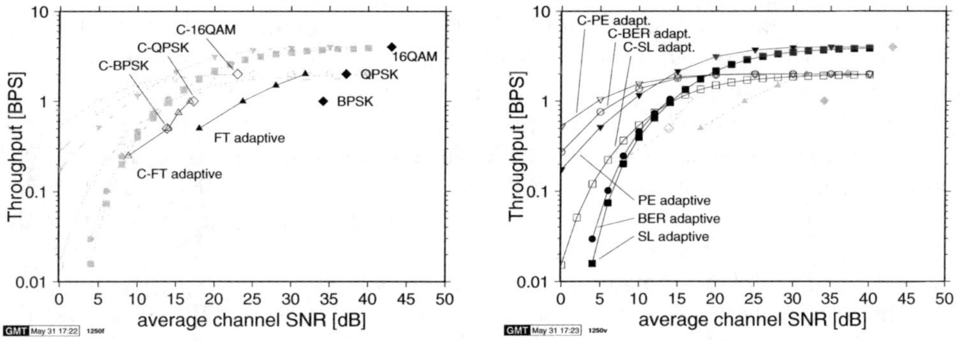

(a) fixed throughput schemes **(b)** variable throughput schemes

Figure 22.29: BPS throughput versus average channel SNR for non-adaptive and adaptive modulation as well as for pre-equalised adaptive techniques, for a data bit error rate of 10^{-4}. Note that for the coded schemes the achieved BER values are lower than 10^{-4}. (a) Fixed-throughput systems: coded (C-) and uncoded BPSK, QPSK, 16-QAM, and fixed-throughput (FT) adaptive modulation. (b) Variable-throughput systems: coded (C-) and uncoded switching level adaptive (SL), target BER adaptive (BER) and pre-equalised adaptive (PE) systems. Note that the separately plotted variable-throughput graph also shows the light grey benchmark curves of the complementary fixed-rate schemes and vice versa

adaptive modems' performance is marked on the graph as diamonds, and it can be seen that the uncoded fixed schemes require the highest channel SNR of all examined transmission methods to achieve a data BER of 10^{-4}. Channel coding employing the advocated turbo coding schemes dramatically improves the SNR requirements, at the expense of half the data throughput. The uncoded fixed-throughput (FT) adaptive scheme, marked by filled triangles, yields consistently worse data throughput than the coded (C-) fixed modulation schemes C-BPSK, C-QPSK and C-16QAM, with its throughput being about half the coded fixed scheme's at the same SNR values. The coded FT (C-FT) adaptive system, however, delivers very similar throughput to the C-BPSK and C-QPSK transmission, and can deliver a BER of 10^{-4} for SNR values down to about 9 dB.

The variable throughput schemes, highlighted in Figure 22.29(b), outperform the comparable fixed throughput algorithms. For high SNR values, all uncoded schemes' performance curves converge to a throughput of 4 bits/symbol, which is equivalent to 16-QAM transmission. The coded schemes reach a maximal throughput of 2 bits/symbol. Of the uncoded schemes, the "data" switching level (SL) and target BER adaptive modems deliver a very similar BPS performance, with the target-BER scheme exhibiting slightly better throughput than the SL adaptive modem. The adaptive modem employing pre-equalisation (PE) significantly outperforms the other uncoded adaptive schemes and offers a throughput of 0.18 BPS at an SNR of 0 dB.

The coded transmission schemes suffer from limited throughput at high SNR values, since the half-rate channel coding limits the data throughput to 2 BPS. For low SNR values, however, the coded schemes offer better performance than the uncoded schemes, with the exception of the "speech" SL adaptive coded scheme, which is outperformed by the uncoded PE adaptive modem. The poor performance of the coded SL scheme can be explained by the lower uncoded target BER of the "speech" scenario, which was 1%, in contrast to the 10% uncoded target BER for the coded BER and PE adaptive schemes. The coded PE adaptive modem outperforms the target-BER adaptive scheme, thanks to its more accurate control of the uncoded BER, leading to a higher throughput for low SNR values.

It is interesting to observe that for the given set of four modulation modes the uncoded PE adaptive scheme is close in performance to the coded adaptive schemes, and that for SNR values of more than 14 dB it outperforms all other studied schemes. It is clear, however, that the coded schemes would benefit from higher-order modulation modes, which would allow these modems to increase the data throughput further when the channel conditions allow. Before concluding this chapter, in the next section let us consider the generic problem of optimum power and bit allocation in the context of uncoded OFDM systems.

22.6 A Fast Algorithm for Near-Optimum Power and Bit Allocation in OFDM Systems [439]

22.6.1 State-of-the-art

In this section the problem of efficient OFDM symbol-by-symbol based power and bit allocation is analysed in the context of highly dispersive time-variant channels. A range of solutions published in the literature is reviewed briefly and Piazzo's [439] computationally efficient algorithm is exposed in somewhat more detail.

When OFDM is invoked over highly frequency selective channels, each subcarrier can be allocated a different transmit power and a different modulation mode. This OFDM symbol-by-symbol based "resource" allocation can be optimised with the aid of an algorithm which, if the channel is time variant, has to be repeated on an OFDM symbol-by-symbol basis. Some of the existing algorithms [83, 440] are mainly of theoretical interest due to their high complexity. Amongst the practical algorithms [426, 427, 439, 441, 442] the Hughes-Hartog algorithm (HHA) [442–444] is perhaps best known, but its complexity is somewhat high, especially for real-time OFDM symbol-by-symbol based applications at high bit rates. Hence the HHA has stimulated extensive research for computationally more efficient algorithms [426,427,439,441,442]. The most efficient appears to be that of Lai *et al.* [442], which is a fast version of the HHA and that of Piazzo [439].

22.6.2 Problem description

Piazzo [439] considered an OFDM system using N subcarriers, each employing a potentially different modulation mode and transmit power. Below we follow the notation and approach proposed by Piazzo [439]. The different modes use different modem constellations and thus carry a different number of bits per subcarrier, ranging from 1 to I bits per subcarrier, corresponding to BPSK and 2^I-ary QAM. We denote the transmit power and the number of bits allocated to subcarrier k ($k = 0, \ldots, N-1$) by p_k and b_k, respectively. If $b_k = 0$, subcarrier k is allocated no power and no bits, hence it is disabled. The total transmit power is $P = \frac{1}{N}\sum_{k=0}^{N-1} p_k$ and the number of transmitted bits per OFDM symbol is $B = \sum_{k=0}^{N-1} b_k$. The i-bit modulation mode is characterised by the function $R_i(S)$, denoting the SNR required at the input of the detector, in order to achieve a target bit error rate (BER) equal to S. Finally, we denote the channel's power attenuation at subcarrier k by a_k, and the power of the Gaussian noise by P_N, so that the SNR of subcarrier k is $r_k = p_k/(a_k \cdot P_N)$.

We consider the problem of minimising the transmit power for a fixed target BER of S and for a fixed number of transmitted bits B per OFDM symbol. We impose an additional constraint, namely that the BER of every carrier has to be equal to S. This constraint simplifies the problem, while producing a system close to the unconstrained optimum system [427,442,444], while [440] considers an unconstrained system. Furthermore, from an important practical point of view, it produces a near-constant BER at the input of the channel decoder, if FEC is used, which maximises the achievable coding gain, since the channel does not become overwhelmed by the plethora of transmission errors, which would be the case for a more bursty error statistics without this constraint. In order to satisfy this constraint, the power transmitted on subcarrier k has to be $p_k = P_N a_k R_{b_k}(S)$, and the total power to be minimised is given by the sum of the N subcarriers' powers across the OFDM symbol:

$$P = \frac{P_N}{N} \sum_{k=0}^{k=N-1} a_k \cdot R_{b_k}(S). \tag{22.6}$$

We now state a property of the optimum system. Namely, in the optimum system if a subcarrier has a lower attenuation than another one - i.e. it exhibits a higher frequency domain transfer function value and hence experiences a higher received SNR - then it must carry at least as many bits as the lower SNR subcarrier. More explicitly:

$$a_k < a_h \Rightarrow b_k \geq b_h. \tag{22.7}$$

The above property in Equation 22.7 can be readily proven. Let us briefly consider a system, which does not satisfy Equation 22.7, where for subcarriers k and h the above condition is violated and hence we have $a_k < a_h$ and $b_k = i_1 < b_h = i_2$. In other words, although the attenuation a_k is lower than a_h, $i_1 < i_2$. Consider now a second system, where the lower attenuation subcarrier was assigned the higher number of bits, i.e. $b_k = i_2$ and $b_h = i_1$. Since the required SNR for maintaining the target BER of S is lower for a lower number of bits, - i.e. we have $R_{i_1}(S) < R_{i_2}(S)$, upon substituting these SNR values in Equation 22.6 we can infer that the second system requires a lower total power P per OFDM symbol for maintaining the target BER S. Thus the first system is not optimum in this sense.

Equation 22.7 states a necessary condition of optimality, which was also exploited by Lai *et al.* in [442], but it can be exploited further, as we will demonstrate below.

From now on, we consider the channel's transfer function or attenuation vector sorted in the order of $a_0 \leq a_1 \leq a_2 \ldots$, which simplifies our forthcoming discussions.

22.6.3 Power and bit allocation algorithm

Piazzo's algorithm [439] solves the above resource allocation problem for the general system by repeteadly solving the problem for a simpler system. Explicitly, the simpler system employs only two modulation modes, those carrying J and $J-1$ bits. This system can be termed the twin-mode system (TMS). On the basis of Equation 22.7 and since the channel's frequency -domain attenuation vector was sorted in the order of $a_0 \leq a_1 \leq a_2 \ldots$, for the optimum TMS (OTMS) the OFDM subcarriers will be partitioned in three groups:

(1) Group J comprises the first or lowest attenuation OFDM subcarriers using a J-bit modulation mode.

(2) Group 0 is constituted by the last or highest attenuation OFDM subcarriers transmitting zero bits.

(3) Group $(J-1)$ hosts the remaining OFDM subcarriers using a $(J-1)$-bit modem mode.

In order to find the OTMS - minimising the required transmit power of the OFDM symbol for a fixed target BER of S and for a fixed number of transmitted bits B per OFDM symbol - we initially assign all the B bits of the OFDM symbol to the highest quality, i.e. lowest attenuation, group J. This of course would be a sub-optimum scheme, leaving the medium quality subcarriers of group $J-1$ unused, since even the highest quality subcarriers would require an excessive SNR, i.e. transmit power, for maintaining the target BER, when transmitting B bits per OFDM symbol. We note furthermore that the above bit allocation may require padding of the OFDM symbol with dummy bits if J is not an integer divisor of B.

Following the above initial bit allocation, Piazzo suggested performing a series of *bit reallocations, reducing the transmit power upon each reallocation.* Specifically, in each power and bit reallocation step we move the $J \cdot (J-1)$ bits allocated to the last, i.e. highest attenuation or lowest quality, $(J-1)$ OFDM subcarriers of group J to group $(J-1)$. For example, if 1 bit/symbol BPSK and 2 bit/symbol 4-QAM are used, then we move $2 \cdot 1 = 2$ bits, which were allocated to the highest attenuation 4-QAM subcarrier to two BPSK modulated subcarriers. The associated trade-off is that while previously the lowest quality subcarrier of class J had to carry 2 bits, it will now be conveying only 1 bit and additionally the highest quality and previously unused subcarrier has to be assigned 1 bit. This reallocation was motivated by the fact that before reallocation the lowest quality subcarrier of class J would have required a higher power for meeting the target BER requirement of S upon carrying 2 bits, than the regime generated by the reallocation step.

In general, for the sake of performing this power-reducing bit reallocation we have to add J subcarriers to group $(J-1)$. Hence we assign the last, i.e. lowest quality, $(J-1)$ OFDM subcarriers of group J and the first, i.e. highest quality, unused subcarrier to group $(J-1)$. Based on Equation 22.6 and upon denoting the index of the last, i.e. lowest quality, subcarrier of group J before reallocation by M_J and the index of the first, i.e. highest quality, unused subcarrier before reallocation by M_0, the condition of successful power reduction after the tentative bit reallocation can be formulated. Specifically, the bit reallocation results in a system using less power, if the sum of the subcarriers' attenuations carrying J bits weighted by their SNR $R_J(S)$ required for the J-bit modem mode for maintaining the target BER of S is higher than that of the corresponding constellation after the above bit reallocation process, when an extra previously unused subcarrier was invoked for transmission. This can be expressed in a more compact form as:

$$R_J(S)\sum_{k=0}^{J-2} a_{M_J-k} > R_{J-1}(S)(a_{M_0} + \sum_{k=0}^{J-2} a_{M_J-k}). \tag{22.8}$$

If Equation 22.8 is satisfied, the reallocation is performed and another tentative reallocation step is attempted. Otherwise the process is terminated, since the optimum twin-mode power and bit allocation scheme has been found.

According to Piazzo's proposition [439] the above procedure can be further accelerated. Since the attenuation vector was sorted, we have $a_{M_J-k} \approx a_{M_J}$ in Equation 22.8. Upon replacing a_{M_J-k} by a_{M_J}, after some manipulations we can reformulate Equation 22.8, i.e. the condition for the modem mode allocation after the bit reallocation to become more efficient as:

$$K_J(S)a_{M_J} - a_{M_0} > 0, \tag{22.9}$$

where $K_J(S) = (J-1)(\frac{R_J(S)}{R_{J-1}(S)} - 1)$ and $K_J(S) > 0$ holds, since $R_J(S) > R_{J-1}(S)$. Piazzo denoted the values of M_J and M_0 after m reallocation steps by $M_J(m)$ and $M_0(m)$. Since initially all the bits were allocated to group J, we have for the index of the last subcarrier of group J at the commencement of the bit reallocation steps $M_J(0) = \lfloor B/J \rfloor - 1$, while for the index of the first unused subcarrier is $M_0(0) = \lfloor B/J \rfloor$, where $\lfloor x \rfloor$ is the smallest integer greater than or equal to x. Furthermore, since in each bit reallocation step the last $(J-1)$ OFDM subcarriers of group J and the first subcarrier of group 0 are moved to group $(J-1)$, after m reallocations we have $M_J(m) = \lfloor B/J \rfloor - 1 - (J-1)m$ and $M_0(m) = \lfloor B/J \rfloor + m$. Upon substituting these values in Equation 22.9 the left-hand side becomes a function of m, namely $f(m) = K_J(S)a_{M_J(m)} - a_{M_0(m)}$. Because the frequency domain channel transfer function's attenuation vector was ordered and since $K_J(S) > 0$, hence it is readily seen that $f(m)$ is a monotonically decreasing function of the reallocation index m. Therefore the method presented above essentially attempts to find the specific value of the reallocation index m, for which we have $f(m) > 0$ and $f(m+1) < 0$. In other words, when we have $f(m+1) < 0$, the last reallocation step resulted in a power increment, not a decrement, hence the reallocation procedure is completed.

The search commences from $m = 0$ and increases m by one at each bit reallocation step. In order to accelerate the search procedure, Piazzo replaced the linearly incremented search by a logarithmic search. This is possible, since the $f(m)$ function is monotonically decreasing upon increasing the reallocation index m. Piazzo [439] stipulated the search range by commencing from the minimum value of the reallocation index m, namely from $m_0 = 0$.

The maximum value, denoted by m_1, is determined by the number of OFDM subcarriers N or by the number of bits B to be transmitted per OFDM symbol, as will be argued below. There are two limitations, which determine the maximum possible number of reallocation steps. Namely, the reallocation steps have to be curtailed when there are no more bits left in the group of subcarriers associated with the J-bit modem mode group or when there are no more unused carriers left after iteratively invoking the best unused carrier from the group of disabled carriers. These limiting factors, which determine the maximum possible number of bit reallocation steps, are augmented further below.

Recall that at the commencement of the algorithm all the bits were assigned to the subcarrier group associated with the J-bit modem mode and hence there were $\lceil B/J \rceil$ subcarriers in group J. Upon reallocating the $J \cdot (J-1)$ bits allocated to the last, i.e. highest attenuation or lowest quality, $(J-1)$ OFDM subcarriers of group J to group $(J-1)$ until no more bits were left in the subcarrier group associated with the J-bit modem mode naturally constitutes an upper limit for the maximum number of reallocation steps m_1, which is given by $\lceil B/J \rceil/(J-1)$. Again, the other limiting factor of the maximum number of bit reallocation steps is the number of originally unused carriers, which was $N - \lceil B/J \rceil)$. Hence the maximum possible number of reallocations is given by $m_1 = min(\lfloor \lfloor B/J \rfloor/(J-1) \rfloor; N - \lfloor B/J \rfloor)$, where $\lceil x \rceil$ is the highest integer smaller than or equal to x.

The accelerated logarithmic search proposed by Piazzo [439] halves the maximum possible range at each bit reallocation step, by testing the value of $f(m)$ at the centre of the range and by updating the range accordingly. Piazzo's proposed algorithm can be summarised in a compact form as follows [439]:

Algorithm 1 $OTMS(B, S, J, N, a_k)$
(1) Initialise $m_0 = 0$, $m_1 = min(\lfloor \lfloor B/J \rfloor/(J-1) \rfloor, N - \lfloor B/J \rfloor)$.
(2) Compute $m_x = m_0 + \lfloor m_1 - m_0 \rfloor/2$.
(3) If $f(m_x) \geq 0$ *let* $m_0 = m_x + 1$; *else let* $m_1 = m_x$.
(4) If $m_1 = m_0 + 1$ *goto step 5); else goto step 2).*
(5) Stop. The number of carriers in group J *is* $N_J = \lfloor B/J \rfloor - m_0 \cdot (J-1)$.

When the algorithm is completed, the value N_J becomes known and this specifies the number of OFDM subcarriers in the group J associated with the J-bit modem mode.

Having generated the optimum twin-mode system, Piazzo also considered the problem of finding the optimum general system (OGS) employing OFDM subcarrier modulation modes carrying $1, ..., I$ bits. The procedure proposed initially invoked Algorithm 1 in order to find the optimumm twin-mode system carrying a total of B bits per OFDM symbol using the I-bit and the $(I-1)$-bit per subcarrier modulation modes. At the completion of Algorithm 1 we know N_I, the number of OFDM subcarriers carrying I bits. These subcarriers are now confirmed. These OFDM subcarriers as well as the associated $I \cdot N_I$ bits can now be eliminated from the resource allocation problem, and the optimum system transmitting the remaining $B - I \cdot N_I$ bits of the remaining $(N - N_I)$ subcarriers can be sought, using subcarrier modulation modes transmitting $(I-1), (I-2), ...$ bits.

Again, Algorithm 1 can be applied to this new system, now using the modulation modes with $(I-1)$ and $(I-2)$ bits per subcarrier and repeating the procedure. After each application of Algorithm 1 a new group of subcarrier is confirmed. Piazzo's general algorithm can be summarised in a compact form as follows [439]:

Algorithm 2 $OGS(B, S, I, N, a_k)$
(1) Initialise $\hat{B} = B$, $\hat{N} = N$, $\hat{a}_k = a_k$, $J = I$.
(2) Perform $OTMS(\hat{B}, S, J, \hat{N}, \hat{a}_k)$ *to compute* N_J.
(3) If $J = 2$, *let* $N_1 = \hat{B} - 2 \cdot N_2$ *and stop.*
(4) Remove the first N_J *carriers from* $\hat{a}_k$, *let* $\hat{B} = \hat{B} - J \cdot N_J$, $\hat{N} = \hat{N} - N_J$, $J = J - 1$ *and goto step 2).*

When the algorithm is completed, the values N_i specifying the number of OFDM subcarriers conveying i bits, become known for all the legitimate modes carrying $i = I, (I-1), ..., 1$ bits per subcarrier. Hence we know the number of bits allocated to subcarrier k $(k = 0, \ldots, N-1)$ expressed in terms of the b_k values as well as the associated minimum power requirements. Hence the system is specified in terms of $p_k = P_N a_k R_{b_k}(S)$. In closing it is worthwile noting that the algorithm can be readily modified to handle the case where the two modes of the twin-mode system carry J and $K < J - 1$ bits.

Having considered the above near-optimum power- and bit allocation algorithm, in the next section we will consider a variety of OFDM systems supporting multiple users and invoking adaptive beam-steering.

22.7 Multi-User AOFDM

M. Münster, T. Keller and L. Hanzo [3]

22.7.1 Introduction

Signal fading as well as co-channel interference are known to have a severe impact on the system performance in multi-cellular mobile environments. *Adaptive modulation* as a method of matching the system to fading induced variations of the channel quality has originally been proposed for single carrier transmission, but its potential was also soon discovered in the context of multicarrier transmissions, with the aim of concentrating the throughput on subcarriers least affected by frequency selective fading [437]. On the other hand, adaptive antenna array techniques have been shown to be effective in reducing co-channel interference at the receiver side [445, 446]. One of the most prominent schemes for performing the combining operation is the *Sample Matrix Inversion* (SMI) technique, which has recently drawn wide interest [132,133,447]. We commence our discussions in the next section with a description of a system amalgamating adaptive modulation and co-channel interference suppression. Initial performance results will be presented in Subsection 22.7.3 assuming perfect knowledge

[3] This section is based on M. Münster, T. Keller and L. Hanzo; Co-Channel Interference Suppression Assisted Adaptive OFDM In Interference Limited Environments, ©IEEE, VTC'99, Amsterdam, NL, 17-19 Sept. 1999

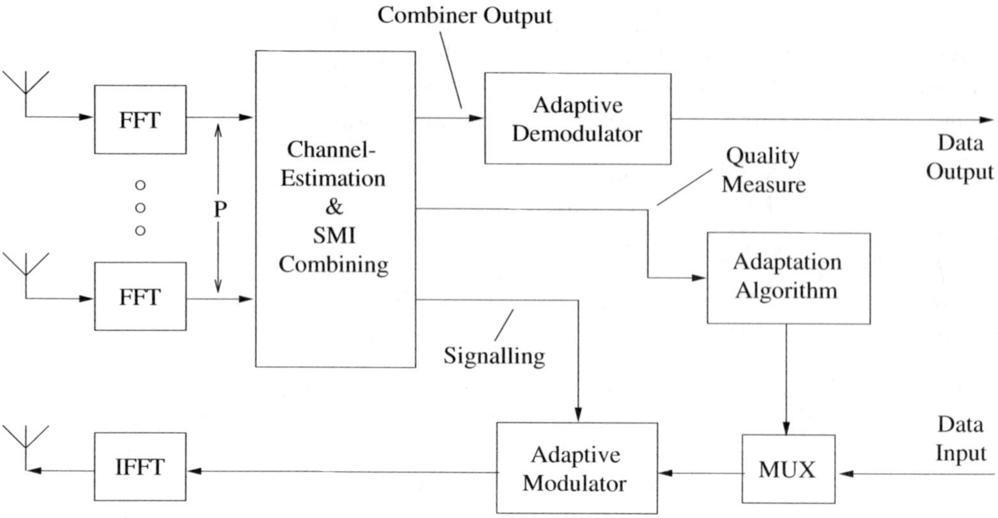

Figure 22.30: Schematic structure of the adaptive transceiver with interference suppression at the receiver

of all channel parameters, whilst in Subsection 22.7.4 the problem of channel parameter estimation will be addressed by means of orthogonal pilot sequences, leading to our conclusions.

22.7.2 Adaptive transceiver architecture

An Overview - The transceiver schematic is shown in Figure 22.30, where the receiver employs a multiple-antenna assisted front end. The signal received by each individual antenna element is fed to an FFT block, and the resulting parallel received OFDM symbols are combined on a subcarrier-by-subcarrier basis. The combining is accomplished on the basis of the weight vector, which has been obtained by solving the Wiener equation, constituting the core of the sample matrix inversion algorithm [445, 446]. After combining the signal is fed into the adaptive demodulator of Figure 22.30, which delivers the output bits in the form of soft-decision information to an optional channel decoder. The demodulator operates in one of a set of four modes, namely 'no transmission', BPSK, QPSK and 16-QAM. Since an interfered channel cannot be considered to constitute a reciprocal system, the modem mode adaptation operates in a closed-loop fashion, where each of the receivers instructs the remote transmitter as to the required set of modulation modes for the next AOFDM symbol, which is necessary for maintaining a given target Bits per Symbol (BPS) performance. On reception of a packet the adaptation algorithm computes the set of modulation modes to be employed by the remote transmitter for the next transmitted AOFDM symbol on the basis of a channel quality measure, namely the subcarrier SNR, which can be estimated by the interference suppression algorithm. The set of requested modulation modes is signalled to the remote receiver along with the next transmitted AOFDM symbol, which is then used by the remote transmitter in its next transmission.

The Signal Model - The $P \times 1$ vector of complex signals, $\mathbf{x}[n,k]$, received by the antenna array in the k-th subcarrier of the n-th OFDM symbol is constituted by a superposition of the independently faded signals associated with the *desired* user and the L *undesired* users plus the Gaussian noise at the array elements:

$$\begin{aligned} \mathbf{x}[n,k] &= \mathbf{d}[n,k] + \mathbf{u}[n,k] + \mathbf{n}[n,k], \quad \text{with} \\ \mathbf{d}[n,k] &= \mathbf{H}^{(0)}[n,k] s_0[k] \\ \mathbf{u}[n,k] &= \sum_{l=1}^{L} \mathbf{H}^{(l)}[n,k] s_l[k], \end{aligned} \tag{22.10}$$

where $\mathbf{H}^{(l)}[n,k]$ for $l = 0, \ldots, L$ denotes the $P \times 1$ vector of complex channel coefficients between the l-th user and the P antenna array elements. We assume that the vector components $H_m^{(l)}[n,k]$ for different array elements m or users l are independent, stationary, complex Gaussian distributed processes with zero-mean and different variance σ_l^2, $l = 0, \ldots, L$. The variable $s_l[n,k]$ - which is assumed to have zero-mean and unit variance - represents the complex data of the l-th user and $\mathbf{n}[n,k]$ denotes the aforementioned $P \times 1$ vector of additive white Gaussian noise contributions with zero mean and variance σ^2 [132].

The SMI Algorithm - The idea behind minimum mean-square error (MMSE) beamforming [446] is to adjust the antenna weights, such that the power of the differential signal between the combiner output and a reference signal - which is characteristic of the desired user - is minimised. The solution to this problem is given by the well-known Wiener equation, which can be directly solved by means of the SMI algorithm in order to yield the optimum weight vector $\mathbf{w}[n,k]$ of dimension $P \times 1$. Once the instantaneous correlation between the received signals, which is represented by the $P \times P$ matrix $\mathbf{R}[n,k]$ - and the channel vector $\mathbf{H}^{(0)}[n,k]$ of the desired user become known, the weights are given by [132, 446, 447]:

$$\mathbf{w}[n,k] = (\mathbf{R}[n,k] + \gamma \mathbf{I})^{-1} \mathbf{H}^{(0)}[n,k], \tag{22.11}$$

where γ represents the so-called diagonal augmentation factor [132]. Assuming knowledge of all channel parameters and the noise variance σ^2, the correlation matrix can be determined by:

$$\begin{aligned} \mathbf{R}[n,k] &\triangleq E_c\{\mathbf{x}[n,k]\mathbf{x}^H[n,k]\} \\ &= \mathbf{R}_d[n,k] + \mathbf{R}_u[n,k] + \mathbf{R}_n[n,k], \quad \text{with} \\ \mathbf{R}_d[n,k] &= \mathbf{H}^{(0)}[n,k]\mathbf{H}^{(0)^H}[n,k] \\ \mathbf{R}_u[n,k] &= \sum_{l=1}^{L} \mathbf{H}^{(l)}[n,k]\mathbf{H}^{(l)^H}[n,k] \\ \mathbf{R}_n[n,k] &= \sigma^2 \mathbf{I}, \end{aligned} \tag{22.12}$$

which is a superposition of the correlation matrices $\mathbf{R}_d[n,k]$, $\mathbf{R}_u[n,k]$ and $\mathbf{R}_n[n,k]$ of the desired and undesired users as well as of the array element noise, respectively.

The combiner output can now be inferred from the array output vector $\mathbf{x}[n,k]$ by means of:

$$y[n,k] = \mathbf{w}^H[n,k]\mathbf{x}[n,k]. \qquad (22.13)$$

The Signal-to-Noise Ratio (SNR) at the combiner output - which is of vital importance for the modulation mode adaptation - is given by [446]:

$$\begin{aligned} SNR &= \frac{E\{|\mathbf{w}^H[n,k]\mathbf{d}[n,k]|^2\}}{E\{|\mathbf{w}^H[n,k]\mathbf{n}[n,k]|^2\}} \\ &= \frac{\mathbf{w}^H[n,k]\mathbf{R}_d[n,k]\mathbf{w}[n,k]}{\mathbf{w}^H[n,k]\mathbf{R}_n[n,k]\mathbf{w}[n,k]} \end{aligned} \qquad (22.14)$$

and correspondingly the Signal-to-Interference+Noise Ratio (SINR) is given by [446]:

$$SINR = \frac{\mathbf{w}^H[n,k]\mathbf{R}_d[n,k]\mathbf{w}[n,k]}{\mathbf{w}^H[n,k](\mathbf{R}_u[n,k]+\mathbf{R}_n)\mathbf{w}[n,k]}. \qquad (22.15)$$

Equation 22.12 is the basis for our initial simulations, where *perfect channel knowledge* has been assumed. Since in a real environment the receiver does not have perfect knowledge of the channel, its parameters have to be estimated, a problem which we will address in Section 22.7.4 on an OFDM symbol-by-symbol basis by means of orthogonal pilot sequences.

The Adaptive Bit-assignment Algorithm - The adaptation performed by the modem is based on the choice between a set of four modulation modes, namely 4, 2, 1 and 0 bit/subcarrier, where the latter corresponds to 'no transmission'. The modulation mode could be assigned on a subcarrier-by-subcarrier basis, but the signalling overhead of such a system would be prohibitive. Hence, we have grouped adjacent subcarriers into 'sub-bands' and assign the same modulation mode to all subcarriers in a sub-band. Note that the frequency domain channel transfer function is typically not constant across the subcarriers of a sub-band, hence the modem mode adaptation will be sub-optimal for some of the subcarriers. The Signal-to-Noise Ratio (SNR) of the subcarriers will be shown to be in most cases an effective measure for controlling the modulation assignment. The modem mode adaptation is hence achieved by calculating in the first step for each sub-band and for all four modulation modes the expected overall sub-band bit error rate (BER) by means of averaging the estimated individual subcarrier BERs. Throughout the second step of the algorithm - commencing with the lowest modulation mode in all sub-bands - in each iteration the number of bits/subcarrier of that sub-band is increased, which provides the best compromise in terms of increasing the number of expected bit errors compared to the number of additional data bits accommodated, until the target number of bits is reached.

The Channel Models - Simulations have been conducted for the indoor Wireless Asynchronous Transfer Mode (WATM) channel impulse response (CIR) of of Section 20.1 [437]. This three-path impulse response exhibits a maximal dispersion of 11 time-domain OFDM samples, with each path faded according to a Rayleigh distribution of a normalised maximal Doppler frequency of $f'_d = 1.235 \cdot 10^{-5}$, where the normalisation interval was the OFDM symbol duration. This model corresponds to the channel experienced by a mobile transmitting at a carrier frequency of 60 GHz with a sampling rate of 225 MHz and travelling at a vehicular velocity of 50 km/h. An alternative channel model, which we considered in our simulations is a Wireless Local Area Network (WLAN) model associated with a seven-path impulse response having a maximal dispersion of 32 samples. However, for this more dispersive and higher Doppler frequency channel adaptive modulation has turned out to be less effective due to its significantly increased normalised Doppler frequency of $f_d' = 3.935 \cdot 10^{-5}$ corresponding to a carrier frequency of 17 GHz, sampling rate of 20 MHz and vehicular velocity of 50 km/h.

22.7.3 Simulation results - perfect channel knowledge

General Remarks - In our initial simulations we assumed that the receiver had perfect knowledge of all channel parameters, which enabled the estimation of the correlation matrix required by the SMI algorithm upon using Equation 22.12. Furthermore, we initially assumed that the receiver was capable of signalling the modulation modes to the transmitter without any additional delay. Throughout our discussions we will gradually remove the above idealistic assumptions. In all simulations we assumed a partitioning of the 512-subcarrier OFDM symbol's total bandwidth into 16 equal-sized 32-subcarrier sub-bands. This has been shown to provide a reasonable compromise between signalling overhead and performance degradation compared to a subcarrier-by-subcarrier based modulation mode assignment.

Two-Branch Maximum–Ratio Combining - Initial simulations were conducted in the absence of co-channel interference. In this scenario the SMI equations take the form of the MMSE maximum-ratio combiner, resulting in a high diversity gain even with the minimal configuration of only two reception elements. Adaptive modulation was performed on the basis of the estimated SNR of each subcarrier, which is given by Equation 22.14. Since due to diversity reception the dramatic fades of the channel frequency response have been mitigated, the performance advantage of adaptive modulation is more modest, as illustrated by Figure 22.31 for the 'transmission frame-invariant' WATM channel, for which the fading profile is kept constant for the OFDM symbol duration, in order to avoid inter-subcarrier interference. For the equivalent simulations in the 'transmission frame-invariant' WLAN channel environment we observed a more distinct performance gain due to adaptive modulation, which is justified by the higher degree of frequency selectivity introduced by the WLAN channel's seven-path impulse response.

SMI Co-Channel Interference Suppression - In these simulations we considered first of all the case of a single dominant co-channel interferer of the same signal strength as the desired user. It is well-known that if the total number of users - whose signals arrive at the antenna array - is less or equal to the number of array elements, the unwanted users are suppressed quite effectively. Hence, for our modulation adaptation requirements we can assume that $SNR \approx SINR$, which enables us to use the algorithm described in Section 22.7.2, on the basis of the SNR estimated with the aid of Equation 22.14. Figure 22.32 illustrates the impact of adaptive modulation in the WATM channel environment under the outlined conditions. At a given SNR the performance gain due to adaptive modulation decreases with an increasing bitrate, since the higher bitrate imposes a more stringent constraint on the modulation

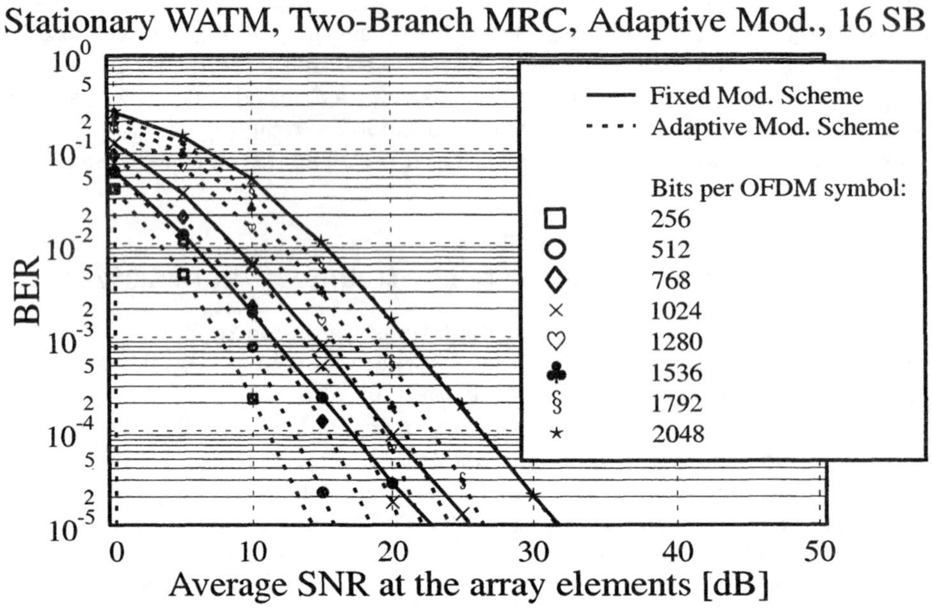

Figure 22.31: BER of 16-sub-band AOFDM modem with *two-branch maximum-ratio combining* in a *'frame-invariant' indoor WATM* environment, assuming *perfect channel knowledge* and *zero-delay signalling* of the modulation modes

mode assignment, invoking a higher number of low-SNR subcarriers. Upon comparing Figure 22.32 and 22.33 we observe that AOFDM attains a significantly higher SNR gain in the presence of co-channel interference, than without interference. As alluded to in the previous Section this is because under co-channel interference the SMI scheme exploits most of its diversity information extracted from the antenna array for suppressing the unwanted signal components, rather than mitigating the frequency domain channel fades experienced by the wanted user. For decreasing values of the Interference-to-Noise Ratio (INR) at the antenna array output, the system performance will gradually approach the performance observed for the MRC system. In order to render our investigations more realistic in our next experiment we allow a continous, i.e. 'frame-variant' fading across the OFDM symbol duration. The system performance corresponding to this scenario is illustrated in Figure 22.33. At low SNRs the observed performance is identical to that recorded in Figure 22.32 for the 'frame-invariant' channel model, whereas at high SNRs we experience a residual BER due to inter-subcarrier interference. Again, for a low number of bits per OFDM symbol the adaptive scheme is capable of reducing the 'loading' of subcarriers with low SNR values, which are particularly impaired by inter-subcarrier interference. Hence AOFDM exhibits a BER improvement in excess of an order of magnitude. So far we

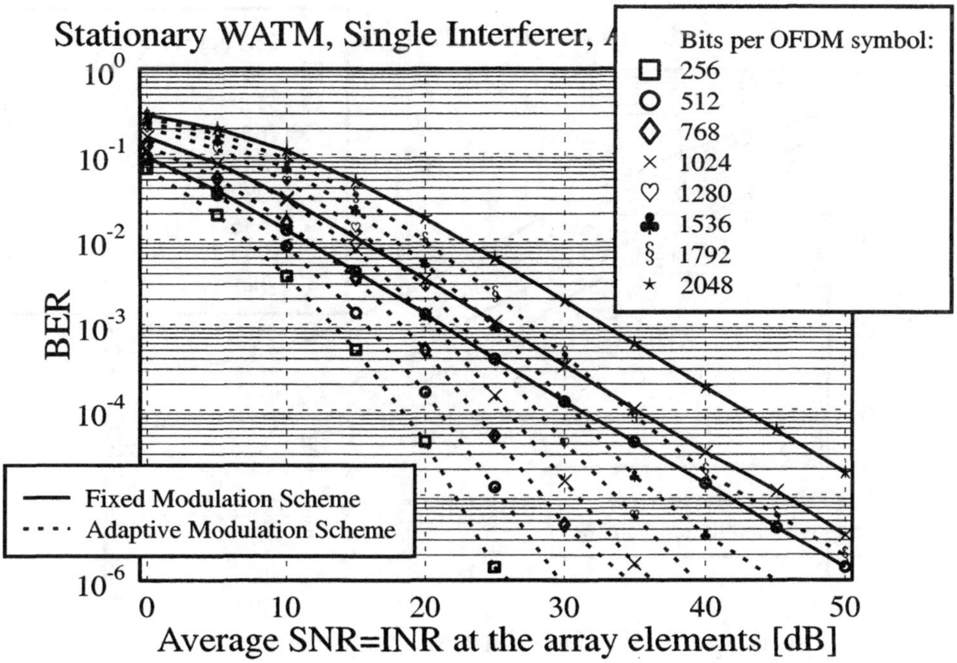

Figure 22.32: BER of 16-sub-band AOFDM modem with *two-branch SMI* and 2 *users* in a *'frame-invariant' indoor WATM* environment, assuming *perfect channel knowledge* and *zero-delay signalling* of the modulation modes

have assumed that the receiver is capable of instantaneously signalling the required modulation modes for the next OFDM symbol to the transmitter. This assumption cannot be maintained in practice. Here we assume a time division duplexing (TDD) system with identical transceivers at both ends of the link, which communicate with each other using adjacent up-link and down-link slots. Hence we have to account for this by incorporating an additional delay of at least one OFDM symbol, while neglecting the finite signal processing delay. Simulation results for this scenario are depicted in Figure 22.34. We observe that the performance gain attained by adaptive modulation is reduced compared to that associated with the zero-delay assumption in Figure 22.33. This could partly be compensated for by channel prediction. When employing a higher number of array elements, the performance gain achievable by adaptive modulation will mainly depend on the number of users and their signal strength. If the number of users is lower, than the number of array elements, or if the interferers are predominantly weak, the remaining degrees of freedom for influencing the array response are dedicated by the SMI scheme to providing diversity for the reception of the wanted user and hence adaptive modulation proves less effective. If the

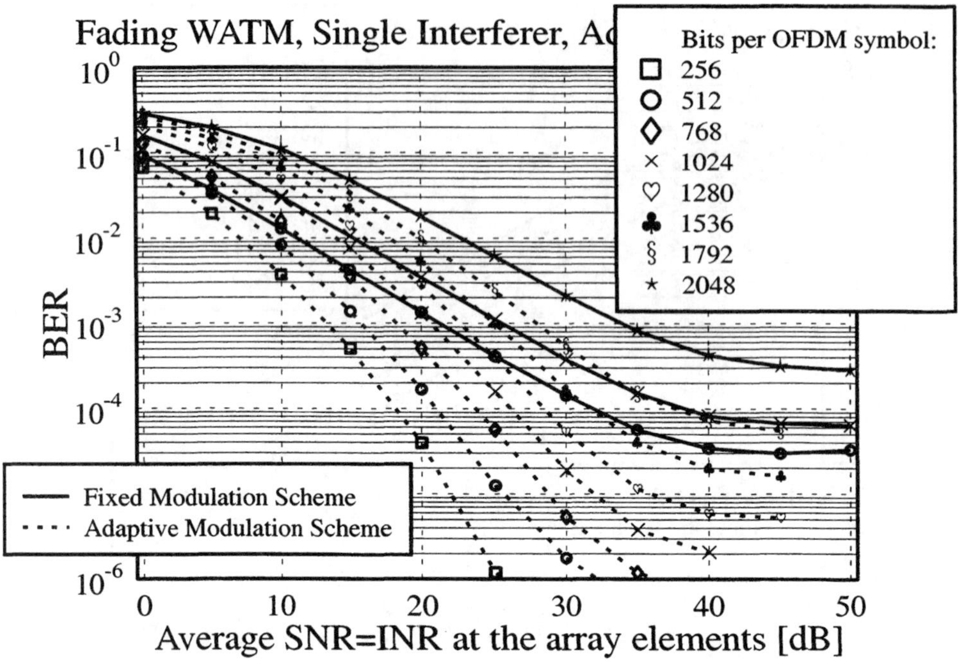

Figure 22.33: Performance results of Figure 22.32 repeated for a *fading indoor WATM* environment

number of users exceeds the number of array elements, the system becomes incapable of suppressing the undesired users effectively, resulting in a residual BER at high SNRs due to the residual co-channel interference. Since for a relatively high number of users the residual interference exhibits Gaussian-noise like characteristics, the SINR given by Equation 22.15 could be a suitable measure for performing the modulation mode assignment. By contrast, for a low number of interferers it is difficult to predict the impact on the system performance analytically. A possible approach would be to use the instantaneous number of errors in each sub-band (e.g. at the output of a turbo decoder) as a basis for the modulation assignment, which constitutes our future work. Let us now consider the issues of channel parameter estimation.

22.7.4 Pilot-based channel parameter estimation

System Description - Vook and Baum [447] have proposed SMI parameter estimation for OFDM by means of orthogonal reference sequences carrying pilot slots, which are transmitted over several OFDM symbol durations. This principle can also be applied on an OFDM symbol-by-symbol basis, as required for adaptive modulation. Upon invoking the idea of pilot based channel estimation by means of sampling

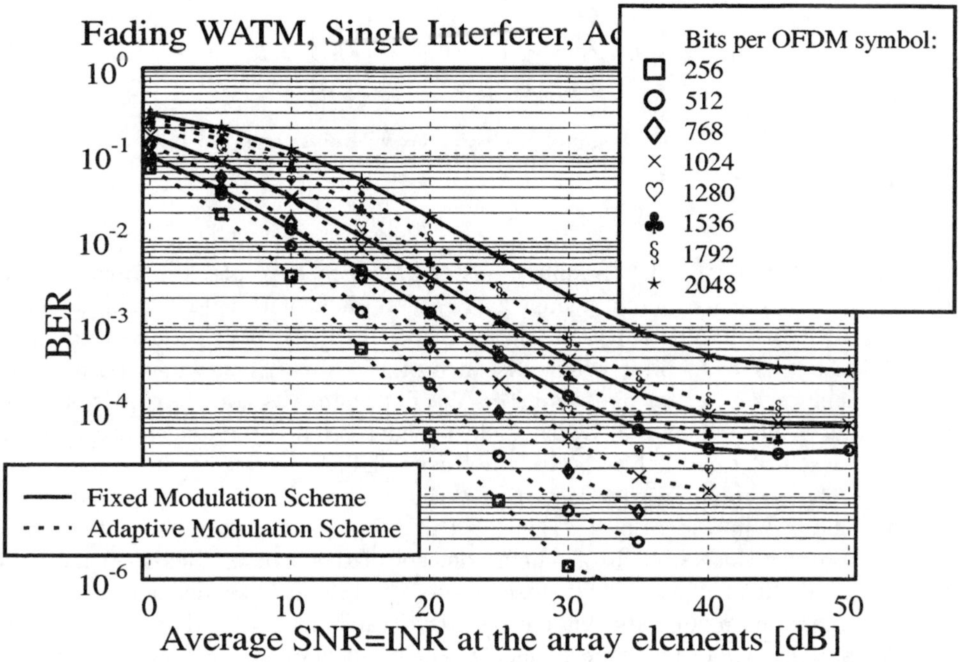

Figure 22.34: Performance results of Figure 22.33 repeated for *one OFDM symbol delayed signalling*

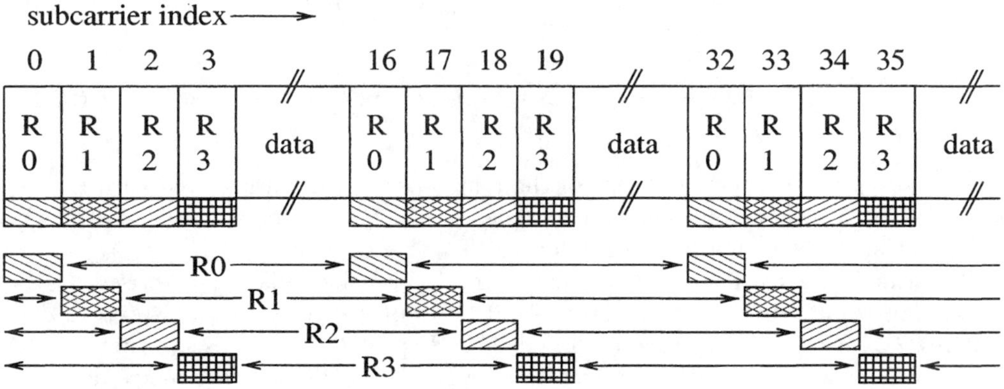

Figure 22.35: Pilot arrangement in each OFDM symbol for a *reference length of* 4 *bit* and a *group distance of* 16 *subcarriers*; interpolation is performed between pilots associated with the same bit position within the reference sequence

code/bit	0	1	2	3
0	1	1	1	1
1	1	1	-1	-1
2	1	-1	-1	1
3	1	-1	1	-1

Table 22.4: Orthogonal Walsh codes with a length of 4 bit

and low-pass interpolating the channel transfer function, we replace each single pilot subcarrier by a group of pilots, which carries a replica of the user's unique reference sequence. This is illustrated in Figure 22.35 for a reference sequence having a length of 4 bit, and for a pilot group distance of 16 subcarriers, which corresponds to the frequency required for sampling the WATM channel's transfer function. The corresponding 4-bit orthogonal Walsh code based reference sequences are listed in Table 22.4. Each of these 4 bits is assigned using BPSK to one of the 4 pilots in a pilot-group. The complex signal received by the m-th antenna in a pilot subcarrier at absolute index k and local index i within the reference sequence is constituted by a contribution of all users, each of which consists of the product of the Walsh code value associated with the user at bit position i of the reference sequence of Table 22.4 and the complex channel coefficient between the transmitter and the m-th antenna. MMSE lowpass interpolation is performed between all pilot symbols of the same relative index i within the k-spaced pilot blocks - as seen in Figure 22.35 - in order to generate an interpolated estimate of the reference for each subcarrier. An estimate of the channel vector $\hat{\mathbf{H}}^{(0)}[n,k]$ and the correlation matrix $\hat{\mathbf{R}}[n,k]$ for the k-th subcarrier of the n-th OFDM symbol is then given by [445–447]:

$$\hat{\mathbf{H}}^{(0)}[n,k] = \frac{1}{N}\sum_{i=0}^{N-1} r^{(0)^*}(i)\mathbf{x}_{LP}[n,k](i) \tag{22.16}$$

$$\hat{\mathbf{R}}[n,k] = \frac{1}{N}\sum_{i=0}^{N-1} \mathbf{x}_{LP}[n,k](i)\mathbf{x}_{LP}^{H}[n,k](i), \tag{22.17}$$

where $r^{(0)}(i)$ denotes the i-th value of the reference sequence associated with the desired user, $\mathbf{x_{LP}}[n,k](i)$ represents the low-pass interpolated received signal at sequence position i and N denotes the total reference length.

Simulation Results - The performance of this scheme is characterised by the simulation results presented in Figure 22.36. Compared to the results presented in Figure 22.34 for 'perfect channel knowledge', in Figure 22.36 we observe that besides the reduced range of supported bitrates there is an additional performance degradation, which is closely related to the choice of the reference length. Specifically, there is a reduction in the number of useful data subcarriers due to the pilot overhead, which reduces the 'adaptively' exploitable diversity potential. Secondly, a relatively short reference sequence results in a limited accuracy of the estimated channel parameters - an effect which can be partly compensated for by a technique referred to as diagonal

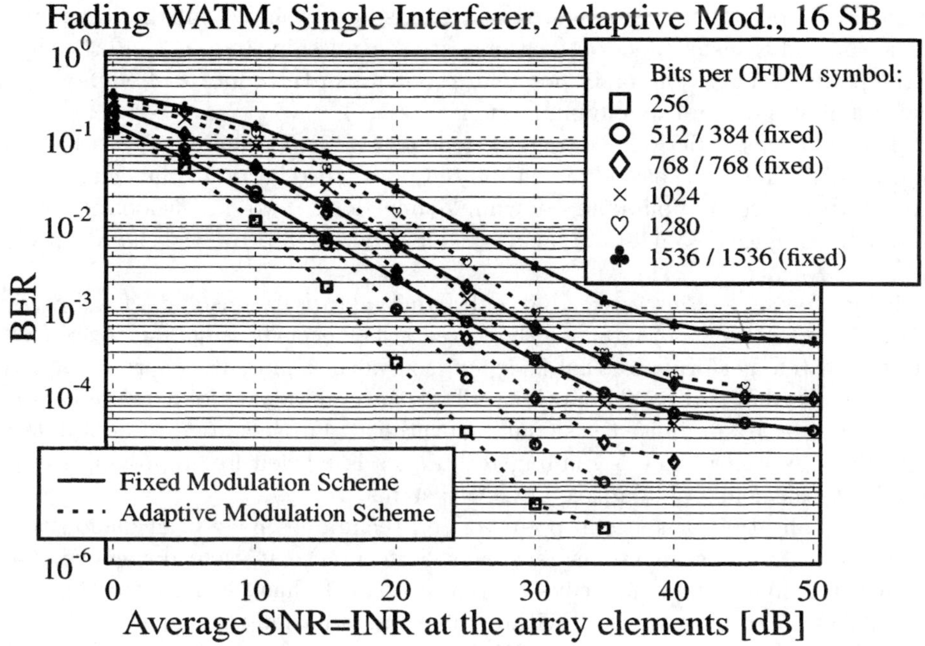

Figure 22.36: BER of 16-sub-band adaptive OFDM modem with *two-branch SMI* and 2 *users* in a *fading indoor WATM* environment, with *pilot based channel parameter estimation* and *one OFDM symbol delayed signalling* of the modulation assignment using a *diagonal loading* of $\gamma = 1.0$

loading [132]. However, the effect of short reference sequences becomes obvious for a higher number of antenna elements, since more signal samples are required, in order to yield a reliable estimate of the correlation matrix. Hence our scheme proposed here is attractive for a scenario having 2-3 reception elements, where the interference is due to 1-2 dominant interferers and an additional Gaussian noise like contribution of background interferers, which renders the SINR of Equation 22.15 to be an effective measure of channel quality. **In conclusion** the proposed adaptive array assisted AOFDM scheme resulted typically in an order of magnitude BER reduction due to employing adaptive modulation.

22.8 Summary

A range of adaptive modulation and spectral predistortion techniques have been presented in this chapter, all of which aim to react to the time and frequency dependent channel transfer function experienced by OFDM modems in fading time dispersive channels. It has been demonstrated that by exploiting the knowledge of the channel transfer function at the transmitter, the overall system performance can be increased substantially over the non-adaptive case. It has been pointed out that the prediction of the channel transfer function for the next transmission time slot and the signalling of the parameters are the main practical problems in the context of employing adaptive techniques in duplex communications.

The channel prediction accuracy is dependent on the quality of the channel estimation at the receiver, as well as on the temporal correlation of the channel transfer function between the uplink and downlink time slots. Two-dimensional channel estimation techniques [133,448] can be invoked in order to improve the channel prediction at the receivers.

It has been demonstrated that sub-band adaptivity instead of subcarrier-by-subcarrier adaptivity can significantly decrease the necessary signalling overhead, with a loss of system performance that is dependent on the channel's coherence bandwidth. We have seen that sub-band adaptivity allows the employment of blind detection techniques in order to minimise the signalling overhead. Further work into blind detection algorithms as well as new signalling techniques is needed for improving the overall bandwidth efficiency of adaptive OFDM systems.

Pre-equalisation or spectral predistortion techniques have been demonstrated to significantly improve an OFDM system's performance in time dispersive channels, while not increasing the system's output power. It has been shown that spectral predistortion can integrate well with adaptive modulation techniques, improving the system's performance significantly. We have seen in Figure 22.1 that a data throughput of 0.5 bits/symbol has been achieved at 0 dB average channel SNR with a BER of below 10^{-4}.

An adaptive power and bit allocation algorithm was also highlighted and the performance of AOFDM in a beam-forming assisted multi-user scenario was characterised.

Chapter 23

Block-Coded Adaptive OFDM [1] [2]

T. Keller, T-H. Liew, L. Hanzo

23.1 Introduction

23.1.1 Motivation

We have seen in Chapter 22 that adaptive modulation techniques can be used in order to achieve a given target level of bit error rate (BER) in a duplex communications link by choosing the appropriate modulation modes for the instantaneous channel conditions. The various adaptive modem mode signalling regimes were portrayed in Figure 22.1. employing recursive systematic Convolutional (RSC) codes as the constituent codes in turbo coding in conjunction with a turbo decoder allowed us to dramatically improve the effective data throughput for a given target bit error rate of 10^{-4} at low SNR values. The performance comparison in Figure 22.29 showed that the observed throughput for the coded, target BER adaptive modem was higher than in the uncoded case for SNR values up to 18 dB. For higher SNR values, however, the redundant information overhead of the FEC limits the system's effective data throughput if only a limited set of modulation modes is employed by the adaptation algorithm. In the case of the half-rate convolutional coding based turbo code employed in Section 22.2.7, half the data bandwidth was unnecessarily absorbed at high SNR values.

[1]This chapter is based on T. Keller, T-H. Liew and L. Hanzo: Redundant Residual Number System Based OFDM Transmission, submitted to IEEE JSAC, 1999

In this chapter, we will investigate the employment of variable rate channel coding schemes within an OFDM transmission system, in order to improve the system's throughput at high SNR values. Analogously to the adaptive modulation modes discussed in Chapter 22, in addition to the modem modes, in this chapter the code rate is also adapted in response to the time and frequency dependent estimated channel conditions. We note, however, that this variable code rate scheme can operate along with time-invariant non-adaptive modulation or in conjunction with additional adaptive OFDM (AOFDM) modulation techniques, as will be demonstrated here.

Similarly to the adaptive modulation schemes discussed in Chapter 22, adaptive coding relies on the same fundamental principles of channel quality estimation, parameter adaptation and signalling of the modem parameters employed. Like adaptive modulation, it can operate in open and closed adaptation loops, which were portrayed in Figure 22.29. In this chapter, we will concentrate on the upper bound performance study of adaptive error correction coding for OFDM modems, and assume perfect channel estimation and signalling.

23.1.2 Choice of error correction codes

The behaviour of the adaptive modems in conjunction with long-block-length convolutional coding based turbo coding has been studied in Chapter 22. We have seen that the error correction performance of the turbo decoder was high, resulting in output bit error rates of below 10^{-4} for input bit error rates of up to 1%. The disadvantage of this RSC-based turbo codec was the long block length of 2000 bits, which prevented code rate adaptation on a short-term basis. Radically shortening the interleaving block length of turbo codes results in significantly reduced performance, as does additional puncturing of the code word for code rate adaptation.

Similarly to the adaptive modulation algorithms, the adaptive error correction codec has to be able to vary its code rate according to the frequency dependent channel transfer function observed for the OFDM system in a time dispersive channel. Ideally, the error correction capability of the code would be adjustable for each data bit's expected BER independently. For our experiments, short block length codes of less than 100 bits per code word were employed, in order to allow flexible adaptation of code parameters, while delivering reasonable error protection for the data bits.

Two different block codes have been investigated for adaptive OFDM transmission, namely variable-length redundant residue number system (RRNS) based codes [449, 450] and turbo BCH codes [451, 452].

23.2 Redundant Residue Number System Codes

Residue number system (RNS) [449, 450] based algorithms have been studied in the context of digital filtering, spectral analysis, correlation and matrix operations as well as in image processing [453–456]. Until quite recently RRNSs have only been proposed for fault tolerant processing of signals, for example in digital arithmetics. RRNSs represent each operand with the aid of a set of residues. The residues are generated as the remainders upon dividing the operand to be represented with the aid of the residues by each of the so-called moduli of the RRNS, which have to be relative primes, i.e. do not have a common divisor. Amongst others the following two advantages accrue for RRNS-based processing [457]: they have the ability to use carry-free arithmetic, since all residue-based operations can be processed independently of each other. This is amenable to high-speed parallel processing. A related RRNS property is the lack of ordered significance among residue digits, which implies that as long as a sufficient number of residues was retained, in order to unambiguously represent the results of the computations, any erroneous residue digit can be discarded without affecting the result.

Error detection and correction algorithms based on the RNS have been proposed by Szabo *et al.* [449], and by Watson and Hastings [450], which exploited the properties of the redundant residue number system (RRNS). More recently, a computationally efficient procedure was described in [458] for correcting a single error. In [459], the procedure was extended to correcting double errors as well as simultaneously correcting single errors and detecting multiple errors. Efficient soft-decision multiple error correcting algorithms were suggested in [460]. Furthermore, a RNS-based M-ary modulation scheme has been proposed and analysed in [461], while an RRNS-based CDMA system was the topic of [462].

RRNS(n, k) codes are akin to the well-known family of Reed-Solomon (RS) codes, and both families achieve the so-called maximum minimum distance of $d = n - k + 1$, provided that the moduli of the RRNS code obey certain conditions. It can also be readily shown that the RRNS code's weight distribution can be approximated by the weight distribution of RS codes if all the moduli assume values close to their average value [463–465]. For further details concerning the construction, coding theory, encoding and decoding of RRNS codes the interested reader is referred to the literature [460–465].[3] Based on the above arguments a similar coding performance is achieved by an identical rate RS code and RRNS code, provided that they both use the same number of bits per symbol or bits per residue. However, their encoding and decoding algorithms are distinctly different [460], requiring further research into their implementation, coding theory and application. Since in [460] the authors have developed a soft-decision decoding algorithm, here we favoured RRNS codes, with the intention of stimulating further research in this interesting novel field. Typically there exists a larger variety of RRNS codes for a given code rate and code word length than in the family of RS codes.

The RRNS codes employed in our investigations are systematic, which means that k of the n code residues contain the original data bits, and the additional $(n - k)$ redundant residues can be employed for error correction at the decoder. The error correction capability of the code is $t = \lfloor \frac{n-k}{2} \rfloor$ residues [460].

The code rate, and accordingly the error correction capability of the code, can be readily varied by transmitting only a fraction of the generated redundant residues. If the channel conditions are favourable, then only the systematic information-bearing residues are transmitted, resulting in a unity rate code with no added redundancy and no error correction capability. Upon transmitting two redundant residues with the data bits, the resulting code can correct one residue error for a code rate of $\frac{n}{n+2}$. More of the redundant residues can be transmitted, lowering the code rate and improving

[3] A range of related, but at the time of writing unpublished RRNS-coding oriented papers by the authors can be found under http://www-mobile.ecs.soton.ac.uk/lly/

the code's error resilience at the cost of a lower effective information throughput, when the channel quality degrades.

In our investigations RRNS codes employing 8 bits per residue have been chosen. Three or six systematic information-bearing residues – corresponding to 24 or 48 useful data bits per code word – and up to six redundant residues have been employed. The code parameters for these codes are shown in Table 23.1. As can be seen from the

Code	(5, 3)	(7, 3)	(9, 3)	(8, 6)	(10, 6)	(12, 6)
Sys. residues	3	3	3	6	6	6
Red. residues	2	4	6	2	4	6
Correction capability t	1	2	3	1	2	3
Code rate	0.6	0.43	0.33	0.75	0.6	0.5
Data bits per word	24	24	24	48	48	48
Red. bits per word	16	32	48	16	32	48

Table 23.1: RRNS codes employed in our investigations, each using 8-bit residues

table, the code rates vary from 0.33 to 0.75. The uncoded case, corresponding to a $(3, 3)$ code, is not shown in the table.

23.2.1 Performance in an AWGN channel

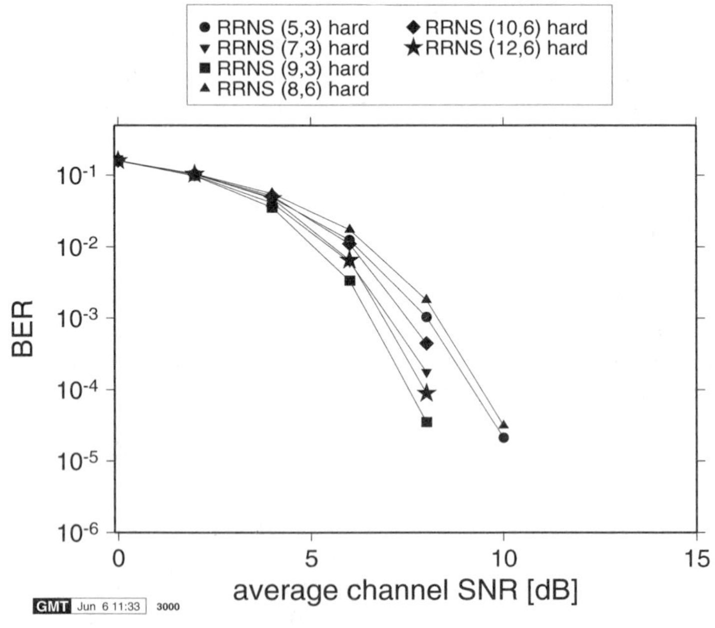

Figure 23.1: BER performance for RRNS-coded QPSK transmission in AWGN channel using hard decision decoding

The BER performance of an OFDM system employing QPSK and the RRNS codes of Table 23.1 is depicted in Figure 23.1. It can be seen that the relative BER performance of the different codes is largely in line with their respective code rates. The (9,3) code, with a code rate of 0.33, exhibits the strongest error correction properties; the (8,6) code, with a code rate of 0.75, is the weakest code of the set. Comparing the performance of the (5,3) code with that of the (10,6), code both having a code rate of 0.6, shows that the longer code exhibits a superior performance. The (12,6) code, having a code rate of 0.5, outperforms the shorter (7,3) code for SNR values in excess of 6 dB.

23.2.1.1 Performance in a fading time dispersive channel

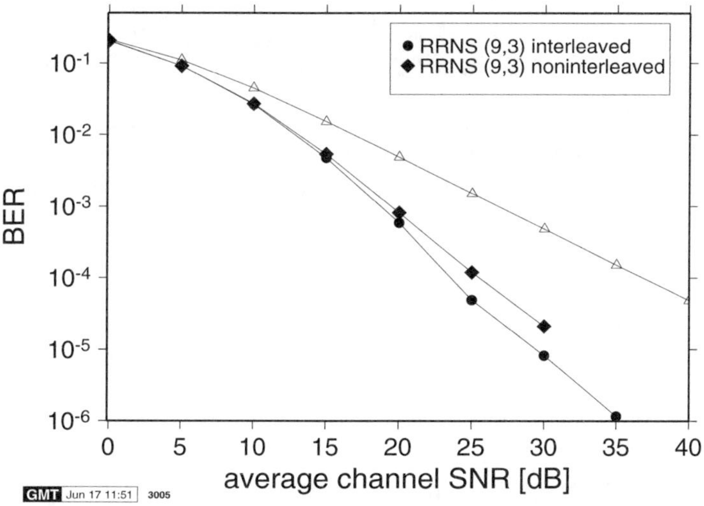

Figure 23.2: BER performance and throughput in bits per symbol for the RRNS-coded OFDM system in the Rayleigh fading shortened WATM channel of Section 20.1.1.2, with 512-FFT QPSK OFDM transmission

Figure 23.2 portrays the BER performance of a RRNS-coded 512-subcarrier modem using a 512-point FFT-based (512-FFT) OFDM modem employing QPSK transmission over the Rayleigh fading short WATM channel described in Section 20.1.1.2. The associated BER curves for uncoded, coded non-interleaved and coded interleaved transmission are shown. The interleaver employed in acquiring the interleaved results was a shortened block interleaver of the structure shown in Figure 23.3. This interleaver structure was chosen since it is sufficiently flexible for varying numbers of residues if adaptive modulation is invoked.

We note that residue-based interleaving has a better performance than bit interleaving, since bit interleaving would increase the probability of residue errors due to spreading bursts of erroneous bits across residues. Since the RRNS decoding algorithm is symbol based, the increased residue error rate would degrade the system's

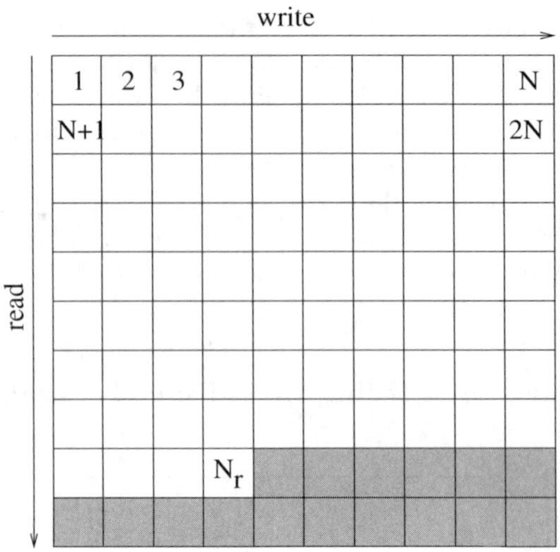

Figure 23.3: Structure of the interleaver employed for the ARRNS system. The interleaver is a square $N \times N$ interleaver for N_r residues, with $N = \lceil \sqrt{N_r} \rceil$; only the first N_r positions are used, where N_r is the number of residues in the OFDM symbol

performance.

It can be seen from Figure 23.2 that the coded schemes deliver a gain of about 12 dB in SNR terms. Interleaving of the residues within the OFDM symbol improves the BER performance by about 2 dB.

23.2.1.2 Adaptive RRNS-coded OFDM

Analogously to the adaptive modulation schemes discussed in Chapter 22, the code rate adaptation reacts to the time- and frequency-varying channel conditions experienced in a duplex link. Each station exploits the channel quality information extracted from the last received OFDM symbol for determining the coding parameters of the next transmitted frame.

Although the longer codes exhibit better error correction properties than the shorter codes investigated, we will concentrate on the group of $(9,3)$, $(7,3)$, $(5,3)$ codes and on uncoded $(3,3)$ transmission for the variable code rate application, since a short code word length allows for increased flexibility in the adaptation to the channel conditions. These codes exhibit an error correction capability of $t = 3, 2, 1$, and 0 residues per code word, respectively. Correspondingly, four error correction coding modes c are defined, which are shown in Table 23.2.

The choice of the coding mode for each code word in the AOFDM symbol is determined on the basis of the estimated channel transfer function. As discussed for the adaptive modulation scheme of Section 22.2.4.2, the predicted bit error probabilities p_e are calculated for all bits to be transmitted in an OFDM symbol, based on the estimated subcarrier SNR and the modulation mode to be employed. More

Mode c	0	1	2	3
n_c	3	5	7	9
k_c	3	3	3	3
t_c	0	1	2	3
R	1	0.6	0.43	0.33

Table 23.2: RRNS coding modes used for the code rate adaptation employing 8-bit residues

explicitly, the expected overall bit error probability for the given OFDM symbol is computed by averaging the individual subcarrier BERs for all legitimate modulation modes M_n, yielding $\bar{p}_e(M_n) = 1/N_s \sum_j p_e(\gamma_j, M_n)$ where γ_j is the subcarrier SNR for $j = 1, \ldots, 512$. The specific subcarrier modem mode allocation with the highest bits per symbol (BPS) throughput, whose estimated BER is lower than the required value is then confirmed. Currently, however, only QPSK/OFDM is used, which simplifies the associated bit allocation and BER computation procedure. AOFDM will be the topic of Section 23.2.2. This algorithm allows the direct adjustment of the desired maximum BER.

If OFDM is to be employed in conjunction with adaptive RRNS coding, then the number of bits per OFDM symbol and the mapping of bits to subcarriers can change from one OFDM symbol to the next. Exlicitly, this implies using a variable bitrate, near-constant BER regime. Hence the RRNS coding scheme adaptation algorithm operates on the basis of the estimated BER, rather than directly relying on the estimated channel transfer function. Once the vector of estimated bit error probabilities $p_e(n)$ for the specific number of bits N_b - which has to be conveyed by the OFDM symbol - is known, the total number of bits to be transmitted is split into blocks of K bits, where K is the number of bits per residue. Again, the error correction capability of the code in each RRNS code word is a given number of residues, not bits. Hence, as argued before, interleaving of bits would increase the residue error rate at the decoder's input and lower the system's performance.

From the bit error probability $p_e(n)$ of bit n, $n = 1 \ldots N_b$ - where N_b is the number of bits allocated to the OFDM symbol - the estimated residue error rate $p_r(r)$ for residue index r - which is defined as the proportion of residues in error - for the $N_r = \lfloor N_b/K \rfloor$ number of residues in the OFDM symbol can be calculated as:

$$p_r(r) = 1 - \left(\prod_{n=0}^{K-1} (1 - p_e(r \cdot K + n)) \right). \tag{23.1}$$

The remaining $N_b - K \cdot N_r$ data bits of the OFDM symbol that are not allocated to any residue are filled with padding bits and hence contain no useful data. The mapping of the residues with index r to the RRNS code words is based on the estimated residue error probabilities $p_r(r)$. The interleaver $I(r)$ of Figure 23.3 is used to map the stream of residues to the residue positions in the transmitted OFDM symbol. The interleaver function used in our experiments is defined more explicitly below.

A received code word of the codec mode c_w is irrecoverable if more than t_c of the received residues are in error. The RRNS code word error probability p_w for word w can be calculated as:

$$p_w(w) = p(R_r(w) > t_{c_w}) = 1 - P(R_r(w) \leq t_{c_w}), \tag{23.2}$$

where $R_r(w)$ is the number of residue errors in code word w, and $p_w(w)$ can be calculated from the residue error probabilities $p_r(r)$ as:

$$p_w(w) = 1 - p\left[R_r(w) = 0\right] - p\left[R_r(w) = 1\right] - \ldots - p\left[R_r(w) = t_{c_w}\right]. \tag{23.3}$$

Upon elaborating further:

$$p(R_r(w) = 0) = \prod_{r=0}^{n_{c_w}-1} (1 - p_r(I(r_{0,w} + r))) \tag{23.4}$$

$$p(R_r(w) = 1) = \sum_{r=0}^{n_{c_w}-1} p_r(I(r_{0,w} + r) \prod_{s=0,s\neq r}^{n_{c_w}-1} (1 - p_r(I(r_{0,w} + s))) \tag{23.5}$$

$$= p(R_r(w) = 0) \cdot \sum_{r=0}^{n_{c_w}-1} \frac{p_r(I(r_{0,w} + r))}{1 - p_r(I(r_{0,w} + r))} \tag{23.6}$$

$$p(R_r(w) = 2) = \frac{1}{2} \cdot p(R_r(w) = 0) \cdot \sum_{r=0}^{n_{c_w}-1} \left[\frac{p_r(I(r_{0,w} + r))}{1 - p_r(I(r_{0,w} + r))} \cdot \sum_{s=0,s\neq r}^{n_{c_w}-1} \frac{p_r(I(r_{0,w} + s))}{1 - p_r(I(r_{0,w} + s))} \right] \tag{23.7}$$

$$p(R_r(w) = 3) = \frac{1}{3!} \cdot p(R_r(w) = 0) \cdot \sum_{r=0}^{n_{c_w}-1} \left[\frac{p_r(I(r_{0,w} + r))}{1 - p_r(I(r_{0,w} + r))} \cdot \sum_{s=0,s\neq r}^{n_{c_w}-1} \left[\frac{p_r(I(r_{0,w} + s))}{1 - p_r(I(r_{0,w} + s))} \cdot \sum_{t=0,t\neq q,r}^{n_{c_w}-1} \frac{p_r(I(r_{0,w} + t))}{1 - p_r(I(r_{0,w} + t))} \right] \right], \tag{23.8}$$

where $r_{0,w}$ is the index of the first residue in code word w.

The code rate adaptation algorithm calculates the word error probability $p_w(w)$ for the RRNS code word index w for the lowest-power codec mode of Table 23.2, $c = 0$. If the word error probability exceeds a certain threshold, i.e. $p_w(w) > \alpha$ for $c = 0$, then the next stronger coding mode $c = 1$ is selected, and the word error probability is evaluated again. If the new RRNS code word error probability exceeds the threshold α, then the next codec mode is evaluated, until the estimated RRNS code word error probability falls below the threshold α, or until the highest power codec mode is selected.

The parameter α is supplied to the algorithm and it can be used to control the adaptation process, similarly to the target BER in the adaptive modulation algorithm of Section 22.2.4.2.

The effect of different settings of the parameter α on the effective data throughput and BER of a QPSK modem in an AWGN channel is shown in Figure 23.4. The

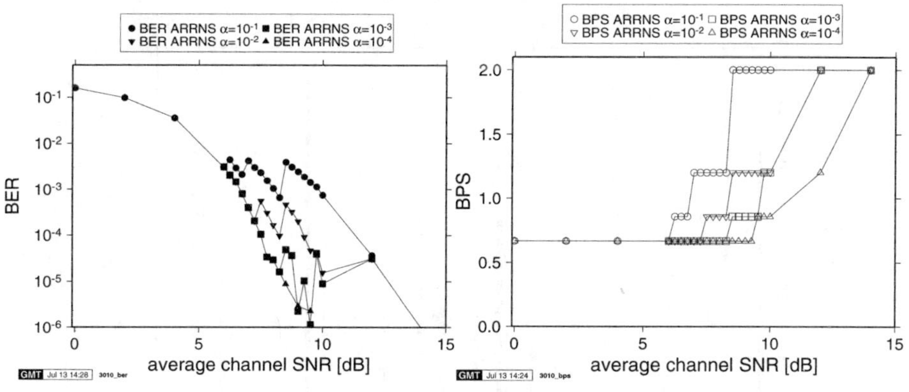

(a) BER (b) BPS throughput

Figure 23.4: BER performance and throughput in bits per symbol for the adaptive RRNS (ARRNS) coded OFDM system in AWGN channels, with QPSK transmission in the subcarriers

word error rate (WER) threshold α was varied from 10^{-1} up to 10^{-4}, and it can be seen that the system performance varies only in a limited range of SNR values, approximately between 7 and 12 dB, with the WER α. The three RRNS-coded BER curves of Figure 23.1 are easily identified in this figure as segments of the adaptive scheme's BER performance curve together with the BER curve of uncoded QPSK. The different codec modes are easily identifiable in this figure, since in the AWGN channel the channel quality is time invariant, and therefore the same codec mode is chosen for all transmitted RRNS code words at a given SNR point. The choice of the RRNS codec mode clearly depends on the value of the WER α. For all four values of α, the system performance is identical for SNR values up to 6 dB. For higher SNR values, it can be observed that the WER α influences the codec mode switching algorithm. For $\alpha = 10^{-1}$, the codec mode is switched to the next lower-power higher-rate mode of Table 23.2 at SNR values of 6.25, 7 and 8.5 dB, resulting in a BER of about 0.7% after each reconfiguration as seen in Figure 23.4. The maximal throughput of 2 bits/symbol is reached at an SNR of 8.5 dB by the $\alpha = 10^{-1}$ curve. For lower values of α, the codec mode switching is more conservative hence lower bit error rates and throughputs are observed. For $\alpha = 10^{-4}$ the modem employs codec mode 3 of Table 23.2 for SNR values of up to 9.5 dB, reaching mode 0 at 14 dB SNR.

Over fading time dispersive channels the ARRNS codec mode is adapted to the time and frequency variant channel conditions. In order to demonstrate the RRNS code allocation process, in Figure 23.5 we captured the expected subcarrier SNR versus the subcarrier index for one specific OFDM symbol. It can be seen that the channel SNR varies by about 18 dB across the set of subcarriers, which will result in a varying BER across the OFDM symbol. In our illustrative example a specific OFDM symbol was selected that contained at least one RRNS code word corresponding to

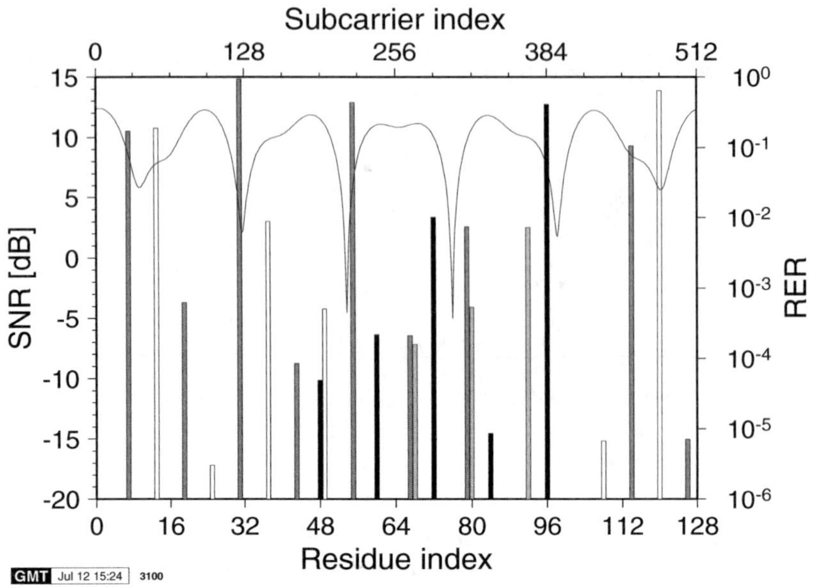

Figure 23.5: Predicted subcarrier SNR (continuous line) for an OFDM symbol and residue error probabilities (vertical bars) for four selected ARRNS code words for a 512-subcarrier OFDM modem employing QPSK in the short WATM channel of Figure 20.3. The average SNR is 10 dB, and $\alpha = 10^{-1}$. Key: light grey = (3, 3) code at indices of 68, 80, 92; black = (5, 3) at 48, 60, 72, 84, 96; white = (7, 3); dark grey (9, 3)

each of the RRNS codec modes of Table 23.2. In order to augment our understanding, in this example the residues of only one RRNS code word per codec mode is shown in the figure, corresponding to a total of $3+5+7+9 = 24$ residues, signified by vertical bars in the figure. Each vertical bar corresponds to one transmitted residue, and the shading of each bar relates it to one of the four ARRNS code words. The position of a bar gives the residue index r, while its height signifies the expected residue error rate (RER) $p_r(r)$ evaluated from Equation 23.1 for each of the residues of the four chosen ARRNS code words, which depends on the subcarrier SNR experienced at index r.

The channel model used for the fading experiments was described in Section 20.1.1.2. The word error probability threshold α was set to 10^{-1}.

The four code words were chosen so that one of each codec mode of Table 23.2 is demonstrated: the light grey bars mark the position and the residue error probabilities for an uncoded (mode 0) (3, 3) RRNS code word. The positions of the residues in the transmitted OFDM symbol are determined by the residue-based interleaver function. It can be seen that, because of the good channel quality in the subcarriers used for transmission, all of the three residues of the uncoded code word exhibit low RER values. The word error probability for this code word is mainly dominated by the highest RER residue transmitted at index 92, which exhibits an error probability of about $7.2 \cdot 10^{-3}$. The resulting word error probability due to all three residues is $7.9 \cdot$

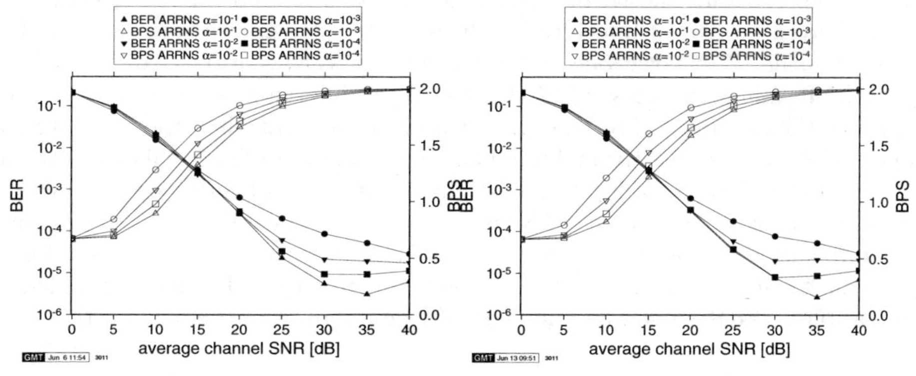

(a) non-interleaved **(b)** interleaved

Figure 23.6: BER and BPS throughput versus average channel SNR for ARRNS-coded 512-subcarrier OFDM transmission employing QPSK over the shortened WATM channel of Figure 20.3. The stipulated WER values were $\alpha = 10^{-1}, 10^{-2}, 10^{-3}$ and 10^{-4}

10^{-3}, which is below the threshold of $\alpha = 10^{-1}$, and therefore uncoded transmission was chosen by the algorithm.

The black bars in Figure 23.5 mark the residues for a mode 1 (5, 3) code word, with five transmitted residues, while the white marked residues form a mode 2 (7, 3) code word, consisting of seven transmitted residues. The first of this code word's residues at residue index $r = 0$ is not visible on the figure, as its expected RER is $2.1 \cdot 10^{-7}$.

The dark grey bars of Figure 23.5, represent a mode 3 (9, 3) RRNS code word with 9 transmitted residues. The word error probability for this mode 3 RRNS code word is 11.5%, which exceeds the set threshold. Since there is no stronger code in the set of coder modes of Table 23.2, the highest possible code rate was chosen.

Figure 23.6 shows the BER performance and data throughput of a 512-subcarrier ARRNS-coded OFDM modem employing QPSK in the SWATM channel of Figure 20.3. Figure 23.6(a) depicts the BER and BPS performance for the modem employing no residue interleaving, while for Figure 23.6(b) a residue interleaver obeying the structure introduced in Figure 23.3 was employed. This comparison is relevant and necessary, since it is not intuitive as to whether residue interleaving results in performance improvements. This is because the residue interleving disperses the bursty channel errors across the OFDM subcarriers, which is expected to improve the coding performance of conventional fixed rate coding. We, however, proposed adaptive RRNS-coding, in order to combat the bursty error distribution across the OFDM subcarriers, which was designed to counteract the frequency selective fading and its bursty errors. Hence interleaving and adaptive RRNS coding may disadvantageously interfere with each other, necessitating a comparison of the interleaved and non-interleved results.

It can be seen that the BER performance is fairly similar for the interleaved and non-interleaved modems. Specifically, although the non-interleaved modem slightly outperforms the interleaved one in BER terms for WERs of $\alpha = 10^{-3}$ and for $\alpha = 10^{-4}$ at SNR values in excess of 25 dB, the BER difference is not significant. Upon comparing the achieved throughput, however, it is clear that the non-interleaved modem offers an average throughput benefit of about 0.1 bit/symbol for all target WERs α in the SNR region up to 25 dB. Since the range of code rates is limited, the BER performance at low values of SNR cannot be significantly lowered. Similarly to the adaptive modulation systems of Chapter 22, transmission blocking would have to be introduced in order to guarantee a target bit error rate. Let us now consider the combination of ARRNS coding and AOFDM modulation in the next section.

23.2.2 ARRNS/AOFDM transceivers

In this section we will demonstrate that upon combining adaptive modulation with adaptive coding, the low SNR performance can be dramatically improved by amalgamating transmission blocking with adaptive error correction coding. We have advocated here the target BER adaptive modulation algorithm of Section 22.2.4.2 due to its high performance and easy adjustability to different target bit error rates.

The transmission parameter adaptation is performed in two steps. First the modulation modes are allocated to the subcarriers according to the algorithm outlined in Section 22.2.4.2. Recall that for AOFDM the expected overall bit error probability for the given OFDM symbol was computed by averaging the individual subcarrier BERs for all legitimate modulation modes M_n, yielding $\bar{p}_e(M_n) = 1/N_s \sum_j p_e(\gamma_j, M_n)$ where γ_j is the subcarrier SNR for $j = 1, \ldots, 512$. The specific subcarrier modem mode allocation with the highest bits per symbol (BPS) throughput, whose estimated BER is lower than the required value is then confirmed. According to this step, the number of bits N_b transmitted in the next OFDM symbol and their estimated bit error probabilities $p_e(n)$ become known. On the basis of this, the code rate adaptation algorithm calculates the residue error rates $p_r(r)$ from Equation 23.1, constructs the interleaver $I(r)$ for the correct number of residues and invokes the appropriate codec modes for the RRNS code words, as outlined above.

Figure 23.7 gives an overview over the ARRNS/AOFDM system's BER and throughput performance over the short WATM channel of Figure 20.3. Two target BER values have been stipulated, both with and without interleaving of the transmitted residues. Figure 23.7(b) and 23.7(a) portray the system's performance if a target BER of 5% is assumed for the adaptive modulation with and without interleaving, respectively. As has been observed in Chapter 22, the uncoded BER, represented by the grey curves, is lower than the AOFDM target BER, which is due to the operation of the AOFDM algorithm. It can be seen that the coded BER is below 1% for all simulated modem configurations, and that the SNR gain is much higher than for fixed QPSK transmission in Figure 23.2. The BER performance is limited, however, by the limited error correction capability of the RRNS (9, 3) mode when the SNR is very low.

Lowering the adaptive modulation's target BER to 0.5%, the BER of the system can be influenced over the whole SNR range by varying the WER α. Figure 23.7(d)

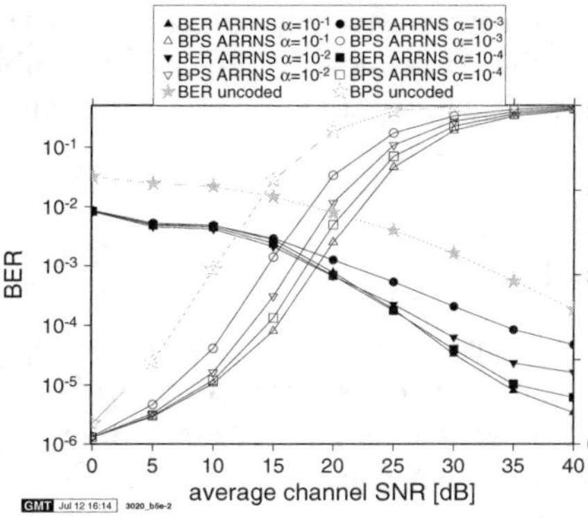

(a) target BER 5% non-interleaved

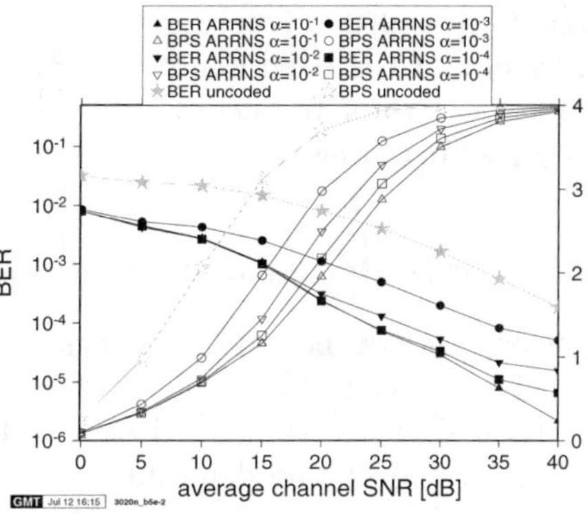

(b) target BER 5% interleaved

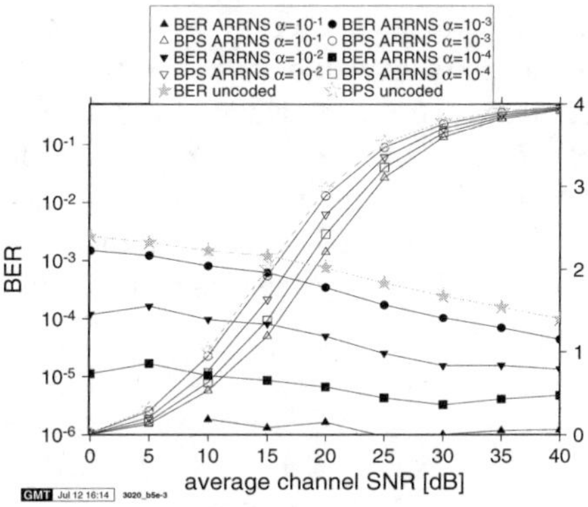

(c) target BER 0.5% non-interleaved

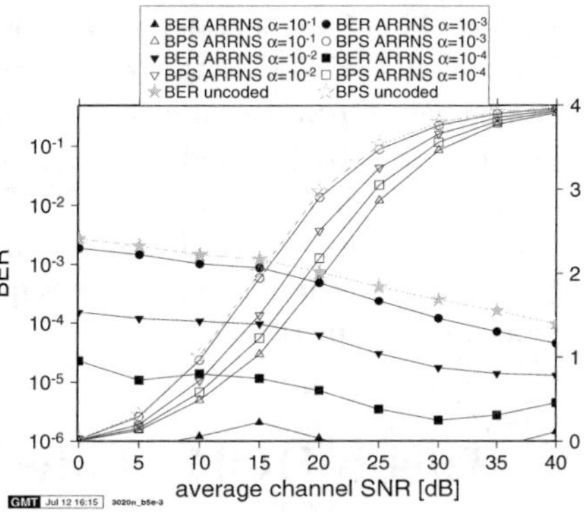

(d) target BER 0.5% interleaved

Figure 23.7: BER and BPS throughput versus average channel SNR for hard decision ARRNS-coded 512-subcarrier OFDM transmission employing adaptive modulation over the shortened WATM channel of Figure 20.3: (a, b) uncoded adaptive modulation target BER 5%, (c, d) uncoded target BER 0.5%, (a, c) non-interleaved, (b, d) interleaved. The stipulated WER values were $\alpha = 10^{-1}, 10^{-2}, 10^{-3}$ and 10^{-4}. The light grey curves show the uncoded BER and BPS throughput and 23.7(c) depict the corresponding BER and BPS throughput. For a WER of $\alpha = 10^{-1}$ the achieved BER is better than $2 \cdot 10^{-3}$, while for $\alpha = 10^{-2}$ a BER of $2 \cdot 10^{-4}$ is never exceeded and for $\alpha = 10^{-3}$ a BER of $2 \cdot 10^{-5}$.

Comparing the interleaved performance with the non-interleaved results, it can be seen that the BER performance of comparable modems is fairly similar. The throughput is slightly higher, however, for the non-interleaved systems, demonstrating the efficiency of the hard decision ARRNS/AOFDM schemes in terms of combating the bursty errors of frequency selective fading. Let us now consider the effects of invoking soft decisions.

23.2.3 Soft decision RRNS decoding

The performance of the ARRNS codes can be increased by soft decision decoding [460]. Previously, the ARRNS decoder assumed that the outputs of the demodulator were binary hard decision values. However, the transceiver is capable of exploiting soft outputs provided by the demodulator at the receiver. Soft decoding of the ARRNS codes can be implemented by combining the classic Chase algorithm [466] with the hard decision ARRNS decoder.

Figure 23.9 shows the soft decision decoded performance of the AOFDM/ARRNS system. Comparison with Figure 23.7 shows an improved BER performance for the soft decoder. This is especially significant for the target bit error rate of 5%, where the soft decoded AOFDM/ARRNS system achieves BER values of below 0.3% at 0 dB SNR. Since the adaptation algorithm is unchanged, the same throughput is observed for the hard and soft decoded systems. A BER of below 10^{-4} was registered for the 0.5% target BER system for a WER of $\alpha = 10^{-2}$. Under these circumstances the interleaved system exhibits a lower throughput and worse BER performance than the non-interleaved system. Let us now compare the performance results to those of turbo BCH codecs in the next section.

23.3 Turbo BCH Codes

As an alternative to the above ARRNS codes, we have investigated the employment of turbo BCH codes for the adaptive modem. In Section 22.2.7, half-rate convolutional turbo codes have been employed in the adaptive modem. However, the convolutional constituent codes can also be replaced by block codes [451, 452], which have been shown for example by Hagenauer [467, 468] to perform impressively even at near-unity coding rates. Generally, block codes are appropriate for high code rates and convolutional codes for lower code rates, typically below 2/3, as argued by Hagenauer *et al.* [467]. Turbo block codes have since been studied in a variety of applications [469–472].

For our adaptive coding system, a turbo (T) code with constituent $BCH(31, 26)$ coders was advocated. The parity bits of the two constituent codes were not punctured, resulting in a $TBCH(36, 26)$ code. The resulting code rate was therefore 0.72. The turbo interleaver employed was a 4×13 bit interleaver, leading to a block size of 52 uncoded or 72 coded bits.

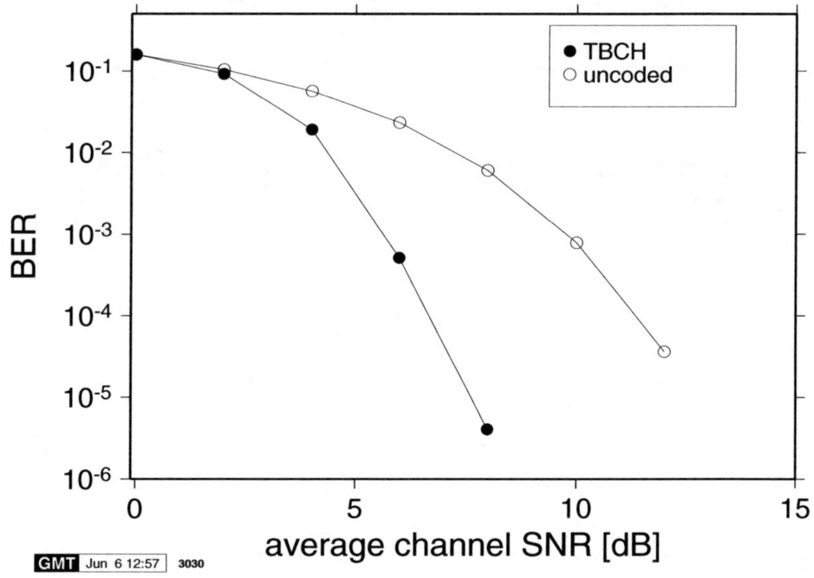

Figure 23.8: BER performance of the 0.72 rate $TBCH(36, 26)$ coded OFDM system over the AWGN channel, employing QPSK transmission in the subcarriers using 6 iterations and a 4×13 turbo interleaver

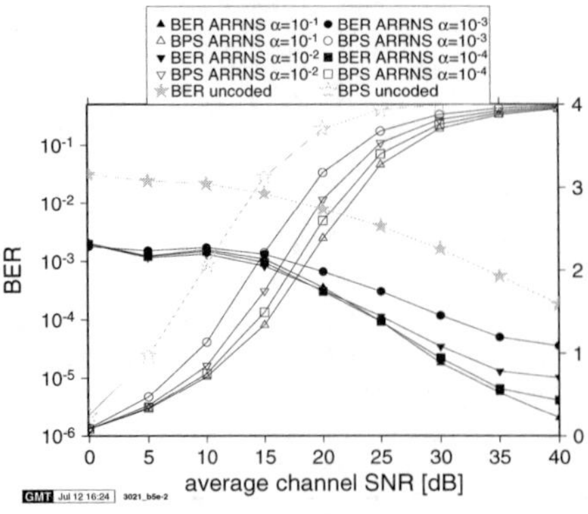

(a) target BER 5% non-interleaved

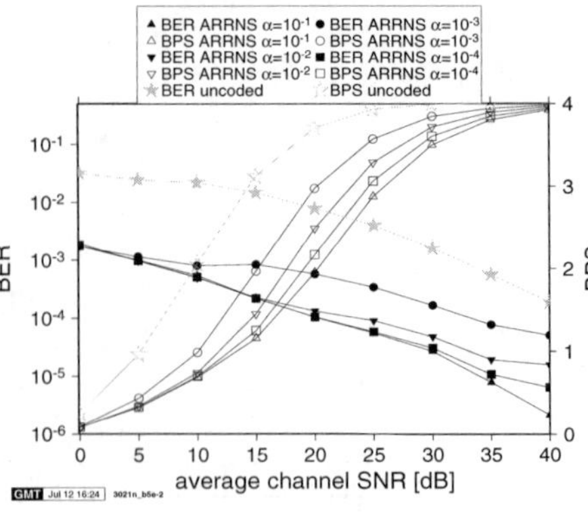

(b) target BER 5% interleaved

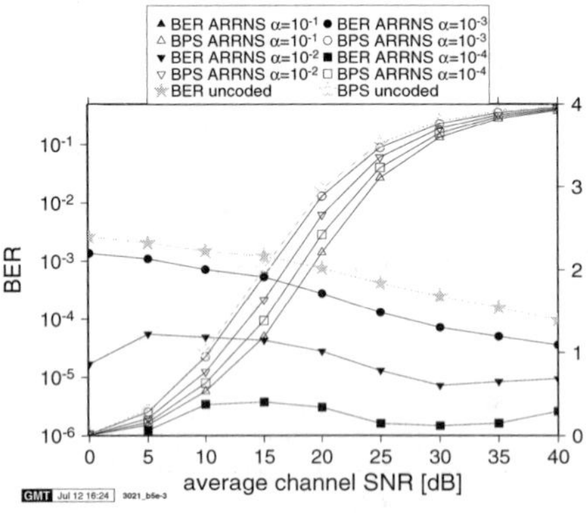

(c) target BER 0.5% non-interleaved

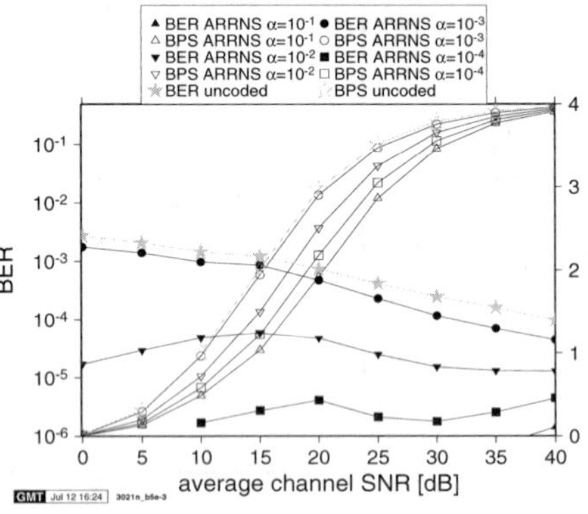

(d) target BER 0.5% interleaved

Figure 23.9: BER and BPS throughput versus average channel SNR for soft decision ARRNS-coded 512-subcarrier OFDM transmission employing adaptive modulation over the shortened WATM channel of Figure 20.3. (a, b) - uncoded adaptive modulation target BER 5%, (c, d) uncoded target BER 0.5%, (a, c) non-interleaved, (b, d) interleaved. The stipulated WER values were $\alpha = 10^{-1}, 10^{-2}, 10^{-3}$ and 10^{-4}. The light grey curves show the uncoded BER and BPS throughput. The corresponding hard decision results are plotted in Figure 23.7

Figure 23.8 shows the BER performance of the TBCH code over an AWGN channel with QPSK transmission using 6 iterations and a 4×13 turbo interleaver. Comparison with the AWGN performance of the hard decision decoded RRNS codes in Figure 23.1 reveals that the TBCH code outperforms all the investigated RRNS codes. The SNR gain of the 0.72 rate TBCH code at a BER of 10^{-4} is about 4.5 dB, which is 0.8 dB better than that of the $1/3$ rate $(9, 3)$ RRNS code. We note, however, that the RRNS coding performance can be improved by invoking turbo coding. This issue is the

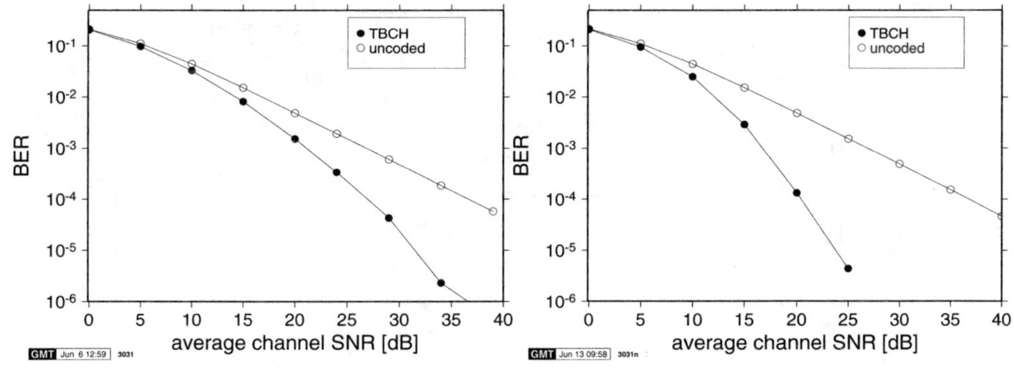

(a) non-interleaved **(b)** interleaved

Figure 23.10: BER versus average channel SNR for the 0.72 rate $TBCH(36, 26)$ coded 512-subcarrier OFDM transmission employing QPSK over the short WATM channel of Figure 20.3 with and without channel interleaving, using 6 iterations and a 4×13 turbo interleaver

subject of further research.

For transmission over the short WATM channel of Figure 20.3, a block channel interleaver obeying the structure depicted in Figure 23.3 was employed. In the case of the 0.72 rate $TBCH(26, 26)$ coder, the interleaving was performed bit-by-bit, instead of the residue-based interleaving of the RRNS system. Figure 23.10 shows the BER performance of the studied TBCH code over the short WATM channel of Figure 20.3. It can be seen that channel interleaving of the TBCH-coded bits significantly improves the performance of the codec. For a BER of 10^{-4} a SNR gain of 10 dB was recorded for the non-interleaved, while an SNR gain of 16.5 dB was observed for the interleaved system.

Figure 23.11 gives a comparison between the 0.72 rate TBCH performance of Figure 23.10 and the long block length half-rate turbo convolutional code of Figure 22.13 in E_b/N_0 terms. It can be seen that the TBCH code with additional channel interleaving performs only about 1 dB worse in E_b/N_0 than the convolutional turbo code at a BER of 10^{-4}.

23.3.1 Adaptive TBCH coding

Unlike the ARRNS code rate adaptation algorithm, which employed four code rates, the proposed TBCH rate adaptation regime operates as a simple on/off switch. If the channel quality is lower than a threshold, then N_d data bits are encoded into N_c coded bits and transmitted in the bit positions allocated for the code word. If the channel quality is sufficiently high for uncoded transmission, then N_c data bits are directly mapped into the allocated bit positions.

Since binary BCH codes are employed, the estimated bit error rate in each code

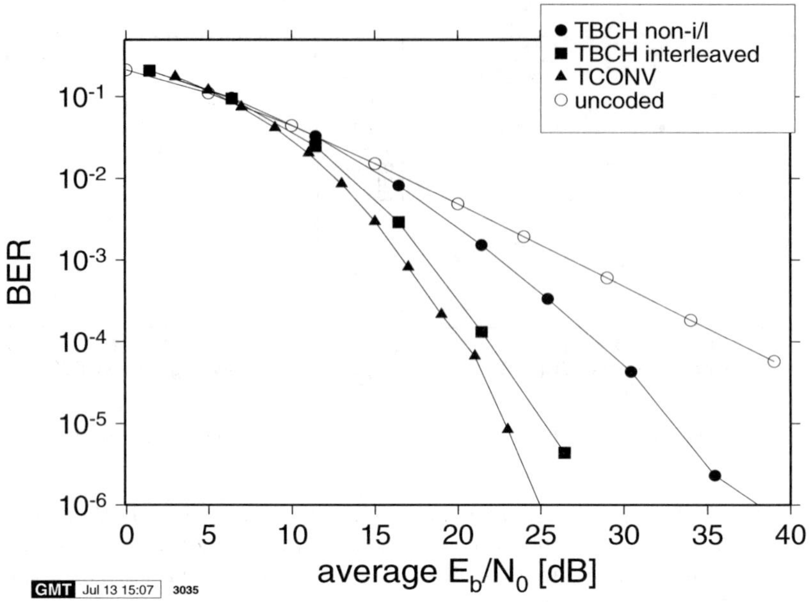

Figure 23.11: Comparison of the BER versus E_b/N_0 performance of the 0.72 rate $TBCH(36, 26)$ coded with and without channel interleaver and the 1/2 rate turbo convolutional coded OFDM system with 2000-bit turbo interleaver of Section 22.2.7 (Figure 22.13) employing QPSK transmission over the short WATM channel of Figure 20.3

word can be used directly as the adaptation criterion. Based on the estimated bit error probability $p_e(n)$, the uncoded bit error probability in the code word w of length N_c can be calculated as:

$$\bar{p}_e(w) = \frac{1}{N_c} \cdot \sum_{n=0}^{N_c} p_e\left(I\left(n + n_0(w)\right)\right), \tag{23.9}$$

where $I()$ is the channel interleaver, and $n_o(w)$ is the index of the first bit of the code word w.

The adaptation algorithm compares the estimated BER in the code word with a bit error rate threshold β, and if $\bar{p}_e(w) \leq \beta$ then N_c data bits are transmitted in the word w. Otherwise, coded transmission is selected.

Figure 23.12 shows the BER and throughput of the ATBCH modem employing QPSK over the short WATM channel of Figure 20.3 for BER thresholds of $\beta = 10^{-2}, 10^{-3}$ and 10^{-4}. It can be seen that for both the interleaved and non-interleaved cases the BER curves follow the trends found in Figure 23.10 for low SNR values before the adaptation algorithm gradually increases the average code rate towards unity, approaching in a throughput of 2 BPS. Comparing the throughput of the interleaved and non-interleaved modems, it can be seen that the non-interleaved scheme yields

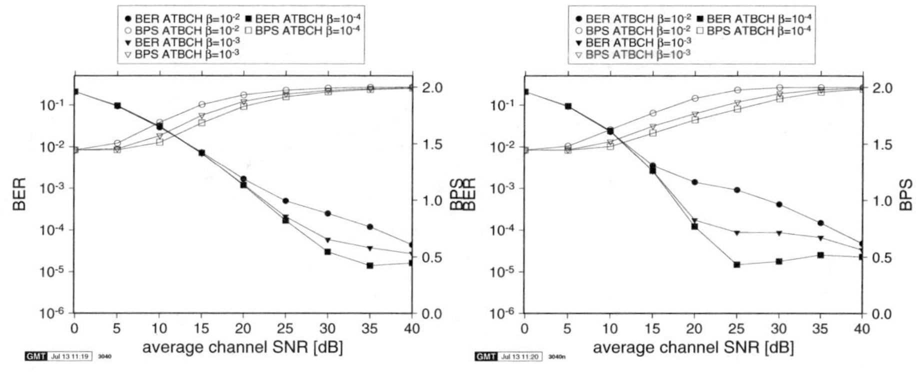

(a) non-interleaved

(b) interleaved

Figure 23.12: BER and BPS throughput versus average channel SNR for adaptive $TBCH(36, 26)$ coded 512-subcarrier OFDM transmission employing QPSK over the short WATM channel of Figure 20.3 with and without channel interleaving

a slightly higher throughput across the SNR range. Let us now combine ATBCH coding with AOFDM in the next section.

23.3.2 Joint ATBCH/AOFDM algorithm

Like in the ARRNS/AOFDM system of Section 23.2.2, the ATBCH code has been combined with the adaptive modulation scheme selection regime discussed in Section 22.2.4.2. The BER and BPS throughput performance of the ATBCH/AOFDM system is determined by the combination of the target BER for the adaptive modulation scheme and the BER threshold β for the code rate adaptation. Figure 23.13 depicts the resulting system performance for uncoded AOFDM target BER values of 10% and 1%, and for ATBCH activation threshold BER values of $\beta = 10^{-2}, 10^{-3}$ and 10^{-4}. For the uncoded AOFDM target BER of 10%, neither the interleaved nor the non-interleaved modems can reduce the data BER to acceptable levels for low SNR values. The interleaved system shows a better BER performance, but exhibits lower throughput than the non-interleaved system. For an uncoded AOFDM target BER of 1%, the ATBCH decoder can efficiently reduce the data BER below 10^{-4} for $\beta = 10^{-4}$. This meets the requirements of the adaptive modem's BER performance, as set out in Chapter 22.

23.4 Signalling

Signalling the codec modes employed for the code words of each OFDM symbol requires a significant amount of bandwidth, analogously to the modulation mode signalling for AOFDM. The effects of erroneous mode detection at the receiver are the

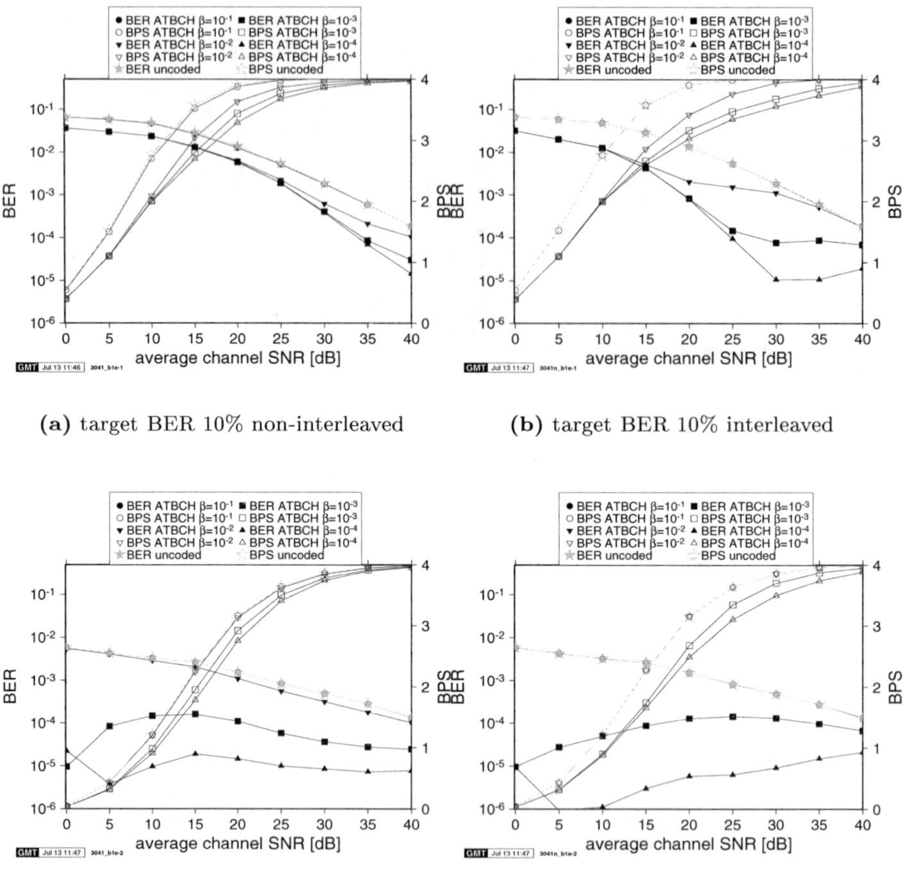

(a) target BER 10% non-interleaved

(b) target BER 10% interleaved

(c) target BER 1% non-interleaved

(d) target BER 1% interleaved

Figure 23.13: BER and BPS throughput versus average channel SNR for ATBCH-coded 512-subcarrier AOFDM over the shortened WATM channel of Figure 20.3: (a, b) - uncoded AOFDM target BER 10%, (c, d) target BER 1%, (a, c) non-interleaved (b, d) - interleaved. The stipulated BER threshold values were $\beta = 10^{-1}, 10^{-2}, 10^{-3}$ and 10^{-4}. The light grey curves show the uncoded BER and BPS throughput

same as for the modulation mode signalling, resulting in loss of synchronisation in the data stream.

Blind detection of the codec mode at the transmitter in the case of open-loop coding adaptation could be employed for longer block length codes, by tentatively decoding the received data sequence and exploiting soft outputs of the decoder as a reliability measure. If decoding of the received sequence increases the reliability measure for the received block's data bits, then the corresponding coder mode is chosen. However, the short block lengths, necessitated by the adaptation algorithms discussed in this chapter make it impossible to use blind detection of the codec mode based on the decoder's soft outputs.

The two different adaptive error correction schemes discussed in this chapter both require signalling of the codec mode for each of the transmitted code words, but the scenarios are different. The ARRNS scheme, which employs four codec modes with code words of variable length between 24 and 72 bits, requires a signalling scheme similar to that of the AOFDM scheme discussed in Section 22.2.6.1.

The ATBCH scheme, with only two different codec modes and a fixed block length, is easier to signal. Depending on the required mode detection error rate, and assuming that the achieved uncoded BER is the uncoded AOFDM target BER, then the number of signalling bits necessary for majority voting signalling can be calculated. If a target BER of 1% is assumed, then a 5-bit signalling sequence results in a detection error rate of 10^{-5}. If 7 signalling bits are used, then a DER of $3.5 \cdot 10^{-7}$ can be achieved.

23.5 Comparison of Coded Adaptive OFDM Schemes

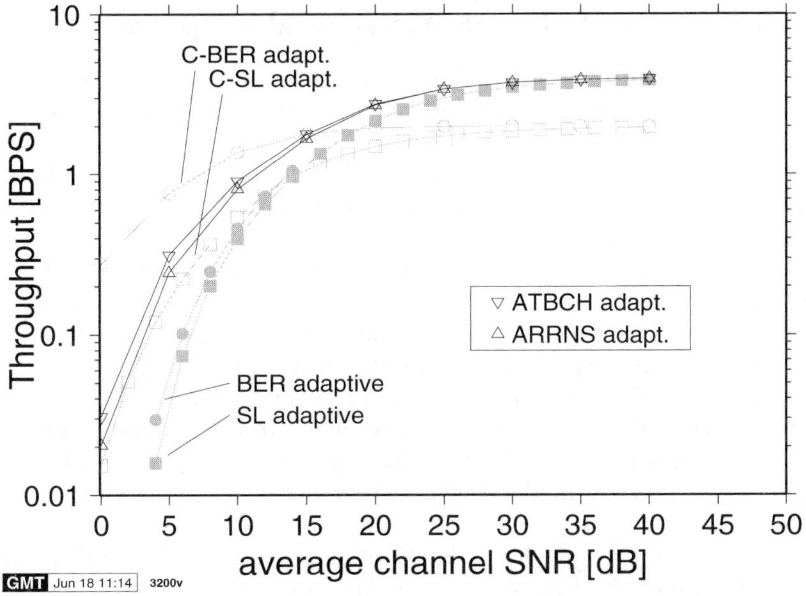

Figure 23.14: BPS throughput versus average channel SNR for AOFDM for a maximum data bit error rate of 10^{-4}. The light grey curves present the performance of the variable-throughput adaptive modulation schemes Figure 22.29 with and without convolutional turbo coding. The-variable throughput systems include convolutionally turbo coded (C-) and uncoded switching level adaptive (SL) and target BER adaptive (BER) systems, as well as the joint adaptive RRNS/AOFDM and ATBCH/AOFDM systems. The short WATM channel of Figure 20.3 was employed

Figure 23.14 shows the throughput of the various adaptive transmission systems studied for a target data BER of 10^{-4}. The lightly shaded curves represent the variable throughput systems' performance graphs from Figure 22.29. The ARRNS/AOFDM and ATBCH/AOFDM systems employed no interleaving and had an AOFDM target BER of 1%. The ARRNS adaptation algorithm used a WER threshold of $\alpha = 10^{-2}$, while the ATBCH system used a data BER threshold of $\beta = 10^{-4}$. The pre-equalisation results from Figure 22.29 were omitted in this figure. It should be noted that the above performance figures do not take into account the signalling overhead required for adaptive modulation and adaptive code rates and hence constitute the upper bound performance of the system.

Let us now in the next section summarise our discourse on various OFDM schemes, before providing an indepth discussion on various system design aspects of QAM- and OFDM-based Digital Video Broadcast Systems in our last chapter.

23.6 Summary, Conclusions and Further OFDM Research

23.6.1 Summary of the OFDM-related chapters

In the last six chapters we have discussed the implementational, algorithmic and performance aspects of orthogonal frequency division multiplexing in predominantly duplex mobile communications environments.

Following a rudimentary introduction to OFDM in Chapter 18 in Chapter 19 we have further studied the structure of an OFDM modem and we have investigated the problem of the high peak-to-mean power ratio observed for OFDM signals, and that of clipping amplification caused by insufficient amplifier back-off. We have investigated the BER performance and the spectrum of the OFDM signal in the presence of clipping, and we have seen that for an amplifier back-off of 6 dB the BER performance was indistinguishable from the perfectly amplified case. We have investigated the effects of quantisation of the time-domain OFDM signal. The effects of phase noise on the OFDM transmission was studied, and two phase noise models were suggested. One model was based on white phase noise, only relying on the integrated phase jitter, while a second model used coloured noise, which was generated from the phase noise mask.

In Chapter 20 we have studied OFDM transmissions over time dispersive channels. The spectrum of the transmitted frequency-domain symbols is multiplied with the channel's frequency domain channel transfer function, hence the amplitude and phase of the received subcarriers is distorted. If the channel is varying significantly during each OFDM symbol's duration, then additional inter-subcarrier interference occurs, affecting the modem's performance. We have seen the importance of channel estimation on the performance of coherently detected OFDM, and we have studied two simple pilot-based channel estimation schemes. Differentially detected modulation can operate without channel estimation, but exhibits lower BER performance than coherent detection. We have seen that the signal-to-noise ratio is not constant across the OFDM symbol's subcarriers, and that this translates into a varying bit error probability across the different subcarriers.

The effects of timing- and frequency errors between transmitter and receiver have been studied in Chapter 21. We have seen that a timing error results in a phase rotation of the frequency domain symbols, and possibly inter-OFDM-symbol interference, while a carrier frequency error leads to inter-subcarrier-interference. We have suggested a cyclic postamble, in order to suppress inter-OFDM-symbol interference for small timing errors, but we have seen that frequency errors of more than 5% of the subcarrier distance lead to severe performance losses. In order to combat this, we have investigated a set of frequency- and timing-error estimation algorithms. We have suggested a time-domain based joint time- and frequency-error acquisition algorithm, and studied the performance of the resulting system over fading time dispersive channels.

Based on the findings of Chapter 20, we have investigated adaptive modulation techniques to exploit the frequency diversity of the channel. Specifically, in Chapter 22, three adaptive modulation algorithms were proposed and their performance was investigated. The issue of signalling was discussed, and we have seen that adaptive OFDM systems require a significantly higher amount of signalling information than adaptive serial systems. In order to limit the amount if signalling overhead, a sub-band adaptive scheme was suggested, and the performance trade-offs against a subcarrier-by-subcarrier adaptive scheme were discussed. Blind modulation mode detection schemes were investigated, and combined with an error correction decoder. We have seen that by combining adaptive modulation techniques with a strong convolutional turbo channel codec significant system throughput improvements were achieved for low SNR values. Finally, frequency-domain pre-distortion techniques were investigated in order to pre-equalise the time-dispersive channel's transfer function. We have seen that by incorporating pre-distortion in adaptive modulation significant throughput performance gains were achieved compared to adaptive modems without pre-equalisation.

We have seen in Chapter 22 that, although channel coding significantly improves the achievable throughput for low SNR values in adaptive modems, with increasing average channel quality the coding overhead limits the system's throughput. In order to combat this problem, in Chapter 23 we have investigated coding schemes which offer easily adjustable code rates. RRNS and BCH codes were employed to implement an adaptive coding scheme allowing us to adjust the code rate across the subcarriers of an OFDM symbol. Combinations of adaptive coding and the adaptive modulation techniques introduced in Chapter 22 were studied, and we have seen that a good compromise between the coded and uncoded transmission characteristics can be found. Let us now in the next subsection draw some conclusions.

23.6.2 Conclusions concerning the OFDM chapters

(1) Based on the implementation-oriented characterisation of OFDM modems, leading to a real-time testbed implementation and demonstration at 34 Mbps we concluded that OFDM is amenable to the implementation of high bit rate wireless ATM networks, which is underlined by the recent ratification of the HIPERLAN II standard.

(2) The range of proposed joint time- and frequency synchronisation algorithms efficiently supported the operation of OFDM modems in a variety of propagation environments, resulting in virtually no BER degradation in comparison to the perfectly synchronised modems. For implementation in the above-mentioned 34 Mbps real-time testbed simplified versions of these algorithms were invoked.

(3) Symbol-by-symbol adaptive OFDM substantially increases the BPS throughput of the system at the cost of moderately increased complexity. It was demonstrated in the context of an adaptive real-time audio system that this increased modem throughput can be translated in improved audio quality at a given channel quality.

(4) The proposed blind symbol-by-symbol adaptive OFDM modem mode detection algorithms were shown to be robust against channel impairments in conjunction with twin-mode AOFDM. However, it was necessary to combine it with higher-complexity channel coding based mode detection techniques, in order to maintain sufficient robustness, when using quadruple-mode AOFDM.

(5) The combination of frequency-domain pre-equalisation with AOFDM resulted in further performance benefits at the cost of a moderate increase in the peak-to-mean envelope flluctuation and system complexity.

(6) The combination of adaptive RRNS and turbo BCH FEC coding with AOFDM provided further flexibility for the system, in order to cope with hostile channel conditions, in particular in the low SNR region.

23.6.3 Suggestions for further OFDM research

In our invstigations we have occasionally assumed perfect knowledge of the channel's frequency-domain transfer function at the receiver and a co-channel interference-free environment. Clearly, these assumptions do not apply to a cellular communications system, and the adaptive schemes in the suggested form will not operate optimally in a co-channel interference limited cellular environment. Interference will generally render the channel non-reciprocal, and will make the channel estimation more arduous. The problem of the non-reciprocal channel could be circumvented by using a closed-loop adaptation system, as it was suggested in Chapter 22. Combining antenna arrays, interference cancellation or joint-detection techniques combined with space-time coding and channel estimation techniques in conjunction with the adaptive OFDM ideas of Chapters 22 and 23 may result in an adaptive OFDM scheme suitable for cellular systems. These issues constitute our current research.

Let us now in our last chapter provide an in-depth systems design example in the context of two recently standardised QAM- and OFDM-based Digital Video Broadcast (DVB) systems, whilst providing a research-oriented outlook to their turbo-coded versions, which further improves their performance.

Chapter 24

Digital Terrestrial and Satellite-based Video Broadcasting to Mobile and Stationary Receivers

C.S. Lee, L. Hanzo, T. Keller, S. Vlahoyiannatos

24.1 DVB-T for Mobile Receivers[1]

24.1.1 Background and motivation

Following the standardisation of the pan-European digital video broadcasting (DVB) systems, we have begun to witness the arrival of digital television services to the home. However, for a high proportion of business and leisure travellers it is desirable to have access to DVB services while on the move. Although it is feasible to access these services with the aid of dedicated DVB receivers, these receivers may also find their way into the laptop computers of the near future. These intelligent laptops may also become the portable DVB receivers of wireless in-home networks.

In recent years three DVB standards have emerged in Europe for terrestrial [140], cable-based [473] and satellite-oriented [474] delivery of DVB signals. The more hostile propagation environment of the terrestrial system requires concatenated

[1]This section is based on C. S. Lee, T. Keller and L. Hanzo: OFDM-based turbo-coded hierarchical and non-hierarchical terrestrial mobile digital video broadcasting, in *IEEE Trans. Broadcasting*, March 2000, ©2000 IEEE. Personal use of this material is permitted. However, permission to reprint/republish this material for advertising or promotional purposes or for creating new collective works for resale or redistribution to servers or lists, or to refuse any copyrighted component of this work in other works must be obtained from the IEEE.

Reed–Solomon (RS) [475, 476] and rate-compatible punctured convolutional coding (RCPCC) [475, 476] combined with orthogonal frequency division multiplexing (OFDM) based modulation, which was detailed throughout the previous chapters. By contrast, the more benign cable and satellite based media facilitate the employment of multilevel modems using up to 256-level quadrature amplitude modulation (QAM). These schemes are capable of delivering high definition video at bit rates of up to 20 Mbit/s in stationary broadcast-mode distributive wireless scenarios.

Recently, there have been a range of DVB system performance studies in the literature [477–480]. Against this background, in this contribution we have proposed turbo coding improvements to the terrestrial DVB system [140] and investigated its performance under hostile mobile channel conditions. We have also studied various video bitstream partitioning and channel coding schemes both in the so-called hierarchical and non-hierarchical transceiver modes to be discussed during our further discourse and compared their performance.

The rest of this section is organised as follows. A brief overview of the enhanced turbo-coded and standard DVB terrestrial scheme is presented in Section 24.1.2, while the channel model is described in Section 24.1.3. Following this, the performance of the improved DVB terrestrial system employing the so-called non-hierarchical and hierarchical format [140] is examined in a mobile environment in Sections 24.1.4 and 24.1.6, before our conclusions and future work areas are presented in Section 24.1.7.

24.1.2 DVB terrestrial scheme

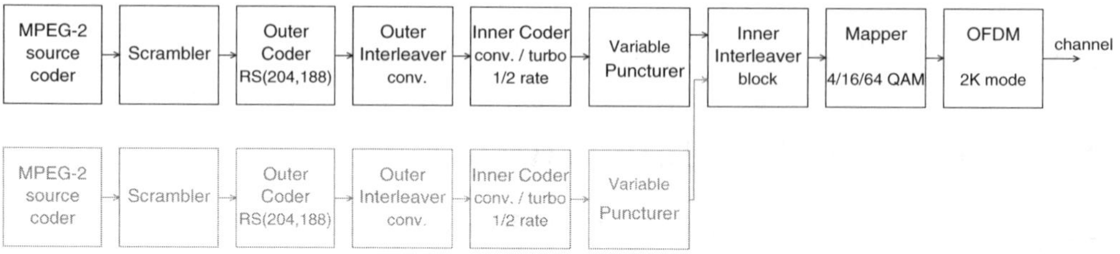

Figure 24.1: Schematic of the DVB terrestrial transmitter functions

The block diagram of the DVB terrestrial (DVB-T) transmitter [140] is shown in Figure 24.1, which consists of an MPEG-2 video encoder, channel coding modules and an OFDM modem [481]. The bitstream generated by the MPEG-2 encoder is packetized into frames of 188-byte long. The video data in each packet is then randomized by the scrambler of Figure 24.1. The specific details concerning the scrambler have not been included in this chapter, since these may be obtained from the DVB-T standard [140].

Due to the poor error resilience of the MPEG-2 video codec, powerful concatenated channel coding is employed. The concatenated channel codec of Figure 24.1 comprises a shortened Reed–Solomon (RS) outer code and an inner convolutional encoder. The 188-byte MPEG-2 video packet is extended by the Reed–Solomon encoder [475, 476] with parity information to facilitate error recovery in order to form a 204-byte packet.

The Reed–Solomon decoder can then correct up to 8 erroneous bytes for each 204-byte packet. Following this, the RS-coded packet is interleaved by a convolutional interleaver and further protected by a half-rate inner convolutional encoder using a constraint length of 7 [475,476].

Furthermore, the overall code rate of the concatenated coding scheme can be adapted by variable puncturing which supports code rates of 1/2 (no puncturing) as well as 2/3, 3/4, 5/6 and 7/8. The parameters of the convolutional encoder are summarised in Table 24.1.

Convolutional Coder Parameters	
Code rate	1/2
Constraint length,	7
n	2
k	1
Generator polynomials (octal format)	171, 133

Table 24.1: Parameters of the $CC(n, k, K)$ convolutional inner encoder of the DVB-T modem

If only one of the two branches of the transmitter in Figure 24.1 is utilised, the DVB-T modem is said to be operating in its non-hierarchical mode. In this mode, the modem can have a choice of QPSK, 16-QAM or 64-QAM modulation constellations, all of which were considered in detail throughout our previous chapters.

A second video bitstream can also be multiplexed with the first one by the inner interleaver, when the DVB modem is in its so-called hierarchical mode [140]. The choice of modulation constellations in this mode is between 16-QAM and 64-QAM. We shall be employing this transmission mode, when the standard data partitioning scheme of [140] is used to split the incoming MPEG-2 video bitstream into two video bit-protection classes with one class having a higher grade of protection or priority than the other one. Recall from Chapter 5 that 16-QAM exhibited two different-integrity subchannels, while 64-QAM had three different-BER subchannels. Hence the higher priority video bits will be mapped to the MSBs of the modulation constellation points and the lower priority video bits to the LSBs of the QAM-constellation. For 16-QAM and 64-QAM, the two MSBs of each 4-bit or 6-bit QAM symbol will contain the more important video data. The lower priority video bits will then be mapped to the lower significance 2 bits and 4 bits of 16-QAM and 64-QAM, respectively, where the BER differences amongst the different subchannels were around a factor 2-3 in Chapter 5 over AWGN channels.

These QPSK, 16-QAM or 64-QAM symbols are then distributed over the OFDM carriers, as discussed in the previous chapters. The parameters of the OFDM system are presented in Table 24.2.

Besides implementing the standard DVB-T system as a benchmark, we have improved the system by replacing the convolutional coder with a turbo codec [433,482]. The turbo codec's parameters used in our investigations are displayed in Table 24.3. The block diagram of the turbo encoder is shown in Figure 24.2. The turbo encoder is constructed of two component encoders. Each component encoder is a half-rate

OFDM Parameters	
Total number of subcarriers	2048 (2K mode)
Number of effective subcarriers	1705
OFDM symbol duration T_s	224 μs
Guard interval	$T_s/4 = 56\mu$s
Total symbol duration (inc. guard interval)	280 μs
Consecutive subcarrier spacing $1/T_s$	4464 Hz
DVB channel spacing	7.61 MHz
QPSK and QAM symbol period	7/64 μs

Table 24.2: Parameters of the OFDM module used in the DVB-T modem [140]

convolutional encoder, whose parameters are listed in Table 24.3. The two component encoders are used to encode the same input bits, although the input bits of the second component encoder are interleaved before encoding. The output bits of the two component codes are punctured and multiplexed, in order to form a single output bitstream. The component encoder used here is known as a half-rate recursive systematic convolutional encoder (RSC) [483]. It generates one parity bit and one systematic output bit for every input bit. In order to provide an overall coding rate of $R = 1/2$, half the output bits from the two encoders must be punctured. The puncturing arrangement used in our work is to transmit all the systematic bits from the first encoder and every other parity bit from both encoders [484]. We note here that one iteration of the turbo decoder involves two so-called Logarithmic Maximum A-Posteriori (log-MAP) [485] decoding operations, which we repeated for the 8 iterations. Hence, the total turbo decoding complexity is about 16 times higher than a constraint length $K = 3$ constituent convolutional decoding. Therefore the turbo decoder exhibits a similar complexity to the $K = 7$ convolutional decoder.

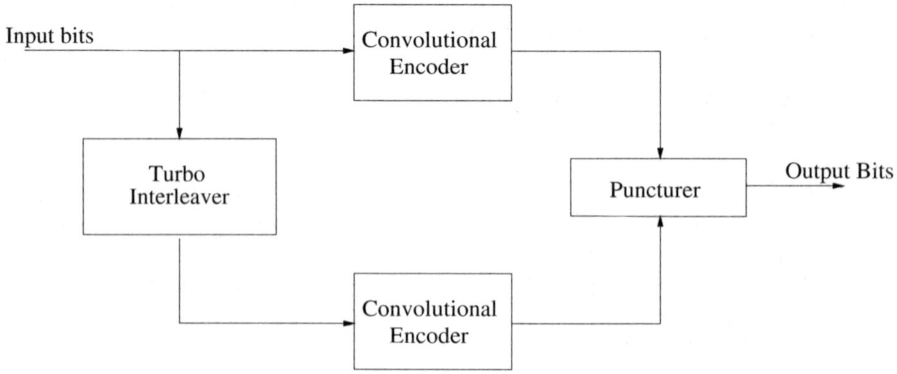

Figure 24.2: Block diagram of turbo encoder

In this section, we have given an overview of the standard and enhanced DVB-T

Turbo Coder Parameters	
Turbo code rate	1/2
Input block length	17 952 bits
Interleaver type	Random
Number of turbo decoder iterations	8
Turbo Encoder Component Code Parameters	
Component code encoder type	Recursive Systematic Convolutional (RSC)
Component code decoder type	log-MAP [485]
Constraint length	3
n	2
k	1
Generator polynomials (octal format)	7, 5

Table 24.3: Parameters of the inner turbo encoder used to replace the DVB-T system's $K = 7$ convolutional encoder of Table 24.1 (RSC: recursive systematic code)

system, which we have used in our experiments. Readers interested in further details of the DVB-T system are referred to the DVB-T standard [140]. The performance of the standard DVB-T system and the turbo-coded system is characterised in Section 24.1.4 and 24.1.6 for non-hierarchical and hierarchical transmissions, respectively. Let us now briefly consider the multipath channel model used in our investigations.

24.1.3 Terrestrial broadcast channel model

The channel model employed in this study was the 12-path COST 207 [486] hilly terrain (HT) type impulse response, with a maximal relative path delay of 19.9 μs. This channel was selected in order to provide a worst-case propagation scenario for the DVB-T system employed in our study.

In the system characterised here, we have used a carrier frequency of 500 MHz and a sampling rate of 7/64 μs. Each of the channel paths was faded independently obeying a Rayleigh fading distribution, according to a normalised Doppler frequency of 10^{-5} [475]. This corresponds to a worst-case vehicular velocity of about 200 km/h. The unfaded impulse response is depicted in Figure 24.3. For the sake of completeness we note that the standard COST 207 channel model was defined in order to facilitate the comparison of different GSM implementations [475] under identical conditions. The associated bit rate was 271 kbit/s, while in our investigations the bit rate of DVB-quality transmissions can be as high as 20 Mbit/s, where there is a higher number of resolvable multipath components within the dispersion-range considered. However, the performance of various wireless tranceivers is well understood by the research community over this standard COST 207 channel and hence its employment is beneficial in benchmarking terms. Furthermore, since the OFDM modem has 2048 subcarriers, the subcarrier signalling rate is effectively 2000-times lower than our

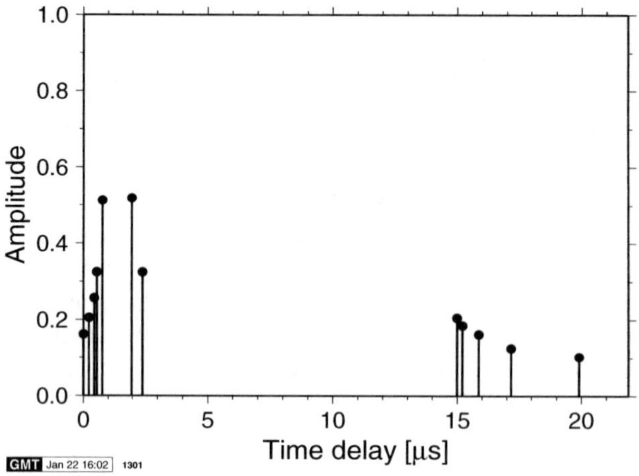

Figure 24.3: COST 207 hilly terrain (HT) type impulse response

maximum DVB-rate of 20 Mbit/s, corresponding to 10 kbit/s. At this sub-channel rate, the individual sub-channel can be considered nearly frequency-flat. In summary, in conjunction with the 200 km/h vehicular speed the investigated channel conditions constitute a pessimistic scenario.

Let us now consider the performance of the so-called non-hierarchical DVB-T scheme in the next section. By contrast, the hierarchical system's performance will be characterised in the next but one section. To be more explicit concerning the hierarchical versus non-hierarchical modes suffice to say here that on the basis of the different sensitivity of the transmitted video bits the hierarachical mode allows us to implement un-equal error protection schemes, as it was detailed for example in [487].

In a low complexity implementation, two protection classes may be envisaged. The higher priority class would contain all the important and error sensitive MPEG2 video synchronisation and header information and some of the more important low frequency variable-length coded Discrete Cosine Transform (DCT) coefficients [488]. The lower priority class would then contain the remaining less important, higher frequency variable-length coded DCT coefficients. Again, the performance of this partitioning process will be characterised in the context of the hierarchical DVB [140] transmission scheme [487] in Section 24.1.6. Let us, however, first consider the architecture of the investigated DVB system in the next section.

24.1.4 Non-hierarchical OFDM DVB performance

In this section we shall elaborate on our findings when the convolutional code used in the standard non-hierarchical DVB scheme [140] is replaced by a turbo code. We will invoke a range of standard-compliant schemes as benchmarks. The 704×576-pixel HDTV-resolution "Football" video sequence was used in our experiments. The MPEG-2 decoder employs a simple error concealment algorithm to fill in missing portions of the reconstructed image in the event of decoding errors. The concealment algorithm will select the specific portion of the previous reconstructed image, which corresponds to the missing portion of the current image, in order to conceal the errors.

In Figure 24.4(a) and (b) the bit error rate (BER) performance of the various modem modes in conjunction with our diverse channel coding schemes are portrayed over stationary, narrowband additive white Gaussian noise channels (AWGN), where the turbo codec exhibits a significantly steeper BER reduction in comparison to the convolutionally coded arrangements.

Specifically, comparing the performance of the various turbo and convolutional codes for QPSK and 64-QAM at a BER of 10^{-4}, the turbo code exhibited an additional coding gain of about 2.24 dB and 3.7 dB respectively, when using half-rate codes in Figure 24.4(a) and (b). Below we will now quantify the associated video performance in terms of the so-called peak signal–to–noise ratio (PSNR), which is defined as follows:

$$\text{PSNR} = 10\ log_{10} \frac{\sum_{n=0}^{N} \sum_{m=0}^{M} 255^2}{\sum_{n=0}^{N} \sum_{m=0}^{M} \Delta^2}, \qquad (24.1)$$

where Δ is the difference between the uncoded pixel value and the reconstructed pixel value. The variables M and N refer to the dimension of the image. The maximum possible 8-bit represented pixel luminance value of 255 was used in Equation 24.1 in order to mitigate the PSNR's dependence on the video material used. The average PSNR is then the mean of the PSNR values computed for all the images constituting the video sequence.

The associated PSNR versus channel signal–to–noise ratio (SNR) graphs in Figure 24.7 demonstrate that approximately 2 dB and 3.5 dB lower channel SNRs are required in conjunction with the rate 1/2 turbo codec for QPSK and 64-QAM, respectively, than for convolutional coding, in order to maintain high reconstructed video quality. The term unimpaired as used in Figure 24.7 and Figure 24.8 refers to the condition, where the PSNR of the MPEG-2 decoder's reconstructed image at the receiver is the same as the PSNR of the same image generated by the local decoder of the MPEG-2 video encoder, corresponding to the absence of channel – but not MPEG-2 coding – impairments.

Comparing the BER performance of the 1/2 rate convolutional decoder in Figure 24.5(a) and the so-called log-MAP [485] turbo decoder using eight iterations in Figure 24.5(b) for QPSK modulation over the worst-case fading mobile channel of Figure 24.3, we observe that at a BER of about 10^{-4} the turbo code provided an additional coding gain of 6 dB in comparison to the convolutional code. By contrast, for 64-QAM using similar codes, a 5 dB coding gain was observed at this BER.

Similar observations were also made with respect to the average PSNR versus channel signal–to–noise ratio (SNR) plots of Figure 24.8. For the QPSK modulation mode and a 1/2 coding rate, the turbo code required an approximately 5.5 dB lower channel SNR for maintaining near-unimpaired video quality than the convolutional code.

Comparing Figure 24.5(a) and Figure 24.6(a), we note that the Reed–Solomon decoder becomes effective in lowering the bit error probability of the transmitted data further below the BER threshold of 10^{-4}. From these figures we also observe that the rate-3/4 convolutional code is unsuitable for transmission over the highly dispersive

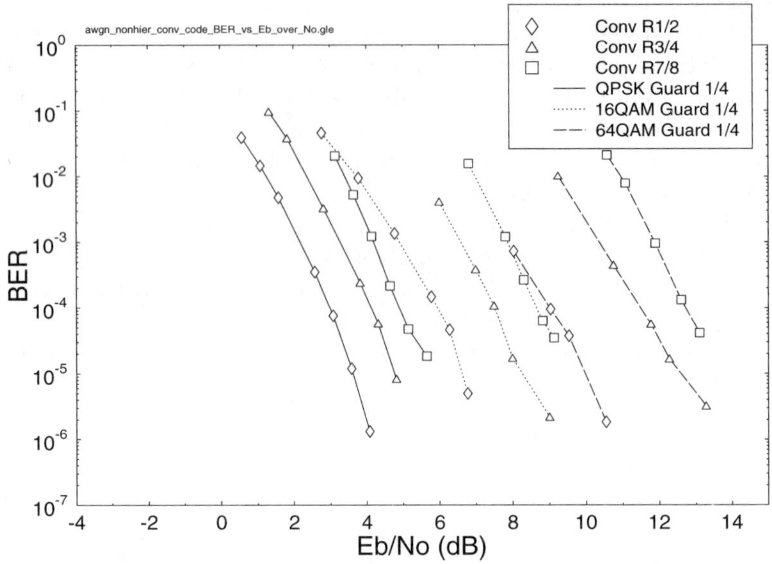

(a) Convolutional Code

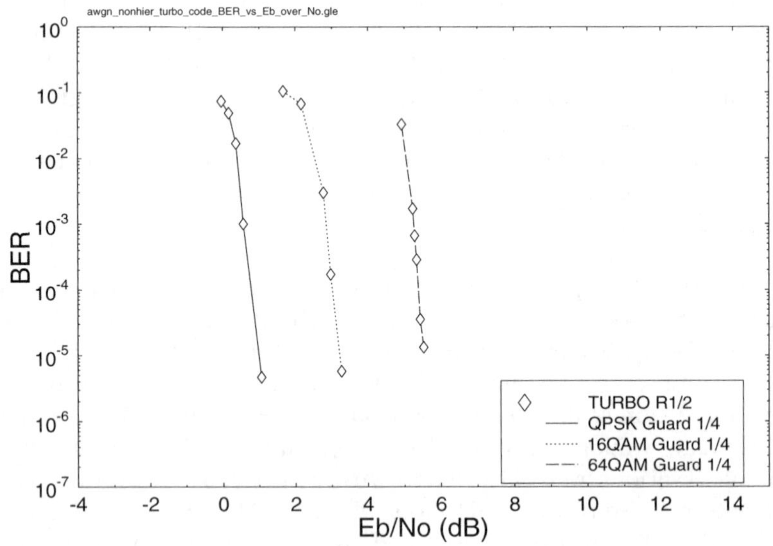

(b) Turbo Code

Figure 24.4: BER after (a) convolutional decoding and (b) turbo decoding for the DVB-T scheme over stationary non-dispersive **AWGN** channels for **non-hierarchical transmission** [487], ©IEEE, 2000, Lee, Keller, Hanzo

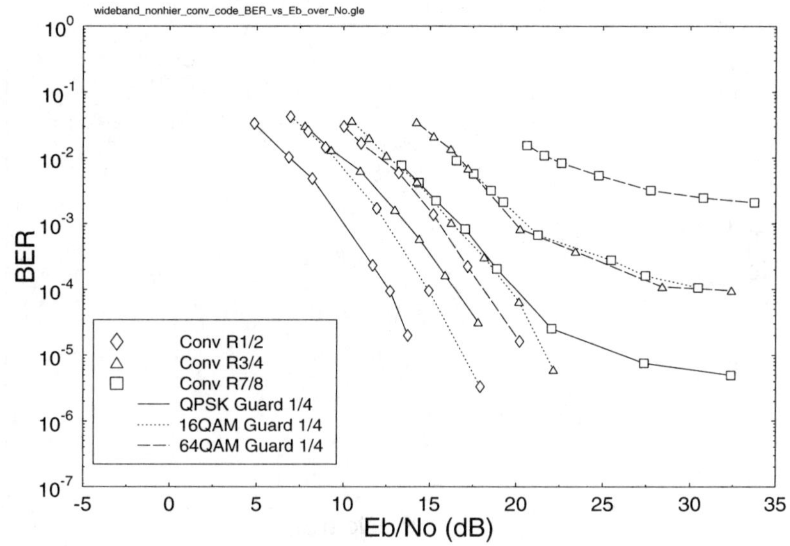

(a) Convolutional Code

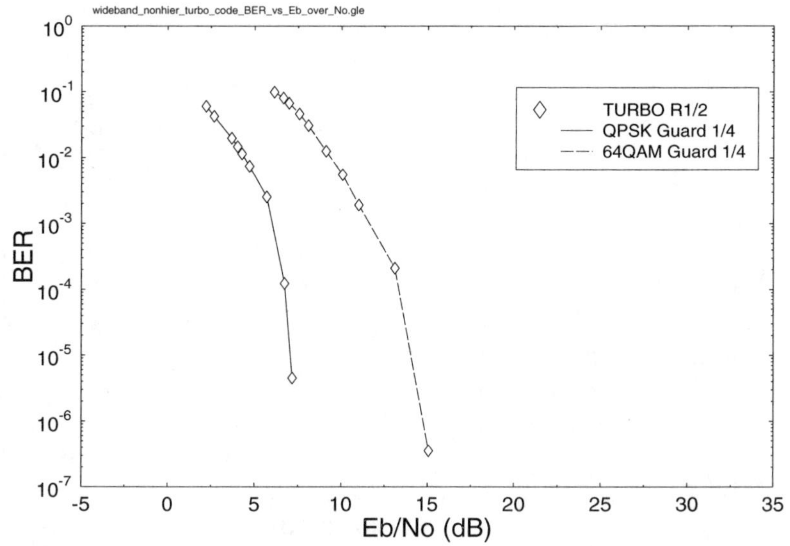

(b) Turbo Code

Figure 24.5: BER after (a) convolutional decoding and (b) turbo decoding for the DVB-T scheme over the **wideband fading channel** of Figure 24.3 for **non-hierarchical transmission** [487], ©IEEE, 2000, Lee, Keller, Hanzo

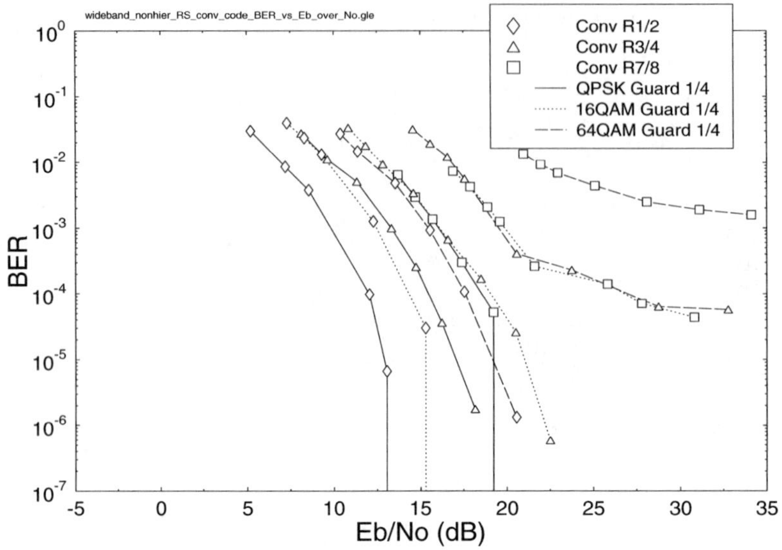

(a) RS and Convolutional Code

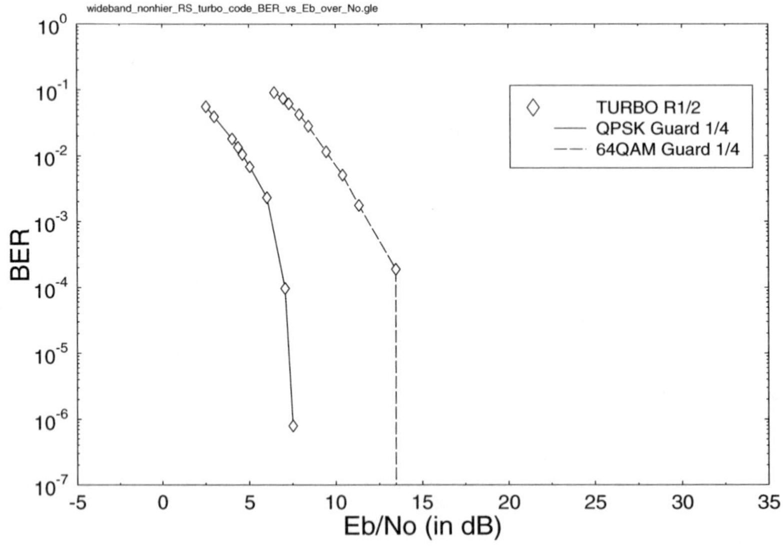

(b) RS and Turbo Code

Figure 24.6: BER after (a) RS and convolutional decoding and (b) RS and turbo decoding for the DVB-T scheme over the **wideband fading channel** of Figure 24.3 for **non-hierarchical transmission** [487], ©IEEE, 2000, Lee, Keller, Hanzo

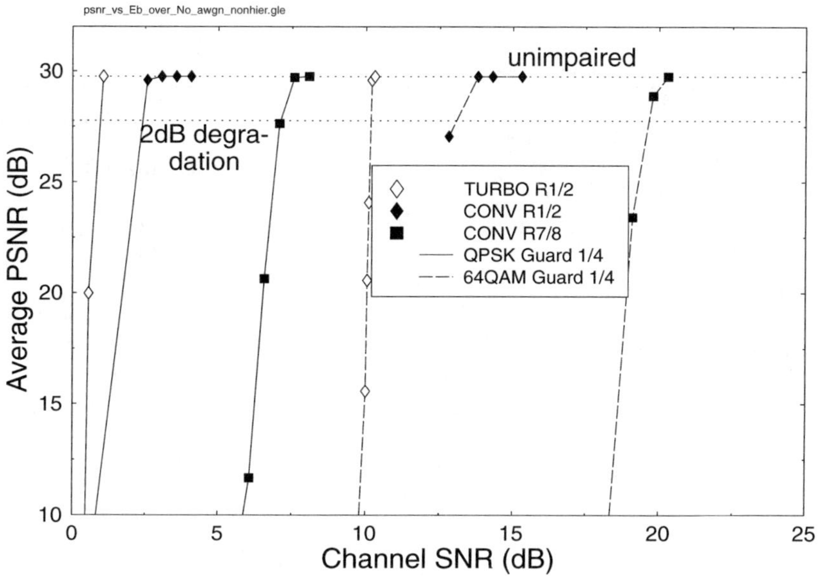

Figure 24.7: Average PSNR versus channel SNR of the DVB scheme [140] over non-dispersive **AWGN** channels for **non-hierarchical transmission** [487], ©IEEE, 2000, Lee, Keller, Hanzo

hilly terrain channel used in this experiment, when 64-QAM is employed. When the rate-7/8 convolutional code is used, both the 16-QAM and 64-QAM schemes perform poorly. As for the QPSK modulation scheme, a convolutional code rate as high as 7/8 can still provide a satisfactory performance after Reed–Solomon decoding.

In conclusion, Tables 24.4 and 24.5 summarise the system performance in terms of the channel SNR (CSNR) required for maintaining less than 2 dB PSNR video degradation. It was observed that, at this PSNR degradation, decoding errors were still perceptually unnoticeable to the viewer due to the 30 frame/s refresh rate, although the typical still frame shown in Figure 24.9 in this scenario exhibits some degradation. It is important to underline once again that the $K = 3$ turbo code and the $K = 7$ convolutional code exhibited comparable complexities. The higher performance of the turbo codec facilitates for example the employment of turbo-coded 16-QAM at a similar channel SNR, where convolutional-coded QPSK can be invoked. This in turn allows us to double the bit rate within the same bandwidth and hence improve the video quality. In the next section, we first outline the philosophy of the MPEG2 video data partitioning scheme, assisting in invoking unequal video bit protection. Then in the next but one section we shall present the results of our investigations employing the DVB-T system [140] in a hierarchical transmission scenario.

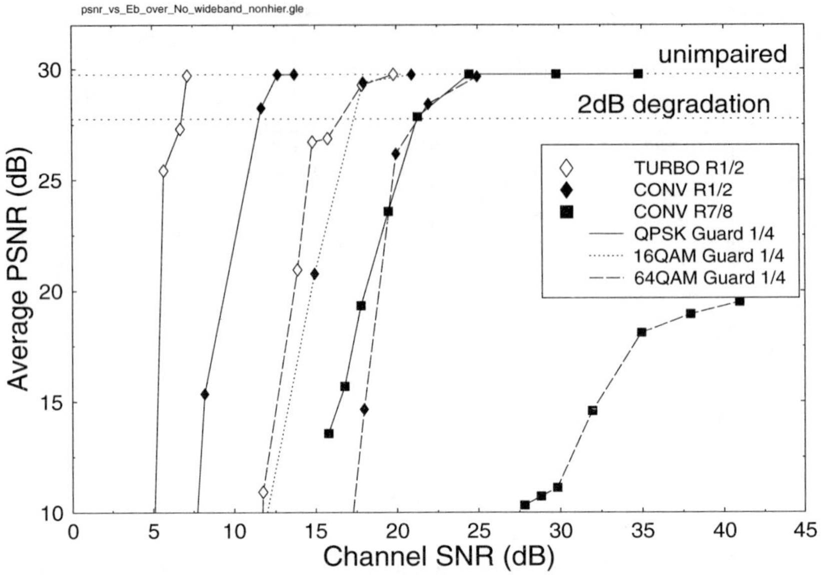

Figure 24.8: Average PSNR versus channel SNR of the DVB scheme [140] over the **wideband fading channel** of Figure 24.3 for **non-hierarchical transmission** [487], ©IEEE, 2000, Lee, Keller, Hanzo

Mod.	Code	CSNR (dB)	E_b/N_0	BER
QPSK	Turbo (1/2)	1.02	1.02	6×10^{-6}
64-QAM	Turbo (1/2)	9.94	5.17	2×10^{-3}
QPSK	Conv (1/2)	2.16	2.16	1.1×10^{-3}
64-QAM	Conv (1/2)	12.84	8.07	6×10^{-4}
QPSK	Conv (7/8)	6.99	4.56	2×10^{-4}
64-QAM	Conv (7/8)	19.43	12.23	3×10^{-4}

Table 24.4: Summary of the **non-hierarchical** performance results over non-dispersive **AWGN** channels tolerating a PSNR degradation of 2 dB. The BER measure refers to BER after Viterbi or turbo decoding

Figure 24.9: Frame 79 of the "Football" sequence, which illustrates the visual effects of minor decoding errors at a BER of 2.10^{-4} after convolutional decoding. The PSNR degradation observed is approximately 2 dB. The sequence was coded using a rate 7/8 convolutional code and transmitted emplying QPSK modulation [487], ©IEEE, 2000, Lee, Keller, Hanzo

Mod.	Code	CSNR (dB)	E_b/N_0	BER
QPSK	Turbo (1/2)	6.63	6.63	2.5×10^{-4}
64-QAM	Turbo (1/2)	15.82	11.05	2×10^{-3}
QPSK	Conv (1/2)	10.82	10.82	6×10^{-4}
64-QAM	Conv (1/2)	20.92	16.15	7×10^{-4}
QPSK	Conv (7/8)	20.92	18.49	3×10^{-4}

Table 24.5: Summary of the **non-hierchical** performance results over **wideband fading channels** tolerating a PSNR degradation of 2 dB. The BER measure refers to BER after Viterbi or turbo decoding

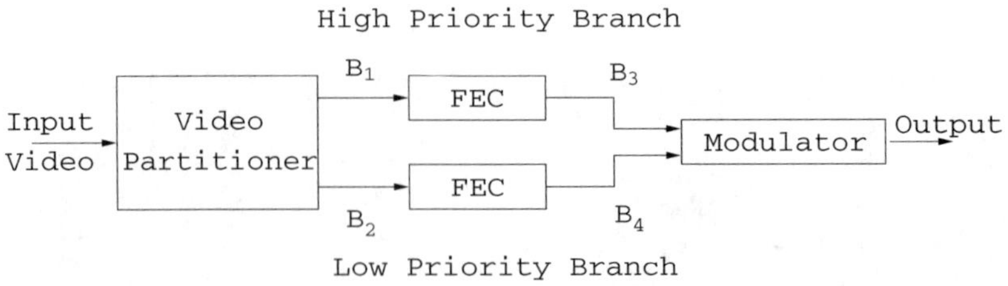

Figure 24.10: Video partitioning for the DVB-T system operating in hierarchical mode

24.1.5 Video data partitioning scheme

The philosophy of the hierarchical transmission mode is that the natural BER difference of a factor 2 to 3 found in Chapter 5 for the 16-QAM modem is exploited for providing unequal error protection for the FEC-coded video streams B3 and B4 of Figure 24.10. If the sensitivity of the video bits requires a different BER ratio between the B3 and B4 streams, the choice of the FEC codes protecting the video streams B1 and B2 of Figure 24.10 can be appropriately adjusted to equal out or to augment these differences. The FEC module of Figure 24.10 represents the concatenated coding system, consisting of a Reed–Solomon codec [475] and a convolutional codec [475]. The modulator can invoke both 16-QAM and 64-QAM [403]. We shall now use an example to illustrate the choice of the various partitioning ratios summarised in Table 24.6.

Modulation	Conv. Code Rate (High Priority)	Conv. Code Rate (Low Priority)	Ratio (High - B1 : Low - B2)
16-QAM	1/2	1/2	1 : 1
	1/2	2/3	3 : 4
	1/2	3/4	2 : 3
	1/2	5/6	3 : 5
	1/2	7/8	4 : 7
	2/3	1/2	4 : 3
64-QAM	1/2	1/2	1 : 2
	1/2	2/3	3 : 8
	1/2	**3/4**	**1 : 3**
	1/2	5/6	3 : 10
	1/2	7/8	2 : 7
	2/3	1/2	2 : 3

Table 24.6: The bit rate partitioning ratios based on the modulation mode and code rates selected for the DVB-T hierarchical mode. The line in bold corresponds to our worked example

We shall assume that 64-QAM is selected and the high and low priority video partitions employ rate 1/2 and 3/4 convolutional codes, respectively. This scenario is portrayed in the third line of the 64-QAM section of Table 24.6. We do not have to take the Reed–Solomon code rate into account, since both partitions invoke the same Reed–Solomon codec. Based on these facts and upon referring to Figure 24.10, the input bit rates B_3 and B_4 of the modulator must be in the ratio 1:2, since the two MSBs of the 64-QAM constellation are assigned to the high priority video partition and the remaining four bits to the low priority video partition.

At the same time, the ratio of B_3 to B_4 is related to the ratio of B_1 to B_2 with the FEC redundancy taken into account, requiring

$$\begin{aligned}
\frac{B_3}{B_4} &= \frac{2\times B_1}{\frac{4}{3}\times B_2} \stackrel{64-QAM}{=} \frac{1}{2}\\
&= \frac{3}{2}\cdot\frac{B_1}{B_2} \stackrel{64-QAM}{=} \frac{1}{2}\\
\frac{B_1}{B_2} &= \frac{1}{2}\times\frac{2}{3}\\
&= \frac{1}{3}.
\end{aligned} \tag{24.2}$$

If, for example, the input video bit rate to the data partitioner module is 1 Mbit/s, the output bit rate of the high and low priority partition would be $B_1 = 250$ kbit/s and $B_2 = 750$ kbit/s, respectively, according to the ratio indicated by Equation 24.2.

Let us consider here the 16-QAM modem and refer to the equally split rate 1/2 convolutional coded high and low priority scenario as Scheme 1. Furthermore, the 16-QAM rate 1/3 convolutional coded high priority data and rate 2/3 convolutional coded low priority data based scenario is referred to here as Scheme 2. Lastly, the 16-QAM rate 2/3 convolutional coded high priority data and rate 1/3 coded low priority data based partitioning scheme is termed as Scheme 3. We then programmed the partitioning scheme of Figure 24.10 for maintaining the required splitting ratio B_1/B_2, as seen in Table 24.7. This was achieved by continuously adjusting the so-called Priority Break Point (PBP) governing the corresponding split of the bitstream into the priority or protection classes using the corresponding alforithms of [487]. The 704×576-pixel "Football" high definition television (HDTV) video sequence was used in these investigations. Having outlined the operation of the data partitioning

	Modulation	Conv. Code Rate (High Prior. – B1)	Conv. Code Rate (Low Prior. – B2)	Ratio (High : Low) (B1 : B2)
Scheme 1	16-QAM	1/2	1/2	1 : 1
Scheme 2	16-QAM	1/3	2/3	1 : 2
Scheme 3	16-QAM	2/3	1/3	2 : 1

Table 24.7: Summary of the three schemes employed in our investigations into the performance of the data partitioning scheme. The FEC-coded high priority video bitstream B3, as shown in Figure 24.10, was mapped to the high priority 16-QAM sub-channel, while the low priority B4-stream to the low priority 16-QAM sub-channel

scheme, let us now characterise its performance in the context of the overall system in the next section.

24.1.6 Hierarchical OFDM DVB performance

Below we will invoke the above DVB-T hierarchical scheme in a mobile broadcasting scenario. We shall also demonstrate the improvements which turbo codes offer when replacing the convolutional code in the standard scheme. Hence the convolutional codec in both the high and low priority partitions was replaced by the turbo codec. We have also investigated replacing only the high priority convolutional codec with the turbo codec, pairing the 1/2 rate turbo codec in the high priority partition with the convolutional codec in the low priority partition. Again, the "Football" sequence was used in these experiments. Partitioning was carried out using the schematic of Figure 24.10 as well as the detailed partitioning algorithms of [487]. The FEC-coded high priority video partition B3 of Figure 24.10 was mapped to the higher integrity 16-QAM or 64-QAM sub-channel. By contrast, the low priority partition B4 of Figure 24.10 was directed to the lower integrity 16-QAM or 64-QAM sub-channel. Lastly, no specific mapping was required for QPSK, since it exhibits no sub-channels. We note, however, that further design trade-offs become feasible, when reversing the above mapping rules. Indeed, this is necessary for example in conjunction with Scheme 2 of Table24.7, since the high number of bits in the low priority partition render it more sensitive than the high priority partition. Again, according to Chapter 5 the 16-QAM sub-channels exhibit a factor of 2 to 3 BER difference, which improves the robustness of the 'reverse-mapped' Scheme 2 of Table 24.7.

Referring to Figure 24.11 and comparing the performance of the 1/2 rate convolutional code and turbo code at a BER of 10^{-4} for the low priority partition, the turbo code, employing 8 iterations, exhibited a coding gain of about 6.6 dB and 5.97 dB for 16-QAM and 64-QAM, respectively. When the number of turbo decoding iterations was reduced to 4, the coding gains offered by the turbo code over that of the convolutional code were 6.23 dB and 5.7 dB for 16-QAM and 64-QAM respectively. We observed that by reducing the number of iterations to 4 halved the associated complexity, but the turbo code exhibited a coding loss of only about 0.37 dB and 0.27 dB in comparison to the 8-iteration scenario for 16-QAM and 64-QAM, respectively. Hence the computational complexity of the turbo codec can be halved by sacrificing only a small amount of coding gain. The substantial coding gain provided by turbo coding is also reflected in the PSNR versus channel SNR graphs of Figure 24.13. In order to achieve transmission with very low probability of error, Figure 24.13 demonstrated that approximately 5.72 dB and 4.56 dB higher channel SNRs are required by the standard scheme compared to the scheme employing turbo coding, when using 4 iterations in both partitions. We have only shown the performance of turbo coding for the low priority partition in Figures 24.11(b) and 24.12(b), since the turbo or convolutional-coded high priority partition was received with very low probability of error after Reed–Solomon decoding for the range of SNRs used.

We also observed that the rate 3/4 and rate 7/8 convolutional codes in the low priority partition were unable to provide sufficient protection to the transmitted video bits, as it becomes evident from Figures 24.11(a) and 24.12(a). In these high coding rate scenarios, due to the presence of residual errors even after the Reed–Solomon decoder, the decoded video exhibited some decoding errors, which is evidenced by the flattening of the PSNR versus channel SNR curves in Figure 24.13(a), before reaching the error-free PSNR.

A specific problem when using the data partitioning scheme in conjunction with the high priority partition being protected by the rate 1/2 code and the low priority partition protected by the rate 3/4 and rate 7/8 codes was that when the low priority partition data was corrupted, the error-free high priority data available was insufficient for concealing the errors, as discussed in Section 24.1.5. We have also experimented with the combination of rate 2/3 convolutional coding and rate 1/2 convolutional coding, in order to protect the high and low priority data, respectively. From Figure 24.13(a) we observed that the performance of this 2/3 rate and 1/2 rate combination approached that of the rate 1/2 convolutional code in both partitions. This was expected, since now more data can be inserted into the high priority partition. Hence, in the event of decoding errors in the low priority data, we had more error-free high priority data that could be used to reconstruct the received image.

Our last combination investigated involved using rate 1/2 turbo coding and convolutional coding for the high and low priority partitions, respectively. Comparing Figures 24.14 and 24.13(a), the channel SNRs required for achieving unimpaired video transmission were similar in both cases. This was expected, since the turbo-convolutional combination's video performance is dependent on the convolutional code's performance in the low priority partition.

Lastly, comparing Figures 24.13 and 24.8, we found that the unimpaired PSNR condition was achieved at similar channel SNRs for the hierarchical and non-hierarchical schemes, suggesting that the data partitioning scheme had not provided sufficient performance improvements in the context of the mobile DVB scheme to justify its added complexity. Again, this was a consequence of relegating a high proportion of video bits to the low integrity partition.

24.1.7 Conclusions and future work

In this contribution, we have investigated the performance of a turbo-coded DVB system in a mobile environment. A range of system performance results were presented based on the standard DVB-T scheme as well as on an improved turbo-coded scheme. The convolutional code specified in the standard system was replaced by turbo coding, which resulted in a substantial coding gain of around 5 dB. It is important to underline once again that the $K = 3$ turbo code and the $K = 7$ convolutional code exhibited comparable complexities. The higher performance of the turbo codec facilitates, for example, the employment of turbo-coded 16-QAM at a similar SNR, where convolutional-coded QPSK can be invoked. This in turn allows us to double the video bit rate within the same bandwidth and hence to improve the video quality. We have also applied data partitioning to the MPEG-2 video stream to gauge its efficiency in increasing the error resilience of the video codec. However, from these experiments we found that the data partitioning scheme did not provide substantial improvements compared to the non-partitioned video transmitted over the non-hierarchical DVB-T system. Our future work will be focused on extending this DVB-T system study to incorporate various types of channel models, as well as on investigating the effects of different Doppler frequencies on the system. Further work will also be dedicated to trellis coded modulation (TCM) and turbo trellis coded

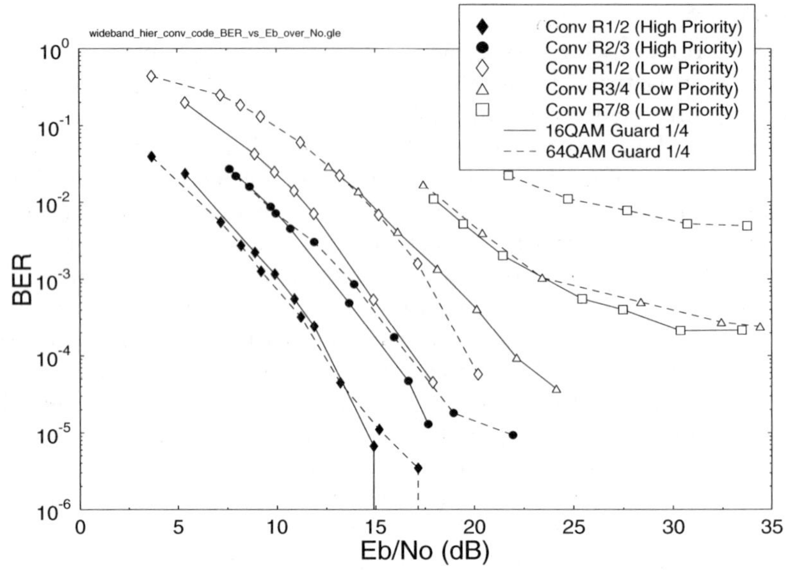

(a) Convolutional Code

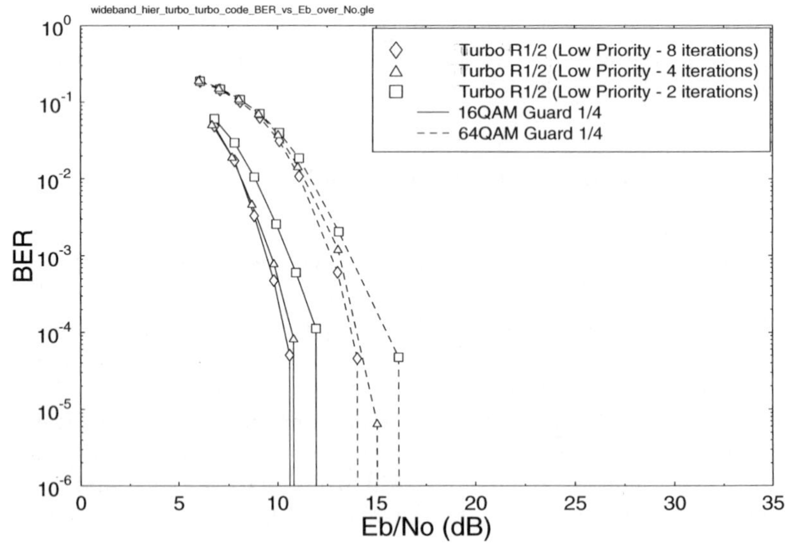

(b) Turbo Code

Figure 24.11: BER after (a) convolutional decoding and (b) turbo decoding for the **DVB-T hierarchical scheme** over the **wideband fading channel** of Figure 24.3 using the schematic of Figure 24.10. In (b), the BER of the turbo or convolutional-coded high priority partition is not shown [487], ©IEEE, 2000, Lee, Keller, Hanzo

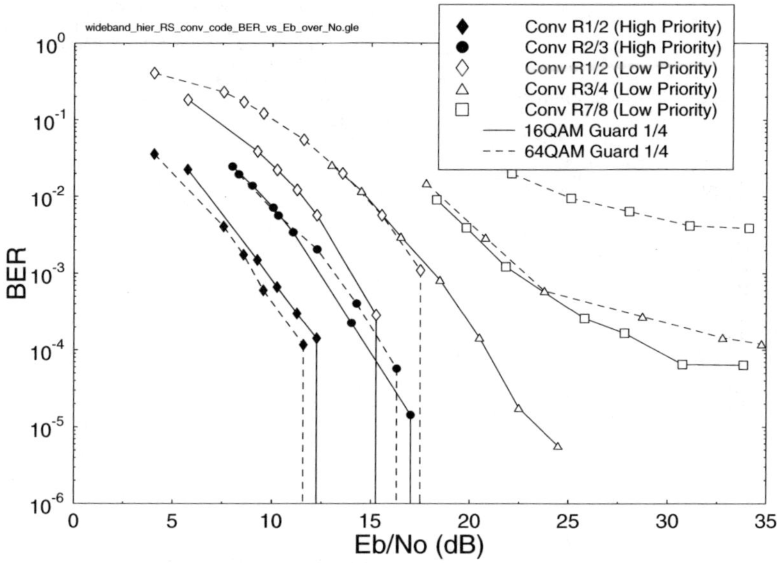

(a) RS and Convolutional Code

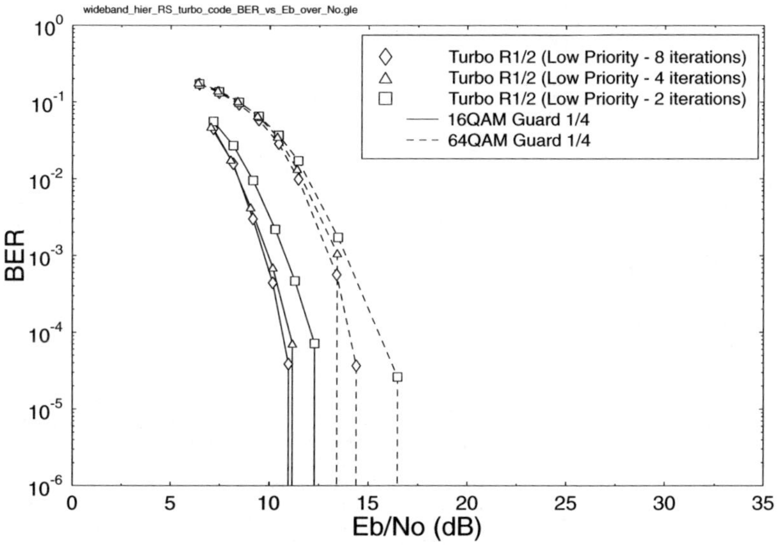

(b) RS and Turbo Code

Figure 24.12: BER after (a) RS and convolutional decoding and (b) RS and turbo decoding for the **DVB-T hierarchical scheme** over the **wideband fading channel** of Figure 24.3 using the schematic of Figure 24.10. In (b), the BER of the turbo or convolutional-coded high priority partition is not shown [487], ©IEEE, 2000, Lee, Keller, Hanzo

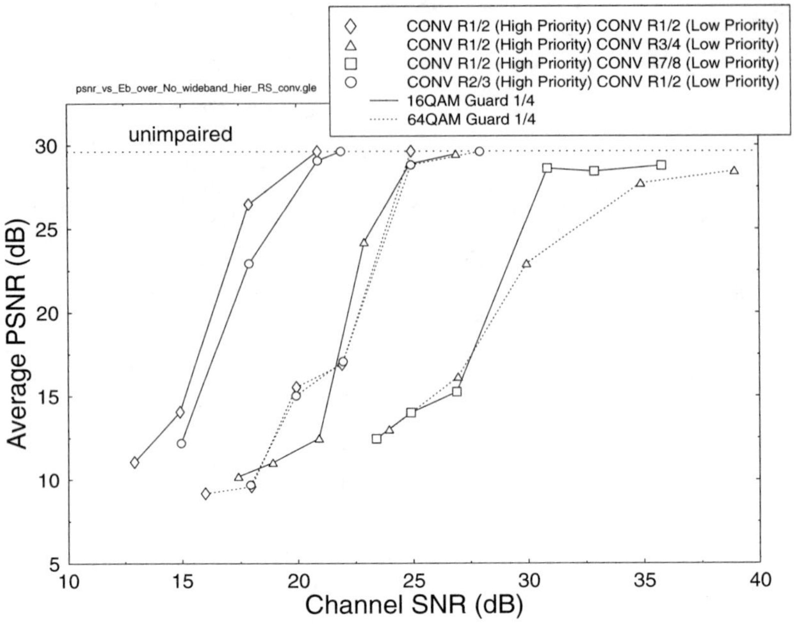

(a) Standard scheme

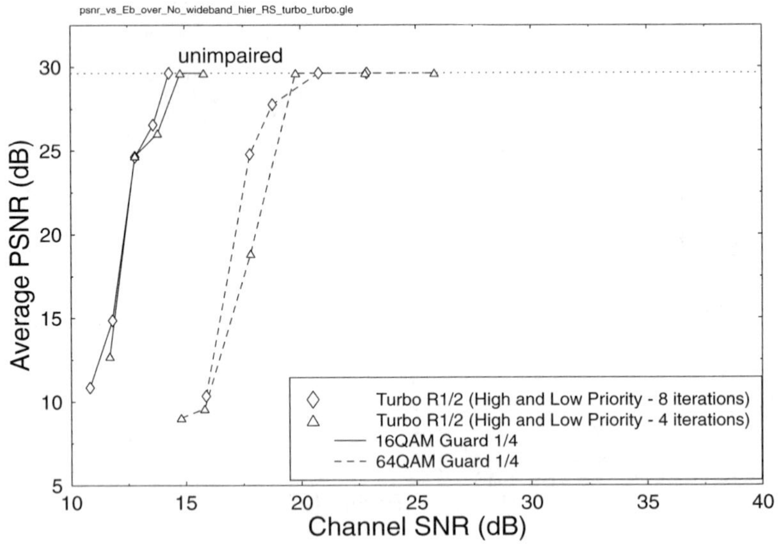

(b) Turbo code in both partitions

Figure 24.13: Average PSNR versus channel SNR for (a) standard DVB scheme [140] and (b) system with turbo coding employed in both partitions, for transmission over the **wideband fading channel** of Figure 24.3 for **hierarchical transmission** using the schematic of Figure 24.10 [487], ©IEEE, 2000, Lee, Keller, Hanzo

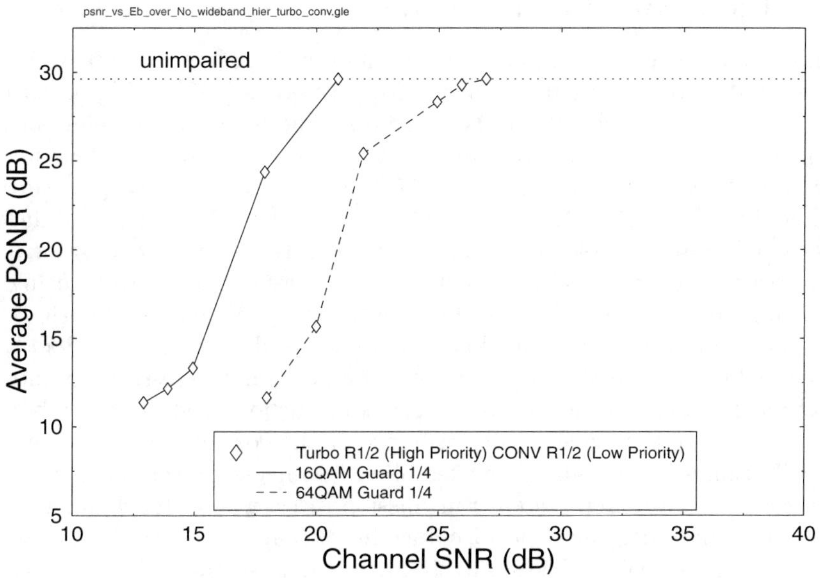

Figure 24.14: Average PSNR versus channel SNR of the DVB scheme, employing turbo coding in the high priority partition and convolutional coding in the low priority partition, over the **wideband fading channel** of Figure 24.3 for **hierarchical transmission** using the schematic of Figure 24.10 [487], ©IEEE, 2000, Lee, Keller, Hanzo

modulation (TTCM) based OFDM. The impact of employing various types of turbo interleavers on the system performance is also of interest. A range of further wireless video communications issues are addressed in [489, 490]. Let us now consider a variety of satellite-based turbo-coded blind-equalised multilevel modulation-assisted video broadcasting schemes.

24.2 Satellite-based Video Broadcasting [2]

24.2.1 Background and motivation

As mentioned at the commencement of this chapter, in recent years three harmonised DVB standards have emerged in Europe for terrestrial [140], cable-based [473] and satellite-oriented [474] delivery of DVB signals. The dispersive wireless propagation environment of the terrestrial system characterised in the previous section requires concatenated Reed–Solomon (RS) [475, 476] and rate compatible punctured convolutional coding (RCPCC) [475, 476] combined with OFDM based modulation. The satellite-based system employs the same concatenated channel coding arrangement, as the terrestrial scheme, while the cable-based system refrains from using concatenated channel coding, opting for RS coding only. The performance of both of the latter schemes can be improved upon invoking blind-equalised multi-level modems, although the associated mild dispersion or linear distortion does not necessarily require channel equalisation. However, since we propose invoking turbo-coded 4-bit/symbol 16-QAM in order to improve the system's performance at the cost of increased complexity, in this contribution we additionally invoked blind channel equalisers. This is further justified by the associated high video transmission rates, where the channel's dispersion may become a more dominant performance limitation.

Lastly, as in the DVB-T system of the previous section, the video codec used in all three systems is the Motion Pictures Expert Group's MPEG-2 codec. These standardisation activities were followed by a variety of system performance studies in the open literature [491–494]. Against this background, similarly to the enhanced DVB-T system of the previous section, we suggested turbo coding improvements to the satellite-based DVB system [474]. Here we present the results of our performance studies for the proposed system under dispersive channel conditions in conjunction with a variety of blind channel equalisation algorithms. The transmitted power requirements of the standard system employing convolutional codecs can be reduced upon invoking more complex, but more powerful turbo codecs. Alternatively, the standard quaternary or 2-bit/symbol system's bit error rate (BER) versus signal-to-noise ratio (SNR) performance can almost be matched by a turbo-coded 4-bit/symbol 16-QAM scheme, while doubling the achievable bit rate within the same bandwidth and hence improving the associated video quality. This is achieved at the cost of an increased system complexity.

The remainder of this section is organised as follows. A succinct overview of the turbo-coded and standard DVB satellite scheme is presented in Section 24.2.2, while our channel model is described in Section 24.2.3. A brief summary of the blind equaliser algorithms employed is presented in Section 24.2.4. Following this the performance of the improved DVB satellite system is examined for transmission over a dispersive two-path channel in Section 24.2.5, before our conclusions and future

[2] This section is based on C.S. Lee, S. Vlahoyiannatos and L. Hanzo: Satellite based turbo-coded, blind-equalised 4-QAM and 16-QAM digital video broadcasting, *IEEE Trans. Broadcasting*, March 2000 [316] ©2000 IEEE. Personal use of this material is permitted. However, permission to reprint/republish this material for advertising or promotional purposes or for creating new collective works for resale or redistribution to servers or lists, or to refuse any copyrighted component of this work in other works must be obtained from the IEEE.

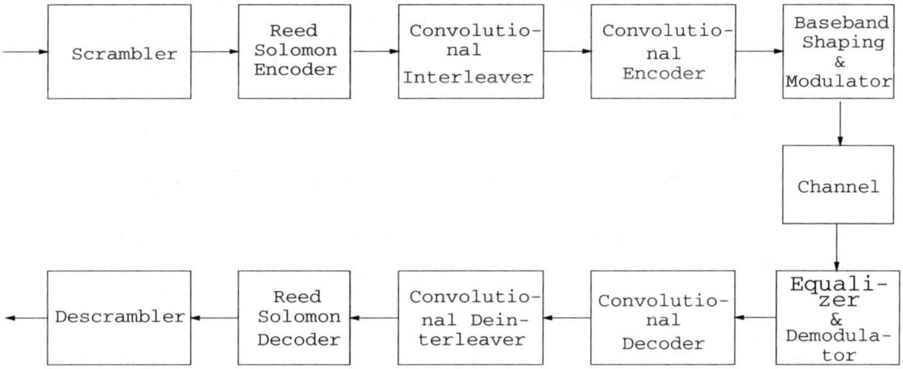

Figure 24.15: Schematic of the DVB satellite system

work areas are presented in Section 24.2.6.

24.2.2 DVB satellite scheme

The block diagram of the DVB satellite (DVB-S) system [474] is shown in Figure 24.15, which is composed of a MPEG-2 video encoder (not shown in the diagram), channel coding modules and a quadrature phase shift keying (QPSK) modem. The bitstream generated by the MPEG-2 encoder is packetized into frames of 188-byte long. The video data in each packet is then randomized by the scrambler. Similarly to the DVB-T scheme, the details concerning the scrambler have not been included in this chanter, since these may be found in the DVB-S standard [474].

Due to the poor error resilience of the MPEG-2 video codec, powerful concatenated channel coding is employed. The concatenated channel codec comprises a shortened Reed–Solomon (RS) outer code and an inner convolutional encoder. The 188-byte MPEG-2 video packet is extended by the Reed–Solomon encoder [475,476] with parity information to facilitate error recovery to form a 204-byte packet. The Reed–Solomon decoder can then correct up to 8 erroneous bytes for each 204-byte packet. Following this, the RS-coded packet is interleaved by a convolutional interleaver and further protected by a half-rate inner convolutional encoder with a constraint length of 7 [475, 476].

Furthermore, the overall code rate of the concatenated coding scheme can be adapted by variable puncturing, not shown in the figure, which supports code rates of 1/2 (no puncturing) as well as 2/3, 3/4, 5/6 and 7/8. The parameters of the convolutional encoder are summarised in Table 24.8.

In addition to implementing the standard DVB-S system as a benchmark, we have improved the system's performance with the aid of a turbo codec [433,482]. The block diagram of the turbo encoder is shown in Figure 24.16. The turbo encoder is constructed of two component encoders. Each component encoder is a half-rate convolutional encoder, whose parameters are listed in Table 24.9. The two component encoders are used to encode the same input bits, although the input bits of

Convolutional Coder Parameters	
Code rate	1/2
Constraint length	7
n	2
k	1
Generator polynomials (octal format)	171, 133

Table 24.8: Parameters of the $CC(n, k, K)$ convolutional inner encoder of the DVB-S modem

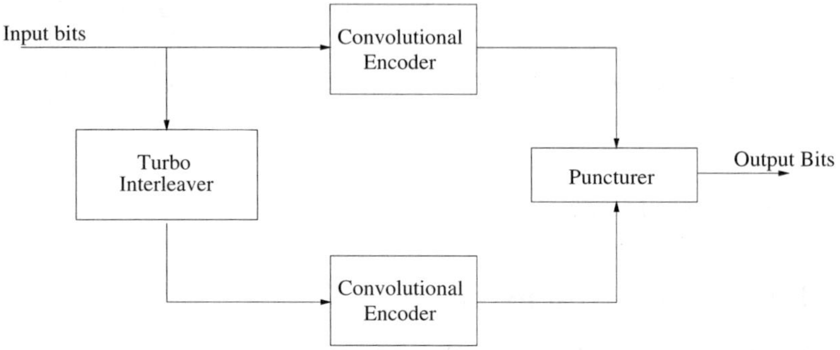

Figure 24.16: Block diagram of turbo encoder

the second component encoder are interleaved before encoding. The output bits of the two component codes are punctured and multiplexed, in order to form a single output bitstream. The component encoder used here is known as a half-rate recursive systematic convolutional encoder (RSC) [483]. It generates one parity bit and one systematic output bit for every input bit. In order to provide an overall coding rate of one half, half the output bits from the two encoders must be punctured. The puncturing arrangement used in our work is to transmit all the systematic bits from the first encoder and every other parity bit from both encoders.

Readers interested in further details of the DVB-S system are referred to the DVB-S standard [474]. The performance of the standard DVB-S system and the performance of the turbo-coded system are characterised in Section 24.2.5. Let us now briefly consider the multipath channel model used in our investigations.

24.2.3 Satellite channel model

The DVB-S system was designed to operate in the 12 GHz frequency band (K-band). Within this frequency band, tropospheric effects such as the transformation of electromagnetic energy into thermal energy due to induction of currents in rain and ice crystals lead to signal attenuations [495,496]. In the past 20 years, various researchers have concentrated their efforts on attempting to model the satellite channel, typically within a land mobile satellite channel scenario. However, the majority of the work

Turbo Coder Parameters	
Turbo code rate	1/2
Input block length	17 952 bits
Interleaver type	Random
Number of turbo decoder iterations	8
Turbo Encoder Component Code Parameters	
Component code encoder type	Convolutional Encoder (RSC)
Component code decoder type	log-MAP [485]
Constraint length	3
n	2
k	1
Generator polynomials (octal format)	7, 5

Table 24.9: Parameters of the inner turbo encoder used to replace the DVB-S system's convolutional coder (RSC: recursive systematic code)

conducted for example by Vogel and his colleagues [497–500] concentrated on modelling the statistical properties of a narrowband satellite channel in lower frequency bands, such as the 870 MHz UHF band and the 1.5 GHz L-band.

However, our high bit rate DVB satellite system requires a high bandwidth, hence the video bitstream is exposed to dispersive wideband propagation conditions. Recently, Saunders *et. al.* [501,502] have proposed the employment of multipath channel models to study the satellite channel, although their study was concentrated on the L-band and S-band only.

Due to the dearth of reported work on wideband satellite channel modelling in the K-band, we have adopted a simpler approach. The channel model employed in this study was the two-path (nT)-symbol spaced impulse response, where T is the symbol duration. In our studies we used $n = 1$ and $n = 2$ (Figure 24.17). This corresponds to a stationary dispersive transmission channel. Our channel model assumed that the receiver had a direct line-of-sight with the satellite as well as a second path caused by a single reflector probably from a nearby building or due to ground reflection. The ground reflection may be strong, if the satellite receiver dish is only tilted at a low angle.

Based on these channel models, we studied the ability of a range of blind equaliser algorithms to converge under various path delay conditions. In the next section we provide a brief overview of the various blind equalisers employed in our experiments, noting that the readers who are mainly interested in the system's performance may proceed directly to our performance analysis section, Section 24.2.5. Recall, furthermore that an indepth discourse on blind channel equalisation was provided in Chapter 7 and the next section offers only a rudimentary overview of the topic.

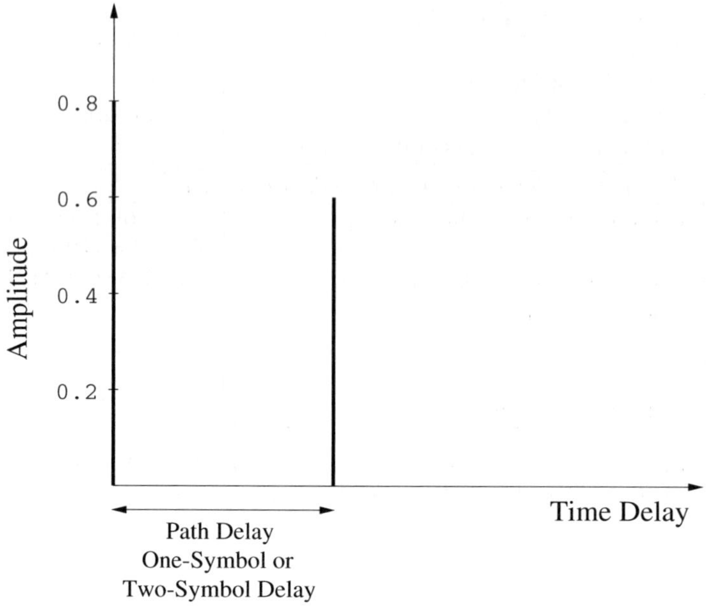

Figure 24.17: Two-path satellite channel model with either a one-symbol or two-symbol delay.

24.2.4 The blind equalisers

In this section the blind equalisers used in the system are presented. The following blind equalisers have been studied:

1) The modified constant modulus algorithm (MCMA) [6]

2) The Benveniste–Goursat algorithm (B-G) [7]

3) The stop-and-go algorithm (S-a-G) [8]

4) The per-survivor processing (PSP) algorithm [223].

We will now briefly introduce these algorithms.

First we define the variables that we will use:

$$\mathbf{y}(n) = [y(n+N_1), \ldots, y(0), \ldots, y(n-N_2)]^T \tag{24.3}$$

$$\mathbf{c}^{(n)} = [c_{-N_1}, \ldots, c_o, \ldots, c_{N_2}]^T \tag{24.4}$$

$$z(n) = \left(\mathbf{c}^{(n)}\right)^T \mathbf{y}(n) = \mathbf{y}^T(n)\mathbf{c}^{(n)} \tag{24.5}$$

where $\mathbf{y}(n)$ is the received symbol vector at time n, containing the $N_1 + N_2 + 1$ most recent received symbols, while N_1, N_2 are the number of equaliser feedback and feedforward taps, respectively. Furthermore, $\mathbf{c}^{(n)}$ is the equaliser tap vector, consisting of the equaliser tap values and $z(n)$ is the equalised symbol at time n, given by the convolution of the received signal with the equaliser's impulse response, while $()^T$ stands for matrix transpose. Note that the variables of Equations 24.3–24.5 assume complex values, when multilevel modulation is employed.

The **modified CMA** (MCMA) is an improved version of Godard's well-known *constant modulus algorithm (CMA)* [5]. The philosophy of the CMA is based on forcing the magnitude of the equalised signal to a constant value. In mathematical terms the CMA is based on minimising the cost function:

$$J^{(CMA)} = E\left[\left(|z(n)|^2 - R_2\right)^2\right], \tag{24.6}$$

where R_2 is a suitably chosen constant and $E[]$ stands for the expectation. Similarly to the CMA, the MCMA, which was proposed by Wesolowsky [6], forces the real and imaginary parts of the complex signal to the constant values of $R_{2,R}$ and $R_{2,I}$ respectively, according to the equaliser tap update equation of [6]:

$$\begin{aligned} \mathbf{c}^{(n+1)} = \mathbf{c}^{(n)} - \lambda \cdot \mathbf{y}^*(n) \cdot \{&Re[z(n)] \cdot ((Re[z(n)])^2 - R_{2,R}) \\ &+ jIm[z(n)] \cdot ((Im[z(n)])^2 - R_{2,I})\}, \end{aligned} \tag{24.7}$$

where λ is the step size parameter and the $R_{2,R}$, $R_{2,I}$ constant parameters of the algorithm are defined as:

$$R_{2,R} = \frac{E\left[(Re[a(n)])^4\right]}{E\left[(Re[a(n)])^2\right]} \tag{24.8}$$

$$R_{2,I} = \frac{E\left[(Im[a(n)])^4\right]}{E\left[(Im[a(n)])^2\right]}, \tag{24.9}$$

where $a(n)$ is the transmitted signal at time instant n.

The **Benveniste–Goursat** (B-G) algorithm [7] is an amalgam of the Sato's algorithm [4] and the decision directed (DD) algorithm of Chapter 7. Strictly speaking, the decision directed algorithm is not a blind equalisation technique, since its convergence is highly dependent on the channel.

This algorithm estimates the error between the equalised signal and the detected signal as:

$$\epsilon^{DD}(n) = z(n) - \hat{z}(n), \tag{24.10}$$

where $\hat{z}(n)$ is the receiver's estimate of the transmitted signal at time instant n. Similarly to the DD algorithm's error term, the Sato error [4] is defined as:

$$\epsilon^{Sato}(n) = z(n) - \gamma \cdot csgn(z(n)), \tag{24.11}$$

where γ is a constant parameter of the Sato algorithm, defined as:

$$\gamma = \frac{E\left[(Re\left[a(n)\right])^2\right]}{E\left[|Re\left[a(n)\right]|\right]} = \frac{E\left[(Im\left[a(n)\right])^2\right]}{E\left[|Im\left[a(n)\right]|\right]} \tag{24.12}$$

and $csgn(x) = sign(Re\{x\}) + jsign(Im\{x\})$ is the complex sign function. The B-G algorithm combines the above two error terms into one:

$$\epsilon^G(n) = k_1 \cdot \epsilon^{DD}(n) + k_2 \cdot |\epsilon^{DD}(n)| \cdot \epsilon^{Sato}(n), \tag{24.13}$$

where the two error terms are suitably weighted by the constant parameters k_1 and k_2 in Equation (24.13). Using this error term, the B-G equaliser updates the equaliser coefficients according to the the following equaliser tap update equations [7]:

$$\mathbf{c}^{(n+1)} = \mathbf{c}^{(n)} - \lambda \cdot \mathbf{y}^*(n) \cdot \epsilon^G(n). \tag{24.14}$$

In our investigations, the weights were chosen as $k_1 = 1$, $k_2 = 5$, so that the Sato error was weighted more heavily, than the DD error.

Again, the **stop-and-go** (S-a-G) algorithm [8] is a variant of the decision directed algorithm of Chapter 7, where at each equaliser coefficient adjustment iteration, the update is enabled or disabled depending on whether the update is likely to be correct. The update equations of this algorithm are given by [8]

$$\mathbf{c}^{(n+1)} = \mathbf{c}^{(n)} - \lambda \cdot \mathbf{y}^*(n) \cdot [f_{n,R} \cdot Re\{\epsilon^{DD}(n)\} + jf_{n,I} \cdot Im\{\epsilon^{DD}(n)\}] \tag{24.15}$$

where * stands for the complex conjugate, $\epsilon^{DD}(n)$ is the decision directed error as in Equation (24.10) and the binary functions $f_{n,R}$, $f_{n,I}$ enable or disable the update of the equaliser according to the following rule. If the sign of the Sato error (the real or the imaginary part independently) is the same as the sign of the decision directed error, then the update takes place, otherwise it does not.

In mathematical terms, this is equivalent to [8]:

$$f_{n,R} = \begin{cases} 1 & \text{if } sgn(Re[\epsilon^{DD}(n)]) = sgn(Re[\epsilon^{Sato}(n)]) \\ 0 & \text{if } sgn(Re[\epsilon^{DD}(n)] \neq sgn(Re[\epsilon^{Sato}(n)]) \end{cases} \tag{24.16}$$

$$f_{n,I} = \begin{cases} 1 & \text{if } sgn(Im\{\epsilon^{DD}(n)\}) = sgn(Im\{\epsilon^{Sato}(n)\}) \\ 0 & \text{if } sgn(Im\{\epsilon^{DD}(n)\}) \neq sgn(Im\{\epsilon^{Sato}(n)\}). \end{cases} \tag{24.17}$$

For a blind equaliser, this condition provides us with a measure of the probability of the coefficient update being correct.

The **PSP algorithm** [223] is based on employing convolutional coding and hence it is a trellis-based sequence estimation technique, in which the channel is not known a priori. Hence an iterative channel estimation technique is employed, in order to estimate the channel jointly with the modulation symbol. In this sense, an initial channel is used and the estimate is updated at each new symbol's arrival.

In our case the update was based on the *least mean squares (LMS)* estimates, according to the following channel-tap update equations [223]:

$$\hat{\mathbf{h}}^{(n+1)} = \hat{\mathbf{h}}^{(n)} + \lambda \cdot \hat{\mathbf{a}}^*(n) \cdot \left(y(n) - \hat{\mathbf{a}}^T(n)\hat{\mathbf{h}}^{(n)} \right), \tag{24.18}$$

where $\hat{\mathbf{h}}^{(n)} = (\hat{h}_{-L_1}^{(n)}, \ldots, \hat{h}_o^{(n)}, \ldots, \hat{h}_{L_2}^{(n)})^T$ is the estimated (for one surviving path) channel tap vector at time instant n, $\hat{\mathbf{a}}(n) = (\hat{a}(n+L_1), \ldots, \hat{a}(0), \ldots, \hat{a}(n-L_2))^T$ is the associated estimated transmitted symbol vector and $y(n)$ is the actually received symbol at time instant n.

Each of the surviving paths in the trellis carries not only its own signal estimation, but also its own channel estimation. Moreover, convolutional decoding can take place jointly with this channel and data estimation procedure, leading to improved bit error rate (BER) performance. Summary of the various equalisers' parameters is given in Table 24.10.

	Step size λ	No. of equalizer taps	Initial tap vector
Benveniste–Goursat	5x10^{-4}	10	$(1.2, 0, \cdots, 0)$
Modified CMA	5x10^{-4}	10	$(1.2, 0, \cdots, 0)$
Stop-and-go	5x10^{-4}	10	$(1.2, 0, \cdots, 0)$
PSP (1 sym delay)	10^{-2}	2	$(1.2, 0)$
PSP (2 sym delay)	10^{-2}	3	$(1.2, 0, 0)$

Table 24.10: Summary of the equaliser parameters used in the simulations. The tap vector $(1.2, 0, \cdots, 0)$ indicates that the first equaliser coefficient is initialised to the value 1.2, while the others are initialised to 0

Having described the components of our enhanced DVB-S system, let us now consider the overall system's performance.

24.2.5 Performance of the DVB satellite scheme

In this section, the performance of the DVB-S system was evaluated by means of simulations. Two modulation types were used, namely the standard QPSK and the enhanced 16-QAM schemes. The channel model of Figure 24.17 was employed. The first channel model had a one-symbol second-path delay, while in the second one the path-delay corresponded to the period of two symbols. The average BER versus SNR per bit performance was evaluated after the equalisation and demodulation process, as well as after Viterbi [475] or turbo decoding [482]. The SNR per bit or E_b/N_o is defined as follows:

$$\text{SNR per bit} = 10\ log_{10} \frac{\bar{S}}{\bar{N}} + \delta \tag{24.19}$$

where $\bar{S}$ is the average received signal power, $\bar{N}$ is the average received noise power and δ, which is dependent on the type of modulation scheme used and channel code rate (R), is defined as follows:

$$\delta = 10\ log_{10} \frac{1}{\text{R} \times \text{Bits per modulation symbol}}. \tag{24.20}$$

Our results are further divided into two subsections for ease of discussion. First, we will present the system performance over the one-symbol delay two-path channel in Section 24.2.5.1. Next, the system performance over the two-symbol delay two-path channel is presented in Section 24.2.5.2. Lastly, a summary of the system performance is provided in Section 24.2.5.3.

24.2.5.1 Transmission over the symbol-spaced two-path channel

The linear equalisers' performance was quantified and compared using QPSK modulation over the one-symbol delay two-path channel model of Figure 24.18. Since all the equalisers have similar BER performance, only the modified CMA results are shown in the figure.

The equalised performance over the one symbol-spaced channel was inferior to that over the non-dispersive AWGN channel. However, as expected, it was better than without any equalisation. Another observation for Figure 24.18 was that the different punctured channel coding rates appeared to give slightly different bit error rates after equalisation. This was because the linear blind equalisers required uncorrelated input bits in order to converge. However, the input bits were not entirely random when convolutional coding was used. The consequences of violating the zero-correlation constraint are not generally known. Nevertheless, two potential problems were apparent. Firstly, the equaliser may diverge from the desired equaliser equilibrium [218]. Secondly, the performance of the equaliser is expected to degrade, owing to the violation of the randomness requirement, which is imposed on the input bits in order to ensure that the blind equalisers will converge.

Since the channel used in our investigations was static, the first problem was not encountered. Instead, the second problem was what we actually observed. Figure 24.19 quantifies the equalisers' performance degradation due to convolutional coding. We can observe a 0.1 dB SNR degradation, when the convolutional codec creates correlation among the bits for this specific case.

The average BER curves after Viterbi or turbo decoding are shown in Figure 24.20(a). In this figure, the average BER over the non-dispersive AWGN channel after turbo decoding constitutes the best case performance, while the average BER of the one-symbol delay two-path MCMA-equalised rate 7/8 convolutionally coded scenario exhibits the worst case performance. Again, in this figure only the modified CMA was featured for simplicity. The performance of the remaining equalisers was characterised in Figure 24.20(b). Clearly, the performance of all the linear equalisers investigated was similar.

It is observed in Figure 24.20(a) that the combination of the modified CMA blind equaliser with turbo decoding exhibited the best SNR performance over the one-symbol delay two-path channel. The only comparable alternative was the PSP algorithm. Although the performance of the PSP algorithm was better at low SNRs, the associated curves cross over and the PSP algorithm's performance became inferior

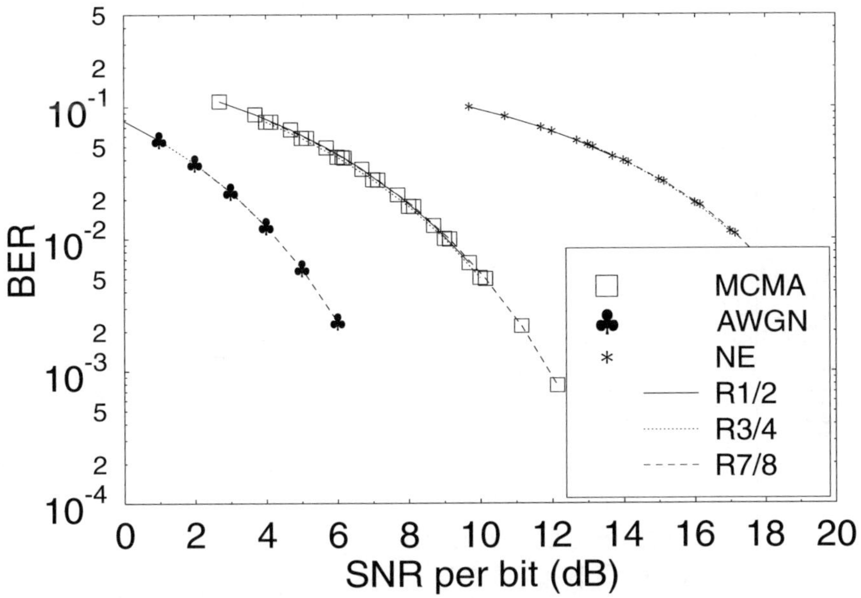

(a) After equalisation and demodulation

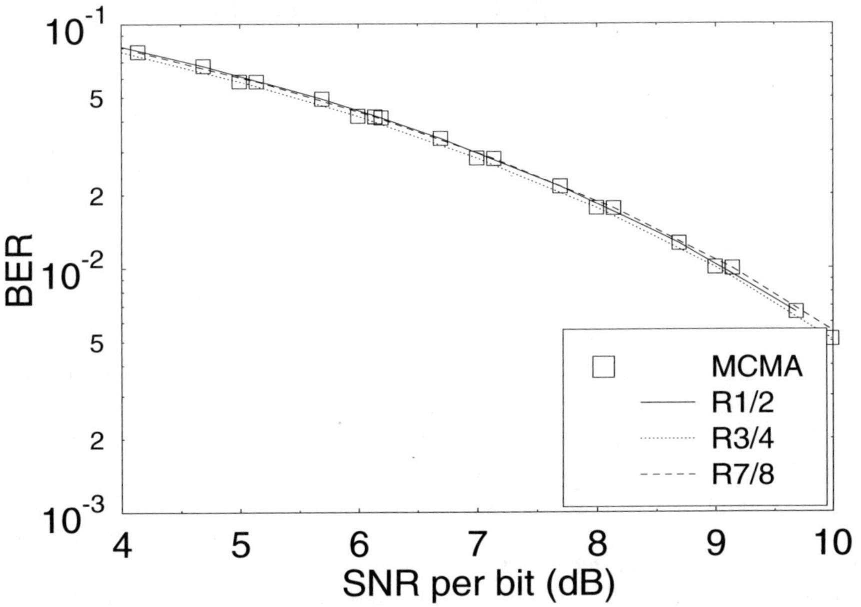

(b) Same as (a) but enlarged in order to show performance difference of the blind equaliser, when different convolutional code rates are used.

Figure 24.18: Average BER versus SNR per bit performance after equalisation and demodulation employing **QPSK** modulation and **one-symbol delay channel** (**NE** = non-equalised; **MCMA** = modified constant modulus algorithm) [316], ©IEEE, 2000, Lee, Vlahoyiannatos, Hanzo

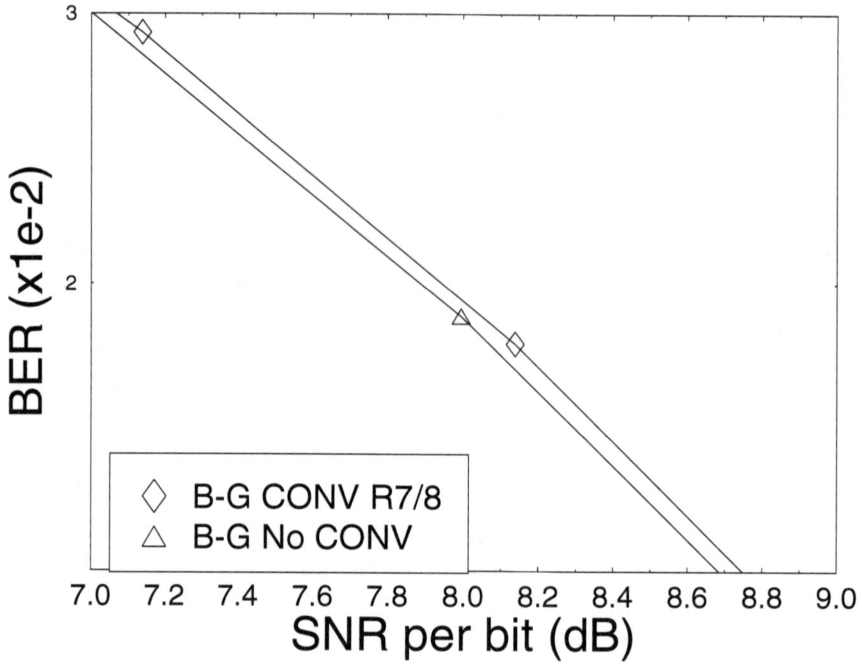

Figure 24.19: Average BER versus SNR per bit performance after equalisation and demodulation employing **QPSK** modulation and the **one-symbol delay two-path channel** of Figure 24.17, for the Benveniste–Goursat algorithm, where the input bits are random (No CONV) or correlated (CONV 7/8) as a result of convolutional coding having a coding rate of 7/8 [316], ©IEEE, 2000, Lee, Vlahoyiannatos, Hanzo

below the average BER of 10^{-3}. Although not shown in Figure 24.20, the Reed–Solomon decoder, which was concatenated to either the convolutional or the turbo decoder, became effective, when the average BER of its input was below approximately 10^{-4}. In this case, the PSP algorithm performed by at least 1 dB worse in the area of interest, which is at an average BER of 10^{-4}.

A final observation in the context of Figure 24.20(a) is that when convolutional decoding was used, the associated E_b/N_o performance of the rate 1/2 convolutional coded scheme appeared slightly inferior to that of the rate 3/4 and the rate 7/8 scenarios beyond certain E_b/N_o values. This was deemed to be a consequence of the fact that the 1/2 rate encoder introduced more correlation into the bitstream than its higher rate counterparts and this degraded the performance of the blind channel equalisers, which performed best, when fed with random bits.

Having considered the QPSK case, we shall now concentrate on the enhanced system, which employed 16-QAM under the same channel and equaliser conditions.

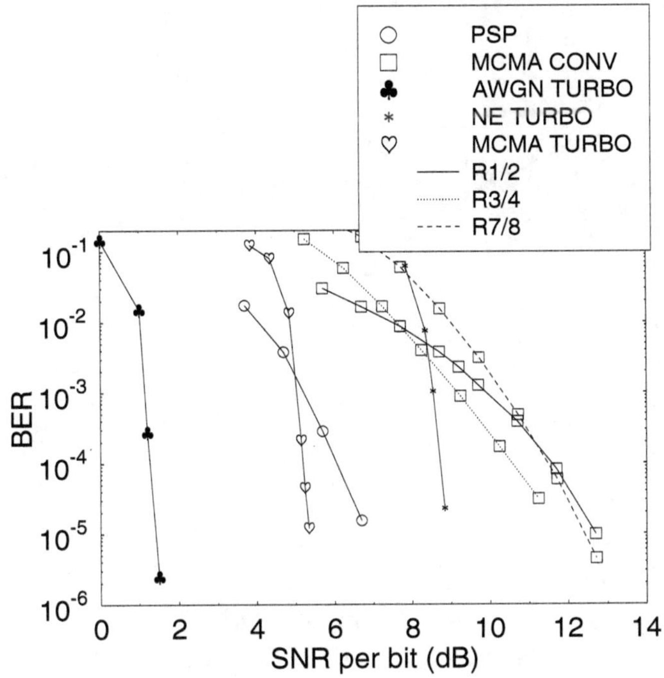

(a) PSP and linear equalisers

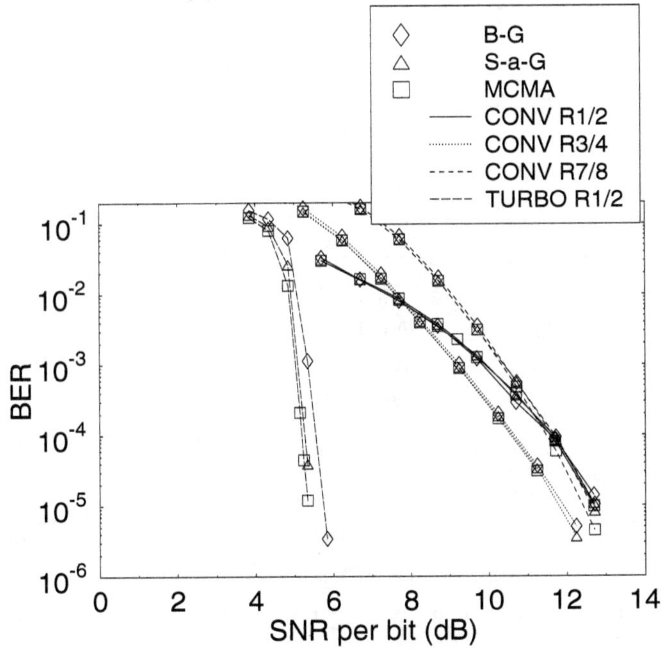

(b) Linear equalisers only

Figure 24.20: Average BER versus SNR per bit performance after convolutional or turbo decoding for **QPSK** modulation and **one-symbol delay channel** (**NE** = non-equalised; **B-G** = Benveniste–Goursat; **S-a-G** = stop-and-go; **MCMA** = modified constant modulus algorithm; **PSP** = per-survivor processing) [316], ©IEEE, 2000, Lee, Vlahoyiannatos, Hanzo

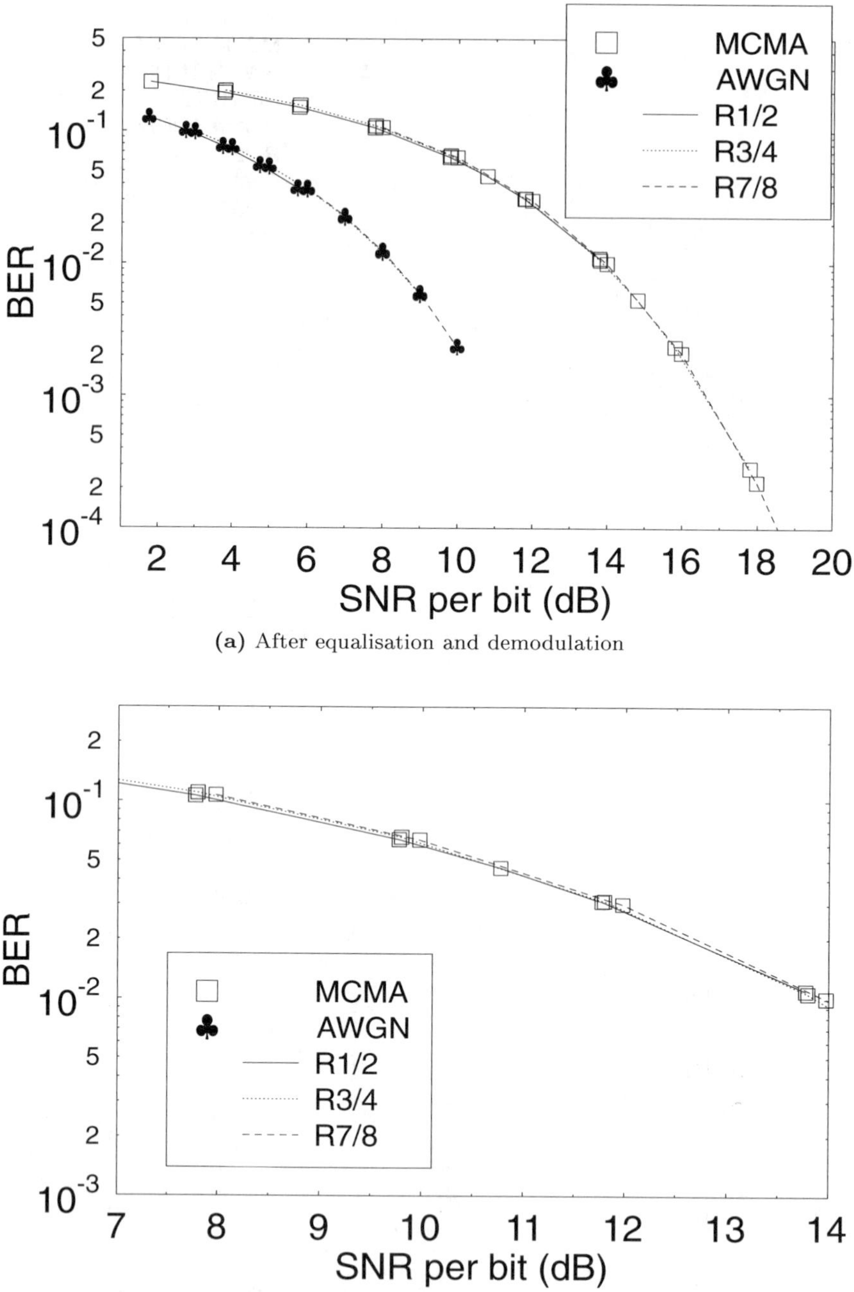

(a) After equalisation and demodulation

(b) Same as (a) but enlarged in order to show performance difference of the blind equaliser, when different convolutional code rates are used.

Figure 24.21: Average BER versus SNR per bit after equalisation and demodulation for **16-QAM** over the **one-symbol delay two-path channel** of Figure 24.17 (**MCMA** = modified constant modulus algorithm) [316], ©IEEE, 2000, Lee, Vlahoyiannatos, Hanzo

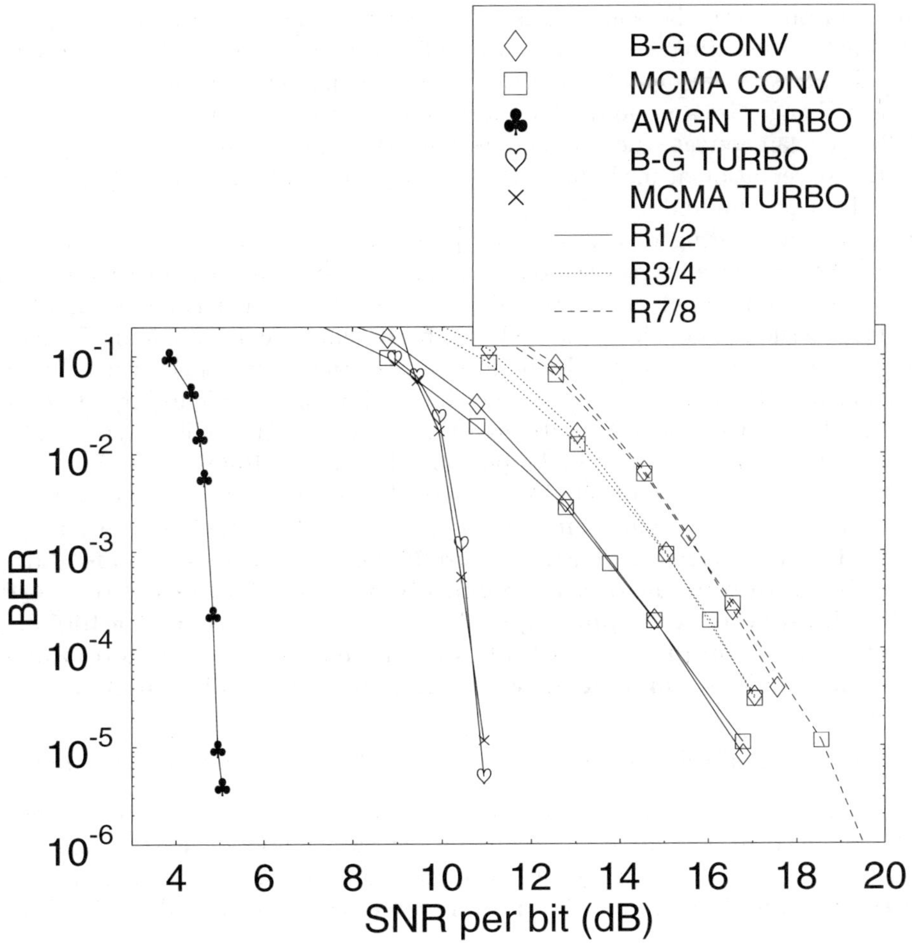

Figure 24.22: Average BER versus SNR per bit after Viterbi or turbo decoding for **16-QAM** over the **one-symbol delay two-path channel** of Figure 24.17 (**B-G** = Benveniste–Goursat; **S-a-G** = stop-and-go; **MCMA** = modified constant modulus algorithm; **PSP** = per-survivor processing) [316], ©IEEE, 2000, Lee, Vlahoyiannatos, Hanzo

In Figure 24.21 and Figure 24.22, the performance of the DVB system employing 16-QAM is presented. Again, for simplicity, only the modified CMA results are given. In this case the ranking order of the different coding rates followed our expectations more closely in the sense that the lowest coding rate of 1/2 was the best performer, followed by rate 3/4 codec, in turn followed by the least powerful rate 7/8 codec.

The stop-and-go algorithm has been excluded from these results, since it does not converge for high SNR values. This happens because the equalisation procedure is only activated when there is a high probability of correct decision directed equaliser update. In our case, the equaliser is initialised far from its convergence point and hence the decision directed updates are unlikely to be correct. In the absence of noise this leads to the update algorithm being permanently de-activated. If noise is present though, then some random perturbations from the point of the equaliser's initialization can activate the stop-and-go algorithm and can lead to convergence. We made this observation at medium SNR values in our simulation study. For high SNR values, the algorithm did not converge.

It is also interesting to compare the performance of the system for the QPSK and 16-QAM schemes. When the one-symbol delay two-path channel model of Figure 24.17 was considered, the system was capable of supporting the use of 16-QAM with the provision of an additional SNR per bit of approximately 4–5 dB. This observation was made by comparing the performance of the DVB system when employing the modified CMA and the half-rate convolutional or turbo code in Figure 24.20 and Figure 24.22 at a BER of 10^{-4}. Although the original DVB satellite system only employs QPSK modulation, our simulations had shown that 16-QAM can be employed equally well for the range of blind equalisers that we have used in our work. This allowed us to double the video bit rate and hence to substantially improve the video quality. **The comparison of Figures 24.20 and 24.22 also reveals that the extra SNR requirement of approximately 4–5 dB of 16-QAM over QPSK can be eliminated by employing turbo coding at the cost of a higher implementational complexity. This allowed us to accommodate a doubled bit rate within a given bandwidth, which improved the video quality.**

24.2.5.2 Transmission over the two-symbol delay two-path channel

In Figure 24.23 (only for the Benveniste-Goursat algorithm for simplicity) and Figure 24.24 the corresponding BER results for the two-symbol delay two-path channel of Figure 24.17 are given for QPSK. The associated trends are similar to those in Figures 24.18 and 24.20, although some differences can be observed, as listed below:

- The "cross over point" is where the performance of the PSP algorithm becomes inferior to the performance of the modified CMA in conjunction with turbo decoding. This point is now at 10^{-4}, which is in the range where the RS decoder guarantees an extremely low probability of error.

- The rate 1/2 convolutional decoding is now the best performer, when convolutional decoding is concerned, while the rate 3/4 scheme exhibits the worst performance.

Finally, in Figure 24.25, the associated 16-QAM results are presented. Notice that the stop-and-go algorithm was again excluded from the results. Furthermore, we observe a high performance difference between the Benveniste–Goursat algorithm and the modified CMA. In the previous cases we did not observe such a significant difference. The difference in this case is that the channel exhibits an increased delay spread. This illustrated the capability of the equalisers to cope with more widespread multipaths, while keeping the equalizer order constant at 10. The Benveniste–Goursat equaliser was more efficient than the modified CMA in this case.

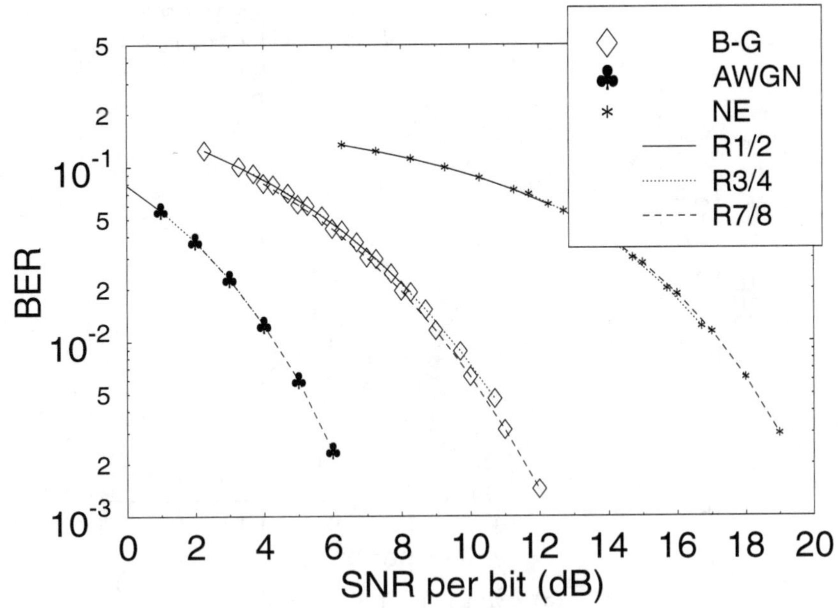

(a) After equalisation and demodulation

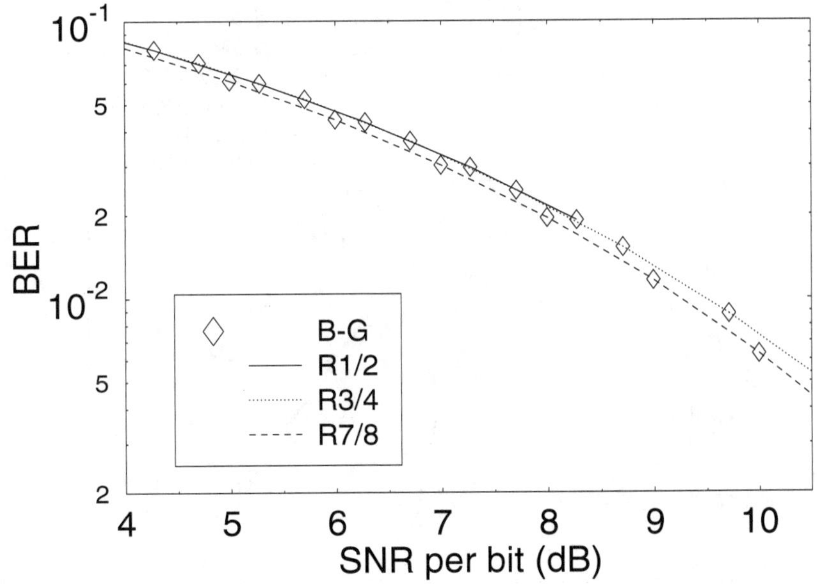

(b) Same as (a) but enlarged in order to show performance difference of the blind equaliser, when different convolutional code rates are used.

Figure 24.23: Average BER versus SNR per bit performance after equalisation and demodulation for **QPSK** modulation over the **two-symbol delay two-path channel** of Figure 24.17 (**B-G** = Benveniste–Goursat) [316], ©IEEE, 2000, Lee, Vlahoyiannatos, Hanzo

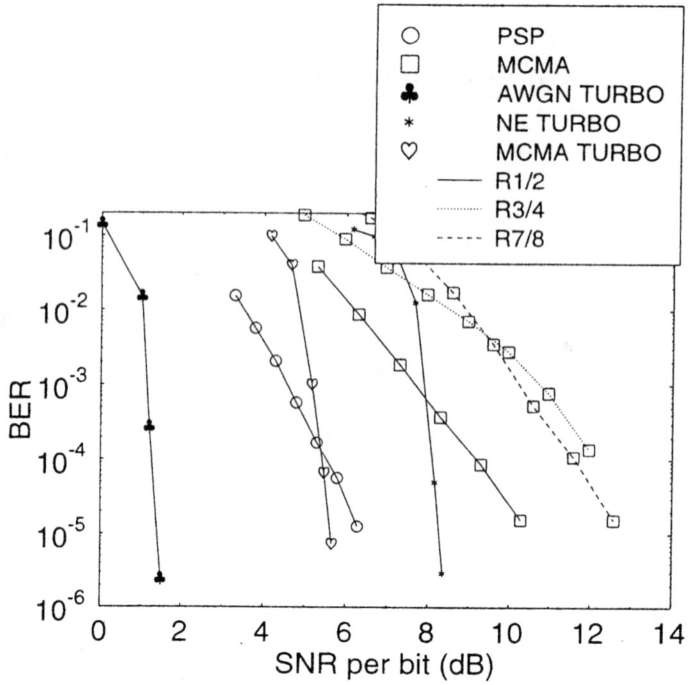

(a) PSP and linear equalisers

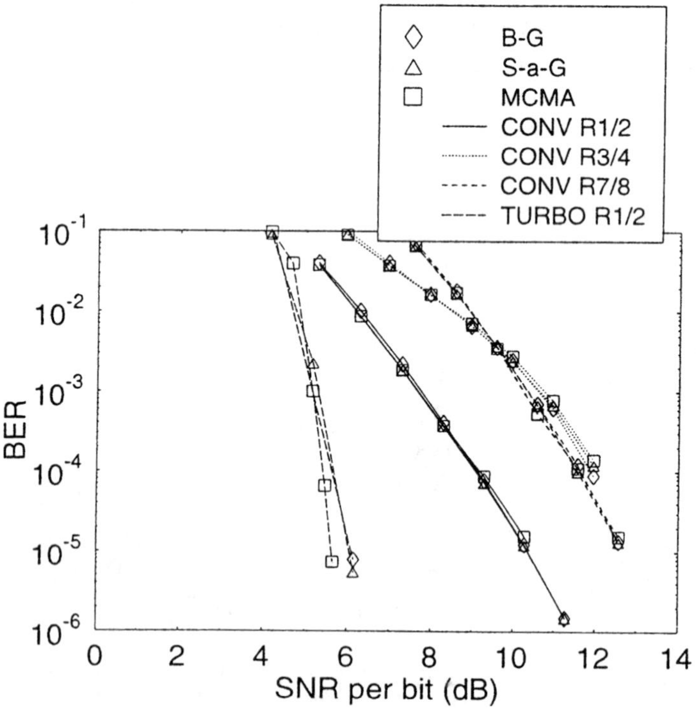

(b) Linear equalisers only

Figure 24.24: Average BER versus SNR per bit performance after convolutional or turbo decoding for **QPSK** modulation over the **two-symbol delay two-path channel** of Figure 24.17 (**B-G** = Benveniste–Goursat; **S-a-G** = stop-and-go; **MCMA** = modified constant modulus algorithm; **PSP** = per-survivor processing) [316], ©IEEE, 2000, Lee, Vlahoyiannatos, Hanzo

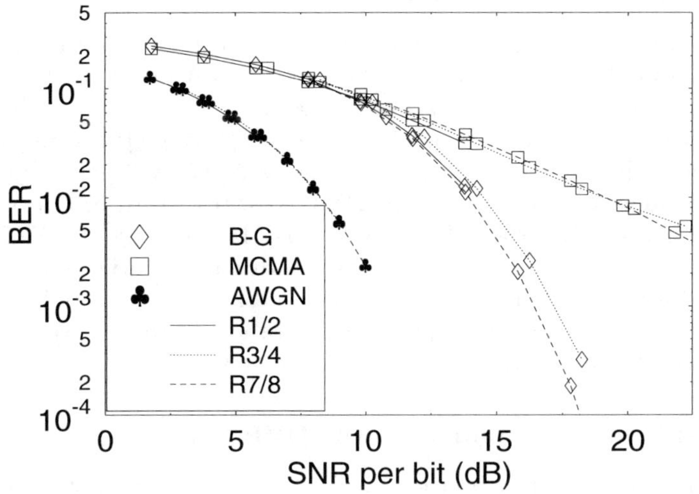

(a) After equalisation and demodulation

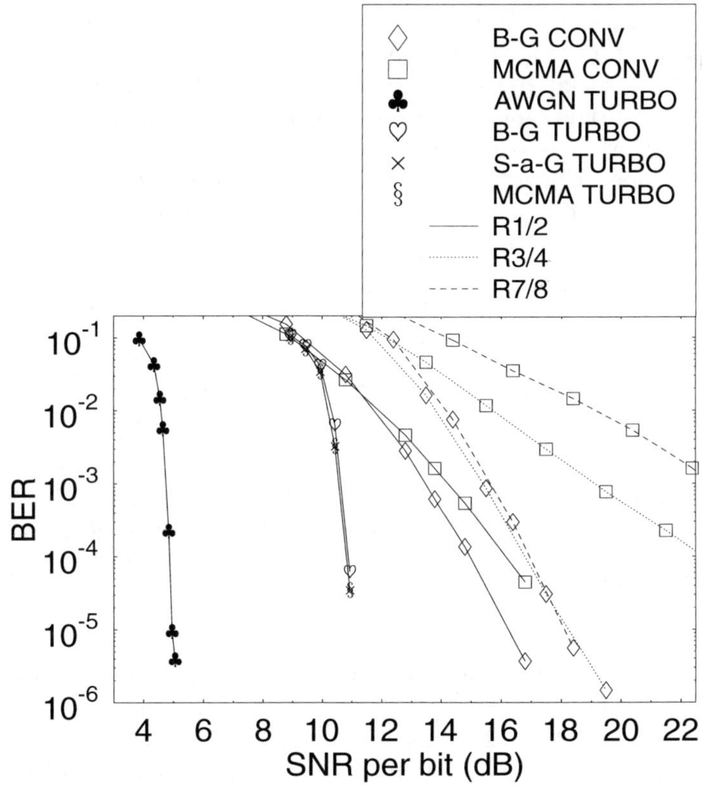

(b) After viterbi or turbo decoding

Figure 24.25: Average BER versus SNR per bit performance (a) after equalisation and demodulation and (b) after Viterbi or turbo decoding for **16-QAM** over the **two-symbol delay two-path channel** of Figure 24.17 (**B-G** = Benveniste–Goursat; **S-a-G** = stop-and-go; **MCMA** = modified constant modulus algorithm; **PSP** = per-survivor processing) [316], ©IEEE, 2000, Lee, Vlahoyiannatos, Hanzo

It is interesting to note that in this case the performance of the different coding rates was again in the expected order: the rate 1/2 scheme is the best, followed by the rate 3/4 scheme and then the rate 7/8 scheme.

If we compare the performance of the system employing QPSK and 16-QAM over the two-symbol delay two-path channel of Figure 24.17, we again observe that **16-QAM can be incorporated into the DVB system, if an extra 5 dB of SNR per bit is affordable in power budget terms. Here, only the B-G algorithm is worthwhile considering out of the three linear equalisers of Table 24.10.** This observation was made by comparing the performance of the DVB system when employing the Benveniste–Goursat equalizer and the half-rate convolutional coder in Figure 24.24 and Figure 24.25.

24.2.5.3 Performance summary of the DVB-S system

Table 24.11 provides an approximation of the convergence speed of each blind equalisation algorithm of Table 24.10. It is clear that PSP exhibited the fastest convergence, followed by the Benveniste–Goursat algorithm. In our simulations the convergence was quantified by observing the slope of the BER curve, and finding when this curve was reaching the associated residual BER, implying that the BER has reached its steady–state value. Figure 24.26 gives an illustrative example of the equaliser's convergence for 16-QAM. The stop-and-go algorithm converges significantly slower than the other algorithms, which can also be seen from Table 24.11. This happens because, during start-up, the algorithm is deactivated most of the time, an effect which becomes more severe with increasing QAM order.

	B-G	MCMA	S-a-G	PSP
QPSK 1 sym	73	161	143	0.139
QPSK 2 sym	73	143	77	0.139
16-QAM 1 sym	411	645	1393	
16-QAM 2 sym	359	411	1320	

Table 24.11: Equaliser convergence speed (in miliseconds) measured in the simulations, given as an estimate of time required for convergence when 1/2 rate puncturing is used (x sym = x-symbol delay two-path channel and x can take either the value 1 or 2)

Figure 24.27 portrays the corresponding reconstructed video quality in terms of the average PSNR versus channel SNR (CSNR) for the one-symbol delay and two-symbol delay two-path channel models of Figure 24.17. Table 24.12 and Table 24.13 provide a summary of the DVB satellite system's performance tolerating a PSNR degradation of 2 dB, which was deemed to be nearly imperceptible in terms of subjective video degradations. The average BER values quoted in the tables refer to the average BER achieved after Viterbi or turbo decoding. The channel SNR is quoted in association with the 2 dB average video PSNR degradation, since the viewer will begin to perceive video degradations due to erroneous decoding of the received video around this threshold.

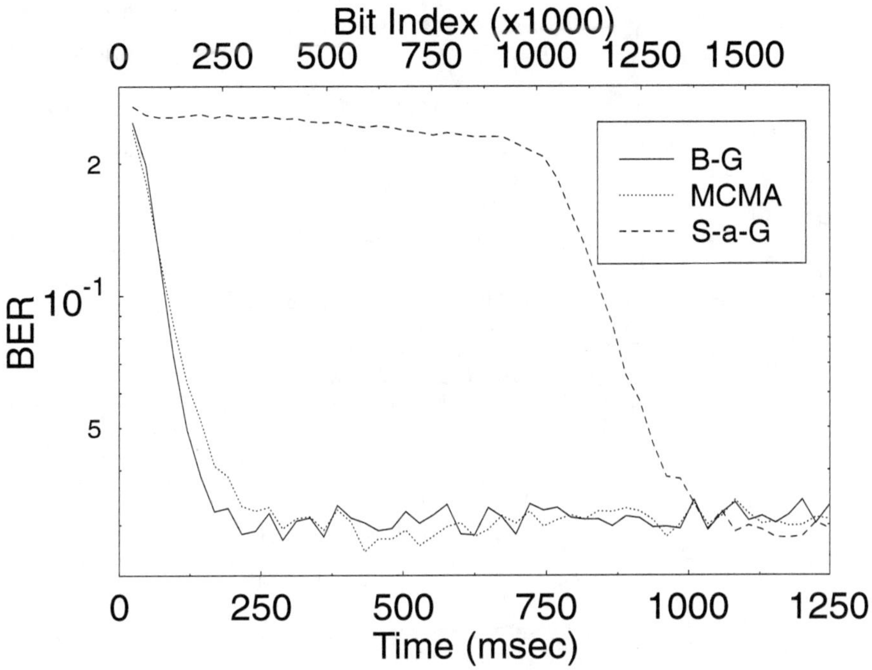

Figure 24.26: Learning curves for 16-QAM, one-symbol delay two-path channel at SNR = 18 dB [316], ©IEEE, 2000, Lee, Vlahoyiannatos, Hanzo

Mod.	Equaliser	Code	CSNR (dB)	E_b/N_0
QPSK	PSP	R=1/2	5.3	5.3
QPSK	MCMA	Turbo (1/2)	5.2	5.2
16-QAM	MCMA	Turbo (1/2)	13.6	10.6
QPSK	MCMA	Conv (1/2)	9.1	9.1
16-QAM	MCMA	Conv (1/2)	17.2	14.2
QPSK	MCMA	Conv (3/4)	11.5	9.7
16-QAM	MCMA	Conv (3/4)	20.2	15.4
QPSK	B-G	Conv (7/8)	13.2	10.8
16-QAM	B-G	Conv (7/8)	21.6	16.2

Table 24.12: Summary of performance results over the dispersive one-symbol delay two-path AWGN channel of Figure 24.17 tolerating a PSNR degradation of 2 dB

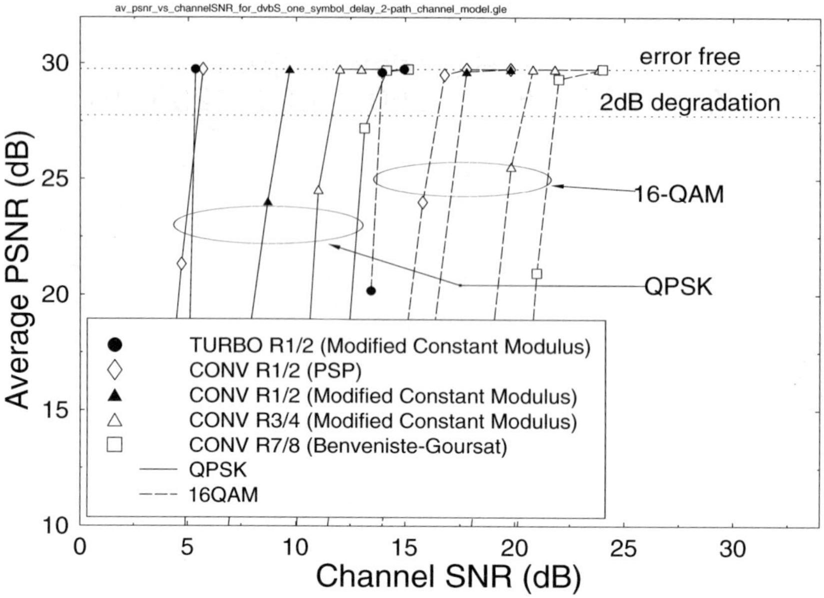

(a) One-symbol delay two-path channel model

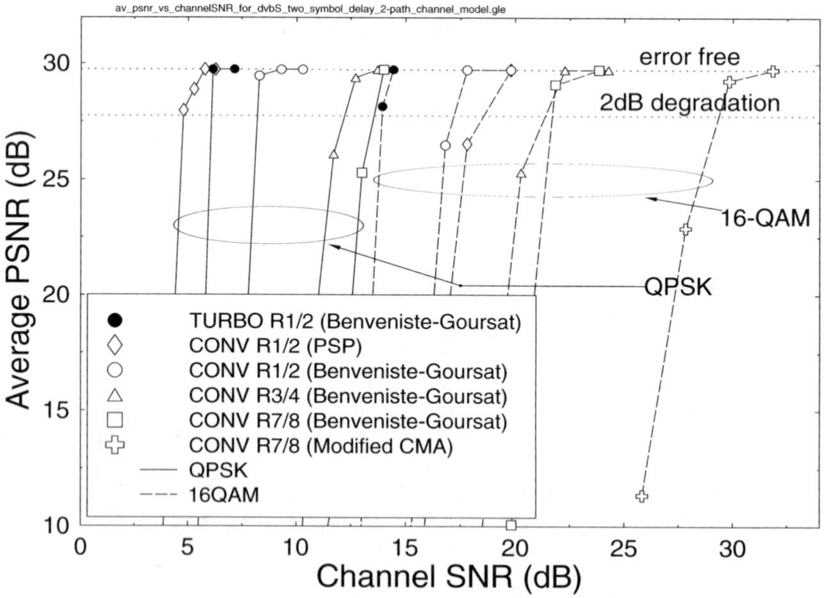

(b) Two-symbol delay two-path channel model

Figure 24.27: Average PSNR versus channel SNR for (a) the one-symbol delay two-path channel model and (b) the two-symbol delay two-path channel model of Figure 24.17 at a video bit rate of 2.5 Mbit/s using the "Football" video sequence [316], ©IEEE, 2000, Lee, Vlahoyiannatos, Hanzo

Mod.	Equaliser	Code	CSNR (dB)	E_b/N_0
QPSK	PSP	R=1/2	4.7	4.7
QPSK	B-G	Turbo (1/2)	5.9	5.9
16-QAM	B-G	Turbo (1/2)	13.7	10.7
QPSK	B-G	Conv (1/2)	8.0	8.0
16-QAM	B-G	Conv (1/2)	17.0	14.0
QPSK	B-G	Conv (3/4)	12.1	10.3
16-QAM	B-G	Conv (3/4)	21.1	16.3
QPSK	B-G	Conv (7/8)	13.4	11.0
16-QAM	MCMA	Conv (7/8)	29.2	23.8

Table 24.13: Summary of performance results over the dispersive two-symbol delay two-path AWGN channel of Figure 24.17 tolerating a PSNR degradation of 2 dB

Mod.	Equaliser	Code	E_b/N_0
QPSK	PSP	R=1/2	6.1
QPSK	MCMA	Turbo (1/2)	5.2
16-QAM	MCMA	Turbo (1/2)	10.7
QPSK	MCMA	Conv (1/2)	11.6
16-QAM	MCMA	Conv (1/2)	15.3
QPSK	MCMA	Conv (3/4)	10.5
16-QAM	MCMA	Conv (3/4)	16.4
QPSK	B-G	Conv (7/8)	11.8
16-QAM	B-G	Conv (7/8)	17.2

Table 24.14: Summary of system performance results over the dispersive one-symbol delay two-path AWGN channel of Figure 24.17 tolerating an average BER of 10^{-4}, which was evaluated after Viterbi or turbo decoding but before RS decoding

Tables 24.14 and 24.15 provide a summary of the SNR per bit required for the various system configurations. The BER threshold of 10^{-4} was selected here, since at this average BER after Viterbi or turbo decoding, the RS decoder becomes effective, guaranteeing near error-free performance. This also translates into near unimpaired reconstructed video quality.

Finally, in Table 24.16 the QAM symbol rate or baud rate is given for different puncturing rates and for different modulation schemes, based on the requirement of supporting a video bit rate of 2.5 Mbit/sec. We observe that the baud rate is between 0.779 and 2.73 MBd, depending on the coding rate and the number of bits per modulation symbol.

Mod.	Equaliser	Code	E_b/N_0
QPSK	PSP	R=1/2	5.6
QPSK	B-G	Turbo (1/2)	5.7
16-QAM	B-G	Turbo (1/2)	10.7
QPSK	B-G	Conv (1/2)	9.2
16-QAM	B-G	Conv (1/2)	15.0
QPSK	B-G	Conv (3/4)	12.0
16-QAM	B-G	Conv (3/4)	16.8
QPSK	B-G	Conv (7/8)	11.7
16-QAM	MCMA	Conv (7/8)	26.0

Table 24.15: Summary of system performance results over the dispersive two-symbol delay two-path AWGN channel of Figure 24.17 tolerating an average BER of 10^{-4}, which was evaluated after Viterbi or turbo decoding but before RS decoding

Punctured Rate	4-QAM Baud Rate (MBd)	16-QAM Baud Rate (MBd)
1/2	2.73	1.37
3/4	1.82	0.909
7/8	1.56	0.779

Table 24.16: The channel bit rate for the three different punctured coding rates and for the two modulations

24.2.6 Conclusions and future work

In this contribution, we have investigated the performance of a turbo-coded DVB system in a satellite broadcast environment. A range of system performance results was presented based on the standard DVB-S scheme, as well as on a turbo-coded scheme in conjunction with blind-equalised QPSK/16-QAM. **The convolutional code specified in the standard system was substituted with turbo coding, which resulted in a substantial coding gain of approximately 4–5 dB. We have also shown that 16-QAM can be utilised instead of QPSK if an extra 5 dB SNR per bit gain is added to the link budget. This extra transmitted power requirement can be eliminated upon invoking the more complex turbo codec, which requires lower transmitted power for attaining the same performance as the standard convolutional codecs.**

Our future work will be focused on extending the DVB satellite system to supporting mobile users for the reception of satellite broadcast signals. The use of blind turbo equalisers will also be investigated in comparison to conventional blind equalisers. Further work will also be dedicated to trellis coded modulation (TCM) and turbo trellis coded modulation (TTCM) based orthogonal frequency division multiplexed (OFDM) and single-carrier equalised modems. The impact on the system performance by employing various types of turbo interleavers and turbo coded is also of interest. A range of further wireless video communications issues are addressed in [489].

Glossary

8-DPSK	8-Phase Differential Phase Shift Keying
ACF	autocorrelation function
ADC	Analog–to–Digital Converter
ADM	adaptive delta modulation
ADPCM	Adaptive Differential Pulse Coded Modulation.
AGC	Automatic Gain Control
AM-PM	amplitude modulation and phase modulation
AWGN	Additive White Gaussian Noise
BCH	Bose-Chaudhuri-Hocquenghem, A class of forward error correcting codes (FEC)
BCM	block code modulation
BER	Bit error rate, the fraction of the bits received incorrectly
BPF	Bandpass Filter
BPSK	Binary Phase Shift Keying
BS	A common abbreviation for Base Station
C	Channel Capacity
CCITT	Now ITU, standardisation group
CD	Code Division, a multiplexing technique where signals are coded and then combined, in such a way that they can be separated using the assigned user signature codes at a later stage.

CISI	controlled inter-symbol interference
CT	cordless telecommunications
DAB	Digital Audio Broadcasting
DECT	A Pan-European digital cordless telephone standard.
DFT	Discrete Fourier Transform
DTX	discontinuous transmission
EQ	Equaliser
FD	Frequency Division, a multiplexing technique, where different frequencies are used for each communications link.
FDD	Frequency-Division Duplex, a multiplexing technique, where the forward and reverse links use a different carrier frequency.
FDM	Frequency Division Multiplexing
FEC	Forward Error Correction
FFT	Fast Fourier Transform
FIR	Finite Impulse Response
FSK	Frequency Shift Keying
GF	Galois field
GMSK	Gaussian Mean Shift Keying, a modulation scheme used by the Pan-European GSM standard by virtue of its spectral compactness.
GSM	A Pan-European digital mobile radio standard, operating at 900MHz.
ICI	Inter–Channel Interference
IFFT	Inverse Fast Fourier Transform
IMD	Intermodulation Distortion
ISI	Inter Symbol Interference, Inter Subcarrier Interference
LOS	Line–Of–Sight
LPF	low pass filter
LSB	least significant bit
LSF	Least Squares Fitting

LTP	long term predictor
MDI	multi-dimensional interference
MLSE	maximum likelihood sequence estimator
MOS	mean opinion score
MS	A common abbreviation for Mobile Station
MSK	minimum shift keying
NLA	non-linear amplification
NLF	non-linear filtering
OFDM	Orthogonal Frequency Division Multiplexing
OOB	out of band
OQAM	offset quadrature amplitude modulation
OQPSK	offset quadrature phase shift keying
OSWE	one-symbol window equaliser
PAM	pulse amplitude modulation
PCM	pulse code modulation
PCN	Personal Communications Network
PD	phase detector
PDF	Probability Density Function
PLL	phase locked loop
PR	PseudoRandom
PRMA	Packet Reservation Multiple Access, a statistical multiplexing arrangement contrived to improve the efficiency of conventional TDMA systems, by detecting inactive speech segments using a voice activity detector, surrendering them and allocating them to subscribers contending to transmit an active speech packet.
PSAM	Pilot symbol assisted modulation, a technique where known symbols (pilots) are transmitted regularly. The effect of channel fading on all symbols can then be estimated by interpolating between the pilots
PSD	Power Spectral Density

PSK	Phase Shift Keying
QAM	Quadrature Amplitude Modulation
QPSK	Quaternary Phase Shift Keying
RF	radio frequency
RPE	regular pulse excited
RSSI	Received Signal Strength Indicator, commonly used as an indicator of channel quality in a mobile radio network.
SEGSNR	segmental signal-to-noise ratio
SIR	Signal to Interference ratio
SNR	Signal to Noise Ratio, noise energy compared to the signal energy
TC	Trellis Coded
TCM	trellis code modulation
TDD	Time-Division Duplex, a technique where the forward and reverse links are multiplexed in time.
TDMA	Time Division Multiple Access
TTIB	transparent tone in band
TWT	travelling wave tube
UHF	ultra high frequency
VAD	voice activity detection
VCO	voltage controlled oscillator
WN	white noise

Bibliography

[1] S. Stapleton and F. Costescu, "An adaptive predistorter for a power amplifier based on adjacent channel emissions," *IEEE Tr. on VT*, vol. 41, pp. 49–57, February 1992.

[2] S. S. L.D. Quach, "A post-distortion receiver for mobile communications," *IEEE Tr. on VT*, vol. 42, pp. 604–616, November 1993.

[3] S. Haykin, *Adaptive Filter Theory*. Prentice Hall, 1996.

[4] Y. Sato, "A method of self–recovering equalization for multilevel amplitude–modulation systems," *IEEE Transactions on Communications*, vol. COM–23, pp. 679–682, June 1975.

[5] D. Godard, "Self–recovering equalization and carrier tracking in two–dimensional data communication systems," *IEEE Transactions on Communications*, vol. COM–28, pp. 1867–1875, November 1980.

[6] K. Wesolowsky, "Analysis and properties of the modified constant modulus algorithm for blind equalization," *European Transactions on Telecommunication*, vol. 3, pp. 225–230, May–June 1992.

[7] M. Goursat and A. Benveniste, "Blind equalizers," *IEEE Transactions on Communications*, vol. COM–28, pp. 871–883, August 1984.

[8] G. Picchi and G. Prati, "Blind equalization and carrier recovery using a "stop–and–go" decision–directed algorithm," *IEEE Transactions on Communications*, vol. COM–35, pp. 877–887, September 1987.

[9] J. McGeehan and A. Bateman, "Phase-locked transparent tone in band (TTIB): A new spectrum configuration particularly suited to the transmission of data over SSB mobile radio networks," *IEEE Transactions on Communications*, vol. COM-32, no. 1, pp. 81–87, 1984.

[10] J. Cavers, "The performance of phase locked transparent tone in band with symmetric phase detection," *IEEE Trans. on Comms.*, vol. 39, pp. 1389–1399, September 1991.

[11] P. Martin and A. Bateman, "Practical results for a modem using linear mobile radio channels," in *Proceedings of IEEE Vehicular Technology Conference (VTC'91)*, (St. Louis, MO, USA), pp. 386–392, IEEE, 19–22 May 1991.

[12] J. Cavers, "An analysis of pilot symbol assisted modulation for rayleigh fading channels," *IEEE Transactions on Vehicular Technology*, vol. 40, pp. 686–693, Nov 1991.

[13] C. Cahn, "Performance of digital phase modulation communication systems," *IRE Trans Comms*, vol. CS-7, pp. 3–6, May 1959.

[14] C. Cahn, "Combined digital phase and amplitude modulation communication system," *IRE Trans Comms*, vol. CS-8, pp. 150–155, Sept 1960.

[15] J. Hancock and R. Lucky, "Performance of combined amplitude and phase modulated communications system," *IRE Trans Comms*, vol. CS-8, pp. 232–237, December 1960.

[16] C. Campopiano and B. Glazer, "A coherent digital amplitude and phase modulation scheme," *IRE Trans. Commun. Systems*, vol. CS-10, pp. 90– 95, 1962.

[17] R. Lucky and J. Hancock, "On the optimum performance of n-ary systems having two degrees of freedom," *IRE Trans Comms*, vol. CS-10, pp. 185–192, June 1962.

[18] R. Lucky, J. Salz, and E. Weldon, *Principles of Data Communication.* New York: McGraw-Hill, 1968.

[19] J. Salz, J. Sheenhan, and D. Paris, "Data transmission by combined AM and PM," *Bell Systems Technical Journal*, vol. 50, pp. 2399–2419, September 1971.

[20] E. Ho and Y. Yeh, "Error probability of a multilevel digital system with intersymbol interference and gaussian noise," *Bell Systems Technical Journal*, vol. 50, pp. 1017–1023, March 1971.

[21] G. Foschini, R. Gitlin, and S. Weinstein, "Optimization of two-dimensional signal constellations in the presence of gaussian noise," *IEEE Trans Comms*, vol. COM-22, pp. 28–38, January 1974.

[22] C. Thomas, M. Weidner, and S. Durrani, "Digital amplitude-phase keying with m-ary alphabets," *IEEE Trans Comms*, vol. COM-22, pp. 168–180, February 1974.

[23] M. Simon and J. Smith, "Carrier synchronization and detection of QASK signal sets," *IEEE Trans Comms*, vol. COM-22, pp. 98–106, February 1974.

[24] M. Simon and J. Smith, "Offset quadrature communications with decision feedback carrier synchronization," *IEEE Trans Comms*, vol. COM-22, pp. 1576–1584, October 1974.

[25] J. Smith, "Odd-bit quadrature amplitude-shift keying," *IEEE Trans Comms*, vol. COM-23, pp. 385–389, March 1975.

[26] K. Miyauchi, S. Seki, and H. Ishio, "New techniques for generating and detecting multilevel signal formats," *IEEE Trans Comms*, vol. COM-24, pp. 263–267, February 1976.

[27] W. Weber, "Differential encoding for multiple amplitude and phase shift keying systems," *IEEE Trans Comms*, vol. COM-26, pp. 385–391, March 1978.

[28] P. Dupuis, M. Joindot, A. Leclert, and D. Soufflet, "16 QAM modulation for high capacity digital radio system," *IEEE Trans Comms*, vol. COM-27, pp. 1771–1781, December 1979.

[29] I. Horikawa, T. Murase, and Y. Saito, "Design and performance of a 200mbit/s 16 QAM digital radio system," *IEEE Trans Comms*, vol. COM-27, pp. 1953–1958, December 1979.

[30] V. Prabhu, "The detection efficiency of 16-ary QAM," *Bell Systems Technical Journal*, vol. 59, pp. 639–656, April 1980.

[31] D. Morais and K. Feher, "NLA-QAM: A method for generating high power QAM signals through non-linear amplification," *IEEE Trans Comms*, vol. COM-30, pp. 517–522, March 1982.

[32] T. Hill and K. Feher, "A performance study of NLA 64-state QAM," *IEEE Trans Comms*, vol. COM-31, pp. 821–826, June 1983.

[33] K. Feher, "Digital communications - satellite/earth station engineering," in *Digital communications - satellite/earth station engineering* [181], pp. 118–120.

[34] K.-T. Wu and K. Feher, "256-QAM modem performance in distorted channels," *IEEE Trans Comms*, vol. COM-33, pp. 487–491, May 1985.

[35] P. Mathiopoulos and K. Feher, "Performance evaluation of a 512-QAM system in distorted channels," *Proc Pt F*, vol. 133, pp. 199–204, April 1986.

[36] M. Borgne, "Comparison of high level modulation schemes for high capacity digital radio systems," *IEEE Trans Comms*, vol. COM-33, pp. 442–449, May 1985.

[37] M. Shafi and D. Moore, "Further results on adaptive equalizer improvements for 16 QAM and 64 QAM digital radio," *IEEE Trans Comms*, vol. COM-34, pp. pp59–66, January 1986.

[38] Y. Saito and Y. Nakamura, "256 QAM modem for high capacity digital radio system," *IEEE Trans Comms*, vol. COM-34, pp. 799–805, August 1986.

[39] A. Rustako, L. Greenstein, R. Roman, and A. Saleh, "Using times four carrier recovery in M-QAM digital radio receivers," *IEEE J-SAC*, pp. 524–533, April 1987.

[40] C.-E. Sundberg, W. Wong, and R. Steele, "Logarithmic PCM weighted QAM transmission over gaussian and rayleigh fading channels," *IEE Proc Pt. F*, vol. 134, pp. 557–570, October 1987.

[41] R. Steele, C.-E. Sundberg, and W. Wong, "Transmission of log-PCM via QAM over Gaussian and Rayleigh fading channels," *IEE Proc.*, vol. 134, Pt. F, pp. 539–556, October 1987.

[42] L. Hanzo, R. Steele, and P. Fortune, "A subband coding, BCH coding and 16-QAM system for mobile radio speech communication," *IEEE Trans. on Vehicular Technology*, vol. 39, pp. 327–340, November 1990.

[43] H. Sari and S. Moridi, "New phase and frequency detectors for carrier recovery in PSK and QAM systems," *IEEE Trans Comms*, vol. COM-36, pp. 1035–1043, September 1988.

[44] J.-I. Chuang, "The effects of time-delay spread on QAM with non-linearly switched filters in a portable radio communications channel," *IEEE Trans Comms*, vol. 38, pp. 9–13, February 1989.

[45] J. Matthews, "Cochannel performance of 16-level QAM with phase locked TTIB/FFSR processing," *IEE colloquium on multi-level modulation*, March 1990.

[46] P. Huish and G. Richman, "Increasing the capacity and quality of digital microwave radio," *IEE colloquium on multi-level modulation*, March 1990.

[47] W. Webb and R. Steele, "16-level circular QAM transmissions over a rayleigh fading channel," *IEE colloquium on multi-level modulation*, March 1990.

[48] E. Issman and W. Webb, "Carrier recovery for 16-level QAM in mobile radio," *IEE colloquium on multi-level modulation*, March 1990.

[49] W. Peterson and E. Weldon Jr, *Error correcting codes*. MIT. Press, 2nd ed., August 1972. ISBN: 0262160390.

[50] W. Webb and R. Steele, "Equaliser techniques for QAM transmissions over dispersive mobile radio channels," *IEE Proc Pt. I*, vol. 138, pp. 566–576, December 1991.

[51] W. Webb, "QAM, the modulation scheme for future mobile radio communications?," *IEE Electronics & Communications Journal*, vol. 4, pp. 1167–176, August 1992.

[52] W. Webb, "Modulation methods for PCNs," *IEEE Communications magazine*, vol. 30, pp. 90–95, December 1992.

[53] R. Steele and W. Webb, "Variable rate QAM for data transmission over Rayleigh fading channels," in *Proceeedings of Wireless '91*, (Calgary, Alberta), pp. 1–14, IEEE, 1991.

[54] K. Feher, "Modems for emerging digital cellular mobile systems," *IEEE Tr. on VT*, vol. 40, pp. 355–365, May 1991.

[55] M. Iida and K. Sakniwa, "Frequency selective compensation technology of digital 16-QAM for microcellular mobile radio communication systems," in *Proceedings of IEEE VTC '92*, (Denver, Colorado, USA), pp. 662–665, IEEE, 10-13 May 1992.

[56] R. Castle and J. McGeehan, "A multilevel differential modem for narrowband fading channels," in *Proceedings of IEEE VTC '92*, (Denver, Colorado, USA), pp. 104–109, IEEE, 10-13 May 1992.

[57] D. Purle, A. Nix, M. Beach, and J. McGeehan, "A preliminary performance evaluation of a linear frequency hopped modem," in *Proceedings of IEEE VTC '92*, (Denver, Colorado, USA), pp. 120–124, IEEE, 10-13 May 1992.

[58] Y. Kamio and S. Sampei, "Performance of reduced complexity DFE using bidirectional equalizing in land mobile communications," in *Proceedings of IEEE VTC '92*, (Denver, Colorado, USA), pp. 372–376, IEEE, 10-13 May 1992.

[59] S. S. T. Nagayasu and Y. Kamio, "Performance of 16-QAM with decision feedback equalizer using interpolation for land mobile communications," in *Proceedings of IEEE VTC '92*, (Denver, Colorado, USA), pp. 384–387, IEEE, 10-13 May 1992.

[60] E. Malkamaki, "Binary and multilevel offset QAM, spectrum efficient modulation schemes for personal communications," in *Proceedings of IEEE VTC '92*, (Denver, Colorado, USA), pp. 325–328, IEEE, 10-13 May 1992.

[61] Z. Wan and K. Feher, "Improved efficiency CDMA by constant envelope SQAM," in *Proceedings of IEEE VTC '92*, (Denver, Colorado, USA), pp. 51–55, IEEE, 10-13 May 1992.

[62] H. Sasaoka, "Block coded 16-QAM/TDMA cellular radio system using cyclical slow frequency hopping," in *Proceedings of IEEE VTC '92*, (Denver, Colorado, USA), pp. 405–408, IEEE, 10-13 May 1992.

[63] P. Kenington, R. Wilkinson, and J. Marvill, "Broadband linear amplifier design for a PCN base-station," in *Proceedings of IEEE Vehicular Technology Conference (VTC'91)*, (St. Louis, MO, USA), pp. 155–160, IEEE, 19–22 May 1991.

[64] R. Wilkinson et al, "Linear transmitter design for MSAT terminals," in *Proc. of 2nd Int. Mobile Satellite Conference*, June 1990.

[65] S. Stapleton, G. Kandola, and J. Cavers, "Simulation and analysis of an adaptive predistorter utilizing a complex spectral convolution," *IEEE Tr. on VT*, vol. 41, pp. 387–394, November 1992.

[66] A. Wright and W. Durtler, "Experimental performance of an adaptive digital linearized power amplifier," *IEEE Tr. on VT*, vol. 41, pp. 395– 400, November 1992.

[67] M. Faulkner and T. Mattson, "Spectral sensitivity of power amplifiers to quadrature modulator misalignment," *IEEE Tr. on VT*, vol. 41, pp. 516–525, November 1992.

[68] S. Sampei and T. Sunaga, "Rayleigh fading compensation for QAM in land mobile radio communications," *IEEE Transactions on Vehicular Technology*, vol. 42, pp. 137–147, May 1993.

[69] T. Sunaga and S. Sampei, "Performance of multi-level QAM with post-detection maximal ratio combining space diversity for digital land-mobile radio communications," *IEEE Tr. on VT*, vol. 42, pp. 294–301, August 1993.

[70] F. Adachi and M. Sawahashi, "Performance analysis of various 16 level modulation schemes under rayleigh fading," *Electronics Letters*, vol. 28, pp. 1579–1581, November 1992.

[71] R. Chang, "Synthesis of band-limited orthogonal signals for multichannel data transmission," *BSTJ*, vol. 46, pp. 1775–1796, December 1966.

[72] M. Zimmermann and A. Kirsch, "The AN/GSC-10/KATHRYN/ variable rate data modem for HF radio," *IEEE Trans. Commun.Techn.*, vol. CCM–15, pp. 197–205, April 1967.

[73] S. Weinstein and P. Ebert, "Data transmission by frequency division multiplexing using the discrete fourier transform," *IEEE Trans. Commun. Techn.*, vol. COM–19, pp. 628–634, October 1971.

[74] L. Cimini, "Analysis and simulation of a digital mobile channel using orthogonal frequency division multiplexing," *IEEE Transactions on Communications*, vol. 33, pp. 665–675, July 1985.

[75] M. Alard and R. Lassalle, "Principles of modulation and channel coding for digital broadcasting for mobile receivers," *EBU Review, Technical No. 224*, pp. 47–69, August 1987.

[76] *Proc. 1st Int. Symp.,DAB*, (Montreux, Switzerland), June 1992.

[77] A. Peled and A. Ruiz, "Frequency domain data transmission using reduced computational complexity algorithms," in *Proceedings of International Conference on Acoustics, Speech, and Signal Processing, ICASSP'80*, vol. 3, (Denver, Colorado, USA), pp. 964–967, IEEE, 9–11 April 1980.

[78] B. Hirosaki, "An orthogonally multiplexed QAM system using the discrete fourier transform," *IEEE Trans. Commun.*, vol. COM-29, pp. 983–989, July 1981.

[79] H. Kolb, "Untersuchungen über ein digitales mehrfrequenzverfahren zur datenübertragung," in *Ausgewählte Arbeiten über Nachrichtensysteme*, no. 50, Universität Erlangen-Nürnberg, 1982.

[80] H. Schüssler, "Ein digitales Mehrfrequenzverfahren zur Datenübertragung," in *Professoren-Konferenz, Stand und Entwicklungsaussichten der Daten und Telekommunikation*, (Darmstadt, Germany), pp. 179–196, 1983.

[81] K. Preuss, "Ein Parallelverfahren zur schnellen Datenübertragung Im Ortsnetz," in *Ausgewählte Arbeiten über Nachrichtensysteme*, no. 56, Universität Erlangen-Nürnberg, 1984.

[82] R. Rückriem, "Realisierung und messtechnische Untersuchung an einem digitalen Parallelverfahren zur Datenübertragung im Fernsprechkanal," in *Ausgewählte Arbeiten über Nachrichtensysteme*, no. 59, Universität Erlangen-Nürnberg, 1985.

[83] I. Kalet, "The multitone channel," *IEEE Tran. on Comms*, vol. 37, pp. 119–124, February 1989.

[84] B. Hirosaki, "An analysis of automatic equalizers for orthogonally multiplexed qam systems," *IEEE Transactions on Communications*, vol. COM-28, pp. 73–83, January 1980.

[85] L. Hanzo, R. Salami, R. Steele, and P. Fortune, "Transmission of digitally encoded speech at 1.2 kbaud for PCN," *IEE Proceedings, Part I*, vol. 139, pp. 437–447, August 1992.

[86] P. Fortune, L. Hanzo, and R. Steele, "On the computation of 16-QAM and 64-QAM performance in rayleigh-fading channels," *IEICE Tr. on Comms.*, vol. E75-B, pp. 466–475, June 1992.

[87] R. Stedman, H. Gharavi, L. Hanzo, and R. Steele, "Transmission of subband-coded images via mobile channels," *IEEE Tr. on Circuits and Systems for Video Technology*, vol. 3, pp. 15–27, Feb 1993.

[88] X. Lin, L. Hanzo, R. Steele, and W. Webb, "A subband-multipulse digital audio broadcasting scheme for mobile receivers," *IEEE Tr. on Broadcasting*, vol. 39, pp. 373–382, December 1993.

[89] W. Webb, R. Steele, J. Cheung, and L. Hanzo, "A packet reservation multiple access assisted cordless telecommunications scheme," *IEEE Transactions on Veh. Technology*, vol. 43, pp. 234–245, May 1994.

[90] L. Hanzo, W. Webb, R. Salami, and R. Steele, "On QAM speech transmission schemes for microcellular mobile PCNs," *European Transactions on Communications*, pp. 495–510, Sept/Oct 1993.

[91] L. Hanzo, J. Streit, R. Salami, and W. Webb, "A low-rate multi-level voice/video transceiver for personal communications," *Wireless Personal Communications, Kluwer Academic Publishers*, vol. 2, no. 3, pp. 217–234, 1995.

[92] L. Hanzo, R. Stedman, R. Steele, and J. Cheung, "A mobile speech/video/data transceiver scheme," in *Proceedings of IEEE VTC '94*, (Stockholm, Sweden), pp. 452–456, IEEE, June 8-10 1994.

[93] L. Hanzo, X. Lin, R. Steele, and W. Webb, "A mobile hi-fi digital audio broadcasting scheme," in *Proceedings of IEEE VTC '94*, (Stockholm, Sweden), pp. 1035–1039, IEEE, June 8-10 1994.

[94] J. Woodard and L. Hanzo, "A dual-rate algebraic CELP-based speech transceiver," in *Proceedings of IEEE VTC '94*, vol. 3, (Stockholm, Sweden), pp. 1690–1694, IEEE, June 8-10 1994.

[95] J. Streit and L. Hanzo, "A fractal video communicator," in *Proceedings of IEEE VTC '94*, (Stockholm, Sweden), pp. 1030–1034, IEEE, June 8-10 1994.

[96] P. Bello, "Selective fading limitations of the KATHRYN modem and some system design considerations," *IEEE Trabsactions on Communications Technology*, vol. COM–13, pp. 320–333, September 1965.

[97] E. Powers and M. Zimmermann, "A digital implementation of a multichannel data modem," in *Proc. of the IEEE Int. Conf. on Commun.*, (Philadelphia, USA), 1968.

[98] R. Chang and R. Gibby, "A theoretical study of performance of an orthogonal multiplexing data transmission scheme," *IEEE Trans. Commun. Techn.*, vol. COM–16, pp. 529–540, August 1968.

[99] B. Saltzberg, "Performance of an efficient parallel data transmission system," *IEEE Trans. Commun. Techn.*, pp. 805–813, December 1967.

[100] K. Fazel and G. Fettweis, eds., *Multi-carrier spread-spectrum.* Kluwer, 1997. p260, ISBN 0-7923-9973-0.

[101] F. Classen and H. Meyr, "Synchronisation algorithms for an ofdm system for mobile communications," in *Codierung für Quelle, Kanal und Übertragung*, no. 130 in ITG Fachbericht, (Berlin), pp. 105–113, VDE–Verlag, 1994.

[102] F. Classen and H. Meyr, "Frequency synchronisation algorithms for ofdm systems suitable for communication over frequency selective fading channels," in *Proceedings of IEEE VTC '94*, (Stockholm, Sweden), pp. 1655–1659, IEEE, June 8-10 1994.

[103] S. Shepherd, P. van Eetvelt, C. Wyatt-Millington, and S. Barton, "Simple coding scheme to reduce peak factor in QPSK multicarrier modulation," *Electronics Letters*, vol. 31, pp. 1131–1132, July 1995.

[104] A. Jones, T. Wilkinson, and S. Barton, "Block coding scheme for reduction of peak to mean envelope power ratio of multicarrier transmission schemes," *Electronics Letters*, vol. 30, pp. 2098–2099, 1994.

[105] D. Wulich, "Reduction of peak to mean ratio of multicarrier modulation by cyclic coding," *Electronics Letters*, vol. 32, pp. 432–433, 1996.

[106] S. Müller and J. Huber, "Vergleich von OFDM–Verfahren mit reduzierter Spitzenleistung," in *2. OFDM–Fachgespräch in Braunschweig*, 1997.

[107] M. Pauli and H.-P. Kuchenbecker, "Neue Aspekte zur Reduzierung der durch Nichtlinearitäten hervorgerufenen Außerbandstrahlung eines OFDM–Signals," in *2. OFDM–Fachgespräch in Braunschweig*, 1997.

[108] T. May and H. Rohling, "Reduktion von Nachbarkanalstörungen in OFDM–Funkübertragungssystemen," in *2. OFDM–Fachgespräch in Braunschweig*, 1997.

[109] D. Wulich, "Peak factor in orthogonal multicarrier modulation with variable levels," *Electronics Letters*, vol. 32, no. 20, pp. 1859–1861, 1996.

[110] H. Schmidt and K. Kammeyer, "Adaptive Subträgerselektion zur Reduktion des Crest faktors bei OFDM," in *3. OFDM Fachgespräch in Braunschweig*, 1998.

[111] R. Dinis and A. Gusmao, "Performance evaluation of OFDM transmission with conventional and 2-branch combining power amplification schemes," in *Proceeding of IEEE Global Telecommunications Conference, Globecom 96*, (London, UK), pp. 734–739, IEEE, 18–22 Nov 1996.

[112] R. Dinis, P. Montezuma, and A. Gusmao, "Performance trade-offs with quasi-linearly amplified OFDM through a 2-branch combining technique," in *Proceedings of IEEE VTC '96*, (Atlanta, GA, USA), pp. 899–903, IEEE, 1996.

[113] R. Dinis, A. Gusmao, and J. Fernandes, "Adaptive transmission techniques for the mobile broadband system," in *Proceeding of ACTS Mobile Communication Summit '97*, (Aalborg, Denmark), pp. 757–762, ACTS, 7-10 October 1997.

[114] B. Daneshrad, L. Cimini Jr., and M. Carloni, "Clustered-OFDM transmitter implementation," in *Proc. of IEEE International Symposium on Personal, Indoor, and Mobile Radio Communications (PIMRC'96)*, (Taipei, Taiwan), pp. 1064–1068, IEEE, 15-18 October 1996.

[115] M. Okada, H. Nishijima, and S. Komaki, "A maximum likelihood decision based nonlinear distortion compensator for multi-carrier modulated signals," *IEICE Trans. On Communications*, vol. E81B, no. 4, pp. 737–744, 1998.

[116] R. Dinis and A. Gusmao, "Performance evaluation of a multicarrier modulation technique allowing strongly nonlinear amplification," in *Proc. ICC 1998*, pp. 791–796, IEEE, 1998.

[117] T. Pollet, M. van Bladel, and M. Moeneclaey, "BER sensitivity of PFDM systems to carrier frequency offset and wiener phase noise," *IEEE Transactions on Communications*, vol. 43, pp. 191–193, February/March/April 1995.

[118] H. Nikookar and R. Prasad, "On the sensitivity of multicarrier transmission over multipath channels to phase noise and frequency offset," in *Proc. of IEEE International Symposium on Personal, Indoor, and Mobile Radio Communications (PIMRC'96)*, (Taipei, Taiwan), pp. 68–72, IEEE, 15-18 October 1996.

[119] W. Warner and C. Leung, "OFDM/FM frame synchronization for mobile radio data communication," *IEEE Transactions on Vehicular Technology*, vol. 42, pp. 302–313, August 1993.

[120] H. Sari, G. Karam, and I. Jeanclaude, "Transmission techniques for digital terrestrial tv broadcasting," *IEEE Communications Magazine*, pp. 100–109, February 1995.

[121] P. Moose, "A technique for orthogonal frequency division multiplexing frequency offset correction," *IEEE Trans. on Communications*, vol. 42, pp. 2908–2914, October 1994.

[122] K. Brüninghaus and H. Rohling, "Verfahren zur Rahmensynchronisation in einem OFDM-System," in *3. OFDM Fachgespräch in Braunschweig*, 1998.

[123] F. Daffara and O. Adami, "A new frequency detector for orthogonal multicarrier transmission techniques," in *Proceedings of IEEE Vehicular Technology Conference (VTC'95)*, (Chicago, USA), pp. 804–809, IEEE, July 15–28 1995.

[124] M. Sandell, J.-J. van de Beek, and P. Börjesson, "Timing and frequency synchronisation in OFDM systems using the cyclic prefix," in *Proceedings of International Symposium on Synchronisation*, (Essen, Germany), pp. 16–19, 14–15 Dec 1995.

[125] N. Yee, J.-P. Linnartz, and G. Fettweis, "Multicarrier CDMA in indoor wireless radio networks," in *PIMRC'93*, pp. 109–113, 1993.

[126] A. Chouly, A. Brajal, and S. Jourdan, "Orthogonal multicarrier techniques applied to direct sequence spread spectrum CDMA systems," in *Proceedings of the IEEE GLOBECOM '93*, (Houston, USA), pp. 1723–1728, November 1993.

[127] G. Fettweis, A. Bahai, and K. Anvari, "On multi-carrier code division multiple access (MC-CDMA) modem design," in *Proceedings of IEEE VTC '94*, (Stockholm, Sweden), pp. 1670–1674, IEEE, June 8-10 1994.

[128] K. Fazel and L. Papke, "On the performance of convolutionally-coded CDMA/OFDM for mobile communication system," in *PIMRC'93*, pp. 468–472, 1993.

[129] R. Prasad and S. Hara, "Overview of multicarrier CDMA," *IEEE Communications Magazine*, pp. 126–133, December 1997.

[130] B.-J. Choi, E.-L. Kuan, and L. Hanzo, "Crest–factor study of MC–CDMA and OFDM," in *Proceeding of VTC'99 (Fall)*, vol. 1, (Amsterdam, Netherlands), pp. 233–237, IEEE, 19–22 September 1999.

[131] Y. Li and N. Sollenberger, "Interference suppression in OFDM systems using adaptive antenna arrays," in *Proceeding of Globecom'98*, (Sydney, Australia), pp. 213–218, IEEE, 8–12 Nov 1998.

[132] Y. Li and N. Sollenberger, "Adaptive antenna arrays for OFDM systems with cochannel interference," *IEEE Trans. on. Comms.*, vol. 47, pp. 217–229, Feb 1999.

[133] Y. Li, L. Cimini, and N. Sollenberger, "Robust channel estimation for OFDM systems with rapid dispersive fading channels," *IEEE Trans. on. Comms.*, vol. 46, pp. 902–915, Apr 1998.

[134] C. Kim, S. Choi, and Y. Cho, "Adaptive beamforming for an OFDM sytem," in *Proceeding of VTC'99 (Spring)*, (Houston, Texas, USA), IEEE, 16–20 May 1999.

[135] L. Lin, L. Cimini Jr., and J.-I. Chuang, "Turbo codes for OFDM with antenna diversity," in *Proceeding of VTC'99 (Spring)*, (Houston, Texas, USA), IEEE, 16–20 May 1999.

[136] M. Münster, T. Keller, and L. Hanzo, "Co–channel interference suppression assisted adaptive OFDM in interference limited environments," in *Proceeding of VTC'99 (Fall)*, vol. 1, (Amsterdam, Netherlands), pp. 284–288, IEEE, 19–22 September 1999.

[137] F. Mueller-Roemer, "Directions in audio broadcasting," *Jnl Audio Eng. Soc.*, vol. 41, pp. 158–173, March 1993.

[138] G. Plenge, "DAB - a new radio broadcasting system - state of development and ways for its introduction," *Rundfunktech. Mitt.*, vol. 35, no. 2, 1991.

[139] ETSI, *Digital Audio Broadcasting (DAB)*, 2nd ed., May 1997. ETS 300 401.

[140] ETSI, *Digital Video Broadcasting (DVB); Framing structure, channel coding and modulation for digital terrestrial television*, August 1997. EN 300 744 V1.1.2.

[141] P. Chow, J. Tu, and J. Cioffi, "A discrete multitone transceiver system for HDSL applications," *IEEE journal on selected areas in communications*, vol. 9, pp. 895–908, August 1991.

[142] P. Chow, J. Tu, and J. Cioffi, "Performance evaluation of a multichannel transceiver system for ADSL and VHDSL services," *IEEE journal on selected areas in communications*, vol. 9, pp. 909–919, August 1991.

[143] K. Sistanizadeh, P. Chow, and J. Cioffi, "Multi-tone transmission for asymmetric digital subscriber lines (ADSL)," in *Proc. ICC93*, pp. 756–760, IEEE, 1993.

[144] ANSI, *ANSI/T1E1.4/94-007, Asymmetric Digital Subscriber Line (ADSL) Metallic Interface.*, August 1997.

[145] A. Burr and P. Brown, "Application of OFDM to powerline telecommunications," in *3rd International Symposium On Power-Line Communications*, (Lancaster), 30 March – 1 April 1999.

[146] M. Deinzer and M. Stoger, "Integrated PLC-modem based on OFDM," in *3rd International Symposium On Power-Line Communications*, (Lancaster), 30 March – 1 April 1999.

[147] R. Prasad and H. Harada, "A novel OFDM based wireless ATM system for future broadband multimedia communications," in *Proceeding of ACTS Mobile Communication Summit '97*, (Aalborg, Denmark), pp. 757–762, ACTS, 7-10 October 1997.

[148] C. Ciotti and J. Borowski, "The AC006 MEDIAN project — overview and state–of–the–art," in *Proc. ACTS Summit '96*, (Granada, Spain), pp. 362–367, 27–29 November 1996.

[149] J. Borowski, S. Zeisberg, J. Hübner, K. Koora, E. Bogenfeld, and B. Kull, "Performance of OFDM and comparable single carrier system in MEDIAN demonstrator 60GHz channel," in *Proceeding of ACTS Mobile Communication Summit '97*, (Aalborg, Denmark), pp. 653–658, ACTS, 7-10 October 1997.

[150] M. D. Benedetto, P. Mandarini, and L. Piazzo, "Effects of a mismatch in the in–phase and in–quadrature paths, and of phase noise, in QDCPSK-OFDM modems," in *Proceeding of ACTS Mobile Communication Summit '97*, (Aalborg, Denmark), pp. 769–774, ACTS, 7-10 October 1997.

[151] T. Rautio, M. Pietikainen, J. Niemi, J. Rautio, K. Rautiola, and A. Mammela, "Architecture and implementation of the 150 Mbit/s OFDM modem (invited paper)," in *IEEE Benelux Joint Chapter on Communications and Vehicular Technology, 6th Symposium on Vehicular Technology and Communications*, (Helsinki, Finland), p. 11, Oct 12–13 1998.

[152] J. Ala-Laurila and G. Awater, "The magic WAND — wireless ATM network demondtrator system," in *Proceeding of ACTS Mobile Communication Summit '97*, (Aalborg, Denmark), pp. 356–362, ACTS, 7-10 October 1997.

[153] J. Aldis, E. Busking, T. Kleijne, R. Kopmeiners, R. van Nee, R. Mann-Pelz, and T. Mark, "Magic into reality, building the WAND modem," in *Proceeding of ACTS Mobile Communication Summit '97*, (Aalborg, Denmark), pp. 775–780, ACTS, 7-10 October 1997.

[154] E. Hallmann and H. Rohling, "OFDM-Vorschläge für UMTS," in *3. OFDM Fachgespräch in Braunschweig*, 1998.

[155] "Universal mobile telecommunications system (UMTS); UMTS terrestrial radio access (UTRA); concept evaluation," tech. rep., ETSI, 1997. TR 101 146.

[156] W. Jakes, ed., *Microwave Mobile Communications*. John Wiley and Sons, 1974. ISBN 0-471-43720-4.

[157] W. Lee, *Mobile cellular communications*. New York: McGraw Hill, 1989.

[158] R. Steele, ed., *Mobile Radio Communications*. IEEE Press-Pentech Press, 1992.

[159] D. Parsons, *The mobile radio propagation channel.* London: Pentech Press, 1992.

[160] D. Greenwood and L. Hanzo, "Characterisation of mobile radio channels," in Steele [158], ch. 2, pp. 92–185.

[161] J. Cheung, *Adaptive Equalisers for Wideband TDMA Mobile Radio.* PhD thesis, Dept. of Electronics and Computer Science, Univ. of Southampton, UK, 1991.

[162] R. Steele and V. Prabhu, "Mobile radio cellular structures for high user density and large data rates," *Proc. of the IEE*, pp. 396–404, August 1985. Pt F.

[163] R. Steele, "The cellular environment of lightweight hand-held portables," *IEEE Communications Magazine*, pp. 20–29, July 1989.

[164] J. Proakis, *Digital Communications.* McGraw-Hill Book Company, 1987.

[165] K. Bullington, "Radio propagation at frequencies above 30 mc/s," *Proc. IRE 35*, pp. 1122–1136, 1947.

[166] R. Edwards and J. Durkin, "Computer prediction of service area for VHF mobile radio networks," *Proc IRE 116 (9)*, pp. 1493–1500, 1969.

[167] W. Webb, "Sizing up the microcell for mobile radio communications," *IEE Electronics and communications Journal*, vol. 5, pp. 133–140, June 1993.

[168] M. Hata, "Empirical formula for propagation loss in land mobile radio," *IEEE Trans. on Vehicular Technology*, vol. 29, pp. 317–325, August 1980.

[169] Y. Okumura, E. Ohmori, T. Kawano, and K. Fukuda, "Field strength and its variability in VHF and UHF land mobile service," *Review of the Electrical Communication Laboratory*, vol. 16, pp. 825–873, September-October 1968.

[170] E. Green, "Radio link design for microcellular systems," *British Telecom Technology Journal*, vol. 8, pp. 85–96, January 1990.

[171] G. O. A. Rustako, N. Amitay and R. Roman, "Propagation measurements at microwave frequencies for microcellular mobile and personal communications," *Proc. of 39th IEEE VTC*, pp. 316–320, 1989.

[172] J. Kiebler, "The design and planning of feeder links to broadcasting satellites," *IEEE J-SAC*, vol. SAC-3, pp. 181–185, January 1985.

[173] C. Loo, "A statistical model for a land mobile radio satellite link," *IEEE Tr.on VT*, vol. VT-34, pp. 122–127, August 1985.

[174] C. Loo, "Digital transmission through a land mobile satellite channel," *IEEE Tr. on Comms.*, vol. 38, pp. 693–697, May 1990.

[175] E. Lutz, D. Cygan, M. Dippold, F. Dolainsky, and W. Papke, "The land mobile satellite communications channel - recording, statistics and channel model," *IEEE Tr. on VT.*, vol. 40, pp. 375–386, May 1991.

[176] J. Hagenauer, F. Dolainsky, E. Lutz, W. Papke, and R. Schweikert, "The maritime satellite communication channel – channel model, performance of modulation and coding," *IEEE J. on Selected Areas in Commum.*, vol. 5, pp. 701–713, May 1987.

[177] C. Loo, "Measurements and models of a land mobile satellite channel and their application to MSK signals," *IEEE Trans. Vehicular Tech.*, vol. VT-35, pp. 114–121, August 1987.

[178] H. Nyquist, "Certain factors affecting telegraph speed," *Bell System Technical Journal*, p. 617, April 1928.

[179] H. Raemer, *Statistical communication theory and applications.* Englewood Cliffs, New Jersey: Prentice Hall, Inc., 1969.

[180] Y. Chow, A. Nix, and J. McGeehan, "Analysis of 16-APSK modulation in AWGN and rayleigh fading channel," *Electronic Letters*, vol. 28, pp. 1608–1610, November 1992.

[181] K. Feher, ed., *Digital communications - satellite/earth station engineering.* Prentice Hall, 1983.

[182] N. Kingsbury, "Transmit and receive filters for QPSK signals to optimise the performance on linear and hard limited channels," *IEE Proc*, vol. 133, pp. 345–355, July 1986. Pt.F.

[183] B. Sklar, *Digital communications - Fundamentals and Applications.* Prentice Hall, 1988.

[184] M. Schwartz, *Information Transmission, Modulation and Noise.* McGraw-Hill, 1990.

[185] K. Feher, ed., *Advanced Digital Communications: Systems and Signal Processing.* Prentice-Hall, 1987.

[186] A. Saleh and D. Cox, "Improving the power-added efficiency of FET amplifiers operating with varying envelope signals," *IEEE Tr. on Microwave Theory Tech.*, vol. MTT-31, pp. 51–56, 1983.

[187] D. Green, "Characterisation and compensation of nonlinearities in microwave transmitters," *IEEE Tr. on Microwave Theory Tech.*, vol. MTT-30, pp. 213–217, 1982.

[188] F. Casadevall, "The LINC transmitter," *RF Design*, pp. 41–48, February 1990.

[189] Y. Akaiwa and Y. Nagata, "Highly efficient digital mobile communications with a linear modulation method," *IEEE JSAC*, vol. SAC-5, pp. 890–895, June 1987.

[190] D. H. A. Bateman and R. Wilkinson, "Linear transceiver architectures," in *Proc. IEEE Veh. Tech. Conf.*, pp. 478–484, 1988.

[191] A. Wright and W. Duntler, "Experimental performance of an adaptive digital linearised power amplifier," *IEEE VT*, vol. 41, pp. 395–400, November 1992.

[192] S. Stapleton and L. Quach, "Reduction of adjacent channel interference using postdistortion," in *Proceedings of IEEE VTC '92*, (Denver, Colorado, USA), pp. 915–918, IEEE, 10-13 May 1992.

[193] J. Namiki, "An automatically controlled predistorter for multilevel quadrature amplitude modulation," *IEEE Tr. Comm.*, vol. COM-31, pp. 707–712, May 1983.

[194] T. Nojima and T. Konno, "Cuber predistortion linearizer for relay equipment in the 800 MHz band land mobile telephone system," *IEEE Tr. on Veh. Tech.*, vol. VT-34, pp. 169–177, November 1985.

[195] P. M. M. Nannicini and F. Oggioni, "Temperature controlled predistortion circuits for 64 QAM microwave power amplifiers," *IEEE Microwave Theory Tech. Dig.*, pp. 99–102, 1985.

[196] Y. Nagata, "Linear amplification technique for digital mobile communications," in *Proceedings of IEEE Vehicular Technology Conference (VTC'89)*, (San Francisco, CA, USA), pp. 159–164, IEEE, 1–3 May 1989.

[197] A. Saleh and J. Salz, "Adaptive linearization of power amplifiers in digital radio systems," *Bell Systems Tech. Journal*, vol. 62, pp. 1019–1033, April 1983.

[198] B. Bunday, *Basic Optimisation Methods.* London: Edward Arnold, 1984.

[199] S. Stapleton and F. Costesu, "An adaptive pre-distortion system," in *Proceedings of IEEE VTC '92*, (Denver, Colorado, USA), pp. 690–693, IEEE, 10-13 May 1992.

[200] L. Franks, "Carrier and bit synchronization — a tutorial review," *IEEE Transactions on Communications*, vol. COM-28, pp. 1107–1121, August 1980.

[201] R. Ziemer and R. Peterson, *Digital Communications and Spread Spectrum System.* New York: Macmillan Publishing Company, 1985.

[202] L. Franks, "Synchronisation subsystems: Analysis and design," in Feher [181], ch. 7.

[203] A. Carlson, *Communication Systems.* McGraw-Hill, 1975.

[204] I. Wassell, *Digital mobile radio communication.* PhD thesis, University of Southampton, 1991.

[205] R. Cupo. and R. Gitlin, "Adaptive carrier recovery systems for digital data communications receivers," *IEEE J-SAC*, vol. 7, pp. 1328–1339, December 1989.

[206] W. Lindsey and M. Simon, "Carrier synchronisation and detection of polyphase signals," *IEEE Trans. Comms.*, pp. 441–454, June 1972.

[207] J. Smith, *Modern communications circuits.* McGraw Hill, 1986.

[208] M. Woodbury, "Inverting modified matrices," tech. rep., Statistical Research Group, Princeton University, Princeton, N.J., U.S.A. Mem.Rep. 42.

[209] B. Picinbono, "Adaptive signal processing for detection and communication," in *Communication Systems and Random Process Theory* (J. Skwirzinsky, ed.), Alphen aan den Rijn, The Netherlands: Sijthof and Noordhoff, 1978.

[210] K. Murota and K. Hirade, "GMSK modulation for digital mobile radio telephony," *IEEE Trans. Commun.*, vol. 29, pp. 1044–1050, July 1981.

[211] L. Lopes, "GSM radio link simulation," tech. rep., University research in Mobile Radio, 1990. IEE Colloquium.

[212] J. Anderson, T. Aulin, and C. Sundberg, *Digital phase modulation.* Plenum Press, 1986.

[213] ETSI, *Digital Video Broadcasting (DVB); Framing structure, channel coding and modulation for 11/12 GHz Satellite Services*, August 1997. ETS 300 421.

[214] ETSI, *Digital Video Broadcasting (DVB); Framing structure, channel coding and modulation for digital terrestrial television*, August 1997. ETS 300 744.

[215] ETSI, *Digital Video Broadcasting (DVB); Framing structure, channel coding and modulation for cable systems*, December 1997. ETS 300 429.

[216] A. Benveniste, M. Goursat, and G. Ruget, "Robust identification of a nonminimum phase system: Blind adjustment of a linear equalizer in data communications," *IEEE Transactions on Automatic Control*, vol. AC-25, pp. 385–399, June 1980.

[217] G. Foschini, "Equalizing without altering or deleting data," *AT&T Technical Journal*, vol. 64, pp. 1885–1911, October 1985.

[218] Z. Ding, R. Kennedy, B. Anderson, and R. Johnson, "Ill-convergence of Godard blind equalizers in data communications systems," *IEEE Transactions on Communications*, vol. COM-39, pp. 1313–1327, September 1991.

[219] S. Bellini, "Bussgang techniques for blind equalisation," in *Proceedings of the IEEE Global Telecommunications Conference*, (Houston, USA), pp. 1634–1640, December 1986.

[220] J. Bussgang, "Cross–correlation functions of amplitude-distorted Gaussian signals," *MIT Research Laboratory Technical Report*, no. 216, 1952.

[221] N. Seshadri, "Joint data and channel estimation using blind Trellis search techniques," *IEEE Transactions on Communications*, vol. COM-42, pp. 1000–1011, February-April 1994.

[222] D. Forney, "Maximum-likelihood sequence estimation of digital sequences in the presence of intersymbol interference," *IEEE Transactions on Information Theory*, vol. IT–18, pp. 363–378, May 1972.

[223] A. Polydoros, R. Raheli, and C. Tzou, "Per–survivor processing: a general approach to MLSE in uncertain environments," *IEEE Transactions on Communications*, vol. COM–43, pp. 354–364, February–April 1995.

[224] A. Polydoros and K. Chugg, "MLSE for an unknown channel - Part I: Optimality considerations," *IEEE Transactions on Communications*, vol. COM-44, pp. 836–846, July 1996.

[225] K. Chugg and A. Polydoros, "MLSE for an unknown channel - Part II: Tracking performance," *IEEE Transactions on Communications*, vol. COM-44, pp. 949–958, August 1996.

[226] C. Antón-Haro, J. Fonolossa, and J. Fonolossa, "Blind channel estimation and data detection using hidden Markov models," *IEEE Transactions on Signal Processing*, vol. 45, pp. 241–247, January 1997.

[227] H. Cirpan and M. Tsatsanis, "Blind receivers for nonlinearly modulated signals in multipath," *IEEE Transactions on Signal Processing*, vol. 47, pp. 583–586, February 1999.

[228] L. Favalli, A. Mecocci, and P. Savazzi, "Blind MLSE equalizer with fuzzy metric calculation for mobile radio environments," *Electronics Letters*, vol. 33, pp. 1841–1842, October 1997.

[229] K. Chugg, "Acquisition performance of blind sequence detectors using per-survivor processing," in *Proceedings of the 1997 47th IEEE Vehicular Technology Conference*, pp. 539–543, May 1997.

[230] K. Chugg, "Blind acquisition characteristics of PSP-based sequence detectors," *International Journal on Selected Areas in Communications*, vol. 16, pp. 1518–1529, October 1998.

[231] E. Baccarelli and R. Cusani, "Combined channel estimation and data detection using soft statistics for frequency selective fast-fading digital links," *IEEE Transactions on Communications*, vol. COM-46, pp. 424–427, April 1998.

[232] S. Chen and Y. Wu, "Maximum likelihood joint channel and data estimation using genetic algorithms," *IEEE Transactions on Signal Processing*, vol. 46, pp. 1469–1473, May 1998.

[233] L. Tong, G. Xu, and T. Kailath, "A new approach to blind identification and equalization of multipath channels," in *Proceedings of the 25th Asilomar Conference*, (Pacific Grive, Canada), pp. 856–860, 4–6 November 1991.

[234] E. Mulines, J. Cardoso, and S. Mayrargue, "Subspace methods for the blind identification of multichannel fir filters," *IEEE Transactions on Signal Processing*, vol. 43, pp. 516–525, February 1995.

[235] M. Tsatsanis and G. Giannakis, "Transmitter induced cyclostationarity for blind channel equalization," *IEEE Transactions on Signal Processing*, vol. 45, pp. 1785–1794, July 1997.

[236] A. Chevreuil, F. Desbouvries, A. Gorokhov, P. Loubaton, and C. Vignat, "Blind equalization in the presence of jammers and unknown noise: Solutions based on second-order cyclostationary statistics," *IEEE Transactions on Signal Processing*, vol. 46, pp. 259–263, January 1998.

[237] A. Chevreuil and P. Loubaton, "Blind second–order identification of FIR channels: Forced cyclostationarity and structured subspace method," *IEEE Signal Processing Letters*, vol. 4, pp. 204–206, July 1997.

[238] M. Tsatsanis and G. Giannakis, "Subspace methods for blind estimation of time-varying FIR channels," *IEEE Transactions on Signal Processing*, vol. 45, pp. 3084–3093, December 1997.

[239] Z. Ding, "Matrix outer-product decomposition method for blind multiple channel identification," *IEEE Transactions on Signal Processing*, vol. 45, pp. 3053–3061, December 1997.

[240] G. Giannakis and E. Serpedin, "Blind identification of ARMA channels with periodically modulated inputs," *IEEE Transactions on Signal Processing*, vol. 46, pp. 3099–3104, November 1998.

[241] G. Giannakis, "Filterbanks for blind channel identification and equalization," *IEEE Signal Processing Letters*, vol. 4, pp. 184–187, June 1997.

[242] R. Heath Jr. and G. Giannakis, "Exploiting input cyclostationarity for blind channel identification in OFDM systems," *IEEE Transactions on Signal Processing*, vol. 47, pp. 848–856, March 1999.

[243] H. Wong and J. Chambers, "Two-stage interference immune blind equaliser which exploits cyclostationary statistics," *Electronics Letters*, vol. 32, pp. 1763–1764, September 1996.

[244] H. Liu, G. Xu, L. Tong, and T. Kailath, "Recent developments in blind channel equalization: From cyclostationarity to subspace," *Signal Processing*, vol. 50, pp. 83–99, April 1996.

[245] Y. Hua, H. Yang, and W. Qiu, "Source correlation compensation for blind channel identification based on second order statistics," *IEEE Signal Processing Letters*, vol. 1, pp. 119–120, August 1994.

[246] Z. Ding, "Characteristics of band-limited channels unidentifiable from second-order cyclostationary statistics," *IEEE Signal Processing Letters*, vol. 3, pp. 150–152, May 1996.

[247] J. Xavier, V. Barroso, and J. Moura, "Closed-form blind channel identification and sourse separation in SDMA systems through correlative coding," *International Journal on Selected Areas in Communications*, vol. 16, pp. 1506–1517, October 1998.

[248] X. Wang and H. Poor, "Blind equalization and multiuser detection in dispersive CDMA channels," *IEEE Transactions on Communications*, vol. 46, pp. 91–103, January 1998.

[249] X. Wang and H. Poor, "Blind joint equalization and multiuser detection for DS-CDMA in uknowon correlated noise," *IEEE Transactions on Circuits and Systems II: Analog and Digital Signal Processing*, vol. 46, pp. 886–895, July 1999.

[250] J. Zhu, Z. Ding, and X.-R. Cao, "Column–anchored zeroforcing blind equalization for multiuser wireless FIR channels," *International Journal on Selected Areas in Communications*, vol. 17, pp. 411–423, March 1999.

[251] H. Zeng and L. Tong, "Blind channel–estimation using the second–order statistics algorithms," *IEEE Transactions on Signal Processing*, vol. 45, pp. 1919–1930, August 1997.

[252] D. Hatzinakos and C. Nikias, "Blind equalization using a tricepstrum based algorithm," *IEEE Transactions on Communications*, vol. COM-39, pp. 669–682, May 1991.

[253] D. Boss, K. Kammeyer, and T. Petermann, "Is blind channel estimation feasible in mobile communication systems ?; a study based on GSM," *International Journal on Selected Areas in Communications*, vol. 16, pp. 1479–1492, October 1998.

[254] T. Endres, S. Halford, C. Johnson, and G. Giannakis, "Blind adaptive channel equalization using fractionally-spaced receivers: A comparison study," in *Proceedings of the Conference on Information Sciences and Systems*, (Princeton, USA), 20–22 March 1996.

[255] C. Johnson Jr. and B. Anderson, "Godard blind equalizer error surface characteristics: White, zero–mean binary source case," *International Journal of Adaptive Control and Signal Processing*, vol. 9, pp. 301–324, July–August 1995.

[256] L. Tong and S. Perreau, "Analysis of a nonparametric blind equalizer for discrete-valued signals," *Proceedings of the IEEE*, vol. 86, pp. 1951–1968, Mar 1996.

[257] J. Proakis, *Digital Communications.* McGraw Hill, 3rd ed., 1995.

[258] A. Nandi, *Blind Estimation using Higher-Order Statistics.* Glasgow: Kluwer Academic Publishers, 1999.

[259] C. Becchetti, A. Cocco, and G. Jacovitti, "Performance comparison of second order based blind equalizers in data communication channels," in *Proceedings of the 1997 13th International Conference on Digital Signal Processing, DSP. Part 1 (of 2)*, vol. 1, (Santorini, Greece), pp. 147–150, 2–4 July 1997.

[260] M. Kristensson and B. Ottersten, "Asymptotic comparison of two blind channel identification algorithms," in *Proceedings of the 1997 1st IEEE Signal Processing Workshop on Signal Processing Advances in Wireless Communications, SPAWC'97*, pp. 361–364, 16–18 April 1997.

[261] J. Altuna and B. Mulgrew, "Comparison of cyclostationary blind equalization algorithms in the mobile radio environment," *International Journal of Adaptive Control and Signal Processing*, vol. 12, pp. 267–282, May 1998.

[262] K. Skowratananont and J. Chambers, "Comparison of blind channel estimation and equalisation techniques for a fading environment," in *Proceedings of the 1998 6th IEE Conference on Telecommunications*, no. 451, (Edinburgh, UK), pp. 27–31, 21 March – 2 April 1998.

[263] J. Shynk, P. Gooch, G. Krishnamurthy, and C. Chan, "Comparative performance study of several blind equalization algorithms," in *Proceedings of SPIE - The International Society for Optical Engineering*, vol. 1565, (San Diego, USA), pp. 102–117, 22–24 July 1991.

[264] T. Schirtzinger, X. Li, and W. Jenkins, "Comparison of three algorithms for blind equalization based on the constant modulus error criterion," in *Proceedings of the 1995 International Conference on Acoustics, Speech, and Signal Processing. Part 2 (of 5)*, (Detroit, USA), pp. 1049–1052, 9–12 May 1995.

[265] T. Endres, S. Halford, C. Johnson Jr., and G. Giannakis, "Simulated comparisons of blind equalization algorithms for cold start-up applications," *International Journal of Adaptive Control and Signal Processing*, vol. 12, pp. 283–301, May 1998.

[266] L. Hanzo, W. Webb, and T. Keller, *Single- and Multi-carrier Quadrature Amplitude Modulation.* John Wiley-IEEE Press, April 2000.

[267] L. Hanzo and W. Webb, *Modern Quadrature Amplitude Modulation - Principles and Applications for Fixed and Wireless Channels.* IEEE Press - John Wiley, 1994.

[268] S. Haykin, *Blind Deconvolution.* Prentice Hall, 1 ed., 1994.

[269] D. Lainiotis, S. Katsikas, and S. Likothanassis, "Optimal seismic deconvolution," *Signal Processing*, vol. 15, pp. 375–404, December 1988.

[270] D. Huang and F. Gustafsson, "Sufficient output conditions for identifiability in blind equalization," *IEEE Transactions on Communications*, vol. 47, pp. 191–194, February 1999.

[271] L. Tong, G. Xu, and T. Kailath, "Blind identification and equalization based on second-order statistics: a time domain approach," *IEEE Transactions on Information Theory*, vol. COM-40, pp. 380–389, December 1994.

[272] S.-C. Pei and M.-F. Shih, "Fractionally spaced blind equalization using polyperiodic linear filtering," *IEEE Transactions on Communications*, vol. 46, pp. 16–19, January 1998.

[273] K. Dogancay and R. Kennedy, "Least squares approach to blind channel equalization," *Signal Processing*, vol. 58, pp. 63–78, April 1997.

[274] T. Endres, C. Johnson, and M. Green, "Robustness to fractionally-spaced equalizer length using the constant modulus criterion," *IEEE Transactions on Signal Processing*, vol. 47, pp. 544–548, February 1999.

[275] J. LeBlanc, I. Fijalkow, and C. Johnson Jr., "CMA fractionally spaced equalizers: Stationary points and stability under iid and temporally correlated sources," *International Journal of Adaptive Control and Signal Processing*, vol. 12, pp. 135–155, March 1998.

[276] M. Magarini, A. Spalvieri, and G. Tartara, "Asymptotic analysis of stabilisation technique for the blind fractionally spaced equaliser," *Electronics Letters*, vol. 32, pp. 1947–1948, October 1996.

[277] C. Papadias and D. Slock, "Fractionally spaced equalization of linear polyphase channels and related blind techniques based on multichannel linear prediction," *IEEE Transactions on Signal Processing*, vol. 47, pp. 641–654, March 1999.

[278] V. Yang and D. Jones, "A vector constant modulus algorithm for shaped constellation equalization," *IEEE Signal Processing Letters*, vol. 5, pp. 89–91, April 1998.

[279] O. Shalvi and E. Weinstein, "New criteria for blind deconvolution of nonminimum phase systems (channels)," *IEEE Transactions on Information Theory*, vol. IT-36, pp. 312–321, March 1990.

[280] O. Shalvi and E. Weinstein, "Super-exponential methods for blind deconvolution," *IEEE Transactions on Information Theory*, vol. 39, pp. 504–519, March 1993.

[281] H. Chiang and C. Nikias, "Adaptive deconvolution and identification of nonminimum phase FIR systems based on cumulants," *IEEE Transactions on Automatic Control*, vol. 35, pp. 36–47, January 1990.

[282] D. Boss and K.-D. Kammeyer, "Blind GSM channel estimation," in *Proceedings of the 1997 47th IEEE Vehicular Technology Conference*, (Phoenix, USA), pp. 1044–1048, 4–7 May 1997.

[283] J. Choi, I. Song, and R. Park, "Some convergence properties of Godard's quartic algorithm," *Signal Processing*, vol. 56, pp. 313–320, February 1997.

[284] Z. Ding, R. Johnson, and R. Kennedy, "On the (non)existence of undesirable equilibria of Godard blind equalizers," *IEEE Transactions on Signal Processing*, vol. 40, pp. 2425–2432, October 1992.

[285] Y. Li, K. Liu, and Z. Ding, "Length -and cost- dependent local minima of unconstrained blind channel equalizers," *IEEE Transactions on Signal Processing*, vol. 44, pp. 2726–2735, November 1996.

[286] Z. Ding, R. Kennedy, B. Anderson, and R. Johnson Jr., "Local convergence of the Sato blind equalizer and generalizations under practical constraints," *IEEE Transactions on Information Theory*, vol. 39, pp. 129–144, January 1993.

[287] Z. Ding and R. Kennedy, "On the whereabouts of local minima for blind adaptive equalizers," *IEEE Transactions on Circuits and Systems II: Analog and Digital Signal Processing*, vol. 39, pp. 119–123, February 1992.

[288] Z. Ding and R. Johnson Jr., "On the nonvanishing stability of undesirable equilibria for FTR Godard blind equalizers," *IEEE Transactions on Signal Processing*, vol. 41, pp. 1940–1944, May 1993.

[289] Y. Li and Z. Ding, "Convergence analysis of finite length blind adaptive equalizers," *IEEE Transactions on Signal Processing*, vol. 43, pp. 2120–2129, September 1995.

[290] H. Zeng, L. Tong, and C. Johnson, "Relationships between the constant modulus and Wiener receivers," *IEEE Transactions on Information Theory*, vol. 44, pp. 1523–1539, July 1998.

[291] P. Regalia and M. Mboup, "Undermodeled equalization: A characterization of stationary points for a family of blind criteria," *IEEE Transactions on Signal Processing*, vol. 47, pp. 760–770, March 1999.

[292] M. Gu and L. Tong, "Geometrical characterizations of constant modulus receivers," *IEEE Transactions on Signal Processing*, vol. 47, pp. 2745–2756, October 1999.

[293] Y. Li and K. Liu, "Static and dynamic convergence behaviour of adaptive blind equalizers," *IEEE Transactions on Signal Processing*, vol. 44, pp. 2736–2745, November 1996.

[294] V. Weerackody, S. Kassam, and K. Laker, "Convergence analysis of an algorithm for blind equalization," *IEEE Transactions on Communications*, vol. 39, pp. 856–865, June 1991.

[295] W. Lee and K. Cheun, "Convergence analysis of the stop-and-go blind equalization algorithm," *IEEE Transactions on Communications*, vol. 47, pp. 177–180, February 1999.

[296] Y. Li and Z. Ding, "Global convergence of fractionally spaced Godard (CMA) adaptive equalizers," *IEEE Transactions on Signal Processing*, vol. 44, pp. 818–826, April 1996.

[297] Z. Ding, "On convergence analysis of fractionally spaced adaptive blind equalizers," *IEEE Transactions on Signal Processing*, vol. 45, pp. 650–657, March 1997.

[298] J. J. Shynk and C. K. Chan, "Performance surfaces of the constant modulus algorithm based on a conditional gaussian model," *IEEE Transactions on Signal Processing*, vol. 41, pp. 1965–1969, May 1993.

[299] S. Douglas, A. Cichocki, and S. Amari, "Fast-convergence filtered regressor algorithms for blind equalisation," *Electronics Letters*, vol. 32, pp. 2114–2115, November 1996.

[300] C. Papadias and D. Slock, "Normalized sliding window constant modulus (CM) and decision-directed algorithms: a link between blind equalization and classical adaptive filtering," *IEEE Transactions on Signal Processing*, vol. 45, pp. 231–235, January 1997.

[301] J. Anderson and S. Mohan, "Sequential coding algorithms: A survey and cost analysis," *IEEE Transactions on Communications*, vol. COM-32, pp. 1689–1696, February 1984.

[302] Z. Xie, C. Rushforth, R. Short, and T. Moon, "Joint signal detection and parameter estimation in multiuser communications," *IEEE Transactions on Communications*, vol. 41, pp. 1208–1216, August 1993.

[303] A. Papoulis, *Probability, Random Variables, and Stochastic Processes*. McGraw-Hill International Editions, 2. ed., 1984.

[304] S. Haykin, *Communications Systems*. John Willey & Sons, 2. ed., 1994.

[305] B. Noble and J. Daniel, *Applied Linear Algebra*. Englewood Cliffs, New Jersey, USA: Prentice–Hall International Editions, 3. ed., 1986.

[306] D. Hatzinakos, "Blind equalization based on prediction and polycepstra principles," *IEEE Transactions on Communications*, vol. 43, pp. 178–181, February–April 1995.

[307] D. Hatzinakos, "Blind equalization using decision feedback prediction and tricepstrum principles," *Signal Processing*, vol. 36, pp. 261–276, April 1994.

[308] A. Bessios and C. Nikias, "POTEA: the power cepstrum and tricoherence equalization algorithm," *IEEE Transactions on Communications*, vol. 43, pp. 2667–2671, November 1995.

[309] A. Petropulu and C. Nikias, "Blind deconvolution of coloured signals based on higher–order cepstra and data fusion," *IEE Proceedings, Part F: Radar and Signal Processing*, vol. 140, pp. 356–361, December 1993.

[310] G. Kechriotis, E. Zervas, and E. Manolakos, "Using recurrent neural networks for adaptive communication channel equalization," *IEEE Transactions on Neural Networks*, vol. 5, pp. 267–278, March 1994.

[311] S. Amari and A. Cichocki, "Adaptive blind signal processing - neural network approaches," *Proceedings of the IEEE*, vol. 86, pp. 2026–2048, October 1998.

[312] C. You and D. Hong, "Nonlinear blind equalization scheme using complex-valued multilayer feedforward neural networks," *IEEE Transactions on Neural Networks*, vol. 9, pp. 1442–1455, November 1998.

[313] Y. Fang and T. Chow, "Blind equalization of a noisy channel by linear neural network," *IEEE Transactions on Neural Networks*, vol. 10, no. 4, pp. 918–924, 1999.

[314] S. Choi. and A. Cichocki, "Cascade neural networks for multichannel blind deconvolution," *Electronics Letters*, vol. 34, pp. 1186–1187, June 1998.

[315] S. Mo and B. Shafai, "Blind equalization using higher order cumulants and neural network," *IEEE Transactions on Signal Processing*, vol. 42, pp. 3209–3217, November 1994.

[316] L. H. C.S. Lee, S. Vlahoyiannatos, "Satellite based turbo-coded, blind-equalised 4-qam and 16-qam digital video broadcasting," *IEEE Transactions on Broadcasting*, vol. , p. , March 2000.

[317] J. Massey, "Coding and modulation in digital communications," in *Proc of International Zurich Seminar on Digital Communications 1994*, (Zurich, Switzerland), Mar 1974.

[318] H. Imai and S. Hirakawa, "A new multi-level coding method using error correcting codes," *IEEE Transactions on Information Theory*, vol. 23, pp. 371–377, May 1977.

[319] G. Ungerboeck, "Channel coding with multilevel/phase signals," *IEEE Transactions on Information Theory*, vol. IT-28, pp. 55–67, Jan 1982.

[320] G. Ungerboeck, "Treliis-coded modulation with redundant signal sets part 1: Introduction," *IEEE Communications Magazine*, vol. 25, pp. 5–11, Feb 1987.

[321] E. Biglieri and M. Luise, "Coded modulation and bandwidth-efficient transmission," in *Proc. of the Fifth Tirrenia Intern. Workshop*, (Elsevier, Netherlands), 8–12 Sept 1991.

[322] "Special issue on coded modulation," *IEEE Comms. Magazine*, vol. 29, December 1991.

[323] E. Biglieri, D. Divsalar, P. McLane, and M. Simon, *Introduction to trellis coded modulation with applications.* New York: MacMillan Publishing Co., 1991.

[324] G. Forney Jr, R. Gallager, G. Lang, F. Longstaff, and S. Qureshi, "Efficient modulation for band-limited channels," *IEEE Journal on Selected Areas in Communications*, vol. 2, pp. 632–647, Sept 1984.

[325] C. Shannon, *Mathematical Theory of Communication.* University of Illinois Press, 1963.

[326] J. Wozencraft and R. Kennedy, "Modulation and demodulation for probabilistic coding," *IEEE Trans. Inf. Theory*, vol. IT-12, pp. 291–297, 1966.

[327] J. Wozencraft and I. Jacobs, *Principles of communications engineering.* J.Wiley, 1965.

[328] R. Blahut, *Theory and practice of error control codes.* Addison-Wesley, 1983. ISBN 0-201-10102-5.

[329] E. Berlekamp, *Algebraic Coding Theory.* McGraw-Hill, New York, 1968.

[330] W. Peterson, *Error correcting codes.* Cambridge, Mass, USA: MIT. Press, 1st ed., 1961.

[331] A. Michelson and A. Levesque, *Error control techniques for digital communication.* J. Wiley and Sons, 1985.

[332] K. Wong and L. Hanzo, "Channel coding," in Steele [158], ch. 4, pp. 347–488.

[333] M. Oeder, "Rotationally invariant trellis codes for mPSK modulation," in *Proc. IEEE ICC'85*, (Chicago, USA), pp. 552–556, June 1985.

[334] L.-F. Wei, "Trellis-coded modulation with multidimensional constellations," *IEEE Trans Inf.Theory*, vol. IT-33, pp. 483–501, July 1987.

[335] *International Consultative Committee for Telephone and Telegraph Recommendations.* Geneva. V.29 - V.33.

[336] L. Wei, "Rotationally-invariant convolutional channel coding with expanded signal space, part i and ii," *IEEE Tr. on Selected Areas in Comms*, vol. SAC-2, pp. 659–686, September 1984.

[337] K. Shanmugam, *Digital and Analog Communications Systems.* New York, USA: John Wiley, 1979.

[338] W. Lee, *Mobile communications engineering.* McGraw Hill Book Co., 1982.

[339] I. Gradshteyn and I. Ryzhik, *Table of integrals, series and products.* Academic Press, 1980.

[340] G. Saulnier and W. Raffety, "Pilot-aided modulation for narrowband satellite communications," in *Proc. Mobile Satellite Conf*, pp. 329–336, 1988.

[341] A. Bateman and J. McGeehan, "Feedforward transparent tone in band for rapid fading protection in multipath fading," in *IEE Int.Conf. Comms.*, vol. 68, pp. 9–13, 1986.

[342] A. Bateman and J. McGeehan, "The use of transparent tone in band for coherent data schemes," in *IEEE Int. Conf. Comms*, (Boston, Mass, USA), 1983.

[343] A. Bateman, G. Lightfoot, A. Lymer, and J. McGeehan, "Speech and data transmissions over a 942MHz TAB and TTIB single sideband mobile radio system," *IEEE Transactions on Vehicular Technology*, vol. VT-34, pp. 13–21, Feb 1985.

[344] A. Bateman and J. McGeehan, "Data transmissions over UHF fading mobile radio channels," *Proc IEE Pt.F*, vol. 131, pp. 364–374, 1984.

[345] J. McGeehan and A. Bateman, "A simple simultaneous carrier and bit synchronisation system for narrowband data transmissions," *Proc. IEE, Pt.F*, vol. 132, pp. 69–72, 1985.

[346] J. McGeehan and A. Bateman, "Theoretical and experimental investigation of feedforward signal regeneration," *IEEE Trans. Veh. Tech*, vol. VT-32, pp. 106–120, 1983.

[347] A. Bateman, "Feedforward transparent tone in band: Its implementation and applications," *IEEE Trans. Veh. Tech*, vol. 39, pp. 235–243, August 1990.

[348] M. Simon, "Dual pilot tone calibration technique," *IEEE Trans. Veh. Tech.*, vol. VT-35, pp. 63–70, May 1986.

[349] M. Fitz, "A dual-tone reference digital demodulator for mobile communications," *IEEE Tran. Veh. Tech.*, vol. VT-42, pp. 156–166, May 1993.

[350] S. Gamnathan and K. Feher, "Pilot tone aided QPRS systems for digital audio broadcasting," *IEEE Trans. on Broadcasting*, vol. 38, pp. 1–6, March 1992.

[351] F. Davarrin, "Mobile digital communications via tone calibration," *IEEE Trans. Veh. Tech.*, vol. VT-36, pp. 55–62, May 1987.

[352] J. Cavers, "Performance of tone calibration with frequency offset and imperfect pilot filter," *IEEE Tran. Veh. Tech.*, vol. 40, pp. 426–434, May 1991.

[353] D. Esteban and C. Galand, "Application of quadrature mirror filters to split band voice coding scheme," in *Proceedings of International Conference on Acoustics, Speech, and Signal Processing, ICASSP'77*, (Hartford, Conn, USA), pp. 191–195, IEEE, 9–11 May 1977.

[354] J. Johnston, "A filter family designed for use in quadrature mirror filter banks," in *Proceedings of International Conference on Acoustics, Speech, and Signal Processing, ICASSP'80*, (Denver, Colorado, USA), pp. 291–294, IEEE, 9–11 April 1980.

[355] J. Lodge and M. Moher, "Time diversity for mobile satellite channels using trellis coded modulations," in *IEEE Global Telecommun. Conf.*, (Tokyo, Japan), 1987.

[356] M. Moher and J. Lodge, "TCMP – a modulation and coding strategy for rician fading channels," *IEEE Journal on Selected Areas in Communications*, vol. 7, pp. 1347–1355, December 1989.

[357] S. Sampei and T. Sunaga, "Rayleigh fading compensation method for 16-QAM in digital land mobile radio channels," in *Proceedings of IEEE Vehicular Technology Conference (VTC'89)*, (San Francisco, CA, USA), pp. 640–646, IEEE, 1–3 May 1989.

[358] J. Cavers, "Pilot symbol assisted modulation in fading and delay spread," in *Proceedings of IEEE VTC '93*, (Secaucus, NJ, USA), pp. 13–16, IEEE, May 18-20 1993.

[359] M. F. J.P. Seymour, "Improved carrier synchronisation techniques for mobile communications," in *Proceedings of IEEE VTC '93*, (Secaucus, NJ, USA), pp. 901–904, IEEE, May 18-20 1993.

[360] AT&T Information Services, *A trellis coded modulation scheme that includes differential encoding for 9600 bit/sec full-duplex,two-wire modems*, August 1983. CCITT SG XVII.

[361] R. Salami, L. Hanzo, R. Steele, K. Wong, and I. Wassell, "Speech coding," in Steele [158], ch. 3, pp. 186–346.

[362] K. Larsen, "Short convolutional codes with maximal free distance for rate 1/2, 1/3 and 1/4," *IEEE Trans. Info. Theory*, vol. IT-19, pp. 371–372, May 1973.

[363] K. Wong, L. Hanzo, and R. Steele, "Channel coding for satellite mobile channels," *International Journal on Satellite Communications*, vol. 7, pp. 143–163, July–Sep 1989.

[364] P. Ho, J. Cavers, and J. Varaldi, "The effect of constellation density on trellis coded modulation in fading channels," in *Proceedings of IEEE VTC '92*, (Denver, Colorado, USA), pp. 463–467, IEEE, 10-13 May 1992.

[365] S. Fechtel and H. Meyr, "Combined equalisation, decoding and antenna diversity combining for mobile personal digital radiotransmission using feedforward synchronisation," in *Proceedings of IEEE VTC '93*, (Secaucus, NJ, USA), IEEE, May 18-20 1993.

[366] R. Bultitude and G. Bedal, "Propagation characteristics on microcellular urban mobile radio channels at 910MHz," *IEEE J-SAC*, vol. 7, pp. 31–39, January 1989.

[367] R. Steele, "Deploying personal communications networks," *IEEE Comms. Magazine*, pp. 12–15, September 1990.

[368] R. Bultitude, S. Mahmoud, and W. Sullivan, "A comparison of indoor radio propagation characteristics at 910MHz and 1.75GHz," *IEEE J-SAC*, vol. 7, pp. 20–30, January 1989.

[369] H. Harmuth, *Transmission of Information by Orthogonal Time Functions*. Berlin: Springer Verlag, 1969.

[370] H. Harmuth, "On the transmission of information by orthogonal time functions," *AIEE*, July 1960.

[371] H. Harmuth, "Die orthogonalteilung als verallgemeinerung der zeit- und frequenzteilung," *AEÜ*, vol. 18, pp. 43–50, 1964.

[372] D. Saha and T. Birdsall, "Quadrature-quadrature phase shift keying," *IEEE Trans. Comms.*, vol. 37, pp. 437–448, May 1989.

[373] H. Landau and H. Pollak, "Prolate spheroidal wave functions...," *BSTJ*, vol. 41, pp. 1295–1336, July 1962.

[374] W. Lee, "Spectrum efficiency in cellular," *IEEE Tr. on Vech. Tech*, vol. 38, pp. 69–75, May 1989.

[375] J. Williams, L. Hanzo, R. Steele, and J. Cheung, "A comparative study of microcellular speech transmission schemes," *IEEE Tr. on Veh. Technology*, vol. 43, pp. 909–925, Nov 1994.

[376] R. Steele, J. Williams, and L. Hanzo, "Speech performance of adaptive transceivers for PCN," in *IEE Colloquium on Advanced channel coding and modulation*, (London), 16th November 1992.

[377] J. Williams, *Simulation of digital modulation and demodulation*. PhD thesis, University of Southampton, 1992. Ph.D. mini-thesis.

[378] R. Steele and J. Williams, "Third generation PCN and the intelligent multimode portable," *IEE Electronics and Comm. Journal*, vol. 5, pp. 147–156, June 1993.

[379] P. Vary and R. Sluyter, "MATS-D speech codec: Regular-pulse excitation LPC," in *Proc. of the Nordic Seminar on Digital Land Mobile Radio Communications (DMRII)*, (Stockholm, Sweden), pp. 257–261, October 1986.

[380] P. Vary and R. Hoffmann, "Sprachcodec für das europäische Funkfernsprechnetz," *Frequenz 42 (1988) 2/3*, pp. 85–93, 1988.

[381] J. Schur, "Über potenzreihen, die im innern des einheitskreises beschränkt sind," *Journal für die reine und angewandte Mathematik, Bd 14*, pp. 205–232, 1917.

[382] N. Kitawaki, M. Honda, and K. Itoh, "Speech-quality assessment methods for speech coding systems," *IEEE Communications Magazine*, vol. 22, pp. 26–33, October 1984.

[383] A. Gray and J. Markel, "Distance measures for speech processing," *IEEE Transactions on ASSP*, vol. 24, no. 5, pp. 380–391, 1976.

[384] N. Kitawaki, H. Nagabucki, and K. Itoh, "Objective quality evaluation for low-bit-rate speech coding systems," *IEEE Journal on Selected Areas in Communications*, vol. 6, pp. 242–249, Feb. 1988.

[385] E. Bacs and L. Hanzo, "A simple real-time adaptive speech detector for SCPC systems," in *Proc of ICC'85*, (Chicago, USA), pp. 1208–1212, May 1985.

[386] D. Freeman, G. Cosier, C. Southcott, and I. Boyd, "The voice activity detector for the pan-european digital cellular mobile telephone service," in *Proc. of ICASSP'89*, (Glasgow, UK), pp. 369–372, 23-26 May 1989.

[387] S. Hansen, "Voice activity detection (VAD) and the operation of discontinuous transmission (DTX) in the GSM system," in *Proc. of Digital Cellular Radio Conference*, (Hagen, Germany), pp. 2b/1–2b/14, October 12-14 1988.

[388] L. Hanzo, W. Webb, R. Salami, and R. Steele, "On speech transmission schemes for microcellular mobile PCNs," *European Transactions on Communications*, vol. 4, pp. 495–510, Sept./Oct 1993.

[389] W. Webb, L. Hanzo, R. Salami, and R. Steele, "Does 16-QAM provide an alternative to a half-rate GSM speech codec ?," in *Proceedings of IEEE Vehicular Technology Conference (VTC'91)*, (St. Louis, MO, USA), pp. 511–516, IEEE, 19–22 May 1991.

[390] O. Wasem, D. Goodman, C. Dvorak, and H. Page, "The effect of waveform substitution on the quality of PCM packet communications," *IEEE Tr. ASSP*, vol. 36, pp. 342–348, March 1988.

[391] S. Nanda, D. Goodman, and U. Timor, "Performance of PRMA: A packet voice protocol for cellular systems," *IEEE Tr. on VT*, vol. 40, pp. 584–598, August 1991.

[392] D. Goodman, "Cellular packet communications," *IEEE Trans Comms*, vol. 38, pp. 1272–1280, August 1990.

[393] L. Hanzo, J. Cheung, R. Steele, and W. Webb, "Performance of PRMA schemes via fading channels," in *Proceedings of IEEE VTC '93*, (Secaucus, NJ, USA), pp. 913–916, IEEE, May 18-20 1993.

[394] K. Wong, L. Hanzo, and R. Steele, "A sub-band codec with embedded reed-solomon coding for mobile radio speech communication," in *Proceedings of ICCS 1988, Singapore*, 31 October – 3 November 1988.

[395] H. Kolb Private Communications.

[396] J. Lindner Private Communications.

[397] D. Schnidman, "A generalized nyquist criterion and an optimum linear receiver for a pulse modulation system," *BSTJ*, pp. 2163–2177, November 1967.

[398] W. V. Etten, "An optimum linear receiver for multiple channel digital transmission systems," *IEEE Trans. Comm.*, vol. COM-23, pp. 828–834, August 1975.

[399] A. Kaye and D. George, "Transmission of multiplexed PAM signals over multiple channel and diversity systems," *IEEE Trans. Comm. Techn.*, vol. COM-18, pp. 520–525, October 1970.

[400] M. Aaron and D. Tufts, "Intersymbol interference and error probability," *IEEE Trans. Inform. Theory*, vol. IT-12, pp. 26–34, January 1966.

[401] D. Tufts, "Nyquist's problem: The joint optimization of transmitter and receiver in pulse amplitude modulation," *Proc. IEEE*, vol. 53, pp. 248–259, March 1965.

[402] H. Schüssler, *Digitale Systeme zur Signalverarbeitung.* Berlin, Heidelberg, and New York: Springer Verlag, 1974.

[403] W. Webb and L. Hanzo, *Modern Quadrature Amplitude Modulation: Principles and Applications for Wireless Communications.* IEEE Press-Pentech Press, 1994. ISBN 0-7273-1701-6.

[404] R. O'Neill and L. Lopes, "Performance of amplitude limited multitone signals," in *Proceedings of IEEE VTC '94*, (Stockholm, Sweden), IEEE, June 8-10 1994.

[405] X. Li and L. Cimini, "Effects of clipping and filtering on the performance of OFDM," in *Proceedings of IEEE VTC '97*, (Phoenix, Arizona, USA), pp. 1634–1638, IEEE, 4–7 May 1997.

[406] A. Garcia and M. Calvo, "Phase noise and sub–carrier spacing effects on the performance of an OFDM communications system," *IEEE Communications Letters*, vol. 2, pp. 11–13, January 1998.

[407] W. Robins, *Phase Noise in signal sources*, vol. 9 of *IEE Telecommunication series.* Peter Peregrinus Ltd., 1982.

[408] C. Tellambura, Y. Guo, and S. Barton, "Equaliser performance for HIPERLAN in indoor channels," *Wireless Personal Communications*, vol. 3, no. 4, pp. 397–410, 1996.

[409] T. Ojanperä, M. Gudmundson, P. Jung, J. Sköld, R. Pirhonen, G. Kramer, and A. Toskala, "FRAMES: - hybrid multiple access technology," in *Proceedings of IEEE ISSSTA'96*, (Mainz, Germany), pp. 334–338, IEEE, Sept 1996.

[410] M. Failli, "Digital land mobile radio communications COST 207," tech. rep., European Commission, 1989.

[411] J. Torrance and L. Hanzo, "Comparative study of pilot symbol assisted modem schemes," in *Proceedings of IEE Conference on Radio Receivers and Associated Systems (RRAS'95)*, (Bath, UK), pp. 36–41, IEE, 26–28 September 1995.

[412] K. Fazel, S. Kaiser, P. Robertson, and M. Ruf, "A concept of digital terrestrial television broadcasting," *Wireless Personal Communications*, vol. 2, pp. 9–27, 1995.

[413] J. Kuronen, V.-P. Kaasila, and A. Mammela, "An all-digital symbol tracking algorithm in an OFDM system by using the cyclic prefix," in *Proc. ACTS Summit '96*, (Granada, Spain), pp. 340–345, 27–29 November 1996.

[414] M. Kiviranta and A. Mammela, "Coarse frame synchronization structures in OFDM," in *Proc. ACTS Summit '96*, (Granada, Spain), pp. 464–470, 27–29 November 1996.

[415] Z. Li and A. Mammela, "An all digital frequency synchronization scheme for OFDM systems," in *Proceedings of IEEE International Symposium on Personal, Indoor and Mobile Radio Communications, PIMRC'97*, (Marina Congress Centre, Helsinki, Finland), pp. 327–331, IEEE, 1–4 Sept 1997.

[416] J. Bingham, "Method and apparatus for correcting for clock and carrier frequency offset, and phase jitter in multicarrier modems." U.S. Patent No. 5206886, April 27 1993.

[417] T. de Couasnon, R. Monnier, and J. Rault, "Ofdm for digital tv broadcasting," *Signal Processing*, vol. 39, pp. 1–32, 1994.

[418] P. Mandarini and A. Falaschi, "Sync proposals." MEDIAN Design Note, January 1996.

[419] T. Keller and L. Hanzo, "Orthogonal frequency division multiplex synchronisation techniques for wireless local area networks," in *Proc. of IEEE International Symposium on Personal, Indoor, and Mobile Radio Communications (PIMRC'96)*, vol. 3, (Taipei, Taiwan), pp. 963–967, IEEE, 15-18 October 1996.

[420] H. Matsuoka, S. Sampei, N. Morinaga, and Y. Kamio, "Adaptive modulation system with variable coding rate concatenated code for high quality multi-media communications systems," in *Proceedings of IEEE VTC '96*, (Atlanta, GA, USA), pp. 487–491, IEEE, 1996.

[421] S.-G. Chua and A. Goldsmith, "Variable-rate variable-power mQAM for fading channels," in *Proceedings of IEEE VTC '96*, (Atlanta, GA, USA), pp. 815–819, IEEE, 1996.

[422] D. Pearce, A. Burr, and T. Tozer, "Comparison of counter-measures against slow Rayleigh fading for TDMA systems," in *IEE Colloquium on Advanced TDMA Techniques and Applications*, (London, UK), pp. 9/1–9/6, IEE, 28 October 1996. digest 1996/234.

[423] V. Lau and M. Macleod, "Variable rate adaptive trellis coded QAM for high bandwidth efficiency applications in rayleigh fading channels," in *Proceedings of IEEE Vehicular Technology Conference (VTC'98)*, (Ottawa, Canada), pp. 348–352, IEEE, May 1998.

[424] J. Torrance and L. Hanzo, "Latency and networking aspects of adaptive modems over slow indoors rayleigh fading channels," *IEEE Tr. on Veh. Techn.*, vol. 48, no. 4, pp. 1237–1251, 1998.

[425] J. Torrance, L. Hanzo, and T. Keller, "Interference aspects of adaptive modems over slow rayleigh fading channels," *IEEE Tr. on Veh. Techn.*, vol. 48, pp. 1527–1545, Sept 1999.

[426] A. Czylwik, "Adaptive OFDM for wideband radio channels," in *Proceeding of IEEE Global Telecommunications Conference, Globecom 96*, (London, UK), pp. 713–718, IEEE, 18–22 Nov 1996.

[427] P. Chow, J. Cioffi, and J. Bingham, "A practical discrete multitone transceiver loading algorithm for data transmission over spectrally shaped channels," *IEEE Trans. on Communications*, vol. 48, pp. 772–775, 1995.

[428] J. Torrance, *Adaptive Full Response Digital Modulation for Wireless Communications Systems.* PhD thesis, Dept. of Electronics and Computer Science, Univ. of Southampton, UK, 1997.

[429] K. Miya, O. Kato, K. Homma, T. Kitade, M. Hayashi, and T. Ue, "Wideband CDMA systems in TDD-mode operation for IMT-2000," *IEICE Trans. on Comms.*, vol. E81-B, pp. 1317–1326, July 1998.

[430] O. Kato, K. Miya, K. Homma, T. Kitade, M. Hayashi, and M. Watanabe, "Experimental performance results of coherent wideband DS-CDMA with TDD scheme," *IEICE Trans. on Comms.*, vol. E81-B, pp. 1337–1344, July 1998.

[431] T. Keller and L. Hanzo, "Blind-detection assisted sub-band adaptive turbo-coded OFDM schemes," in *Proceeding of VTC'99 (Spring)*, (Houston, Texas, USA), pp. 489–493, IEEE, 16–20 May 1999.

[432] J. Torrance and L. Hanzo, "Optimisation of switching levels for adaptive modulation in a slow Rayleigh fading channel," *Electronics Letters*, vol. 32, pp. 1167–1169, 20 June 1996.

[433] C. Berrou, A. Glavieux, and P. Thitimajshima, "Near shannon limit error-correcting coding and decoding: Turbo codes," in *Proceedings of the International Conference on Communications*, (Geneva, Switzerland), pp. 1064–1070, May 1993.

[434] L. Bahl, J. Cocke, F. Jelinek, and J. Raviv, "Optimal decoding of linear codes for minimising symbol error rate," *IEEE Transactions on Information Theory*, vol. 20, pp. 284–287, March 1974.

[435] T. Keller, M. Muenster, and L. Hanzo, "A burst–by–burst adaptive OFDM wideband speech transceiver." submitted to IEEE JSAC, 1999.

[436] T. Keller, J. Woodard, and L. Hanzo, "Turbo-coded parallel modem techniques for personal communications," in *Proceedings of IEEE VTC '97*, (Phoenix, Arizona, USA), pp. 2158–2162, IEEE, 4–7 May 1997.

[437] T. Keller and L. Hanzo, "Adaptive orthogonal frequency division multiplexing schemes," in *Proceeding of ACTS Mobile Communication Summit '98*, (Rhodes, Greece), pp. 794–799, ACTS, 8–11 June 1998.

[438] C. Shannon, "Communication in the presence of noise," *Proceedings of the I.R.E.*, vol. 37, pp. 10–22, January 1949.

[439] L. Piazzo, "A fast algorithm for near-optimum power and bit allocation in ofdm systems." to appear in Electronics Letters, December 1999.

[440] T. Willink and P. Wittke, "Optimization and performance evaluation of multicarrier transmission," *IEEE Transactions on Information Theory*, vol. 43, pp. 426–440, March 1997.

[441] R. Fischer and J. Huber, "A new loading algorithm for discrete multitone transmission," in *Proceeding of IEEE Global Telecommunications Conference, Globecom 96*, (London, UK), pp. 713–718, IEEE, 18–22 Nov 1996.

[442] S. Lai, R. Cheng, K. Letaief, and R. Murch, "Adaptive trellis coded mqam and power optimization for ofdm transmission," in *Proceeding of VTC'99 (Spring)*, (Houston, Texas, USA), IEEE, 16–20 May 1999.

[443] D. Hughes-Hartogs, "Ensemble modem structure for imperfect transmission media." U.S Patents Nos. 4,679,227 (Jul. 1988) 4,731,816 (Mar. 1988) and 4,833,796 (May 1989).

[444] J. Bingham, "Multicarrier modulation for data transmission: an idea whose time has come," *IEEE Communications Magazine*, pp. 5–14, May 1990.

[445] L. Godara, "Applications of antenna arrays to mobile communications, part II: Beam-forming and direction-of-arrival considerations," *Proceedings of the IEEE*, vol. 85, pp. 1193–1245, Aug 1997.

[446] J. Litva and T.-Y. Lo, *Digital Beamforming in Wireless Communications.* Artech House Publishers, Inc., Boston, London, 1996.

[447] F. Vook and K. Baum, "Adaptive antennas for OFDM," in *Proceedings of IEEE Vehicular Technology Conference (VTC'98)*, vol. 2, (Ottawa, Canada), pp. 608–610, IEEE, May 1998.

[448] Y. Li, "Pilot-symbol-aided channel estimation for OFDM in wireless systems," in *Proceeding of VTC'99 (Spring)*, (Houston, Texas, USA), IEEE, 16–20 May 1999.

[449] N. Szabo and R. Tanaka, *Residue Arithmetic and Its Applications to Computer Technology.* New York: McGraw-Hill Book Company, 1967.

[450] R. Watson and C. Hastings, "Self-checked computation using residue arithmetic," *Proceedings of the IEEE*, vol. 54, pp. 1920–1931, December 1966.

[451] R. Pyndiah, "Iterative decoding of product codes: Block turbo codes," in *Proceedings of the International Symposium on Turbo Codes & Related Topics*, (Brest, France), pp. 71–79, Sept 1997.

[452] P. Adde, R. Pyndiah, O. Raoul, and J.-R. Inisan, "Block turbo decoder design," in *Proceedings of the International Symposium on Turbo Codes & Related Topics*, (Brest, France), pp. 166–169, Sept 1997.

[453] W. Jenkins and B. Leon, "The use of residue number system in the design of finite impulse response filters," *IEEE Transactions on Circuits Systems*, vol. CAS-24, pp. 191–201, April 1977.

[454] M. Soderstrand, "A high-speed, low-cost, recursive digital filter using residue number arithmetic," *Proceeding IEEE*, vol. 65, pp. 1065–1067, July 1977.

[455] M. Soderstrand and E. Fields, "Multipliers for residue number arithmetic digital filters," *Electronic Letters*, vol. 13, pp. 164–166, March 1977.

[456] M. Soderstrand, W. Jenkins, and G. Jullien, *Residue Number System Arithmetic: Modern Applications in Digital Signal Processing.* New York, USA: IEEE Press, 1986.

[457] E. Claudio, G. Orlandi, and F. Piazza, "A Systolic Redundant Residue Arithmetic Error Correction Circuit," *IEEE Transactions on Computers*, vol. 42, pp. 427–432, April 1993.

[458] H. Krishna, K.-Y. Lin, and J.-D. Sun, "A coding theory approach to error control in redundant residue number systems - part I: Theory and single error correction," *IEEE Transactions on Circuits and Systems-II: Analog and Digital Signal Processing*, vol. 39, pp. 8–17, January 1992.

[459] J.-D. Sun and H. Krishna, "A coding theory approach to error control in redundant residue number systems - part II: Multiple error detection and correction," *IEEE Transactions on Circuits and Systems-II: Analog and Digital Signal Processing*, vol. 39, pp. 18–34, January 1992.

[460] T. Liew, L.-L. Yang, and L. Hanzo, "Soft-decision redundant residue number system based error correction coding," in *Proceeding of VTC'99 (Fall)*, (Amsterdam, Netherlands), pp. 2974–2978, IEEE, 19–22 September 1999.

[461] L.-L. Yang and L. Hanzo, "Residue number system arithmetic assisted m-ary modulation," *IEEE Communications Letters*, vol. 3, pp. 28–30, Feb. 1998.

[462] L.-L. Yang and L. Hanzo, "Performance of residue number system based DS-CDMA over multipath fading channels using orthogonal sequences," *ETT*, vol. 9, pp. 525–536, November - December 1998.

[463] H. Krishna, K.-Y. Lin, and J.-D. Sun, "A coding theory approach to error control in redundant residue number systems - part I: theory and single error correction," *IEEE Trans. Circuits Syst.*, vol. 39, pp. 8–17, January 1992.

[464] J.-D. Sun and H. Krishna, "A coding theory approach to error control in redundant residue number systems - part II: multiple error detection and correction," *IEEE Trans. Circuits Syst.*, vol. 39, pp. 18–34, January 1992.

[465] H. Krishna and J.-D. Sun, "On theory and fast algorithms for error correction in residue number system product codes," *IEEE Trans. Comput.*, vol. 42, pp. 840–852, July 1993.

[466] D. Chase, "A class of algorithms for decoding block codes with channel measurement information," *IEEE Trans. on Info. Theory*, vol. IT-18, pp. 170–182, January 1972.

[467] J. Hagenauer, E. Offer, and L. Papke, "Iterative decoding of binary block and convolutional codes," *IEEE Transactions on Information Theory*, vol. 42, pp. 429–445, March 1996.

[468] H. Nickl, J. Hagenauer, and F. Burkett, "Approaching shannon's capacity limit by 0.27 dB using simple hamming codes," *IEEE Communications Letters*, vol. 1, pp. 130–132, Sept 1997.

[469] T. Liew, C. Wong, and L. Hanzo, "Block turbo coded burst-by-burst adaptive modems," in *Proceeding of Microcoll'99, Budapest, Hungary*, pp. 59–62, 21-24 March 1999.

[470] B. Yeap, T. Liew, J. Hamorsky, and L. Hanzo, "Comparative study of turbo equalisers using convolutional codes and block-based turbo-codes for GMSK modulation," in *Proceeding of VTC'99 (Fall)*, (Amsterdam, Netherlands), pp. 2974–2978, IEEE, 19–22 September 1999.

[471] M. Yee, T. Liew, and L. Hanzo, "Radial basis function decision feedback equalisation assisted block turbo burst-by-burst adaptive modems," in *Proceeding of VTC'99 (Fall)*, (Amsterdam, Netherlands), pp. 1600–1604, IEEE, 19–22 September 1999.

[472] C. Wong, T. Liew, and L. Hanzo, "Turbo coded burst by burst adaptive wideband modulation with blind modem mode detection," in *Proceeding of ACTS Mobile Communication Summit '99*, (Sorrento, Italy), pp. 303–308, ACTS, June 8–11 1999.

[473] ETSI, *Digital Video Broadcasting (DVB); Framing structure, channel coding and modulation for cable systems*, December 1997. EN 300 429 V1.2.1.

[474] ETSI, *Digital Video Broadcasting (DVB); Framing structure, channel coding and modulation for 11/12 GHz Satellite Services*, August 1997. EN 300 421 V1.1.2.

[475] R. Steele and L. Hanzo, eds., *Mobile Radio Communications.* IEEE Press-John Wiley, 2 ed., 1999.

[476] A. Michelson and A. Levesque, *Error Control Techniques for Digital Communication.* Wiley-Interscience, 1985.

[477] S. O'Leary and D. Priestly, "Mobile broadcasting of DVB-T signals," *IEEE Transactions on Broadcasting*, vol. 44, pp. 346–352, September 1998.

[478] W.-C. Lee, H.-M. Park, K.-J. Kang, and K.-B. Kim, "Performance analysis of viterbi decoder using channel state information in COFDM system," *IEEE Transactions on Broadcasting*, vol. 44, pp. 488–496, December 1998.

[479] S. O'Leary, "Hierarchical transmission and COFDM systems," *IEEE Transactions on Broadcasting*, vol. 43, pp. 166–174, June 1997.

[480] L. Thibault and M. Le, "Performance evaluation of COFDM for digital audoo broadcasting Part I: parametric study," *IEEE Transactions on Broadcasting*, vol. 43, pp. 64–75, March 1997.

[481] P. Shelswell, "The COFDM modulation system: the heart of digital audio broadcasting," *Electronics & Communication Engineering Journal*, vol. 7, pp. 127–136, June 1995.

[482] C. Berrou and A. Glavieux, "Near optimum error correcting coding and decoding: turbo codes," *IEEE Transactions on Communications*, vol. 44, pp. 1261–1271, October 1996.

[483] S. Wicker, *Error Control Systems for Digital Communication and Storage.* Prentice Hall, 1994.

[484] A. Barbulescu and S. Pietrobon, "Interleaver design for turbo codes," *IEE Electronic Letters*, pp. 2107–2108, December 1994.

[485] P. Robertson, E. Villebrun, and P. Hoeher, "A comparison of optimal and sub-optimal MAP decoding algorithms operating in the log domain," in *Proceedings of the International Conference on Communications*, pp. 1009–1013, June 1995.

[486] M. Failli, "Digital land mobile radio communications COST 207," tech. rep., European Commission, 1989.

[487] C. Lee, T. Keller, and L. Hanzo, "Turbo-coded hierarchical and non-hierarchical mobile digital video broadcasting," *IEEE Transaction on Broadcasting*, March 2000.

[488] B. Haskell, A. Puri, and A. Netravali, *Digital Video: An Introduction To MPEG-2.* Digital Multimedia Standards Series, Chapman & Hall, 1997.

[489] L. Hanzo, P. Cherriman, and J. Streit, "Video compression and communications over wireless channels: From second to third generation systems, WLANs and beyond." Book in preparation[3].

[490] H. Gharavi and L. Hanzo, eds., *Proceedings of the IEEE*, vol. 87, October 1999.

[491] G. Reali, G. Baruffa, S. Cacopardi, and F. Frescura, "Enhancing satellite broadcasting services using multiresolution modulations," *IEEE Transactions on Broadcasting*, vol. 44, pp. 497–506, December 1998.

[492] Y. Hsu, Y. Chen, C. Huang, and M. Sun, "MPEG-2 spatial scalable coding and transport stream error concealment for satellite TV broadcasting using Ka-band," *IEEE Transactions on Broadcasting*, vol. 44, pp. 77–86, March 1998.

[493] L. Atzori, F. D. Natale, M. D. Gregario, and D. Giusto, "Multimedia information broadcasting using digital TV channels," *IEEE Transactions on Broadcasting*, vol. 43, pp. 383–392, December 1997.

[494] W. Sohn, O. Kwon, and J. Chae, "Digital DBS system design and implementation for TV and data broadcasting using Koreasat," *IEEE Transactions on Broadcasting*, vol. 44, pp. 316–323, September 1998.

[495] J. Griffiths, *Radio Wave Propagation and Antennas - An Introduction.* Prentice Hall, 1987.

[496] M. Karaliopoulos and F.-N. Pavlidou, "Modelling the land mobile satellite channel: a review," *Electronics and Communication Engineering Journal*, vol. 11, pp. 235–248, October 1999.

[497] J. Goldhirsh and W. Vogel, "Mobile satellite system fade statistics for shadowing and multipath from roadside trees at UHF and L-band," *IEEE Transactions on Antennas and Propagation*, vol. 37, pp. 489–498, April 1989.

[498] W. Vogel and J. Goldhirsh, "Multipath fading at L band for low elevation angle, land mobile satellite scenarios," *IEEE Journal on Selected Areas in Communications*, vol. 13, pp. 197–204, February 1995.

[3]For detailed contents please refer to http://www-mobile.ecs.soton.ac.uk

[499] W. Vogel and G. Torrence, "Propagation measurements for satellite radio reception inside buildings," *IEEE Transactions on Antennas and Propagation*, vol. 41, pp. 954–961, July 1993.

[500] W. Vogel and U. Hong, "Measurement and modelling of land mobile satellite propagation at UHF and L-band," *IEEE Transactions on Antennas and Propagation*, vol. 36, pp. 707–719, May 1988.

[501] S. Saunders, C. Tzaras, and B. Evans, "Physical statistical propagation model for mobile satellite channel," tech. rep., European Commission, 1998.

[502] S. Saunders, *Antennas and Propagation for Wireless Communication Systems Concept and Design.* John Wiley & Sons, 1999.

Index

Symbols

Q

R

S

T

Author Index

C

D

E

F

G

H

L

M

N

O

P

Q

R

S

T

U

V

W

X

Y

Z